Springer Series in

CHEMICAL PHYSICS 102

Springer Series in

CHEMICAL PHYSICS

The purpose of this series is to provide comprehensive up-to-date monographs in both well established disciplines and emerging research areas within the broad fields of chemical physics and physical chemistry. The books deal with both fundamental science and applications, and may have either a theoretical or an experimental emphasis. They are aimed primarily at researchers and graduate students in chemical physics and related fields.

Please view available titles in *Springer Series in Chemical Physics* on series homepage http://www.springer.com/series/676

Gheorghe Duca

Homogeneous Catalysis with Metal Complexes

Fundamentals and Applications

With 103 Figures

Springer

Gheorghe Duca
Academy of Sciences of Moldova
Chisinau, Moldova

Series Editors

Springer Series in Chemical Physics ISSN 0172-6218
ISBN 978-3-642-24628-9 ISBN 978-3-642-24629-6 (eBook)
DOI 10.1007/978-3-642-24629-6
Springer Heidelberg New York Dordrecht London

Library of Congress Control Number: 2012936482

Printed on acid-free paper

Springer is part of Springer Science+Business Media (www.springer.com)

This book is dedicated to the memory of
Prof. A.Ya. Sychev and Prof. A.P. Purmal

Preface

In the six chapters of the present book are considered both theoretical problems of the functioning of homogeneous catalysts, including the role of transition metal ions (mainly, iron, copper, and manganese) and their complex compounds, and practical aspects of such catalysis in various fields, involving normal biochemical and pathological (diseases) processes, oxyacids and natural polyphenols oxidation, in environmental processes (atmosphere, water), in the chemical industry, and in food chemistry. Naturally, the description of the homogeneous catalysis role in these applied fields does not pretend to be comprehensive. The purpose is different—to show, using certain examples, the influence, role, and mechanisms of catalytic reactions with metal complexes participation, in the processes taking place in the most important applied fields.

Another aim was to attract attention and raise the interest in scientific research in this rapidly developing field of catalysis (homogeneous catalysis with metal complexes) in various fields and to show the unlimited possibilities in both theoretical and applied aspects.

The third intention was to illustrate the possibilities of a rapidly developing new chemical concept—evolution catalysis, on model systems as the examples (mimetic method), and to reveal the practicability and prospect of such an approach in many applied fields.

It is an important task, in order to comprehend and to outline, from the enormous amount of experimental and theoretic material, certain basic principles of redox homogeneous catalysis, and especially, the mechanisms of processes, as soon as their understanding makes it possible to control and manage them, changing their direction, rate, selectivity, etc., within the required parameters.

In addition, in this book (in certain cases, deliberately) contradictory affirmations and conclusions are often given and compared, with the aim of bringing about the vivid perception of critical attitude of readers regarding the explanation of results, proposed by one or other researcher (or author). It was also considered desirable to show, very briefly, in certain cases (Chap. 1) some methodical approaches to the clarification of the mechanisms of catalytic processes. These approaches are often used by researchers but are rarely described. This part would seem to be useful for young researchers beginning work in the field.

The old proverb says that "a lot of knowledge brings a lot of grief." Grief, because it was impossible to explain everything in detail, that process mechanisms can be treated in a different way, and, very often, they have been revealed, at best, only at a satisfactory qualitative level, and that model systems cannot always explain real processes in the environment, in the human organism, or in real technologies. But this grief inspires action and encourages the researchers to penetrate deeper into the essence of the studied processes to put their knowledge at the service of practical activities, not just to satisfy their curiosity.

I cannot say often enough that this book has apeared as a result of long-term and successful collaboration with Prof. Yu. Scurlatov from Russia.

I would like to thank my colleagues with whom I have collaborated: Dr. V. Isac, Dr. M. Gonta, Dr. V. Covaliov, Dr. V. Gladchi, Dr. O. Covaliova, Dr. L. Romanciuc, Prof. M. Duca (Republic of Moldova), as well as Dr. S. Travin and Dr. E. Shtamm (Russian Federation).

Chisinau Gheorghe Duca

Contents

Chapter 1
Introduction

1.1 Types of Homogeneous Catalysis with Metal Complexes, Their Advantages and Drawbacks

The analysis of one of the most important materials—from both the experimental and theoretical point of view—in the history of chemistry since Boyle's times has allowed us to systematize and reduce it to four conceptual systems: composition doctrine (approximately since 1660), structural chemistry (approximately since 1800), chemical process doctrine (after 1880) and evolutionary chemistry (approximately since 1960). Once we start to consider the nature of the substance, the following hierarchy of its material carriers can be observed: (1) atoms of chemical elements; (2) molecules of chemical compounds; (3) systems of reacting substances, including using of catalysts and (4) highly-organized catalytic systems (enzymes) as a result of evolutionary development of living substances [1].

Therefore higher and higher consecutive levels of chemical knowledge can be distinguished, each subsequent level incorporating the material of lower levels (conceptual systems) in the transformed form. Such an approach serves as the original guide in the strategic development of scientific and technical progress in chemical production.

Nowadays, the fundamental science is not only the basis of current chemical production—it also promotes other fundamental levels (on the new conceptual basis), the creation of quite new production, characterized by sharp increases in labour productivity and the enhanced quality of products obtained. In this regard, the third and fourth conceptual systems —the doctrine of chemical processes and evolutionary chemistry—have been developing very rapidly in last few decades, promising really fantastic results in the fields of both fundamental and applied research.

In these two conceptual systems the problems of redox homogeneous catalysis with transition metals coordination compounds (metal complexes) occupy a very important place.

G. Duca, *Homogeneous Catalysis with Metal Complexes*, Springer Series in Chemical Physics 102, DOI 10.1007/978-3-642-24629-6_1,

Unlike other forms, homogeneous catalysis with metal complexes has some peculiarities [2]:

1. The possibility of revealing the mechanism of catalyst action, its composition and the identification of its intermediate forms on the qualitative, semi-quantitative and sometimes quantitative levels.
2. The ability to obtain practically unlimited amounts of catalytically active and specifically acting systems.
3. Improvement of catalyst specificity and directed change of its activity at the expense of medium parameter variations (nature and number of metal ions, ligands, character of solvent, etc.).
4. Use of the greater part or even of all the molecules of catalyst in catalytic activity, which is impossible in the case of classical heterogeneous catalyst.
5. Decrease in rigidity of process conditions.
6. Comparative ease of the creation of continuous processes.

A characteristic feature of homogeneous catalysis with metal complexes is its clearly expressed practical orientation. As early as the 1940s and 1950s, a series of organic synthesis directions appeared (syntheses with metals carbonyls, catalytic polymerization on Ziegler-Natta systems, cyclohexane oxidation into cyclohexanone, butane into acetic acid, etc.). The number of works on fermentology increased at the same time and research on immobilized enzymes was started. The rapid growth in the amount of work on the thermodynamics of complex formation in solutions at that time promoted studies devoted to the role of coordination in catalysis (approximately since the 1960s).

Homogeneous catalysis today is at the stage of rough development. Although the leading position in the practical use of catalytic phenomena historically belongs to heterogeneous catalysis, and its application is quite considerable from the practical aspect, there are still considerable problem with understanding catalytic processes mechanisms. The existing theories of heterogeneous catalysis mainly have a qualitative character and can only explain some catalytic processes.

The situation with homogeneous catalysis is somewhat different. Its practical applications are still considerably behind those of the heterogeneous version. This is illustrated by the fact that systematic study of homogeneous catalysis at the mechanism level has only been active since the 1960s. Therefore it is not surprising that heterogeneous catalysis has occupied the cardinal position in the chemical industry and that the applications of homogeneous catalysis (in spite of rapid progress) have not yet reached the appropriate level. Though heterogeneous catalysis in its practical applications has built up a good start, theoretical understanding of this phenomenon did not progress as quickly. Some homogeneous-catalysed reaction mechanisms have became clear after 5–10 years study, whereas for some heterogeneous reactions this has taken up to 50 years research.

This can apparently be explained by the fact that, up to the early 1960s, the greatest successes have come in equilibria studies of homogeneous systems (specifically regarding complex compounds of transition metal solutions). By then it became possible to apply various modern physical methods (especially spectroscopy) to detect the intermediate compounds. The simplicity (absence of a border surface) and reproducibility of homogeneous systems was also an important factor.

On the other hand, fast successes in the study of homogeneous reactions mechanisms will help to resolve certain problems in heterogeneous catalysis (for instance, mechanisms of hydrogenization in solution and on the catalyst surface are often similar, although there are also differences). In fact, within approximately the two last decades, great theoretical successes have been achieved in this area, and because of these the mechanisms of some reactions can be described semi-quantitatively and even quantitatively (although quite rarely).

In the field of homogeneous catalysis with metal complexes, certain successes were achieved in the explanation of catalytic activity during the transition from activated metal ion to metal coordination compounds with various ligands. Thus, it became possible to show that the essential importance involves a change in the redox potential during the transition from activated metal ion to metal complexes. Thus, the addition of EDTA to the system $Fe^{3+}{}_{aq}/Fe^{2+}{}_{aq}$ EDTA changes the potential from 0.771 V to 0.143 V, thus greatly reducing the initiation rate of H_2O_2 decomposition. Proceeding from the different lability of the coordination sphere of a metal ion, one can explain the different process mechanisms (for example, H_2O_2 decomposition in the presence of Fe^{3+} complex and Mn^{2+} ions, isoelectronic with it, although quantum-chemical explanations of this difference have appeared lately).

Nowadays the prospects of homogeneous catalysis with metal complexes are fascinating. Its significant advantage is that it frequently provides a way for a reaction which is either extremely difficult without the catalyst or cannot be realized by other means. As such reactions proceed frequently under mild conditions, they result in high selectivity products. Homogeneous catalysts are also potentially more effective than heterogeneous, as in the former the reacting substrate interacts with each molecule of catalyst but in the latter only with superficial atoms or ligands of catalyst.

Varying the ratio between the concentration of ligand and metal it is possible to create homogeneous catalytic systems in which all metal atoms are active and act as catalysts. In its turn this leads to the possibility to use them in much lower concentrations than in the case of heterogeneous catalysts, thus economizing on compounds of rare and precious metals.

Comparative peculiarities of homogeneous and heterogeneous catalysis are given in Table 1.1, and their types are described in details in certain papers [2–4].

The application of homogeneous catalysis in industry, despite its obvious profitability, is complicated by various practical difficulties. For example, it is often rather difficult to regenerate and to extract the catalyst from the reactionary medium, and to separate it from the reaction products.

To solve this problem, another method of catalyst preparation can be used which in essence is binding the transitionmetal complex in a properly prepared solid

Table 1.1 Comparison of homogeneous and heterogeneous catalysis

Parameters	Catalysis	
	Homogeneous	Heterogeneous
Active centers	All the metal atoms	Surface atoms only
Concentration	Low	High
Structure	Definite	Indefinite
Stoichiometry	Definite	Indefinite
Reaction conditions	Mild	Strict (high T° and pressure)
Activity	High	Low
Determination of catalyst composition	Rather complicated	Easy
Stability	Low	High
Catalyst regeneration	Complicated	Easy

matrix. The matrix should have functional groups (ligands) which are bound with the dissolved metal complex:

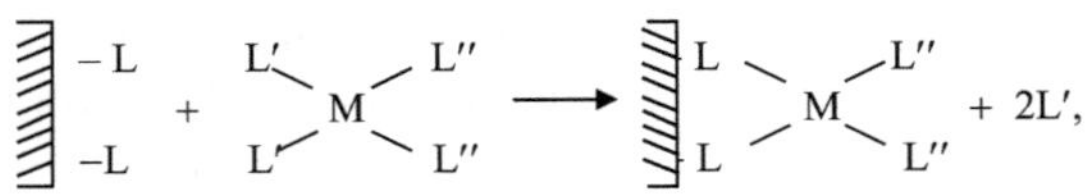

where L, L′, L″ are various ligands.

In this way, the advantages of homogeneous and heterogeneous catalysts are combined, whereas the differences between them reduce considerably. Therefore, it becomes clear that there is no principal difference between the mechanisms of homogeneous and heterogeneous catalysis in many reactions. For example, in the case of olefin and diene catalytic polymerization, irrespective of whether the reaction proceeds in the solution or on the surface, the mechanism is the same—incorporation of monomer occurs along the bond "transition metal–carbon." The difference is in the method of active center formation but not in the mechanism of their action.

There is no generally accepted name for such catalysts. The most widely used are "heterogenized," "hybridized," "immobilized homogeneous," and "exact" catalysts. The last name indicates that all the metal ions can be included in the composition of active centers, and thus the equality between metal ion and active center concentrations is observed. Unlike traditional heterogeneous catalysts, the role of carrier is not reduced only to active component deduction on the surface. While the catalyst is binding with the surface, the significant change in its chemical nature occurs. In addition it becomes possible to preserve the catalyst in the dispersed form and thus to prevent its aggregation.

The opportunity to create heterogenized catalysts by fixation of metal complexes on a substrate (polymeric carrier) is very important [5, 6].

Enzymatic catalysis and the possibility of its modeling is no less interesting. Currently, it has found broad industrial applications [7]. The main difference between enzymatic catalysis and the other catalysis types is singularity of the enzyme active centre structure (caused in particular by its albumen part) not present in other cases, for example in homogeneous solutions of low molecular substances—complex compounds of transition metals, though the latter can also be components or analogues of active centres. This is because in enzymes globula the activation of molecular groups occurs (for example, metal coordination compounds), which in the usual state are not very active catalysts.

The structure of active centres and influence of surrounding albumen determine the unique properties of enzymes: (1) their high catalytic efficiency (sometimes exceeding non-biological catalysts 10^{10}–10^{13} times), (2) their exceptional selectivity and (3) their ability to respond on the regulating interaction of small changes in properties and medium structure [8].

Immobilized enzymes are used today when working at higher temperatures (higher than 333 K) and for the convenience of their introduction and extraction from the reaction medium. Immobilized enzymes are those connected with some carrier. Immobilization consists of: (1) covalent joining of enzyme to polymer carrier (cellulose, glass, polystyrol, polyaminoacids, etc.); (2) covalent joining of enzyme molecules to each other with the help of added polyfunctional reagent; (3) enzyme incorporating into three-dimensional network of gel of polymer; (4) enzyme adsorption on water insoluble carriers (for example ionites); and (5) enzyme incorporation into semipermeable microcapsulas or polymeric fibers. In these ways a considerable increase in enzyme stability can be reached. Besides, it can be easily separated from the products of catalysed reaction and thus catalytic processes can be stopped at the required stage, and the catalyst can be used in a flow regime or repeatedly. Immobilized enzymes have rather large advantages in their use in medical applications since it is possible to avoid the allergic reaction of organisms.

As a rule, immobilization increases the effective lifetime and temperature stability of enzymes, although there can be exceptions.

1.2 Perspectives of the Industrial Use of Homogeneous Catalysis with Metal Complexes

High activity, specificity and possibility to modify metal complex catalysts—these are the features that would determine their wide industrial applications.

High activity can be achieved due to the change in ligand nature, fastening of an active centre on the appropriate carrier, creation of the multi-centered catalysis structure, selection of solvent and certain process conditions, etc.

Because of the high efficiency of homogeneous catalysts, it became possible to develop a highly effective technological process of olefin production by ethylene polymerization. Synthesis of acetic acid is possible by methanol carbonylation (mild conditions, rhodium complexes as catalysts, many times exceeding the activity of cobalt or acid catalysts, which makes them competitive in spite of the high cost of rhodium). The hydrogenase analogues were created via complex homogeneous and heterogeneous catalytic systems of aromatic nitrocompound hydration (high rate, mild conditions). Thus it became possible to realize the continuous hydration of aromatic nitrocompounds to corresponding amines such as aniline, toluidine and chloraniline. Earlier, such processes were carried out under the hard conditions of high temperature and hydrogen pressure, which often resulted in low yields of main product, required complicated equipment and made the process expensive.

Until now, various complex catalysts and catalytic system types have been developed that make it possible to hydrate the benzene into cyclohexane under mild conditions and at a high rate (sometimes exceeding the corresponding values for heterogeneous catalysts).

Another factor of homogeneous catalysis with metal complexes is its specificity. It can be achieved by variation of nature and number of coordinated ligands, carriers and solvent. So the process of tetralene production by naphthalene hydration was developed using the highly-selective homogeneous complex catalyst, its selectivity being tens of times higher than that of skeleton nickel one, usually applied in these instances.

The use of enzymes and especially immobilized ones is varied. Thus, catalase is used in cheese production as H_2O_2 arrests decomposition after the cold sterilization, in the production of foamy rubber and foamy cement, and also to accelerate fur and feather bleaching.

In the early 1970s it became possible to carry out the covalent binding of amyloglucosidase (a fermentation that ensures starch transformation into glucose–starch sugaring) with porous glass. A multitonnage process using immobilized enzymes is the transformation (conversion) of D,L-glucose (obtained from corn starch) into D,L-fructose, in which the sugar contents is approximately 1.5 times higher. For these purposes the immobilized enzyme glucoisomerase is used. Suitable equipment with the capacity of 250,000 tonnes and more per year was developed in the USA.

Using of immobilized enzymes makes it possible to transform cane sugar into a glucose–fructose mixture. The economic benefit of such a transformation is quite considerable.

The opportunity to develop the synthesis of heterogenized low-temperature and efficient catalysts of ammonium and nitrogen-containing compounds from the molecular nitrogen under ordinary pressure, using the principles of immobilized enzymes and heterogenized metal complexes, is attractive.

Existing and possible basic areas for the use of homogeneous catalysis with metal ions and complexes in various reactions are extensive, as can clearly be seen from Scheme 1.1.

The field of homogeneous catalytic processes with metal complex participation is comprehensive and universal. Therefore, it is necessary to put certain limitations on subsequent consideration of the experimental and theoretical data discussed in this book.

We should consider only the redox catalytic processes in the liquid phase (mainly in polar solvents—basically in water—and in nonpolar solvents). This is because of the exceptional importance of redox processes in relatively simple chemical systems, in the reactions proceeding in living organisms, in environmental processes (in atmosphere, water and soil), and in industrial technologies (especially in the food-processing industries).

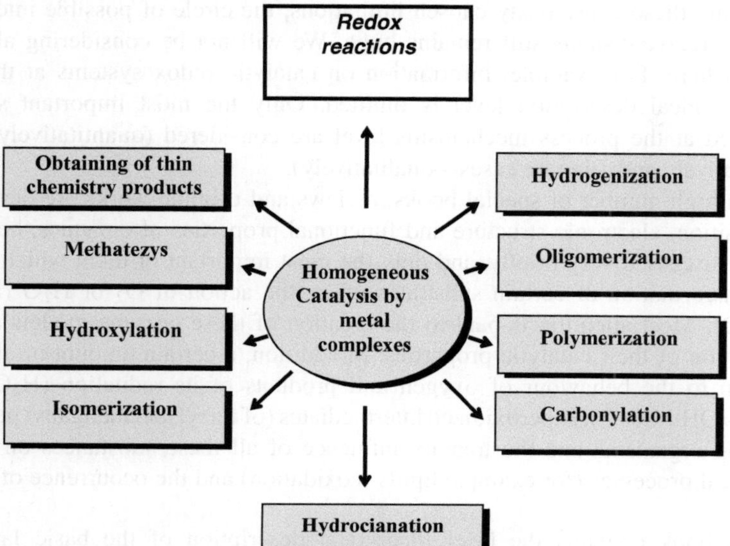

Scheme 1.1 Applications of homogeneous catalysis with the transition metal complexes

Only the ions or complex compounds of iron, manganese, cobalt, copper (and in part those of zinc and molybdenum) at different oxidation degrees will be considered, these metal compounds being necessary for living organisms catalysing redox transformations.

Many are involved as catalysts in the environmental redox processes. For example, iron or manganese compounds are catalysts of the redox reactions proceeding in the atmosphere with the formation of so-called "acid rain." They—and also copper compounds—are responsible for catalytic processes occurring worldwide in oceans, seas, rivers and water basins. These elements' compounds are used as catalysts in various chemical technologies, and also in spontaneous processes, associated with raw (food) or prepared materials.

Many substances can act as oxidizers in these processes. Only those catalytic processes will be considered in which dioxygen or products of its reduction are used as oxidants, in particular hydrogen peroxide, H_2O_2, as these oxidizers are unique with regard to their biological activity, environmentally friendly behaviour and adaptability to production. To a smaller degree, catalytic systems will also be considered where hydroperoxides ROOH related to hydrogen peroxide, iodozyl benzene and some other oxidants are used.

The solvent in such systems is usually water, but also possibly certain organic substances. The temperature conditions of these processes are close to standard ones ($t = 25°C$, $p = 1$ atm), although there can be small deviations from these conditions.

Despite these consciously chosen limitations, the circle of possible interesting catalytic redox systems still remains huge. We will not be considering all these systems here. For example, information on catalytic redox systems at the phenomenological description level is omitted. Only the most important systems examined at the process mechanisms level are considered (quantitatively, semi-quantitatively and—in rare cases—qualitatively).

As a high number of special books, reviews and original works are devoted to composition, electronic structure and functional properties of enzymes, here they will be discussed very briefly, and only the most important of them which realize the transformation of certain substrates under the action of O_2 or H_2O (ROOH, C_6H_5IO). Most attention is paid to the creation of these enzyme models and the description of their catalytic properties. In addition, a certain amount of attention is given to the behaviour of oxygen and products of its reduction (H_2O_2, free radicals OH·, HO^{2-}), superoxidized intermediates (of ferryl and manganyl particles) in living organisms and the toxicity influence of all these substances on certain biological processes (for example lipids reoxidation) and the occurrence of certain diseases.

This book contains the brief theoretical description of the basic laws and mechanisms of the redox reactions with O_2 and H_2O_2 in association with water compartments, catalysed by transition metal coordination compounds. There is also a detailed discussion of these processes and their applications in various fields of industrial chemistry, the food-processing industry, biochemistry, medicine, analytical chemistry, ecological chemistry and other areas.

Nowadays it is mainly the compounds of twelve elements (Ti, V, Mn, Fe, Co, Zn, Mo, Rh, Pd, W, Cu and Ni) that are used as catalysts in industry. Among them, the extremely interesting and well studied catalysts are the coordination compounds of manganese, iron and copper. Not only was the composition of these complexes defined in the solutions, but also their stability constants and the formal kinetics of various substrates redox transformations were studied. The mechanisms of these processes were revealed on a good qualitative (and sometimes on semi-qualitative and even quantitative) level, and quantum-chemical calculations were made, confirming the probability of the occurrence of certain reactive intermediates. All this creates the basis for further optimisation of important reactions.

Consideration of other metal complexes, other than those of manganese, iron, copper and cobalt, as redox transformations catalysts is not expedient here for the following reasons: (1) some of them are considered in detail in certain papers; (2) experimental material and redox transformations mechanisms are often given in literature only on the descriptive level.

Besides, as there is a great variety of catalytic redox reactions, some model of these processes should be chosen, selecting those that will reveal the general regularities and mechanisms of such processes. As oxidizers in such models the particles were chosen which are the products of four-electron reduction of the oxygen molecule:

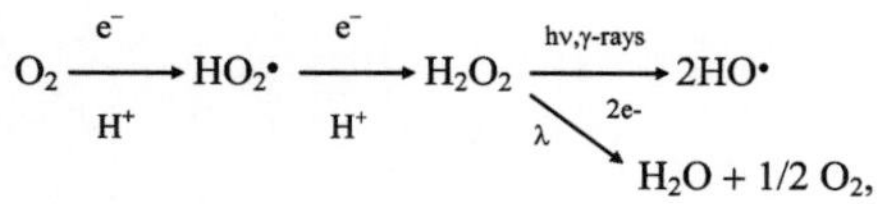

i.e., O_2, H_2O_2, $HO_2^{\bullet}$, $OH^{\bullet}$ and also ferryl ($Fe^{IV} = O^{2+}$) and manganyl ($Mn^{IV} = O^{2+}$) ions or their complexes with ligands. The oxidizer in most cases was hydrogen peroxide. The substrates were carbonic acids, alcohols, aldehydes, dyes, etc. The ligands of metal coordination compounds were carbonic acids, alcohols, aldehydes, dyes, etc. Thus, the summary scheme of processes to be considered further can be presented as

$$M^{z+}L_n + O_x + SH_2 \rightarrow P_i$$

where $M^{z+}L_n$ represents copper, iron, manganese, cobalt and sometimes other elements coordination compounds, O_x the oxidizer H_2O_2, ROOH, O_2, $HO_2^{\bullet}$, $OH^{\bullet}$, etc., SH_2 substrates (more often organic compounds), and P_i the reaction products.

Even such self-restriction of the reactions considered results in the necessity to describe only the main material from the huge experimental and theoretical information available from the scientific literature and from data obtained by the authors. Therefore plenty of interesting experimental data are deliberately not given in this book After a brief theoretical statement of basic laws and mechanisms of the redox catalytic processes, the main regularities and mechanisms and their role, and the possibility of their use are shown in biochemistry, medicine, environmental chemistry, analytical chemistry (catalymetry) and, for practical purposes, in various areas of industrial and food chemistry.

References

1. K.W. Boyack, K. Börner, R. Klavans, Scientometrics **79**(1), 45–60 (2009)
2. W.N.M. van Leeuwen, *Piet Homogeneous Catalysis*, XIII (Springer, Heidelberg, 2004) p. 407
3. G. Swiegers, *Mechanical Catalysis: Methods of Enzymatic, Homogeneous, and Heterogeneous Catalysis* (Wiley, New York, 2008) p. 320
4. B. Cornils, W.A. Herrmann, *Applied Homogeneous Catalysis with Organometallic Compounds: Developments* (Wiley-VCH,Weinheim, 2002), 1450
5. A. Corma, H. Garcia, F.X. Llabres i Xamena, Chem. Rev. **110**(8), 4606–4655 (2010)
6. S. Ogasawara, S. Kato, J. Am. Chem. Soc. **132**(13), 4608–4613 (2010)
7. H.E. Schoemaker, D. Mink, M.G. Wubbolts. Science **299**(5613) 1694–1697 (2003)
8. R.H. Crabtree, *Green Catalysis* (Wiley-VCH, Weinheim, 2009) p. 1082

Chapter 2
Homogeneous Redox Catalysis with Transition Metal Compounds in Oxide and Peroxide Systems

2.1 General Principles of Catalytic Redox Reactions with Metal Compounds

Catalytic reactions are usually subdivided into homogeneous, heterogeneous, and fermentative, which is conditioned not only by the existence of one- or many-phase catalytic systems but also most probably, as it seemed earlier, by the completely distinguished mechanisms of their action. As a result, the theories describing the conduct of systems related to these three groups should be different [1]. However, the development of metal–complex catalysis acts as a bridge, and with its help it is possible to connect all these three types of catalysis. The hope still exists that in the future a general theory of catalysis will be created. Indeed, during the last few decades many experimental and theoretical data appeared confirming that coordination compounds of several transition metals which are used in catalysis of certain reactions in homogeneous, heterogeneous, and enzymatic conditions act in accordance with the same (or close) mechanisms. This is especially specific for homogeneous and heterogeneous catalysis by metal complexes [2]. The difference in the rate and selectivity of the same reaction course in homogeneous, heterogeneous, and fermentative catalysis is caused by a number of specific features such as the degree of optimum spatial arrangement of reagents [3].

With the ions' transition from solvated transition metal to their complex compounds with definite ligand (L), substantial growth of catalytic reaction rate is often observed. Such a situation is stipulated by the fact that ligands coordination, for example, in an equatorial plane, and central metal ion entry in internal coordination sphere, and also substrate and oxidizer entry into an axial position, create favorable opportunities of electron rearrangements within the coordination sphere of the metal ion. Ligands, accepting or giving back a part of their electron density to the central atom of the metal, thus reduce the compensating influence of substrate and oxidizer. Therefore, the total energy of their interaction accompanied by the breakage of old links and formation of new ones changes comparatively little (original buffer effect of ligands producing the possibility of power expenses

G. Duca, *Homogeneous Catalysis with Metal Complexes*, Springer Series in Chemical Physics 102, DOI 10.1007/978-3-642-24629-6_2,

compensation for bonds breakage by 80–90%). This effect of power compensation frequently brings about the essential contribution to the reaction velocity increase on introduction of the ligands into the catalytic system, although other effects (which will be considered further in Sect. 2.3) also have a strong influence on the efficiency of the catalytic system.

Substrate and oxidizer penetration into the inner coordination sphere of the catalyst–metal complex and subsequent redistribution of electron density in this sphere, resulting in catalytic redox transformation of initial substances into products, provides the essential inner-sphere mechanism of the reaction course. Electron transfer according to such a mechanism is much faster than in the case of the outer sphere mechanism due to the formation of an intermediate in which electron transfer is facilitated [4].

Another possible process is initial substances' transformation according to an outer-sphere mechanism. In this case, reacting substrates and oxidizer do not enter the inner coordination sphere of the metal, overlapping of electron environments of the metal complex and reaction of substances is insignificant, the coordination of latter with metal being realized by means of hydrogen or donor–acceptor bonds' formation. Though the energy of such bonds is insignificant (8–30 kJ/mol), several of them can be formed and even the insignificant reduction of the activation energy (of the order of 10 kJ/mol) can cause a tenfold increase in the catalytic process.

Numerous attempts were undertaken in order to classify various reactions and the transition metal compounds catalyzing them. As a basis of such classifications, different attributes were considered and each of them was limited, and, therefore, they will not be discussed here. Let us consider only the scheme of the metal with ligands and substrates interaction in the solution (Scheme 2.1).

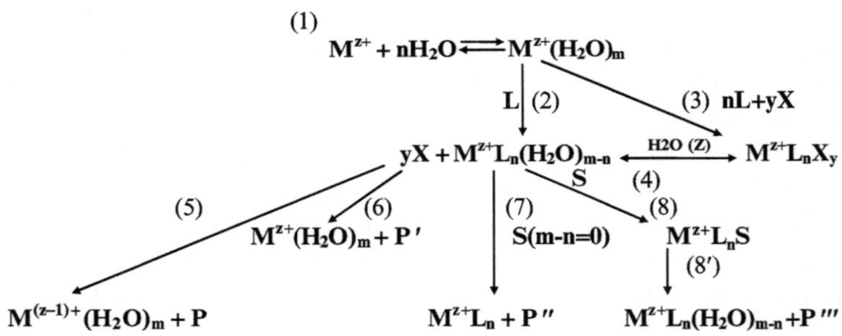

Scheme 2.1 Interaction of metal with ligands and substrates in the solution

Here, L is generally a polydentate ligand, X an acidogroup or other ligand, S a substrate, Z a solvent, P reaction product, and $m - n = 0$ for inner-sphere electron transfer.

Brief explanatory notes to Scheme 2.1:

1. Hydration (solvation) occurs when a metal salt is introduced into the solvent.
2. Formation of coordination compound of transition metal by a replacement reaction.

3. Formation of coordination compound with mixed ligands.
4. Reaction of hydration (solvation).
5. and 6. Under the influence of the metal ion, coordinated ligands are transformed into the reaction products P and P′, with or without change of the metal ion's oxidation number.
7. Case of saturated coordination complex with outer-sphere electron transfer, resulting in products P″ formation.
8. 8′. Reaction products P‴ form as a result of unsaturated coordination metal complex interaction with substrate, formation of an unstable intermediate between them and its subsequent disintegration.

Each of these routes is a set of many stages (mechanism of reaction). A rather complex picture of catalysis by complex compounds can be distinguished from this scheme. To simplify the picture of catalytic processes and to provide their detailed elaboration, only one group of processes—the ionic redox reactions—will be considered further, these being of fundamental importance in chemistry, biochemistry, bioinorganic chemistry, medicine, environmental chemistry, analytical chemistry, and technological processes.

In general, the majority of redox transformations can be presented in the form [5]

$$DH_2 + A \rightleftarrows D + AH_2.$$

Thus, redox reactions represent a general result of two H atoms or two electrons and two protons transfer from the electron donor DH_2 (various reducers, of which H_2 is the most elementary) to the electrons acceptor A (various oxidizers, the most elementary of which is O_2). It was considered earlier that such a simultaneous two-electron transfer represents the reaction of the redox transformation:

$$DH_2 \rightleftarrows DH^- + H^+,$$

$$DH^- + A \rightleftarrows (DHA)^- \rightarrow D + AH^- (\text{ion-molecular mechanism}),$$

$$AH^- + H^+ \rightleftarrows AH_2.$$

However, during the period 1921–1931, Michaelis has put forward the hypothesis that such an ion–molecular mechanism is not general. Redox transformations can also occur as a result of a one-electron transfer:

$$DH_2 \rightleftarrows DH^- + H^+, \quad A^- + H \rightleftarrows AH,$$

$$DH^- + A \rightarrow DH + A^-, \quad D^- + AH \rightarrow D + AH^-,$$

$$DH \rightleftarrows D^- + H^+, \quad AH^- + H^+ \rightleftarrows AH_2 (\text{ion-radical mechanism}).$$

When the reactions of one-electron transfer take place, the realization of the radical–chain mechanism, including the stages of chain initiation (i), continuation (1, 2) and breakage (o), is possible:

$$\text{i. } DH^- + A \rightarrow DH + A^-, \quad A^- + H^+ \rightleftarrows AH,$$

$$\text{1. } AH + DH^- \rightarrow AH^- + DH, \quad DH \rightleftarrows D^- + H^+,$$

$$\text{2. } D^- + A \xrightarrow{H+} AH,$$

$$\text{o. } AH + AH \rightarrow A_2H_2, \quad \text{(radical-chain mechanism)}$$

$$\text{o}'. AH + D^- \rightarrow AH^- + D.$$

These three processes essentially cover the whole set of ionic redox transformations.

The mechanisms of redox reaction catalysis by variable valency metal ions can be summarized as follows [6]:

1. Ion-molecular mechanism of intermediate complex formation:

$$DH_2 \rightleftarrows DH^- + H^+,$$

$$DH^- + M^{2+} + A \rightleftarrows DH^-M^{2+}A \rightarrow M^{2+} + AH^- + D,$$

$$AH^- + H^+ \rightleftarrows AH_2;$$

2. Ion-radical mechanism of compensating reactions:

$$DH_2 \rightleftarrows DH^- + H^+,$$

$$DH^- + M^{2+} \rightarrow DH + M^+, \quad D^- + M^{2+} \rightarrow D + M^+,$$

$$M^+ + A \rightarrow M^{2+} + A^-, \quad M^+ + AH \rightarrow M^{2+} + AH^-,$$

$$AH^- + H^+ \rightleftarrows AH_2;$$

3. Radical-chain mechanism of compensating reactions:

$$DH_2 \rightleftarrows DH^- + H^+,$$
$$DH^- + M^{2+} \rightarrow M^+ + DH \text{ – initiation;}$$

$$\left.\begin{array}{l} M^+ + A \rightarrow M^{2+} + A^-, \\ AH + DH^- \rightarrow AH^- + DH, \\ D^- + M^{2+} \rightarrow D + M^+, \end{array}\right\} \text{ – chain continuation;}$$

$$\left.\begin{array}{l} M^+ + DH \rightarrow M^{2+} + DH^-, \\ AH + AH \rightarrow A_2H_2, \text{etc.} \end{array}\right\} \text{ – chain breakage;}$$

4. Mixed mechanism—reaction rate is stipulated by the rates of ion-molecular and ion-radical or radical-chain mechanisms.

Mechanisms of such kinds of redox transformations are frequently rather complex and can be different, depending on the redox pair nature (electron donor–acceptor), catalyst (metal ion or its complex), conditions of process course, solvent, etc. Therefore, it is very important to choose as simple a reaction as possible as a model of redox transformation. The most suitable model appears to be the catalytic disintegration of H_2O_2 for several reasons (simple molecule, the products of its disintegration do not complicate the reaction, the course of reaction can easily be watched by the disappearance of hydrogen peroxide and molecular oxygen evolving, the substrate itself can be either oxidizer or reducer, intermediate reactive particles can be identified, and their concentration in many cases can be measured, etc.).

These and other reasons which have resulted in the choice of the model reaction as

$$2H_2O_2 \xrightarrow[\text{catalyst}]{} 2H_2O + O_2$$

are described in [6]. Besides, it seems to be the only and the simplest reaction that can take place depending on conditions in accordance with all the basic mechanisms of redox transformations and it has been most thoroughly investigated all, qualitatively as well as quantitatively. It is expected that, with transition from this reaction to other redox transformations, the basic regularities will be retained. Therefore, this redox process model will be widely used; it is also important for understanding many of the processes taking place in living organisms or those observed in nature around us.

Metal ions of variable valency are efficient catalysts of redox transformations. In the case when metal ions provide stationary catalytic regimes, interactions are inevitably carried out in the system which determines stationary concentrations of the metal ion in various oxidation states:

$$\begin{array}{l} 1.\, M^{2+} + DH^- \rightarrow M^+ + DH, \\ 2.\, M^+ + A \rightarrow M^{2+} + A^-, \\ \hline DH^- + A \rightarrow DH + A^-. \end{array}$$

Before giving the summary of metal ions influence, as catalysts, on redox processes, it is important to compare proton catalysis with metal ions catalysis, and to reveal the differences. In the line $H^+ - M^{z+} - ML_n{}^{z+}$ (n, number of ligands L), the proton is an elementary and effective catalyst of many reactions. Its efficiency is the result of an absence of electrons and a very small proton radius, only equal to about 10^{-5} nm. The radius of other cations is of the order of about 10^{-1} nm, i.e., about four times the proton order magnitude. Due to its very small radius, the proton exerts a high polarization action on adjacent molecules and ions. For the same reason, the transition from one particle to another is especially simple. Thus, there are no appreciable changes in the arrangement of electron environments of a particle and essential steric difficulties do not arise. Therefore, its catalytic action is not selective or specific and it catalyzes many reactions due to polarization of certain bonds of the transformed molecules.

Peculiarities of metal ions and their differences from the proton consist in their ability to do the following:

1. To coordinate with themselves various particles (substrates, ligands), forming reaction complexes with different stability.

2. To be in various oxidation states and to have a varying coordination number and the ability to change it.
3. To bring together the reacting partners resulting in their favorable arrangement, playing thus a role not only as certain matrix but also as bridge that can provide the electron transfer from one reacting component to another.
4. To act in common with the reaction partners in removing the symmetry interdiction observed in the case of noncatalytic reactions.

These differences between acid catalysis and metal ions (or complexes) catalysis can also be explained by using quantum–chemical notions [7,8]. The proton (in the case of Bronsted's acids) delivers the low-level vacant 1s-orbital in catalytic systems, at the expense of which structural-flexible electron-deficiency system reagent–catalyst (proton) appears, thus facilitating the formation of bridge transition structures. In the case when the base is the catalyst, the latter delivers, in contrast, high-level undivided pairs of electrons, forming, with the reagent, the electron-excessive system, thus facilitating bond dissociation. The peculiarity of metal–complex catalysis is the fact that the catalyst contains both low-level vacant orbitals (p-orbitals, for example), and high-level occupied orbitals (d-orbitals, for example), which can interact with occupied and vacant orbitals of reagents, respectively, thus weakening or completely breaking off certain bonds in the reagent. Such interaction, on one hand, determines structural flexibility of metal–complex systems, and on the other represents an opportunity of the broadest properties modification of the catalyst itself by the replacement of one ligand with another.

In most cases, the redox processes in metal–complex catalysis proceed in a very complex way and experimental identification of high-reactive particles thus formed (such as free radicals, metal compounds in unusual oxidationstates, intermediate compounds, etc.) is rather difficult and frequently simply impossible. The quantum–chemical approach in such cases, at least with respect to the calculation of one or another elementary act initiation, can give valuable information.

From the numerous quantum–chemical treatments available, regarding the charge transfer on reagent coordination to catalyst, and its application to homogeneous catalysis with metal complexes at different stages of such research, we shall mention a few [6,9–12]. It was suggested in these works that, on catalyst–reagent bond formation, due to the charge transfer, common molecular orbitals with redistributed electron cloud will be formed in the system (joint Fermi level F, ε_F), thus facilitating the initiation of the catalyzed reaction. The charge transfer is being realized due to the alignment of Fermi levels of initial catalyst and reagent with general Fermi level formation in this system and is completely determined by Fermi levels difference of system ingredients. The value of charge transfer Δq is connected with the so-called catalytic capacity C_K (i.e., the opportunity of accepting electrons without essential change of energy) by the following expression:

$$C_K = \frac{\partial q_K}{\partial \varepsilon_{FK}}.$$

Therefore, it is possible to calculate energetic levels of MO and ψ-function of molecular complex MO, those of reagent, catalyst–reagent coordination compound, active particles (free radicals, intermediate compounds), direction and size of charge transferred from the catalyst to the reagent and vice versa, possible redistribution of electron cloud in the reagent, resulting in the end in its transformation. The MO LCAO method in its half-empirical variant (Malliken–Vosburg–Gelmholtz method with charges self-coordination on atoms) was used for calculations.

2.2 Intermediate Compounds, O_2 Molecule Activation, and Free Radicals in the Catalysis of Oxidation Processes

2.2.1 Nature of Forming Intermediate Compounds

An intermediate compound is a substance which exists during a certain time ranging from several oscillatory periods to several seconds or minutes and which reacts selectively with certain substances. An intermediate compound can be more or less thermodynamically stable in comparison with the initial substances. Various cases of potential energy (V) changes in instances of formation of intermediate compounds of different stability are given in Fig. 2.1.

In case (2) with the appropriate choice of conditions (for example, with decreased temperature) the intermediate compound can be extracted and identified. In case (3), in contrast, the intermediate compound is characterized by the minimum on the potential curve. It is less stable from a thermodynamic point of view than the initial substances and, kinetically, is more inclined to be disintegrated. The higher the minimum position on the potential curve, the less the stationary concentration and lifetime of the intermediate compound. In this case the intermediate compound is more difficult to isolate (extract). It can be determined in some cases only in

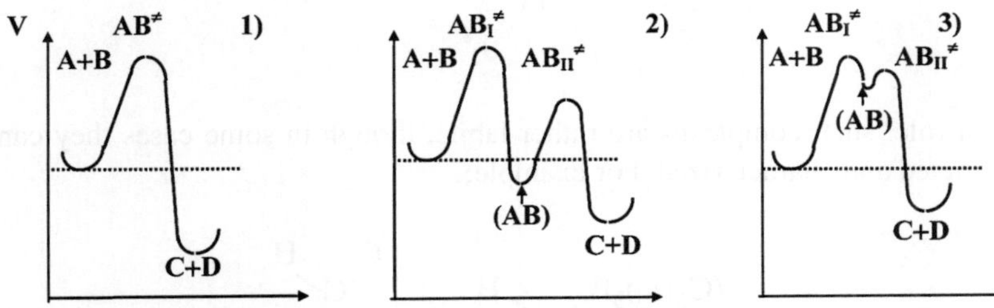

Fig. 2.1 Potential energy change during intermediate compound (AB) formation from A and B and final products C and D via the first and the second transition states ($AB_I^{\neq}$ and $AB_{II}^{\neq}$) for the following cases (1)—intermediate compound will not be formed; (2)—formed intermediate compound is more stable than the initial components; (3)—intermediate compound is less stable than the initial components

solution. It can exist in solution in sufficient quantities to be discovered (for example spectrophotometrically), when the minimum on the curve is not situated too high.

The nature of intermediate compound catalyst–substrate can be different. It can be complex with the complete or partial charge transfer (CCT or PCT); binding via the hydrogen bonds; "onium" complex ($R{-}Cl + AlCl_3 \leftrightarrow R{-}Cl\ldots AlCl_3$). Rather frequently, it is the complex with CCT. Thus, the formation of both π- and σ-complexes is possible:

1. In the case of π-complex formation, the following occur: (a) occupied π-orbitals of substrate (ligand) gain donor properties; (b) empty p- and d-orbitals of the metal get acceptor properties:

$$(M \xleftarrow[\cdots\cdots\cdots\cdots]{\pi\text{-electron}} SH_2).$$

   Figure 2.2 shows both (a) and (b) cases. Such mutual bonding caused by interaction of occupied and free orbitals of substrate and metal (catalyst) results in a number of consequences such as weakening of C–C bonds and increase of intermediate compounds stability with the reduction of the unsaturated substrate's π-electron density and increase of d-electron density of the metal.
2. In the case of σ-complex formation between σ-containing orbitals of substrate and d-orbitals of the metal ion, weakening of the A–B bond takes place with its further breakage (Fig. 2.3).

   In a similar way, not only the breakage of such a bond as H–H or haloid–haloid and haloid–H takes place, but also the breakage of C–haloid, C–H, and even C–C bonds proceeds easily enough.
3. Unlike the most Bronsted or Lewis acids, the ions of transition metals or their complexes can simultaneously join two identical or different substrate molecules by means of σ- or π-bonding to the same center. Thus, a very favorable configuration for inner-molecular transfer appears:

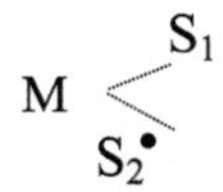

   As a rule, such complexes are rather labile, though in some cases they can be extracted and characterized. For example:

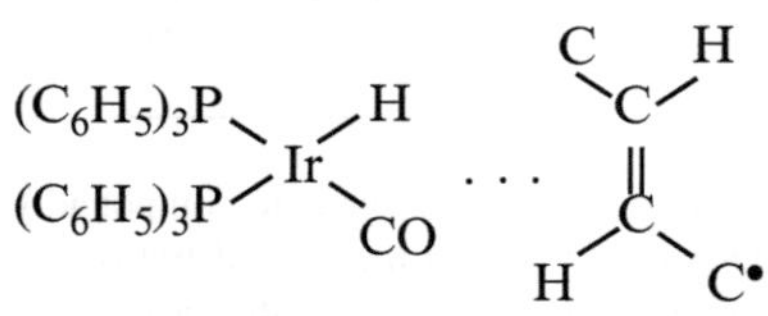

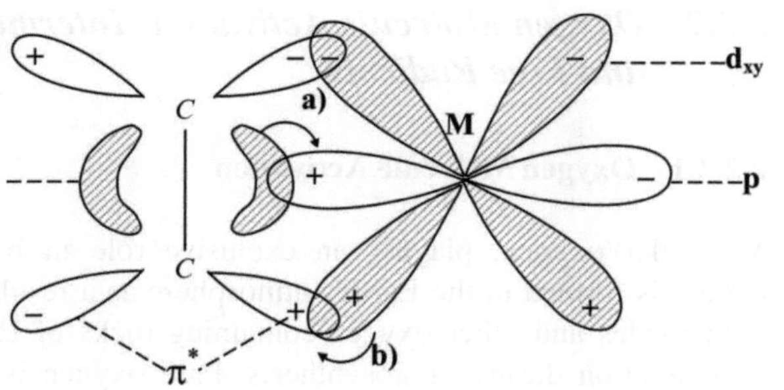

Fig. 2.2 π-Complex formation

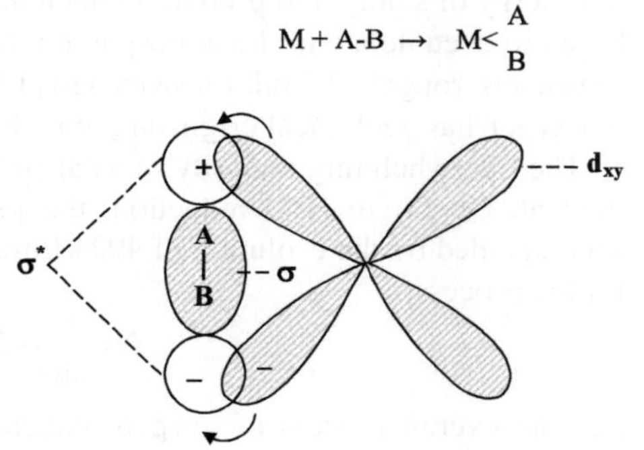

Fig. 2.3 σ-Complex formation

4. π-Complexes of transition metals often turn into σ-complexes easily enough. For example, if in Fig. 2.2 the p-orbital of metal is filled with the electrons passing from the ligand, the substrate π-electrons cannot pass to the metal and the π-complex will turn into a σ-complex.
5. The peculiarity of intermediate compounds formation with metal ion (complex) is also the opportunity of not only coordination number but also the oxidation degree of the latter changing. Thus, transition metal ions frequently act as catalysts in redox reactions. The intermediate compounds of SH_2 with ML which are rapidly formed or disintegrated, should include the labile metal ions. Another distinctive feature of intermediate compounds formation is the need for free coordination places present in the initial metal complex to undergo further binding with the substrate. For Cr(II, III), Mn(II, III), and Fe(II, III) ions the coordination number is most often equal to 6; for Co(II), Ni(II), and Cu(II), 4 or 6; and for Cu(I), 2 or 4. The number of free coordination places depends both on the strength of L with M (from K) interaction, and on the ratio between ligand and metal ion in the solution.

Intermediate compound formation is a fast reaction, although its mechanism, as a rule, is unknown, and its disintegration is a slower reaction. The intermediate compounds can be identified by various methods: spectrophotometry, eletron parametric resonance, kinetic data, etc.

2.2.2 *Oxygen Molecule Activation: Intermediate Species, and Free Radicals*

2.2.2.1 Oxygen Molecule Activation

Molecular oxygen, playing an exclusive role in both animate and inanimate nature, is formed in the Earth's atmosphere as a result of solar energy interaction with oxides and other oxygen-containing rocks of Earth crust and solar energy accumulation during photosynthesis. Free oxygen is too chemically reactive to appear on Earth without the photosynthetic action of living organisms, which use the energy of sunlight to produce elemental oxygen from water. Elemental O_2 only began to accumulate in the atmosphere after the evolutionary appearance of these organisms, roughly 2.5 billion years ago [13]. Thus, in the modern atmosphere 70% of oxygen has geological origin and only 30% photosynthetic.

The overwhelming majority of vital processes takes place using oxygen as an oxidant. On O_2 to H_2O reduction, the joining of four electrons to O_2 occurs, accompanied by the evolution of 492 kJ/mol of free energy ($\Delta G = -492\,\text{kJ/mol}$) for the process:

$$O_2 \xrightarrow[4H^+]{4e^-} 2H_2O,$$

i.e., the overall process of oxygen reducing to water is thermodynamically very favorable. However, the first stage of this process, one-electron O_2 reduction to H_2O, is thermodynamically disadvantageous ($\Delta G = +16.3\,\text{kJ/mol}$) and consequently, under normal conditions, oxygen is inert in relation to many organic compounds [14]. The evolution of huge amounts of energy with O_2 reducing to H_2O in one act is possible only in the case of x complex fermentative reactions. For example, O_2 interaction with hydrocarbons (RH),

$$RH + O_2 \rightarrow R^\bullet + HO_2^\bullet$$

is a strongly endothermic reaction ($\Delta H_{HO2} - \Delta H_{RH} = 200\text{–}390 = -190\,\text{kJ/mol}$). Therefore in living organisms oxygen cannot directly oxidize hydrocarbons, aminoacids, steroids, and other compounds.

In the long chain of reactions from O_2 to H_2O,

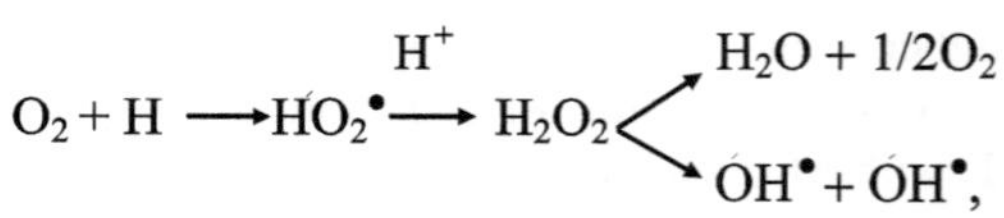

H_2O_2 is the most stable intermediate product.

Transition from O_2 to H_2O_2 (two-electron reduction) is accompanied by the evolution of 151 kJ/mol of energy, while the subsequent transition from H_2O_2 to H_2O evolves 341 kJ/mol of free energy. So H_2O_2 is a stronger two-electron oxidant than O_2. However, it should be noted that there are not many examples of one-stage two-electron reactions of organic substances oxidation by oxygen and hydrogen peroxide. Such processes can proceed only in the presence of catalysts—ions of some metals.

From the practical point of view, processes taking place with self-oxidation by oxygen are undesirable (for example in a number of food industry branches); however, a wider use of molecular oxygen as oxidant is expected in the very near future, as in normal conditions oxygen is rather inert (its apparent activity as an oxidant of some part of a substrate is caused simply by the presence of microimpurities of metal ions activating the molecular oxygen). Thus, the molecular oxygen needs to be activated, i.e., to be transformed from basic triplet state into the bound or exited (singlet) state [14]. H, R, M^{z+}, and the surface of the reaction vessel can act as bond partners.

To understand the problem of oxygen activation better, it is necessary to consider its electron structure.

Electron Structure of the Oxygen Molecule

The scheme of O_2 energy levels in the MO LCAO method is presented below.

The bond between oxygen atoms is realized by one two-electron σ-bond $(3\sigma_g)^2$ and by two three-electron π-bonds, $(1\ \pi_u)^2$ and $(1\ \pi_g{}^*)^1$, each of them consisting of two binding and one loosening electrons. As two of the least stable bound electrons are located on degenerated π*-orbitals, their spins are parallel ($S = 1$) and the O_2 molecule is in the triplet $(2S + 1)$ state $^3\Sigma_g$. This is the lowest state in relation to the energy (according to Hund's rule), i.e., it is the basic state, and is characterized by the most stable electrons arrangement in O_2. In this state the O_2 molecule is paramagnetic and quite often acts as a free radical.

The first exited state $^1\Delta_g$ will be formed with the transition of the second $\pi_g{}^*$-electron onto the orbital of the first $\pi_g{}^*$-electron (i.e., both $\pi_g{}^*$-electrons occupy the same orbital). In this case, obviously, $S = 0$ and $2S + 1 = 1$, i.e., the first excited state is the singlet one Δ_g (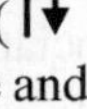). Excitation energy $^3\Sigma_g \rightarrow ^1\Delta_g$ reaches 102 kJ/mol. This state is rather durable and is used in organic synthesis.

The second excited state—when two $\pi_g{}^*$-electrons occupy different $\pi_g{}^*$-orbitals ($\pi_{g\,x}{}^*$ and $\pi_{g\,y}{}^*$)—is also possible, but their spins are antiparallel. Transition to this excited and also singlet state $^1\Sigma_g$ (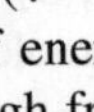) from the basic one $^3\Sigma_g$ is accompanied by the consumption of 159 kJ/mol of energy. Singlet oxygen is the general name of oxygen in the excited state, although frequently only $^1\Delta_g$ is understood by this name. Other excited states of O_2 are also known, but they will not be considered here as they do not take part in normal chemical reactions. In the $^1\Delta_g$ condition the bond in O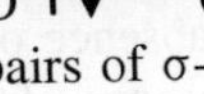O is almost similar to that in ethylene C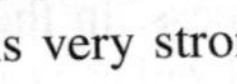C, as it is formed by two pairs of σ- and π-electrons. Electron affinity is very strong at the excited

Table 2.1 Characteristics of O–O bonds

Frequency	Calculated bond order = number of binding/loosening electrons/2	Frequency of valency fluctuations (sM^{-1})	Dissociation energy (kJ/mol)	Bond length (Å)	Electron affinity (EV)
O_2	2	1,555	489	1.21	0.43
$O_2(^1\Delta_g)$	2				
$O_2^{\bullet-}$	1.5	1,089	396	1.32–1.35 (for instance, in KO_2–1.28)	
O_2^{2-}	1	880 (H_2O_2)	215	1.49 (BaO_2)	

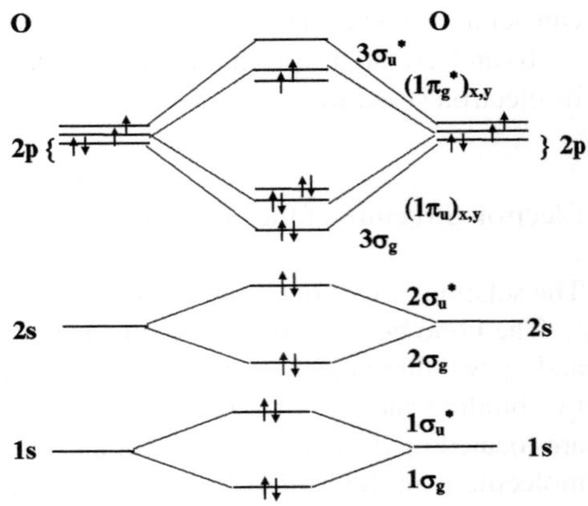

Fig. 2.4 Scheme of O_2 energy levels

state of oxygen $^1\Delta_g$. Almost all the reactions of oxygen in the excited state ($^1\Delta_g$) are related to its interaction with molecules containing π-electrons (aromatic and aliphatic unsaturated molecules). Usually peroxides are formed thus, which, being unstable, lead to the formation of more stable secondary products of reaction.

The electrons joining to the O_2 molecule result in filling π*-orbitals, and bond O–O order, frequency of valent fluctuation, and dissociation energy will be reduced, while the bonds lengths will be increased (Table 2.1).

Besides, as for electron isolation from O_2, the energy of 12 eV is needed, the O_2 particle thus being very unstable and consequently rare. Due to the positive affinity of electrons to O_2 (+0.43 eV), $O_2^{\bullet-}$ will be formed frequently and easily, including in living organisms. For the same reason, O_2 easily joins on to metal ions.

As can be seen from Fig. 2.4, the O_2 molecule in its basic state is bi radical (O_2 ↑↑) and therefore it easily interacts with many organic radicals $R^\bullet$, forming various particles such as: $ROO^\bullet$, ROOH, and $RO^\bullet$; i.e., for hydrocarbons autoxidation oxygen activation occurs without the unprofitable stage of electron transfer ($H + O_2 \rightarrow HO_2$). Naturally, the initiation stage of organic radicals' formation should precede this process. In the absence of such initiation (in the absence of organic radicals) the molecular oxygen can be activated only in the presence of catalysts, mainly transition valency metals and complexes, whose role is reduced either to O_2 activation or $R^{\bullet}$ generation.

2.2.2.2 Oxygen Interaction with Metal Ion and Substrate

In spite of the fact that many oxidation reactions are exothermal, few substances interact at normal temperatures with oxygen without catalysts, generally metal compounds.

With vast amounts of research being conducted into liquid-phase hydrocarbons' oxidation mechanisms in nonpolar media, less work is devoted to the detailed study of oxidation by oxygen in aqueous medium. The mechanism of oxygen activation in water solutions, and the role of metal ions and their concentrations, is not always clear.

The following mechanisms of organic substrate oxidation by molecular oxygen are possible in principle in the presence of metal ions: "cyclic," "activation," and "induced." In the "cyclic" mechanism the SH_2 substrate oxidation occurs on its interaction with metal (metal complex) as catalyst, and the role of O_2 is reduced only to catalyst reoxidation:

$$2M^{2+} + SH_2 \underset{W_1}{\longrightarrow} 2M^{+} + S + 2H^{+}, \tag{2.1}$$

$$2M^{+} + O_2 \xrightarrow[W_2]{2H^{+}} 2M^{2+} + 2H_2O_2, \tag{2.2}$$

where W_1 and W_2 are reaction rates (2.1) and (2.2).

Whether the interaction of O_2 with M^+ or with $M^{2+}SH_2$ takes place, it is apparently dependent on the reaction conditions. This mechanism usually takes place in the case of metal ions with not very high redox potential and easily oxidizing substrates. The higher (M^{2+}/M^{+}), the more stable the M^+ state, and thus, the higher the oxidation rate of substrate according to the first reaction (2.1) and the lower the rate of M^{2+} regeneration on O_2 interaction with M^+ according to the second reaction (2.2). In this case (2.2) is the limiting stage, when the reaction rate does not depend on substrate concentration and is proportional to O_2 concentration.In contrast, on low φ values the reaction rate does not depend on O_2 concentration and is defined by SH_2 and M^{2+} concentrations.

An example of a cyclic process is the ascorbic acid oxidation by oxygen in the presence of copper chloride and the compounds of Cu(II) and Fe(III). Schematically, the electrons transfer from π-orbitals of ascorbic acid oxygen via d-orbitals of metal on vacant π^*-loosening oxygen orbitals is shown in Fig. 2.5.

The presence of sufficiently stable oxidation states of metal $M^{(z-1)+}$, such as the cases of copper [Cu(II)–Cu(I)] and iron [Fe(III)–Fe(II)], facilitates the electron transfer from M^{z+} ion to O_2. Therefore, it becomes clear that ions such as Zn(II), Mn(II), Co(II), and Ni(II), which have no lower stable oxidation states, are inactive as catalysts in this reaction. This "cyclic" mechanism of oxidation prevails when the

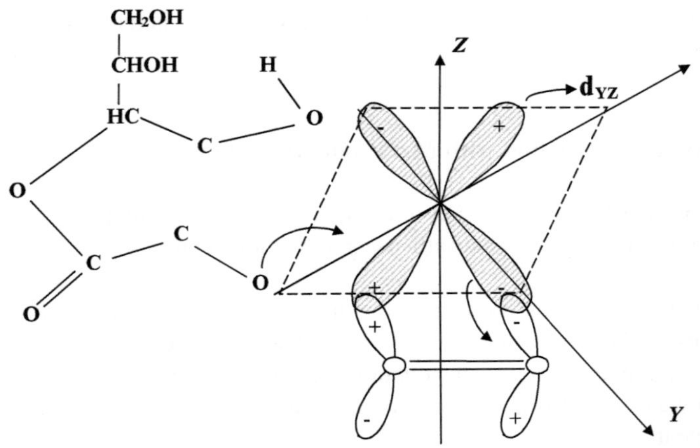

Fig. 2.5 Schematic representation of electrons transfer stage on the ascorbic acid oxidation with the molecular oxygen in the presence of Cu(II) and Fe(III) compounds

substrate possesses strong reducing properties, thus easily coordinating as a ligand with metal ion or complex and transferring the electron to it.

Cyclic two-electron catalytic processes of substrate oxidation and reduction of O_2 up to H_2O_2 by joint action of electron pair of metals, for example "light-blue oxidases," in which the active center consists of an even number of metal ions (for example, in lactase), are also possible:

$$Cu(II)_2 + \text{diphenol} \rightarrow Cu(I)_2 + \text{quinone} + 2H^+,$$

$$Cu(I)_2 + O_2 \rightarrow Cu(II)_2 + H_2O_2.$$

Such a process is also possible with one-stage two-electron oxidation of substrate (alcohols into aldehydes), for example, in galactosidase, containing one Cu^+ ion in its active center:

$$Cu^+ + O_2 \rightarrow Cu^{3+} + H_2O_2,$$

$$Cu^{3+} + RCH_2OH \rightarrow Cu^+ + RCHO + 2H^+.$$

An example of "cyclic" mechanism of substrate oxidation with two-electron transfer is the Walker process of acetylene production from ethylene:

$$(C_2H_4 + 1/2O_2 \xrightarrow{PdCl_2} CH_3CHO)$$

The difference consists only of the stage (2.3), i.e., in the fact that Pd reoxidation is easily done by the reduction of Cu^{2+} ion, and only then is Cu^+ reoxidized by molecular oxygen:

$$C_2H_4 + PdCl_2 + H_2O \rightarrow CH_3CHO + 2HCl + Pd.$$
$$Pd + 2CuCl_2 \rightarrow PdCl_2 + 2CuCl. \quad (2.3)$$
$$2CuCl + 2HCl + \frac{1}{2}O_2 \rightarrow 2CuCl_2 + H_2O.$$

Interaction of ethylene with palladium ion goes through the formation of monomer $[C_2H_4PdCl_2(H_2O)]$ or dimer $[C_2H_4PdCl_2]$ π-complexes (depending on the conditions).

Alcohol's oxidation into appropriate aldehydes runs in general in a similar way [15]:

$$RCH_2OH + PdCl_2 \rightarrow RCHO + Pd + 2HCl$$

with subsequent Pd reoxidation by $CuCl_2$, and then with CuCl reoxidation by oxygen. Actually this reaction proceeds, apparently, in a more complicated way.

The same "cyclic" mechanism may be responsible for the reactions of oxidation states, taking place on account of hydrogen ion detachment from two substrate molecules:

$$2RH + M^{z+} \rightarrow R - R + M^{(z-2)+} + 2H^+,$$
$$M^{(z-2)+} + 2H^+ + 1/2O_2 \rightarrow M^{z+} + H_2O,$$

The overall reaction: $2RH + 1/2O_2 \rightarrow R - R + H_2O.$

An example of such a reaction is the interaction of aromatic compounds with Pd(II) salts [15]:

$$2C_6H_6 + PdX_2 \rightarrow C_6H_5 - C_6H_5 + Pd + 2HX.$$

In the presence of oxygen the reaction becomes catalytic, since Pd is reoxidized by oxygen to Pd^{2+}. The reaction of phenols combination proceeds similarly.

In the "activation" process, the role of metal ion consists in O_2 activation by its coordination with metal. Thus the O_2 orbitals' completion and composition changes, corresponding to the electron excitation of O_2. Such excitation occurs only at the expense of coordination interaction, without using outside energy, that prepares the oxygen molecule for subsequent interaction with substrate:

$$M^+ + O_2 \rightarrow (M^+O_2) \rightarrow \text{(bound activated oxygen)}(M^+O_2) + SH_2 \rightarrow S + H_2O_2.$$

(For example, in colorless copper-containing oxidases copper ion exists in the unique number and form of Cu^{2+}.) Thus, the triple complex metal ion–O_2–substrate will be formed, and the essence of catalysis by metal ion consists, apparently, in the formation of a bridge inside this complex. Using this bridge, two electrons and two hydrogen atoms are "synchronously" transferred from SH_2 substrate to O_2 oxidizer, thus passing the thermodynamically unadvantageous stage of one-electron transfer. Therefore, the bound oxygen in the triple complex is in the intermediate condition between the anion radical $O_2^{-\bullet}$ and the singlet oxygen $1\Delta_g$ or $1\Sigma_g$, and the metal ion—in the reduced form (for example, phenylidenamine oxidation in the presence of copper ions).

The "induction" processes appear due to external free radicals inducing, for example, in ascorbic acid simultaneous oxidation by oxygen and hydrogen peroxide. Hydrogen peroxide takes part in $OH^{\bullet}$ radicals induction, and then the chain is continued:

$$OH^{\bullet} + SH_2 \rightarrow R_I^{\bullet}H$$

Radical of OH joining to SH_2

$$R_I^{\bullet}H + O_2 \rightarrow R_IHO_2$$

$$R_IHO_2 + CuSH^{+} \rightarrow Cu^{2+} + OH^{\bullet} + 2S$$

Generation of free radicals in the solution is frequently assisted by the transition metal salts, which catalyze the decomposition of hydroperoxides formed in the system, yielding free radicals.

In the absence of hydroperoxides, the role of metal ion can be as follows: electrons transfer from substrate to metal:

$$RH + M^{z+} \xrightarrow{e} RH^{+} + M^{(z-1)+}$$

$$RH^{+} \rightarrow R^{\bullet} + H^{+}$$

hydrogen detachment from substrate (in the presence of O_2):

$$M^{z+} + O_2 \rightarrow M^{(z+1)+}O_2^{\bullet-}$$

$$M^{(z+1)+}O_2^{\bullet-} \xrightarrow{RH} M^{z+} - O - O - H + R^{\bullet}$$

and also generation of free radicals by the direct transfer of electrons from metal to substrate:

$$M^{z+}X \rightarrow M^{(z-1)+} + X^{\bullet}$$

$$X^{\bullet} + RH \rightarrow HX + R^{\bullet}$$

2.2.2.3 Ligand Properties of Oxygen

As the ligand protonation constant value is a relative measure of its capacity to form the coordination bond, the absence of O_2 protonized forms testifies to the extremely low σ-donor capacity of O_2, i.e., oxygen cannot be the donor of undivided pairs of electrons, which could take part in the formation of coordination compound. However, due to the low-level vacant π*-loosening orbitals, it is very easy to accept the electrons from the completed t_{2g}-orbitals of metal ion when the ionization potential of the latter is less than O_2 affinity to electron. Thus, the molecular oxygen is poor σ-donor and good π-acceptor.

The strength of $M{-}O_2$ bonds depends on the metal ion capacity for transferring the electrons from filled molecular orbitals of L_nM complex (or metal ion) onto the loosening orbitals of oxygen molecules.

Any factor increasing the electron density on the metal ion (for example, its preliminary interaction with ligands possessing strongly expressed donor properties) also increases the bond strength $M{-}O_2$ and weakens the O–O bond. Thus, for example, the electronegativity reducing of σ-donor ligand X (Cl > Br > I) in the complex IrX $(CO)(PPh_3)_2(O_2)$ increases the strength of the $Ir{-}O_2$ bond. Therefore, the increase of O–O bond length and the reduction of convertibility of oxygen with iridium bouding takes place.

The scheme of inner complex electron transfer on oxygen coordination with metal ions can be presented as follows:

$$M^+ + O_2 \rightleftarrows M^+O_2 \rightleftarrows MO_2{}^+ \rightleftarrows M^{2+}O_2^{\bullet -} \rightleftarrows M^{2+} + O_2^{\bullet -}$$

Necessary mutual orientation of t_{2g}-orbitals of metal ion and oxygen π^*-orbitals is provided in M^+O_2, when electron transfer is possible. $MO_2{}^+$ is a complex with PCT while $M^{2+}O_2^{\bullet -}$ is an anion radical [$O_2^{\bullet -}$ is a typical ligand: $pK_d(HO_2) = 4.88$].

If M^+ is a one-electron donor, oxygen is monodentate in the metal–O_2 complex (electron density transfer from M is relatively little) and bond $M{-}O{-}O^-$ makes an angle of 120–150° (distance in $O{-}O^-$ is 1.3 A, $\nu \approx 1{,}100$–$1{,}150\,\mathrm{sm}^{-1}$ and higher), i.e., the state of the oxygen atom corresponds to sp^2-hybrid orbitals, and ethyl-like reactive forms such as $O_2(^1\Delta q)$, $O_2^{\bullet -}$ and O_2^{2-} with sp^2-hybridization of orbitals can play certain part. Similar bond type ($M{-}O{-}O^-$) can be found in hemoglobin, myoglobin, and in some other compounds. This form of M and O_2 bond is called Poling's structure. In such a monodentate model the π-bond in O_2 remains localized (thus, the electron transfer is realized from the d-orbital of the metal into the π^*-orbital of oxygen). Metal ions should have a good electron-donor ability, and therefore metals in low oxidation degree provide sufficient electron transfer and complex formation with O_2.

It becomes clear from the aforesaid that the catalysis of oxidation by molecular oxygen is facilitated in the case of processes, resulting in M^+ formation proceeding in the system (for example, when the substrate itself has strong reduction properties). Complexes with porphyrin, phthalocyanine, and other ligands with plain four-site coordination are single-nuclear complexes of this type. Coordinated oxygen is close to $O_2^{\bullet -}$. Linear bond M–O–O has not been found up to now.

If M^+ is a two-electron donor (electron density transfer from the metal is great), the oxygen usually acts as a bidentate ligand and both oxygen atoms in O_2 are located equivalently around the metal. Thus, delocalization of oxygen's π-electrons takes place on metal with σ-bond formation (Fig. 2.6).

Such triangular coordination is characteristic for transition metals with low oxidation degree (Pt^0, Pd^0, Ni^0, Ir^{+1}, Rh^{+1}, etc.), for example, $Pt(PPh_3)_2O_2$, $Pd(PPh_3)_2O_2$, $IrCl(PPh_3)_2O_2$, etc. In $Pt(PPh_3)_2O_2$, the condition of oxygen is close to $O_2{}^{2-}$.

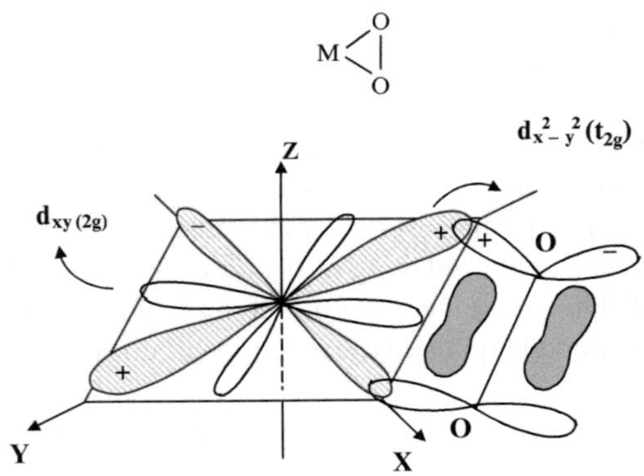

Fig. 2.6 Electron distribution in 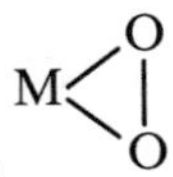 complex: filled orbitals are shown as *shaded*, unfilled ones are *unshaded*

In the complex 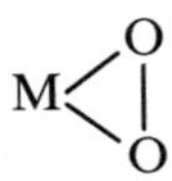 named the Griffith's structure, bound oxygen can be formally considered as existing in the form of $O_2{}^{2-}$, i.e., as peroxide anion. Bond length O–O in such compounds reaches 1.4–1.6 A ($\nu \approx 750–950\,cm^{-1}$).

On two ions, M^+ participation in coordination μ-peroxocomplex can be formed M-O-O-M,, in which oxygen is in the condition close to O_2^- or O_2^{2-}. Quantum–chemical calculations using Huckel's method for molecular orbitals were performed regarding the fragment $M–O_2–M$ (M from Ti to Ni) have given certain information on the character of O_2 binding (for example, filling of bond $M–O_2$ is maximal in the middle of the third-complex's line). The calculation involved has shown that not only completely filled d-orbitals can take part in oxygen activation but also partially populated ones, that electron density value can serve the measure of activation, that metal d-orbitals can interact with molecular orbitals situated higher than 1 π_g-orbital. Due to O_2 coordination to ion or metal complex of a, b, c type

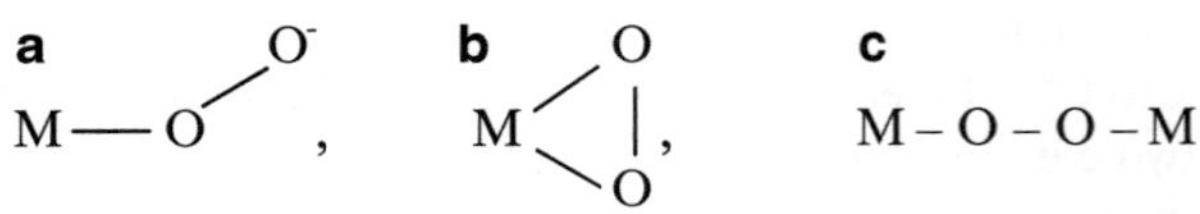

it can be in condition from $O_2{}^-$ up to $O_2{}^{2-}$, i.e., in more reactionary condition in relation to many substrates than before the coordination.

Only for those metal ions which can interact with O_2 is the following inequality correct: $\varphi(M^{3+}/M^{2+})$, $\varphi(M^{2+}/M^+)$, $\varphi(M^{3+}/M^+) < \varphi(O_2/H_2O_2) = 0.69\,V$. Therefore, iron ions, for example, cannot interact with $O_2[\varphi(Fe^{3+}/Fe^{2+}) = 0.771\,V]$, and only iron complexes with redox potential value less than 0.69 V

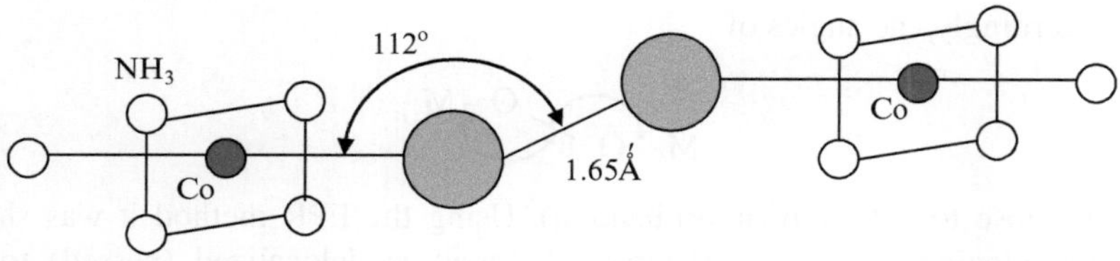

Fig. 2.7 Cobalt ammonia μ-peroxocomplex

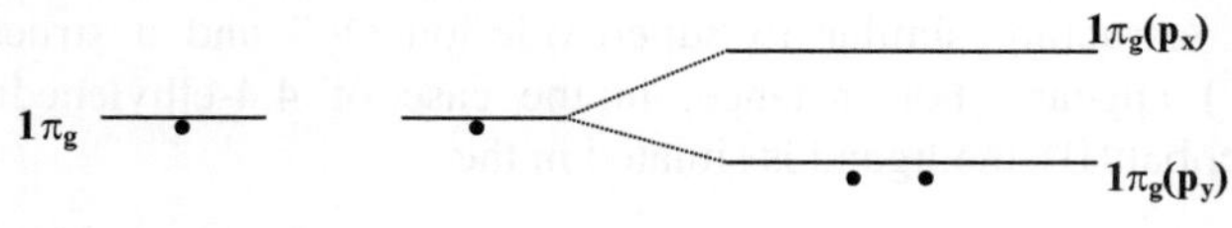

Fig. 2.8 Removal of two 1 π_g MO-orbitals of O_2 degeneration

interact with O_2 and catalyze the reactions of oxidation by oxygen. It was shown that with redox potential reducing $\varphi(M^{2+}/M^{+})$ the rate of self-oxidation of M^+ increases, and the epikinetic order observed changes from 2 to 1, which is connected with the disintegration of intermediately forming peroxocomplex $MOOM^{2+}$.

Oxygen as a ligand will form complex compounds with many metals belonging to the VIII collateral (secondary) group (Ru, Os, Rh, Ir), especially in the presence of other ligands such as triphenylphosphine or similar. But the most numerous are its coordination compounds both with one-nuclear and more often with two-nuclear cobalt(II) complexes, for example, with the following configuration (Fig. 2.7):

$$2[Co(NH_3)_6]^{2+} + O_2 \rightleftarrows [(NH_3)_5CoO_2Co(NH_3)_5]^{4+} + 2NH_3.$$

Two-nuclear complexes will be formed by a two-step mechanism:

$$Co^{II} + O_2 \underset{K_{-1}}{\overset{K_1}{\rightleftarrows}} Co^{II}O_2,$$

$$Co^{II}O_2 + Co^{II} \underset{K_{-2}}{\overset{K_1}{\leftrightarrows}} Co^{II}O_2Co^{II}.$$

Under the influence of the metal ion, the degeneration of two 1 π_g MO-orbitals of O_2, on which there were two unpaired electrons, is removed. Thus, the energy of these orbitals' splitting is much higher than the energy of two electrons pairing, due to the fact that they occupy a lower 1 $\pi_g(p_y)$–orbital (Fig. 2.8).

Thus, under the influence of the metal, the electron configuration of O_2 will be similar to that existing in ethylene:

and, accordingly, the angles of

become close to 120° (sp^2-hybridization). Using the EPR method it was shown that the unpaired $e_g(d_{x^2-y^2})$-electron of cobalt is delocalized (passed) to the oxygen molecule (into the loosening 1 π_g*-orbital), thus weakening the O–O bond and lengthening it. Similar reduction of the oxygen molecule practically turns it into astructure similar to superoxide-ion O_2^- and a structure similar to $Co^{III}(O_2^-)$ appears. For instance, in the case of 4,4-ethylenedinitrilo-di-2-pentanonatocobalt(II), the ligand is situated in the

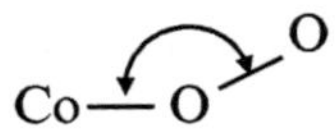

plane, the angle is 1.26 Å, and the bond length of O–O reaches 1.26 Å (close to the bond length in O_2).

The redistribution of electron density on the formation of a four-centered molecular orbital in $[(NH_3)_5Co–O–O–Co(NH_3)_5]^{4+}$ or in similar complexes is shown in Fig. 2.9.

With two electrons transferring from two cobalt ions to an oxygen molecule, a structure will be formed similar to peroxide-ion O_2^{2-}, and the distance O–O increases even more than in the peroxide ion (1.65 Å instead of 1.29 Å in O_2).

In similar complexes of transition metals with oxygen, π-binding is the main factor. The higher the electron density on the metal, the more favorable are the conditions for the formation of the complex. The increase of metal electron density can be achieved by suitable choice of ligand. Thus, it was shown that if phosphines with various donor capacity are taken as ligands $[(C_6H_4Cl)_3P < (C_6H_5)_3P < (CH_3C_6H_4)_3P) < (CH_3OC_6H_4)_3P)]$, the better the donor, the faster the O_2 binding, and the more stable the oxygen complex.

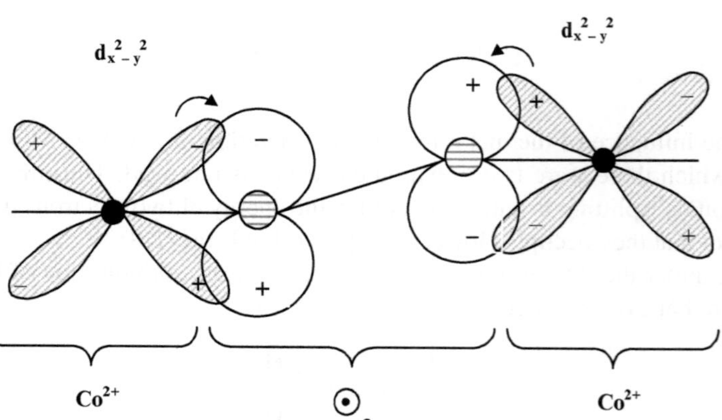

Fig. 2.9 Distribution of the electron density in cobalt ammonia μ-peroxocomplex

Therefore, along with O_2 coordination, its activation occurs, and consequently in such condition it can take part in various oxidation reactions. The majority of these reactions have radical-chain character and thus more often oxidation is not specific, as many other by-products of the reaction will be formed. Specificity of oxidation reaction, apparently, should be increased if both O_2 and substrate are coordinated around one and the same metal ion.

For better understanding of O_2 condition in various natural complex compounds, its convertibility in these compounds, and also possible creation of synthetic oxygen carriers, numerous research projects on model systems, especially using iron compounds, have been carried out.

2.2.2.4 Reactions of Oxygen Reduction Products (OH, O^-, HO_2), Peroxyl- and Oxy Radicals

Consecutive four-electron reduction of O_2 to H_2O proceeds with the formation of certain free radicals and relatively stable molecules (H_2O_2):

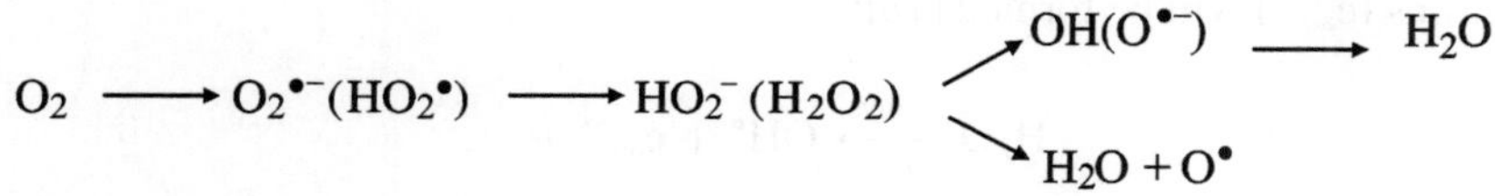

The reactions of H_2O_2 with metal ions, their complexes, and various substrates will be considered later.

$^1O_2$ is a singlet oxygen in $^1\Delta_g$ condition. At usual temperatures, conditions, and an oxygen pressure of 2 atm, its lifetime is 72 min [16]. It can be obtained by electrical discharge in gaseous oxygen, with the help of substances photosensibilizing the oxygen oxidation (illumination by light at the presence of methylene blue, Bengalia rose, other sensitizers): by decomposition of hydrogen peroxide in the presence of NaClO ($H_2O_2 + OCl^- \rightarrow {}^1O_2 + H_2O + Cl$) and by other methods. Especially convenient are photosensibilizing reactions of oxidation. $^1O_2$ obtained by one method or another possesses very strong affinity to electrons. Almost all reactions of O_2, known nowadays are related to its interaction with the system of π-electrons. And the rate of such interaction is higher, as is the electron density of the unsaturated bond. As an example, such a reaction can produce the so-called ox-reaction with an olefin containing allyl hydrogen:

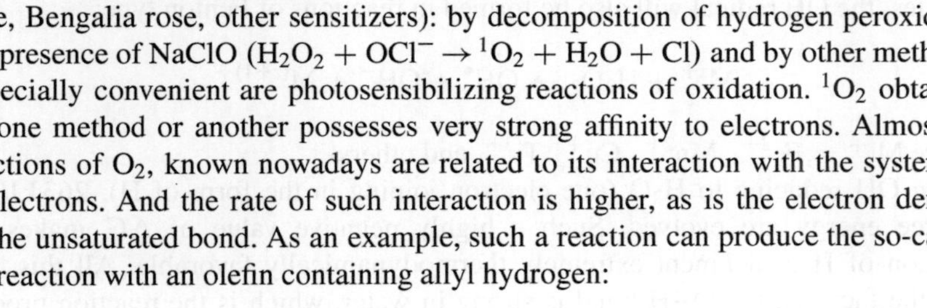

or 1,2-joining to a double bond with high electron density:

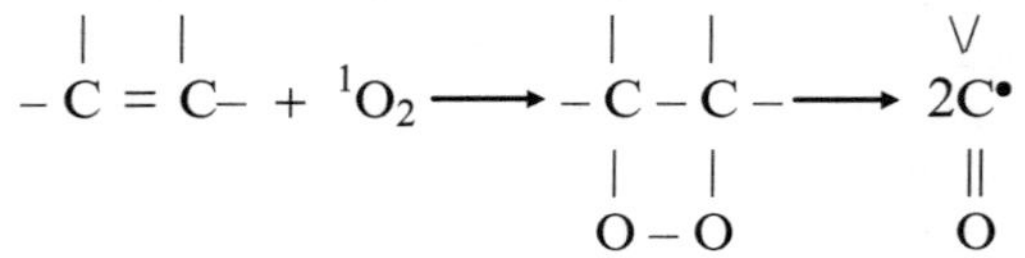

As $^1O_2$ oxidizes the organic substances, it is neutralized in the organism by beta carotene, greasy amines, etc., which transform it into $^3O_2(^3\Sigma_g)$ [17].

Reactions of OH Radicals

As a result of one-electron reduction of H_2O_2 (rather weak oxidizer) during photochemical ($H_2O_2 + h\gamma$) or catalytic ($Fe^{2+} + H_2O_2$) decomposition of hydrogen peroxide, and also as a result of water radiolysis ($H_2O \xrightarrow{\gamma\text{-rays}}$), the OH radical—the strongest oxidizer in aqueous solutions—will be formed.

In the latter case, among other primary particles, OH radicals and hydrated electrons (e_{aq}^-) will be formed [18]:

$$H_2O \xrightarrow{\gamma\text{-rays}} OH^\bullet + e_{aq}^- + \ldots$$

Hydroxyl radicals will also be produced during hydrated electrons' interaction with nitrogen oxide or hydrogen peroxide:

$$e_{aq}^- + N_2O + H_2O \rightarrow OH^\bullet + OH^- + N_2,$$

$$e_{aq}^- + H_2O_2 \rightarrow OH^\bullet + OH^-.$$

Besides, the OH radical will also be formed in reactions of Fenton type:

$$M^{z+} + H_2O_2 \rightarrow OH^\bullet + OH^- + M^{(z+1)+},$$

where $M^{z+} = Fe^{2+}$, Mn^{2+}, Cu^+, Ti^{3+}, and others.

On OH reducing to H_2O (one electron joining in the form of H), 263 kJ/mol of free energy are evolved. Such a highly negative value of ΔG makes the reaction of H detachment extremely thermodynamically favorable. All this, and also the fact that the O–H bond is strong in water, which is the reaction product ($H + OH \rightarrow H_2O$) and reaches 499 kJ/mol, causes the oxidation of almost each organic compound by the OH radical [19]. As ΔG has a highly negative value, any thermodynamically possible way of organic substance oxidation can take place, resulting in the formation of a variety of oxidation products (oxidant is not specific), which is in many cases undesirable.

Basically, the high potential for reaction of OH is caused by small values of the activation energy of hydroxyl radical interaction with organic compounds.

Thus, for example, on OH interaction with benzene and its derivatives, it reaches only 0.34–2.35 kJ/mol. For many other reactions with OH radical participation, the activation energy is within the limits of 4–21 kJ/mol. OH radicals interaction with subsequent joining of the OH-group to organic compounds is called the reaction of hydroxylation [20]. The first stage of this reaction with aromatic compounds is, apparently, the formation of intermediate oxycyclohexadienyl radicals:

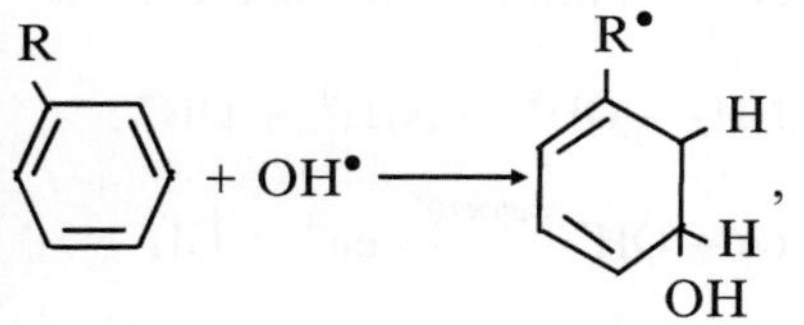

but not hydrogen detachment with phenol radical formation as was thought earlier.

Having a high redox potential, the OH radical is the most powerful due to its oxidizing capacity, interacting with the majority of organic partners with rate constants close to those of diffusion. As a rule, OH radical oxidizes organic molecules by the mechanism of H atom detachment, with the formation of water molecules [21]:

$$-\overset{|}{\underset{|}{C}}-H + OH^{\bullet} \longrightarrow -\overset{|}{\underset{|}{C}}^{\bullet} + H_2O$$

or, in general: $RH + OH^{\bullet} \rightarrow R^{\bullet} + H_2O$.

The oxidation of alcohols can serve as a typical example of such reactions. Usually, the most poorly connected hydrogen atom is detached with high rate constants, i.e., that situated in the α -position relative to the alcohol group, for example:

$$C_2H_5OH + OH^{\bullet} \xrightarrow{k=1.2\times10^9\,M^{-1}\,s^{-1}} CH_3CH{-}OH{+}H_2O.$$

On tertiary butyl alcohol oxidation, only β-alcohol radicals can be formed:

$$CH_3-\overset{\displaystyle CH_3 \atop |}{\underset{\displaystyle | \atop CH_3}{C}}-OH + OH^{\bullet} \xrightarrow{\kappa = 2{,}5\cdot 10^8\ M^{-1}c^{-1}} (CH_3)_2CH_2C-OH + H_2O.$$

The same values have rate constants of H-atoms detachment from saturated hydrocarbons ($\approx 10^8\ M^{-1}\ s^{-1}$).

Still, the constants of OH interaction with carboxyl groups are lower, since in this case oxidation occurs by the mechanism of electrons, instead of H-atom acceptance.

Such reactions are less thermodynamically favorable as soon as hydroxide ions are formed instead of water:

$$CH_3COO^- + OH^\bullet \xrightarrow{\approx 5\times 10^7} CH_3COO^\bullet + OH^-.$$

The oxidation of amines proceeds similarly (in accordance with the mechanism of electrons' acceptance) with rather low rate constants ($M^{-1}\ s^{-1}$):

$$NH_3 + OH^\bullet \rightarrow NH_3{}^+ + OH^-,$$

$$en + OH^\bullet \xrightarrow{\approx 5\times 10^7} en^+ + OH^-.$$

OH radicals' adhesion alongside the double bond occurs with high rate constants [21]:

$$CH_2{=}CH-CH_2-OH + OH \xrightarrow{1{,}2\cdot 10^9} CH_3-CH(OH)-CH_2-OH,$$

$$PhOH + OH \xrightarrow{\approx 5\cdot 10^9}$$ 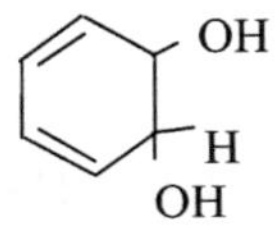

In the systems containing hydrogen peroxide and transition metal ion, the following reactions should be taken into account [19]:

$$H_2O_2 + OH^\bullet \xrightarrow{1.7\times 10^7} HO_2^\bullet + H_2O,$$

$$HO_2^\bullet + OH^\bullet \xrightarrow{7\times 10^8} O_2^{\bullet -} + H_2O.$$

as well as OH reactions with metal ions.

The OH radical interacts with activated metal ions by following the mechanism of H-atom detachment from coordinated water molecule:

$$M^{z+}(H_2O)_n + OH^\bullet \rightarrow M^{z+}(H_2O)_{n-1}OH + H_2O \rightarrow M^{(z+1)+}(H_2O)_{n-1}OH^-.$$

Therefore, such reactions proceed with similar rate constants ($\approx 10^8\,M^{-1}\ s^{-1}$), irrespective of the central ion nature.

The OH radical interacts with complex metal ions by direct transfer of electrons from the central atom, and the rate constants can be essentially higher than in the case of metal ion with coordinated water molecule oxidation, for example:

$$Fe(CN)_6^{4-} + OH^\bullet \xrightarrow{\approx 10^{10}} Fe(CN)_6^{3-} + OH^-.$$

In general, OH radicals appear not only in metal–peroxide systems but also in many redox systems, for example, water radiolysis, photolysis of aqueous solutions of many aromatic compounds, reactions of HClO acid and other halogen-containing oxidizers. The following rule can be formulated: for the reactions in aqueous medium including formation of active intermediates of oxidizing nature, there is an opportunity for hydroxyl radical formation in the secondary reaction of interaction with water (or its dissociated form) [6]:

$$H_2O + OX \rightarrow OH^{\bullet} + H^+ + OX^-,$$

$$OH^- + OX \rightarrow OH^{\bullet} + OX^-,$$

where OX is an intermediate of oxidizing nature.

To date a significant amount of information about the hydroxyl radicals' potential of reaction has been obtained. The most direct way of obtaining data related to the reaction rate constants with $OH^{\bullet}$ participation is pulse radiolysis of water solutions. Reliable data on rate constants can also be obtained by competitive methods with the use of dyes having high reactivity toward OH and high extinction coefficients, facilitating the analytical procedure of the dye concentration's determination.

Methods of OH radicals identification in real redox systems have also been developed well enough. High reactivity of hydroxyl radicals directly entails their small stationary concentration, and also the impossibility of their detection by EPR methods [22]. Therefore, all the methods that can prove OH presence in solution, and can determine their quantity, are indirect, i.e., they are based on kinetic displays.

The main groups of OH detection methods are the following:

1. Methods based on various sorts of radical traps (of maleic acid type), i.e., on OH radical transformation into less reactive radicals with subsequent tracing by EPR methods [19].
2. Methods based on hydroxyl radicals using compounds with double bond, such as allyl alcohol, phenol. The difference between this group of methods and the previous one is that the whole EPR spectrum is not registered, but only changes in kinetics or the main reaction stoichiometry on acceptor introduction into the reaction medium. This group of methods is especially convenient and gives the most distinct results for the systems in which radical-chain processes take place.

These methods are used in cases when it is more convenient to study the kinetics of the acceptor consumption instead of kinetic characteristics of basic reaction. The well-known example is paranitrosodimethylaniline (PNDMA) used as OH radicals acceptor.

3. Methods based on determination of reactivity "spectrum" of an active intermediate particle and its comparison with values known from the literature. These methods are applied rather seldom because of significant labour inputs; however, they give quite reliable results.

A version of these methods is the establishing of oxidation reaction products distribution by their composition.

4. Methods based on the capacity of OH radicals to initiate the vinyl polymerization chain processes are to be mentioned separately. To establish strictly the radical character of the reaction, the analysis of edge groups of polymers is to be done (namely, by IR or NPR spectroscopy methods). To establish qualitatively the radical character of the reaction, the visual way of polymerization monitoring is quite suitable. A well-known example of such a method involves methyl-methacrylate polymerization in the systems generating OH radicals. Using methylmethacrylate for mechanisms diagnostics in water solution is convenient, when the monomer has good solubility in water, whereas the polymer will form a separate phase.

The reactivity of the OH radical in alkaline media is toreceive special mention. At $pH > 12$, acid–basic dissociation of hydroxide radicals proceeds: $OH^{\bullet} \leftrightarrows O^{-} + H^{+}$ ($pK = 11.85$). As pK of the $OH^{\bullet}$ radical is lesser than that of water, $O^{-} + SH \rightarrow OH^{-} + S$ transition is less thermodynamically favorable than the transition $OH^{\bullet} + SH \rightarrow H_2O + S$. Because of this, the lesser activity of the hydroxyl radical ionized form can be explained. However, in metal-containing systems the dissociation of OH practically does not play any role, since usually metal ion hydrolysis essentially limits the working range to lower pH values.

Chemically, $O^{\bullet-}$ can be exposed to the same reactions as the OH radical. However, its negative charge considerably reduces its electrophilic capacity and, therefore, the reactions of one-electron oxidation as well as the reactions of joining will proceed less effectively. The reactions of the hydrogen atom will proceed more often and in this sense it is more selective than $OH^{\bullet}$.

One of major radicals, especially in biological systems, generated in the presence of H_2O_2 is the superoxide radical [18, 20, 21, 23]. Its occurrence is connected to hydrogen peroxide reactions with various oxidizers:

$$H_2O_2 + OH^{\bullet} \rightarrow HO_2^{\bullet} + H_2O,$$

$$H_2O_2 + M^{2+} \rightarrow HO_2^{\bullet} + M^{+} + H^{+}.$$

The $HO_2^{\bullet}$ radical can also be generated directly by hydrogen atom interaction with the oxygen molecule: $H + O_2 \rightarrow H_2O$ with rate constant $k = 2 \times 10^{10} M^{-1} s^{-1}$.

Unlike the hydroxyl radical, the dissociation of superoxide radical proceeds within the working pH limits:

$$HO_2^{\bullet} \rightleftarrows O_2^{\bullet-} + H^{+}, \quad pK = 4.88.$$

The $O_2^{\bullet-}$ radical can be obtained on activated electrons interaction with O_2 [23]: $e_{aq}^{-} + O_2 \rightarrow O_2^{\bullet-}$ with rate constant $2 \times 10^{10} M^{-1} s^{-1}$.

Superoxide can also be generated by the reaction of electron transfer from the CO_2 radical to O_2 [24]:

$$CO_2^{\bullet -} + O_2 \xrightarrow{k=2\times10^9 M^{-1}s^{-1}} O_2^{\bullet -} + CO_2,$$

or by means of α-hydroxy-peroxyl radical transformation:

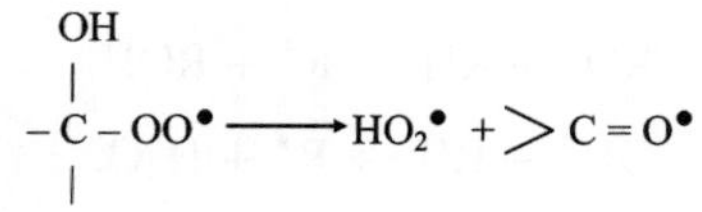

The superoxide anion radical in dissociated form is a relatively weak oxidant and thus can react mainly with metal ions:

$$M^{+} + O_2^{\bullet -} \xrightarrow{(H^{+})} M^{2+} + HO_2^{-}.$$

Usually the rate constant of such reactions is defined by the lability of the coordination sphere of the metal ion. On the other hand, the O_2^{-} radical can easily be exposed to oxidation up to molecular oxygen:

$$M^{2+} + O_2^{-} \rightarrow M^{+} + O_2,$$

$$O_2^{\bullet -} + \text{oxidant} \rightarrow O_2 + \text{product}.$$

Reactions of such type are used as a method to determine quantitatively the presence of superoxide radicals in solution:

$$C(NO_2)_4 + O_2^{\bullet -} \xrightarrow{k=2\times10^9 M^{-1} s^{-1}} C(NO_2)_3^{-} + NO_2 + O_2 \; (\lambda = 350 \text{ nm}, \varepsilon = 15{,}000).$$

The superoxide radical (HO_2) in its protonized form is a strong oxidant. HO_2 radical can detach H atoms from saturated hydrocarbons and alcohols. The rate constant of such a reaction is—four to seven orders less than that of similar ones with OH radical.

The reactions of superoxide radicals can play an essential role in the mechanism of organic substrate oxidation [23], especially when the generation of more reactive particles is for whatever reasons improbable. Such a situation occurs, for example, in pesticides oxidation in natural conditions. The basic channels of organic matter destruction will be thus the reactions of hydrolysis and direct interaction with molecular oxygen:

$$RH + H_2O \rightarrow \text{products}, \tag{2.4}$$

$$RH + O_2 \rightarrow R^{\bullet} + HO_2^{\bullet}. \tag{2.5}$$

In view of the low rate of initiation [reaction (1.5)] and absence of channels of radicals linear destruction, chain development is inevitable:

$$R^\bullet + O_2 \rightarrow RO_2^\bullet,$$
$$RO_2^\bullet + RH \rightarrow R^\bullet + ROOH,$$
$$ROOH \rightarrow RO^\bullet + OH^\bullet,$$
$$RO^\bullet + RH \rightarrow R^\bullet + ROH,$$
$$OH^\bullet + RH \rightarrow R^\bullet + H_2O.$$

Breakage:

$$\left.\begin{array}{l} RO_2^\bullet + RO_2^\bullet \rightarrow \\ HO_2^\bullet + HO_2^\bullet \rightarrow \\ R^\bullet + HO_2^\bullet \rightarrow \end{array}\right\} \text{products.}$$

The superoxide ion $O_2^{\bullet -}$ has an unpaired electron in molecular orbital $(1\,\pi_g{}^*)_y$, and, therefore, the distance between nuclei grows and becomes equal to 1.32–1.35 Å and the bond is weakened. For example, in oxyhemoglobin the charge transfer proceeds from metal ion to O_2 accompanied by O–O bond length increase. The state of oxygen here is close to $O_2^{\bullet -}$. Besides, in living organisms, the formation of $O_2^{\bullet -}$ was proved to proceed in oxidase and xanthynogenase reactions.

Reacting with H^+, $O_2^{\bullet -}$ is exposed to disproportioning:

$$2O_2^{\bullet -} + 2H^+ \rightarrow H_2O_2 + O_2.$$

In general, $O_2^{\bullet -}$ is a stronger oxidizer than HO_2. Therefore, a higher interaction rate constant becomes clear:

$$O_2^{\bullet -} + HO_2^\bullet \xrightarrow{1\times10^8\,M^{-1}\,s^{-1}} O_2 + HO_2{}^- \,(H_2O_2)$$

in comparison with:

$$2HO_2^\bullet \xrightarrow{9\times10^5\,M^{-1}\,s^{-1}} O_2 + H_2O_2.$$

Submarines and spaceships are provided with oxygen, formed in this reaction by addition of water (H^+OH^-) to solid KO_2 ($K^+O_2{}^-$).

As already mentioned, $O_2{}^-(HO_2)$ activity is rather insignificant (in comparison with OH). It can act both as oxidant and reducer and can take part in nucleophylic reactions of replacement ($RX + O_2{}^- \rightarrow RO_2 + X^-$), joining to a double bond, $(R - \underset{|}{C} = \underset{|}{C} - R + 2O_2{}^- \longrightarrow 2RCOO^-)$, reactions of dehydrogenization, $(- CH_2 - + O_2 \longrightarrow -\underset{|}{C}H- + HOO^-)$,

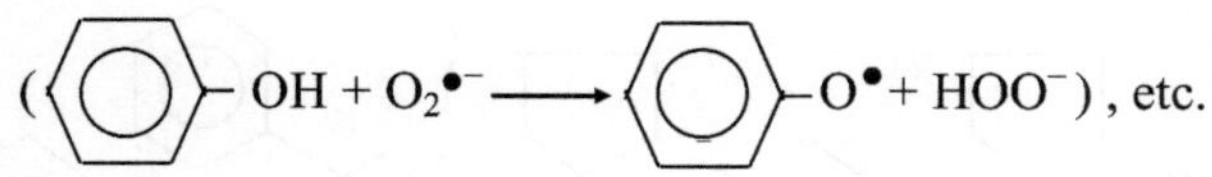

etc., and others.

The last reaction proceeds slowly in the case of monophenols, but—in the case of catechin, hydroquinone, and other polyvalent phenols—rapidly. Unlike singlet oxygen, O_2^- is a nucleophilic agent. In general, O_2^- possesses rather good reactivity both as oxidant and as reducer, and therefore it easily reduces organic compounds with acceptor properties. Thus, for example, on its interaction with quinone, the anion radical will be obtained:

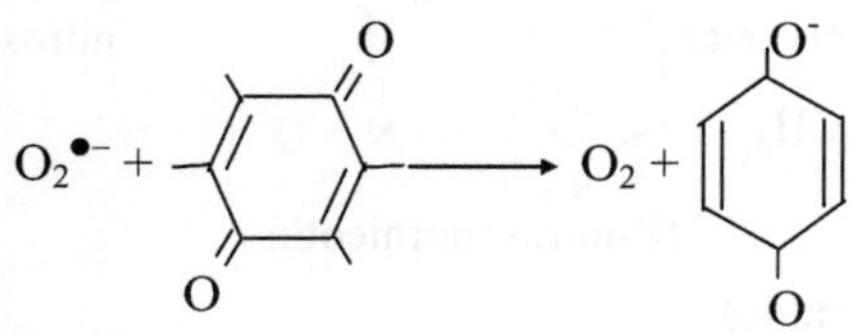

In the presence of $O_2^{\bullet-}$, tetrazolium and tetramethane salts will be reduced rapidly. These sensitive reactions are used for $O_2^{\bullet-}$ detection. The higher the alkalinity of solvent, the more stable $O_2^{\bullet-}$ becomes.

All the reactive particles discussed above may be arranged, according to their activity, as follows: $OH^\bullet > O^{\bullet-} > O > HO_2^\bullet$. Oxygen atoms are obtained as a rule by ozone photolysis and HO_2^- by electrolytic reduction of O_2. The activity of the dissociated form of $O^{\bullet-}$ is approximately 3–4 times lower, that of atomic oxygen O about 300 times lower, and that of $HO_2^\bullet$ 4–7 times lower than the activity of the hydroxyl radical OH.

The products of O_2 reduction can display both harmful and useful action.

Harmful Action

1. Smog: represents a mixture of polycyclic hydrocarbons (PCH) and other substances [25]:

$$\mathrm{PCH} \xrightarrow{h\nu} \underset{\text{sensitizer}}{\mathrm{PCH}^*} + {}^3O_2 \rightarrow \mathrm{PCH} + {}^1O_2$$

   Along with $^1O_2$, $OH^\bullet$, $HO_2^\bullet$, and O, other reactive particles may also be present in photochemical smog.
2. Radiation sickness: on irradiation by γ-rays $H_2O \rightarrow OH^\bullet$ and other active radicals.
3. Smoking: tobacco smoke contains many electron donors, which form with O_2 superoxide radical $O_2^{\bullet-}$. It contains such carcinogens as benzapyrene, and also highly dangerous mutagenic products of proteins and aminoacids pyrolysis. Besides, it contains the whole family of tobacco alkaloids, which interact with nitrogen oxides (NO) and are transformed into strong mutagenic nitrozocompounds [26]:

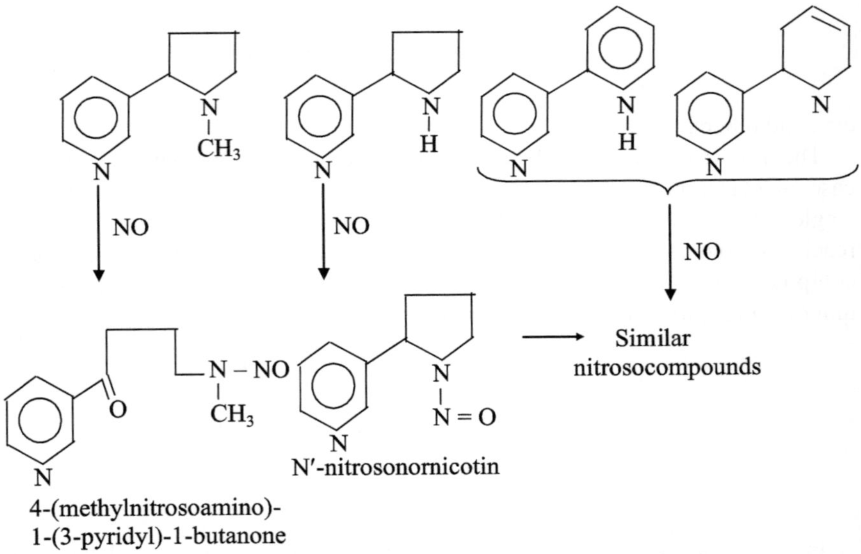

All these compounds influence the DNA of cells. OH radicals also affect them. As a result, DNA is damaged which leads to malignant cells instead of healthy ones. Therefore, the risk of lung cancer is four times higher for smokers than for nonsmoking people.

4. At the end of the electron transporting chain in an organism, O_2 is usually restored by "peaceful" enzyme—cytochromoxidase acting in accordance with the "wholesale" 4-electron mechanism (breath and photosynthesis). If normal exchange in healthy organisms takes place, $O_2^{\bullet-}$ is usually formed by xanthynogenase. Considering secondary methods of oxygen exchange, all the same superoxides (xanthynoxidases, etc.) will be formed: in mitochondria $O_2^{\bullet-}$, H_2O_2 (they suppress ATP synthesis), in cell walls of plants, in the nuclei of tumor cells. How are the organisms protected from $O_2^{\bullet-}$? For this, superoxidedismutase (SOD) is used: $O_2^- + O_2^- + 2H^+ \rightarrow H_2O_2 + O_2$. There are no superoxidedismutases in anaerobic organisms and, therefore, they do not support O_2^- and they are not protected against it. Smokers artificially introduce O_2 into their system, thus destroying it.

Useful Action of Superoxides

1. $O_2^{\bullet-}$ in liver destroys toxic substances. It also participates in collagen formation for wound healing.
2. Neonatal jaundice is treated by light: O_2^- formed oxidizes the excess of yellow bilirubine.

3. Bacteria produce $O_2^{\bullet -}$, thus regulating the density of bacteria colonies.
4. Phagocytosis—microbes are grasped by special cells (reactionary oxygen radicals). The presence of $O_2^{\bullet -}$ in the outside walls of plants represents a barrier for pathogenic microorganisms' penetration.
5. Products of O_2 reduction are used to resolve problems of environmental protection: the natural water of rivers and lakes contains 3×10^{-6} to 3×10^{-5} mol/L H_2O_2, 2×10^{-13} to 2×10^{-12} mol/L $^1O_2$, and 10^{-17} mol/L and more $OH^\bullet$, $O_2^{\bullet -}$ will also be formed.

There are various ways of peroxide radicals formation:

– By interaction of carbon radicals with molecular oxygen:

$$-\overset{|}{\underset{|}{C}}{}^\bullet + O_2 \longrightarrow -\overset{|}{\underset{|}{C}}-O-O^\bullet;$$

– By the reactions of transition metal ions with molecular oxygen (peroxyl complex will be formed):

$$M^{z+} + O_2 \rightarrow MOO^{z+};$$

– By metal ions interactions with hydroperoxides (ROOH):

$$ROOH + M^{(z+1)+} \rightarrow ROO + H^+ + M^{z+}.$$

Carbon peroxyl radicals formation proceeds quickly ($k \approx 10^9$–$10^{10}\,M^{-1}\,s^{-1}$), while the joining of oxygen molecules to thiyl radicals (RS) occurs much slower ($k \approx 10^7$–$10^8\,M^{-1}\,s^{-1}$).

Peroxide radicals take part in the set of reactions, including those of hydrogen atom separation from the carbon bond C–H:

$$-\overset{|}{\underset{|}{C}}-O-O^\bullet + H-\overset{|}{\underset{|}{C}}- \xrightarrow{\kappa < 10^3\,M^{-1}\,s^{-1}} -\overset{|}{\underset{|}{C}}-OOH + -\overset{|}{\underset{|}{C}}{}^\bullet,$$

transforming into hydroperoxides although with rather low rate constants. That is why two peroxyl radicals interaction is possible with the formation of mostly unstable tetroxide:

$$2 >COO^\bullet \longrightarrow >COOOOC<,$$

which can be disintegrated in various ways:

$$R_2CHOOOOCHR_2$$

$$R_2CHOH + R_2C=O + O_2 \leftarrow R_2CHOOOOCHR_2 \rightarrow 2R_2C=O + H_2O$$

or $R_2CHOOOOCHR_2 \rightarrow O_2 + 2R_2CHO^{\bullet}$. Interaction of the latter radicals can result in peroxides formation:

$$2RCHO^{\bullet} \rightarrow R_2CHOOCHR_2$$

or in disproportioning:

$$2RCHO^{\bullet} \rightarrow R_2C = O + R_2CHOH.$$

Halogenated peroxyl radicals such as, for example, the 3-chloromethyl peroxyl radical, will rapidly oxidize the electron donors (D) (tyrosine, methionine, vitamins C and E):

$$CCl_3OO + D \xrightarrow{K \approx 10^8\,M^{-1}\,c^{-1}} CCl_3OO^- + D^{\bullet +}$$

and hence they play an important role in the metabolism of hydrocarbons halogen derivatives.

Oxyradicals can be produced in different ways [24]:

$$R{-}OOH + M^{z+} \rightarrow RO^{\bullet} + OH^- + M^{(z+1)+},$$
$$R{-}OH + OH^{\bullet} \rightarrow RO^{\bullet} + H_2O,$$

and manymore.

Oxyradicals usually have oxidizing properties and easily detach the hydrogen atom:

$$RO^{\bullet} + {>}CH-OH \longrightarrow ROH + {>}C^{\bullet}-OH^{\bullet}$$

This reaction proceeds rather quickly and consequently only trace amounts of oxyradicals $RO^{\bullet}$ can be detected in the starting medium.

Phenomenal oxyradicals can also act as oxidizers. Free semiquinones are oxyradicals existing in acid–base equilibrium:

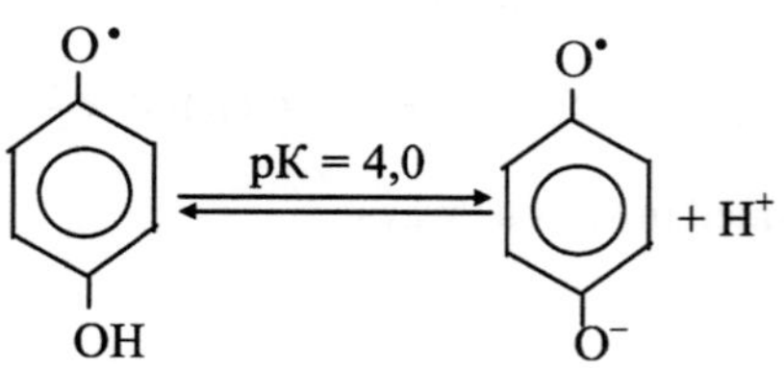

(QH•) (Q•⁻)

They act as reducers which are easily oxidized up to the appropriate quinones (Q), for example, in interaction with tetranitromethane (TNM):

$$Q^{\bullet -} + C\,(NO_2)_4 \xrightarrow{K \approx 10^8\,M^{-1}\,c^{-1}} C(NO_2)_3{}^- + NO_2 + Q.$$

Unpaired electrons of oxyradicals can be delocalized (it is confirmed by EPR-measurements), this fact being expressed as a resonance of two structures:

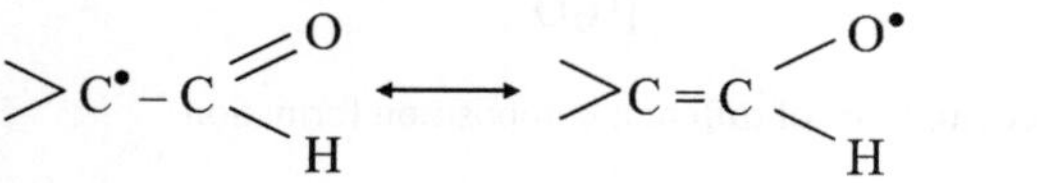

Such delocalized oxyradicals are called "crypto"-oxyradicals. Although these radicals can act chemically as oxygen radicals, in the physical sense they are not the same [24].

2.2.3 *Peroxocomplexes of Some Metals*

In the systems $M^{z+}{}_{aq} - H_2O_2$ and $M^{z+}L_n - H_2O_2$, the innersphere electron transfer is often preceded by intermediate formation of active peroxocompounds.

2.2.3.1 Peroxocomplexes of Manganese

Intermediate compound $Mn^{3+} \cdot HO_2$ was detected by using the stop–flow technique on $\lambda = 470\,nm$. In general, the formation of numerous peroxocomplexes of various compositions is possible: $L_nMH_2O_2{}^{z+}$, $L_nMHO_2{}^{(z-1)+}$, $L_nMO_2{}^{(z-2)+}$, $L_nM(H_2O_2) \cdot (HO_2{}^-)^{(z-1)+}$, $L_nM(H_2O_2)_2{}^{z+}$, and $L_nM\,(HO_2{}^-)^{z+}$. For instance, in the system Mn^{2+}–phen–H_2O_2 a peroxocomplex will be formed at $\lambda = 360\,nm$ having the composition $Mn^{2+} : H_2O_2 = 1 : 1$ [27]:

$$[\text{Mnphen}_2]^{2+} + H_2O_2 \rightarrow [\text{Mnphen}_2HO_2]^+ + H^+,$$

$$K_{eq.} = \frac{[\text{Mnphen}_2HO_2{}^+]\,[H^+]}{[\text{Mnphen}_2{}^{2+}]\,[H_2O_2]} = \frac{[\text{Mnphen}_2HO_2{}^+] \cdot K_{H_2O_2}}{[\text{Mnphen}_2]\,[HO_2{}^-]}$$

$$= K_{\text{st.ofPCinH}_2\text{O}_2},\ \text{as}\ K_{H_2O_2} = \frac{[HO_2{}^-]\,[H^+]}{[H_2O_2]}.$$

Indirect proof of peroxocomplex (PC) formation is the change of the reaction order by H_2O_2 in the course of the catalase process (Fig. 2.10).

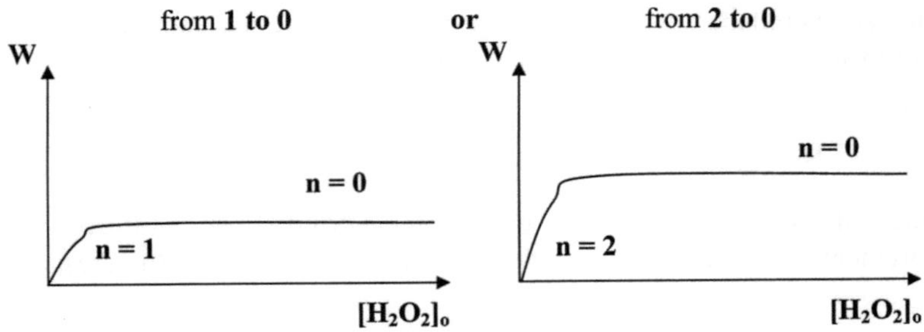

Fig. 2.10 Peroxocomplexes of different composition formation

When the stability constant of the peroxocomplex ($K_{st.PC}$) cannot be defined directly from the experimental data, it can sometimes be approximated using correlation diagrams of the following type:

$$\frac{\log K_{MOH^+}}{\log K_{MHO^{2+}}} = \frac{\log K_{LmMOH^+}}{\log K_{LmMHO^{2+}}}$$

if all the values except of $K_{LmMHO}{}^{2+}$ are known.

Under certain conditions, diperoxocomplexes of various compositions, active and inactive, in the catalysis of H_2O_2 decomposition can be formed from monoperoxocomplexes.

Determination of the dependence of peroxocomplex concentration on the parameters influencing its formation (such as initial concentrations of M^{z+}, H_2O_2, HO_2^-, L, etc.) and thermodynamic consideration of system makes it possible to define the PC composition, and the comparison of the latter with the process rate, to reveal its role in catalysis.

2.2.3.2 Peroxocomplexes of Iron

Formation of iron(III) peroxocomplex in acid medium containing $Fe(HO_2)^{2+}$ and its participation in catalytic disintegration of H_2O_2 has been proved by spectrophotometric methods [28–32].

Alcohols' oxidation up to aldehydes and acids can proceed through the formation of intermediate peroxocomplex $Fe(H_2O_2)^{3+}$:

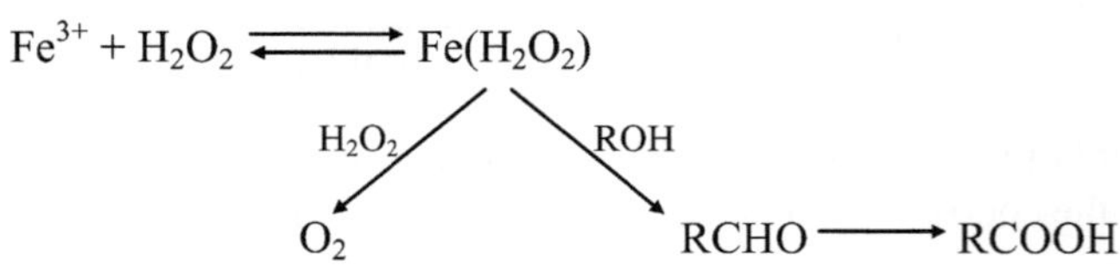

Coordination of various ligands (for example, α, α′-dipyridyl or 1,10-phenanthroline) to Fe^{2+} ions in the presence of H_2O_2 results in the formation of intermediate iron(II) peroxocompound according to the following scheme:

$$[FeL]^{2+} + H_2O_2 \xrightarrow{\text{rapidly}} [FeL \cdot H_2O_2]^{2+}$$

$$\xrightarrow{\text{slowly}} [FeL \cdot (H_2O_2)]^{2+} \xrightarrow{\text{limiting stage}} [FeL]^{3+} + OH^- + OH^\bullet$$

In the system Fe^{3+}–EDTA–H_2O_2, various peroxocomplexes of the following composition can be formed $[Fe\ EDTA(OH)HO_2]^{3-}$ [33].

2.2.3.3 Peroxocomplexes of Cobalt

Apart from simple cobalt peroxide $[Co(O)_2]$ [34], there exist two-nuclear cobalt peroxocompounds [35] which are unstable and are decomposed in water solutions. Thus, single-nuclear intermediate complexes, such as $[(H_3N)_5Co(HO_2)]^{2+}$, will be formed. The system Co^{2+}–glycin (Gl)–H_2O_2 generates the mixture of stable red–violet complexes: cobalt(III) *cis*- and *trans*-hydroxobis (glycinate) and unstable cobalt(III) μ-peroxotetra(glycin) $[GlCo(O_2)CoGl]$. In the system Co^{2+}–EDTA–H_2O_2 two-nuclear peroxocomplex will also be formed [36]. Using α, α′-dipyridyl (dipy), 1,10-phenanthroline (phen), and pyridine (Py) as ligands, formation of such peroxocomplexes—$[Co(Dipy)(HO_2)]^{2+}$, $[Co(Phen)(HO_2)]^{2+}$, $[Co(Py)(HO_2)]^{2+}$—was detected at $\lambda \approx 330$–400 nm [37, 38].

2.2.3.4 Peroxocomplexes of Copper

Copper peroxide compounds of $Cu(O_2)H_2O$, $Cu(OH)(HO_2)$, $Cu(O_2)(HO_2)$, $Cu^{2+}LHO_2{}^-$, and $\left[Cu^{2+}LHO_2{}^-\right]_2$ composition were obtained long ago. Peroxocomplexes forming in the systems Cu^{2+}–Phen–H_2O_2 and Cu^{2+}–Dipy–H_2O_2 were considered in detail [39, 40]. The complexes involved are formed by the following reactions:

$$[CuPhen]^{2+} + HO_2{}^- \rightleftarrows \left[Cu^{2+}Phen\,(HO_2{}^-)\right],$$

$$2\left[Cu^{2+}Phen\,(HO_2{}^-)\right] \rightleftarrows [CuPhen\,(HO_2{}^-)]_2\,.$$

Monomer of peroxocomplex is an active intermediate compound in the catalysis of H_2O_2 decomposition [6].

Stability data on certain metal peroxocomplexes are given in Table 2.2.

Table 2.2 Stability constants of transition metals belonging to the first period peroxocomplexes

Peroxocomplex	Stability constant[a], (L/mol)	Determination method	Reference
$[Fe(HO_2)]^{2+}$	2×10^9	Spectrophotometrical	[41]
$[Cu(HO_2)]^+$	5×10^6	Calculated	[42]
$[Mn(Phen)_2(HO_2)]^+$	3.6×10^2	Spectrophotometrical	[43]
$[Fe(EDTA)(OH)(HO_2)]^{3-}$	1×10^4	Spectrophotometrical	[44]
$[Fe(EDTA)(O_2)]^{3-}$	2.4×10^4	Spectrophotometrical	[33]
$[CoPhen(HO_2)]^{2+}$	4.3×10^3	Spectrophotometrical	[38]
$[CoDipy(HO_2)]^{2+}$	3×10^2	Spectrophotometrical	[38]
$[CoPy(HO_2)]^{2+}$	5×10^7	Spectrophotometrical	[38]
$[CuPhen(HO_2)]^+$	6.2×10^6	Spectrophotometrical	[43]
$[Cu(NH_3)_3(HO_2)]^+$	2×10^2	Calculated	[42]
$[CuDipy(HO_2)]^+$	3×10^5	Calculated	[42]
$[CuDipy(HO_2)]$	1×10^9	Calculated	[42]
$[Cu(Dipy)_2(HO_2)]^+$	1×10^3	Calculated	[42]
$[CuGlycin(HO_2)]^+$	3×10^5	Calculated	[42]
$[CuGystatin(HO_2)]^+$	4×10^4	Calculated	[42]
$[CuImidazol(HO_2)]^+$	8.9×10^9	Calculated	[45]
$[Cu(Imidazol)_2(HO_2)]^+$	2.3×10^9	Calculated	[45]
$[Cu(Imidazol)_3(HO_2)]^+$	2.0×10^3	Calculated	[45]

[a]Has been determined in water solutions

2.3 Theoretical Aspects of Catalysis with Metal Complexes

2.3.1 Influence of Complex Formation on Metal Ion Reactivity and Catalytic Processes Mechanisms (Role of Coordination in Catalysis)

The complete theory allowing one to predict the change of reactivity of transition metal ions (manganese, iron, and others) due to complex formation with ligands of various nature and structure has not yet been elaborated. However, it is known that complex formation of metal ion with ligand can change its hydrolytical stability, electrochemical properties, cause a change of electron transfer character (complexes with partial electron transfer, one- or multi-electron transfer), and change the lability of metal coordination sphere and the mechanism of redox reactions [6].

Revealing the mechanisms of redox catalysis with metal ions makes it possible to outline the basic ways of complex formation's influence on the catalytic properties of metal ions of transition valency.

2.3.1.1 Effect of Metal Ions' Hydrolytic Stability Increase

With the pH increase of water solution of M^{2+} salt, part of the M^{2+} ions will pass into the form of hydroxocomplex MOH^+:

$$M^{2+} + H_2O \xrightarrow{K_{hydr}} MOH^+ + H^+.$$

The consequence of M^{2+} transition into MOH^+ will be the decrease of redox potential, since usually $K_{M^{2+}(OH)} > K_{M^{+}(OH^{-})}$, and, therefore,

$$\varphi_{[M^{2+}(OH^{-})/M^{+}(OH^{-})]} = \varphi_{(M^{2+}/M^{+})} - RT\ln\frac{K_{M^{2+}(OH_{-})}}{K_{M^{+}(OH_{-})}}.$$

Therefore, the rate of reaction:

$$M^{2+} + DH^{-} \rightarrow M^{+} + DH^{\bullet}$$

will decline, as the oxidation state M^{2+} will be stabilized and the effective rate constant χ will reduce.

If metal hydroxoform has not yet been formed (because of low pH value), then

$$W = \chi\left[M^{2+}\right]_0 \cdot [DH^{-}]$$

and up to pH < $pK_{diss}DH_2$ the rate will grow because of $[DH^{-}]$ increase from DH_2.

For characteristic complex forming metals, such as Fe(II) and Cu(II), the limit of their thermodynamic stability is defined by pH value from 3 to 5.

When adding a salt solution of the bidentate ligand LH_2 to the metal, at low pH, the dominant forms in the solution are LH_2, LH^{-}, and M^{2+}. With increase of pH and ligand excess, the share of other complexes will increase: LH_2, LH^{-}, and M^{2+} which at high enough pH value may be hydrolyzed, forming the hydroxocomplexes $LMOH^{+}$, $LM(OH)_2$, and $L_2M(OH)_2$, and two nuclear (polynuclear) complexes:

$$2LMOH^{+} \rightleftarrows L_2M_2(OH)_2^{2+}.$$

Colloidal or residual $M(OH)_2$ will be formed in the presence of a ligand at higher pH values than in its absence. Obviously, the higher the stability constant of M^{2+} with L, the higher the pH values that promote the residual metal hydroxide formation. It means that the pH of a solution can be increased to higher values and the solution homogenity may be kept. But at higher pH, the $[DH^{-}]$ will be increased, resulting in a considerable increase of the reaction rate $W = \chi[M^{2+}] \cdot [DH^{-}]$. Therefore, the first factor of influence of complex formation on metal ions' catalytic properties is the preservation of the system's homogenity at higher pH by the prevention of metal ions' aggregation and their exclusion from the reaction area.

2.3.1.2 Change of Metal Electrochemical Properties Due to Complex Formation

Due to complex formation, the redox potential of the metal ion will be changed. Varying the number and nature of ligands makes it possible to change its value within wide limits ($\approx -1/+1.5$ eV), leaving the nature of M^{2+} constant.

Usually the initial stage of catalysis is a stage of substrate activation, resulting in the formation of reactionary substrate particles characterized by low equilibrium constants. In general, the activation stage can be presented as follows:

$$\text{I.}\quad LM^{2+}DH_2 \underset{k_2}{\overset{k_1}{\rightleftarrows}} LM^{+} + DH + H^{+} \quad (\Delta G_{\mathrm{I}}^{\mathrm{o}})\,.$$

Further, the active substrate particle $DH^{\bullet}$ (electron donor) will interact with LM^{2+} with high rate constant:

$$\text{II.}\quad LM^{2+} + DH^{\bullet} \rightarrow LM^{+} + H^{+} + D \quad (\Delta G_{\mathrm{II}}^{\mathrm{o}})\,.$$

With the substrate activation stage to be repeated again and again, another A oxidant is to be introduced into the system (electron acceptor) transforming LM^{+} into LM^{2+}:

$$\text{IIa.}\quad LM^{+} + A \rightarrow LM^{2+} + A^{-} \quad (\Delta G_{\mathrm{IIa}}^{\mathrm{o}})\,.$$

Thus, to estimate the influence of complex formation on the metal ion's catalytic properties, not only the reaction of substrate DH^{-} activation reaction (I) should be considered but also the catalytic cycle of reactions:

$$\text{I.}\quad LM^{2+} + DH_2 \rightarrow LM^{+} + DH^{\bullet} + H^{+}\ (\Delta G_{\mathrm{I}}^{\mathrm{o}})$$

$$\text{IIa.}\quad LM^{+} + A \xrightarrow{H^{+}} LM^{2+} + AH^{\bullet} \quad (\Delta G_{\mathrm{IIa}}^{\mathrm{o}})$$

$$DH_2 + A \rightarrow DH^{\bullet} + AH^{\bullet} \quad \Delta G_{\mathrm{a}}^{\mathrm{o}}$$

Here, $\Delta G_{\mathrm{a}}^{\mathrm{o}} = \Delta G_{\mathrm{I}}^{\mathrm{o}} + \Delta G_{\mathrm{IIa}}^{\mathrm{o}}$ is a change of free energy describing brutto-process. $\Delta G_{\mathrm{a}}^{\mathrm{o}}$ is defined by the nature of the substrate pair (DH_2 + A), and $\Delta G_{\mathrm{I}}^{\mathrm{o}}$ and $\Delta G_{\mathrm{IIa}}^{\mathrm{o}}$ values by the choice of catalyst. Thus, the reversible process appears, yielding the catalyst regeneration and its recurrence again and again:

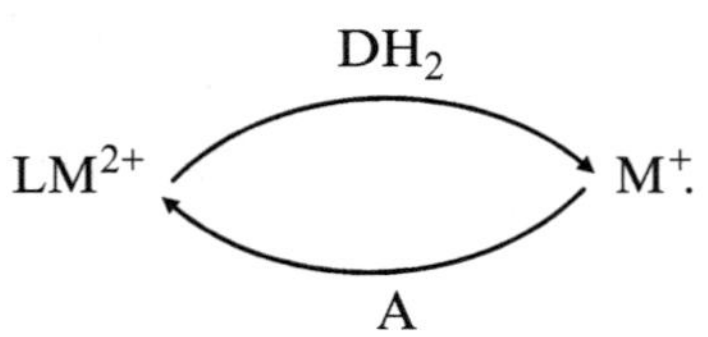

Hence, for each substrate pair, an optimal complex catalyst could be found by varying the ligand surrounding the metal ion.

The most active catalyst will be that which equally rapidly oxidizes DH_2 and reduces A by reactions I and IIa. This can be achieved by altering the ligand surrounding the central metal ion.

2.3.1.3 Change of Electron Transfer Character (One- or Many-Electron) on Complex Formation

In a number of cases of complex formation, the stabilization of metal oxidation degree will be observed to differ from the initial one by two units. It can be done either by two consecutive one-electron transfers:

$$M^{z+} + DH^- \rightarrow M^{(z-1)+} + DH^\bullet$$

$$M^{(z-1)+} + DH^- \rightarrow M^{(z-2)+} + DH^\bullet$$

or as a result of simultaneous two-electron transfer:

$$M^{(z-2)+} + A \xrightarrow{H^+} M^{z+} + AH^-,$$

i.e., in the case of reaction without substrate radicals' formation. This case was considered as an example of the reaction mechanism in the system Fe^{3+}–TETA–H_2O_2, in which a two-electron transfer stage appears [46]:

$$\text{Fe(II)TETA} + H_2O_2 \rightarrow \text{Fe(IV)TETA}.$$

In general, in the case of a multistage process described by a number of one-electron stages, redox potentials of each stage are unequal and the presence of oxidizer (or reducer) is needed in order to perform the process, with a "power" equal to the most difficult energy stage (for example, the stage IIa to provide the reversible redox cycle).

If many electron processes may be realized in one stage, their redox potential, being the average of the potentials of each one-electron stage, will be lower than that necessary for the most difficult stage.

Thus it is possible to use a weaker oxidizer (reducer) and to save energy (as used for the work of metal enzymes). In this case, the catalyst can play a role of original "switch" of a one-electron mechanism into a multielectron one, accepting from the electron donor DH_2 the electrons one by one and giving some of them back to the substrate (acceptor A).

When such a switching of the one-electron mechanism into a multielectron one occurs, the redox potential of the catalyst (connected with the change of its oxidation number) must not be essentially changed. This is assured by the fact that the catalyst represents a cluster or multinuclear complex compound including some weakly bound metal ions, able to participate in the electrons exchange. Reacting with an external reducer, such a catalyst can accept electrons one by one, not significantly changing its redox potential. Then in the reaction with the substrate, which is in the coordination sphere of the complex, the transfer of two or several electrons occurs simultaneously. Thus, substrate radicals are not generated, and final products will be formed at once:

$$DH_2 + A \xrightarrow{2e^-} D + AH_2.$$

Thus, the selectivity of the redox reaction will be sharply increased. As the multinuclear complex compounds canvary considerably, the routes of reactions can be different.

2.3.1.4 Complex Formation Can Change the Lability of the Metal Coordination Sphere

The lability of the metal coordination sphere represents the replacement rate of a solvent molecule by one substrate molecule. Lability of a metal ion is characterized by the rate constant of the above-mentioned replacement. When the aqua ion of metal interacts with the substrate in accordance with the outer-sphere electron transfer mechanism:

$$\overset{\curvearrowleft\ \bar{e}}{M^{z+}{}_{\cdot aq} \ldots SH_2}$$

the increase of the metal coordination sphere lability as the result of complex formation with the ligand can change the mechanism into the inner sphere one, and lead to an increase of the effective rate constant:

$$LM^{z+}{}_{aq} + SH_2 \rightarrow LsM^{z+} \cdot SH_2 + aq.$$

An example of the same sort is the coordination sphere lability change with the transition from $Fe^{3+}{}_{aq}$ to $Fe^{3+}EDTA_{aq}$. The theory of complex formation influence on the lability of metal ion coordination sphere is poorly developed.

2.3.1.5 Complex Formation Influence on the Change of Redox Reaction Mechanisms

The mechanism of redox reactions can be changed with the transition from $M^{3+}{}_{aq}$ ion catalysis to catalysis by LM^{z+} complex.

The simplest mechanism is the ion–molecular one:

$$DH_2 \rightleftarrows DH^- + H^+$$

1. $DH^- + M^+ \rightleftarrows (DHM)^{(z-1)+} \rightarrow (DM)^{(z-2)+} + H^+ \rightarrow D + M^{(z-2)+},$

2. $M^{(z-2)+} + A \rightleftarrows (MA)^{(z-2)+} \overset{H^+}{\rightleftarrows} (MAH)^{(z-1)+} \rightarrow M^{z+} + AH^-,$

$$AH^- + H^+ \rightleftarrows AH_2.$$

It can be presented either as a sequence of two fast one-electron transfers, or as one two-electron transfer. This mechanism, which does not include substrate radicals in catalytic cycles of transformations, represents the greatest technological interest, since it is the mechanism of selective catalysis. Complex formation of metal ion with appropriate ligands favors the display of this mechanism, as it either brings two or several metal ions together via the ligand, thus creating the conditions of collective and simultaneous two- or many-electrons transfer, or changes the electron density on the metal, so that two electron transfers can easily proceed.

Other mechanisms are radical and radical-chain ones, when metal oxidation number changes by one unit. With such one-electron transfer, the redox reaction proceedswith some difficulty.

A large number of reactions of metal ions of different oxidation numbers with the initial DH_2 and A, and also reactions between free radicals and metal ions, are chemically possible [6]:

1. $M^{2+} + DH^{-} \rightarrow M^{+} + DH^{\bullet}$,
2. $M^{+} + A \rightarrow M^{2+} + A^{\bullet -}$,
3. $DH^{\bullet} + DH^{\bullet} \rightarrow D + DH_2$,
4. $AH^{\bullet} + AH^{\bullet} \rightarrow A + AH_2$,
5. $DH^{\bullet} + AH^{\bullet} \rightarrow D + AH_2$,
6. $DH^{\bullet} + A \rightarrow D + AH^{\bullet}$,
7. $M^{2+} + DH^{\bullet} \rightarrow M^{+} + D + H^{+}$,
8. $M^{+} + AH^{\bullet} \rightarrow M^{2+} + AH^{-}$,
9. $M^{2+} + A^{\bullet -} \rightarrow M^{+} + A$,
10. $M^{+} + DH^{\bullet} \rightarrow M^{2+} + DH^{-}$,
11. $AH^{\bullet} + DH_2 \rightarrow AH_2 + DH^{\bullet}$.

Stages 4, 5, 8, and 11 result in the formation of the reaction product $AH_2(AH^{-})$, and only in stage 11 does the regeneration of the active intermediate particle $DH^{\bullet}$ proceed, and due to it the radical-chain mechanism can be realised.

With the transition from $M^{z+}{}_{aq}$ to $M^{z+}L$, the electron density value on M^{z+} will be changed. If these changes are not significant, they will promote one-electron transfer and the opportunity of radical or radical-chain mechanism. Therefore, the results of quantum-chemical calculations [for example, made by the method of MO LCAO in the approximation of Mulliken–Wolfsberg–Helmholtz (MWH) of the stage of catalytic redox reaction's initiation] sometimes help to find not only the direction and size of charge transfer, but also the effects of the ratio between HOMO and lowest free molecular orbital (LFMO) levels of the catalyst and substrate and the possible course of the process (see Chap. 4). These are major factors of complex formation influence on the metal-catalyst ion.

2.3.2 *Qualitative Model of the Theory of PCT Complexes*

As was previously shown, complex formation in catalytic redox systems and processes plays an extremely important role. As a rule, a result of complex formation is charge transfer. On the analysis of charge transfer processes, the inner-sphere reactions of electron exchange have been considered more often. There are little data available about the kinetics of inner-sphere electron transfer, though these processes represent the greatest interest for catalytic redox reactions. An attempt was undertaken to create, at least on a qualitative level, the theoretical model of an inner-sphere electron transfer reaction in polar medium, taking as an example the general consideration of a bimolecular redox reaction with intermediate complex formation [47]:

$$A + D^{e-} \rightleftarrows (AD)^{e-} \rightarrow A^{e-} + D,$$

where A is the acceptor, D^{e-} the donor of PCT, and $(AD)^{e-}$ the intermediate complex with PCT.

In inner-sphere reactions, the products occur not as a result of proper electron transfer but as a result of dissociation of the above-mentioned intermediate complex. During electron transfer over distances shorter than those of Van-der Waals ($\approx <$ 5A), the dielectric permeability of solvent decreases and the role of Coulomb's interaction increases. The reduction of distance R in inner-sphere transfer results in the decrease of solvent reorganization energy E_S. Thus, the quantitative model of inner-sphere electron transfer has been proposed.

The limiting case when $E_s \rightarrow 0$ was considered, when the intermediate complex formed can be disintegrated either into to initial particles, being in the basic state N, or to electron transfer products only on complex dissociation from the electron-exited state E:

$$A + D^{e-} \rightleftarrows \left[(AD)_N{}^{e-} \rightleftarrows (AD)_E{}^{e-}\right] \rightleftarrows A^{e-} + D.$$

An intermediate complex $(AD)^{e-}$ is represented thus as the "superposition" of two states:

$$(AD)^{e-} = (AD^{e-} \leftrightarrow A^{e-}D).$$

The stages of two-electron transfer were considered similarly. In this case, in contrast to inner-sphere transfer, whereas for inner-sphere reactions the probability of electron transfers between the terms corresponds to the states without transfer AD^{2e-} and two-electron charge transfer AD^{2e} is close to one. The structure of the formed complex can also be represented as a superposition of the states A and D differing by two electrons:

$$(AD)^{2e-} = (AD^{2e-} \leftrightarrow A^{2e-}D).$$

Similarly, as well as for one-electron transfer, the dissociation of this complex from two-electron exited state occurs at two-electron transfer:

$$A + D^{2e-} \rightleftarrows \left[(AD)_N{}^{2e-} \rightleftarrows (AD)_E{}^{2e-}\right] \rightleftarrows A^{2+} + D.$$

Further simplification of the model involves the assumption that spontaneous transitions between basic- and electron-exited complex states proceed much faster than complex dissociation into the initial reagents or the products of electron transfer. The basic postulates of this model of inner-sphere electron transfer have been formulated. The existence of several conformation states of the initial products and the products of electron transfer is possible, this fact opening an opportunity for the interpretation of properties and reactivity of metal complex ions in various conformations and redox transformations. The consequences of this qualitative model of inner-sphere electron transfer have been formulated.

If a third particle takes part in the interaction with intermediate complex, triple (or multicomponent) charge transfer complexs can be formed.

This model was applied to the inner-sphere transfer reactions with metal complex ions participation. Thus inner-sphere reactions of metal ions in reduced form (M^+) with two-electron acceptor (A) or metal ions in oxidized form (M^{2+}) or superoxidized one (M^{3+}) with two-electron donor (D^{2-}) can be of two types—with the resulting one and two-electron transfer:

$$M^+ + A \overset{k_f}{\rightleftharpoons} (C_N \rightleftharpoons C_E) \overset{k_d'}{\rightleftharpoons} M^{2+} + A^-,$$

$$M^+ + A \overset{k_d}{\rightleftharpoons} (C_N \rightleftharpoons C_E) \overset{k_f'}{\rightleftharpoons} M^{3+} + A^{2-},$$

$$M^{2+} + D^{2-} \rightleftharpoons (C_N \rightleftharpoons C_E) \rightleftharpoons M^+ + D^{\cdot -},$$

$$M^{3+} + D^{2-} \rightleftharpoons (C_N \rightleftharpoons C_E) \rightleftharpoons M^+ + D,$$

where C_N and C_E are basic and electron-exited states of the appropriate intermediate charge transfer complex. The expression for rate constant of the resulting electron transfer is as follows:

$$k_{el.} = \frac{k_f \cdot k_f' \cdot K_{el.}}{k_f + k_f' \cdot K_{el.}},$$

where $K_{el.}$ is the equilibrium constant of the given redox reaction.

It is shown in this model that the binary intermediate complexes can be of four types—with inner-sphere one- and two-electron transfer, with superposition of structures:

$$(MA)_I{}^+ = (M^+A \leftrightarrow M^{2+}A^-),$$

$$(MA)_{II}{}^+ = (M^+A \leftrightarrow M^{3+}A^{2-}),$$

$$MD = (M^{2+}D^{2-} \leftrightarrow M^+D^{\bullet -}),$$

$$MD^+ = (M^{3+}D^{2-} \leftrightarrow M^+D).$$

With these complexes, interaction with redox ligand or with other metal ions, the formation of triple charge transfer complexes of various types is possible, and also of binuclear charge transfer ones.

Creation of a qualitative theoretical model allows the final result to be obtained (for example, the rate constant of the resulting electron transfer $K_{el.}$) without performing difficult (and frequently inexact) quantum-chemical calculations, this fact being important and practically valuable. Within the framework of this model, it is possible to discuss the various phenomena of metal-complex redox catalysis.

Using various experimental data obtained earlier, the proposed theoretical model of inner-sphere electron transfer was confirmed. For example, in the case of the reaction proceeding with intermediate formation of binuclear charge transfer complex, the resulting process can be represented as follows:

$$(\mathrm{CuDFH})^+ + \mathrm{Cu}^{2+} \rightarrow 2\mathrm{Cu}^+ + \mathrm{DK} + \mathrm{H}^+,$$

where DFH_2 is dihydroxyfumaric acidand DK diketosuccinic acid. The experimental values of rate constant of this reaction make 10.5 L/mol s. Calculated value $k_{\mathrm{el.}} = k_{\mathrm{f}}' \exp(-\Delta G_{\mathrm{el}}/RT) \approx 15$ l/mol s is in accordance with the experimental data. Numerous confirmations of the model and its various aspects were obtained on the study of elementary interaction mechanisms of O_2 and H_2O_2 with metal ions in the reduced form and ions in the oxidized form with hydrogen peroxide, ascorbic acid, its analog—dihydroxyfumaric acid—and also catechins [48].

All the aforesaid shows that the model and the concept of PCT can be successfully applied for the analysis of different aspects of metal complex ions' redox transformations.

2.4 General Principles of Redox Catalysis Mechanisms Determination

2.4.1 *Possible Mechanisms of Catalytic Redox Systems $Fe^{z+}{}_{aq}$–H_2O_2 and $Fe^{z+}L_n$–H_2O_2*

The display of any function of metal is more or less dependent on the electron structure of the metal ion, and on its change during coordination with ligand or substrate. The detailed consideration of the results of system $Fe^{2+}{}_{aq}$–H_2O_2 (Fenton's reagent) and L_nFe^{2+}–H_2O_2 study, and also the mechanisms of the appropriate processes is available in the form of a critical review [6, 10, 48]. According to these and other data of kinetic regularities, study of H_2O_2 decomposition in the system Fe^{2+}–H_2O_2 made it possible to determine the rate constants of elementary stages and acid-basic equilibria of various reactions in the system Fe^{2+}–H_2O_2 (Table 2.3) [49], but also has enabled development of the various schemes of H_2O_2 catalytic disintegration mechanisms.

General experimental data [10], partially given in Table 2.3 and Scheme 2.2 [49], made it possible not only to establish possible elementary stages of H_2O_2 decomposition by Fe^{2+} aqua ions but also their sequence depending on the experimental conditions.

A. Processes with Fe^{2+} participation

1. $Fe^{2+} + H_2O_2 \longrightarrow Fe^{3+} + OH\cdot + OH^-$,
2. $Fe^{2+} + H_2O_2 \longrightarrow FeO^{2+} + H_2O$,
3. $Fe^{2+} + H_2O_2 \longrightarrow Fe^{4+} + 2OH^-$,
4. $FeOH^{+} + H_2O_2 \longrightarrow Fe^{3+} + OH\cdot + 2OH^-$,
5. $FeOH^{+} + H_2O_2 \longrightarrow FeO^{2+} + H_2O + OH^-$,
6. $(H_2O)_5Fe(H_2O)^{2+} + H_2O_2 \rightleftharpoons (H_2O)_5Fe(H_2O_2)^{2+} + H_2O$,
7. $(H_2O)_5Fe(H_2O_2)^{2+} \longrightarrow (H_2O)_4(OH^-)FeO^{2+} + H_2O$,
8. $(H_2O)_5Fe(H_2O_2)^{2+} \xrightarrow{-H^+} (H_2O)_5(OH^-)Fe^{3+} + OH\cdot$,
9. $(H_2O)_5Fe(H_2O_2)^{2+} + Fe^{2+}(H_2O)_6 \xrightarrow{-H^+} 2(H_2O)_5Fe^{2+}(OH^-) + H_2O$,
10. $Fe^{2+} + HO_2^- + H_2SO_4 \longrightarrow Fe\genfrac{}{}{0pt}{}{\diagup SO_4H}{\diagdown OOH} + H^+$,
11. $Fe^{2+}SO_4 + H_2O_2 \longrightarrow FeSO_4(H_2O_2)$,
12. $Fe\genfrac{}{}{0pt}{}{\diagup SO_4H}{\diagdown OOH} + H_2O_2 \longrightarrow Fe\genfrac{}{}{0pt}{}{\diagup SO_4H}{\diagdown OH} + H_2O + O_2$,
13. $Fe^{2+} + OH\cdot \longrightarrow Fe^{3+} + OH^-$,
14. $Fe^{2+} + HO_2\cdot \longrightarrow Fe^{3+} + HO_2^-$,
15. $Fe^{2+}(H_2O_2)(HO_2) \rightleftharpoons Fe^{2+}(HO_2\cdot)(HO^+) + OH^-$,
16. $Fe^{2+}(HO_2\cdot)(HO^+) \rightleftharpoons Fe^{3+}(HO_2\cdot)(HO\cdot)$,
17. $[Fe_2(OH)_2]^{2+} + H_2O_2 \longrightarrow [Fe_2(OH)_2]^{4+} + 2OH^-$;

B. Processes with Fe^{3+} participation

18. $Fe^{3+} + HO_2\cdot \longrightarrow Fe^{2+} + H^+ + O_2$,
19. $Fe^{3+} + O_2\cdot^- \longrightarrow Fe^{2+} + O_2$;

C. Processes with $OH\cdot$, $HO_2\cdot$ ($O_2\cdot^-$) participation (besides those given above)

20. $OH\cdot + H_2O_2 \longrightarrow HO_2\cdot + H_2O$,
21. $HO_2\cdot + H_2O_2 \longrightarrow O_2 + H_2O + OH\cdot$;

D. Processes with Fe^{4+} participation

22. $FeO^{2+} + H_2O_2 \longrightarrow Fe^{2+} + O_2 + H_2O$,
23. $Fe^{4+} + OOH^- \longrightarrow Fe^{2+} + O_2 + H^+$,
24. $Fe^{4+} + HO_2^- \rightleftharpoons FeOOH^{3+} \longrightarrow Fe^{3+} + HO_2\cdot$,
25. $(H_2O)_4(OH^-)Fe^{2+}O + H_2O \longrightarrow (H_2O)_4(OH^-)_2Fe^{3+} + OH\cdot$,
26. $(H_2O)_4(OH^-)Fe^{2+}O + Fe^{2+} \longrightarrow 2Fe^{3+}(OH^-)(H_2O)_5$,
27. $Fe^{4+} + Fe^{2+} \longrightarrow 2Fe^{3+}$,
28. $FeO^{2+} + H_2O_2 \longrightarrow HO_2\cdot + OH^- + Fe^{3+}$;

E. Acid-alkali equilibrium

29. $Fe^{2+} + H_2O \longrightarrow FeOH^+ + H^+$,
30. $2Fe^{2+} + 2H_2O \longrightarrow Fe(OH)_2^{2+} + 2H^+$,
31. $H_2O_2 \longrightarrow HO_2^{2-} + H^+$,
32. $HO_2\cdot \longrightarrow H^+ + O_2\cdot^-$,
33. $(H_2O)_5Fe^{2+}(H_2O_2)^{2+} \longrightarrow (H_2O)_4(OH^-)Fe(H_2O_2)^{2+} + H^+$.

Scheme 2.2 Reactions related to H_2O_2 disintegration process in the system $Fe^{2+}{}_{aq}$–H_2O_2 [10]

Table 2.3 Rate constants of elementary stages and acid–basic equilibrium in the system $Fe^{2+}{-}H_2O_2$

Reaction	Constanta[a]	References
$Fe^{2+} + H_2O_2 \rightarrow Fe^{3+} + OH^{\bullet} + OH^-$	68	[48]
	76.5	[49]
	62	[50]
	53	[51,52]
	58	[53]
	66	[54]
	79	[27]
	49.5	[55]
	73	[56]
	76	[57]
	56.2	[58]
	41.7	[59]
	61.7	[60]
	60.3	[61]
	50.1	[62,63]
$Fe^{2+} + OH^{\bullet} \rightarrow Fe^{3+} + OH^-$	2.6×10^8	[64]
	5.1×10^8	[65]
	3×10^8	[57,66]
	5×10^8	[67]
	3.2×10^8	[67–69]
	2.5×10^8	[70]
	2.7×10^8	[71]
	3.3×10^8	[72]
$Fe^{2+} + HO_2^{\bullet} \rightarrow Fe^{3+} + HO_2^-$	7.2×10^5	[73]
	1.5×10^6	[74]
$Fe^{3+} + OH^{\bullet} \rightarrow Fe^{4+} + OH^-$	7.9×10^7	[10,70]
$OH^{\bullet} + H_2O_2 \rightarrow HO_2^{\bullet} + H_2O$	4.5×10^7	[64]
	1.7×10^7	[65]
	3×10^7	[70]
$OH^{\bullet} + OH^{\bullet} \rightarrow H_2O_2$	5.3×10^9	[57]
	5.02×10^9	[75]
	8×10^9	[64]
	6×10^7	[76]
$OH^{\bullet} + O_2^{\bullet -} \rightarrow OH^- + O_2$	$1 \times 10^{1}0$	[10,70]
$OH^{\bullet} + HO_2^- \rightarrow H_2O + O_2$	1.4×0^{10}	[10,70]
$HO_2^{\bullet} + HO_2^{\bullet} \rightarrow H_2O_2 + O_2$	1.9×10^6	[77]
	2.2×10^6	[78]
	8.1×10^5	[79]
$O_2 \cdot^- + O_2 \cdot^- \xrightarrow{H^+} O_2 + HO_2^-$	< 0.35	[67]
$HO_2^{\bullet} + O_2^{\bullet -} \rightarrow HO_2^- + O_2$	8.5×10^7	[67]
$HO_2^{\bullet} + H_2O_2 \rightarrow H_2O + O_2 + OH^{\bullet}$	0.2	[10]
	3.7	[10]
$(H_2O)_4(OH^-)Fe^{2+}(H_2O_2) \rightarrow (H_2O)_4(OH^-)FeO^{2+} + H_2O$	3×10^5	[80]
$(H_2O)_5Fe(H_2O_2)^{2+} \rightarrow (H_2O)_5(OH^-)Fe^{3+} + OH^{\bullet}$	2.8×10^3	[80]

(continued)

Table 2.3 (continued)

Reaction	Constanta[a]	References
$(H_2O)_5Fe(H_2O_2)^{2+} + Fe^{2+}(H_2O)_6 \rightarrow 2(H_2O)_5$ $Fe^{3+}(OH^-) + H_2O$	3×10^5	[80]
$Fe^{2+} + O_2^{\bullet-} \xrightarrow{H^+} Fe^{3+} + HO_2^-$	7.3×10^5	[73]
$Fe^{3+} + H_2O_2 \rightarrow Fe^{2+} + HO_2^\bullet + H^+$	9.1×10^{-7}	[81]
$Fe^{3+} + H_2O_2 \rightleftarrows FeHO_2^{2+} + H^+$	2×10^{-3}	[82]
$Fe^{3+}HO_2^- \rightarrow Fe^{2+} + HO_2^\bullet$	2×10^{-3}	[83]
	1.6×10^{-3}	[48]
$FeHO_2^{2+} + FeOH^{2+} \rightarrow 2Fe^{2+} + H_2O + O_2$	5	[48]
$Fe^{3+} + HO_2^\bullet \rightarrow Fe^{2+} + H + O_2$	7.3×10^5	[73]
	2.1×10^6	[84]
	3.3×10^5	[58]
$Fe^{3+} + O_2^{\bullet-} \rightarrow Fe^{2+} + O_2$	4×10^8	[76]
	1.29×10^9	[84, 85]
	1.1×10^9	[83]
	5×10^7	[58]
	1.9×10^9	[74]
$H_2O_2 \rightleftarrows HO_2^- + H^+$	2.63×10^{-12}	[86]
$HO_2^\bullet \rightleftarrows H^+ + O_2^{\bullet-}$	3.55×10^{-5}	[87]
$HO_2^\bullet + H^+ \rightleftarrows H_2O_2^+$	$(3.16/3.98) \times 10^{-5}$	[77, 78]
$OH \rightleftarrows O^{\bullet-} + H^+$	1.02×10^{-12}	[88]
$Fe^{2+} + OH^- \rightleftarrows FeOH^+$	$K = 3.63 \times 10^5$	[89]
$Fe^{3+} \rightleftarrows FeOH^{2+} + H^+$	2×10^{-3}	[90]
	1.4×10^{-3}	[91]
	1.83×10^{-3}	[92]
$(H_2O)_5Fe^{2+}(H_2O) + H_2O_2 \rightleftarrows (H_2O)_5Fe(H_2O_2)^{2+} + H_2O$	1.8×10^{-2}	[80]
$(H_2O)_5Fe(H_2O_2)^{2+} \rightleftarrows (H_2O)_4(OH^-)Fe(H_2O_2)^{2+} + H^+$	6×10^{-5}	[80]
$Fe^{2+} + HO_2^\bullet \rightleftarrows FeHO_2^{2+}$	2.1×10^6	[84]

[a]Rate constants are given in L/mol s, disintegration constants in s^{-1}, and equilibrium constants in mol/L

The following set of sequential stages in the system $Fe^{2+}{}_{aq}$–H_2O_2 is considered to be generally accepted (stage numbering is in accordance with Scheme 2.2) [93].

1. $Fe^{2+} + H_2O_2 \rightarrow Fe^{3+} + OH^\bullet + OH^-$,
13. $Fe^{2+} + OH^\bullet \rightarrow Fe^{3+} + OH^-$,
20. $OH^\bullet + H_2O_2 \rightarrow HO_2^\bullet + H_2O$,
14. $Fe^{2+} + HO_2^\bullet \rightarrow Fe^{3+} + HO_2^-$,
18. $Fe^{3+} + HO_2^\bullet \rightarrow Fe^{2+} + H^+ + O_2$,
19. $Fe^{3+} + O_2^{\bullet-} \rightarrow Fe^{2+} + O_2$.

Such a set of stages will be realized under certain conditions: $0.5 < [(H_2O_2)/(Fe^{2+})] < 200$, $pH > 2$. The process of H_2O_2 disintegration proceeds according to the ion-radical cyclic mechanism up to the ratio $[H_2O_2]/[Fe^{2+}] \approx 3 \times 10^6$ (the equality of rates of initiation and chain continuation stages is characteristic for this mechanism), and in the case $[H_2O_2]/[Fe^{2+}] > 3 \times 10^6$—according to the radical-chain mechanism [only under these conditions that is it possible to reveal precisely the stages of chain initiation reaction (1), continuation (20, 21), and breakage (18 and/or 19)].

In the Ruff's system ($Fe^{3+}{}_{aq}$–H_2O_2)H_2O_2, decomposition, catalyzed by Fe^{3+} aqua-ions, proceeds by radical-chain mechanism in acid medium at pH ≤ 2 (this and other mechanisms are considered in detail in [10]). Basic experimental proofs of this mechanism are based on (1) inhibiting action of various organic substances [42, 48, 74, 83, 94–102], (2) generation in the given system of $OH^{\bullet}$ radicals [100, 101, 103–110] and/or O_2^{-} [42, 83, 94–96, 102, 110], and (3) experimental data proving the presence of one-electron transfer stages which include the redox cycle: $Fe^{3+} \Leftrightarrow Fe^{2+}$ [42,48,105,111]. The supporters of ion-molecular mechanism of H_2O_2 disintegration in Ruff's system [41, 82, 112–119] give as the main proof the data on various intermediate compounds formation and in particular that of peroxocomplex $FeHO_2^{2+}$ [84, 118–121]. However, the formation of such an intermediate does not contradict (as shown in [10]) the process running by the radical-chain mechanism.

It was proved qualitatively and quantitatively [42] that, from all possible reactions of H_2O_2 decomposition in Ruff's system, only a definite sequence of stages is actually realized, as presented in Scheme 2.2. This scheme is considered to be generally accepted for the description of H_2O_2 decomposition mechanism by Fe^{3+} aqua-ions (radical-chain mechanism).

The nature of ligand interaction with the transition metal ion renders great influence on the mechanism of H_2O_2 decomposition that can be seen from Table 2.4 (Table 2.4 is taken from [49]).

Ligands coordination with transition metal ions (in particular, with Fe^{3+} ions) results in many cases in the acceleration of H_2O_2 decomposition [10,42,46,132–137]. The greatest activity in H_2O_2 disintegration reaction is displayed by iron(III) cations with triethyleneamine (so-called inorganic catalase) [6].

The mechanisms of H_2O_2 decomposition by coordinative compounds of Fe(III) are various: radical chain (I), ion molecular (II), and chain nonradical (III). The general scheme of mechanisms in the systems Fe^{3+}(Fe^{2+})–ligand–H_2O_2 is given in Scheme 2.3 (the scheme and explanatory material are taken from [6] and [49]).

Such a mechanism is realized in the systems Fe^{3+}–Dipy–H_2O_2 [10, 122, 137], Fe^{3+}–EDTA–H_2O_2 [138, 139], Fe^{3+}–H_2LH–H_2O_2 [10, 57], Fe^{3+}–acidic chromium dark-blue–H_2O_2 [48], Fe^{3+}–ethylenediamine–H_2O_2 [140], Fe^{3+}–glutamic acid–H_2O_2, and some other systems [141, 142].

In cycle II, basic processes do not include the reactions with the participation of $OH^{\bullet}$ and $O_2^{-\bullet}$ radicals. The stage of two-electron transfer takes place:

$$L'Fe^{2+} \underset{+2e^-}{\overset{-2e^-}{\rightleftarrows}} L'Fe^{4+}.$$

This mechanism is realized, for example, in the case of H_2O_2 decomposition by coordination compounds of Fe^{3+} with triethylenetetramine [10, 46, 137] and hystidine [10, 126, 143–145].

In cycle III, ionic- and diperoxocomplexes will be formed. The oxidation degree of iron(III) compounds is not changed and $OH^{\bullet}$ and $HO_2^{\bullet}$ ($O_2^{\bullet-}$) radicals are

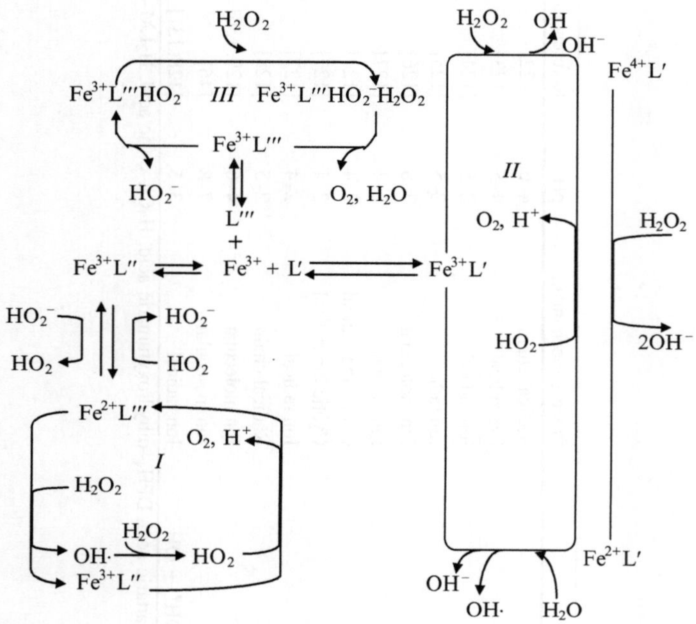

Scheme 2.3 Generic mechanisms of hydrogen peroxide decomposition with Fe(III) coordination compounds

not generated into the reaction medium. Such a mechanism is observed on H_2O_2 decomposition by Fe^{3+} complexes with $4', 4'', 4'''$-tetrasulfophthalocyanine [131], N, N'-bis-(2-methylenepyridyl)-ethylenediamine [146], and tetrapyridyl [146].

Complex formation of various ligands with Fe^{2+} results in the increase of its activity in the reactions of hydrogen peroxide disproportioning (if pH value changes thus to the favorable one), and the process of homogeneous H_2O_2 disintegration can proceed in weakly acid and neutral media. The oxidizing agent of Fenton's type can be obtained by introducing the ligand into the system $Fe^{2+}-H_2O_2$ for the organic substances hydroxylation in weakly acid—and in some cases neutral—media. The nature of ligand and pH of the medium considerably changes the mechanism of H_2O_2 disproportioning in the systems Fe^{2+}–ligand–H_2O_2. In many cases, mainly in weakly acid media (Table 2.4), on complex catalyst interaction with H_2O_2, one-electron transfer (from catalyst to substrate) takes place, and $OH^\bullet$ radicals are generated in the reaction medium (ion radical or radical-chain mechanism of H_2O_2 decomposition is realized).

Generation of $OH^\bullet$ radical does not always proceed by means of the elementary stage of H_2O_2 disintegration. Hydrogen peroxide can form intermediate compounds of $L_nFe \cdot H_2O_2^{2+}$ type, or others, which, depending on conditions, will disintegrate with the formation of $OH^\bullet$ radical, or the process will proceed via the formation of compounds of ferryl-ions type with generation, at the end, of the same $OH^\bullet$ radicals.

Table 2.4 Mechanisms of Fe^{2+} complexes interaction with H_2O_2

Systema[a]	Elementary mechanism of interaction with $\mathbf{H_2O_2}$	Process mechanism	pH	References
Fe^{2+}–Dipy–H_2O_2	$FeDipy_n{}^{2+} + H_2O_2 \rightarrow FeDipy_n{}^{3+} + OH^{\bullet} + OH$	Radical chain	4–6	[122]
Fe^{2+}–Phen–H_2O_2	$FePhen^{2+} + H_2O_2 \rightleftarrows FePhenH_2O_2{}^{2+} \rightarrow FePhen$	Ion radical	4–5	[123]
	$FePhen_n{}^{2+} + H_2O_2 \rightarrow FePhen_n{}^{3+} + OH + OH^-$	Ion radical	2–5	[124]
Fe^{2+}–His–H_2O_2	$Fe(His)_2 + H_2O_2 \rightarrow Fe(His)_2{}^+ + OH^{\bullet} + OH^-$	Ion radical	2–5	[125]
	$Fe^{2+}(His)_2 + H_2O_2 \rightarrow Fe^{4+}(His)_2 + 2OH^-$	Ion molecular	7–8	[126]
Fe^{2+}–H_4T–H_2O_2	$H_2TFe^{2+} + H_2O_2 \rightarrow H_2Tfe^{3+} + OH^{\bullet} + OH^-$	Ion radical	2–4	[127]
Fe^{2+}–DFH_4–H_2O_2	$FeDFH_3{}^+ + H_2O_2 \rightarrow FeDFH_3{}^{2+} + OH^{\bullet} + OH^-$	Cyclic ion radical	2–4	[127]
	$FeDFH_2 + H_2O_2 \rightarrow FeDFH_2 + OH^{\bullet} + OH^-$	Cyclic ion radical	2–4	[127]
Fe^{2+}–H_3Cit–H_2O_2	$FeHCit + H_2O_2 \rightarrow FeHCit^+ + OH^{\bullet} + OH^-$	Ion radical	2–4	[127]
Fe^{2+}–H_2LM–H_2O_2	$Fe^{2+}HLM + H_2O_2 \rightarrow Fe^{4+}HLM + 2OH^-$	Radical chain	2–5	[128]
Fe^{2+}–EDTA–H_2O_2	$EDTAFe^{2+} + H_2O_2 \rightarrow EDTAFeO^{2+}$	Ion molecular	4–6	[129, 130]
Fe^{2+}–TETA–H_2O_2	$TETAFe^{2+} + H_2O_2 \rightarrow TETAFe^{4+} + 2OH^-$	Ion molecular	7–8	[46]
Fe^{2+}–CN^-–H_2O_2	$[Fe(CN)_5H_2O]^{3-} + H_2O_2 \rightarrow\rightarrow [Fe(CN)_5(H_2O)]^{2-} + OH^{\bullet} + OH^-$	Ion radical	3–5	[128, 131]

[a] Notations: Dipy—α, α′-dipyridyl, Phen—1,10-phenanthroline, His—gystidin, H_4T–tartaric acid, DFH_4–dihydroxyfumaric acid, H_3Cit–citric acid, H_2LM–lumomagnezon, EDTA–ethylenediamintetraacetic acid, TETA–triethylenetetramine

However, from the experimental data available at present, one cannot definitely conclude when this may occur.

In certain cases, especially in neutral media, the process mechanism includes the stages of two-electron transfer. It can be considered that ligands influence on ions catalytic activity in the reactions of hydrogen peroxide disproportionation is frequently connected to the initial redox potential. Therefore, to reveal the major factors influencing the catalytic activity of a number of Fe^{2+} complexes and that of other transition metals in the reactions of H_2O_2 disintegration, first of all it is necessary to find out the character of interrelation between the values of redox potential and catalytic activity of Fe^{2+} complexes using various nitrogen- and oxygen-containing compounds as ligands.

In the case of introduction of the ligand into the system $Fe^{2+}-H_2O_2$ the following types of its influence on the process are possible:

1. The reducing capacity of Fe^{2+} ions considerably increases in the forming complex, which results in essential enhancing of the reaction rate of catalyst interaction with H_2O_2. At the same time, such ligands stabilize the oxidation state of Fe^{3+}, and Fe^{3+} complexes will not react with hydrogen peroxide (or the peroxide particles radicals). In such cases, the oxidation of complex-bound Fe^{2+} into Fe^{3+} actually occurs in the system Fe^{2+}–ligand–H_2O_2 and the disproportioning process thus ceases[1] (i.e., the catalysis of H_2O_2 disintegration should not be considered).
2. Another possible variant of Fe^{2+} complexes interaction with H_2O_2 is that the direct reaction
   (a) $LFe^{2+} + H_2O_2 \rightarrow LFe^{3+} + OH^{\bullet} + OH^-$ proceeds with extremely slow rate, and the reverse one (b) $LFe^{2+} + HO_2- \rightarrow LFe^{2+} + HO_2\bullet$ will be realized rather effectively. In such systems, the process of H_2O_2 catalytic disproportioning is limited by the stage (a) and practically does not take place.
3. The rate of direct and reverse reactions will be increased, i.e., redox cycles will be formed in the system:

$$Fe^{2+} \overset{H_2O_2}{\rightleftarrows} LFe^{3+}.$$

The scheme of the aforementioned process mechanism involves many stages, and only in such cases is catalytic disproportioning of H_2O_2 by Fe^{2+} coordination compounds realized effectively.

Due to the fact that the choice of quantitative characteristic parameters of the above-mentioned ligands having an influence on H_2O_2 disproportioning by Fe^{2+} aqua ions represents a significant research and practical interest (it opens the way of

[1] In the consideration in question, the possibility of one-electron transfers realization is taken into account only, since just such mechanism has been discovered in weakly acid media. At the same time some conclusions are also acceptable for the cases of two-electron transfers stage realization.

target choice of a ligand for catalyst obtaining with certain properties), the problem involved is considered in detail below.

2.4.2 *Stages of Revealing Catalytic Redox Reactions Mechanism (Methodic Aspect*[2]*)*

The occurrence of numerous new intermediate particles along with the initial substances and the reaction products during the course of catalytic reactions brings about great experimental and calculation difficulties in the revealing of the reaction mechanism. Generation of each of these intermediate particles (free radicals, intermediate compounds, complexes with changed oxidation degree of the metal, substrate radicals, and other compounds) is characterized by the equilibrium constant of elementary rate constant. The set of all these data is necessary for revealing on a quantitative level (that practically is seldom possible) or even on a good qualitative level (there are many more examples available) the mechanism of the catalytic reaction. Actually, in order to study the mechanisms of catalytic processes it is necessary to determine the structure, stability constant of metal complexes-catalysts, separate equilibrium constants, initial substances concentration of the reacting system during process running, oxidation degree of metal ion, oxidation degree of complex compound, concentrations of final substances, nature of intermediate compounds formed, intermediate particles (free radicals), oxidation degree of metal in these intermediate compounds, bond character change in substrate on complex formation with catalyst, and other parameters. Hence, the scheme of revealing redox reaction mechanisms occurring in the natural environment or in model systems, with transition metals complex compounds as catalysts, is given in Scheme 2.4.

Brief decoding of the six-stage scheme is as follows:

1. Thermodynamic study includes the revealing of structure and stage stability constants of metal complex compounds as catalysts, determination of each possible equilibrium which can take place in the reaction medium, calculation of various complexes' structures, and revealing the composition of the most catalytically active metal complexes.
2. Study of formal kinetics including determining of the reaction orders by separate components and recording the expression for experimental rate as the function of the parameters determining it (W_{exp}.).
3. Defining and studying intermediate compounds and particles which will be formed during the reactions. This is the central and frequently the most difficult stage in revealing a reaction mechanism. Thus the change of oxidation degree of central metal ion of complex compound is revealed, whether the various

[2]This section is intended basically for the beginning researchers in the field of study of homogeneous catalytic redox reactions mechanisms.

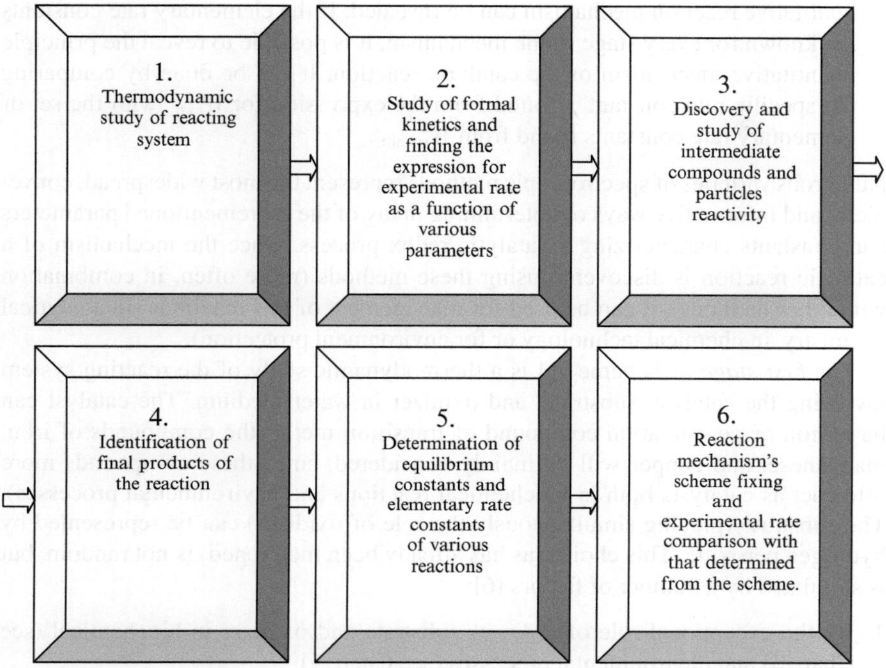

Scheme 2.4 Succession of revealing mechanisms of oxidation–reduction reaction

intermediate compounds will be formed as the result of catalyst interaction with substrate and oxidant (reducer), appropriate structures and interaction constants are determined, and various reactionary particles (free radicals) which can be formed from initial substrates or oxidizers (reducers) are identified. The sequence of their formation and concentration is defined, as well as interactions with substrate and other parameters.

4. Identification of final reaction products, their composition, concentration (output), and their influence on conditions of reaction realization (temperature, nature of solvent, ratio between the initial components), catalyst modification (for example, change of ligand nature), etc.
5. Definition of equilibrium constants and elementary rate constants of catalytic process' separate stages—necessary condition for the elimination of reactions which do not proceed when processes run in the natural environment or under the conditions of an experiment simulating a natural process.
6. On the basis of all previous experimental and calculation data, the probable scheme of a catalytic reaction is to be being considered and the rate constant may be found using it ($W_{schem.}$). Bodenstain–Semenov's method of stationary concentrations can be used for this, for example. At this final stage of research, the proposed mechanism scheme is checked on $W_{exp.}$ and $W_{schem.}$ comparison. If the elementary rate constant of at least one mechanism stage is absent, only the

qualitative reaction mechanism can be revealed. If the elementary rate constants are known for every stage of the mechanism, it is possible to reveal the principle quantitative mechanism of the catalytic reaction. It can be done by comparing the specific rate constant χ found from the expression for $W_{\text{exp.}}$ with the set of elementary rate constants found from $W_{\text{schem.}}$.

Numerous varieties of spectroscopic methods represent the most widespread, convenient, and informative ways of determining many of the aforementioned parameters and constants characterizing a catalytic redox process. Once the mechanism of a catalytic reaction is discovered using these methods (more often, in combination with other methods), it can be used for management of this reactions (in analytical chemistry, in chemical technology or for environment protection).

The first stage of Scheme 2.4 is a thermodynamic study of the reacting system involving the catalyst, substrate, and oxidizer in water medium. The catalyst can be an ion or coordination compound of transition metal (the compounds of iron, manganese, and copper will be mainly considered, since these compounds more often act as catalysts both in biochemical reactions and environmental processes). The substrate (playing simultaneously the role of oxidizer) can be represented by hydrogen peroxide. This choice (as has already been mentioned) is not random, but is stipulated by a number of factors [6]:

1. By the exceptional role of H_2O_2 as substrate and oxidizer in biochemical (see Chap. 3) and environmental processes (see Chap. 5)
2. By the relative simplicity of hydrogen peroxide
3. By the catalytic disintegration of H_2O_2 to environmentally friendly molecules of water and oxygen or to the reactionary radicals $HO_2^\bullet$ and $OH^\bullet$
4. By the opportunity of realization (depending on conditions) of basic (ion-molecular, radical, radical-chain) mechanisms of homogeneous redox catalysis in the presence of transition metal compounds (as catalysts)
5. By the ease of the concentration's determination for both initial hydrogen peroxide and products of its transformation.

Thus, the initial reacting system represents the following:

$$M^{z+}{}_{aq}\left(M^{z+}L_n\right) - H_2O_2.$$

In the solution containing a metal ion and a ligand, the equilibrium is established:

$$M^{z+} + nL \rightleftarrows M^{z+}L \rightleftarrows M^{z+}L_2 \rightleftarrows \ldots \rightleftarrows M^{z+}L_n\,(I)$$

The problem arises of finding out which of these forms of complexes is catalytically active (or the most catalytically active)? With this purpose it is necessary to determine the possible composition of metal complexes forming in solution, their preliminary stability constants, and then with the help of these constants to calculate the shares of the complexes formed ($\alpha_i = [ML_i]/C_M$) and to compare their increase or decrease with process rate changing.

To determine the composition of metal complexes formed in solution, one of the variants of Ostromyslensky–Job's method of continuous changes may be used, in which optical density is the property which varies with change of metal ion/ligand ratio (from 0 to 100%) [6]. The optical density maximum at a particular wavelength, while the metal ion/ligand ratio varies, corresponds to the composition of the complex compound.

This method is used only when one complex is formed in the system. In the case of metal complexes of various compositions being formed in the system, a similar spectrophotometric study should be performed at various wavelengths, corresponding to the absorption of complexes of various compositions.

There are many experimental and calculation methods for the determination of formed metal complexes' consecutive stability constants which are well described in many works [147–151].

The second stage of Scheme 2.4 is a study of formal kinetics and finding the expression for experimental rate ($W_{\mathrm{exp.}}$) as a function of various parameters (usually $[M^{2+}]$, [L], $[H_2O_2]$, [S], $[H^+]$). As a rule, kinetic methods of determination of the order of substances taking part in the reaction and of the effective rate constant are used.

The third stage of Scheme 2.4 is fixation and research of intermediate compounds and reactionary particles. In a catalytic redox reaction with O_2 or H_2O_2 participation in the natural environment (in the atmosphere, water, or soil) various intermediate particles are formed during its intermediate stages, including oxygenated metal complexes, peroxocomplexes, metal complexes in which metals are in higher or lower oxidation degrees than the initial metal ions, free radicals of oxidizer ($HO_2^{\bullet-}$ or $OH^{\bullet}$ radicals), and substrate free radicals. Therefore, to reveal the stages of such catalytic processes, i.e., its mechanism, these intermediate formations, their concentration, and the sequence of their occurrence in reactionary system should be identified, where possible. Intermediate stages of the general process frequently pass very quickly, and are rather reactive while forming compounds. It means that they cannot always be identified by the usual experimental methods. Frequently they are detected using special methods of fast reactions study (for example stop flow method, etc.), or their presence is proved by some indirect method (for example, by kinetic methods).

An intermediate compound is a substance which exists for from several oscillatory periods to several seconds or minutes and interacts selectively with certain substrates. Intermediate compounds formed can be thermodynamically more or less stable than the initial substance.

Intermediate compounds can be identified by various methods: spectrophotometrically, EPR, according to kinetic data, etc. To determine the structure and reactivity of these compounds, different methods can be successfully used: electron spectroscopy, Roentgen and infrared (IR) spectroscopy, EPR methods.

The study of transition metals' oxygenated complexes can be considered as an example of these methods' utilization. The reaction of the formation of oxygenated complex can be represented as follows:

$$L_nM^{z+} + O_2 \rightleftarrows M^{z+}L_n \cdot O_2,$$

where $n = 1$ or 2. On such coordination, part of the electron density is transferred from the central metal ion to O_2 and, depending on the value of such transfer in the oxygenated complex, O_2 gets the properties of superoxide- or peroxide-ion.

The formation of an oxygenated complex can be proved by spectrophotometric method. So, on manganese chloride and *n*-tripropylphosphine dissolving in tetrahydrofuran, the violet oxygenated manganese complex will be reversibly formed, where the manganese has the oxidation number 3+ and the oxygen state is close to superoxide-ion $O_2^{\bullet-}$. Reversibility of this complex was confirmed by the possibility of realization of numerous oxygenation–deoxygenation cycles.On the basis of electron, Roentgen electron and IR-spectroscopy, as well as EPR methods, it was proved that all the ions of 3d-transition metals in low oxidation degrees will form coordination compounds with suitable ligands capable of being bound reversibly with O_2 [152]. With the help of Roentgen electron spectroscopy methods, it is possible to determine the distribution of the electron density in the active center of 3d-transition metal complex with O_2 and oxidation degree of metal in oxygenated complex. Thus, complete or partial electron transfer from metal ion to oxygen molecule is observed.

Mass spectroscopy methods offer the opportunity to investigate the kinetics and to calculate activation energy of thermal deoxygenation of coordination compounds [153].

Electron spectroscopy gives the information on charge transfer strips (metal–oxygen), metal–ligand, for example, in the case of oxygenated co-complexes with various nitrogen-donor and also other ligands. Strips referring can be performed starting from the simplified scheme of molecular orbitals (MO) for one-nuclear oxygenated complex. Spectrophotometric methods enable one to identify oxygenated metal complexes and to obtain some data on their structures.

The results obtained by EPR allow one to draw qualitative conclusions about the electron structure of oxygenated metal complexes, in particular to distinguish low-spin one-nuclear oxygenated complexes from diamagnetic and paramagnetic two-nuclear complexes, as super thin splitting of the first complexes strongly differs from that of the second ones.

IR results enable the distinguishing of one-nuclear oxygenated complexes from appropriate two-nuclear ones, as the frequency of valent fluctuations O–O for the former usually lies in the range of 1000–1160 sm^{-1}, and for the latter in the range of 790–910 sm^{-1}. In two-nuclear oxygenated complexes these latter frequencies are close to the appropriate values describing the peroxide-ion. Thus, the IR-spectra provide important information on the character of an electron structure of the O_2 molecule in oxygenated metal complexes.

Similar oxygenated complexes catalyze many reactions of organic substrates oxidation by oxygen, as oxygen is already in an activated state in the complex. So, for example, an oxygenated complex with *o*-phenanthroline or diethyl amine catalyzes the reaction of hydrazine oxidation [152].

Many reactions of 3d transition metals oxygenate complexes in solution or on matrixes, which realize substrates oxidation, proceed catalytically. Therefore, oxygen activation through the formation of oxygenated metal complex and catalytic oxidation of various organic substances is rather widespread in the processes, for example, of fresh water self-purification under the action of photocatalysis and with the participation of air oxygen.

Another ecologically important reaction is the catalytic oxidation of SO_2 by oxygen in the presence of complex compounds of cobalt(II) and manganese(III) with the derivatives of phthalocyanine and porphyrin,respectively.

Using metal porphyrins ($Me^{z+}P$) as catalysts of S(IV) oxidation to S(VI) in the treatment of emitted industrial gases has shown good prospects for the given method [154, 155]. Pulse microsecond radiolysis with spectral registration of intermediate products makes it possible to watch the intermediate particles ion radical generation and their subsequent transformation [156]. The reaction of SO_2 catalytic oxidation by oxygen was studied in the presence of catalysts—cobalt tetrasulfophthalocyanine [Co(II)TSP] and manganese tetratolylporphyrin [Mn(III)TTP]. Intermediate compounds in these systems were determined spectrophotometrically. The stage of initiation is the reducing of $M^{z+}P$, O_2, or $M^{z+}P\ldots O_2$. The catalyst ($M^{z+}P$) promotes the co-ordination and direct O_2 reduction to peroxide ions O_2^{2-}, and also generation of substrate active form (radicals $HSO_3^{\bullet}$, being the initiators of S(IV) self-oxidation).

The spectral electrochemical method allows one to prove the formation of a number of free radicals and intermediate particles appearing in Scheme 2.4. In the absence of catalyst Co(II)TSP, the O_2 spectrum has maximum adsorption at $\lambda = 250\,\text{nm}$. Introduced 1×10^{-5} M Co(II)TSP into the system made the adsorption maximum shift by 260 nm and correspond to the formation of intermediate superoxide complex Co(II)TSP...O_2^{-}. Somewhat earlier, the formation of Co(II)TSP structures was registered spectrophotometrically. Therefore, the following reaction is possible:

$$\text{Co(II)TSP}\ldots O_2 + e^-_{\text{solv.}} \rightarrow \text{Co(II)TSP}\ldots O_2^{\bullet -}.$$

Further, the superoxide complex was turned into the peroxide one:

$$\text{Co(II)TSP}\ldots O_2^{\bullet -} \rightarrow \text{Co(III)TSP}\ldots O_2^{2-}$$

with $\lambda_{\max} = 350$ and 670 nm, characteristic for complex Co(II)TSP. Similar superoxide and peroxide complexes were registered spectrophotometrically. If another catalyst—Mn(III)TTP—is introduced into the system, oxo-, superoxide, and peroxide complexes Mn(III)TTP...O_2, Mn(III)TTP...$O_2^{-\bullet}$ and Mn(IV)TTP ...O_2^{2-} will be formed, respectively. Further, spectrophotometry showed the formation of complex SO_2...Co(II)TSP...O_2 if more SO_2 is added to the system. Thus, SO_2 interacts with peroxocomplex till sulfate ion formation according to two probable mechanisms:

$$\mathrm{Co(III)TSP}\ldots\mathrm{O}_2^{2-} + \mathrm{SO}_2 \rightarrow \mathrm{Co(III)TSP} + \mathrm{SO}_4^{2-},$$

$$\mathrm{Co(III)TSP}\ldots\mathrm{O}_2^{2-} + \mathrm{HSO}_3^{-} \rightarrow \mathrm{Co(III)TSP} + \mathrm{SO}_4^{2-} + \mathrm{OH}^{-},$$

$$\mathrm{Co(III)TSP}\ldots\mathrm{O}_2^{2-} + \mathrm{SO}_3^{2-} \xrightarrow{\mathrm{H_2O}} \mathrm{Co(III)TSP} + \mathrm{SO}_4^{2-} + 2\mathrm{OH}^{-}$$

or

$$\mathrm{SO}_2\ldots\mathrm{Co(II)TSP}\ldots\mathrm{O}_2 + \mathrm{e}^{-}_{\mathrm{solv.}} \rightarrow \mathrm{SO}_2\ldots\mathrm{Co(II)TSP}\ldots\mathrm{O}_2^{-} \rightarrow$$

$$\rightarrow \mathrm{SO}_2\ldots\mathrm{Co(III)TSP}\ldots\mathrm{O}_2^{2-} \rightarrow \mathrm{Co(III)TSP} + \mathrm{SO}_4^{2-}.$$

Thus, the pathway complexity of the environmentally important reaction of SO_2 oxidation by oxygen to SO_4^{2-} (H_2SO_4) in the presence of catalyst Co(II)TSP [and also Mn(III)MP] is proved by spectrophotometrical data of intermediate oxo-, superoxo-, and peroxocomplexes' as well as free radicals' formation and destruction.

Hydrogen peroxide is another oxidizer frequently participating in natural catalytic and noncatalytic processes. Its activation (as well as that of O_2) can also be carried out by metal coordination to complex with the formation of a more or less stable peroxocomplexes. As an example, peroxocomplexes may be taken, intermediate substances being formed during H_2O_2 catalytic disintegration:

$$\mathrm{Mn}^{3+} + \mathrm{HO_2}^{-} \rightarrow \mathrm{Mn}^{3+}\cdot\mathrm{HO_2}^{-} \rightarrow \mathrm{Mn}^{2+} + \mathrm{HO_2}\cdot$$

Intermediate compound $Mn^{3+} \cdot HO_2^{-}$ was detected spectrophotometrically using the technique of stop–flow at $\lambda = 470$ nm. In the general case, the formation of numerous peroxocomplexes of different composition is possible in solution: $L_nMH_2O_2^{z+}$, $L_nMHO_2^{(z-1)+}$, $L_nMO_2^{(z-2)+}$, $L_nM(H_2O_2)(HO_2^{\bullet -})^{(z-1)+}$, and $L_nM(H_2O_2)^{2z+}$. For example, in the system Mn^{2+}–phen–H_2O_2 yellow–green peroxocomplex will be formed at $\lambda = 360$ nm having the structure $Mn^{2+} : H_2O_2 = 1 : 1$:

$$[\mathrm{Mnphen_2}]^{2+} + \mathrm{H_2O_2} \rightarrow [\mathrm{Mnphen_2HO_2}]^{+} + \mathrm{H}^{+}.$$

Other intermediate particles which appear in the course of redox catalytic reactions in natural processes can be metal ions or complexes in other oxidation degrees than the initial metal ion compounds (catalysts). Spectrophotometry or EPR methods are often used for their detection and the determination of their concentration.

So, for example, Fe^{2+} ions from the bottom sediments are again oxidized by oxygen which is contained in water. In the case of their binding in complexes, the latter (that is even more effective) can participate in the generation of intermediate particles, which then will oxidize the pollutants. Thus, iron ions and complexes play an important role in natural water self-purification (see Chap. 5). The mechanism of O_2 activation by iron coordination compounds includes the formation of intermediate oxygen complex [10] and subsequent generation of hydrogen peroxide, free $O_2^{\bullet}$ and $OH^{\bullet}$ radicals, and iron compounds with oxidation number 3+.

$$LFe^{2+} + O_2 \rightleftharpoons (LFeO_2)^{2+} \xrightarrow{H^+} Fe^{2+} + L + H_2O_2$$
$$(LFeO_2)^{2+} \xrightarrow{2Fe^{2+}} 2LFe^{3+} + H_2O_2$$
$$(LFeO_2)^{2+} \xrightarrow{2H^+} LFe^{3+} + O_2^-$$

$$LFe^{2+} + H_2O_2 \rightleftharpoons (FeH_2O_2)^{2+} \xrightarrow{H^+} Fe^{2+} + L + H_2O$$
$$(FeH_2O_2)^{2+} \xrightarrow{2Fe^{2+}} 2Fe^{3+} + H_2O$$
$$(FeH_2O_2)^{2+} \xrightarrow{2H^+} LFe^{3+}OH^- + OH.$$

Fe^{2+} ions (or compounds) can be identified and determined quantitatively on the basis of the specific reaction of Fe^{2+} with *o*-phenanthroline.

Fe^{3+} ions concentration can also be determined spectrophotometrically in the presence of dye (xylenol orange) or ammonia rodanin. Thus, for example, in the case of ammonia rodanin dye, the compound will be formed with $\lambda_{max} = 550$ nm and $\varepsilon_{550} = 2.66 \times 10^4$ L/mol s.

In the case of a catalytic process, the fixation of other oxidation degrees of metal compound (catalyst) in comparison with the initial state is not always an easy task, and—sometimes—impracticable. These difficulties become stronger when the catalytic system activity increases and, as a consequence, the necessity of catalyst content reduces. Nevertheless, sometimes the formation of higher oxidation degrees of metal can be detected using spectrophotometrical methods. So, for some systems, Mn(II)–L–H_2O_2, the formation of Mn(III) and Mn(IV) compounds can be detected using this method. Various agents (complexons), benzydine, pyriphosphate, and gluconate can be used for Mn(III) detection.

The possibility of Mn(III) and Mn(IV) stabilization in the form of their gluconate complexes in alkaline medium is of peculiar interest [157–159]. Such complexes can be detected spectrophotometrically—Mn(IV) gluconate complex has the adsorption peak at 280 nm with $\varepsilon > 10^4$ L/mol s, i.e., it is possible to work with a manganese concentration of about 10^{-4} g-ion/L).

However, for example, in the case of catalytic process with manganese compounds' participation (especially in the environment), the detection of various manganese oxidation numbers is not always a simple task and is not always possible. Thus, in the case of carbonates in natural water (in sufficient concentration) and of hydrogen peroxide in neutral medium (pH 7–8), the system Mn^{2+}–HCO_3^-–H_2O_2 will be formed, in which carbonate complexes of manganese $Mn(HCO_3)_2$ are very effective catalysts of H_2O_2 decomposition. Thus, free $O_2^{-\bullet}$ and $OH^\bullet$ radicals will be formed, oxidizing the organic pollutants. This process is one of the possible ways of water reservoirs self-purification. For manganese detection in higher oxidation degrees of oxidation than 2+, gluconate is added to the system Mn^{2+}–H_2O_2 (neutral medium). In these conditions, unlike those specified above in alkaline medium [152–159] the formation of manganese(III) gluconate complex ($\lambda_{max} = 227$ nm) proceeds slowly.

When the pH value increases to 8, a "hump" appears at 280 nm, along with a spectrum of manganese(III) gluconate complex, which indicates the beginning of manganese(IV) gluconate complex formation.

In the case of a more complex system Mn^{2+}–HCO_3^-–H_2O_2, it is also possible to identify a gluconate complex of Mn(III) and to study the kinetics of its formation [160].

However, the direct detection of Mn(IV) gluconate complex in neutral medium is impossible, though this does not mean that this complex cannot appear. It can be rather quickly recombined with manganese(II) gluconate complex:

$$\text{LMn(IV)} + \text{LMn(II)} \rightarrow 2\text{LMn(III)}.$$

The difficulties of high metal oxidation degrees detection are proportional to their reactivity in redox catalytic processes in solutions. As an example, the problem of detection of assumed rather reactive particle—ferryl-ion $Fe^{iv}O^{2+}$—in which iron oxidation degree is 4+ can be used (see Chap. 3).

Radicals of oxidizer or substrate are widespread reactive intermediate particles in redox processes. When hydrogen peroxide or oxygen is the oxidizer in such processes, $HO_2^{-\bullet}$ and $OH^{\bullet}$ radicals can be generated in the catalytic system (as already shown above), playing very important roles both in biochemical and atmospheric processes, and also in the reactions occurring in natural and waste waters.

For the detection and quantitative definition of free radicals in catalytic redox reactions, the method of inhibitors is used more often [6]. An inhibitor (In) is a substance, whose addition into the reaction medium at a concentration much less than the concentration of substrate results in the essential increasing of the reaction rate. The action of inhibitors is reduced to the reaction with active intermediate particles (in particular with free radicals), chain carriers in the radical-chain mechanism. Interaction of In with free radical $R^{\bullet}$ is the only way of In's influence on radical-chain noncatalytic process rate.

In the case of catalysis of redox transformation in solutions by metal ions the inhibition of the process may proceed in different ways. One of these ways is that inhibitors are the acceptors of free radicals and metal ions with lower oxidation numbers than the initial metal ions. For instance, in the case of catalysis by M^{2+} ions in one-electron transfer mechanism:

$$M^{2+} + DH^- \rightarrow M^+ + DH^{\bullet},$$

and further

$$M^+ + A \rightarrow M^{2+} + A^{\bullet -}.$$

Fig. 2.11 Kinetic curves character in radical–chain reaction mechanism

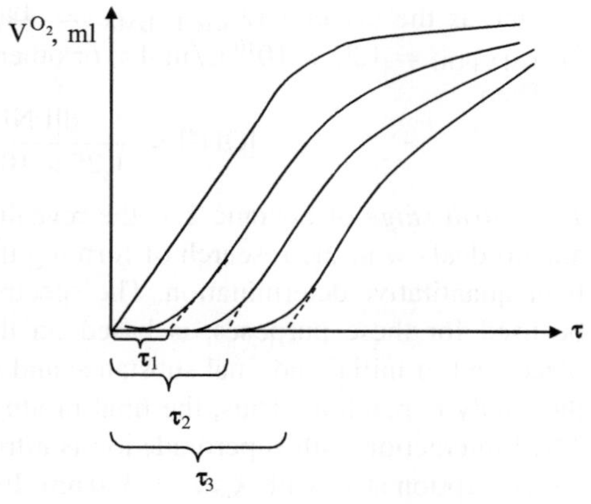

The conclusion of the radical reaction mechanism can be made if M^+ or free radicals are detected.

The method of inhibitors can be used for deciding about the radical-chain mechanism of redox catalysis. In this case, (1) well-defined induction period τ appears, (2) inhibitor consumption in the induction period occurs with constant rate equal to the inhibition rate W_i ($W_i = W_{\text{initiation}}$), (3) increasing of the induction period is done proportionally to the concentration of inhibitor introduced, and (4) after the inhibitor's consumption the initial rate is restored. These attributes of the radical-chain mechanism are well illustrated in Fig. 2.11.

Here,

$$W_{\text{I}} = \tilde{n} \cdot \frac{[\text{In}]_0}{\tau},$$

where $\tilde{n}$ is the number of radicals interacting with one molecule of inhibitor (usually = 1 or 2).

The expression

$$W_{\text{I}} = f\left(C_{\text{M}}, C_{\text{L}}, C_{\text{SH2}}, C_{\text{H}_+}, \text{etc.}\right)$$

in comparison with W_{exp} gives valuable information on the process mechanism.

With the help of spectrophotometric methods, it is possible to determine the presence of $HO_2^{\bullet}$ ($O_2^{\bullet -}$) radical since the inhibitor interacting with the free radical generates a colored compound. For example, if superoxide ion-radical $O_2^{-\bullet}$ is formed in the system, the specific inhibitor can be TNM [$C(NO_2)_4$]:

$$\text{C}\,(\text{NO}_2)_4 + \text{O}_2^{\bullet -} \xrightarrow{k=1.9\times 10^9\ \text{l/mol s}} \text{C(NO}_2)_3{}^- + \text{O}_2 + \text{NO}_2.$$

The nitrophorm [ion $C(NO_2)_3^-$] formed has an electron adsorption strip at $\lambda_{\max} =$ 350 nm.

In the case when $OH^{\bullet}$ radicals will be formed in the reaction, the spectrophotometric method can also be used for their detection, and paranitrosodimethylaniline as the detector ($\lambda_{max\ PNDMA} = 440\,nm$, $\varepsilon_{440} = 3.4 \times 10^4\,L/mol\ s$ and $k_{PNDMA+OH} = 1.25 \times 10^{10}\,L/mol\ s$) or other detectors.

Then

$$[OH^\bullet] = \frac{d[PNDMA]/dt}{1.25 \times 10^{10}[PNDMA]}.$$

The fourth stage of Scheme 2.4 (the revealing of catalytic redox reactions mechanism) deals with the research of forming the final products' chemical nature and their quantitative determination. The spectrophotometric method, which can also be used for these purposes, is based on the difference in the adsorption strips placement of initial and final substance and on the change of their intensity during the catalytic reaction. Thus, the final product of the reaction considered above of TNM interaction with superoxide ion is nitrophorm ion, which is characterized by the adsorption strip with $\lambda_{max} = 350\,nm$. Its formation and accumulation kinetics can be monitored spectrophotometrically.

The fifth stage of Scheme 2.4 is the determination of equilibrium constants and elementary rate constants.

To get the quantitative characteristics of certain stages of the mechanism, it is important to know the elementary rate constants of free radicals' interaction with all reacting substrates (K_{R+S}). There are many such constants in the literature, determined by various methods, and in particular by the method of competing acceptors (MCA). For example, the rate constant of $OH^\bullet$ radicals interaction with any substrate $S(K_{OH+S})$ can be defined by this method by the comparison of experimentally obtained constants with those available from the literature. For this purpose, the number of chemically various substrates-acceptors (S) can be chosen for which the constants of interaction with $OH^\bullet$ are known. It is necessary to investigate their influence on the PNDMA oxidation process in the system Mn^{2+}–HCO_3^-–H_2O_2–PNDMA–S. PNDMA consumption during the reaction can be followed spectrophotometrically.

The addition of S in certain concentrations results in the clear braking of PNDMA oxidation rate caused by the competition of S and PNDMA for $OH^\bullet$ radicals. Calculation of K_{OH+S} values is carried out according to the expression:

$$\frac{1}{\Delta C_{PDNMA}} = A\left(1 + \frac{k_3[H_2O_2] + k_4[HCO_3^-]}{k_1[PNDMA]} + \frac{k_2[S]}{k_1[PNDMA]}\right),$$

where ΔC_{PNDMA} is PNDMA concentration change during H_2O_2 catalytic decomposition in the presence of certain amount of competing acceptor S and A is the empirical coefficient identical to studied substances under constant experimental conditions. It can be determined as a portion on an axis of ordinates of dependence $1/\Delta C_{PNDMA}$ from [S]/[PNDMA]—the angle of inclination is equal to $A \cdot k_2/k_1$. In the given expression it is necessary to take into account the consumption of $OH^\bullet$ radicals not only in reactions:

$$OH^{\bullet} + PNDMA \xrightarrow{\kappa_1} \text{decolorizing} \left(\kappa_1 = 1.25 \times 10^{10}\ \text{l/mol s}\right), \quad (2.6)$$

$$OH^{\bullet} + S \xrightarrow{k_2} \text{products.} \quad (2.7)$$

If other substances are present in the water solution besides PNDMA, the $OH^{\bullet}$ consumption will proceed in many ways, and the appropriate number of rate constants and substrates concentration should generally appear in the formula for K_{OH+S} (in this case, k_2) calculation. Additional ways of OH consumption in this system will be:

$$OH^{\bullet} + H_2O_2 \xrightarrow{k_3} HO_2^{\bullet} + H_2O \left(k_3 = 1.7 \times 10^{7}\ \text{l/mol s}\right), \quad (2.8)$$

$$OH^{\bullet} + HCO_3{}^{-} \xrightarrow{k_4} CO_3^{\bullet -} + H_2O \left(k_4 = 1.5 \times 10^{7}\ \text{l/mol s}\right). \quad (2.9)$$

Hence, for K_{OH+S} determination at neutral pH, the system $Mn^{2+}-HCO_3{}^{-}-H_2O_2$–PNDMA–S can be used. Proceeding from methodical reasons, as an example, the opportunity to determine the elementary rate constant of k_{OH+S} has been explained in enough detail.

The sixth stage of Scheme 2.4 describing the revealing of a redox reaction mechanism consists in drawing up the probable scheme of the catalytic reaction mechanism, reaction rate calculation ($W_{calc.}$) (using the method of stationary concentration), and $W_{exp.}$ and $W_{calc.}$ rates comparison. In the case of concurrence of a specific rate constant χ with a set of elementary rate constants, and also concurrence of interacting components by the orders, it can be considered that the reaction mechanism was proved quantitatively. Concurrence of $W_{exp.}$ and $W_{calc.}$ only by orders of interacting components testifies that the mechanism is correct only qualitatively.

Further examples of reactions that can proceed in the environment are given on a model level. Thus, in waste waters from textile manufactures residual dyes can be present. The problem of dyes removal or destruction in waste water is encountered all over the world. Besides, the reactions of oxidation by Ruff's system ($Fe^{3+}-H_2O_2$) of organic dyes are important for the development of kinetic methods of determination of iron micro amounts in weakly acid water. Taking as an example the destruction (oxidation) of a dye—acid chrome dark blue (ACDB)—by hydrogen peroxide in the presence of Fe(III) compounds, the method of revealing the mechanism of this process will be shown. If this mechanism is known, it is possible to operate this reaction. Certainly, in this ($Fe^{3+}-H_2O_2$–ACDB) and in other similar systems, the data obtained by other methods (mainly volumometric, sometimes IR-spectroscopic and EPR) are also used along with the spectrophotometric method. Nevertheless, the adsorption electron spectra frequently provide a large (or essential) contribution into the revelation of reaction mechanisms.

Oxidation rate in this system is founded on the basis of formal kinetics [10]:

$$W_{exp.}{}^{ACDB} = \chi \cdot [Fe^{3+}]_0 \cdot \frac{[ACDB]_0 \cdot [H_2O_2]_0}{(1 + m \cdot [ACDB]_0) \cdot (1 + n \cdot [H_2O_2]_0)},$$

Initiation

i. $Fe^{3+} + ACDB \rightleftharpoons Fe^{3+}ACDB \xrightarrow{\kappa_i} Fe^{2+} + R_1$ ($\kappa_i = 20$ l/mol·s),

Chain continuation

1. $Fe^{2+} + H_2O_2 \longrightarrow Fe^{3+} + OH^- + OH\cdot$ ($\kappa_1 \approx 53$ l/mol·s),
2. $OH\cdot + H_2O_2 \longrightarrow HO_2\cdot + H_2O$ ($\kappa_2 = 3{\cdot}3{\cdot}10^7$ l/mol·s),
3. $OH\cdot + KXTC \longrightarrow R_2$ ($\kappa_3 = 1{\cdot}10^9$ l/mol·s),
4. $OH\cdot + Fe^{2+} \longrightarrow Fe^{2+}OH^-$ ($\kappa_4 = 3{\cdot}10^8$ l/mol·s),
5. $Fe^{3+} + HO_2^- \longrightarrow Fe^{2+} + HO_2\cdot$ ($\kappa_5 = 2{\cdot}1{\cdot}10^6$ l/mol·s),
6. $R_2 + Fe^{3+} \longrightarrow Fe^{2+} + R$ ($\kappa_6 = 10^8$ l/mol·s),

Chain breakage

7. $Fe^{2+} + HO_2\cdot \longrightarrow Fe^{3+} + HO_2^-$ ($\kappa_7 = 7.2{\cdot}10^5$ l/mol·s).

Scheme 2.5 Electrons transfer stage on the ascorbic acid oxidation with the molecular oxygen in the presence of Cu(II) and Fe(III) compounds

where m and n are constant values. By the method of inhibition (as described above, using spectrophotometric methods), it was found that TNM and PNDMA brake the reaction of ACDB oxidation, i.e., $O_2^{-\bullet}$ and $OH^\bullet$ radicals are generated in the studied system. On introduction into the reaction medium of TNM and PNDMA simultaneously, PNDMA is consumed first, and only later the TNM. This testifies that $OH^\bullet$ radicals are generated in the system first, and $HO_2^\bullet(O_2^{-\bullet})$ will be formed in later stages. ACDB possesses reducing properties and at its interaction with Fe^{3+}, Fe^{2+} will be formed. The scheme of reaction is as follows (Scheme 2.5):

Such a scheme is confirmed by the following experimental data obtained by the spectrophotometric method [10]:

- ACBD effectively interacts with Fe^{3+}, reducing it to Fe^{2+}.
- First $OH^\bullet$ and then $HO_2^\bullet$ radicals will be formed in the system.
- ACDB is oxidized by $OH^\bullet$ radicals competitively with H_2O_2.
- Fe^{2+} ions interact with H_2O_2, instead of the complex Fe^{2+} with ACBD that follows from the study of systems Fe^{3+}–ACDB–H_2O_2(I) and Fe^{2+} – ACDB–H_2O_2(II). In system (II), the initial oxidation rate of ACDB is almost twice as much, as in the case of system (I).
- The number of elementary rate constants has been calculated.

According to the above-mentioned mechanism scheme, using the method of stationary concentrations, the following expression can be obtained:

$$W_{cal}^{ACBD} = \left(\frac{\kappa_1 \cdot \kappa_3}{\kappa_i}\right)^{1/2} \cdot [Fe^{3+}]_0 \cdot [H_2O_2]_0^{0.5} \cdot [ACBD]_0{}^{0.5},$$

which is in qualitative accordance with W_{exp}. [10].

Knowledge of catalytic redox reactions' mechanisms makes it possible to choose optimal conditions for the analysis of different substances (metal ions, inorganic and organic substances), including those in natural objects. For analytical scopes, the spectroscopic methods in catalymetry are especially suitable.

2.5 Mechanisms of Oxide and Peroxide Systems Redox Transformations Catalysis with Transition Metal Compounds

2.5.1 *Mechanisms of O_2 Transformation in Oxide Catalytic Systems*

A detailed consideration of mechanisms of dioxygen transformation (activation) in solutions was given briefly in Sect. 2.2.2, and also in [129]. Here the focus will be on separate aspects of this problem and on examples of O_2 transformation mechanisms catalyzed by manganese, iron, cobalt, and copper complexes.

The most favorable conditions for ML_n interaction with O_2 are provided in the following cases [129]: first, in tetrahedral or cubic symmetry configuration—exited states of metal complex with electron configuration d^5–d^7; second, in D_{4h} symmetry—strong tetragonal distortion of octahedron (coordination number = 6), formation of square prism structure (coordination number = 6) or square pyramids (coordination number = 5), in the case when energy of one of t_{2g}-orbitals (d_{xy}) increases, and the energy of one of e_g-orbitals (d_{z^2}) is lowered (in case of ions with electron configuration d^3–d^{10}—high-spin, and ions d^5–d^{10}—low-spin). Other types of symmetry distortions can also be efficient (for example, for ions with few d-electrons). Besides, the formation of a complex of MO_2^+ (or LMO_2^+) type is also influenced by thermodynamic factors. Mechanisms of interaction of transition metals compounds of the first line with O_2 are considered below.

The mechanism of Cu^+ compounds interaction with O_2 in the system Cu^+–L–O_2 (L = 2,2′-dipyridyl) was studied in some detail in weakly acid, neutral, and alkaline media. The most complete mechanism of O_2 interaction with Cu^+L_2 (L = 2,2′-dipyridyl) is given in [80, 161] (Scheme 2.6), where indexes I and II refer to structures MO^{2+} according to Griffith and Pauling, respectively. The reaction rate in acid and alkaline media is limited by the rate of $L_2\left(CuO_2^+\right)_I$ formation, and in neutral media by the rate of $L_2\left(CuO_2^+\right)_{II}$. Complex $L_2\left(CuO_2^+\right)_I$ is an oxygen complex of "peroxide" type with equivalent O atoms:

$L_2(CuO_2^+)_I = L_2$ 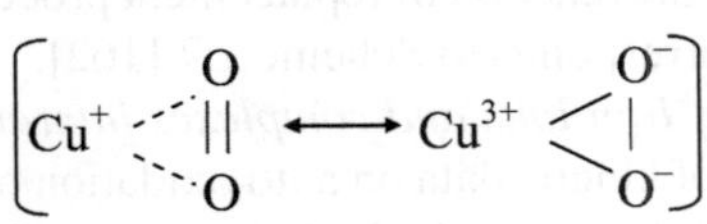 ,

and complex

$$Cu^{+}L_2 + O_2 \underset{3{,}7\cdot 10^{6},\ s^{-1}}{\overset{10^{5},\ M^{-1}s^{-1}}{\rightleftarrows}} L_2(CuO_2^{+})_I,$$

$$L_2(CuO_2^{+})_I + H^{+} \xrightarrow{\sim 10^{10},\ M^{-1}s^{-1}} L_2(CuHO_2)^{2+}\ (\longrightarrow Cu^{2+}L_2 + HO_2\cdot),$$

$$L_2(CuO_2^{+})_I + OH^{-} \underset{10^{9},\ s^{-1}}{\overset{3\cdot 10^{10},\ M^{-1}s^{-1}}{\rightleftarrows}} L(OH^{-})(CuO_2^{+})_I + L,$$

$$L(OH^{-})(CuO_2^{+})_I \xrightarrow[(OH^{-})]{6\cdot 10^{4},\ s^{-1}} LCu^{2+}(OH^{-})_2 + O_2\cdot^{-}$$

$$L(OH^{-})(CuO_2^{+})_I \xrightarrow[(L)]{7\cdot 10^{2},\ s^{-1}} L_2Cu^{III}O^{+} + HO_2^{-},$$

$$L_2(CuO_2^{+})_I \underset{1{,}5\cdot 10^{5},\ s^{-1}}{\overset{6\cdot 10^{5},\ s^{-}}{\rightleftarrows}} L_2(CuO_2^{+})_{II},$$

$$L_2(CuO_2^{+})_{II} + Cu^{+}L_2 \xrightarrow[(2H^{+})]{\sim 10^{9},\ M^{-1}s^{-1}} 2Cu^{2+}L_2 + H_2O_2,$$

$$L_2(CuO_2^{+})_{II} + L \xrightarrow{\sim 10^{8},\ M^{-1}s^{-1}} Cu^{2+}L_2 + O_2\cdot^{-} + L,$$

$$L_2(CuO_2^{+})_{II} \xrightarrow{5\cdot 10^{4},\ s^{-1}} Cu^{2+}L_2 + O_2\cdot^{-},$$

$$Cu^{+}L_2 + HO_2\cdot(O_2\cdot^{-}) \xrightarrow[-H^{+}]{} Cu^{2+}L_2 + HO_2^{-},$$

$$Cu^{2+}L_2 + O_2\cdot^{-}(HO_2\cdot) \xrightarrow[-H^{+}]{} Cu^{+}L_2 + O_2,$$

Scheme 2.6 Mechanism of O_2 interaction with $Cu^{+}L_2$

$$L_2(CuO_2^{+})_{II} = L_2\left[Cu^{+} \cdots O{=}O \longleftrightarrow Cu^{2+} - O{-}O^{-}\right],$$

in which oxygen has only partially passed into the form $O_2^{-\bullet}$ radical (complex of "superoxide" type). Cu^{+} aqua ion can similarly form with oxygen complexes of types I and II, and rate constants of its formation (and also that of oxygenated dipyridyl complexes) are given in [129]. When an additional ligand [OH^{-}, EDTA, malonic acid (H_2A), etc.] is introduced into the system where complexes of types I or II are formed, the reaction of replacement proceeds and the mechanism of $Cu^{+}L_2$ autooxidation corresponds to Scheme 2.7 [162].

Mechanism of iron ions and complexes interaction with O_2 was also studied in detail. The sum of kinetic data on auto oxidation of YFe^{2+} (Y = EDTA tetra anion) can be expressed, for example, by Scheme 2.8 [129], where x is oxidation product of coordinated EDTA and B^{-} is anion of iron(II) salt. Thus the high value of the

$$L_2(CuO_2^{+})_I + A^- \underset{\sim 10^9,\ M^{-1}s^{-1}}{\overset{4\cdot 10^9,\ M^{-1}s^{-1}}{\rightleftarrows}} L(A^-)(CuO_2^{+})_I + L,$$

$$L_2(CuO_2^{+})_I + HA^- \underset{10^9,\ M^{-1}s^{-1}}{\overset{1{,}3\cdot 10^9,\ M^{-1}s^{-1}}{\rightleftarrows}} L(AH^-)(CuO_2^{+})_I + L,$$

$$L_2(CuO_2^{+})_{II} + A^- \xrightarrow{2{,}1\cdot 10^9,\ M^{-1}s^{-1}} L(A^-)(CuO_2^{+})_{II} + L,$$

$$L(A^-)(CuO_2^{+})_{I,\ II} \longrightarrow LCuA + O_2{\cdot}^-,$$

$$L(AH^-)(CuO_2^{+})_I \longrightarrow LCuA + HO_2{\cdot}.$$

Scheme 2.7 Mechanism of $Cu^{+}L_2$ auto-oxidation

$$1.\ YFe^{2+} + O_2 \underset{10^6,\ c^{-1}}{\overset{2{,}3\cdot 10^3,\ M^{-1}s^{-1}}{\rightleftarrows}} Y(FeO_2)^{2+},$$

$$2.\ Y(FeO_2)^{2+} + H^+ \xrightarrow{\sim 10^{10},\ M^{-1}c^{-1}} Y(FeHO_2)^{3+}\ (\longrightarrow YFe^{3+} + HO_2{\cdot}),$$

$$3.\ Y(FeO_2)^{2+} + YFe^{2+} \xrightarrow[2H^+]{\sim 10^{10},\ M^{-1}s^{-1}} 2YFe^{3+} + H_2O_2,$$

$$4.\ Y(FeO_2)^{2+} + B^- \xrightarrow{\sim 10^{10},\ M^{-1}s^{-1}} Y(FeO_2)^{2+}\ldots B^- \begin{cases} \longrightarrow YFe^{2+} + O_2 + B^- \\ \longrightarrow YFe^{4+}B^- + H_2O_2, \end{cases}$$

$$5.\ YFe^{4+}B^- + YFe^{2+} \xrightarrow{\sim 10^{10},\ M^{-1}s^{-1}} 2YFe^{3+} + B^-,$$

$$6.\ YFe^{4+}B^- \longrightarrow XFe^{3+} + B^-,$$

Scheme 2.8 Mechanism of YFe^{2+} auto-oxidation

rate constant of $Y(FeO_2)^{2+}$ peroxide type complex interaction with the second iron ion (stage 3) is observed.

Complexes of superoxide type with $Co^{2+}(d^7)$ [163–165] and $Fe^{2+}(d^6)$ [166] with salicylideneimine, porphyrins, and other macrocyclic ligands were obtained, this fact testified by angles M–O–O in these complexes (118–138°), O–O bond lengths (1.26–1.3 Å) and IR-spectra frequencies ($\nu = 1,100$–$1,150\ \text{sm}^{-1}$). On Co^{2+} oxygenation, oxygen complexes of only superoxide type will be formed which on Co^{2+} ions excess in water solutions will form binuclear peroxocomplexes. Mechanism of oxygenation of Co^{2+} complexes in water solutions can be expressed as:

$$L_5Co^{2+}X + O_2 \rightleftarrows L_5CoO_2^{2+} + X,$$

$$L_5CoO_2^{2+} + XCo^{2+}L_5 \overset{X}{\rightleftarrows} (CoOOCo)^{2+} \xrightarrow[2H^+]{Y} 2L_5Co^{3+}Y + H_2O_2,$$

where X is a bridge ligand.

According to the PCT concept, binuclear oxygen-containing complexes can be presented as follows:

$$CoOOCo^{2+} = \left[Co^{+} \cdots O=O \cdots Co^{2+} \longleftrightarrow Co^{2+} - O^{-}-O^{-} - Co^{2+} \right]$$

Similar two-nuclear peroxocomplexes will form natural O_2 carriers—gemetrin $(Fe^{2+})_2$ and gemocyanin $(Cu^{2+})_2$ [167]. Complexes of the same type were also obtained in the case of Mn^{2+} (d^5) compounds with phthalocyanine [168].

The general scheme of typical formation mechanism (for different metal ions and complexes) and further transformations of oxygen complexes of metals in the absence of oxidized substrates under the influence of solvent, ions H^+ and OH^-, free ligand (L), modifying ligand (BH), or bridge one (X–) can be summarized as Scheme 2.9 [129].

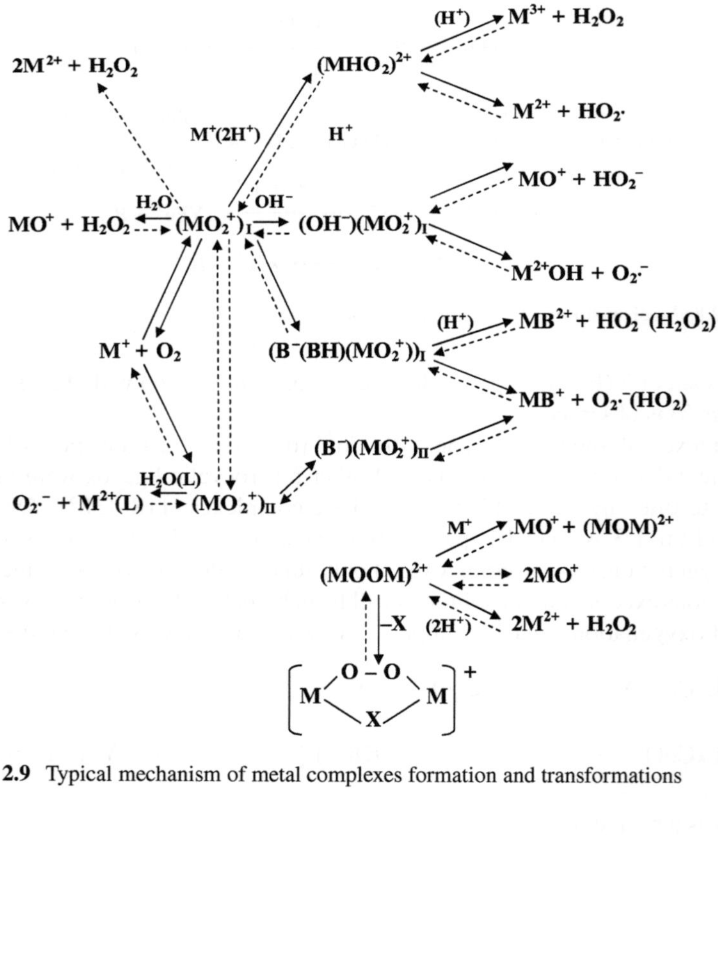

Scheme 2.9 Typical mechanism of metal complexes formation and transformations

Considering the reactivity of oxygen complexes of metals, the symmetry of surrounding ligands should be taken into account (hence the presence of potential-vacant places in the coordination sphere), the charge of metal ion (provides the certain value of oxygen redox potential in the composition of complex or facilitates the breakage of O–O bond), and lability of metal ion coordination sphere (in the reactions of ligand replacement by substrate and triple complex formation of partial carry electron transfer and its subsequent disintegration on the products of one- and two-electron transfers).

Oxygen metal complexes of superoxide type can take part in the following reactions [129]:

1. Outer-sphere electron transfer

$$(MO_2{}^+)_{II} + RH \rightarrow M^{2+} + HO_2{}^- + R^\bullet;$$

2. Inner-sphere redox transformations.

$$(MO_2{}^+)_{II} + RH \rightleftharpoons (MO_2RH)^+ \rightarrow \begin{cases} M^{2+} + HO_2{}^- + R^\bullet \\ M^+ + HO_2\cdot + R^\bullet \\ M^+ + H_2O_2 + P \end{cases}$$

Oxygen complexes of peroxide type can also be transformed by two redox directions:

1. By the reactions of inner-sphere (one- or two-electron) oxidation of coordinated redox-ligand.
2. By the reactions resulting in O–O bond breakage on substrate (RH) coordination:

$$(MO_2{}^+)_I + RH \rightarrow MO^+ + ROH.$$

Similar rather important reactions of catalytic hydroxylation by metal oxygen complexes of numerous organic compounds will be considered in detail in Chap. 3 (hydroxylation by monogenases and by their models).

As a model hydroxylating system on the basis of O_2, the Udenfrind's system should bementioned:

$$Fe^{3+}\left(\text{or}\,Fe^{2+}\right) - EDTA - O_2\ AK\,(\text{or}\,DH_2)\,.$$

Here AK is ascorbic acid and DH_2 any suitable reducer, which hydroxylates the aromatic compounds, olefines, and saturated hydrocarbons (pH 3.5–6.5) according to the following reaction:

$$RH + DH_2 + O_2 \rightarrow ROH + D + H_2O.$$

However, the mechanism of processes in this system and similar ones is rather complicated, the stages realized depending on the ligand chosen, properties of oxidized substrate, conditions of hydroxylation realization, etc. Literary data on the mechanism in Udenfrind's system are rather inconsistent, it is insufficiently developed even on a qualitative level, and therefore it will not be considered further (the statement of the versions of mechanisms in system Udenfrind's system can be found in [129, 169–171]. It should be mentioned that one of such versions is the assumption that an active intermediate particle in hydroxylating Udenfrind's system is an oxygen complex of the peroxide type—$Y(FeO_2)^{2+}$ [129].

2.5.2 Mechanisms of H_2O_2 Decomposition in the Peroxide Catalytic Systems

2.5.2.1 Active Intermediate Particles and Methods of Their Revelation

Despite the 100-year's history of the study of H_2O_2 catalytic decomposition in the presence of transition metal's ions or complexes, the interest in this research area does not die away. It occurs not only because hydrogen peroxide is an intermediate product of metabolic catalytic transformations in living organisms, takes part in many atmospheric processes (for example, in the formation of "acid rains"), largely causes natural waters' self-purification, undergoes numerous transformations on the realization of various industrial processes and many others (for details see Chaps. 3–6), but also because our knowledge in understanding that elementary catalytic redox act needs to become deeper all the time [129]. Revealing the nature of the primary active intermediate particles in the reactions of electron transfer, i.e., discovering the number of electrons transferred in the elementary act (deeper insight into the "elementary act" itself) is the dominating factor. With changing the notion of the elementary act, the study of H_2O_2 catalytic decomposition is continually being renewed.

It is known that hydrogen peroxide is dual by nature: it is a strong two-electron oxidizer in its undissociated form and an electron donor in the ionized condition. Thus, on the one hand it reduces systems containing ions of transition metals, on the other it generates a number of free radicals ($OH^{\bullet}$, $HO_2{}^{\bullet}$), intermediate particles of ferryl or manganyl type, and intermediate compounds of various composition.

On metal compound interaction in oxidized form M^{2+} (here and later M^{2+} is metal ion or metal complex in the oxidized form, with M^{+} in the reduced form) with hydrogen peroxide the peroxocomplex will be formed, undergoing acid-alkaline dissociation and then decomposition:

$$M^{2+} + H_2O_2 \rightleftarrows (MH_2O_2)^{2+} \rightleftarrows H^{+} + MHO_2{}^{+} \rightleftarrows M^{+} + HO_2\cdot$$

Peroxocomplex $MHO_2{}^{+}$, possessing reducing properties, either dissipates on the products of one-electron transfer, or, interacting with the second metal ion, is oxidized to O_2 or free radicals.

In the case of H_2O_2 interaction with the reduced metal form M^{+}, two different mechanisms have been discussed. The first was identified by Gaber and Vilsteter in 1931 [172]:

$$M^{+} + H_2O_2 \rightarrow M^{2+} + OH^{-} + OH^{\bullet},$$

$$M^{+} + OH^{\bullet} \rightarrow M^{2+}OH^{-}.$$

The second was identified by Bray and Gorin in 1932 [173]:

$$Fe^{2+} + H_2O_2 \rightarrow FeO^{2+} + H_2O,$$

$$FeO^{2+} + Fe^{2+} \xrightarrow{H^{+}} Fe^{3+} + Fe\,(OH)^{2+}$$

with intermediate formation of ferryl-ion $Fe^{IV} = O^{2+}$ (this situation from the quantum-chemical aspect will be considered in Chap. 3).

Because of high reactivity, with both $OH^{\bullet}$ radicals and particles of ferryl-ion type, their stationary concentration in the system is very low and consequently the presence of these particles is usually judged by the occurrence of their interaction product with suitable acceptor. For $OH^{\bullet}$ radical and ferryl ion, one such acceptor is maleic acid which forms, with $OH^{\bullet}$ Radicals, the sticking radical by the double bond (R_1).

With ferryl ion, the carboxyl radical (R_2) will be formed:

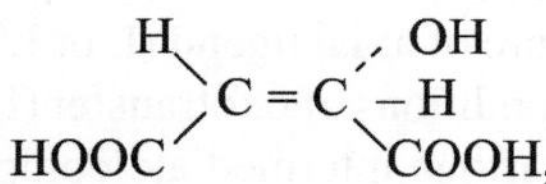

With ferryl ion, carboxyl radical (R_2) will be formed:

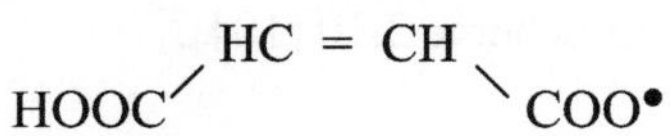

R_1 and R_2 are much longer living particles and they can be detected by EPR method in flow conditions.

The other method of intermediate particle detection is based on the analysis of the reaction products structure (their ratio) that, in particular, was used for the proof of the ferryl particle's occurrence in biochemical processes.

Intermediates of ferryl ions and complexes in H_2O_2 decomposition catalysis were considered in detail in [93]. Here these data will only be briefly used for the experimental confirmation of such high-reactive particles formation. There are experimental data according to which $OH^{\bullet}$ radicals are generated in Fenton's system at low pH, and at higher pH values in the presence of suitable ligand-stabilizing high oxidation degree iron, not the OH radical; the reaction:

$$L_nFe^{2+} + H_2O_2 \rightarrow \left(L_nFe^{IV} = O\right)^{2+} + H_2O$$

involves the less reactive ferryl complex $(L_nFe^{IV} = O)^{2+}$ (although acting as oxidant). The proof of either $OH^{\bullet}$ radicals or ferryl particles' formation can serve the peculiarities of transformation products distribution of a number of substrates. In acid media, the distribution is close to that taking place on $OH^{\bullet}$ radicals generated by radiolysis. In neutral and alkaline media, this distribution is more stereospecific, i.e., in the latter case ferryl iron presumably represents the oxidant [57, 174–178]. For ferryl particles' detection and their differentiation from $OH^{\bullet}$ radicals, various experimental methods were used [179, 180]: pulse radiolysis, stop flow methods, generation of Fe^{IV} compounds by Fe(III) salts or complexes oxidation with free $OH^{\bullet}$ radicals, and others. On cyclohexanol oxidation by hydrogen peroxide in the presence of Fe(II) salts in acetonitryl, radioselectivity of products was observed, this fact being explained by the generation of ferryl particles [181]. The same ferryl complex was discovered in reactions of H_2O_2 with iron(III) porphyrin in water solution [182] and with complex Fe^{II}EDTA [183]. The formation of ferryl complexes was confirmed with the help of Raman spectroscopy and other methods involving H_2O_2 interaction with some enzymes such as peroxidases and catalases [184–187].

Ferryl group formation experimentally was also proved in model systems, simulating peroxide oxidation of lipids [188]. The investigated mechanisms of olefins epoxidation by hydrogen peroxide and hydroperoxides involve the formation of oxoporphyrin π radical-cation $P^{+}Fe^{IV} = O$ [189, 190] as substrates oxidant. It was proved by intensive strip occurrence in a resonant Raman spectrum at $\lambda = 852\,\mathrm{sm}^{-1}$ (~15 K) [191]. The ferryl group's electron configuration ($\sigma^2\pi^4$) proves that there is a double bond Fe = O in it [192].

The source of oxygen, nature of axial ligand (L or L′), production method, nature of solvent, and temperature conditions of oxotransfer (frequently below −50 °C) can influence the formation of various intermediate compounds [193]. Ferryl particle $Fe^{IV}O$ cannot be detected in water solutions or extracted in the form of appropriate porphyrin complex; it is easier to discover it in a nonwater medium. Various methods for its formation are given in Scheme 2.10 [104].

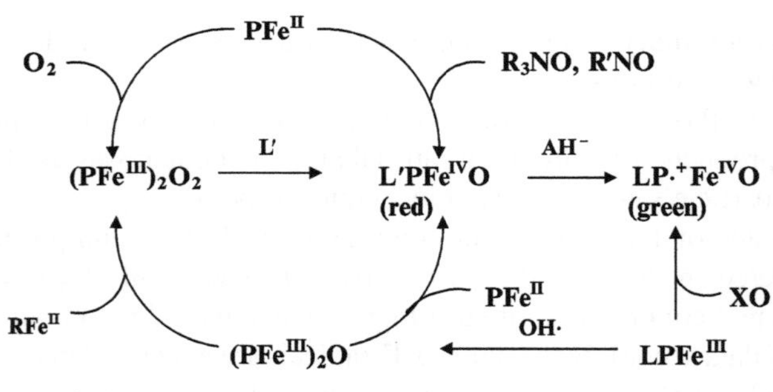

Scheme 2.10 Methods of the ferryl particle formation

Here, PFe^{II} is the porphyrin complex with Fe^{II}, $L'PFe^{IV}O$ the complex with additional ligand (L′, methylimidazole or pyridine), $LP^{\bullet +}Fe^{IV}O$ the same with $L = ClO_4^-$, $LPFe^{III}$ the complex of Fe(III), $(PFe^{III})_2O$ the binuclear oxocomplex of Fe(III), $(PFe^{III})_2O_2$ the μ-peroxodimer, AH^- an electron acceptor, and XO an oxidizer (R_3NO or R′NO)].

In the processes realized by the enzymes, the formation of oxoiron(IV) porphyrin cation radical is presumed. In particular, generation of such a particle is supposed to take place in the reactions of alkanes hydroxylation, carried out by cytochrome P-450 [194, 195]. Scheme 2.11 covers this process [196].

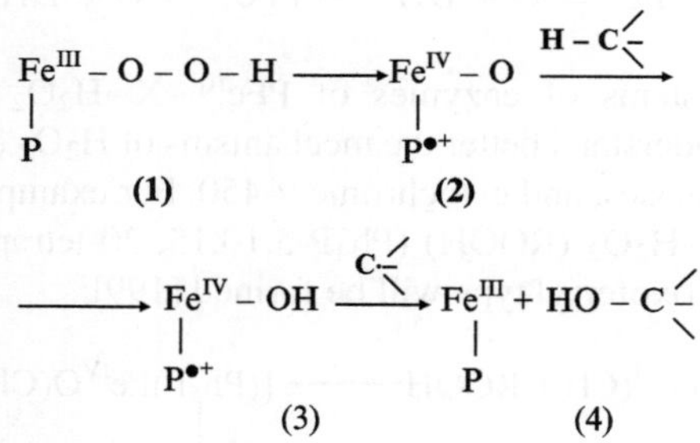

Scheme 2.11 Hydroxylation of alkanes with cytochrome-450

Heterolytical breakage of the O–O bond in the hydroperoxo complex of Fe(I) is assumed in this scheme with the formation of one-electron oxidized iron porphyrin (2) (oxidant). Complex (2) interacting with substrate tears the hydrogen atom from it and forms the hydroxocomplex Fe^{IV} (3), interacting further with the substrate radical which transforms it into the hydroxylated product with the initial ferryl-form (4) of cytochrome P-450 regeneration.

On H_2O_2 interaction with various peroxidases (Per), inner and outer sphere intermediate compounds will be formed, and then the compounds PerO of ferryl type (Fe^V–O or, more precisely, $P^+Fe^{IV}O$) will be the following [197]:

$$Per + H_2O_2 \rightleftarrows Per \cdot H_2O_2 \rightleftarrows (PerH_2O_2) \rightarrow PerO + H_2O.$$

The energy barrier of Fe–O bond formation is insignificant (activation energy of total process is approximately 14.7 kJ/mol). In the presence of reducers (DH^-) or additional electrons, the final product becomes the compound of $Fe^{IV}O$ type which can, in its turn, be reduced to the state Fe(III) of initial peroxidase [93]:

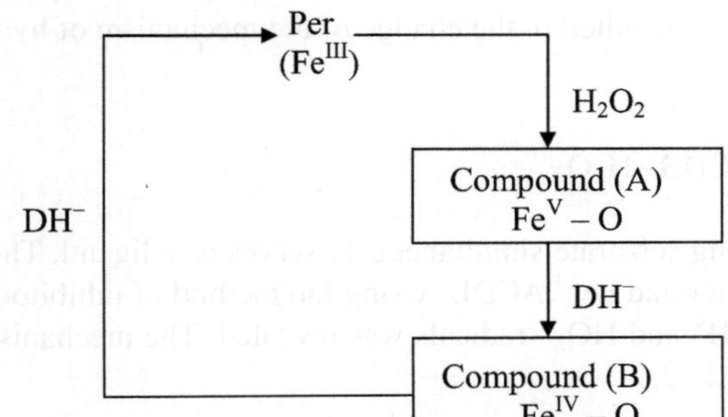

Compounds A and B can compete for the reducer; at low pH this competition is won by the compound B, at high pH (pH 9) by compound A. The A compound is oxoiron(IV) porphyrin π-cation radical $P^{+}Fe^{IV} = O$) [198], and the B compound is a result of one-electron reduction of compound A:

$$P^{\bullet +}Fe^{V} = O + DH^{-} \rightarrow PFe^{IV} = O + DH^{\bullet}$$

Numerous model systems of enzymes of $PFe^{III}{-}X{-}H_2O_2$ (ROOH) type (see Chap. 3) help us to understand better the mechanisms of H_2O_2 or ROOH interaction with peroxidases, catalases, and cytochrome P-450. For example, in the model system $(Ph_4P)Fe(III)Cl{-}H_2O_2$ (ROOH) (Ph_4P-5,10,15, 20-tetraphenyl-porphyrinate) iron porphyrinates of oxoferryl type will be formed [199]:

$$(Ph_4P)Fe^{III}(Cl) + ROOH \longrightarrow [(Ph_4P)Fe^{IV}O(Cl)RO^{\bullet}]$$

$$[(Ph_4P)Fe^{IV}O(Cl)RO^{\bullet}] \rightarrow [(Ph_4P^{\bullet +})Fe^{IV}O(Cl) + RO^{-}$$

$$[(Ph_4P)Fe^{IV}O(Cl)RO^{\bullet}] \rightarrow [(Ph_4P)Fe^{IV}O(Cl) + RO^{\bullet}$$

Particles $(Fe^{IV} = O)^{2+}$ can be turned into $OH^{\bullet}$ radicals in the conditions of catalytic oxidation of some organic substrate (S), for example, in the systems $Fe^{2+}{}_{(aq)}{-}H_2O_2{-}S$, $L_nFe^{2+}{-}H_2O_2{-}S$ and enzyme–Fe(III)–$O_2(H_2O_2)$–S.

2.5.2.2 Mechanisms of H_2O_2 Decomposition by Iron Coordination Compounds

The mechanisms of H_2O_2 decomposition by Fe^{2+} and Fe^{3+} aqua-ions were briefly considered above in Sect. 2.4.1 according to the materials [93], where it was shown that in the system $Fe^{2+}{}_{aq}{-}H_2O_2$ decomposition of hydrogen peroxide is realized by ion-radical cyclic mechanism, and in the system $Fe^{3+}{}_{aq}{-}H_2O_2$ by radical-chain mechanism. Detailed critical consideration of the problems connected with revealing these mechanisms is given in [10]. The role of the ligand coordination to central metal ion in general is considered in Sect. 2.3. Here some examples are given of peroxide systems in which the influence of ligands complex formation with ions Fe^{2+} and Fe^{3+} resulted in the change of the mechanism of hydrogen peroxide decomposition.

System Fe^{2+}–ACDB–H_2O_2

Here, the oxidizing substrate simultaneously serves as a ligand. The catalyst is the coordination compound Fe^{2+}ACDB. Using the method of inhibitors, generation in this system of $OH^{\bullet}$ and $HO_2{}^{\bullet}$ radicals was revealed. The mechanism of process is given in Scheme 2.12.

$$Fe^{2+} + ACDB \rightleftharpoons Fe^{2+}ACDB,$$
$$Fe^{2+}ACDB + H_2O_2 \longrightarrow Fe^{3+}ACDB + OH\cdot + OH^-,$$
$$OH\cdot + H_2O_2 \longrightarrow HO_2\cdot + H_2O,$$
$$OH\cdot + ACDB \longrightarrow \text{products of ACDB oxidation},$$
$$Fe^{3+}ACDB \longrightarrow Fe^{2+} + \text{products of ACDB oxidation},$$
$$Fe^{2+}(Fe^{2+}ACDB) + HO_2\cdot \longrightarrow Fe^{3+}(Fe^{3+}ACDB) + HO_2^-,$$
$$Fe^{3+}ACDB + HO_2\cdot \longrightarrow Fe^{2+}ACDB + H^+ + O_2.$$

Scheme 2.12 Decomposition of hydrogen peroxide in the system Fe^{2+}-ACDB-H_2O_2

The principal process of the mechanism is a cyclic radical one as in the case of system $Fe^{2+}{}_{aq}$–H_2O_2. ACDB here it plays a double role—as substrate and ligand. The latter's influence does not result in mechanism change compared to the case when water is a ligand for Fe^{2+} ion.

System Fe^{2+}–EDTA–H_2O_2

Ferryl complex $YFeO^{2+}$ (Y is deprotonated EDTA) is generated in the system due to the inner-sphere transfer of two electrons from donor π_g-orbital of iron into the vacant σ_u*-orbital of H_2O_2. The scheme of the process is as in Scheme 2.13 [129].

$$YFe^{2+} + H_2O_2 \longrightarrow YFe^{IV}O^{2+},$$
$$YFe^{IV}O^{2+} + YFe^{2+} \longrightarrow 2YFe^{3+},$$
$$YFe^{IV}O^{2+} \longrightarrow XFe^{3+} \text{ (X – product of EDTA oxidation)},$$
$$YFe^{IV}O^{2+} \xrightarrow{H^+} YFe^{3+} + OH\cdot.$$

Scheme 2.13 Decomposition of hydrogen peroxide in the system Fe^{2+}-EDTA-H_2O_2

Hydrogen peroxide in this system is consumed for complex-bound Fe^{2+} ions and EDTA oxidation. The process (pH 4–6) is ion molecular and ceases on complex transition into the inactive form. The peculiarity of $Fe^{4+}(d^4)$ in the complex $YFe^{IV}O^{2+}$ is the high lability of this complex which is one of the reasons for its high reactivity in relation to various substrates and, thus, its short lifetime. In water solution this complex is hydrolyzed and easily dissipates with $OH^{\bullet}$ radicals formation that will be consumed with EDTA oxidation.

System Fe^{3+}–triethylenetetramine (TETA)–H_2O

This system is extremely active in H_2O_2 decomposition and, therefore, complex Fe^{3+}TETA is sometimes called "inorganic catalase." Catalytic H_2O_2 decomposition by complexes Fe^{3+} and TETA does not include free radicals in the main mechanism (the latter just bring complications to the system by ligand oxidation). The process involved proceeds by the mechanism of two-electron transfers, i.e., by ion-molecular mechanism (Scheme 2.14) [46]:

Initiation:
i. $TETA(III)OH^- + H_2O_2 \longrightarrow TETAFe(IV)(OH^-) + OH^- + OH\cdot;$

Chain continuation:
1. $TETAFe(II) + H_2O_2 \longrightarrow TETAFe(IV) + 2OH^-,$
2. $TETAFe(IV) + HO_2^- \longrightarrow TETAFe(II) + O_2 + H^+;$

Chain breakage:
0. $TETAFe(II) + H_2O_2 \longrightarrow TETAFe(III) + OH^- + OH\cdot.$

Scheme 2.14 Decomposition of hydrogen peroxide in the system Fe^{2+}-TETA-H_2O_2

Redox potential ($\varphi_{TETAFe^{3+}/TETAFe^{2+}}$) $= -0.07$ V, i.e., the state of stabilized Fe(III), and the transition Fe(III) into Fe(IV) becomes equally probable for Fe(III) transition into Fe(II). It provides the opportunity for the formation of both Fe(IV) and Fe(II) compounds and two-electron transition between them. Experiments on ascorbate ion introduction into this system confirm the formation of TETA Fe(IV). H_2O_2 decomposition occurs that may be the result of ascorbate-ion interaction with strong electrons acceptor [with TETA Fe(IV) or with $OH^{\bullet}$]. However, with pH increasing it appears that the concentration of $OH^{\bullet}$ decreases, while the oxidation rate of ascorbate ion increases. It testifies that ascorbate ion being introduced into the investigated system, interacts not with $OH^{\bullet}$ radicals but with other strong oxidants which can only be TETA Fe(IV).

2.5.2.3 Mechanisms of H_2O_2 Decomposition by Manganese Coordination Compounds

Coordination compounds of manganese are contained in active centers of manganese-containing catalases and dismutases [200, 201] and in numerous enzymes of redox transformations of organic tricarbonic acids (Krebs cycle) [202] and other enzymes. The unique ability of the manganese ion to change rather easily its oxidation degree from +2 up to +7 on stabilizing certain ligands and oxidizers creates an opportunity for its coordination compounds to be used as efficient catalysts of hydrogen peroxide decomposition in water solutions. The methods and problems of such catalytic processes are described in detail in [203]. Detailed kinetic regularities of these processes running at the presence of catalysts–manganese coordination compounds—are given in [6, 204–224] and other works. The revealed mechanisms of hydrogen peroxide decomposition in certain catalytic systems are given here only briefly and in general form [225].

Coordination compounds of manganese with triethyleneamine [204, 205], diethylenetriamine [214, 216], tetraethylenepentamine [226], α,α'-dipyridyl [221, 223], *o*-phenanthroline [221–223, 227], 2,2′,2″-tripyridyl [227], 2- and 3-aminodiamine [227], 2,9-dimethyl- and 4,7-diphenyl-1,10-phenanthroline [227], ethylenediamine [227], diethanolamine [227], pyridoxalphosphate–ethylenediamine [216, 227], N, N', N''-trimethyldiethylenetriamine [227], acetylacetone [208, 209], gystidin [204], and glycine [217, 218], with a number of aminoacids [219, 220] and some other compounds possess catalase activity. Among all these coordination compounds, bicarbonate complex $Mn(HCO_3)_2$ is extremely active in H_2O_2 decomposition. According to their activity as catalases, manganese(II) complexes can be put in the following order:

$$Mn\,(HCO_3)_2 > MnTrien^{2+} > MnHis_2^{2+} > MnDien_2^{2+} > MnAcac_2^{2+} > MnEn_2^{2+} > MnDipy_2^{2+} > MnPhen_2^{2+} > MnAla_3 > MnVal_3 > MnGly_3 > MnSer_2 > MnAsp_2,$$

where HCO_3^- is the bicarbonate ion, Trien is triethylenetetramine, His histydin, Dien ethylenediamine, Acac acetylacetone, En ethylenediamine, Dipy α,α'-dipyridyl, Phen *o*-phenanthroline, Ala alanine, Val valine, Gly glycine, Ser serine, and Asp asparagine.

In the process of catalytic decomposition of hydrogen peroxide, the following redox transformations of manganese coordination compounds can take place:

$$L_nMn(II) \underset{+2e}{\overset{-2e}{\rightleftarrows}} L_nMn(IV); L_nMn(III) \underset{+e}{\overset{-e}{\rightleftarrows}} L_nMn(IV); L_nMn(III) \overset{+e}{\rightleftarrows} Mn(II).$$

Depending on the specific transformation, the process can proceed by ion-radical (radical-chain or cyclic) or ion-molecular mechanism. The three mechanisms are discussed below.

First, *ion-radical (radical-chain) mechanism* of H_2O_2 decomposition proceeds in the systems $Mn(II)-HCO_3^--H_2O_2$, $Mn(II)-Trien-H_2O_2$, $Mn(II)-His-H_2O_2-H_3BO_3$, $Mn(II)-Phen-H_2O_2$, $Mn(II)-Dipy-H_2O_2$ [204, 208, 210, 213, 222, 223]. Scheme 2.15 of such a mechanism in the most effective system $Mn(II)-HCO_3^--H_2O_2$ (reaction effectively proceeds at pH 7–8 and with $[HCO_3^-]/[Mn^{2+}] \approx 10^5$), can be given as follows [203]:

Initiation:

i_1. $(HCO_3^-)_2Mn(II) + H_2O_2 \longrightarrow (HCO_3^-)_2Mn(IV) + 2OH^-$,

i_2. $(HCO_3^-)_2Mn(IV) + HO_2^- \longrightarrow LMn(III) + O_2\cdot^- + H^+$;

Chain continuation:

1. $(HCO_3^-)_2Mn(IV) + O_2\cdot^- \longrightarrow (HCO_3^-)_2Mn(III) + O_2$,
2. $(HCO_3^-)_2Mn(III) + H_2O_2 \longrightarrow (HCO_3^-)_2Mn(IV) + OH\cdot + OH^-$,
3. $OH\cdot + H_2O_2 \longrightarrow O_2\cdot^- + H^+ + H_2O$,
4. $OH\cdot + HCO_3^- \longrightarrow CO_3\cdot^- + H_2O$,
5. $CO_3\cdot^- + H_2O_2 \longrightarrow HCO_3^- + O_2\cdot^- + H^+$;

Chain breakage:

0. $(HCO_3^-)_2Mn(III) + O_2\cdot^- \longrightarrow (HCO_3^-)_2Mn(II) + O_2$.

Scheme 2.15 Radical-chain mechanism of hydrogen peroxide decomposition in the system $Mn(II)$-HCO_3^-

Stages (4) and (5) in addition generate the initiating particle $O_2^{\bullet-}$, that apparently is one of the reasons for the extremely high efficiency of H_2O_2 disintegration in the presence of complex $Mn(HCO_3)_2$. On this basis the catalytic method of trace amounts of Mn^{2+} ions detection was developed [203].

Second, *ion-radical cyclic mechanism* (Scheme 2.16) [203]—here, $W^{O_2}/W_I = 1$, and manganese compounds in various oxidation degrees and free radicals take part in the basic process. In such systems, Mn(IV) compounds react with $HO_2^{-\bullet}$ by one-electron transfer, and Mn(III) complexes formed are capable of oxidizing hydrogen peroxide. Such a mechanism is observed in the system Mn^{2+}—En—H_2O_2.

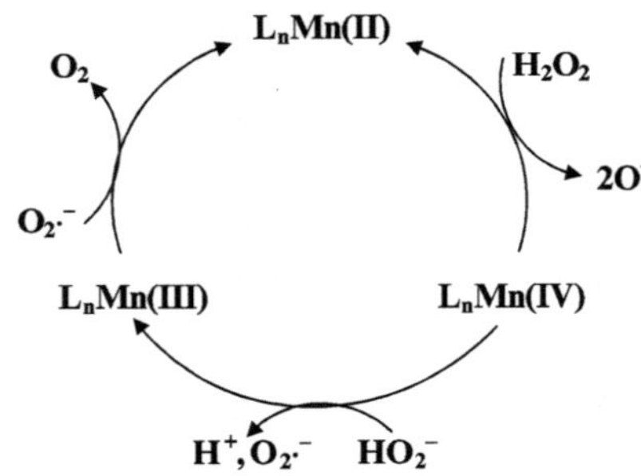

Scheme 2.16 Ion-radical cyclic mechanism of hydrogen peroxide decomposition

Third, *ion-molecular mechanism* [203] of H_2O_2 decomposition differs from the previous one first of all by the fact that free radicals in reactionary volume will not be formed, proved by the absence of process braking by the specific acceptors of $OH^{\bullet-}$ and $O_2^{\bullet}$ radicals (PNDMA and TNM, respectively). The process can proceed through the peroxocomplexes formation and redox reactions realized inside the complex catalyst [6]. In the absence of peroxocomplex formation, the stages of two-electron transfer from the catalyst to H_2O_2 can be realized and, in contrast, the process can also proceed by the following ion-molecular mechanism [203]:

1. $MnL_n + H_2O_2 \rightarrow Mn(IV)L_n + 2OH^-$,
2. $Mn(IV)L_n + HO_2^- \rightarrow Mn(II)L_n + H^+ + O_2$.

The ion-molecular mechanism of H_2O_2 decomposition was defined for the systems Mn(II)–diethyleneamine–H_2O_2 [214, 216], Mn(II)-glycine (alanine, valine, serine, treonine, asparagine)–H_2O_2 [214, 215], and others. For example, in the system Mn(II)—Dien—H_2O_2 inhibitors PNDMA and TNM do not influence the reaction rate of H_2O_2 disintegration which specifies the absence of $OH^{\bullet-}$ and $O_2^{\bullet}$ radicals' generation. The process mechanism consists of stages (1) and (2) (L_n = 2Dien).

The general Scheme 2.17 of redox transformations mechanism of manganese mononuclear complex compounds in the process of H_2O_2 decomposition can be given in the following form [199]:

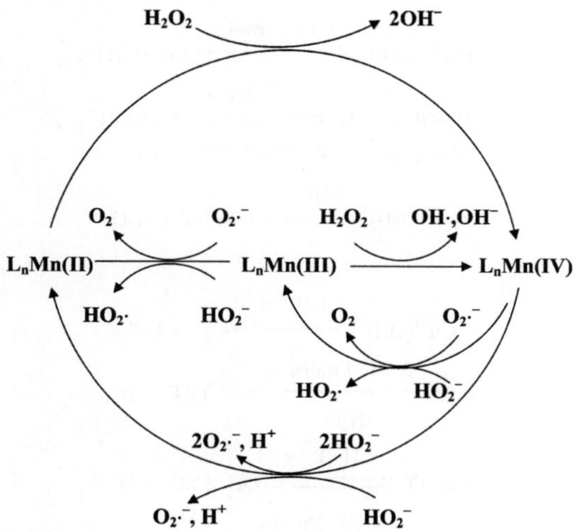

Scheme 2.17 Generic scheme of Mn complexes transformations on hydrogen peroxide decomposition

The three mechanisms discussed above result from this scheme as special cases. Coordination compounds of manganese catalyse H_2O_2 decomposition in neutral and weakly alkaline media ($6 < pH < 11$). Just within these limits of pH values, the formation of catalase-active manganese complexes with various ligands occurs more often. As against the neutral media, in weakly alkaline media manganese hydroxocomplexes as a rule manifest catalase activity. With increase of pH the properties of manganese complexes to realize two-electron transfers and those of carrying out radical-free catalytic decomposition of hydrogen peroxide increase too.

2.5.2.4 Mechanisms of H_2O_2 Decomposition by Copper Ions and Complexes

The detailed review of the experimental proofs of appropriate mechanisms is given in [6, 129]. Here, just the final results of the detailed study are provided.

Hydrogen peroxide can force out water molecules from the labile coordination sphere of Cu^+ ion, forming dihydroxocomplex Cu^{3+} $[Cu(OH)_2{}^+]$, and then, due to acid–basic transformations, not $OH^•$ radical will be generated (Cu(III) formation was also shown in [5, 11]), but cupryl-ion ($Cu^{III}O^+$). In weakly acid and neutral media it is exposed to hydrolysis with the formation of the particle $(OH)CuO^+$. The mechanism of Cu^+ interaction with H_2O_2 is as shown in Scheme 2.18 [228, 229]:

Cu(III) ions —exhibit two to three orders less reactivity toward various substrates than $OH^•$ radicals (i.e., approximately the same as atomic oxygen).

Taking as an example the study of bis-dipyridyl copper complex $(Dipy)_2Cu^+$ with hydrogen peroxide interaction, the ligand influence on the process mechanism has been shown (Scheme 2.19) [129].

$$Cu^+ + H_2O_2 \xrightarrow{5\cdot10^3,\ l/mol\cdot s} Cu^{III}(OH)_2^+,$$

$$Cu(OH)_2^+ + H^+ \overset{10^3,\ l/mol\cdot s}{\rightleftharpoons} CuOH^{2+},$$

$$Cu(OH)^+ + H^+ \rightleftharpoons Cu^{3+},$$

$$Cu^{III}(OH)_2^+ \xrightarrow{3\cdot10^4,\ s^{-1}} Cu^{III}O^+ + H_2O,$$

$$Cu^{III}(OH)^{2+} \xrightarrow{\sim3\cdot10^2,\ s^{-1}} Cu^{2+} + OH\cdot,$$

$$Cu^{2+} \xrightarrow[H_2O]{3\cdot10^4,\ l/mol\cdot s} Cu^{2+}OH^- + H^+,$$

$$Cu^{III}O^+ \overset{H_2O}{\rightleftharpoons} (OH^-)CuO^+ + H^+,$$

$$(OH^-)CuO^+ \xrightarrow[H_2O]{<10^2,\ l/mol\cdot s} Cu^{2+}(OH^-)_2 + OH\cdot$$

$$Cu^{3+} + Cu^+ \xrightarrow{\sim2\cdot10^9,\ l/mol\cdot s} 2Cu^{2+},$$

$$Cu^{3+} + Cu^{3+} \xrightarrow[H_2O + 1/2O_2]{3\cdot10^7,\ l/mol\cdot s} 2Cu^{2+} + H_2O_2.$$

Scheme 2.18 Mechanism of Cu^+ interaction with H_2O_2

$$(Dipy)_2Cu^+ + H_2O_2 \xrightarrow{2\cdot10^3,\ l/mol\cdot s} (Dipy)_2Cu^{III}O^+,$$

$$(Dipy)_2Cu^{III}O^+ + H^+ \xrightarrow{3\cdot10^3,\ l/mol\cdot s} (Dipy)_2CuOH^{2+},$$

$$((Dipy)_2CuOH^{2+} \overset{H^+}{\rightleftharpoons} (Dipy)_2Cu^{3+}),$$

$$(Dipy)_2Cu^{3+}((Dipy)_2CuO^+, (Dipy)_2CuOH^{2+}) + (Dipy)_2Cu^+ \xrightarrow[H^+]{\sim10^{10}} 2(Dipy)_2Cu^{2+},$$

$$(Dipy)_2CuO^+ \xrightarrow[H_2O]{10^4,\ l/mol\cdot s} (Dipy)_2Cu^{2+}OH^- + OH\cdot,$$

$$(Dipy)_2CuOH^{2+} \longrightarrow (Dipy)_2Cu^{2+} + OH\cdot.$$

Scheme 2.19 Effect of ligand on process mechanism on Cu^+ interaction with H_2O_2

From mechanisms comparison of H_2O_2 interaction with Cu^+ and $(Dipy)_2Cu^+$ their certain similarity can be seen, especially in the cases of cupryl ion and cupril complex formation. In the latter case, $OH^\bullet$ radicals' formation by the reactions of water or hydride-ion oxidation is typical.

On alcohol (ROH) introduction into the similar system, "cupryl-ion" $L_2Cu^{III}O^+$ formed will interact with it, generating $RO^\bullet$ and $R_\alpha{}^\bullet$ radicals:

$$L_2Cu^{III}O^+ + ROH \rightarrow RO^\bullet + L_2CuOH^+,$$
$$L_2Cu^{III}O^+ + RHOH \rightarrow R_\alpha + L_2CuOH^+.$$

Hydroxocomplex L_2CuOH^{2+} or unhydrolyzed ion L_2Cu^{3+} will interact with primary alcohol groups forming the appropriate aldehyde:

$$L_2CuOH^{2+}\,(L_2Cu^{3+}) + RCH_2OH \rightarrow Cu^+L_2 + RCHO.$$

For the systems $L{-}Cu^{2+}{-}H_2O_2$ (where L is bidentate ligand, in particular L = α, α′-dipyridyl) the process mechanism is radical-chain and can be given in the following form (Scheme 2.20) [42].

i. $DipyCuOH^+ + HO_2^- \longrightarrow DipyCu^+ + H_2O + O_2^{\cdot -}$,

1. $DipyCu^+ + H_2O_2 \longrightarrow DipyCu^{2+} + OH^- + OH\cdot$,

2. $OH\cdot + H_2O_2 \longrightarrow O_2^{\cdot -} + H^+ + H_2O$,

3. $O_2^{\cdot -} + DipyCu^{2+} \longrightarrow O_2 + DipyCu^+$,

0. Reaction breakage with OH· involvement.

Scheme 2.20 Radical-chain mechanism of H_2O_2 oxidation in the system $L\text{-}Cu^{2+}\text{-}H_2O_2$

2.6 Mechanisms of Organic Substrates (S) Oxidation in Oxide and Peroxide Systems

2.6.1 Mechanisms of Organic Substrates Oxidation by Oxygen [Systems M^{z+} ($M^{z+}L_n$)$-O_2-S$]

Oxygen coordinated with many transition metals compounds is already in the activated state and can be used as oxidizer in catalytic reactions with many organic compounds (S) that are of great practical importance. Thus, the coordination of the O_2 molecule to metal ions or complexes is one of the most effective ways of its activation; the causes of it have already been discussed above (Sect. 2.2). As an example of systems of $M^{z+}(L_nM^+){-}O_2{-}S$ type, those systems in which the catalysts are manganese ions or complex compounds will be considered later. Manganese ions were used as catalyst in the reactions of aldehydes, methylbenzens, methylbenzonitriles, *n*-pentane, acetophenone, 4,8-diamino-1,5-dihydroxynaphthaldehyde of tiosemicarbazone, hydrazine, ascorbic acid, morine, *o*-hydroxyphenylthiourea, and many other organic compounds' oxidation by molecular oxygen. However, fully proved mechanisms of such processes are seldom provided.

As an example of *catalysis by manganese ions*, the system $Mn^{3+}-O_2-RH$ can be taken, where RH signifies alkylaromatic compounds, and Mn^{3+} ions participate in the initiation stage. The process proceeds by radical-chain mechanism, for which the qualitative scheme is given in [230].

Aqua-ions of Mn^{2+} usually catalyze the oxidation of chelate-forming organic substances; manganese–substrate complex will be formed, which oxidizes the substrate inside the coordination sphere by interacting with O_2. Although the mechanisms of similar processes are available in the literature [231, 232], they are not confirmed by sufficient experimental data and can be considered as speculative. However, it is possible to agree with the opinion of these authors, that the substrates' oxidation in such cases is realized by manganese ions in high oxidation degrees [with Mn(III) or Mn(IV) formation] inside the coordination sphere of manganese substrate complex. Proceeding from rather limited information about the kinetic regularities and mechanisms of substrates catalytic oxidation by molecular oxygen in the presence of manganese ions, some assumptions can be made [203]:

- The process can proceed by a radical chain or ion radical (nonchain) mechanism with the reversible redox transformations of catalyst:

$$\mathrm{Mn(III)} \underset{-e}{\overset{+e}{\rightleftarrows}} \mathrm{Mn(II)}.$$

- The substrate should possess chelate-forming capacity and should form metal–substrate complex reacting with oxygen in an inner- or outer-spheric way (more often inside the coordination sphere of such complex with one- or two-electron transfer).

Catalysis by manganese complexes of the reactions of substrates oxidation by molecular oxygen is more effective than by manganese ions and consequently will be considered in more detail. As catalysts for such processes, manganese stearate, some porphyrin, phthalocyanic, phenanthrolinic, dipyridylic, and other manganese complexes are used. At the initial stage the intermediate compound of catalyst and O_2 will be formed; the activation of the latter proceeds, though the mechanism of such activation is not always clear.

The interaction of such oxygenerated metal complexes with substrate proceeds more often via the formation of a triple complex (catalyst–O_2–substrate), disintegration of which results in the reaction products. However, sometimes oxygenerated complex interaction with the substrate is supposed to proceed without the formation of a triple complex. For example, the oxidation of organic substrate (RH) was investigated in the system $MnSt_2-O_2-$8-pentadecanone (St represents stearate) [233]. Caprylic and enantic acids are the main products, the yield of which reaches up to 60%. At the initial stages, the formation of free radicals occurs and the mechanism of their appearance looks like:

$$MnSt_2 + O_2 \rightarrow MnSt_2{}^{\delta+} \ldots O_2{}^{\delta-},$$

$$MnSt_2^{\delta+}\ldots O_2^{\delta-} + RH \longrightarrow \begin{cases} MnSt_2OH + RO^{\bullet} \\ MnSt_2OOH + R^{\bullet} \end{cases}$$

The oxidized form of manganese, reacting with ketone RH, generates free radicals and starts the chain reaction with participation of Mn(II) and Mn(III) at the stage of chain continuation. Thus, the triple complex will not be formed.

Manganese stearate also catalyzes the oxidation of hexadecane by molecular oxygen, and this reaction is autocatalytic [234]. Very frequently for the oxidation of a large number of organic compounds the systems PMn–O_2–reducer ($NaBH_4$) are used, where P is porphyrin or its derivatives [235–240]. So, the most probable process of cyclohexane oxidation in the system TPhPMn–O_2–$NaBH_4$–cyclohexane (where TPhP is tetraphenylporphyrin) in benzene–ethanol solution on room temperature is shown in the following form [239]:

$$TPhPMn(II) + O_2\ TPhPMn(II)\cdot O_2 \xrightarrow{RH} \begin{cases} TPhPMn(III) + O_2^{\bullet-} \\ \text{Oxidation products}^{\bullet} \end{cases}$$

An example of effective catalysts of organic substances' oxidation in water medium are Mn(II) complexes with 1,10-phenanthroline [241–243]. Mn(III) compounds forming in the reaction medium of the system Mn(II)–Phen–O_2–methylethylketone will oxidize the enol form (EnH) of methylethylketone (RH), generating free radicals, and the mechanism of the chain process is shown in Scheme 2.21 [243]; here Mn(III) and Mn(II) are appropriate manganese complexes with 1,10-phenanthroline. On the basis of this scheme, the rate expression was obtained by the method of stationary concentrations complying with the experimentally found expression that confirms the applicability of the offered mechanism.

$$RH \underset{}{\overset{H^+}{\rightleftharpoons}} EnH$$

$$Mn(III) + EnH \longrightarrow Mn(II) + H^+ + R\cdot \text{ (initiation)},$$

$$R\cdot + O_2 \longrightarrow RO_2\cdot,$$

$$RO_2\cdot + Mn(II) \longrightarrow ROO^- + Mn(III),$$

$$ROO^- + H^+ \longrightarrow ROOH,$$

$$ROOH \overset{H^+}{\longrightarrow} CH_3CHO + CH_3COOH \text{ (products)},$$

$$ROOH + Mn(II) \longrightarrow RO^- + OH\cdot + Mn(III),$$

$$RH + OH\cdot \longrightarrow R\cdot + H_2O,$$

$$R\cdot + Mn(III) \longrightarrow \text{Products (breakage)}.$$

Scheme 2.21 Mechanism of organic substrates oxidation in the system Mn(II)-Phen-O_2

In general, the activation of molecular oxygen by ions and complexes of transition metals is well known [129, 152, 167, 244, 245]. The mechanism of O_2 activation includes the stage of the formation of oxygenated intermediate compound of catalyst $^{\bullet}O_2$ type having high oxidizing capacity in relation to various substrates (S). Oxidation of the latter usually proceeds through the formation of intermediate complex of M[Mn(II)...O–O...S] type [246–248]. The mechanisms including the stages of O_2 reducing up to H_2O_2 and Mn(II) compounds oxidation to Mn(III) and even to Mn(IV) are also proved [157, 249, 250].

One of the most detailed investigated processes of organic compound oxidation in water media and in aerobic conditions is the effective photochemical oxidation of leicobase [hydrochloric acid salt of diantipyril-dimethyl-aminophenylmethane (LB) in the system Mn(II)Phen(Dipy)–O_2–LO] [251]. Among plenty of dyes having different complex-forming capacities in relation to manganese (indigocarmine, murexide, alizarine, neutral red, active light blue, *o*-dianisidine, some leico-bases in the systems Mn(II)–Phen(Dipy)–O_2), only leico-bases and *o*-dianisidine will be oxidized efficiently. In anaerobic conditions or in the absence of catalyst, the oxidation of LB does not occur or (in the second case) occurs very slowly, which shows that LB are oxidized by molecular oxygen, which is activated by complexes of manganese(II) with 1,10 phenanthroline (α, α'-dipyridyl). It was found that the complex having the composition of $MnPhen_2^{2+}$ brings the main contribution into the catalytic activity of the aforementioned system. Activating action of ultraviolet light is caused, apparently by the manganese complex and O_2 molecules transition into the excited condition (on system irradiation with light having $\lambda = 300$–340 nm the oxidation rate of LB increases 20–30 times).

Scheme 2.22 of the considered process is represented by the totality of the following stages (L = Phen or Dipy):

$$\mathbf{MnL_2^{2+} + O_2 \rightleftharpoons MnL_2O_2^{2+},}$$

$$\mathbf{MnL_2^{2+} + O_2 \overset{h\nu}{\rightleftharpoons} (MnL_2O_2^{2+})\cdot,}$$

$$\mathbf{MnL_2O_2^{2+}(MnL_2O_2^{2+})\cdot + LB \rightleftharpoons MnL_2O_2^{2+}...LB(MnL_2O_2^{2+}...LB)\cdot,}$$

$$\mathbf{MnL_2O_2^{2+}...LB(MnL_2O_2^{2+}...LB)\cdot \longrightarrow MnL_2 + \text{products of LB oxidation}}$$

Scheme 2.22 Photochemical oxidation of leucobase in the system Mn(II)-Phen(Dipy)-O_2

The oxidation of leicobases proceeds through the formation of a triple intermediate complex. This scheme explains all the experimental data obtained:

- Direct dependence of LB oxidation rate (W^{LB}) on [Mn^{2+}], efficiency of process (beginning from pH 5) due to ligand complex formation with Mn^{2+} ion and change of W^{LB} with the change of concentration in solution of complex with composition MnL_2^{2+}
- Increase of W^{LB} on ultraviolet irradiation
- Participation of O_2 in LB oxidation and its activation due to the formation of compounds of $MnL_2O_2^{2+}$ $\left(MnL_2O_2^{2+}\right)^*$ type

– Outer sphere mechanism of LB oxidation in these systems
– H_2O_2 and $OH^{\bullet -}$, $O_2^{\bullet}$ radicals, and also Mn(III) nonparticipation in the process of LB oxidation

The mechanisms of substrate's oxidation by oxygen can be attributed not only to the outer-sphere but also the inner sphere ones. In more detail, this problem is considered on the examples when catalysts are the coordination compounds of manganese [225].

In alkaline media manganese(II) complexes with such ligands as gluconate and sorbitol (pH 9.5–10.5), having interacted with O_2, practically immediately convert into the appropriate compounds of manganese(IV) (although complexes of manganese(III) can be detected spectrophotometrically), which oxidize various substrates (S) [157, 252]:

1. $L_2Mn(II) + O_2 \rightleftarrows [L_nMn(II)O_2] \rightarrow L_2Mn(IV) + O_2^{2-}$,
2. $L_2Mn(II) + L_2Mn(IV) \rightarrow 2L_2Mn(III)$,
3. $L_2Mn(III) + S \rightarrow L_2Mn(II) + P_1$ (products of one-electron oxidation of S). Probably, in some cases in competition with (2), the following reaction also proceeds:
   2′. $L_nMn(IV) + S \rightarrow L_2Mn(II) + S_2$ (products of two-electron oxidation of S).

Therefore, the cyclic mechanism of substrate oxidation is realized in these systems according to Scheme 2.23.

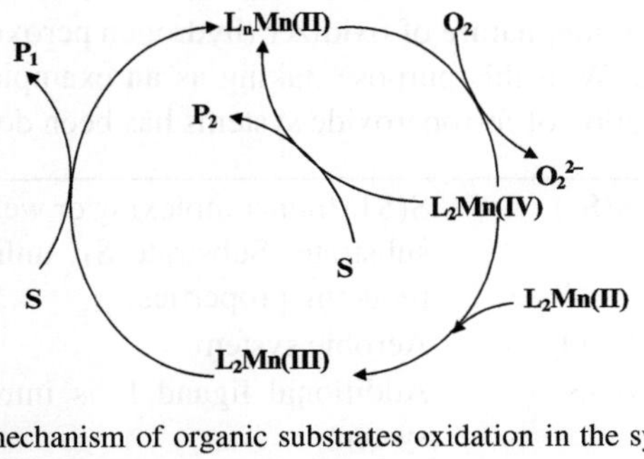

Scheme 2.23 Cyclic mechanism of organic substrates oxidation in the system involving Mn(II) complexes

Summing up the brief consideration of transition metals oxygenated complexes [129], it is to be noted that oxygen complexes of "superoxide" type $(MO_2{}^+)_{II}$ (complex of Pauling type–O_2 has only partially passed into the form of $O_2^{\bullet}$ radical) can participate both in reactions of inner sphere electron transfer (hydrogen atom isolation)

$$(MO_2{}^+)_{II} + RH \rightarrow M^{2+} + HO_2{}^- + R^{\bullet},$$

and inner sphere transformations.

$$(MO_2{}^+)_{II} + RH \rightleftarrows (MO_2RH)^+ \begin{cases} \rightarrow M^{2+} + HO_2{}^- + R^{\bullet} \\ \rightarrow M^{2+} + HO_2{}^{\bullet} + R^{\bullet} \\ \rightarrow M^{+} + H_2O_2 + \text{products.} \end{cases}$$

In the case of oxygen complexes of "peroxide type" $(MO_2{}^+)_I$ (the complex of Griffith's type–O_2 has partially passed into the form O_2^{2-}), the transformations of two types can also proceed: the reactions of inner sphere one-electron (two-electron) oxidation of redox ligand or reactions with O–O peroxide bond breakage with subsequent substrate hydroxylation:

$$(MO_2{}^+)_I + RH \rightarrow \text{Intermediate compound} \rightarrow MO^+ + ROH.$$

The transformations of the latter type are frequently realized in the cells of living organisms by means of monooxygenases. Such reactions will be considered in Chap 3.

2.6.2 *Mechanisms of Organic Substrate's Oxidation by Hydrogen Peroxide [Systems $M^{z+}(M^{z+}L_n)-H_2O_2-S$]*

The results of numerous pieces of research on the mechanisms of organic compounds' oxidation by the system $Fe^{2+} - H_2O_2$ are reflected most fully in Walling's work [57]. However, subsequent development in this direction has shown that the scheme involved does not cover the whole variety of such mechanisms. It was necessary to take into account the nature of the substrate and the possibility of its interaction with metal ion, nature of oxidizer (hydrogen peroxide, ROOH and O_2), and ligand influence. With this purpose, taking as an example the compounds of iron(II), the classification of ferroperoxide systems has been developed [10, 93]:

1. $Fe^{2+}-H_2O_2-S(S_1)$	S(S1), noncomplexing or weakly complexing substrate. Substrate S_1, unlike S, possesses reducing properties
2. $Fe^{2+}-H_2O_2-S-O_2$	Aerobic system
3. $Fe^{2+}-L-H_2O_2-S$	Additional ligand L is introduced into the system
4. $Fe^{2+}-L-H_2O_2-S-O_2$	Aerobic system with additional ligand
5. $Fe^{2+}-H_2O_2-SL$	S_L, substrate and ligand at the same time
6. $Fe^{2+}-H_2O_2-SL-O_2$	Aerobic system with S_L
7. $Fe^{2+}-L-H_2O_2-SL$	System including L and S_L
8. $Fe^{2+}-SL-H_2O_2-S$	System including S_L and S
9. $Fe^{2+}-SL-H_2O_2-S-O_2$	Aerobic system including S_L and S

All these systems can be divided into two groups. In the first group (systems 1–4), oxidizing substrate will not form a complex with Fe^{2+} or will form it very poorly. Thus, overlapping of electron orbitals Fe^{2+} and S will be little. The reaction of oxidation in the systems of this group proceeds by outer-sphere way (outside the first coordination sphere of Fe^{2+}, i.e., within the reaction bulk). In the second group (systems 5–9) the substrate also represents a ligand. Here we can expect the formation of intermediate compounds with substrate and inner sphere electron transfer. These groups reflect only the extreme cases of complex formation properties of ligands. All these systems are considered in detail and characterized in [10,93]. The general scheme of mechanisms of H_2O_2 disintegration and organic substrates oxidation is given in Scheme 2.24.

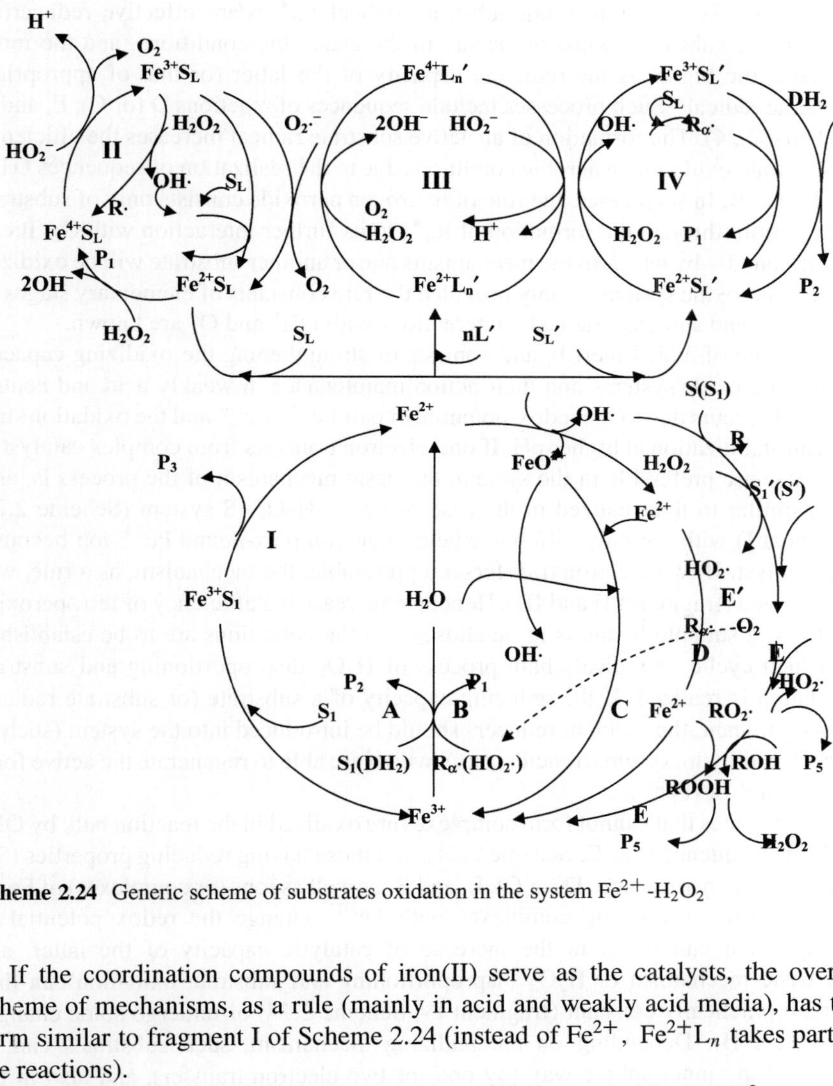

Scheme 2.24 Generic scheme of substrates oxidation in the system Fe^{2+}-H_2O_2

If the coordination compounds of iron(II) serve as the catalysts, the overall scheme of mechanisms, as a rule (mainly in acid and weakly acid media), has the form similar to fragment I of Scheme 2.24 (instead of Fe^{2+}, $Fe^{2+}L_n$ takes part in the reactions).

For simplification, not all possible reactions of generated radicals $OH^{\bullet}$ are given in the fragments II, IV, and I (sequence C) (they are similar to the transformations given in sequence E of fragment 1).

Ferroperoxide systems can be classified according to various attributes: efficiency of its action, complex formation capacity of substrate in relation to Fe^{2+}, the mechanism of H_2O_2 disproportioning, and substrate oxidation.

If substrate introduced (S) and substrate radical formed (R) do not possess the reducing properties, the oxidation process will cease with time and will end after all Fe^{2+} is transformed into Fe^{3+} (sequences D or C will be realized, see Scheme 2.24). If substrate S_1 and/or forming substrate radical $R_\alpha{}^\bullet$ is/are effective reducer(s), quantitative substrate oxidation occurs in the anaerobic conditions, and the more effective the higher is the reducing capacity of the latter (or that of appropriate substrate radical). Such processes include sequences of reactions D (or C), E, and B (Scheme 2.24). The formation of an active substrate radical increases the efficiency of substrate oxidation in aerobic conditions due to the realization of sequences D (or C), E, and B. In such cases, the role of hydrogen peroxide consists only of substrate preparation (through the formation of $R_\alpha{}^\bullet$) to its further interaction with O_2. It can be estimated—by which of the mechanisms one or another substrate will be oxidized by ferroperoxide systems—only provided the rate constants of elementary stages of substrates and substrate radicals' interactions with $OH^\bullet$ and O_2 are known.

The role of added inert ligand consists of strengthening the oxidizing capacity of ferroperoxide systems and their action maintenance in weakly acid and neutral media. It occurs due to the redox-potential of pair Fe^{3+}/Fe^{2+} and the oxidation state of iron stabilization at higher pH. If one-electron transfers from complex catalyst to substrate are preferable in the system, the basic mechanism of the process is, as a rule, similar to that realized in the case of $Fe^{2+}{-}H_2O_2{-}S$ system (Scheme 2.24, fragment I) with the only difference being that complex-bound Fe^{2+} ion becomes the catalyst. If two-electron transfers are preferable, the mechanism, as a rule, will be changed (fragments II and III). Hence, to increase the efficiency of ferroperoxide systems, a suitable ligand is to be chosen and the conditions are to be established at which cyclic or radical-chain process of H_2O_2 disproportioning and substrate oxidation is realized. If the reducing capacity of a substrate (or substrate radical) is insignificant, the efficient reducers should be introduced into the system (such as ascorbic, dihydroxyfumaric acids, etc.), which are able to regenerate the active form of the catalyst (Fe^{2+}).

S substrates that cannot form complexes are oxidized in the reaction bulk by $OH^\bullet$ radicals (sequence D or C, Scheme 2.24), and those having reducing properties (S_1) are oxidized in sequences D (or C), E, and B, as well as ions (or complexes) of Fe^{3+}.

S_1 substrates forming complexes with Fe^{2+}, change the redox potential of catalyst that can result in the increase of catalytic capacity of the latter, and the basic mechanism of H_2O_2 disproportioning and substrate oxidation can thus remain practically constant (fragment IV, Scheme 2.24) or undergo some changes (fragment II). Depending on the realizing mechanism, such substrates can be oxidized the inner sphere way (by one- or two-electron transfer), and also in the reaction bulk by generated radicals $OH^\bullet$ or $O_2{}^{\bullet-}$. If S_L will form complexes only with Fe^{3+}, the mechanism of process will involve the sequence A instead of sequence B.

All the ferroperoxide systems can be divided into two groups with regard to the mechanism of H_2O_2 disproportioning and substrate oxidation: in the first group ion-radical mechanism (cyclic or radical-chain) is realized, this fact being described by fragments I, II, and IV; in the second group—ion-molecular mechanism—fragment III (Scheme 2.24).

The examples of organic substrates' oxidation by Fenton reagent, complicated by participation of oxygen (aerobic conditions), can call on the reactions of ethyl alcohol oxidation [10, 93, 253, 254] (Scheme 2.25).

$$1.\ Fe^{2+} + H_2O_2 \longrightarrow Fe^{3+} + OH\cdot + OH^-,$$
$$2.\ OH\cdot + C_2H_5OH \longrightarrow R_\alpha\cdot + H_2O,$$
$$3.\ R_\alpha\cdot + O_2 \longrightarrow RO_2\cdot,$$
$$4.\ RO_2\cdot \longrightarrow CH_3CHO + HO_2\cdot,$$
$$5.\ RO_2\cdot + ROH \longrightarrow CH_3CH(OOH)OH + R_\alpha\cdot,$$
$$6.\ RO_2\cdot + Fe^{2+} \xrightarrow{H^+} Fe^{3+} + CH_3CH(OOH)OH,$$
$$7.\ HO_2\cdot + C_2H_5OH \longrightarrow R_\alpha\cdot + H_2O_2,$$
$$8.\ HO_2\cdot + Fe^{2+} \xrightarrow{H^+} Fe^{3+} + H_2O_2,$$
$$9.\ HO_2\cdot + Fe^{3+} \longrightarrow Fe^{2+} + O_2 + H^+,$$
$$10.\ CH_3CH(OOH)OH \longrightarrow CH_3CHO + H_2O_2.$$

Scheme 2.25 Oxidation of ethyl alcohol

A similar mechanism is realized on substrates oxidation and in the typical systems $Fe^{2+}{-}S_L{-}H_2O_2{-}O_2$, $L{-}Fe^{2+}{-}H_2O_2{-}S{-}O_2$ with the only difference being that the catalysts in this case serve the coordination compounds of Fe^{2+} with S_L or L accordingly [10].

According to the experimental proofs of process, regarding the route in aerobic conditions as described in Scheme 2.25, the following should be related:

1. Difference in oxidation rates of substrate in aerobic and anaerobic systems
2. Excess of oxidation product formed over the quantity of disproportioned H_2O_2
3. Generation of $OH^\bullet$ radicals, Fe^{2+} concentration decrease and the accumulation of Fe^{3+} during the reaction

Therefore, it can be believed that the mechanism of organic substrates' oxidation in aerobic conditions by $OH^\bullet$ radicals generated by ferroperoxide systems in the cases when active radical (of $R_\alpha{}^\bullet$ type) will be formed is described by Scheme 2.25. Hence, the role of H_2O_2 in the process of aerobic oxidation of organic substrates (particularly alcohols) is reduced to the preparation of the latter for their further oxidation by oxygen (in the case of alcohols—to the transformation of ROH into $R_\alpha{}^\bullet$).

Coordinative compounds of manganese(II) are also effective catalysts of organic compounds oxidation by hydrogen peroxide. Numerous data have been obtained on the systems of Mn(II)–L–H_2O_2–S type from study and appropriate examples of such systems are analyzed in detail in [203, 225]. Only some generalizations are given here.

Catalase-active coordinative compounds of manganese are at the same time peroxidase-active too, caused by the formation of active intermediate particles within the catalase process [$HO_2^{\bullet}$, $OH^{\bullet}$ radicals, manganese(III, IV) compounds] which will oxidize the organic substances. H_2O_2 decomposition in the presence of Mn(II) complexes proceeds by *radical-chain or ion-molecular mechanisms*. Introduction of noncomplexing or weakly complexing substrates do not essentially change *radical-chain mechanisms*, when substrate radicals formed in the reaction are poorly active or have rather strong reduction properties. In this case, the scheme of the peroxidase process differs from the catalase one only by the additional stage of chain breakage:

$$OH^{\bullet} + S \rightarrow P_I(\text{products of substrate oxidation}).$$

O_2 participation in the peroxidase process brings complications into the scheme of the mechanism. Some organic dyes (indigomonosulfate, indigocarmine, indigotetrasulfate, murexide, carmine—stability of the latter with Mn(II) falls into the limits $1<pK<3$) and also alcohols (ethanol, isopropanol, ethyleneglycol, glycerine) in the systems Mn(II)–HCO_3^-–H_2O_2–S. Mn(II)–Trien–H_2O_2–S and Mn(II)–His–H_2O_2–H_3BO_3–S are efficiently oxidized by such radical-chain mechanism with $OH^{\bullet}$ radicals' generation. These substrates' oxidation is done by $OH^{\bullet}$ radicals in the reaction bulk, i.e., outside the first coordination sphere of complex catalyst. Such a process mechanism is "outer-sphere."

In the case of alcohols oxidation (ROH) by $OH^{\bullet}$ radicals, substrate radicals will be formed ($R_{\alpha}^{\bullet}$), having rather strong reducing properties to interact with oxygen in solution (in aerobic conditions). The mechanism of process thus includes the following stages:

$$OH^{\bullet} + ROH \rightarrow R_{\alpha}^{\bullet} + H_2O,$$

$$R_{\alpha}^{\bullet} + O_2 \rightarrow RO_2^{\bullet},$$

$$RO_2^{\bullet} \rightarrow RCHO + HO_2^{\bullet},$$

$$RO_2^{\bullet} + ROH \xrightarrow{2H^+} R_{\alpha}^{\bullet} + RCHO + H_2O_2,$$

$$O_2^{\bullet -} + ROH \rightarrow R_{\alpha}^{\bullet} + HO_2^{-}.$$

Hereby, a certain similarity in mechanisms of alcohols' oxidation by the systems Fe(II)–L–H_2O_2–S and Mn(II)–L–H_2O_2–S is observed. It is known that:

- In the case of $R_\alpha^\bullet$ formation O_2 will also take part in the oxidation process
- The process rate in the presence and absence of O_2 should be different, and the quantity of oxidizing substrate should be significantly higher than that spent for S oxidation by hydrogen peroxide
- The role of H_2O_2 in such systems is reduced to the substrate preparation (in the case of alcohols, their transformation into alkoxyradical $R_\alpha^\bullet$) for its further interaction with O_2
- The mechanism of process can be qualified as outer sphere

If OH radical is not generated in the system Mn(II)–L–H_2O_2–S, substrate oxidation can proceed in the outer-sphere way by the Mn(IV) compounds formed [and possibly, in some cases, by Mn(III) compounds also]. For example, in the system Mn(II)–diethyleamine(tridentate ligand)–H_2O_2–S (some organic dyes: indigocarmine, indigomonosulfonate, indigotetrasulfonate, etc.) Mn(II)$Dien_2$ is a catalytically active saturated complex, and substrates' oxidation is provided in the outer-sphere way by Mn(IV) compounds. The mechanisms of such substrates' oxidation in the systems Mn(II)–ethylenediamine–H_2O_2–S, Mn(II)–1,10-phenanthroline–H_2O_2–S and in many others are similar. As a rule, in such systems the substrate is competitive with hydrogen peroxide for the oxidized form of catalyst, and as a result S is oxidized and the H_2O_2 role is reduced practically to catalyst transformation to higher oxidation degree.

Ion-molecular mechanism is realized in the systems where the substrate can form rather strong complexes with Mn^{2+} ($pK_{stability} \approx 3$–8). Thus ligand-substrate complex or peroxocompound of L_nMn(II)H_2O_2 (HO_2^-) type will be formed and alternate oxidation of complex-bound Mn(II) ion proceeds to Mn(IV) by interaction of the latter with the substrate:

$$L_n\mathrm{Mn(II)S} + H_2O_2 \rightarrow L_n\mathrm{Mn(IV)S} + 2OH^-,$$

$$L_n\mathrm{Mn(IV)S} \rightarrow L_n\mathrm{Mn(II)} + P_i.$$

The role of substrate oxidizer is played by the catalyst's oxidized forms via the formation of mixed ligand-substrate complex and substrate oxidation inside the catalyst coordination sphere (inner-sphere mechanism). Such a mechanism takes place in systems where the organic dyes are used as substrates: lumomagnezone, lumogallione, alizarine, and alizarine C (substrates of S_L type, $pK_{stability}>3$). Such substrates introduction alters the principle mechanism of hydrogen peroxide decomposition. Hence, the specific inhibitors of $OH^\bullet$ and $HO_2^\bullet$ radicals (paranitrosodimethylamine and TNM, accordingly) do not affect the S_L peroxidase oxidation rate in the catalase process. It testifies to the advantageous realization in these systems of not radical-chain and ion-molecular inner sphere mechanisms. The main reason for mechanism change and the altering of catalyst redox potential should be considered because of oxidizing substrate S_L coordination with Mn(II). The general mechanism of substrates of S_L type in the systems Mn(II)–L–H_2O_2–S_L can be expressed by Scheme 2.26, where S_L in Mn(IV)S_L is oxidized yielding the reaction product P, and the reduced form Mn(II) interacting with the excess of S_L, will be repeatedly transformed into LMn(II) S_L.

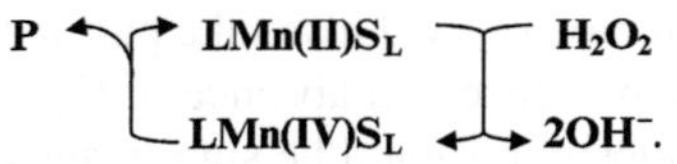

Scheme 2.26 Generic mechanism of substrates oxidation in the systems of Mn(II)-L- H_2O_2-S_L

Among the catalytic processes of hydrocarbons' oxidation by hydrogen peroxide with transition metal ions' participation, the reactions of aromatic compounds' hydroxylation are of special importance. The reactions involved can be carried out in water, nonwater, and mixed solvent in mild conditions [248]. The study of similar reactions with the use of elementary systems of Fenton's type ($Fe^{2+} + H_2O_2$) is important both for the many industrial products obtained (phenol, hydroxyquinone, etc.) and for understanding the mechanisms of this reaction proceeding in living organisms by enzymatic means (cytochrome P-450).

On using Fenton's reagent, forming $OH^{\bullet}$ radicals interact with aromatic substrate and the general way of these transformations, with the formation of numerous final products, looks as follows [255]:

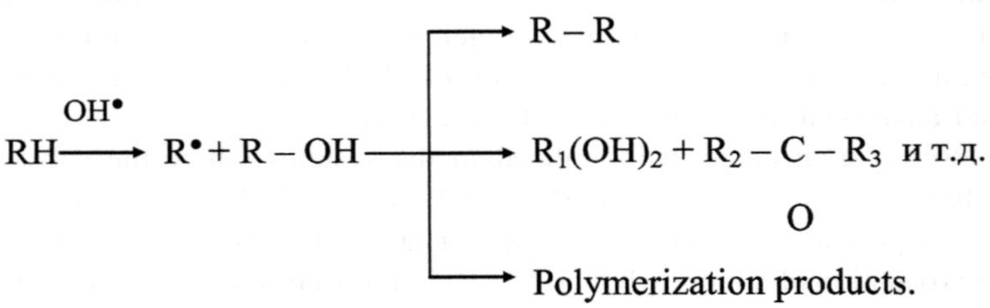

However, numerous pieces of research on benzene, phenol, toluene, and the production of their derivatives by means of Fenton's reagent or similar systems with Cu^{2+}, Co^{2+}, and other transition metals have shown that in water medium these processes proceed with poor selectivity, with small yield, and by radical mechanisms. Such character of the mechanisms was judged, in particular, by small values of NJH-shift (migration of hydrogen isotopes on marked compounds hydroxylation), characteristic of just this kind of mechanism.

On the transition to aprotonic or water-aprotonic solvents, in some cases the selectivity and hydroxylation products' yield somewhat increases. Unusually high values of NJH-shift thus observed, character of products distribution, and other data specify that in aprotonic solvents hydroxylating particle can be not only $OH^{\bullet}$ radical but also another particle. This particle presumably can be ferryl-ion or complex:

$$Fe^{2+} + H_2O_2 \rightarrow Fe^{IV}O^{2+} + H_2O.$$

The particle's generation can result in the increase of selectivity of hydroxylation reaction. However, there are no precise experimental and theoretical confirmations of such capacity that ion may have.

From the metal ions used (Fe, Cu, Zn, Co, Cr, W, etc.), the most active catalyst turned out to be Fe^{3+} ions [255]. Thus, on benzene hydroxylation in a water–acetonitryl medium, maximal activity was observed at pH corresponding to Fe^{3+} ions existence in dimer hydroxoform $[Fe_2(OH)_2]^{4+}$. Despite numerous works on aromatic compounds' hydroxylation, the detailed mechanism of this process is still insufficiently investigated, and the data available are sometimes inconsistent.

One of the versions of unsaturated and saturated compounds' hydroxylation is opening up the opportunity of using a system of Fenton's type with certain updating. Such modified systems have allowed one to carry out reactions of olefins' and saturated compounds' catalytic chlorohydroxylation [256]. In these systems, instead of hypochloric acid (HOCl) as olefin carrier in a noncatalytic process, a modified oxidizer was used (mixture of HOOH/HCl in the ratio 1:1) and as catalysts–complexes $Fe^{II}L_4{}^{2+}$ [L = tetrakis-(triphenylphosphine oxide) ($OPPh_3$) or α, α′-dipyridyl (dipy)]. The reaction was run in acetonitryl (MeCN) or in water acetonitryl (0.2 M $H_2O/MeCN$):

$$RCH = CHR' + (HOOH/HCl) \xrightarrow{Fe^{II}L_4^{2+}} RCH(OH)CH(Cl)R'.$$

Scheme 2.27 showing the mechanism of catalytic chlorohydroxylation of olefins and hydrocarbons is given in [256].

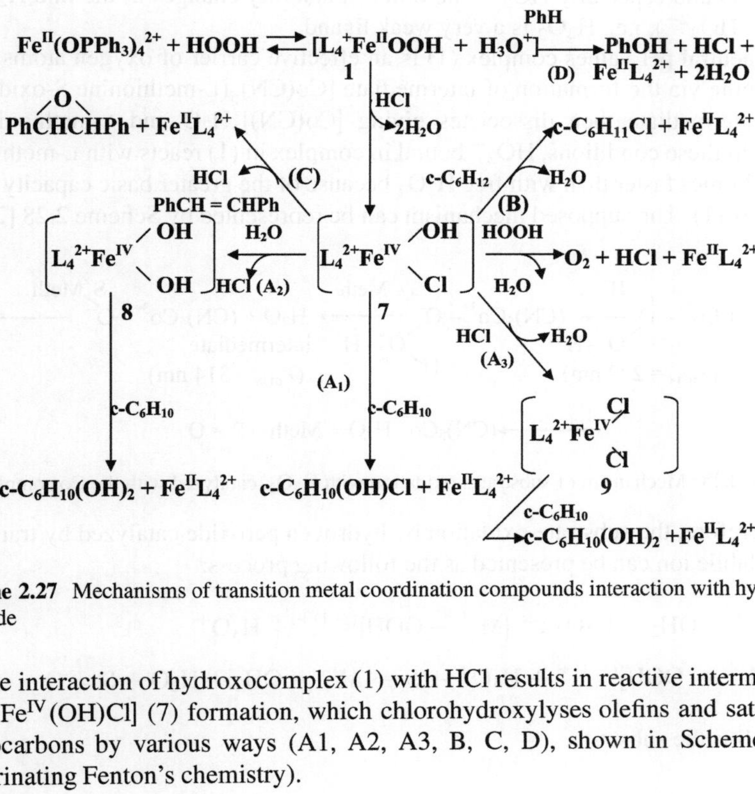

Scheme 2.27 Mechanisms of transition metal coordination compounds interaction with hydrogen peroxide

The interaction of hydroxocomplex (1) with HCl results in reactive intermediate $[L_4{}^{2+}Fe^{IV}(OH)Cl]$ (7) formation, which chlorohydroxylyses olefins and saturated hydrocarbons by various ways (A1, A2, A3, B, C, D), shown in Scheme 2.27 (chlorinating Fenton's chemistry).

That means that use of such an oxidizer as HOOH/HCl with certain catalysts can replace such oxidizers as $Cl_2/HOCl$ in synthetic chemistry. The compound (7) appeared to be more reactive than HOCl (produced in vivo with the help of mieloperoxidase—see Chap. 4) in relation to organic substances in water medium. Therefore, the most interesting assumption appears to be the possibility of formation of reactive intermediates such as (7) for oxyradical damage realization in biochemical systems. However, it is only an assumption.

On various coordinative compounds of transition metals' interaction with hydrogen peroxide, hydroperoxocomplexes of $M^{z+}L_n(OOH^-)$ type will often be formed, which on one hand can act in the reaction of activation in water solutions, and on other to provide the process of oxygen atom transfer to a suitable substrate. The activation of pentacyano-(hydroperoxo)cobaltate(III), $[Co(CN)_5OOH]^{3-}$ (1) in water medium and oxygen atom transfer from it on L-methionine with the formation of L-methionine-*S*-oxide can serve as an example of such process realization [257].

The activation proceeds in acid medium according to the following reactions:

$$[Co(CN)_5OOH]^{3-} + H_3O^+ \rightleftarrows [Co(CN)_5H_2O_2]^{2-} + H_2O,$$

$$[Co(CN)_5H_2O_2]^{2-} + H_2O \rightarrow [Co(CN)_5H_2O]^{2-} + H_2O_2.$$

Hydrogen peroxide coordinated in the complex $[Co(CN)_5H_2O_2]^{2-}$ is more labile than H_2O and especially HO_2^- (the order of stability changes in the line $H_2O_2 < H_2O < HO_2^{2-}$), i.e., H_2O_2 is a very weak ligand.

At neutral pH values complex (1) is an effective carrier of oxygen atoms on L-methonine via the formation of intermediate $[Co(CN)_5$(L-methionine S-oxide)$]^{2-}$. This intermediate then dissociates giving $[Co(CN)]_5H_2O$ and L-methionine-*S*-oxide. In these conditions, HO_2^- bound in complex in (1) reacts with L-methionine 2×10^4 times faster than with free H_2O_2 because of the greater basic capacity of the complex (1). The supposed mechanism can be represented by Scheme 2.28 [257].

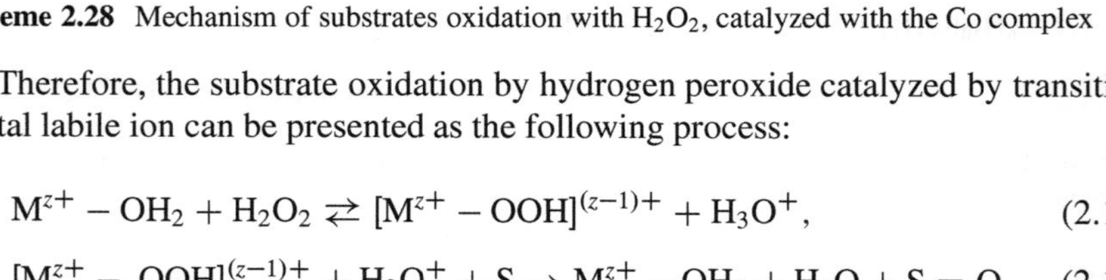

Scheme 2.28 Mechanism of substrates oxidation with H_2O_2, catalyzed with the Co complex

Therefore, the substrate oxidation by hydrogen peroxide catalyzed by transition metal labile ion can be presented as the following process:

$$M^{z+} - OH_2 + H_2O_2 \rightleftarrows [M^{z+} - OOH]^{(z-1)+} + H_3O^+, \quad (2.10)$$

$$[M^{z+} - OOH]^{(z-1)+} + H_3O^+ + S \rightarrow M^{z+} - OH_2 + H_2O + S = O \quad (2.11)$$

where S is the substrate.

In contrast, the metal ion can reduce coordinated HOO– by a one- or two-electron process, generating the new oxidants [258]:

$$[M^{z+} - OOH]^{(z-1)+} \rightarrow [M^{(z+1)+} = O]^{(z-1)+} + OH^{\bullet}, \tag{2.12}$$

$$[M^{z+} - OOH]^{(z-1)+} + H^{+} \rightarrow [M^{(z+2)+} = O]^{z+} + H_2O. \tag{2.13}$$

High medium acidity prevents the reaction (1.10) running, while $OH^{\bullet}$ radical formation on H_2O_2 reduction by uncoordinated metal ion is still possible:

$$H_2O_2 + e^{-} + H^{+} \rightarrow H_2O + OH^{\bullet}. \tag{2.14}$$

It proves to be true that the low pH value in Fenton's system (Fe^{2+}–H_2O_2) favors OH radicals generation in accordance with the reaction at (1.14). At higher pH and high concentration of H_2O_2 [it is favorable for reaction (1.10)], the yield of ferryl particles increases ($Fe^{IV} = O$) [reaction (1.13)] which can proceed in accordance with an inner sphere mechanism [257].

2.7 Ferryl Particles in the Catalysis of Oxidation Processes

2.7.1 Experimental Confirmation of Ferryl Particles Formation

Intermediate compounds of ferryl, manganyl, etc. type play an extremely important role not only in the reactions of catalytic oxidation of various organic compounds in solutions but also in a great variety of enzymatic and nonenzymatic processes in vivo. The detailed review of this problem is given in [93]. But here some theoretical possibilities of the existence of these intermediates are discussed briefly important for further consideration of enzymatic processes mechanisms study and modeling in which intermediates are involved.

Fenton reagent (system Fe^{2+}–H_2O_2) from the moment of its discovery in 1894 has been known as one of the most "strong" liquid-phase oxidisers in acid media. However, Fenton has ascertained at the end of the last century only that the traces of iron(II) sulfate efficiently catalyse the process of tartaric, maleic, and other α-acids' oxidation by hydrogen peroxide. A fundamental study of H_2O_2 decomposition by Fe^{2+} ions performed by Gaber and Weiss in the 1930s, and also the variants of the scheme of the radical-chain process in question made by Barb, Baxendale, etc. [58], have contributed to the development of knowledge about the mechanism of Fenton reagent action. The following stages can be included in the modified scheme (Scheme 2.29).

$$Fe^{2+} + H_2O_2 \longrightarrow Fe^{3+} + OH\cdot + OH^-,$$
$$Fe^{2+} + OH\cdot \longrightarrow Fe^{3+} + OH^-,$$
$$OH\cdot + H_2O_2 \longrightarrow HO_2\cdot + H_2O_2,$$
$$HO_2\cdot + Fe^{3+} \longrightarrow Fe^{2+} + H^+ + O_2,$$
$$Fe^{2+} + HO_2\cdot \longrightarrow Fe^{3+} + HO_2^-.$$

Scheme 2.29 Radical-chain mechanism of H_2O_2 oxidation with Fe^{2+} ions

Such a scheme complies with the experiment and, as a matter of fact, has not undergone essential changes till now. The revealing of the scheme of the H_2O_2 decomposition mechanism by Fe^{2+} ions made it possible to reveal the active point of Fenton's reagent which turned out to be the $OH^{\bullet}$ radical.

As shown earlier (see Sect. 2.5.2.1) in parallel with the concept of $OH^{\bullet}$ radicals formation in the system Fe^{2+}–H_2O_2, the hypothesis of Gray and Gorin [173] has appeared which in essence consists of "ferryl ion" (Fe^{IV} – –O) formation in the given system according to the following reaction:

$$Fe^{2+} + H_2O_2 \rightarrow Fe^{IV} = O + H_2O,$$

and only then, after its hydrolysis, will the hydroxyl radical $OH^{\bullet}$ be generated.

The idea of $Fe^{IV}O^{2+}$ formation was supported by many researchers [80,129,131, 176,259–262], etc., which used the given approach for substantiation of the results obtained with Fenton's reagent action mechanism study.

The possibility of ferryl particle formation within the complexes composition was also considered at pH 7 [263], but not in acid medium in which the system $Fe^{2+}{}_{aq}$–H_2O_2 can only exist. It is supposed that under certain conditions (suitable pH, ligand, solvent, and temperature) the intermediate complex will be formed, and then ferryl ions or ferryl complexes, hydrolysis of these resulting in $OH^{\bullet}$ radical formation:

$$Fe^{2+} + H_2O_2 \overset{-H^+}{\leftrightarrows} Fe^{2+}\,(HO_2{}^-) \overset{H^+}{\longrightarrow} Fe^{IV} = O^{2+} + H_2O \rightarrow Fe^{3+} + OH^- + OH^{\bullet}.$$

If the hydrolysis of the ferryl particle proceeds at high rate, only $OH^{\bullet}$ radicals can be detected experimentally in the system [10]. Suitable ligands can stabilize a high oxidation degree of metal (Fe^{IV}, Mn^{IV}). There are experimental and theoretical proofs that in acid medium only $OH^{\bullet}$ radicals will be formed in Fenton's system, and at higher pH, when iron ion is bound into the complex with a suitable ligand, the possibility of less reactive particle formation $L_nFe^{IV} = O^{2+}$ (which is a more specific oxidizer) arises.

At various pHs the oxidizing particles differing from each other can be formed, following from the peculiarities of certain substrates transformation products distribution. In fact, in acid solutions the distribution of products is approximately the same as on using $OH^{\bullet}$ radicals (generated by electrolysis), while in neutral and basic media it is more stereospecific. In the latter case the oxidizer is not $OH^{\bullet}$ radical and apparently ferryl iron [93, 264].

Such high-oxidized complexes of iron(IV) are similar to each other with their basic parameters ($\lambda_{max} \approx 430$ nm, $\varepsilon = 900 \pm 100$ L/mol s), the time of half-transformation varies from 0.1 to 1 s [165,179,180], and they are more stereospecific in the reactions of oxidation and hydroxylation [265], especially important in biological systems. In general, the ferryl complex is somewhat weaker but is a more selective oxidizer than $OH^{\bullet}$ radical.

Experimental proofs of ferryl-complex generation in redox catalytic reactions are as follows [93]:

1. The results of cyclohexanol oxidation by hydrogen peroxide in the presence of iron(II) salts (product radioselectivity was explained by the occurrence of intermediate compound of ferryl type, transmitting the atom of oxygen) [181]
2. The data on hydrogen peroxide interaction with iron porphyrinate [182]
3. Results of Fe^{II}EDTA complex interaction with H_2O_2 and alcohols, aminoacids and carbonic acids influence on this system [183]
4. Resonance Raman-spectra obtained on plants' peroxidases protonation [184, 185]
5. The data on ferrimyoglobin interaction with H_2O_2 [266]
6. Presence of ferryl group in model systems simulating ligands peroxide oxidation [267]
7. Occurrence of so-called compounds I and II (see Chap. 3) in catalytic cycle of peroxidases and catalases on their interaction with hydrogen peroxide [268]
8. Results of model systems of cytochrome P-450 of $PFe^{III}-H_2O_2(ROOH)$ [199] type study where P is the general mark for porphyrins

Numerous experimental proofs of ferryl particles $Fe^{IV} = O^{2+}$ generation or their complexes $L_nFe^{IV} = O^{2+}$ both in enzymatic processes (compounds I and II) and in their modeling systems have allowed one—at the same time—to establish their high reactivity (only a little less than that of $OH^{\bullet}$ radical) and their large specificity as oxidizers of various organic substances compared to such systems as $Fe^{2+}{}_{aq}-H_2O_2-S$, $L_nFe^{2+}-H_2O_2-S$ and enzyme (or its model)–$O_2(H_2O_2)$–S (where S is an organic substrate). However, despite obvious successes, these particles' identification still remains a difficult problem for the specific system investigation, that causes numerous discussions about the generation of "superoxidized" states of iron or manganese (manganyl particles).

2.7.2 *Quantum-Chemical Methods of Ferryl (Manganyl) Particles Formation Study*

In spite of the fact that in many catalytic systems the mechanism of hydrogen peroxide disintegration by ML_n-catalysts have been investigated on qualitative, semiqualitative, and (in rare cases) quantitative levels, a number of difficulties arise in the explanation of this process mechanism and especially of its initiation stages. It is stipulated by the complexity of this reaction, despite its apparent simplicity:

$$2H_2O_2 \xrightarrow{ML_n} 2H_2O + O_2.$$

Actually the occurrence of numerous new intermediate particles during this reaction brings out great experimental and calculation difficulties with revealing its mechanism. Generation of each of these intermediate particles (intermediate compounds, free radicals, complexes with the changed metal oxidation degree, and other compounds) is characterized by the equilibrium constant or elementary rate constant. The set of all these data related to the mechanism of this catalytic reaction needs to be established on a quantitative level (practically that is seldom possible) or at least at good qualitative one (there are considerably more examples available). In reality, to study the mechanisms of catalytic processes, the composition, stability constants of metal complexes-catalysts, separate equilibrium constants, initial substances concentrations in the reacting medium during the process, metal oxidation degree in the ion or complex compound, nature of intermediate compounds formation, changing of bond character in substrate on complex formation with catalyst, and other parameters should be determined.

On each of these stages, the number of experimental (impossibility of intermediate particles definition) and calculation (some elementary rate constants are unknown) difficulties arise; they are not always possible to overcome in the real system under investigation. Therefore, in many cases supposed mechanisms of H_2O_2 disintegration in catalytic systems, proceeding from the experimental data, are subject to a certain degree of arbitrariness. This especially concerns the initial stage of the mechanism—the initiation stage, on which the further course of process largely depends.

Even in those rare cases when all the above-mentioned experimental and other difficulties are overcome and the mechanism of the catalytic redox process is revealed completely and reliably enough, the information obtained has ascertaining character and does not explain different catalytic capacity of various transition metal ions and ligands' various influence on them, as well as the occurrence of different intermediate particles (for instance, free radicals) at the initiating stage.

Despite approximation of semi-empirical quantum-chemical methods, their use brings the above problems on the electron level. Statements of the molecular orbitals theory (MO) can be used on the initiation stage consideration in the reaction of substrate interaction with the catalyst (complex compound of metal). Revealing the difference in energy levels of boundary MO reagents [mainly those of maximum-filled molecular orbital (MFMO) and LFMO within the pair donor–acceptor] can be useful with the most favorable ways of chemical reactions discussion, as the reaction usually runs between two centers having the greatest difference in electron density on boundary MO.

Quantum-chemical study of the systems manganese(iron) ligand–H_2O_2 were carried out with the purpose of substantiating additionally the possibility of elementary stages realization of H_2O_2, $O_2^{\bullet-}$, $HO_2^{\bullet}$, $OH^{\bullet}$ interaction with certain coordination compounds of manganese(iron) in various oxidation degrees. The systems including H_2O_2 and manganese(iron) where the ligand is bicarbonate ion, triethyleneamine (Trien), ethylenediamine (En), and phenanthroline (phen) were studied in most detail. A quantum-chemical study was performed on the basis of catalytically active complexes calculation [$Mn^{z+}(HCO_3)_2$, Mn^{z+}Trien, $Mn^{z+}En_2$, $Mn^{z+}Phen_2$, $Fe^{2,3+}{}_{aq}$], in which the charge of central ion changes from +2 up to +4, and also that of molecules and particles of H_2O_2, HO_2^-, $OH^{\bullet}$ and $O_2^{\bullet-}$ types. Semi-empirical method MO LCAO in the approximation of MWH was used to make calculations with self-coordination according to the charges of all the atoms present in the system. On the basis of electron structure of catalyst and reagents calculation, the estimation of their relative reactivity was done, which in the Klopman's scheme [269] serves the value of total energy of system catalyst-reagent change on the appropriate transition state formation.

The main results of calculations have allowed one to make the following conclusions [203, 261–272]. Calculated values showing charge transfer to H_2O_2 from Mn(II) complex and rather large for small molecules of H_2O_2 change of energy, testify to the essential O–O bond weakening in H_2O_2 molecule and the possibility of its further breakage. As a result, Mn(II) compounds are transformed into Mn(III) ones and H_2O_2 dissipates yielding $OH^{\bullet}$ and OH^-. However, as the $OH^{\bullet}$ formation in the stage of initiation was not discovered experimentally, for example, in the system Mn^{2+}–HCO_3^-–H_2O_2 (this radical was discovered only at the stage of chain continuation), it was assumed that it, not leaving the coordination sphere of complex catalyst, reacts with Mn(III) compounds forming complex-bound Mn(IV) compounds.

The realization of such a process can be given schematically by the following scheme:

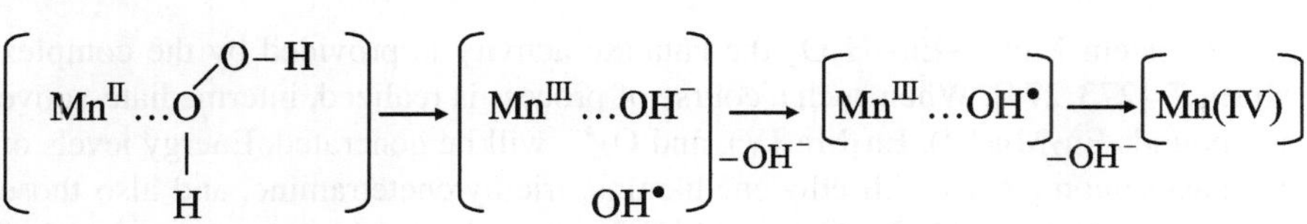

The value of electron density transfer (δ_q) from $HO_2^{\bullet-}$ to manganese(IV) complexes in this system reaches ~ 0.7 V, that makes the realization of one-electron transfer of catalyst to Mn(III) possible. The process can proceed and through the consecutive quick stages of one-electron transfer with practically simultaneous coming of two $O_2^{\bullet-}$ particles into the reaction bulk and regenerated Mn(II) complexes formation. The data are given that testify to the possibility of initiation mechanism realization in the systems considered via the sequence of two stages—i_1 and i_2 in which two electrons transfer is realized from the catalyst to H_2O_2 (i_1) and one electron transfer from H_2O_2 to the catalyst (i_2):

$$i_1.\quad LMn(II) + H_2O_2 \rightarrow LMn(IV) + 2OH^-,$$

$$i_2.\quad LMn(IV) + HO_2^- \rightarrow LMn(III) + O_2^{\bullet-} + H^+.$$

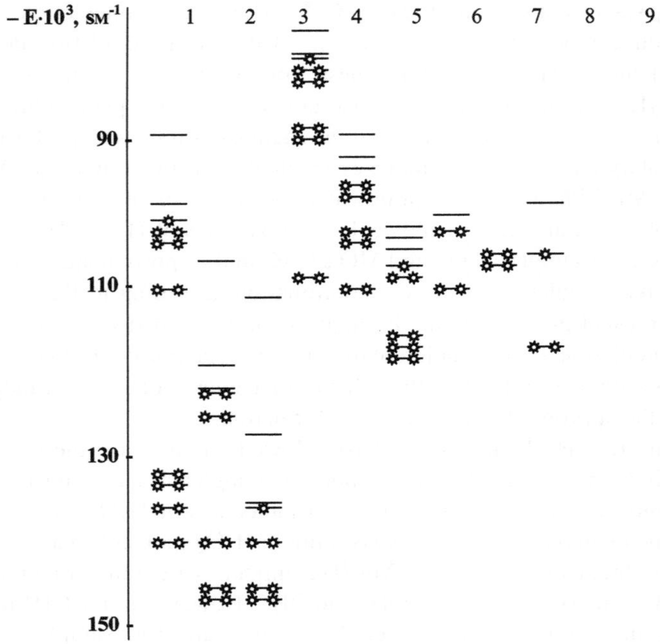

Fig. 2.12 Energy levels of manganese complexes with ethylenediamine: $[MnEn_2(H_2O)]^{2+}$ (1), $[MnEn_2(H_2O)]^{3+}$ (2), $[MnEn_2(H_2O)]^{4+}$ (3), triethylenetetramine $[MnTrien(H_2O)]^{2+}$ (4), $[MnTrien(H_2O)]^{3+}$ (5), $[MnTrien(H_2O)]^{4+}$ (6), and also those of peroxide particles: H_2O_2(7), $O_2^{\bullet-}$ (8), ¡@4;¿ $HO_2^{\bullet}$ (9)

In the system Mn^{2+}–En–H_2O_2 the catalase activity is provided by the complex $MnEn_2^{2+}$ [273, 274]. When such a course of process is realized, intermediate-active compounds $En_2Mn(III)$, $En_2Mn(IV)$, and $O_2^{\bullet-}$ will be generated. Energy levels of manganese complexes with ethylenediamine, triethylenetetramine, and also those of peroxide particles H_2O_2, $O_2^{\bullet-}$ and $HO_2^{\bullet}$ were calculated by quantum-chemical methods MO LCAO (Fig. 2.12) [203].

The results of total energy change in the intermediate state catalyst–oxidizer, and also calculated values of redox capacities and charge transfers from one particle to another on their interaction with common MO formation, are given in [275]. HOMO and LFMO levels of complexes with the charge +2 and those of hydrogen peroxide are located close to each other (they are pseudodegenerated) (Fig. 2.12) which provides maximal electron density transfer between the reagents. On electron density transfer from complex to H_2O_2, the O–O bond is labilized which must result in its breakage with free $OH^{\bullet}$ radicals' formation which has not been detected experimentally [203]. It can be understood if inside the coordination sphere the transfer of one more electron takes place from the central ion of complex onto $OH^{\bullet}$ (without the latter coming out from the coordination sphere) or spontaneous two-electron transfer occurs from complex to H_2O_2.

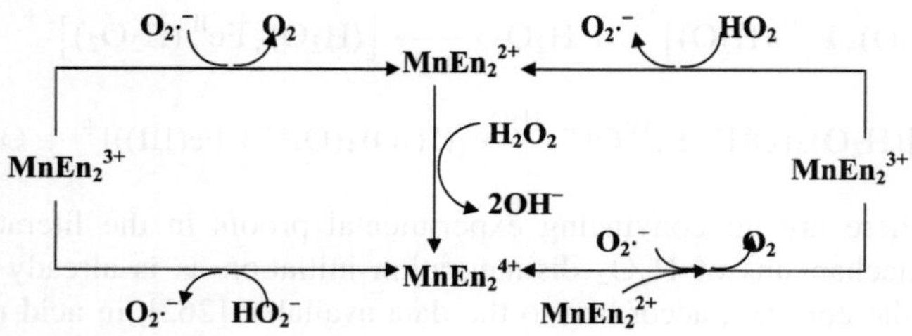

Scheme 2.30 Radical-chain mechanism of H_2O_2 oxidation with Fe^{2+} ions

Summarizing the experimental data and quantum-chemical calculations of Mn(II)–ethylenediamine(En)–H_2O_2 system, the scheme of H_2O_2 disintegration can be presented as follows (Scheme 2.30) [203]:

The analysis of decomposition mechanisms with complexes $Mn(HCO_3)_2$, $MnTrien^{2+}$, $MnPhen_2^{2+}$, and $MnEn_2^{2+}$ testifies that in the first two cases radical-chain mechanism is realized with $OH^{\bullet-}$ and $O_2{}^{\bullet-}$ radicals' generation. In the system Mn(II)–Phen–H_2O_2 the radical-chain process proceeds, but in the reaction bulk only $O_2{}^{\bullet-}$ radicals will be formed. The same radicals are intermediate-active particles on H_2O_2 disintegration by $MnEn_2^{2+}$ complexes, but the mechanism in this case is cyclic ion-radical. The results of quantum-chemical study of the above-mentioned systems made it possible to reveal probable reasons for various sets of consecutive stages' realization in these systems, in spite of the fact that central metal ion in the complexes $MnTrien^{2+}$, $MnEn_2^{2+}$, and $MnPhen_2^{2+}$ is in the same environment (four atoms of nitrogen; another two coordination places are occupied by water molecules), and also to state the certain judgment of rather unequal activity of these complexes (taking as examples $MnTrien^{2+}$ and $MnEn_2^{2+}$) in the reactions of H_2O_2 disintegration, that emphasizes the specificity of the ligand and its influence on catalytic activity of the transition metal ion [203].

The above-mentioned quantum-chemical methods were used to perform the study of stages of H_2O_2 catalytic disintegration initiation in Fenton's system which is one of the most effective liquid-phase oxidizing reagents in acid medium.

It was supposed that the stage of initiation of hydrogen peroxide disintegration in Fenton's system an is elementary one [3]. According to the opposite point of view, $OH^{\bullet-}$ radical and Fe^{3+} ion generation does not serve the unequivocal proof of this stage's simplicity. Probably, as a result of two-electron transfer from the catalyst to substrate, ferryl-ion will be formed first [173]:

$$\left[(H_2O)_5\,Fe^{II}\,(H_2O)\right]^{2+} + H_2O_2 \xrightarrow[-H_2O,-H^+]{} [(H_2O)_5\,Fe^{IV}O]^{2+} \rightarrow$$

$$\xrightarrow{H_2O} [(H_2O)_5Fe(III)(OH^-)]^{2+} + OH^\bullet$$

Later on it was stated [80] that a peroxocomplex is generated first, from which later, due to inner molecular electron transitions, a complex with ferryl ion will be formed, which will then transform into the complex Fe(III) and $OH^\bullet$:

$$\left[(H_2O)_5\,Fe^{II}\,(H_2O)\right]^{2+} + H_2O_2 \xrightarrow[-H_2O]{} \left[(H_2O)_5\,Fe^{II}\,(H_2O_2)\right]^{2+} \xrightarrow[-H^+]{}$$

$$\rightarrow [(H_2O)_4(OH^-)Fe^{IV}O]^+ \xrightarrow{H_2O} [(H_2O)_4(OH^-)_2Fe(III))]^+ + OH^\bullet$$

However, there are no convincing experimental proofs in the literature of the proposed mechanisms of H_2O_2 disintegration initiation as is already mentioned above. On the contrary, according to the data available [262], in acid media $OH^\bullet$ radicals will be formed, and only in neutral and alkaline media (when iron ions are bound in complexes) does the possibility of ferryl complexes occurrence increase. Nevertheless, the idea of ferryl particle formation in ferroperoxide systems has been discussed in many papers [10].

With the purpose of theoretical consideration of ferryl particles initiation possibility in Fenton's system, quantum chemical calculations of $Fe(H_2O)_6{}^{n+}$ electron structure and those of certain hypothetical intermediate particles were made, and also the calculation of all peroxide particles (H_2O_2, HO_2^-, $HO_2^\bullet$, OH) by the method MO LCAO in approximation of MWH [276]. On the basis of these calculations, the self-coordinated values of atoms and orbital fillings were found, and also the values of electron energy levels of corresponding MO (Fig. 2.13).

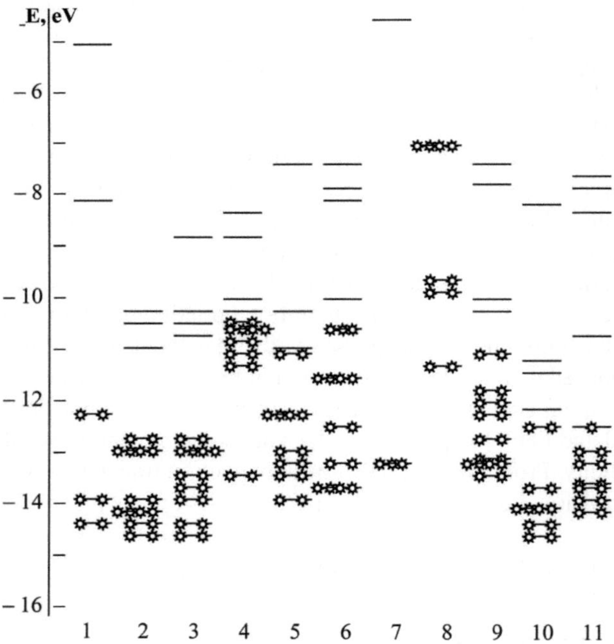

Fig. 2.13 Energy levels of MO of molecules' active zone: 1 – H_2O_2, 2 – $[Fe(H_2O)_6]^{2+}$, 3 – $[Fe^{II}(H_2O)_5(H_2O_2)]^{2+}$, 4 – $[Fe^{II}(OH^-)(H_2O)_4(H_2O_2)]^+$, 5 – $[Fe^{IV}O(H_2O)_4(OH^-)]^+$, 6 – $[Fe(III)(OH)_2(H_2O)_4]^+$, 7 – $OH^\bullet$, 8 – HO_2^-, 9 – $[Fe^{II}(H_2O)_5(HO_2^-)]^+$, 10 – $[Fe^{IV}O(H_2O)_5]^{2+}$, 11 – $[Fe(III)(OH^-)(H_2O)_5]^{2+}$

In addition, quantitative assessment was made of electron transfer from MO complex on the MO of the peroxide particle and the values of charge transferred (δ_q) were determined.

The analysis of complexes energy levels testifies to the low probability of peroxocomplex formation having the composition $[Fe^{II}(H_2O)_5(H_2O_2)]^{2+}$ (set of levels 3) and the impossibility of realization of such inner molecular transformations in which $[Fe^{II}(OH^-)(H_2O)_4(H_2O_2)]^+$ (4) and $[Fe^{IV}O(H_2O)_4(OH^-)]^+$ (5) are generated, for the levels of high-filled molecular orbitals (HFMO) and LFMO of these particles are much higher than the corresponding levels of both $[Fe(H_2O)_6]^{2+}$ (2), and $[Fe^{II}(H_2O)_5(H_2O_2)]^{2+}$ (3). Thus, in highly acid medium the formation of ferryl ion is impossible, and this complies with the experimental data.

However, with the increase in pH the intermediate peroxocomplex having the structure $\left[Fe^{II}(H_2O)_5\,(HO_2{}^-)\right]^+$ can be formed, this fact being testified by the analysis of the nature of HFMO and LFMO levels of $[Fe(H_2O)_6]^{2+}$ (2), $HO_2{}^-$ (8) compounds, and hypothetical complex $\left[Fe^{II}(H_2O)_5\,(HO_2{}^-)\right]^+$ (9), and also charge transfer calculation from Fe^{II} aquacomplex to H_2O_2 and from $HO_2{}^-$ to this aquacomplex (accordingly $\delta_q = -0.05$ and 0.76 V).

The comparison of MO energy levels of peroxocomplex having the composition $[Fe^{II}(H_2O)_5(HO_2{}^-)]^+$ (9) on the one hand, and MO of the complexes $[Fe^{IV}O(H_2O)_5]^{2+}$ (10) and $[Fe(III)(OH^-)(H_2O)_5]^{2+}$ (11) on the other, testifies to the form of the stage-by-stage course of the process:

$$\left[Fe^{II}\,(H_2O)_6\right]^{2+} + H_2O_2 \rightarrow \left[Fe^{II}\,(H_2O)_5\,(HO_2{}^-)\right]^+ \rightarrow$$
$$\rightarrow [Fe^{IV}O(H_2O)_5]^{2+} \xrightarrow{H_2O} [Fe(III)(H_2O)_5(OH^-)]^{2+} + OH^\bullet.$$

Thus, quantum-chemical calculations confirm the possibility of peroxocomplex $[Fe^{II}(H_2O)_5(HO_2{}^-)]^+$ and ferryl-ion $[Fe^{IV}O(H_2O)_5]^{2+}$ formation as intermediate particles. Therefore, on the pH increasing in the system, this possibility is enhanced too (H_2O_2 dissociation increases). It is increased even more with the introduction into the system of suitable ligands (porphyrin, etc.), able to stabilize the high oxidation degree of iron (Fe^{IV}). If the transformations considered proceed at high rate, it is impossible to detect them at the present level of experimental techniques, and only $[Fe(III)(H_2O)_5(OH^-)]^{2+}$ and $OH^\bullet$ radical can be detected in the system. This shows the limitations of the available experimental opportunities and confirms the importance of quantum-chemical calculations.

References

1. G.C. Boreskov, *Heterogeneous Catalysis* (Nova Science, New York, 2003), p. 236
2. W. Herrmann, C. Kohlpaintner Water-soluble ligands, metal complexes, and catalysts: synergism of homogeneous and heterogeneous catalysis. Angew. Chem. **32**(11), 1524–1544 (1993)
3. A. Shilov, Priroda [Nature] **44**, 20 (1979) [Russian]
4. D. Austruc, *Organometallic Chemistry and Catalysis* (Springer, Berlin, 2007), p. 608

5. C.I. Zamaraev, Kinetika i kataliz [Kinet. Catal.] **21**, 36 (1980) [Russian]
6. A.Ya Sychev, *Redox Catalysis with Metal Complexes* (Stiinta, Kishinev, 1976), p. 191 [Russian]
7. A.A. Bagaturiants, Zh. Fiz. Khim. [J. Phys. Chem.] **57**, 1100 (1983) [Russian]
8. J.J. Spivey, K.M Dooley (eds.) Catalysis, vol. 22 (Royal Society of Chemistry, Cambridge, 2010), p. 328
9. A.E. Shilov, Uspekhi khimii [Russian Chem. Rev.] **43**, 863 (1974) [Russian]
10. A.Ya. Sychev, V.G. Isak, *Homogeneous Catalysis with Iron-Containing Compounds* (Stiinta, Kishinev, 1988), p. 216 [Russian]
11. A.Ya. Sychev, S.S. Budnikov, Zh. Fiz. Khim. [J. Phys. Chem.] **44**, 106 (1970) [Russian]
12. A.Ya. Sychev, S.S. Budnikov, S.M. Bardina, Zh. Fiz. Khim. [J. Phys. Chem.] **44**, 2421 (1970) [Russian]
13. NASA Research Indicates Oxygen on Earth 2.5 Billion Years Ago. Release 07-215 [online.] Updated 27 Sept 2007, http://www.nasa.gov/home/hqnews/2007/sep/HQ_07215_Timeline_of_Oxygen_on_Earth.html
14. A. Bakac, Inorg. Chem. **49**, 3584–3593 (2010)
15. E.-I. Negishi, A. de Meijeire (eds.), *Handbook of Organopalladium Chemistry for Organic Synthesis*, vol. 3 (Wiley, New York, 2002)
16. F. Wilkinson, W.P. Helman, A.B. Ross, J. Phys. Chem. Ref. Data **24**(2), 663–677 (1995)
17. A. Telfer, S. Dhami, S.M. Bishop, D. Phillips, J. Barber, Biochemistry **33**(48), 14469–14474 (1994)
18. J.A. La Verne, Radiat. Res. **153**(2), 196–200 (2000)
19. K.B. Storey (ed.), *Functional Metabolism: Regulation and Adaptation* (Wiley, New Jersey, 2004), p. 594
20. C. Knipe, W.E. Watts (eds.), *Organic Reaction Mechanisms* (Wiley, New Jersey, 2005), p. 594
21. V.D. Roshchina, V.V. Roshchina, *Ozone and Plant Cell* (Kluwer Academic, Netherlands, 2004), p. 240
22. B.C. Gilbert, M.J. Davies, D.M. Murphy (eds.), *Electron Paramagnetic Resonance* (Royal Society of Chemistry, Cambridge, 2002), p. 360
23. C. Von Sonntag, *Free-Radical-Induced DNA Damage and Its Repair: A Chemical Perspective* (Springer, Berlin, 2006), p. 523
24. K.D. Asmus *Reactive Oxygen Species in Chemistry, Biology and Medicine* (Plenum, New York, London, 1988), pp. 37–54
25. M. Allaby, R. Garratt, *Fog, Smog, and Poisoned Rain* (Facts On File, New York, 2003), p. 208
26. O. Sterner, *Chemistry, Health and Environment* (Wiley-VCH, Weinheim, 2010), p. 345
27. T. Rigg, W. Taylor, I.J. Weiss, Phys. Chem. **22**, 527 (1954)
28. K. Duerr, J. Olah, R. Davydov, M. Kleimann, J. Li, N. Lang, R. Puchta, E. Hübner, T. Drewello, J.N. Harvey, N. Jux, I. Ivanović-Burmazović, Dalton Trans. **39**, 2049–2056 (2010)
29. E.E. Chufán, K.D. Karlin, J. Am. Chem. Soc. **125**(52), 16160–16161 (2003)
30. M.L. Kremer, Israel J. Chem. **9**, 321 (1971)
31. J.M. Flicstein, J. Catal. **8**, 145 (1967)
32. S.V. Kryatov, V.E. Rybak-Akimova, Chem. Rev. **105**(6), 2175–2226 (2005)
33. C. Walling, M. Kurz, H.J. Schugar, Inorg. Chem. **9**, 932 (1970)
34. J. Cho, R. Sarangi, H.Y. Kang, J.Y. Lee, M. Kubo, T. Ogura, E.I. Solomon, W. Nam, J. Am. Chem. Soc. **132**(47), 16977–16986 (2010)
35. A.E. Kahil, H. Taube, J. Am. Chem. Soc. **74**, 2312–2314 (1952)
36. R.G. Galman, J. Phys. Chem. **65**, 556 (1961)
37. A.Ya. Sychev, S.F. Nedopekina, in *Proceedings of 2nd Conference on Organic Substances Oxidation in Liquid Phase*, Moscow, 1971, p. 26 [Russian]
38. G.A. Shagisultanova, N.P. Glukhova, Zh. Fiz. Khim. [J. Phys. Chem.] **46**, 508 (1972) [Russian]
39. A.Ya. Sychev, B. TKHO, Zh. Fiz. Khim. [J. Phys. Chem.] **44**, 726 (1971) [Russian]
40. Yu.I. Scurlatov, A.P. Purmal, Zh. Fiz. Khim. [J. Phys. Chem.] **43**, 1580 (1969) [Russian]

41. M.F. Sisemore, M. Selke, J.N. Burstyn, J.S. Valentine, Inorg. Chem. **36**(6), 979–984 (1997)
42. A.P. Purmal, Mechanisms of redox catalysis with metal ions in polar media, Ph.D. thesis, Institute of Chemical Physics of the USSR Academy of Sciences, Moscow, 1970, p. 28 [Russian]
43. A.Ya. Sychev, Catalytical properties of some transition metals coordination compounds in catalase, peroxidase and oxidase reactions, Ph.D. thesis, Kishinev, 1970, p. 232 [Russian]
44. A.C. Babko, N.V. Loria, Zh. Neorg. Khim. [J. Inorg. Chem.] **13**, 506 (1968) [Russian]
45. V.S. Sharma, J. Schubert, J. Am. Chem. Soc. **91**, 6291 (1969)
46. T.P. Vorobieva, V.G. Isac, A.P. Purmal, A.Ya. Sychev, Zh. Fiz. Khim. [J. Phys. Chem.] **47**, 1731–1739 (1974) [Russian]
47. Yu.I. Scurlatov, G.G. Duca, D.G. Batir, S.O. Travin, Koord. Khim. [Coord. Chem.] **15**, 291 (1989) [Russian]
48. Yu.N. Kozlov, A.D. Nadejdin, A.P. Purmal, Kinetika i kataliz [Kinet. Catal.] **14**, 141 (1973) [Russian]
49. A.W. Vermilyea, B.M. Voelker, Environ. Sci. Technol. **43**(18), 6927–6933 (2009)
50. F. George, *Catalysis. Study of Homogeneous Processes* (Foreign Literature, Moscow, 1957) p. 183
51. J.H. Baxendale, M.G. Evans, G.S. Park, Trans. Faraday Soc. **42**, 155 (1946)
52. I.A. Kulikov, V.S. Koltunov, V.I. Marchenco, A.S. Milodranova, L.C. Nikitova. Zh. Fiz. Khim. [J. Phys. Chem.] **53**, 647 (1979) [Russian]
53. H. Po, R. Sutin, Inorg. Chem. **7**, 621 (1968)
54. T.L. Conocchioli, E.L. Hamilton, R. Sutin, J. Am. Chem. Soc. **87**, 926 (1965)
55. C.F. Wells, M.A. Salam, Nature (London) **203**, 751 (1964)
56. C.F. Wells, M.A. Salam, Nature (London) **205**, 690 (1965)
57. C. Walling, Acc. Chem. Res. **8**, 125 (1975)
58. W.G. Barb, J.H. Baxendale, P. George, K.R. Hardave, Trans. Faraday Soc. **47**, 462 (1951)
59. F.S. Daition, H.C. Sutton, Trans. Faraday Soc. **49**, 1011 (1953)
60. J.P. Keene, Radiat. Res. **22**, 14 (1964)
61. H. Liu, X.Z. Li, Y.J. Leng, C. Wang, Water Res. **41**(5), 1161–1167 (2007)
62. C.F. Wells, M.A. Salam, Trans. Faraday Soc. **83**, 620 (1967)
63. C.F. Wells, M.A. Salam, J. Chem. Soc. A 24 (1968)
64. H.A. Schwarz, J. Phys. Chem. **66**, 255 (1962)
65. J.H. Baxendale, A.A. Khan, Int. J. Radiat. Phys. Chem. **1**, 11 (1969)
66. C. Von Sonntag, Water Sci. Technol. **58**(5), 1015–1021 (2008)
67. A.K. Picalev, S.A. Cabakchi, *Reactivity of Water Radiolysis Primary Products* (Energoizdat, Moscow, 1982), p. 323 [Russian]
68. A. Stuglik, Z.P. Zagorski, Radiat. Phys. Chem. **17**, 229 (1981)
69. C.K. Duesterberg, S.E. Mylon, T.D. Waite, Environ. Sci. Technol. **42**(22), 8522–8527 (2008)
70. M.A. Anbar, P. Neta, Int. Appl. Radiat. Izotop. **18**, 493 (1967)
71. A.C. Pical, P.Ya. Glazunov, V.I. Spitsyn, Izv. Akad. Nauk. SSSR Ser. Khim. [Proc. Acad. Sci. USSR Chem. Ser.] 401 (1965) [Russian]
72. B. Chutny, Collect. Czech. Chem. Commun. **31**, 358 (1966)
73. J.P. Keene, Radiat. Res. **17**, 14 (1962)
74. A.D. Nadejdin, Yu.I. Kozlov, A.P. Purmal, Zh. Fiz. Khim. [J. Phys. Chem.] **50**, 910 (1976) [Russian]
75. J.K. Thomas, J. Rabini, M.S. Matheson, E.J. Hart, S. Hordon, J. Phys. Chem. **70** 2409 (1966)
76. K. Schested, E. Bjergbakke, O.L. Rasmussen, J. Phys. Chem. **51**, 3159 (1969)
77. G. Czapski, L.M. Dorfman, J. Phys. Chem. **68**, 1169 (1964)
78. G. Czapski, B.H. Bielski, J. Phys. Chem. **67**, 2180 (1963)
79. B.H. Bielski, A.O. Allen, J. Phys. Chem. **81**, 1048 (1977)
80. Yu.I. Scurlatov, Ph.D. thesis, Institute of Chemical Physics USSR Academy of Sciences, Moscow, 1981, p.168
81. D.L. Ingles, Aust. J. Chem. **25**, 87 (1972)
82. J. Lewis, S. Richard, M. Salter, J. Am. Chem. Soc. **85**, 2434 (1963)

83. V.M. Berdnikov, O.S. Juravleva, Kinetika i kataliz [Kinet. Catal.] **14**, 878 (1973) [Russian]
84. G.G. Jayson, I.P. Keene, D.A. Stirhing, A.I. Swallow, Trans. Faraday Soc. **65**, 453 (1969)
85. D. Benar, J. Phys. Soc. **74**, 3209 (1970)
86. V.A. Rabinovich, Z.Ya. Havin, *Short Chemical Reference-Book* (Chemistry Publishing, Moscow, 1977), p. 193 [Russian]
87. J. Rabani, M.A. Mulac, M.S. Matheson, J. Phys. Soc. **69**, 53 (1965)
88. J. Rabani, M.S. Matheson, J. Am. Chem. Soc. **86**, 3175 (1964)
89. Yu.Yu. Lurie, *Reference Book on Analytical Chemistry* (Moscow, Chemistry Publishing, 1979), p. 529 [Russian]
90. R.M. Milburn, W.C. Wosburgh, J. Am. Chem. Soc. **47**, 1352 (1945)
91. K. Mitteilungen, Chemistry **29**, 253 (1975)
92. A.W. Fordnam, Aust. J. Chem. **22**, 1111 (1969)
93. A.Ya. Sychev, V.G. Isac, Uspekhi khimii [Russ. Chem. Rev.] **64**, 1183 (1995)
94. V.A. Plakhotnik, Izv. Akad. Nauk. SSSR Ser. Khim. [Proc. Acad. Sci. USSR Chem. Ser.] 503 (1974) [Russian]
95. B.V. Efremov, V.A. Plakhotnik, Dokl. Akad. Nauk. SSSR [Proc. USSR Acad. Sci.] **206**, 110 (1972) [Russian]
96. B.V. Efremov, V.A. Plakhotnik. Zh. Fiz. Khim. [J. Phys. Chem.] **47**, 2179 (1973) [Russian]
97. B.V. Efremov, V.A. Plakhotnik, *Catalytic Decomposition of H_2O_2 Inhibiting with Tetranitromethane Under the Influence of Fe(III) Ions. I. Initial Stage of Reaction*, vol. 6061 (VINITI, Moscow, 1973) [Russian]
98. W.C. Bray, S. Peterson, J. Am. Chem. Soc. **72**, 1401 (1950)
99. W.S. Andersen, A. Niemann, Acta Chem. Scand. **12**, 896 (1958)
100. A.G. Parts, Nature (London) **168**, 4262 (1951)
101. H.I. Bechanon, A. Goosen, J. Lovelock, J. South Afr. Chem. Inst. **29**, 85 (1976)
102. V.M. Berdnikov, Yu.N. Kozlov, A.P. Purmal, Khim. Vysokikh Energ. [High Energ. Chem.] **3**, 370 (1969) [Russian]
103. C. Walling, A. Goosen, J. Am. Chem. Soc. **73**, 3718 (1951)
104. M.S. Bains, J. Indian Chem. Soc. **53**, 83 (1973)
105. W.G. Barb, J.H. Baxendale, P. George, K.R. Hargrave, Trans. Faraday Soc. **47**, 462–591 (1951)
106. A.M. Brook, L. Castlle, L.J.R. Smith, J. Chem. Soc. Perkin Trans. **2**, 687 (1982)
107. S. Tamagaki, K. Suzuki, H. Okamoto, W. Tagaki, Tetrahedron Lett. **24**, 4847 (1983)
108. M.I. Kolthoff, E.S. Parry, J. Am. Chem. Soc. **73**, 3718 (1951)
109. E.S. Rudakov, L.C. Volkova, V.P. Tretyakov, V.V. Zamashchikov, Kinetika i kataliz [Kinet. Catal.]. **23**, 26 (1982) [Russian]
110. T.P. Vorobieva, V.M. Berdnikov, A.P. Purmal, Kinetika i kataliz [Kinet. Catal.] **11**, 100 (1970)
111. A.P. Rudenko, I.V. Tsybanova, Zh. Fiz. Khim. [J. Phys. Chem.] **45**, 387 (1971) [Russian]
112. M.L. Klemer, Trans. Faraday Soc. **59**, 2535 (1963)
113. M.L. Klemer, J. Catal. **1**, 351 (1962)
114. M.L. Klemer, S. Baer, J. Phys. Chem. **78**, 1919 (1974)
115. P. Jones, R. Kitching, M.L. Tobe, W.F.K. Wynne-Jones, Trans. Faraday Soc. **55**, 79 (1959)
116. D.G. Braun, Int. J. Chem. Kinet. **22**, 463 (1990)
117. M.L. Klemer, Int. J. Chem. Kinet. **17**, 1299 (1985)
118. M.L. Hagget, P. Jones, W.F.K. Wynne-Jones, Discuss. Faraday Soc. **29**, 153 (1960)
119. P. George, Adv. Catal. **4**, 367 (1952)
120. P. George, N. Uri, Quart. Rev. **6**, 186 (1952)
121. M.L. Kremer, G. Stein, Trans. Faraday Soc. **55**, 959 (1959)
122. W.G. Barb, J.H. Baxendale, P. George, K.R. Hargrave, Trans. Faraday Soc. **51**, 935 (1955)
123. G.V. Krauze, *Catalytical Reactions in Liquid Phase* (Nauka, Alma-Ata, 1967), p. 491 [Russian]
124. A.Ya. Sychev, V.G. Isac, N.V. Suen, Indigocarmine catalytic oxidation by hydrogen peroxide in the presence of Fe(II) complexes with 1,10-phenanthroline, Cherkassy, 1980 [Article deposited at the NIITEKHIM, No. 59, in Russian]
125. A.Ya. Sychev, V.G. Isac, N.V. Suen, Khim. Fizika [Chem. Phys.] **2**, 1565 (1983) [Russian]

126. A.Ya. Sychev, V.G. Isac, N.V. Suen, Proc. Acad. Sci. MSSR. Biol. Chem. Ser. **3**, 11 (1983) [Russian]
127. G.G. Duca, Kataliz okisleniya vinnoj i digidroksifumarovoj kislot, Ph.D. thesis, Kishinev, 1979 [Russian]
128. J. Sobkovski, Rocz. Chem. **43**, 1729 (1969)
129. A.Ya. Sychev, S.O. Travin, G.G. Duca, Yu.I. Scurlatov, *Catalytic Reactions and the Protection of Environment* (Stiinta, Kishinev, 1983), p. 271 [Russian]
130. S.O. Travin, Ph.D. thesis, Institute of Physical Chemistry, USSR Academy of Science, Moscow, 1979
131. G. Davis, A.R. Garafalo, Inorg. Chem. **15**, 1101 (1976)
132. P. Benchev, *Complex Formation and Catalytic Activity* (Mir, Moscow, 1975), p. 342 [Russian]
133. C.B. Yatsimyrskii (ed.), *Coordination Compounds Chemistry Review* (Naukova Dumka, Kiev, 1975), p. 225 [Russian]
134. A.E. Shilov, Kinetika i kataliz [Kinet. Catal.] **21**, 26 (1980)[Russian]
135. A.E. Shilov, Redox metal enzymes and their models. Theoretical and methodical aspects. Part II. Chernogolovka, 1982 [Russian]
136. J. Kendlyn, C. Taylor, D. Thompson, *Reactions of transition metals coordination compounds* (Mir, Moscow, 1970) [Russian]
137. J.H. Wang, J. Am. Chem. Soc. **77**, 5715 (1955)
138. J.P. Kachanova, E.L. Kudreavtseva, A.P. Purmal, Zh. Fiz. Khim. [J. Phys. Chem.] **48**, 1449 (1974) [Russian]
139. E.N. RizkolA, O.H. Ei-Shafei, N.M. Guindy, Inorg. Chem. Acta. **57**, 199 (1982)
140. J.P. Kachanova, Yu.N. Kozlov, A.P. Purmal, Zh. Fiz. Khim. [J. Phys. Chem.] **43**, 2580 (1969) [Russian]
141. A.A. Kadirov, E.B. Bekasova, A.G. Muftahov, Moscow, 1985, VINITI, Nr.3115
142. V.G. Isac, N.T. Niung, A.Ya. Sychev, Zh. Fiz. Khim. [J. Phys. Chem.] **61**, 2212 (1987) [Russian]
143. V.G. Isac, A.Ya. Sychev, Proc. Acad. Sci. MSSR. Biol. Chem. Ser. **1**, 56 (1983) [Russian]
144. V.G. Isac, Ph.D. thesis, Kishinev, 1973 [Russian]
145. P. Valdmeier, P. Sigel, Inorg. Chem. Acta. Rev. **5**, 659 (1971)
146. M. Barteri, M. Farinella, B. Pispisa, L. Splendorini, Inorg. Chem. **17**, 3366 (1978)
147. K.B. Yatsymirskii, V.P. Vasilev, *Instability Constants of Complex Compounds* (Pergamon, Oxford, New York, 1960), p. 214 [Translated from Russian]
148. F.J.C. Rossotti, H. Rossotti, *The Determination of Stability Constants and Other Equilibrium Constants in Solutions* (McGraw-Hill, New York, 1961), p. 425
149. H.L. Schläfer, *Komlexbildung in Lösung* (Springer, Berlin, 1961)
150. M. Beck, I. Nadpal, *Study of Complex Formation* (Mir, Moscow, 1989), p. 411 [Russian]
151. Yu.I. Salnikov, A.N. Glebov, F.V. Deviatov, *Multi-Nuclear Complexes in Solution* (Kazan University, Kazan, 1989), p. 288 [Russian]
152. Yu.I. Bratushko, *Coordination Compounds of 3d-Transition Metals with Molecular Oxygen* (Naukova Dumka, Kiev, 1987), p. 168 [Russian]
153. V.A. Pokrovskaya, Yu.I. Bratushko, C.B. Yatsymirskyi, E.N. Korol, Theor. Exp. Chem. **14**, 760 (1978) [Russian]
154. G.E. Zaikov, S.A. Maslov, V.L. Rubailo, *Acid Rains and the Environment* (Chemistry Publishing, Moscow 1991, p. 140. [Russian]
155. L. Khorvat, *Acid Rains* (Stroiizdat, Moscow, 1990), p. 80 [Russian]
156. C.I. Weschler, T.E. Graedel, in *Heterogeneous Atmospheric Chemistry*, ed. by D.R. Schrayer (American Geophysical Union, Washington, 1982), p. 196
157. M.E. Bodini, D.T. Sawyer, J. Am. Chem. Soc. **98**, 8366 (1976)
158. M.E. Bodini, M.A. Willis, T.L. Riechel, D.T. Sawyer, Inorg. Chem. **15**, 1538 (1976)
159. D.T. Sawyer, M.E. Bodini, J. Am. Chem. Soc. **97**, 6588 (1975)
160. A.Ya. Sychev, V.G. Isac, U. Pfanmuller, Zh. Fiz. Khim. [J. Phys. Chem.] **33**, 1351 (1978) [Russian]

161. Yu.I. Scurlatov, S.O. Travin, in *Proceedings of IV Conference on Liquid-Phase Oxidation of Compounds*, Baku, 1979, p. 86 [Russian]
162. Yu.I. Scurlatov, S.O. Travin, A.P. Purmal, Proc. Acad. Sci. USSR Chem. Ser. **10**, 2201 (1979) [Russian]
163. L.A. Rodley, W.T. Robinson, Nature **235**, 438 (1972)
164. R. Boca, H. Elias, W. Haasec, M. Huber, R. Klementa, L. Muller, H. Paulus, I. Svoboda, M. Valko, Inorganica Chim. Acta **278**(2), 127–135 (1998)
165. Y. Sun, K. Chen, L. Jia, H. Li, Phys. Chem. Chem. Phys. **13**, 13800–13808 (2011)
166. J.P. Collman, R.R. Gange, C.A. Reed, T.R. Halbert, G. Lang, W.T. Robinson, J. Am. Chem. Soc. **97**, 1427 (1975)
167. E. Bayer, P. Shretsman, *Structure and Bond* (Mir, Moscow, 1969), p. 273 [Russian]
168. A. Geiss, M. Keller, H. Vahrenkamp, J. Organomet. Chem. **541**(1–2), 441–443 (1997)
169. S.C. Mitchell, Flavin mono-oxygenase (FMO)—the other oxidase. Curr. Drug Metabol. **9**(4), 280–284 (2008)
170. G.A. Hamilton, R.I. Workman, L. Woo, J. Am. Chem. Soc. **86**, 3390–3391 (1964)
171. R.A. Floyd, C.A. Lemis, Biochem **22**, 2645 (1983)
172. F. Haber, R. Willstäter, Anal. Chem. **64**, 2844 (1931)
173. W.S. Bray, M.H. Gorin, J. Am. Chem. Soc. **54**, 2124 (1932)
174. J.D. Rush, W.H. Koppenol, J. Free Radic. Biol. Med. **1**, 281 (1985)
175. M. Grootveld, B. Halliwell, Free Radic. Res. Commun. **1**, 243 (1986)
176. J.T. Gloves, M. Van Der Puy, J. Am. Chem. Soc. **98**, 5290 (1976)
177. H. Sugimoto, D.T. Sawyer, J. Am. Chem. Soc. **106**, 4283 (1984)
178. P.N. Balasubramanian, T.C. Bruice, J. Am. Chem. Soc. **108**, 549 (1986)
179. J.D. Rush, B.H. Bielski, J. Am. Chem. Soc. **108**, 523 (1986)
180. B.H. Bielski, M.J. Thomas, Studies of hypervalent iron in aqueous solutions. 1. Radiation-induced reduction of iron(VI) to iron(V) by CO_2-. J. Am. Chem. Soc. **109**(25), 7761–7764 (1987)
181. J.T. Groves, in *Metal Ion Activation of Dioxygen*, vol. 2, ed. by Th.G. Spiro (Wiley-Interscience, New York, 1980), p. 125
182. T.C. Bruice, M.F. Zipplies, W.A. Lee, Proc. Natl. Acad. Sci. USA **83**, 4646 (1986)
183. J.D. Rush, W.H. Koppenol, J. Biol. Chem. **261**, 6730 (1986)
184. A.J. Sitter, C.M. Reczek, J. Terner, J. Biol. Chem. **260**, 7515 (1985)
185. S. Hashimoto, Y. Tatsuno, T. Kitagawa, Proc. Natl. Acad. Sci. USA **83**, 2417 (1986)
186. R. Makino, T. Uno, Y. Nishimura, T. Jizuka, M. Tsuboi, Y. Ishimura, J. Biol. Chem. **261**, 8376 (1986)
187. J. Terner, C.M. Reczek, A.J. Sitter, in *Oxygen Complexes and Oxygen Activation by Transition Metals*, ed. by A.E. Martell, D.T. Sawyer (Plenum, New York, London, 1988), p. 327
188. M. Tien, B.A. Svingen, S.D. Aust, in *Oxygen and Oxy-Radicals in Chemistry and Biology*, ed. by M.A.J. Rodgers, E.L. Powers (Acadamic, New York, London, 1981), p. 147
189. B. Meunier, Chem. Rev. **92**, 1411 (1992)
190. T.G. Traylor, Sh. Tsuchiya, Y.S. Byen, C.H. Kim, J. Am. Chem. Soc. **115**, 2775 (1993)
191. K. Bajdor, K. Nakamoto, J. Am. Chem. Soc. **106**, 3045 (1984)
192. K. Tatsumi, R. Hoffman, Inorg. Chem. **20**, 3771 (1981)
193. R.H. Holm, Chem. Rev. **87**, 1401 (1987)
194. T.J. McMurry, J.T. Groves, in *Cytochrome P-450: Structure, Mechanism and Biochemistry*, ed. by P.R. Ortiz De Montellano (Plenum, New York, 1986), p. 1
195. J.T. Groves, J. Chem. Educ. **62**, 928 (1985)
196. J.S. Valentine, J.a.N. Burstin, L.O.D. Margerum, in *Oxygen Complexes and Oxygen Activation by Transition Metals*, ed. by A.E. Martell, D.T. Sawyer (Plenum, New York, London, 1988), p. 175
197. Y. Dou, S. Haswell, J. Greenman, J. Wadhawan, Electrochem. Comm, **11**(10), 1976–1981 (2009)

198. H.M. Goff, M.A. Phillippi, A.D. Boersma, A.P. Hansen, in *Electrochemical and Spectrochemical Studies of Biological Redox Compounds*, ed. by K.M. Kadish (American Chemical Society, Washington, 1982), p. 357 (Advances in Chemistry Ser., vol. 201)
199. W.A. Lee, T.C. Bruice, J. Am. Chem. Soc. 107, 513 (1985)
200. S.V. Khangulov, N.V. Voevodskaya, V.V. Barynin et al., Biofizika [Biophysics] **32**, 960 (1987) [Russian]
201. S.V. Khangulov, V.V. Barynin, V.. Melik-Adomyan et al., Bioorg. Khim. [Bioorg. Chem.] **12**, 741 (1986) [Russian]
202. I.A. Chernavin, *Physiology and Biochemistry of Microelements* (Vysshaya Shkola, Moscow, 1970), p. 70 [Russian]
203. A.Ya. Sychev, V.G. Isac, *Manganese Coordination Compounds in Catalysis* (Stiinta, Kishinev, 1990), p. 321 [Russian]
204. A.Ya. Sychev, V.G. Isac, Zh. Fiz. Khim. [J. Phys. Chem.] **47**, 244 (1973) [Russian]
205. A.Ya. Sychev, V.G. Isac, L. Dao Van, Zh. Fiz. Khim. [J. Phys. Chem.] **51**, 363 (1977) [Russian]
206. A.Ya. Sychev, V.G. Isac, L. Dao Van, Zh. Fiz. Khim. [J. Phys. Chem.] **52**, 107 (1978) [Russian]
207. A.Ya. Sychev, V.G. Isac, V.N. Moraru, Zh. Fiz. Khim. [J. Phys. Chem.] **52**, 624 (1978) [Russian]
208. A.Ya. Sychev, V.G. Isac, C.h.T.T. Fyong, Proc. Acad. Sci. MSSR Biol. Chem. Ser. **6**, 66 (1984) [Russian]
209. A.Ya. Sychev, V.G. Isac, C.h.T.T. Fyong, Zh. Fiz. Khim. [J. Phys. Chem.] **58**, 2331 (1984) [Russian]
210. A.Ya. Sychev, V.G. Isac, U. Pfanmuller, Zh. Fiz. Khim. [J. Phys. Chem.] **57**, 760 (1983) [Russian]
211. A.Ya. Sychev, S.S. Budnikov, V.G. Isac, Proc. Acad. Sci. MSSR Biol. Chem. Ser. **2** (1986) [Russian]
212. D.G. Batyr, V.G. Isac, A.A. Kirienco, Koord. Khim. [Coord. Chem.] **10**, 1372 (1984) [Russian]
213. V.G. Isac, A.Ya. Sychev, *Redox Catalysis by Metal Complexes* (Kishinev State University, Kishinev, 1984) [Russian]
214. D.G. Batyr, V.G. Isac, A.A. Kirienco, L.B. Unculita, Koord. Khim. [Coord. Chem.] **9**, 1085 (1983) [Russian]
215. V.G. Isac, A.A. Kirienco, Proc. Acad. Sci. MSSR Biol. Chem. Ser. **4**, 82 (1981) [Russian]
216. D.G. Batyr, V.G. Isac, S.O. Travin, A.A. Kirienco, Koord. Khim. [Coord. Chem.] **12**, 1065 (1986) [Russian]
217. D.G. Batyr, V.G. Isac, S.V. Kilmininov, Yu.Ia. Kharitonov, Koord. Khim. [Coord. Chem.] **13**, 3 (1987) [Russian]
218. D.G. Batyr, V.G. Isac, S.V. Kilmininov, Koord. Khim. [Coord. Chem.] **13**, 367 (1987) [Russian]
219. D.G. Batyr, V.G. Isac, S.V. Kilmininov, Koord. Khim. [Coord. Chem.] **13**, 1371 (1987) [Russian]
220. D.G. Batyr, V.G. Isac, S.V. Kilmininov, Ch.T.T. Fyong, Proc. Acad. Sci. MSSR Biol. Chem. Ser. **6**, 46 (1986) [Russian]
221. Ya.D. Tighineanu, A.Ya. Sychev, V.M. Berdnikov, Zh. Fiz. Khim. [J. Phys. Chem.] **45**, 1723–2533 (1971) [Russian]
222. Ya.D. Tighineanu, Redox catalysis by manganese ions in water solutions, Ph.D. thesis, Moscow, 1984 [Russian]
223. Ya.D. Tighineanu, V.M. Berdnikov, Zh. Fiz. Khim. [J. Phys. Chem.] **59**, 349 (1985) [Russian]
224. G.R.A. Johnson, N.B. Nazhat, R.A. Sandalla-Nazhat. J. Chem. Soc. Chem. Commun. 407 (1985)
225. A.Ya. Sychev, V.G. Isak, Russ. Chem. Rev. **62**, 279 (1993) [Russian]
226. N.M. Reybel, Zh. Fiz. Khim. [J. Phys. Chem.] **40**, 1112 (1966) [Russian]
227. L.S. Davidova, N.N. Plata, T.P. Brin, Zh. Fiz. Khim. [J. Phys. Chem.] **42**, 258 (1968) [Russian]

228. D. Meyerstein, Inorg. Chem. **10**, 2244 (1971)
229. G.V. Buxton, R.M. Sellers, *Compilation of Rate Constants for the Reaction of Metal Ions in Unusual Valency States* (New York, 1978), p. 27
230. N.N. Lebedev, N.G. Digurov, T.V. Buharina, in *Proceedings of VI Conference on Organic Substances Oxidation in Liquid Phase*, vol. 1, Lvov, 1986, p. 114 [Russian]
231. M. Constantini, A. Dromari, M. Joufret et al., J. Mol. Catal. **7**, 98 (1980)
232. J.L. Benet, H.J. Hertog, J.M. Eseriche, R.M. Saez, Analyst **111**, 1325 (1986)
233. O.P. Dmitrieva, V.E. Agabekov, N.I. Mitskecich. Proc. Acad. Sci. USSR **25**, 329 (1981) [Russian]
234. R.V. Kucher, L.I. Opeiga, T.V. Novikova, Proc. Acad. Sci. USSR Ser. B **5**, 37 (1987) [Russian]
235. E.A. Lukasheva, O.P. Dmitrieva, Proc. Acad. Sci. USSR **294**, 1167 (1987) [Russian]
236. A.I. Samokhvalova, A.B. Solovyova, A.L. Chugreeva et al., Proc. Acad. Sci. USSR **284**, 633 (1986) [Russian]
237. E.A. Lukasheva, A.B. Solovyova, L.B. Karmilova, N.S. Enicolopean, Kinetika i kataliz [Kinet. Catal.] **26**, 56 (1985) [Russian]
238. A.B. Solovyova, A.I. Samokhvalova, E.I. Karakozova, Kinetika i kataliz [Kinet. Catal.] **25**, 1080 (1984) [Russian]
239. I. Tabushi, N. Koga, J. Am. Chem. Soc. **101**, 6456 (1979)
240. A.B. Solovyova, T.I. Karakozova, C.A. Bogdanova, Proc. Acad. Sci. USSR **269**, 160 (1983)
241. A.Ya. Sychev, Ya.D. Tighineanu, Zh. Analit. Khim. [J. Analyt. Chem.] **24**, 1842 (1969) [Russian]
242. Ya.D. Tighineanu, *Research in the Field of Pharmacy and Chemistry* (Stiinta, Kishinev, 1975), p. 84 [Russian]
243. V.D. Komissarov, E.T. Denisov, Kinetika i kataliz [Kinet. Catal.] **10**, 513 (1969) [Russian]
244. C.B. Yatsymirskyi (ed.) *Biological Aspects of Coordination Chemistry* (Naukova Dumka, Kiev, 1979) [Russian]
245. A.V. Savitskyi, V.I. Neliubin, Uspekhi khimii [Russ. Chem. Rev.] **44**, 214 (1975) [Russian]
246. E.A. Lukasheva, A.B. Solovyova, A.L. Chugreev, N.S. Enicolopean, Proc. Acad. Sci. USSR **294**, 1167 (1987) [Russian]
247. N.S. Enikolopean, C.A. Bogdanova, C.A. Ascarov, Uspekhi khimii [Russ. Chem. Rev.] **52**, 20 (1983) [Russian]
248. A.B. Lukasheva, A.B. Solovyova et al., Kinetika i kataliz [Kinet. Catal.] 26 (1985) [Russian]
249. A.Ya. Sychev, V.G. Isac, U. Pfanmuller, Zh. Fiz. Khim. [J. Phys. Chem.] **57**, 760 (1983) [Russian]
250. J. Reinado, F. Toribio, D. Perez-Benditat. Analyst **112**, 771 (1987)
251. V.G. Isac, A.Ya. Sychev, L.I. Budarin, M.L. Kazakevich, L.B. Vozian, A.P. Filipov, S.G. Levitchi, F. Zuher, Zh. Fiz. Khim. [J. Phys. Chem.] **67**, 2160–2164 (1993) [Russian]
252. D.T. Richens, C.G. Smith, D.T. Sawyer, Inorg. Chem. **18**, 707 (1979)
253. D.I. Metelitsa, Uspekhi khimii [Russ. Chem. Rev.] **40**, 1175 (1971) [Russian]
254. G.O. Shulpin, D. Attanasio, G.V. Nizova, J. Muzau, L. Siber, in *1st European Congress on Catalysis*, vol. 1, Montpelier, 1993, p. 327
255. A.A. Karakhanov, S.M. Volkov, A.G. Dedov, in *Catalysis—Fundamental and Applied Research*, ed. by O.A. Pertii, V.V. Lunin (State University, Moscow, 1987), pp. 147–172 [Russian]
256. D.T. Sawyer, J.P. Hage, A. Sobkowiak, J. Am. Chem. Soc. **117**, 106 (1995)
257. S.A. Mizza, B. Bocquet, C. Robyr, S. Thomi, A.F. Williams, Inorg. Chem. **35**, 1332 (1996)
258. M. Fontecave, J.L. Pierre, Bull. Soc. Chim. Fr. **130**, 77 (1993)
259. M.L. Kremer, Int. J. Chem. Kinet. **40**(9), 541–553 (2008)
260. A. Luis, J.I. Lombraña, F. Varona, A. Menéndez, Kor. J. Chem. Eng. **26**(1), 48–56 (2009)
261. J.T. Groves, M. Van Der Puy, J. Am. Chem. Soc. **96**, 5274 (1974)
262. D.L. Ingles, Aust. J. Chem. **26**, 1021 (1973)
263. W.H. Koppenol, J.F. Liebman, J. Phys. Chem. **98**, 99 (1984)
264. J.D. Rush, W.H. Koppenol, J. Free Radic. Biol. Med. **1**, 281 (1986)

265. B.H.J. Bielski, in *Oxygen Radicals in Biology and Medicine*, ed. by M.G. Simic, K.A. Taylor, J.F. Ward, C. Von Sonntag (Plenum, New York, London, 1988), p. 123
266. M.L. Kremer, Isr. J. Chem. **21**, 72 (1981)
267. M. Tien, B.F. Svingen, C.D. Aust, in *Oxygen and Oxy-Radicals in Chemistry and Biology*, ed. by M.A.J. Rodgers, E.L. Powers (Academic, New York, 1981), p. 147
268. H.M. Goff, M.A. Fhillipi, A.D. Boersma, A.P. Hansen, in *Electrochemical and Spectrochemical Studies of Biologycal Redox Compounds*, ed. by K.M. Kadish (American Chemical Society, Washington, 1982), p. 357 (Advances in Chemistry Ser., vol. 201)
269. G. Klopman, *Reaction Capacity and Reaction Routes* (Mir, Moscow, 1977) [Russian]
270. A.Ya. Sychev, V.G. Isac, S.S. Budnikov, Proc. Acad. Sci. MSSR Biol. Chem. Ser. **2**, 54 (1986)
271. V.G. Isac, A.Ya. Sychev, *Redox Catalysis by Metal Complexes* (Moldova State University, Kishinev, 1984) [Russian]
272. D.G. Batyr, V.G. Isac, A.A. Kirienco, F.A. Spatar, Koord. Khim. [Coord. Chem.] **13**, 1369 (1987) [Russian]
273. V.G. Isac, A.A. Kirienco, Proc. Acad. Sci. MSSR Biol. Chem. Ser. 81 (1981)
274. D.G. Batyr, F.A. Spatar, A.A. Kirienco, V.G. Isac, Koord. Khim. [Coord. Chem.] **40**, 1372 (1984) [Russian]
275. D.G. Batyr, F.A. Spatar, A.A. Kirienco, V.G. Isac, Koord. Khim. [Coord. Chem.] **13**, 360 (1987) [Russian]
276. A.Ya. Sychev, L.D. Alapini, A.S. Dimoglo, Proc. Acad. Sci. MSSR Biol. Chem. Ser. **3**, 70 (1992)

Chapter 3
Biochemical Processes and Metal Complexes' Role as Catalysts

3.1 Catalytic Influence of Transition Metal Compounds on Biological Processes

In Chap. 2, some theoretical problems of catalysis with transition metal ions or complexes, oxidation of various organic substances with oxidizers such as oxygen and hydrogen peroxide, and also the mechanisms of these processes were considered. This provides the necessary basis for a search of possible fields of homogeneous catalysis with metal complexes' practical use. It can be said without exaggeration that the advances reached in this field during the last 20 years made it possible, to a great degree, to consider in a new way both the processes proceeding in the world around us (atmosphere, water, and soil), and the processes within the sphere of human activity (in the chemical industry, pharmacology, foodstuffs production, biology, medicine, etc.). Having faced in the late 1920s the complex problems resulting from his own activities (environmental problems, chemical processes selectivity and optimization, agricultural problems, etc.), man was forced to appeal more and more to living nature, and in particular to evolutional chemistry. The great successes reached within the last few decades in the field of enzymatic chemistry, revealing the action mechanisms of a number of important enzymes, served as a good basis for numerous attempts on their modeling or analogous systems creation, although less efficient than natural enzymes, but more simple and convenient for practical use.

Advances in electronics and other related fields are dealing, in particular, with the creation of new materials. The synthesis of these materials is often a catalytic process as well.

At last, demographic problems in the twentieth century have sharply aggravated the problem with the increasing population provision demands, primarily food and other essential items. One of the consequences is great attention to revealing the mechanism of nitrogen molecule activation and C–C and C–H bonds breakage in hydrocarbons.

G. Duca, *Homogeneous Catalysis with Metal Complexes*, Springer Series in Chemical Physics 102, DOI 10.1007/978-3-642-24629-6_3,

Further in Chap. 3, only those fields in which metal complexes are used (mainly those of copper, iron, manganese, and cobalt) as appropriate redox biological process catalysts, in which the oxidant is O_2 or its reduction products are $HO_2^{\bullet}$, H_2O_2, $OH^{\bullet}$, etc., will be considered.

3.1.1 Role of Transition Metal Ions in Biological Systems

To provide the normal proceeding of biochemical processes in humans and many other animals and plants, in the main ten ions or metal compounds are needed. Four of them (sodium, potassium, magnesium, and calcium) are metal compounds of the main sub-groups of the periodic table. The character of the bonds they form with ligands is mostly electrostatic. They are united by large mobility in the organisms; therefore, sodium and potassium play an important role in nervous conductivity and other important functions, and magnesium and calcium in muscular activity and many other reactions. However, they are not practically involved in catalytic processes in vivo (besides magnesium in plants), and therefore they will not be considered further. The role of vanadium, chromium, molybdenum, niobium, and cadmium will also not be considered because of their very low concentrations in vivo and less important roles.

Six other elements (iron, nickel, manganese, copper, cobalt, zinc) belong to the transition metals. These elements form more or less strong bonds, often of covalent character, with numerous chemical parts of organisms or substrates and metabolites. They form part of the reaction centers of various enzymes, metal proteins, in formation of a plenty of various coordination compounds with organism's chemical substances, and their transformation products which can be considered as ligands.

Possible ligands with which metal ions interact (for instance, in an enzymes' reaction center) can be: for manganese–carboxylates, phosphates, imidazole (adenosinetriphosphatase, enolase); for iron–porphyrin, imidazole, other ligands, and also the variety of protein parts of enzymes; for cobalt–benzimidazole and sugar carbanion (vitamin B_{12}); for copper—different nitrogen bases ("blue" proteins); and for zinc–imidazole or NH_2-bases (dehydrogenases, carbonic anhydrase).

These metal ions can be catalysts in various model reactions (nonenzymatic). For example, copper ions are catalysts of pyridoxal transformations, glycine ether hydrolysis, acetodicarboxylate decarboxylation, and oxaloacetate decomposition; iron ions–dimethyloxaloacetate decomposition; manganese ions–glycine ether hydrolysis.

To realize catalytic processes in living organisms, "trace" amounts of metal ions are often enough, less than 5×10^{-5} g per g of living substance. About one-third of enzymes studied have in their composition "trace" amounts of metals. Lack of these metals results in various diseases. The need by living organisms was first established for iron as long ago as in the seventeenth century, for copper in 1928, for manganese in 1931, for zinc in 1934, for cobalt in 1935, and for nickel in 1973 [1]. Thus, anemia (because of the lack in iron) has been known for a long time.

The main function of trace metal ions in an organism is taking part in enzymatic activity. Thus, rats fed with food low in Mn^{2+} ions (0.4 mg/kg of weight) compared to normal contents of this ion (50 mg/kg of weight) resulted in significant diminishing in catalytic activity of food arginine disintegration with liver arginase [1].

Manganese ions Mn^{2+} also take part in pyruvate carboxylase enzyme activity, catalyzing the interaction of bicarbonate ion with acetyl coenzyme A, resulting in oxaloacetate formation. Thus, a possible lack of Mn^{2+} ions can be compensated with Mg^{2+} ions. In general, for normal activity of the human organism from 3 to 4 mg of manganese per day are needed. Galactosyltransferases are also dependent on manganese in the human organism. The enzymes dependent on magnesium and manganese, namely, glutathione reductase, glucose-6-phosphate dehydrogenase, isocitrate dehydrogenase, and malic enzymes, are widely represented in reactions. For enzymes such as copper-containing cytochrome oxidase, manganese ions are also necessary; otherwise the enzyme becomes low-active. It seems that an important role is played here by the eventual process of Mn^{2+} transformation into Mn^{3+}, and on this account, reduction of oxidative products of phenol substrates.

Manganese superoxide dismutase (MnSOD) is essential for life as dramatically illustrated by the neonatal lethality of laboratory animals that are deficient in MnSOD. Numerous studies have shown that MnSOD can be induced to protect against pro-oxidant insults resulting from cytokine treatment, ultraviolet light, irradiation, certain tumors, amyotrophic lateral sclerosis, and ischemia/reperfusion. Moreover, overexpression of MnSOD has been shown to protect against pro-apoptotic stimuli as well as ischemic damage. Also, several studies have reported declines in MnSOD activity during diseases or pathological states such as cancer, progeria, asthma, and transplant rejection [2].

In plants, manganese involvement in nitrogen exchange reactions is also important. It can be shown by taking as an example nitrate assimilation by plants, which will then be reduced to nitrite with subsequent formation of hydroxylamine. Manganese participates in the stages of reduction of nitrates to nitrites and hydroxylamine to ammonia. All these processes are catalyzed by manganese- and molybdenum-containing nitrate-, nitrite-, and hydroxylamine reductases.

Detailed steps of manganese involvement in the respiration reactions (O_2 absorbance) are not quite clear. It is supposed [3] that manganese presence activates the number of reactions within the cycle of tri- and dicarboxylic acids, and decarboxylation reactions as well. The presence of manganese can increase demand for oxygen in the process of reduced nicotinamide adenine dinucleotide regeneration, and also increase oxidation and decomposition processes, catalyzed by peroxidase and peptidase, respectively [4].

Manganese influences the whole series of processes proceeding in plants, namely, synthesis of ascorbic and other organic acids, and acceleration of plants development and seeds maturing. Thus, for instance, it activates redox processes in vine plant by increasing many enzymes' activity (ascorbinase, peroxidase, and polyphenoloxidase), increasing the contents of chlorophyll and sugar accumulation, accelerating berries ripening, and enhancing the mean weight of grapes and berries [5].

Influence of lack in manganese and also its toxicity in living organisms is well described, for example, in [6]. With its lack, convulsions are observed, level of catecholamines in cerebrum changes, metabolic processes in bone growth are changed, and many other factors dealing with obesity, atherosclerosis, and other pathologic processes are observed.

Ions and compounds of these and other metals play a part in numerous functions in an organism, as described in [7, 8]. Further, only the catalytic role of metal ions or compounds in processes proceeding in vivo will be considered. Catalytic processes proceeding in vivo can be conditionally separated into two groups—without and with protein involvement. The first group includes the processes of ligand peroxidation, reactions of Haber–Weiss type, influence of some drugs (antibiotics) on ADN, etc., in which transition metal ions or compounds are catalysts. The second group includes mainly enzymatic systems. Enzymes, their structures, and functions are given in great and increasing amounts of the literature; therefore, these problems will be only slightly touched upon when needed to provide clear statements. Main attention will be paid to the mimetic approach—enzymes modeling (imitation) and using data obtained for better understanding of enzymatic systems' functional action mechanisms and practical use of such model systems in industry (see Chap. 7).

The main role of the aforesaid five transition metals, mainly that of iron, manganese, and copper, consists of the realizing of catalytic redox reactions with their participation in both nonenzymatic and enzymatic processes. Besides, these metal compounds can slow down the proceeding of chemical reactions in vivo, providing a blockage of certain active centers, by acting as toxicants and by inhibiting radical chain processes or taking part, equally with enzymatic systems, in electron transport.

In living organisms, the most important is the presence of iron ions which turn part of numerous enzymes and oxygen-binding proteins into specific catalysts. The capacity of iron to exist in several oxidation degrees (II, III, and IV), relatively easy transformation into each other in the presence of the appropriate ligands and oxidizers, and the ability to join molecular oxygen make this element unique for further catalytic activation of oxygen molecules, hydrogen peroxide, and hydroperoxides, and realization of subsequent numerous reactions with free radical formation from these substances. These properties of iron are widely used by electron transport proteins in cells' mitochondria, by hydroxylating enzymes (cytochrome P-450), and by oxygen-transferring proteins (hemoglobin) [9].

Equally with these normal biochemical processes, the redox capacity of iron compounds also represents a source of potential toxicity [10]. Under aerobic conditions, uncontrolled redox transformations can proceed, resulting in H_2O_2, $O_2^{\bullet}$, and $OH^{\bullet}$ radical formation. The last radical is especially toxic, reacting with most organic molecules exclusively efficiently (with rate constants close to diffusion ones). $OH^{\bullet}$ radical attacks the membranes, proteins, nucleic acids, and many other cell components, and therefore, its formation in noticeable amounts in vivo is undesirable.

There is a number of protective means against its formation. The most important is stable control over its absorption, transport, and storage inside the organism. Additional means are protective enzymes (superoxidedismutase, catalase, glutathioneperoxidase, and also vitamins C, E, etc.). However, when iron levels become too high, toxicity is caused by initiation of formation of excessive amounts of free oxygen radicals with iron compounds.

In the case of a healthy person, iron absorption through the gastrointestinal tract is carefully regulated (about 1–2 mg of iron is absorbed every day, and the same amount is lost daily). While heme iron possesses large biological suitability, simple inorganic iron salts and their complexes are poorly absorbed. Endogenic iron-absorbing systems, apparently, accept iron(II), but not iron(III), although reductases bound with membranes are able to reduce iron(III) to iron(II) [11, 12]. A recent study has shown that ascorbate enhances iron uptake into intestinal cells through the formation of $FeCl_3$–ascorbate complex [13].

Iron is transported into human blood by transferrin (molecular weight about 80,000) [14]. Transferrin has iron-joining places with high affinity to iron(III) (stability constant reaches $10^{20}\,M^{-1}$ at pH 7.4) [15]. In the case of healthy people, transferrin concentration reaches about 25–40 μM. In the presence of transferrin, concentration of free iron(III) is less than 10^{-12} M, and the latter is unable to realize $OH^{\bullet}$ radical formation and thus produce toxic damages in cells.

Iron compounds are transported by transferrin through the small intestine and liver to tissues which need iron for normal metabolism, especially to the marrow—the place of hemoglobin synthesis [16].

In transferrin, all the iron-joining places can never be busy (about one-third of them only), which makes it possible, equally with its main function, to be involved in the protection against infections, taking away iron ions from microorganisms which is needed for their growth and reproduction [17, 18].

Provided the level of iron in cells is higher than that needed, its excess is stored in two storage forms: soluble ferritin and insoluble hemosiderin [19]. These storage iron forms are present practically in each cell, but there are especially many of them in liver, spleen, and cerebrum. Its content in human blood is one of the diagnostic indexes for various diseases. Ferritin (molecular weight ≈450,000) contains iron in the form of iron(III) phosphate hydroxide [20]. In such a form, iron(III) is practically not able to generate free oxygen radicals.

It was considered that the main role of ferritin is to store (accumulate) Fe^{3+} ions. However, equal with this function, ferritin can serve as biocatalyst of some redox processes, for instance, with superoxide dismutase [21], and also by ferroxidase (accelerates oxidation of Fe^{2+} to Fe^{3+} ions with oxygen) and peroxidase activity [22]. Extremely important is the fact that iron of ferritin (FERR) is an initiator of lipid peroxidation on account of Fe^{3+} reducing with various reducers to Fe^{2+} ions [23–25].

On ferritin peroxidase activity study, the rates of free radical initiation in the systems FERR–ROOH (H_2O_2)–Ar, where ROOH is the hydroperoxide of *tert*-butyl or cumyl, and Ar is various aromatic amines, namely, 3, 3′, 5, 5′-tetramethylbenzidine (TMB), these being acceptors of radicals of hydroperoxides (peroxides), were determined. Free radicals are formed according to the following interactions of ferritin (shown here as Fe^{3+}) with ROOH (H_2O_2) [22]:

$$Fe^{3+} + ROOH \underset{K_{-1}}{\overset{K_1}{\rightleftharpoons}} [Fe^{3+}\ldots OOR(H)] \quad (1)$$

$$[Fe^{3+}\ldots OOR(H)] \xrightarrow{K_2'} [Fe^{3+}OH] + RO^{\bullet} \quad (2')$$

$$[Fe^{3+}\ldots OOH(H)] \xrightarrow{K_2''} Fe^{2+} + H^+ + RO_2^{\bullet} \quad (2'')$$

$$Fe^{2+} + ROOH \xrightarrow{K_2'''} [FeOH]^{2+} + RO^{\bullet} \quad (2)$$

Active radicals $RO^{\bullet}$ ($OH^{\bullet}$) formed, and to a lesser degree, peroxide radicals $RO_2^{\bullet}$ ($HO_2^{\bullet}$), will oxidize aromatic amines present in the system, thus realizing the peroxidase process. It was thus shown that peroxidase activity of ferritin is comparable with its feroxidase activity [22], which points to the numerous functions of this protein.

The body of an adult person contains 4–5 g of iron, mainly in the form of hemoglobin (66%), and also in the form of myoglobin ($\approx$10%), various iron-containing enzymes, transport protein transferrin in storage proteins (ferritin and hemosiderin), and some other organ, tissue, and intercellular fluids [10]. In the case of some diseases, blood transfusion accompanied with excessive iron introduction is needed. On repeated transfusion, this excessive amount of iron becomes significant, and is accumulated mainly in the liver in the form of ferritin and hemosiderin.

Under these conditions, an excess of iron is present also in low molecular form inside the cells in plasma, which is accompanied by heart, endocrine tonsil, and liver damage with free oxygen radicals. To such diseases which need regular blood transfusion we can add thalassemia (inherited hemolytic anemia in children—degenerative hemoglobin formation) which results in reduction of red cell lifetime and, accordingly, anemia [16]. If excess iron is not removed with a chelate-forming agent such as desferrioxamine, the patients will die.

3.1.2 Catalytic Reactions in the Initiation of Lipid Peroxidation

Catalytic activation of molecular oxygen is a necessary condition of free radicals formation from it. Really, in physiological conditions, oxygen molecule, being triplet ($^3\Sigma_g$), is rather inert with regard to many organic compounds (RH), which, as a rule, are singlet. Such behavior is caused by the Woodworth–Hoffman rule of symmetry conservation, as well as thermodynamic obstacles in one electron joining to O_2 ($\Delta G^o \approx +15\,\mathrm{kJ/mol}$). Reactions of O_2 with most organic compounds

$$RH + O_2 \longrightarrow R^{\bullet} + HO_2^{\bullet}$$
$$RH + O_2 \longrightarrow RH^{+} + O_2^{\bullet-}$$

$$2RH + O_2 \longrightarrow 2R^{\bullet} + H_2O_2$$

are endothermic and thus in physiological conditions proceed extremely slowly without a catalyst (besides some substrates, for instance, flavines [26]).

Thus, necessity in O_2 catalytic activation with the reduced forms generation becomes obvious. The reactions with the reduced forms of oxygen formation (superoxide anion radical $O_2^{\bullet-}$, and perhydroxyl radical $HO_2^{\bullet}$) proceed, for example, in the case of lipid peroxide oxidation. The radicals are initiated with metal compounds (iron) on their interaction with unsaturated aliphatic acids. Other forms of reduced oxygen, such as hydrogen peroxide H_2O_2 (or hydroperoxide ROOH) and $O_2^{\bullet-}$–are sources of free radicals in the presence in vivo of iron ions or compounds (in physiological conditions often in concentrations of μmol/L) in accordance with the Haber–Weiss reaction:

$$Fe^{3+} + H_2O_2(ROOH) \longrightarrow Fe^{2+} + HO_2^{\bullet}(RO_2^{\bullet}) + H^{+}$$
$$Fe^{3+} + O_2^{\bullet-} \longrightarrow Fe^{2+} + O_2$$
$$Fe^{2+} + H_2O_2(ROOH) \longrightarrow Fe^{3+} + OH^{-} + OH^{\bullet}(RO^{\bullet}).$$

Hydrogen peroxide is generated in many biological processes and in the presence of transition metal compounds (especially iron). Such reactions play a significant role in vivo.

Equally, with the reducing superoxide anion radical $O_2^{\bullet-}$ in vivo, other reducers are also present (NADPH, NADH, cysteine, etc.), able to compete with it. Besides, in physiological conditions, the presence of iron complexes with suitable ligand (possessing sufficient donor properties and capacity to stabilize high oxidation degree of iron) and hydrogen peroxide or $OH^{\bullet}$ radical can result in active ferryl particles formation. They can oxidize plenty of organic compounds in a more selective way than hydroxyl radical:

$$LFe^{3+} + OH^{\bullet} \longrightarrow (LFe^{IV}{=}O)^{2+} + OH^{-}\cdot$$

For these and other processes in living organisms, as mentioned above, there are enough iron compounds. However, only a small amount of Fe^{2+} ions is in the free state in vivo, where it is mainly bound with various biological compounds such as ADP and deoxyribonucleic acid (DNA) [27]. Within the system Fe^{2+}–H_2O_2, in the presence of ADP and in higher degree ATP, $OH^{\bullet}$ radical generation increases about 20–50 times compared to that with the lack of these nucleotides (AMP does not accelerate this process). Substitution of Fe^{2+} ion with Fe^{3+} in these nucleotide systems diminishes the efficiency about one order, and addition of the reducer (ascorbic acid), making the system reversible, results in H_2O_2 decomposition to $OH^{\bullet}$ with high yield [28]. Highly reactive hydroxyl radical formed by Haber–Weiss reaction or in another way, in physiological conditions can interact with any organic substances (alkanes, alkens, aromatic compounds, etc.):

$$R^{-}H + OH^{\bullet} \longrightarrow R^{\bullet} + H_2O$$

Iron ions and its complex compounds, such as $Fe^{2+\bullet}ADP$, Fe^{3+}, ascorbate, or some iron-containing proteins (ferritin, hemoglobin, and myoglobin), play an important stimulating role in catalytic lipid oxidation, i.e., in the processes where aliphatic acids are oxidation substrates. Lipids represent the complex of aliphatic acids, especially polyunsaturated:

LipH =

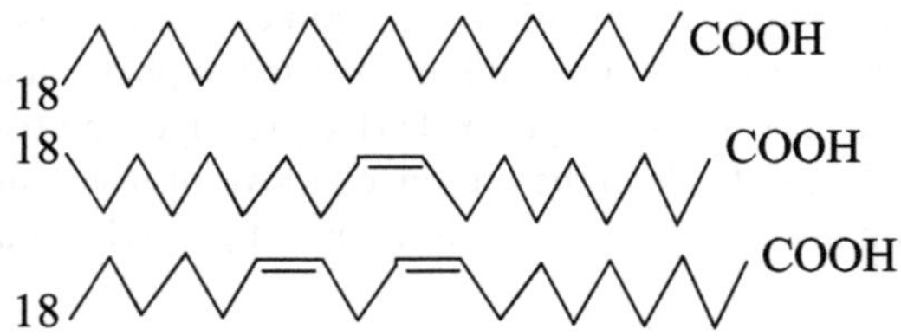

- stearic acid
- oleic acid
- linolic acid

and many other similar acids.

Reduction reactions catalyzed by iron compounds can proceed in vivo under certain conditions:

$$O_2(H_2O_2) \xrightarrow{FeLn} O_2^{\bullet-}, OH^{\bullet}, LFe^{IV}O^{2+}$$

These reactive particles catalytically interact with ligands, forming numerous products and initiating the whole chain of reactions, many of which are rather undesirable for living organism normal functioning:

$$O_2^{\bullet-}, OH^{\bullet}, LFe^{IV}O^{2+} + LipH \xrightarrow{FeL_n} \text{Products (RCHO, RCO, ROH, RCOOH, } R-\underset{\diagdown O \diagup}{CH-CH_2}\text{).}$$

Thus, this process renders noticeable influence on biological membranes and drug toxicity, and brings in its contribution to carcinogenesis and other cases of oxidation stress manifestation. A detailed review regarding many aspects of lipid peroxidation is given in collection [29], as well as in subsequent numerous publications. Here, the catalytic role of iron compounds in lipid peroxidation will be only briefly described.

$OH^{\bullet-}$ radicals generated in one way or another can detach a hydrogen atom from a lipid (LipH) and form a hydrocarbon radical of polyunsaturated aliphatic acid ($Lip^{\bullet}$), which, having rapidly reacted with O_2, will be turned into peroxy radical ($LipO_2^{\bullet}$). The latter, interacting with lipid molecule, will generate lipid hydroperoxide (LipOOH), which is rather stable in vivo. However, the presence of iron compounds (or some other transition metals) results in LipOOH decomposition catalysis [30]. Ligands that are suitable for such iron complexes are phosphates, phosphate ethers (ADP), and porphyrins. These iron complexes may be involved in membranes' lipid peroxidation chains with the formation of alkoxy and peroxy radicals ($LipO^{\bullet}$ and $LipO_2^{\bullet}$, respectively). The scheme of such radical-chain transformations appears as follows [10, 30]:

$$\text{Initiation:}\quad LipH + OH^{\bullet} \longrightarrow Lip^{\bullet} + H_2O$$

$$Lip^{\bullet} + O_2 \longrightarrow LipO_2^{\bullet}$$

$$\text{Continuation:}\quad LipO_2^{\bullet} + LipH \longrightarrow LipOOH + Lip^{\bullet}$$

$$LipOOH + LFe^{2+} \longrightarrow LFe^{3+} + OH^- + LipO^{\bullet}$$

$$LipOOH + LFe^{3+} \longrightarrow LFe^{2+} + H^+ + LipO_2^{\bullet}$$

$$\text{Breakage:}\quad 2LipO_2^{\bullet} \longrightarrow [ROOOOR] \quad \text{non radical products} + O_2^{\bullet}$$

Under relatively low partial O_2 pressure already, it is joined to $Lip^{\bullet}$ with the rate constant close to the diffusion one ($\approx 10^9$ L/mol s). Thus, the main radical in solution is $LipO_2^{\bullet}$, but not $Lip^{\bullet}$ and therefore it is less probable that $Lip^{\bullet}$ will be involved in the breakage stage. Limiting is the stage of hydrogen atom detachment from LipH.

Functions of iron complexes in lipid peroxidation are numerous. They initiate the process, cause catalytic decomposition of hydroperoxides, and generate free radicals ($O_2^{\bullet -}$, $OH^{\bullet}$):

$$LFe^{3+} + O_2 \rightleftarrows \underbrace{[LFe^{2+}O_2 \leftrightarrow LFe^{3+}O_2^{\bullet -}]}_{\text{charge transfer complex}} \rightleftarrows LFe^{3+} + O_2^{\bullet -}$$

$$2O_2^{\bullet -} + 2H^+ \longrightarrow H_2O_2 + O_2$$

$$LFe^{2+} + H_2O_2 \longrightarrow LFe^{3+} + OH^- + OH.$$

Catalytic decomposition of hydroperoxides which are unstable results in the formation of numerous products containing aldehyde, keto, hydroxy, epoxy, and carboxy groups. Aldehyde formation is often the cause of taste deterioration in many foodstuffs (rancid taste of fats and oil). They (namely, 4-hydroxynonenals) are also formed in vivo on lipid peroxidation of liver macrosomes, accumulate in systems NADPH–Fe, and inhibit the synthesis of many substances important for cells [31]. The $O_2^{\bullet -}$ particle itself in the absence of iron ions cannot be lipids' peroxidation initiator [32]; this fact is shown in vivo in the example of such particle's formation in adriamycin redox cycle (antibiotic used in tumor therapy) and in the absence of its interaction with lipids. Only the presence of Fe^{3+} ions stimulates the formation at first of Fe^{2+} and then subsequent reaction of lipid peroxidation—$Lip^{\bullet}$ generation [33]:

$$Fe^{3+} + O_2^{\bullet-} Fe^{2-} + O_2$$

$$LipH + O_2 + Fe^{2+} + H^+ \longrightarrow Lip^{\bullet} + H_2O_2 + Fe^{3+}.$$

The initiator of lipid peroxidation could have been $OH^{\bullet}$ radical [34]. However, there is another point of view. According to this, such an initiator is the particle preceding $OH^{\bullet}$ radical formation, namely, ferryl. Really, $OH^{\bullet}$ radical was not detected with the help of the usual fixators [10]. Ferryl particle formation was also supposed to take place in model processes of microsomal and xanthine oxidase oxidation [35]:

$$ADP - Fe^{2+}O_2 \xrightarrow[-2e-]{2H} ADP - [Fe^{IV}O]^{2+} + H_2O.$$

It was also assumed that these particles were also generated on complex ADP–Fe^{2+} interaction with H_2O_2, proceeding from the fact that introduction of mannitol ($OH^{\bullet}$ remover) into the system ADP–Fe^{2+}–H_2O_2–LipH did not change the rate of lipid oxidation. Similar statements were expressed regarding the ferryl particle formation and its initiating role in lipid peroxidation and in other systems, for instance, in the system Fe^{2+}–diethylenetriaminepentaacetic acid (DETAPAC)–H_2O_2–LipH) [32]. But this point of view needs further confirmation. Fe^{2+}–H_2O_2-mediated lipid peroxidation is stimulated in the presence of $Ni^{2+}$and is inhibited in the presence of Co^{2+} [36].

The lipid peroxidation catalyzed by ferryl particle $LFeO^{2+}$ (or $OH^{\bullet}$) is shown on scheme 3.1 [30].

Of course, it should be considered that there are plenty of other reducers in the cells (ascorbic acid, NADH, and NADPH) which can also generate $OH^{\bullet}$ and play a more important role than $O_2^{\bullet}$ by interaction with $H_2O_2^-$ [37]. Besides, hydrogen peroxide can oxidize complex $ADP \cdot Fe^{2+}$ to $ADP \cdot Fe^{3+}$ (L = ADP), generating $OH^{\bullet}$, although the latter does not initiate lipid peroxidation in these conditions (mannitol–$OH^{\bullet}$ remover, does not inhibit the reaction) [38]. All this testifies to the fact that the problem of lipid peroxide oxidation is still disputable.

It is already clear that lipid peroxidation represents great danger for living organism and, therefore, in vivo concentrations of lipid peroxides as free radical reaction initiators must be low [39]. Nevertheless, protective systems against lipid peroxidation do not possess unlimited possibilities. As a result of various types of violations (such as alcohol introduction into the organism or unfavorable environment, and for other reasons), especially during the aging process, radical chain processes will be increased. This is manifested in particular in a large amount of pigment formation (lipofuscin—old age pigment, cirrhosis), possessing characteristic fluorescence. This seems to occur as a result of peroxidized lipids' interaction with denaturated proteins (see Chap. 4). Thus, in vivo, due to free radical processes with transition metal ions' participation, membrane lipid deterioration can proceed, this fact being connected with living organisms' aging [40].

3.1.3 Ligands Influence of Haber–Weiss Reaction

For the research of catalytic generation of $OH^{\bullet}$ radicals from H_2O_2 or their formation in the process of lipids peroxidation, sufficiently high concentrations of iron compounds (namely, FeADP) are often used, of the order of 50–200 μM, although the concentration of iron available is much less and does not exceed 5 μM in vivo, even in cases of living creatures organs intensive inflammation [41]. Here, rather high amounts of $O_2^{\bullet-}$ and H_2O_2 are accumulated, produced with activated neutrophils, which in the presence of iron compounds can become sources of $OH^{\bullet}$ radicals.

Limitation of deterioration processes can be caused by the reduction of metal catalytic ions supply to the places of $O_2^{\bullet-}$ and H_2O_2 accumulation due to these ions' complex formation, diminishing their activity. Various ligands influence $O_2^{\bullet-}$ ability to reduce Fe^{3+} ions in a different way. Thus, specific ligands desferrioxamine and batophenanthroline inhibit this reaction (Fe^{3+} ions are not reduced to Fe^{2+} ions), and EDTA, ATP, and picolinic acid promote it. The use of specific complex-forming agents which bind Fe^{3+} ions (and thus prevent its reduction) can contribute to oxidative deterioration, suppressing therapeutic treatment of heart, kidney, skin, and other tissues by introducing drugs—specific ligands, and also weakening the process of posttraumatic degenerations in cerebrum and spinal cord [42–44].

If $OH^{\bullet}$ radical is still formed in vivo by the Haber–Weiss reaction, it will act more directly and in a more destructive way than hydroxide radical, generated by radiolysis in solution. In the first case, this radical interacts (because of its high reactivity) with the nearest biological target, situated no more than 5–10 Å from it. In the second case, interaction proceeds in the bulk of solution, and, consequently, at large distances.

Taking into consideration the Haber–Weiss reaction it is possible to understand local-specific deterioration of biological targets, and also other processes realized in vivo. Thus, on hemoglobin oxidation with peroxides, Fe^{2+} ions will be released which, in common with H_2O_2, can act as efficient catalysts of $OH^{\bullet}$ generation [45]. The same iron is released from hemoglobin on oxidation stress. Undamaged hemoglobin is not an efficient catalyst of the Haber–Weiss reaction. However, it is known that hemoglobin's iron can efficiently support bacterial growth (often even with lethal finality) in patients (as on bacterial introduction into the open wound), and also, apparently, it can be used as biological Fenton's reagent (jointly with H_2O_2).

The second example is the role of ferritin (protein of storage iron(III)) in the reactions of lipid peroxidation. Under physiological conditions, cellular reducers such as ascorbate will reduce ferritin's iron(III) to iron(II) which will then be released from the cell's mitochondrial ferritin in the form of phosphate complexes and compounds with ATP and ADP, and take part in lipid peroxidation [46] in accordance with general Scheme 3.1.

Another intercellular reducer—superoxide anion $O_2^{\bullet-}$—can also liberate iron from ferritin in the form of iron(II), which in the same way takes part in lipid peroxidation [47–50]. Pulse radiolysis research of superoxide interaction with

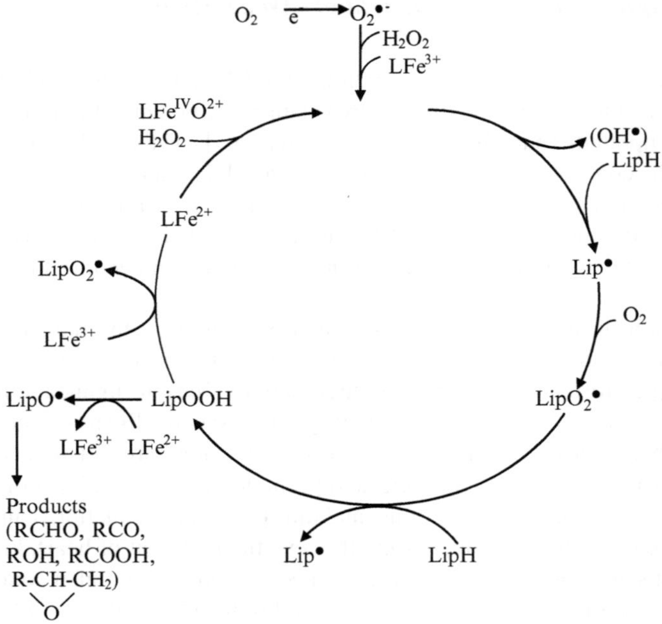

Scheme 3.1 Lipid peroxidation catalyzed by ferryl particle $LFeO^{2+}$ or $OH^{\bullet}$

ferritin made it possible to find the second order rate constant of this reaction, which coincides with the rate constant of the reaction of $O_2^{\bullet-}$ with Fe(III)–EDTA ($k = 3 \times 10^5$L/mol^{-1} s^{-1}) [51,52]. The intracellularly occurring Cu(I)–glutathione complex (Cu(I)–[GSH]$_2$) has the ability to reduce molecular oxygen into superoxide radicals ($O_2^{\bullet-}$). Based on such an ability, we addressed the potential of this complex to generate the redox-active Fe^{2+} species, during its interaction with free Fe^{3+} and with ferritin-bound iron [53].

The aforesaid data lead to the conclusion that reducing cellular agents such as ascorbate or superoxide anion radical interact with ferritin and supply transition metals as catalysts for the Haber–Weiss reaction [54]. As to the possibility of the Haber–Weiss reaction realization on $O_2^{\bullet-}$ interaction with transferrin (iron-transporting protein), discrepancies in the data exist which soon deny such possibility [54].

However, the influence of chelate-forming agents is also to be considered, which can greatly change the rate of the Haber–Weiss reaction. While the complex of Fe(III) with EDTA is rather rapidly reduced with $O_2^{\bullet-}$ to Fe(II), DETAPAC, forming chelates with Fe(III), it is non-active in Haber–Weiss reaction, although this complex is active in Fenton reaction [54] Another ligand, desferal, is an important clinical treatment means against iron excess in an organism. In patients who suffer from iron excess, as a result of leukemia and not connected with transferrin, this iron was found in plasma, accelerating radical reactions in it [55, 56]. Transferrin enters into the cells on endocytosis, and thus the pH value of vacuoles, which contain transferrin, falls. This causes Fe^{3+} ions' release from transferrin protein, which, interacting with superoxide anions $O_2^{\bullet-}$ present in the cell and then with H_2O_2, according to Haber–Weiss reaction, will form highly reactive $OH^{\bullet}$ radicals, deteriorating various targets in the cell.

In reality, these reactions are complex, and dependent on conditions (pH of environment, presence of one or another ligand, suitable temperature, and solvent), first perferryl and ferryl particles may be formed:

$$LFe^{2+} + O_2^{\bullet-} \rightleftarrows [LFe^{3+} + O_2^{\bullet-} \leftrightarrow LFe^{2+}O_2] \leftrightarrows Fe^{3+} + O_2$$

and

$$LFe^{2+} + H_2O_2 \longrightarrow LFe^{IV}O^{2+} + H_2O$$

and then $OH^{\bullet}$ radical

$$LFe^{IV}O^{2+} + H_2O \longrightarrow LFe^{3+} + OH^{-} + OH.$$

In the presence of $O_2^{\bullet-}$, desferral does not transform iron(III) into catalytically active particles (iron(II)) for the Haber–Weiss reaction. Desferral can inhibit membrane lipid peroxidation, initiated with peroxide systems [57]; however, it can act sooner as an electron or a hydrogen atom supplier, but not as an iron remover. Another example, phytinic acid, present in large amounts in each plant, inhibits free radical reactions catalyzed with metal ions [58–60], thus protecting the seeds against oxidative deterioration during storage [61]. The Fe(III) complex with phytinic acid reacts relatively slowly with $O_2^{\bullet-}$, which makes it harder for iron to be a good catalyst in the Haber–Weiss reaction.

Although the majority of chelate-forming agents studied make the reaction $Fe^{3+} + O_2^{\bullet-}$ slower (i.e., this reaction is a rate-determining stage in the Haber–Weiss reaction), it has still not been studied enough in physiological conditions to reach any final conclusions. Nonetheless, the data available help us to understand better the role of iron, copper, $O_2^{\bullet-}$, H_2O_2, ROOH, $RO^{\bullet}$, $RO_2^{\bullet}$, and $OH^{\bullet}$ in the problems of human health connected with oxidation stress [54]. As soon as iron compounds are contained mainly in storage, transport, or functional (enzymes) proteins, it is less probable that iron joined to protein could catalyze $OH^{\bullet}$ radicals formation outside the protein ($OH^{\bullet}$ radicals would immediately attack the protein itself) [62].

Therefore, it follows that real catalysts of $OH^{\bullet}$ radicals formation are iron ions or compounds released from protein [63]. Really, under certain biological conditions, for instance, at low pH in micro-surroundings of activated macrophages in rheumatic inflammation of joints, iron can be released from transferrin [64]. Another cause of metal ions release from protein can be oxidation stress (see Chap. 4) with subsequent formation of destructive $OH^{\bullet}$ radicals [41].

Very often with lack of iron in plants, chlorosis can develop. Resolving this problem is important for agriculture. In contrast, iron excess in an organism is also a serious illness, which needs the use of suitable chelate-forming agents for its binding. The use of desferrioxamine (desferral) as a drug for iron excess removal from an organism is limited for several reasons, namely, its high price, short action time, etc. Therefore, relatively recently on the basis of three 8-hydroxyquinoline subunits, water-soluble tripodal O-TRENSOX has been synthesized [65, 66] which forms a strong complex with Fe^{3+} ion (Fe(III)$^{\bullet}$ O-TRENSOX). In contrast to iron(III) complexes with EDTA or citrate, compound Fe(III)$^{\bullet}$ O-TRENSOX, in the presence of a reducer such as ascorbate, does not induce damage in radicals (this was shown on such a target as DNA) with hydrogen peroxide (the Fenton reaction does not proceed). Complex Fe(III)$^{\bullet}$ O-TRENSOX can prevent (or turn back) iron chlorosis, while O-TRENSOX itself can be used for iron excess removal [66]:

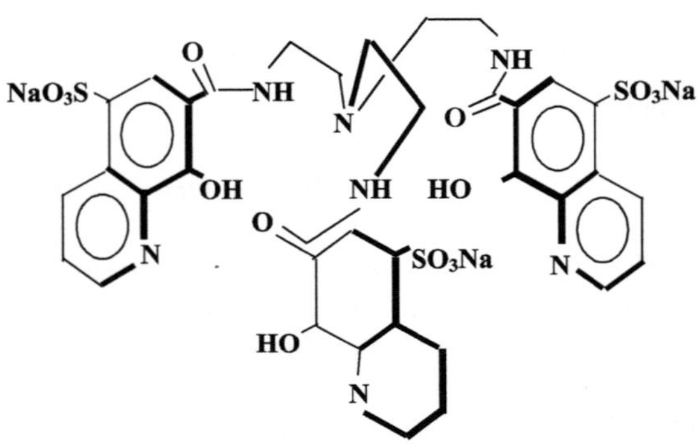

O - TRENSOX

3.1.4 Toxicity of O_2 Reduced Forms with Regard to Biological Systems and Methods of Protection

DNA deterioration in the cells of mammals which are subject to oxidation stress (see Chap. 4) is caused by local-specific formation of OH$^{\bullet}$ radicals [67, 68] which damage DNA [41]. Other particles such as ferryl apparently can also belong, under certain conditions in vivo, to the class of such high-reactive particles deteriorating DNA and other cellular targets [69, 70].

Complex compounds of transition metals not only influence in vivo biological substrates of nonenzymatic systems but also can significantly influence functional activity of enzymes themselves and enzyme chains. An example of the latter is the influence of CO(II, III) and Cu(II) complexes with 1,10-phenanthroline (phen) and its substitutes (2,9-dimethyl-4,7-diphenyl-5,6-dimethyl-5-nitro-5-aminophenanthroline) on the respiratory electron transporting cellular chain [71]. This respiratory chain is localized in cell mitochondria. Its substrates are organic carbonic acids formed due to foodstuff metabolism. Respiration is electron transfer from substrates of respiration to oxygen, proceeding by the chain of conjugated enzymatic reactions. In this respiratory chain, the substrates, coenzymes NADH and ubiquinone, system of cytochromes (b, c + c_1, a + a_3), and O_2, which is reduced to water, are included. The great amount of energy formed is stored in the form of ATP.

Complex compounds of Co(II) with phenyl substitutes in 1,10-phenanthroline significantly activated respiration of mitochondria, inhibited the synthesis of ATP, and degraded the integrity of mitochondrial membrane. Copper complexes of 1, 10-phenanthroline and its substitutes also rendered significant activating influence on mitochondrial respiration. In chemical reactions using respiratory chain ingredients, a similar picture was observed. Biological activity of CO(II, III) and Cu(II) in 1, 10-phenanthroline complexes in the processes of respiration in mitochondria is caused mostly by their catalytic activity in reactions of biological substrates' oxidation (carbonic acids) [71].

The number of other influences of such complexes on biological systems is to be noted. Thus, copper(II) complex with 2,9-dimethylphenanthroline turned out to be an efficient inhibitor of plasmodium and *Mycoplasma gallisepticum* growth—a dangerous pathogen causing different diseases of living organisms, from humans to plants. Thus, inhibition was caused by influence on the respiratory chain [72]. Complex $[Cu(phen)]Cl_2$ possesses strong bactericidal influence on gonococcus [73] and fungicidal action [74]. $[Mn(phen)_2(mal)]\cdot 2H_2O$ and $[Ag_2(phen)_3(mal)]\cdot 2H_2O$ ($malH_2 =$ malonic acid) were also found to exhibit fungistatic and fungicidal activity against *Candida albicans* [75]. Although copper compounds are found in all respiratory tissues [76], too high concentration levels are undesirable. Thus, pathologic high copper concentrations were found in tissues of patients with Wilson's disease and in the cerebral fluid of patients with Parkinson's disease [77]. Living organisms are forced to control copper content, preventing the accumulation of its too high (toxic) concentrations. Otherwise, copper ion excess along with highly reactive oxygen derivatives, biological reducers, and pharmaceutic agents will oxidatively destroy nucleic acids, proteins, and lipids [78, 79].

Apparently, copper toxicity is caused with its catalytic oxidative redox cycle between mono- and bivalent oxidation states according to the following reactions:

$$Cu^+ + O_2 \longrightarrow Cu^{2+} + O_2^{\bullet -} \tag{3.1}$$

$$2O_2^{\bullet -} + 2H^+ \longrightarrow H_2O_2 + O_2 \tag{3.2}$$

$$Cu^+ + H_2O_2 \longrightarrow Cu^{2+} + OH^{\bullet} + OH^-\cdot \tag{3.3}$$

Binding of Cu^+ ions with some biological chelate-forming substances inhibits $OH^{\bullet}$ radicals formation. Thus, introduction of penicillamine and triethylenetetramine (drug used in Wilson's disease's treatment) inhibited free radical formation. It is interesting that in reaction (3.3), an oxidant will be formed which behaves similarly with $OH^{\bullet}$, but it is not excluded that it represents an intermediate of superoxidized copper. Biological ligands such as cysteine, histidine, glutathione, and the above-mentioned penicillamine and triethylenetetramine prevent these highly reactive oxidants' formation [77].

Provided $O_2^{\bullet -}$ radical, much less reactive in relation to many biological targets, is formed *in vivo*, it can be diffused on large enough distances, till it meets Fe(III) complex with biological target (DNA, for instance). Thus, Fe(III) will be reduced to Fe(II). The latter, interacting with H_2O_2, generates $OH^{\bullet}$radical near the biological target, being ligand for Fe(III) and substrate for $OH^{\bullet}$. These $OH^{\bullet}$ radicals formed by local-specific mechanisms act more efficiently and selectively than the same $OH^{\bullet}$ sradicals obtained directly within the cell.

Besides, $OH^{\bullet}$ radical can be involved in epileptic pathogenesis induced with iron(II) and iron(III) compounds. Introduction of iron or heme into the brain of rats caused chronic epileptic nidus. Thus, the level of peroxide radicals, namely, $O_2^{\bullet -}$ and $OH^{\bullet}$, increased significantly in brain, and were detected by EPR methods (using spin-trap techniques) [80].

In its turn, $OH^{\bullet}$ radical is responsible for membrane lipid peroxidation occurrence, in particular, for creatine oxidation; the product of this reaction can be methylguanine (endogenic convulsant causing characteristic epileptic manifestations). Thus, $OH^{\bullet}$ radicals, as well as $O_2^{\bullet -}$, can be involved in epileptic pathogenesis, induced with iron compounds, by its formation through the Haber–Weiss mechanism. The same processes often result in epilepsy with head injuries.

In the same way, phagocytic mechanism of $OH^{\bullet}$ action in cells can be described as well. Phagocytes (neutrophils, eosinophils, and mononuclear phagocytes) are able to generate a large amount of rather reactive oxidation agents ($OH^{\bullet}$, $O_2^{\bullet -}$, etc.) for use in intruded pathogens' deterioration. This remarkable property of phagocytes is increased on the presence in these cells of respiratory destroying oxidases, realizing oxygen reduction to $O_2^{\bullet -}$ with NADH participation:

$$O_2 + \text{NADH} \xrightarrow{\text{oxidase}} O_2^{\bullet -} + \text{NADP}^+ + \text{H}^+.$$

The presence of superoxide dismutases in cells results in the rapid disappearance of $O_2^{\bullet -}$ and H_2O_2 accumulation.

The presence of $O_2^{\bullet -}$ and H_2O_2 creates the possibility for phagocytes to produce great amounts of various oxidants. Thus, in the presence of myeloperoxidase, a strong oxidant may be formed—hypochloric acid [81]:

$$H_2O_2 + Cl + H^+ \xrightarrow{\text{myeloperoxidase}} HOCl + H_2O$$

transforming then into appropriate oxidizing radicals (see Chap. 4). Besides, they can generate $OH^{\bullet}$ radicals using iron and copper compounds as catalyst, apparently, by means of the Haber–Weiss reaction. The same process takes place if H_2O_2 is substituted with alkyl hydroperoxides:

$$O_2^{\bullet -} + \text{ROOH} \xrightarrow{\text{iron or copper compounds}} \text{RO}^{\bullet} + \text{OH}^- + O_2.$$

Possessing the remarkable capacity to generate numerous oxidative particles, phagocytes manage lipid peroxidation in inflammation places and at those parts where the need for such oxidants appears (see Chap. 4). Thus, neutrophils produce $OH^{\bullet}$ radicals which realize hydrogen atom detachment from lipid (LipH) and thus begin the radical reactions of lipid peroxidation considered earlier. Although the mechanism of lipid peroxidation management by phagocytes is not quite clear, it also seems to involve $O_2^{\bullet-}$ radicals [82, 83].

Besides, $OH^{\bullet}$ and $O_2^{\bullet-}$- free radicals, phagocytic systems, as supposed, can also form another form of reactive oxygen particles. Thus, singlet oxygen $^1O_2$ can be formed (it was detected in chloroplasts and in eye crystal thread lens). Besides, at pH<5 (conditions that may be realized in some macrophages), $HO_2^{\bullet}$ radicals can be formed—stronger oxidants than $O_2^{\bullet-}$.

Finally, $OH^{\bullet}$ radical formation, besides phagocytic activity, in H_2O_2 presence in a biological system, can be caused by transition metals' catalytic influence, iron complexes with adenosine, guanosine triphosphates, citrates, and DNA, with the help of Fenton or Haber–Weiss reaction. Metal ion removal from sensitive places by chelate formation with certain ligands can suppress tissue degradation with $OH^{\bullet}$ radicals. The most active in this relation is desferrioxamine:

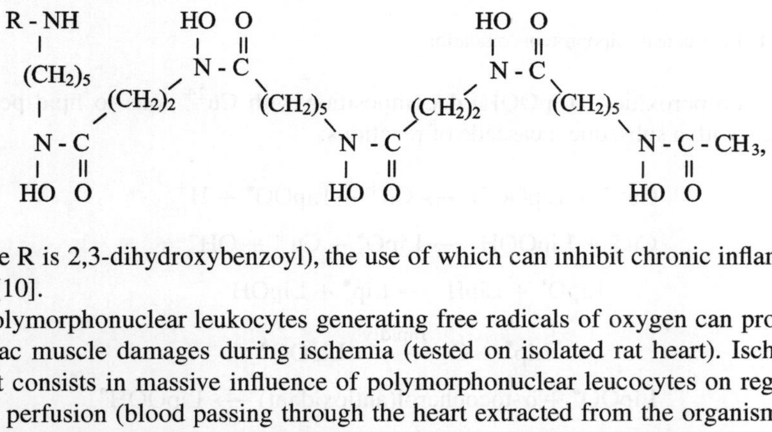

where R is 2,3-dihydroxybenzoyl), the use of which can inhibit chronic inflammation [10].

Polymorphonuclear leukocytes generating free radicals of oxygen can provoke cardiac muscle damages during ischemia (tested on isolated rat heart). Ischemic insult consists in massive influence of polymorphonuclear leucocytes on regional heart perfusion (blood passing through the heart extracted from the organism and placed into the artificial medium) with superoxide radicals formation as part of the phagocytic response (reaction) [84, 85].

For quantitative superoxide radicals determination generated with phagocytes or polymorphonuclear leucocytes, the EPR method was used with 5,5-dimethyl-pyrroline-*N*-oxide as spin trap [86, 87].

Low-density lipoprotein oxidation is related to early atherosclerotic damage (see Chap. 4) [88]. This process is accelerated by the addition of trace amounts of $CuCl_2$ to such lipids (LipH). A simplified scheme of this process in vitro is given in Fig. 3.1 [89]. Here, initiation of a lipid peroxidation chain reaction is shown, as well as antioxidants' influence on the process, secondary and tertiary reactions with lipid hydroperoxides. It is supposed that initiation depends on lipid hydroperoxides (LipOOH) proceeding in low-density lipoproteins. The process begins with these

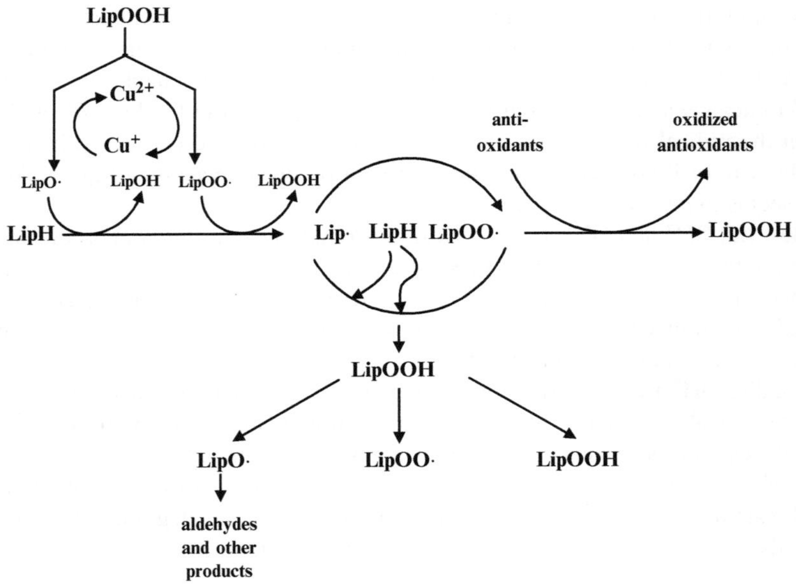

Fig. 3.1 Low-density lipoproteins oxidation

lipid hydroperoxides' (LipOOH) decomposition with Cu^{2+} ions to lipid peroxyl radicals with a subsequent cascade of reactions:

$$Cu^{2+} + \text{LipOOH} \longrightarrow Cu^{+} + \text{LipOO}^{\bullet} + H^{+}$$

$$Cu^{+} + \text{LipOOH} \longrightarrow \text{LipO}^{\bullet} + Cu^{+} + OH^{-}$$

$$\text{LipO}^{\bullet} + \text{LipH} \longrightarrow \text{Lip}^{\bullet} + \text{LipOH}$$

$$\text{Lip}^{\bullet} + O_2 \xrightarrow{\text{rapidly}} \text{LipOO}^{\bullet}$$

$$\text{LipOO}^{\bullet} + \alpha\text{-tocopherol(antioxidant)} \longrightarrow \text{LipOOH}^{\bullet}$$

Thus, a self-maintaining chain reaction occurs as shown in Fig. 3.1.

Destruction of $\text{LipOO}^{\bullet}$ radicals by vitamin E shows low-density lipoprotein protection against oxidation. Only if the antioxidants are completely spent does rapid accumulation of LipOOH proceed. However, the latter being a relatively unstable compound, as a rule, it is subject to secondary and tertiary reactions catalyzed by copper ions. Thus, $\text{LipOO}^{\bullet}$ and $\text{LipO}^{\bullet}$ radicals will be formed which compose the chains. In their turn, $\text{LipO}^{\bullet}$ radicals can attack both proteins (thus, peptide bonds will be disintegrated) and be subject to a β-disintegration reaction (forming aldehydes, joining LipH). Some of these aldehydes are rather reactive (namely, malonaldehyde, 4-hydroxynonenal, and 2,4-alkadienals) and, having interacted with LipH, will then change by various ways, including the formation of cell-toxic substances.

One of the reasons for the absence of a cells' protection against the joint presence of H_2O_2 and Fe^{2+} in them ($OH^\bullet$ radicals are formed) is increased generation of lipid peroxides inactivating some processes, such as synthesis of prostaglandin H synthase (PGHS) [90]. Formation of Fe^{3+} ions along with $OH^\bullet$ radicals on Fenton reaction

$$Fe^{2+} + H_2O_2 \longrightarrow Fe^{3+} + OH^- + OH^\bullet$$

means that iron ions in the ratio Fe^{2+}:$Fe^{3+} = 1:1$, together with H_2O_2, will generate more lipid peroxides than the $OH^\bullet$ radical alone. Really, in this case, perferryl ion $Fe^{2+}OO^\bullet \leftrightarrow Fe^{3+}O_2^{\bullet-}$ can be formed, which can generate additional amounts of lipid peroxides, deteriorating PGHS and, apparently, prostacyclin synthase [91].

Inner cellular damage caused by intercellular H_2O_2 penetrating into the cells will be decreased in the presence of Fe^{2+} in the intercellular liquid. In the presence of H_2O_2 and Fe^{2+} in amounts less than 250 and 500 μM, respectively, cell cytolysis (deterioration) does not proceed. Thus, intracellular iron(II) compounds protect the contents of cells against excessive H_2O_2 penetration into them by its transformation into $OH^\bullet$ radicals which are mainly ruined on their interaction with intracellular substances (proteins, amino acids, buffer salts, or antioxidants) [90].

Thus, general scheme of oxygen toxicity manifestation, represented in [92], can be slightly completed, taking into consideration influence of specific (L′) or non-specific (L) ligands [93] (Fig. 3.2). In Fig. 3.2, electron donor $D^{\bullet-}$ reduces O_2 to $O_2^{\bullet-}$ and H_2O_2. The latter, interacting with L_nFe^{2+}, will form ferryl particle $L_nFe^{IV}O^{2+}$, which, being a strong oxidant, interacts with biological target RH, generating $R^\bullet$. Radical $R^\bullet$, joining O_2, will form $RO_2^\bullet$, which on reaction with RH is turned into hydroperoxide ROOH and $R^\bullet$. Depending on ligand nature (L or L′) and value of redox potential φ, either Haber–Weiss reaction (left part) or reaction of $O_2^{\bullet-}$ dismutation with the formation of H_2O_2 (right part) can be realized.

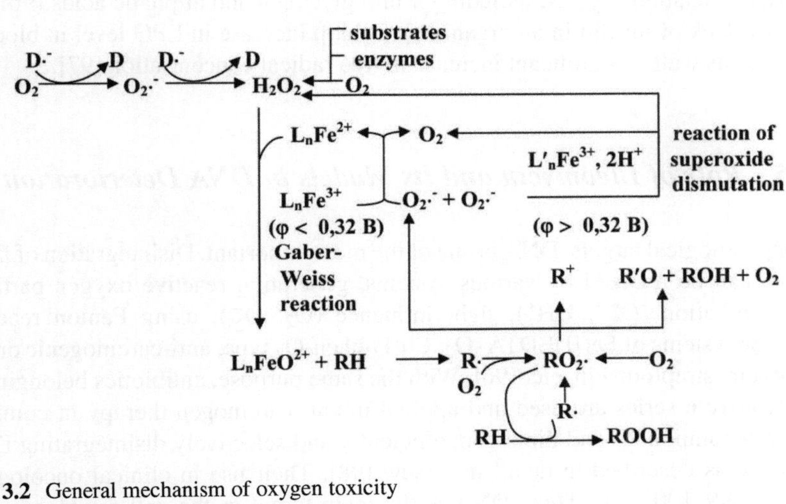

Fig. 3.2 General mechanism of oxygen toxicity

Lipids can also be biological targets of RH type. As soon as many reducers are contained in the cell (namely, ascorbic acid) which can directly reduce Fe(III) to Fe(II) with higher rate than superoxide ion $O_2^{\bullet -}$, determining the role of the latter in oxygen toxicity manifestation, though not excluded, is considered as doubtful. Actually H_2O_2 and $OH^{\bullet}$ radicals, but not $O_2^{\bullet -}$ radical, caused deterioration of phospholipids of cardiomyocyte membranes and lipid peroxidation [94]. Thus, it was shown that H_2O_2 causes relatively large deterioration of membrane phospholipids compared to $OH^{\bullet}$ and $O_2^{\bullet -}$ radicals.

Oxygen toxicity is manifested on monosaccharide autoxidation (namely, glyceraldehyde). Monosaccharide enediols formed are catalytically oxidized by iron compounds generating ketoaldehydes, $O_2^{\bullet -}$, and $OH^{\bullet}$. Oxidation of monosaccharides contributes to hyperglycemia (diabetes) occurrence, i.e., joint oxidation of tissues by means of these highly reactive particles. Such tissue oxidation is often accompanied by secondary processes—atherosclerosis, kidney disease, cataract, and retinal pathology [95].

Thus, monosaccharide autoxidation can contribute to diabetic complication development. Monosaccharide concentration is significantly increased in hyperglycemia and hence, concentration of H_2O_2 and other reactive particles ($O_2^{\bullet -}$, $OH^{\bullet}$) involved in monosaccharide autoxidation. For instance, the cellular toxic particles mentioned above can be generated in erythrocytes and eye crystal lens.

Catalytic oxidation of monosaccharide enediols with iron compounds can be accompanied by oxidation stress in hyperglycemia and atherosclerosis connected with it.

Insulin being one of the main regulators of substance exchange within the organism's cells (exchange of hydrocarbons, lipids, proteins and amino acids, nucleic acids and nucleotides) will decrease glucose level in blood due to the increasing of glucose introduction into cells, intensification of glucose exchange in various metabolic cycles, including it into glycogen and aliphatic acids [96]. On diabetes (lack of insulin in an organism), a sharp increase in LPO level in blood is observed, as well as significant increase in free radical concentration [97].

3.1.5 Role of Bleomycin and Its Models in DNA Deterioration

Among biological targets, DNA is one of the most important. Disintegration of DNA threads can be realized by various systems, generating reactive oxygen particles by γ-irradiation ($O_2^{\bullet -}$, $OH^{\bullet}$), light influence (O_2, O_2^{-}), using Fenton reaction ($OH^{\bullet}$) or systems of Fe(II)EDTA–O_2, Cu(I)-phen-O_2 type, anti-carcinogenic drugs, adramycin, streptomycin, etc. [96]. With the same purpose, antibiotics belonging to the bleomycin series are used and applied in anti-carcinogen therapy in common with iron compounds and dioxygen, efficiently and selectively disintegrating DNA threads as is described in detail in review [98]. Their use in clinical oncology is given in [99, 100], etc. Thus, DNA is the main target in the cell for bleomycins [101–103]. They can disintegrate RNA as well [98]. Bleomycins are glycopeptides

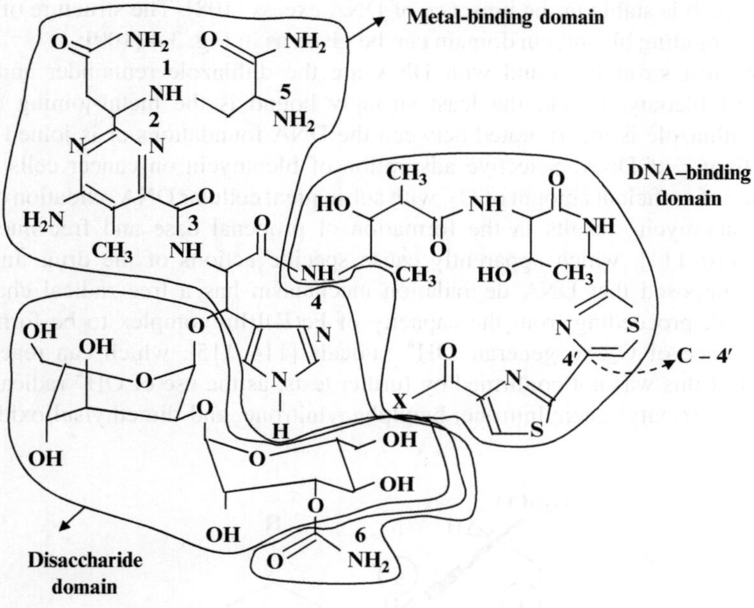

Fig. 3.3 Structure of bleomycin and its domains

with a molecular mass of ≈1,500. The structure of bleomycins (Blm-X) is as shown in Fig. 3.3.

Here, X–$NH(CH_2)_3S^+(CH_3)_2$ (Blm–A_2),
–$NH(CH_2)_3NH(CH_2)_4NH_2$ (Blm–A_5),
–$NH(CH_2)_3SCH_3$ (DMBlm–A_2: dimethylbleomycin),
–NH_2(BlmB_2'), $NH(CH_2)_4NHC(NH)NH_2$ (Blm–B_2),
–$NH(CH_2)_3NH(CH_2)_3N(CH_2)_4$ (Blm–BAPP),

$$-NH(CH_2)_4NH-\underset{H}{\overset{CH_3}{\overset{|}{C}}}H-Ph \quad \text{(peplomycin).}$$

Numbers 1-6 are used to indicate nitrogen atoms coordinated with Fe^{2+} ions and occupying six coordination places in the complex.

Bleomycin can be bound strongly enough with DNA. In the presence of dioxygen, Fe(II)Blm binding with DNA results in the destruction of the latter [104]. Molecule Blm can accomplish various functions, and thus was conventionally divided into several parts (domains). Pyrimidoblaminic acid and β-oxohistidine form a metal-binding domain [105, 106]. The remainder of dithiazole and edge amine X is called DNA-binding domain. Disaccharide promotes O_2 binding for oxidation disintegration of DNA [107]. Equally with Fe(II)Blm complex, complexes Fe(II)Blm, Cu(II)Blm, and Co(III)Blm will also be formed, which can join DNA at different rates. Dioxygen O_2 joins Fe(II)Blm forming a triple complex (Oxygenated

Blm), which is stable in the presence of DNA excess [108]. The structure of metal (copper)-binding bleomycin domain can be given as in Fig. 3.4 [109]:

The most strongly bound with DNA are the dithiazole remainder and edge amine of bleomycin, and the least strongly bound is the metal-joining center. Thus, dithiazole is incorporated between the DNA foundations or is joined to the small fissure of DNA. Selective adsorption of bleomycin on cancer cells in the presence of sufficient amount of O_2, with subsequent cellular DNA oxidation cutting by Fe-bleomycin, results in the formation of propenal base and free nucleonic bases [110, 111], which apparently cause specific actions of the drug. Initially, it was supposed that DNA degradation mechanism has a free radical character [112, 113], proceeding from the capacity of Fe(II)Blm complex to be formed in the presence of O_2, to generate $OH^{\bullet}$ radicals [114, 115], which can react with DNA. But this was not confirmed by further tests, as the use of $OH^{\bullet}$ radical traps (such as *tert*-butyl-phenylnitrone, butylphenylnitrone, and dimethylsulfoxide) did

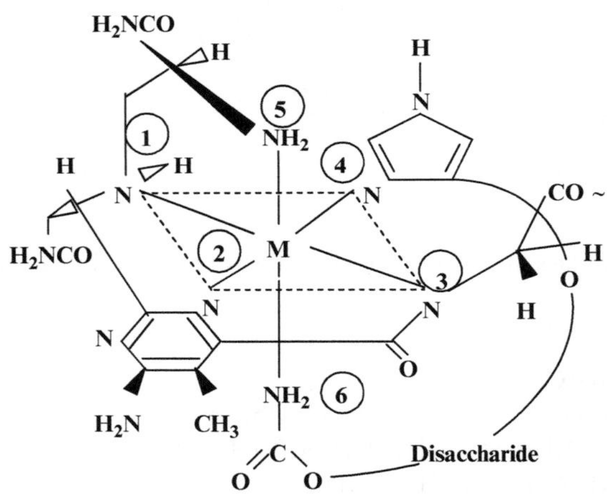

Fig. 3.4 Structure of metal-binding domain of bleomycin

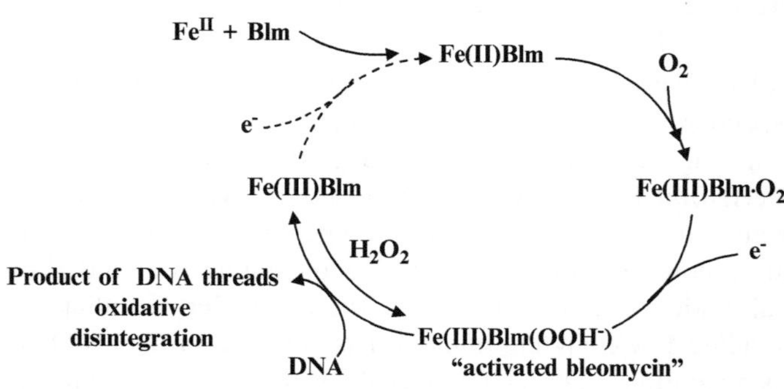

Fig. 3.5 DNA degradation by iron complex with bleomycin

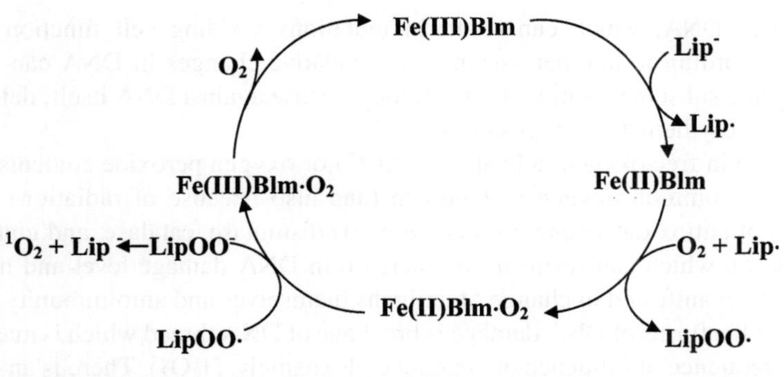

Fig. 3.6 Supposed scheme of lipogenase-like reaction activated with iron(II, III) complexes with Blm

not provide inhibition of DNA disintegration. FeBlm also catalyzes oxo-transfer to olefin substrates in a stereoselective way (such as *cis*-stilbene) in the presence of phenyl hypoiodite (PhIO), H_2O_2, ROOH, and O_2 [116]. These FeBlm properties are unique, since this is the first example of a mononuclear non-hemic system capable of activating oxygen.

Proceeding from the results of electron spectroscopy, EPR, and Mössbauer effect, mechanism of bleomycin influence on DNA is represented as the result of formed "activated bleomycin" which causes oxidative DNA degradation. The scheme of DNA disintegration by iron complex with bleomycin and dioxygen is given in [117]. In a somewhat different, cyclic, form it can be represented as in Fig. 3.5.

An additional electron for "activated bleomycin" formation can come either from the reducer present in solution, or from other Fe(II)Blm molecule as a result of disproportioning. "Activated bleomycin" is sooner low-spin iron (III) peroxo-complex [Fe(III)Blm(HOO^- than ferryl complex in hemic systems according to mass spectrometric data [118, 119] (this has not been revealed definitely).

It can then oxidize DNA and be reactivated anew by hydrogen peroxide. It is not excluded, but has not been proved, that in the presence of electron donors, Fe(III)Blm can be partly reduced to Fe(II)Blm (in Fig. 3.5, it is shown by the dotted thread). The details of DNA oxidative destruction are given in [98, 106, 120–122]. The destruction of DNA and RNA [123] results in disintegration of their lactose phosphate frame in C_4' position and liberation of free bases (propenals) [124].

It can be seen from Fig. 3.5 that hydrogen peroxide influences results in the repeated regeneration of "activated bleomycin" and selective catalytic oxidation of both relatively simple organic compounds and complex ones such as DNA and RNA. Thus, complex of bleomycin with iron(II) acts as "proteinless enzyme." One molecule of this antibiotic is able to catalyze the reduction of 4,7 oxygen molecules and destruct several DNA [88]. In its turn, DNA catalyzes auto-activation of some oxidized metal complexes, thus efficiently protecting itself from deterioration [125].

Oxidative DNA deterioration results in the occurrence of inflammatory autoimmune diseases such as rheumatic arthritis and systematic erythematosus lupus (skin reddening), and also cancer and aging [126]. This is apparently caused by changes in DNA, which can result in mutations yielding cell function change [127]. According to another assumption, oxidative changes in DNA can make it an antigenic substance, stimulating antibody release against DNA itself, detected in the case of erythematosus lupus [128].

Increase in free oxygen radicals ($O_2^{\bullet}$, $OH^{\bullet}$) or oxygen peroxide contents in vivo due to the common oxygen metabolism (and also because of radiation) reduces the level of antioxidative enzymes (superoxide dismutase, catalase, and glutathione peroxidase), which can result in the increase in DNA damage level and mutation possibility, manifested in change of antigens themselves and autoimmunity [129].

One of the forms of DNA damage is breakage of DNA thread which is manifested as a consequence of influence on oxidant cells (namely, H_2O_2). There is an enzyme in the cell—poly-ADP–ribose polymerase, being the part of the system, repairing the damaged DNA. On DNA deterioration, this enzyme will be activated, consuming NAD^+ and ATP. If the degree of DNA damage becomes too high, the amount of NAD^+ and ADP is not enough to repair the damaged DNA that results in cell ruin. Such damage is a "self-destroying response" of cells in the case of extensive DNA damage so that it will be better for the cell to be ruined than endure the appearance of somatic mutations (connected with body), resulting in malignant consequences [129].

Iron compounds joining DNA, in the presence of oxygen oxidants (O_2, H_2O_2), catalyze $OH^{\bullet}$ generation (see [130]) and, therefore, subsequent DNA damage. This statement has been recently confirmed in a study of the complex $[Fe^{II}EDTA]^{2-}$ and H_2O_2 influence on DNA [131]. Thus, proof is given that it is $OH^{\bullet}$ radical and not ferryl complex that is responsible for DNA damage.

One of the numerous products obtained by oxygen radical attack on DNA is 8-hydroxydeoxyguanosine (8-OHdG), which is a mutagen causing increase in frequency of bases separation from DNA.

Many chemotherapeutic drugs, e.g., bleomycin, cause biological effects similar to that from radiation. Bleomycin acts as an antitumor agent due to its capacity of triple complex formation with iron and DNA. This complex can be activated with cytochrome P-450 and, apparently, DNA deterioration also involves its interaction with $OH^{\bullet}$ radicals. This results in the breakage of simple and double DNA thread with the production of free and propenal bases. Their further decomposition results in the formation of malonic aldehyde. By its accumulation, the conclusion is made on the presence of "catalytic" iron in biological liquids [129].

On activation, human granulocytes (leukocytes containing granula in cytoplasm) produce a large quantity of free oxygen radicals in inflammation places in an organism. Granulocyte activation with the tumoral promoter tetra-deconyl phorbol acetate (TPA) causes an increase in 8-OHdG in cell and, hence, breakage of DNA threads. The rate of such thread unwinding is significantly increased within the cells of patients with rheumatoid arthritis compared to normal cells.

Chromosomal deviations of norm in the case of patients with sclerosis are provoked by oxygen radicals inducing DNA deterioration [132]. Blood lymphocytes in the case of patients with autoimmune diseases (rheumatoid arthritis, systematic erythematosus lupus, and Bechets syndrome) turn out to be highly sensitive to the toxic effects of alkylating agents such as *N*-methyl-*N*-nitrosourea, compared to normal patients [133]. All these point out that DNA damage with oxygen radicals can be caused with various environmental mutagens and can play an important role in somatic mutations yielding autoimmune diseases [129].

Like iron(III) interaction with bleomycin (Fe(III)Blm), cobalt salt interaction with bleomycin results in the appropriate complex formation CO^{III}Blm [134, 135]. Cobalt ion joins the metal-binding domain of bleomycin by means of five N-donor primary and secondary amino groups, pyrimidine, and imidazole ring. CO^{III}Blm is a low-spin complex, strongly joins DNA [136], and, unlike Fe(III)Blm complex, does not provoke DNA oxidative damage [137].

However, Co^{III} Blm can disintegrate DNA, provided it is irradiated with visible light [138]. Cobalt salts form several complexes with bleomycin (orange, brown, and green). Three model compounds [139] were revealed and studied, namely $[Co(PMA)(N\text{-methylimidazole})]^{2+}$ (1), $[Co(PMA)(H_2O)]^{2+}$ (2), and $[Co(PMA)Cl]^{+}$ (3), imitating spectrophotometric properties of these three Co^{III}Blm complexes, photodisintegration of DNA by them on their UV irradiation. Under their UV irradiation, in water solutions $OH^{\bullet}$ radicals will be formed (from water) [140] which induces the breakage of DNA threads. The bond of cobalt(III) from these model compounds with DNA is weak (electrostatic), while cobalt(III) with Blm forms a much stronger bond. Because of this, model complexes (1), (2), and (3) do not realize DNA photodisintegration with the same efficiency as Co^{III}Blm. DNA photodisintegration with model complexes (1)–(3) is the result of $OH^{\bullet}$ radical influence. However, model complex $[Co(PMA)]^{2+}$ modification by stronger (covalent) joining of acrydine or bithiazole promotes photodegradation of these complexes, which is rather similar to that of Co^{III}Blm [139].

"Activated bleomycin" [Fe(III)Blm(HOO^-)] seems to be responsible for a serious side effect—lung fibrosis (benign tumor from joint tissue); bleomycin therapy led to marked lung deteriorations which limits the drug usage. This process can be reversed by introducing some antioxidants. Fibroma occurrence induced with bleomycin apparently involves high-active oxidants' influence on the components of cellular membranes. As "activated bleomycin" is a peroxo-compound of iron(III) with bleomycin, this complex easily realized oxo-transfer to the olefin compounds of membranes. Hence, it becomes clear that interaction between cell lipids (olefins) and Fe(III)Blm(HOO^-) complex must result in lung deterioration.

Research in vitro has confirmed that iron ions bound in complex with bleomycin and peplomycin (another member of the bleomycin group) catalyze lipid peroxidation with oxidation products distribution, similar to that produced by lipoxygenase [141]. Also, it was proved that bleomycin causes redox-sensitive activation of phospholipase D [141, 142].

Earlier lipogenase activity and evolving of singlet oxygen ($^1\Delta_g$), accompanying it, were discovered under the influence of Fe(III)Blm and Fe(II)Blm·O_2 complexes (supposing it is Fe(II)Blm· $O_2^{\bullet-}$) on sodium linoleate (as lipid oxidation model) in vitro [143]. Later on, the assumed scheme of such lipogenase activity mechanism and $^1O_2$ formation was proposed [144]. This scheme can be represented in a somewhat simplified and cyclic form as follows (Fig. 3.6):

Complex Fe(III)Blm, interacting with linoleate anion, obtains electron and Fe(II)Blm, and appropriate ($Lip^{\bullet}$) will be formed. Complex of Fe(II)Blm with O_2 forms the oxygenated complex Fe(II)Blm·O_2 (Fe(III)Blm·$O_2^{\bullet-}$). Equally O_2, influencing $Lip^{\bullet}$, will generate a new radical $LipOO^{\bullet}$. The latter, receiving the electron from Fe(II)Blm·O_2, will be transformed into a dissociated form of hydroperoxide linoleate ($LipOO^-$) and complex—into Fe(III)BlmO_2. The oxygenated complex Fe(III)Blm·O_2 on oxygen lacking can once again be transformed into Fe(III)Blm. Singlet oxygen $^1O_2$ generation from hydroperoxide linoleate was proved to proceed, using a chemiluminescent method in visible (633 nm) and its more specific variant—infrared (1,268 nm) fields [144], which confirmed the earlier assumption on its formation [145]. It was also found that Fe(III)Blm complex stimulates $^1O_2$ generation (luminescence appears) in the system Fe(III)Blm–H_2O_2–OCl^- as well, which also catalyzes linoleate (Lip^-) interaction with $^3O_2$, transforming the latter into $^1O_2$.

Naturally, this supposed scheme of lipogenase-like reaction needs knowledge of further details. It is possible that "activated bleomycin complex" Fe(III)Blm(OOH^-) will also be formed in vitro.

Provided Fe(II) is replaced with Co(II), the appropriate complex with bleomycin will not disintegrate DNA even in the presence of O_2 and a reducer. However, if the system Co(II)–Blm–O_2 is irradiated with UV light, it results in a DNA disintegration reaction. It was shown [135] that initially formed complex Co(II)Blm·O_2 will then be transformed into two intermediate CO(III) complexes:

$$\text{Co(II)BIm}\cdot\text{O}_2 \longrightarrow \text{Co(III)BIm}\cdot\text{OOH(green)} + \text{Co(III)BIm}\cdot\text{H}_2\text{O(brown)}$$

which are able to take part in the DNA disintegration reaction under irradiation. Complexes Co(III) (peplomycin)·OOH and Co(III) (peplomycin)·H_2O interaction with dodecanucleotide CGCGAATTCGCG turned out to be specific to the disintegration place—near this nucleotide's C_3 and C_{11}, although disintegration products can be different [111].

Bleomycin is a very complex molecule. To clarify the mechanisms of DNA oxidative and photolytic deterioration by bleomycin complexes with metal ions (iron, copper, cobalt, or zinc), a model ligand PMAH has been synthesized [141]:

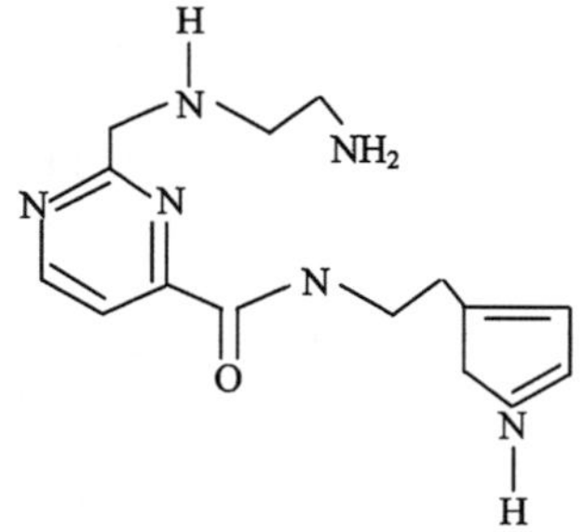

PMAH

which imitates the metal chelate part of bleomycin, and the appropriate complexes $[Fe^{II}(PMA^-)]Cl \cdot MeCN$ (a) and $[Fe(III)(PMA^-)](NO_3)_2 \cdot DMSO$ (b) are analogs of MBlm. Their structure was established by spectral methods. Oxygenated complexes (a) and (b) imitate all spectroscopic and chemical properties of "activated bleomycin." They represent iron(III) peroxo-compounds with PMA^-–$[Fe(III)(PMA^-)(OOH^-)]^+$, which are active intermediates in both DNA destruction and reaction of oxo-transfer. Really, complex (a) catalyzes lipid peroxidation in the presence of O_2 in the same way as Fe(II)Blm [141]. It was also observed that complexes (a) and (b) promote peroxidation of linoleic acid and arachidonic acids with H_2O_2. On such ligand peroxidation, the same products were formed as those in enzymatic peroxidation with lipogenase.

The scheme of lipid peroxidation with model complexes (a) and (b) is given in [141] (Scheme 3.2).

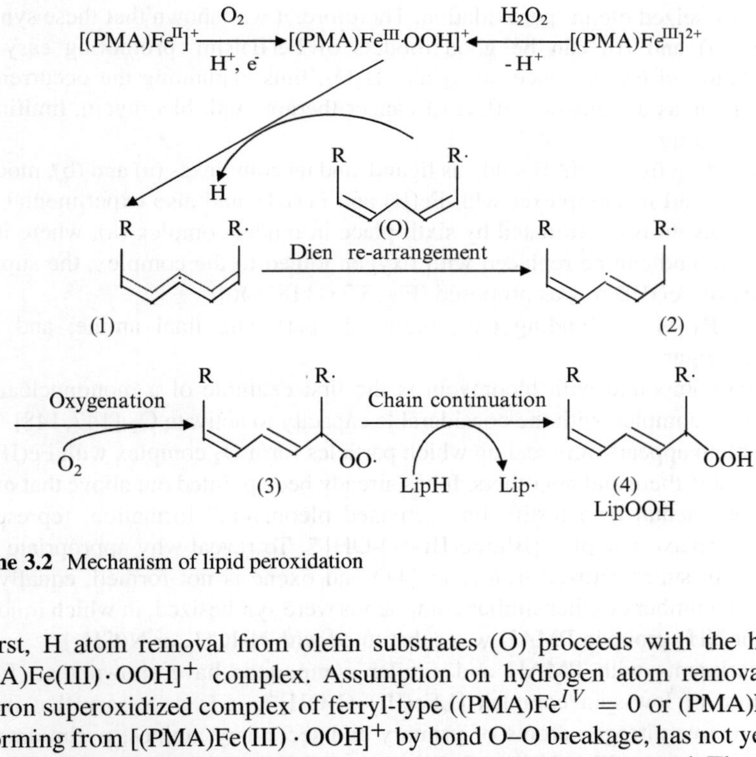

Scheme 3.2 Mechanism of lipid peroxidation

First, H atom removal from olefin substrates (O) proceeds with the help of $[(PMA)Fe(III) \cdot OOH]^+$ complex. Assumption on hydrogen atom removal with oxo-iron superoxidized complex of ferryl-type ($(PMA)Fe^{IV} = 0$ or $(PMA)Fe^{V} = 0$), forming from $[(PMA)Fe(III) \cdot OOH]^+$ by bond O–O breakage, has not yet been proved experimentally, although such an assumption has some ground. Thus, rapid lipid peroxidation is promoted in the presence of O_2 or H_2O_2. In accordance with Scheme 3.2, an H atom is removed from the methylene group of the olefin (linoleic or arachidonic acid) with the formation of C-basic radical (1), then diene radical is generated (2), and with the help of O_2, peroxy radical will be formed (3). The latter interacts with lipid (LipH) and forms lipid hydroperoxide (4).

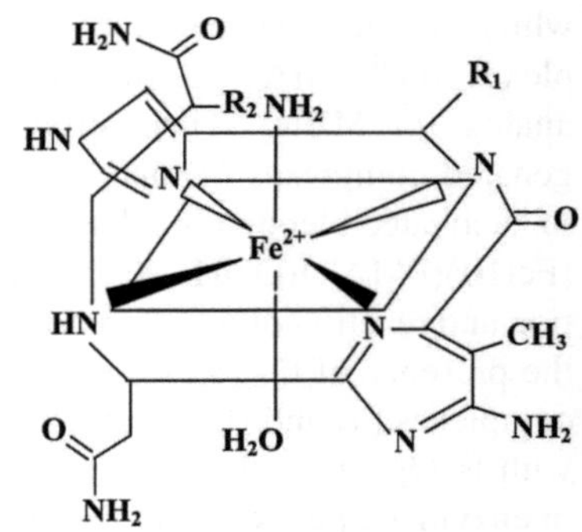

Fig. 3.7 Supposed square-pyramidal, non-hemic structure of Fe(II)Blm in solution

According to Scheme 3.2, the formation of all these products has been proved experimentally. Introduction of α-tocopherol (vitamin E) into the reactive mixture practically seized olefin' peroxidation. Therefore, it was shown that these synthetic analogs (a) and (b) can be good models of Fe(II)Blm, promoting easy lipid peroxidation in the presence of O_2 and H_2O_2, thus explaining the occurrence of lung fibrosis as a main side effect of cancer therapy with bleomycin, limiting the use of this drug.

Proceeding from PMAH study as ligand, and its complexes (a) and (b), modeling bleomycin and its complexes with Fe(II) and Fe(III), and also experimental proof that the solvent is coordinated by sixth place in model complex (a), where it is in labile state and can be replaced with oxygen joined to the complex, the supposed structure of Fe(II)Blm was proposed (Fig. 3.7) [118, 146].

Here, R_1 is the binding part, bithiazole part, and final amine; and R_2 is mannose sugar.

Fe(II) compound with bleomycin is the first example of a mononuclear non-hemic iron complex with the considerable capacity to activate O_2 [147, 148]. Thus, the problem appears of revealing which particles form O_2 complex with Fe(II)Blm, and why it is them and not others. It has already been pointed out above that numerous experimental data testify on "activated bleomycin" formation, representing low-spin peroxo complex $[BlmFe(III){-}O{-}OH]^+$. To reveal why appropriate ferryl complex in superoxidized iron state (4+) and oxene is not formed, equally with PMAH, a number of other similar compounds were synthesized, in which imidazole group or NH_2 group in PMAH was substituted with $NHCH_3$ or $N(CH_3)_2$.

Experiments with PMAH and similar compounds have shown that, for the formation of low-spin complex $[LFe(III){-}O{-}OH]^+$, it is necessary that iron be joined to five nitrogen atoms of primary and secondary amines, pyrimidine, and imidazole rings of PMAH ligand. Like iron–porphyrin complexes [149], in these non-hemic systems, the presence of strong field ligands is needed with spacious π-system in the main plane and that of one axial donor in order to bind and activate O_2. In case of iron porphyrin complexes, a high degree of electron transfer from Fe–O–O to porphyrin π-system provokes the breakage of the O–O bond and the formation of ferryl ($Fe^{IV} = O$) or perferryl ($Fe^{V} = O$) complex [150, 151]. It is supposed that in

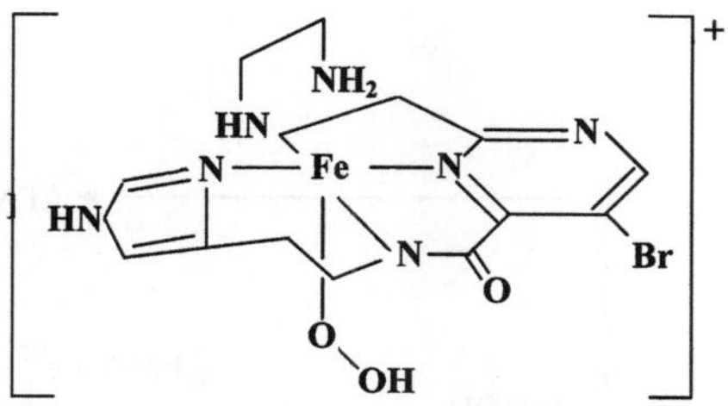

Fig. 3.8 Structure of $[(PMA)Fe(III)–O-OH]^+$ hydroperoxo complex

the case of the Fe(II)–Blm system or model bleomycin analog PMAH, π-electronic density existing in these systems is enough to form the hydroperoxo intermediate Fe(III)($[BlmFe(III)–O-OH]^+$ or $[PMAFe(III)–O-OH]^+$ (Fig. 3.8), but not enough to cause the breakage of the O–O bond like in the porphyrin system [152]:

EPR spectrum ($g = 2.27; 2.18; 1.93$) of $[(PMA)Fe(III)–O-OH]^+$ is identical to that of "activated belomycin." Peroxo complex $[(PMA)Fe(III)–O-OH]^+$ is a very strong oxidant. In methanol solution, it produces $CH_2OH^{\bullet}$ radical by hydrogen atom removal. This radical causes subsequent formation of formaldehyde and $OOCH_3OH^{\bullet}$ radical (established by the EPR method). The peroxo complex involved promotes rapid deterioration of DNA by its sugar part deterioration and propanol state formation. It can also realize stereoselective oxo-transfer to olefin substrates [145], i.e., it possesses the same functional properties as "activated bleomycin."

Initial complex $[Fe(III)(PMA)]^{2+}$ in acetonitrile easily realizes cyclohexane oxidation with *tert*-butyl hydroperoxide (^{t}BuOOH) or hydrogen peroxide (H_2O_2) with cyclohexanone and cyclohexanol formation in the ratio 1:1. This means that functionally this complex acts like the cytochrome P-450 model. Thus, the number of rotations (quantity of moles of products per mole of catalyst per hour) reaches 80. A probable mechanism of cyclohexane oxidation in the presence of ^{t}BuOOH and $[Fe(III)(PMA)]^{2+}$, based on the experimental data obtained, is presented as Scheme 3.3 [153].

As can be seen from Scheme 3.3, the possible mechanism comprises the following stages: intermediate $[(PMA)Fe(III)–O - O^tBu]^+$ formation, homolysis of O–O bond with the formation of two oxidative particles $[(PMA)Fe(III)–O^{\bullet}]^+$ and $^tBuO^{\bullet}$, initiation with $^tBuO^{\bullet}$ radical of H atom detachment from cyclohexane and promoting with $[(PMA)Fe(III)–O^{\bullet}]^+$ radical H atom detachment from ^{t}BuOOH, cyclohexanol and cyclohexanone formation, and intermediate $[(PMA)Fe(III)–O–O^tBu]^+$ formation by $[(PMA)Fe(III)–OH]^+$ interaction with ^{t}BuOOH. The use of ^{t}BuOOH instead of hydrogen peroxide yields the same mechanism of cyclohexane oxidation, but with different product distribution. The intermediate $[(PMA)Fe(III)–O–O^tBu]^+$ has been identified spectrophotometrically under low temperature. In general, $[Fe(III)(PMA)](ClO_4)_2$ complex is an efficient non-hemic catalyst in the reaction of alkane oxidation with *tert*-butyl hydroperoxide at room temperature.

Besides PMAH and its derivatives, other ligands can also be used, forming complexes with the transition metal ions, able to promote DNA disintegration. Such complexes involve $[Cu(phen)_2]^{2+}$ [154, 155] and EDTAFe(II) [156, 157].

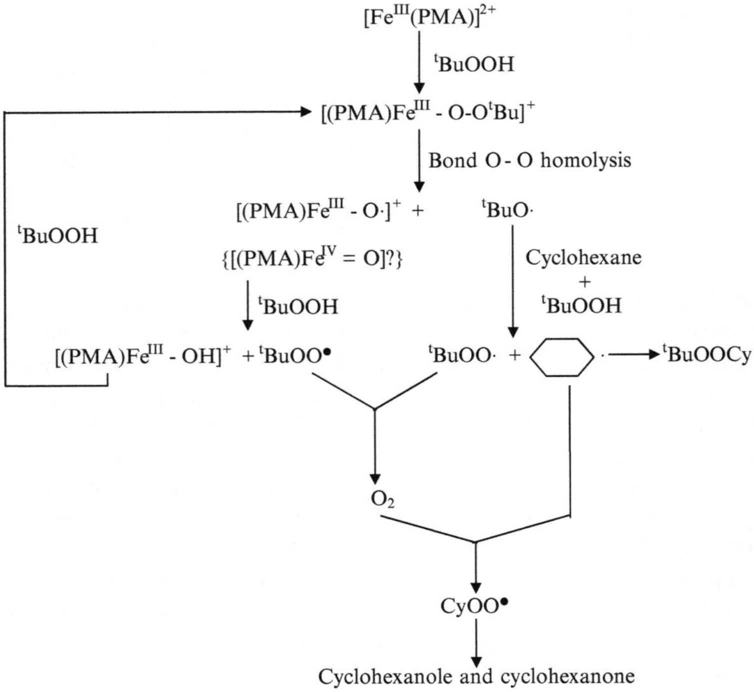

Scheme 3.3 Mechanism of cyclohexane oxidation

Thus, free oxygen radicals are formed attacking the deoxyribose base of DNA. Other complexes need photochemical activation for realization of the same function (namely, $[Cu(phen)_3]^{2+}$) [158].

However, other ligands were synthesized as well, forming complexes with iron(III), able to realize catalytically one-thread oxidative disintegration of plasmid DNA under physiological pH and temperature. Thus, a complex of 1,4,7-trimethyl-1,4,7-triazacyclononane with $FeCl_3$ ($LFeCl_3$) was synthesized, which at low concentrations (~0.5 μM) in the presence of O_2 will efficiently disintegrate DNA. Addition of reducing agents (namely, dithiothreitol) increases this complex efficiency about ten times (for DNA disintegration, only 0.05 μM is needed), and its efficiency becomes closely comparable with that of Fe(II) complex with bleomycin (Fe(II)Blm). $LFeCl_3$ complex can be easily synthesized and is stable in water solutions, which makes it an attractive agent for DNA disintegration [159].

In the process of plasmid DNA disintegration, generated free radicals of oxygen ($OH^•$) or iron-oxo particles react directly with DNA and do not come out into solution, unlike the aforesaid iron complexes with other ligands, and also Fe(II)–EDTA complex (methydiumpropyl) [160–162]. This last complex disintegrates DNA at concentrations similar to the $LFeCl_3$ complex; however, $OH^•$ radicals will be formed fixed in solution. Therefore, in this case, the mechanism of DNA disintegration differs from that for $LFeCl_3$. It is unclear yet whether or not $LFeCl_3$ complex possesses antitumor properties, similar to those found in Fe(II)Blm.

In double DNA threads, breakage can be either single (one thread is broken) or double (both threads are broken). For cell survival, the latter breakage is more pernicious, as in this case, "reparation" (reducing) of DNA is more difficult [163]. Study of the structure of the molecules that can cause such rupture of DNA threads is important both for the appropriate drugs creation against carcinogenic diseases, and for revealing the mechanisms of DNA reparation. With this in mind, complex ((2*S*, 8*R*)-5-amino-2,8-dibenzyl-5-methyl-3,7-diasononandioato) copper(II) has been synthesized and studied (1), able to break orderly two DNA threads, likewise iron complex with bleomycin [164]. Positively charged ammonium group is fixed with the ligand, which can interact with DNA phosphates. Copper(II) is used for hydrogen-detaching active oxygen particles generation. The structure of the complex is as follows [165, 166]:

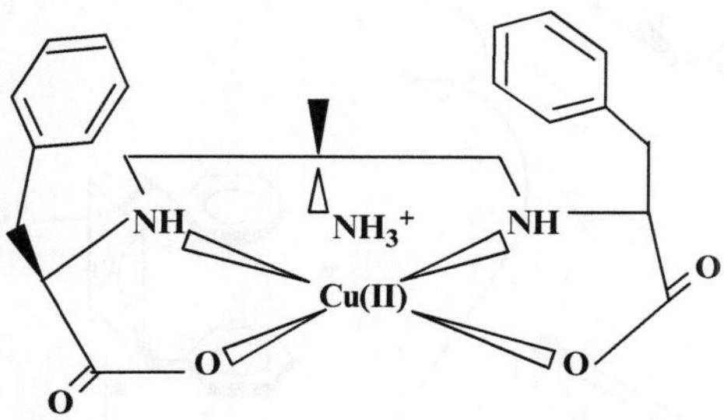

On (1) interaction with DNA in the presence of O_2, $OH^{\bullet -}$ and $O_2^{\bullet -}$ radicals will be formed. In the presence of a reducer (ascorbic acid (AH_2) or others), (1) will generate $OH^{\bullet}$ radicals by Haber–Weiss reaction (5). Appropriate sequence of reactions can be represented by Scheme 3.4.

$$1.\ Cu^{II}L + \bar{e}\ (\text{from reducer}) \longrightarrow Cu^{I}L$$
$$(1)$$
$$2.\ Cu^{I}L + O_2 \rightleftharpoons Cu^{II}L^{\bullet}O_2^{\bullet -}$$
$$3.\ Cu^{II}L^{\bullet}O_2^{\bullet -} \rightleftharpoons Cu^{II}L + O_2^{\bullet -}$$
$$4.\ AH_2 + O_2^{\bullet -} \rightleftharpoons A^{-} + H_2O_2$$
$$5.\ O_2^{\bullet -} + H_2O_2 \longrightarrow O_2 + OH^{-} + OH^{\bullet}.$$

Scheme 3.4 Interaction of DNA with Co complex in the presence of O_2

Clearly, the same processes can be drawn as cyclic Scheme 3.5 [165].

Thus, the mechanism of DNA cutting consists of binding of (1) with DNA, $O_2^{\bullet -}$, and H_2O_2 formation, $OH^{\bullet}$ generation, $OH^{\bullet}$ interaction with deoxyribose, and hydrogen atom removal from it in the $C_4{}'$ position. Interaction of the deoxyribose radicals thus formed yields the products of DNA thread breakage (in particular, propanol base formation). The scheme of DNA thread double rupture is shown in Fig. 3.9 [167].

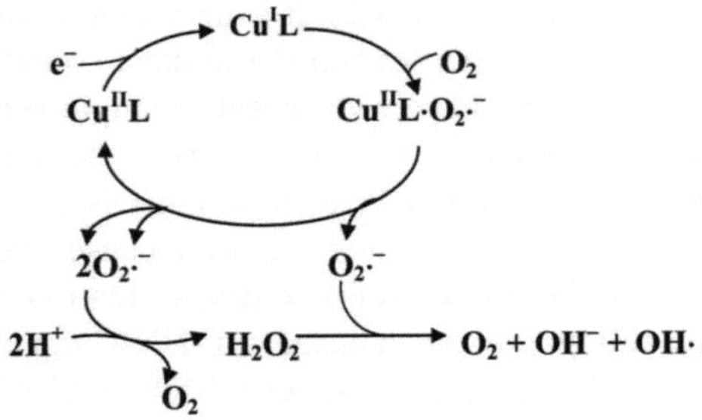

Scheme 3.5 Cyclic scheme of DNA interaction with Co complex

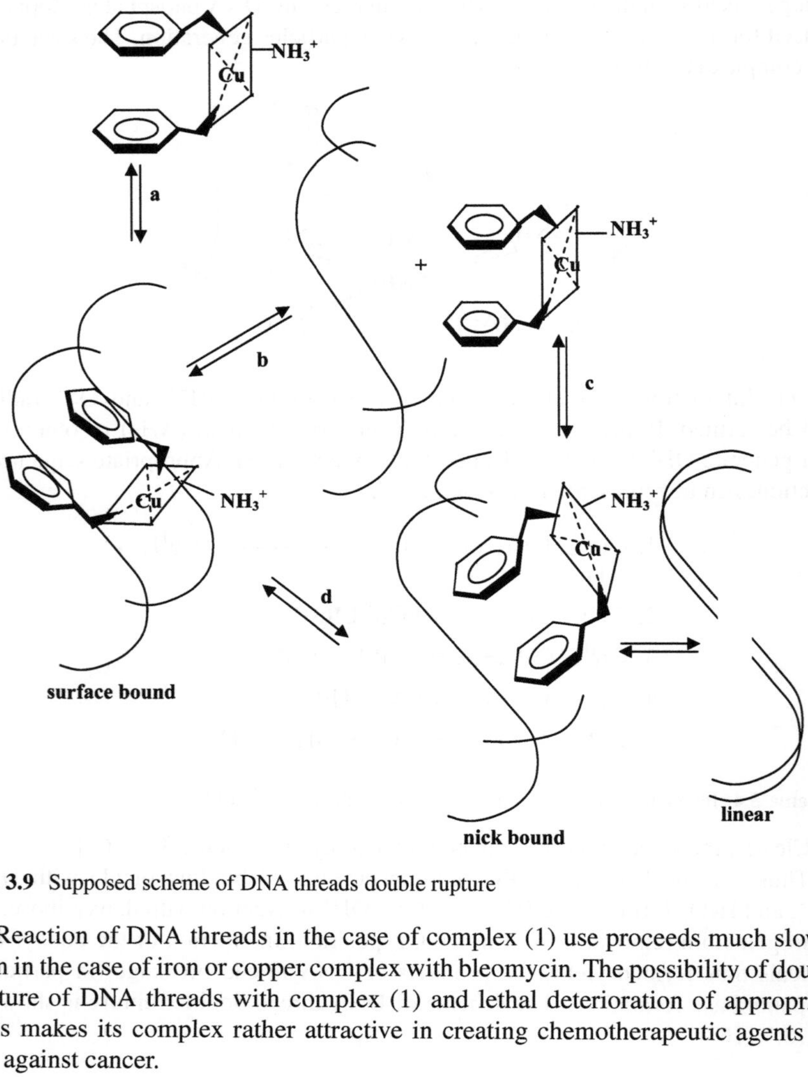

Fig. 3.9 Supposed scheme of DNA threads double rupture

Reaction of DNA threads in the case of complex (1) use proceeds much slower than in the case of iron or copper complex with bleomycin. The possibility of double rupture of DNA threads with complex (1) and lethal deterioration of appropriate cells makes its complex rather attractive in creating chemotherapeutic agents for use against cancer.

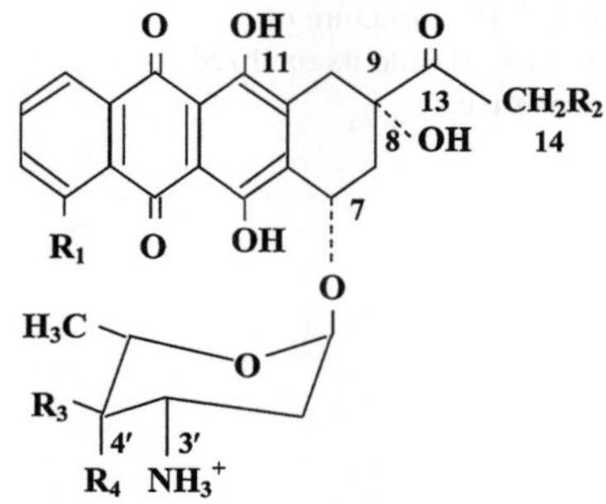

Fig. 3.10 Structure of antitumor anthracyclines. R_i—various remnants of anthracyclines

The clinical use of other antitumor drugs, for instance related to anthracyclines—doxorubicin (DOX, adriamycin), epirubicin, daunorubicin, etc.—is limited by a dose-limiting cardiotoxicity. The latter is connected with oxidative stress which is induced by oxygen free radicals ($O_2^{\bullet-}$ and $OH^{\bullet}$), catalytically generated in vivo by very stable iron(II) complexes with anthracyclines (stability constant of $Fe^{3+}(DOX)_3$ complex is approximately equal to $10^{33}M^{-3}$) [168]. Among anthracyclines, doxorubicin is one of the most efficient antitumor drugs. Its complex with Fe(II) catalyzes the destruction of cell membrane and is incorporated between pair bases of DNA.

The structure of antitumor anthracyclines is presented in Fig. 3.10.

Complex Fe^{3+}–doxorubicin can be subject to auto-reduction to Fe^{2+}–doxorubicin on account of electron transfer from the OH group of α-ketol part of doxorubicin. Thus, generation of $OH^{\bullet}$ radicals proceeds by preliminary formation of $O_2^{\bullet-}$ and H_2O_2 [168]:

$$Fe^{2+} - DOX + O_2 \longrightarrow Fe^{3+} - DOX + O_2^{\bullet-}$$

$$2O_2^{\bullet-} + 2H^+ \longrightarrow H_2O_2 + O_2$$

$$Fe^{2+} - DOX + H_2O_2 \longrightarrow Fe^{3+} - DOX + OH^- + OH^{\bullet}$$

Reducing of Fe^{3+}–DOX complex and lipid peroxidation by means of doxorubicin are realized in parallel with their cardiotoxicity [169]. This is caused, apparently, by $OH^{\bullet}$ radical generation and increased toxicity of anthracyclines such as doxorubicin and epirubicin, as only their complexes with Fe^{3+} (but not of other anthracyclines) are capable of auto-reduction [170].

3.1.6 Intermediates of Ferryl Particles Type in Biochemical Systems

In biochemical processes, oxygen molecule transformation in vivo into more reactive reduced forms and interaction of the latter as oxidant with various endogenous substrates play an important role. In such transformation realization,

Fig. 3.11 Structure of TTPCO(II) and its oxidized derivative

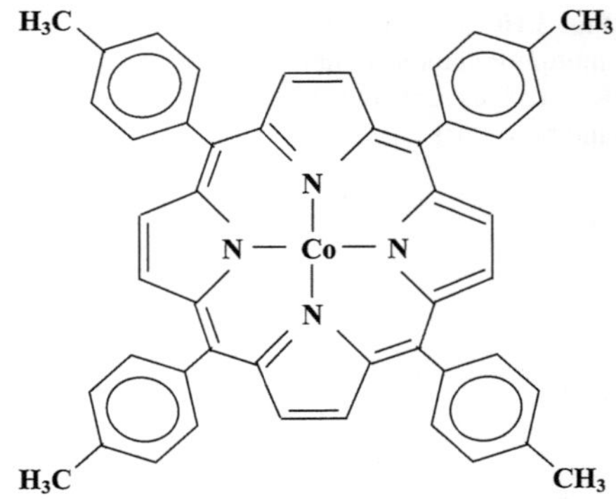

Co(II)TTP

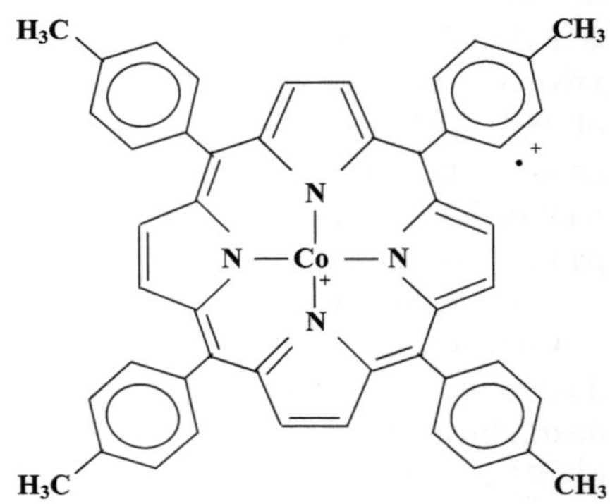

Co(III)TTP$\cdot^{2+}$

metal porphyrins occupy an important place in the metabolic processes of living organism. A high degree of porphyrin cycle conjugation contributes to this, providing relatively easy electron transfer both from porphyrin to central metal, and vice versa. The structure of cobalt tetratolylporphyrin (TTPCo(II)) and its oxidized derivative ($TTP^{\bullet++}$Co(III)) is shown in Fig. 3.11.

Interaction of metal porphyrinates with O_2 results in highly reactive intermediate forms of oxidant generation. The rate formation of such intermediates and their reactivity are affected by many factors: nature of central metal ion, peculiarity of substitutig groups in porphyrin macrocycle, influence of axial ligands' nature, and properties of solvent. Oxidation of metal porphyrinates proceeds through the formation of oxygenated μ-peroxo complex, ferryl intermediate $Fe^{IV} = O$, and, finally, μ-oxydimer:

$$PFe^{II} \xrightarrow{O_2} PFeO_2 \xrightarrow{PFe^{II}} PFe^{III} - O - O - Fe(III)P \longrightarrow PFe^{IV} = O \xrightarrow{PFe^{II}} PFe^{III} - O - Fe(III)P$$

The most general reactions of hydrocarbon (RH), olefin, and aldehyde catalytic transformations can be presented in the following form:

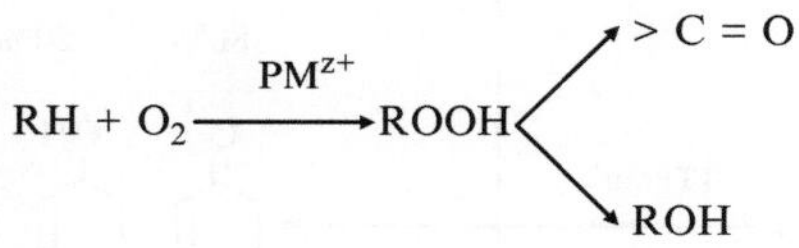

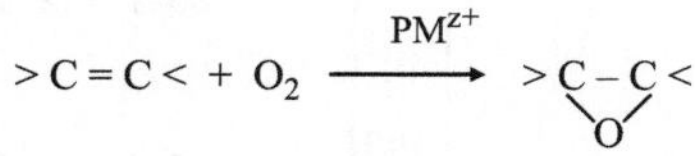

$$RCHO + O_2 \xrightarrow{PM^{2+}} RCOOOH$$

where PM^{z+} metals complex with porphyrin. Catalytic transformation of hydrocarbons can be accompanied by the formation of free radicals, and the mechanism of such processes becomes radical-chain:

Initiation:

$$TPPM^{2+} + O_2 \longrightarrow TPPM^{3+} \cdot O_2^{\bullet -}$$
$$TPPM^{3+} \cdot O_2^{\bullet -} + RCHO \longrightarrow TPPM^{3+} \cdot O_2H + RCO^{\bullet}$$

Chain continuation:

$$RCO^{\bullet} + O_2 \longrightarrow RCOOO^{\bullet}$$
$$RCOOO^{\bullet} + RCHO \longrightarrow RCOOOH + RCO^{\bullet}$$

Breakage:

$$2RCOOO^{\bullet} \longrightarrow \text{Products}$$

Γπe where TTP—tetraphenylporphyrin

Thus, the participation of oxidation metal complex is admitted, although its presence in the reaction medium was not proved [171]. Nevertheless, some oxygenated porphyrin complexes with Mn(II), Fe(II), Cu(I), and other transition metals exist.

Low oxidation potential of Mn(II) and Fe(II) porphyrinates makes them rather unstable in solution, and in the presence of oxygen they are subject to rapid oxidation with the formation of final products with different structures. Irreversible oxidation to dimer forms can also proceed [172].

The efficiency of metal porphyrinates and the processes' course in catalytic oxidation of organic substances with dioxygen depends to a great degree on the absence or presence of an additional reducer (for instance, $NaBH_4$), which creates the necessary conditions for the formation of active intermediates in the course of the reaction. Change of process mechanism can be seen as soon as the various products of oxidation process are obtained in the presence and absence of reducers [173]:

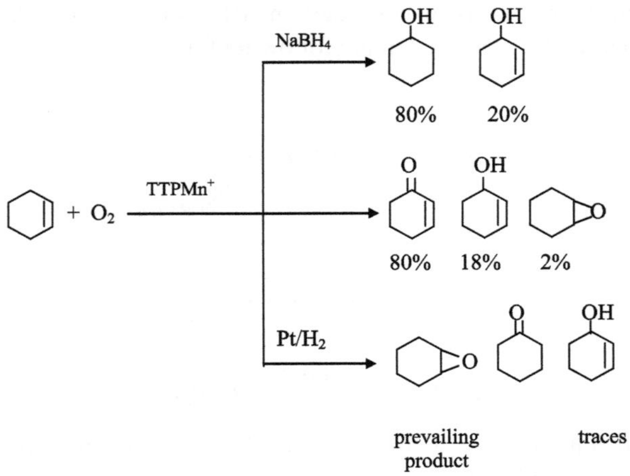

It is supposed that formation of highly oxidized intermediates is complicated in the absence of a reducer, and the reaction course will be different to that in their presence.

On such two-electron oxidants being used as per-acids, phenyl hypoiodite (C_6H_5-IO), sodium hypochlorite (NaOCl), nitrogen oxides, and Mn(III) porphyrinates are transformed into oxo-manganese(V) complexes [161]:

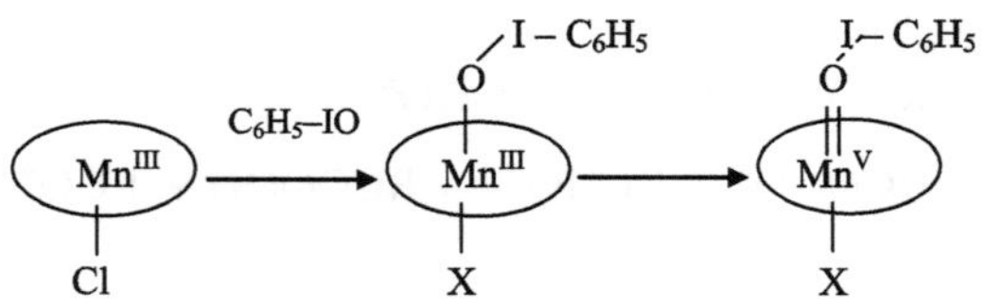

In the same way, iron(III) porphyrinates efficiently realize hydroxylation and epoxidation reactions with these oxidants [173].

Nowadays at least three tens of enzymes (hemic and non-hemic) are known, related to various sources and various classes (peroxidases, catalases, cytochromes P-450, cytochrome oxidases, methane monooxygenases (MMO), etc.), functional activity of which is manifested by means of ferryl intermediates' formation [174].

Ferryl intermediate iron, bound with oxygen, is formally in oxidation degree +5, but it is always present as Fe(IV) = O with one electron isolated from protein, and therefore is represented in the form $P^{\bullet+}$ Fe(IV)=O-ferryl π-cation radical (so-called compound I). Its one-electron reduction generates ferryl particle PFe(IV)=O (Compound II), and two-electron reduction results in initial iron porphyrinate PFe(III). In hydrogen peroxide excess, PFe(IV) = O will be turned into intermediate oxy-compound (compound III), the structure of which can be, apparently, defined as partial charge transfer complex [$Fe^{2+}O_2 \leftrightarrow Fe^{3+}O_2^{\bullet}$] (compound III). The ratio of these compounds looks as in Scheme 3.6.

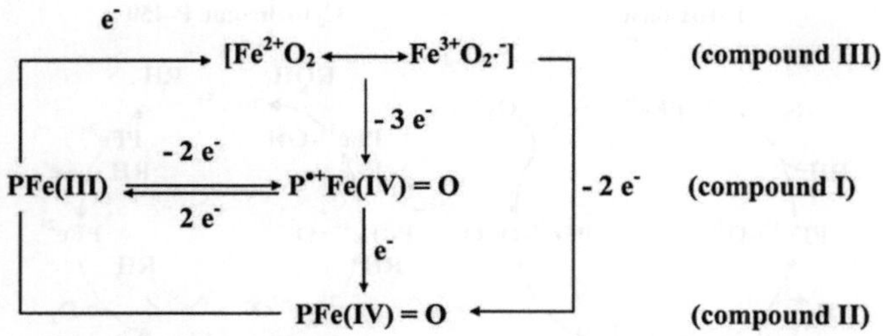

Scheme 3.6 Formation of ferryl intermediates in the functioning of enzymes

These compounds (I–III) are formed as intermediates in the process of various enzymes' functioning. Thus, the difference in mammal peroxidases (myeloperoxidase, lactoperoxidase, saliva peroxidase, eosinophil, and thyroid peroxidases) with regard to substrates oxidation consists in the value of redox potential of the compound I [175]. Myeloperoxidase is used by neutrophils in phagocytosis processes for bactericide HOCl formation from chloride and hydrogen peroxide (see Chap. 4), and neutrophil oxidase transforms compound III into compound I. Prostaglandin H and synthase (peroxidase) form both compound I and compound II [176, 177]. A rather interesting case is the use by catalase of its own substrate—hydrogen peroxide—as oxidant (generating from PFe(III) compound I), and as reducer (regenerating the initial active center of enzyme PFe(III) and evolving O_2). Such behavior is a rather frequent case in catalytic processes, thus illustrating what is known in philosophy as the law of antipodes' unity.

Ferryl intermediates are also generated on oxygenases and oxidases functioning. Comparative representation of substrates transformation mechanisms (without detailed consideration) is given in Fig. 3.12 [174, 178–180].

With all these enzymes' functioning, ferryl particles formation is observed for compound types I and II.

In the case of enzymes with non-hemic active centers, there are no direct spectrophotometric proofs of ferryl particle Fe(IV)=O involvement in the catalytic process, possibly, due to its short lifetime and difficulties in its detection. However, indirect proofs confirm its occurrence in non-hemic enzymes functioning. Oxo-bound two-nuclear iron complex is the active center of these enzymes. The mechanism of its action with ferryl intermediate participation is shown in Fig. 3.13 [181].

Here, X is an unknown group forming bridge bond between two iron atoms and cation radical; $TyrO^\bullet$ is tyrosine radical.

Thus, non-hemic enzymes can fulfill the functions characteristic of hemic proteins, namely methane monoxidase can hydroxylate hydrocarbons like cytochrome P-450, ribonucleotide reductase is capable of oxidizing phenols like peroxidase, etc. Such similarity in functions gives grounds to support the generation of intermediate ferryl particles in the process of non-hemic enzymes activity.

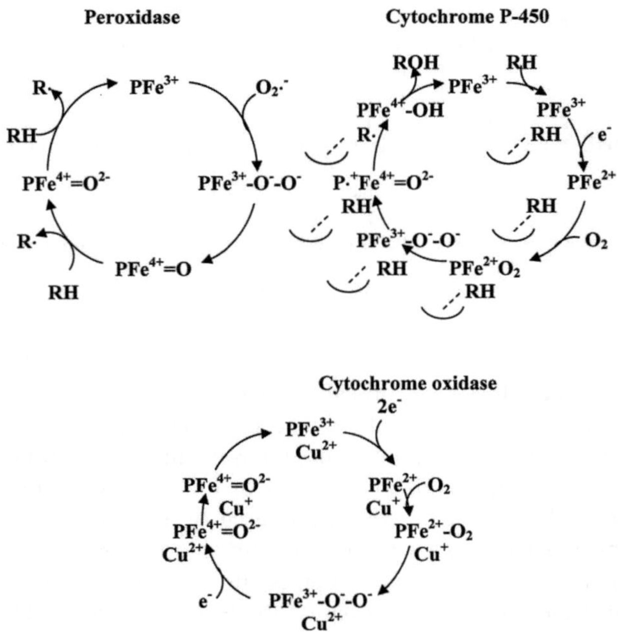

Fig. 3.12 Comparison of approximate reactive mechanisms of peroxidase, cytochrome P-450, and cytochrome oxidase (without details)

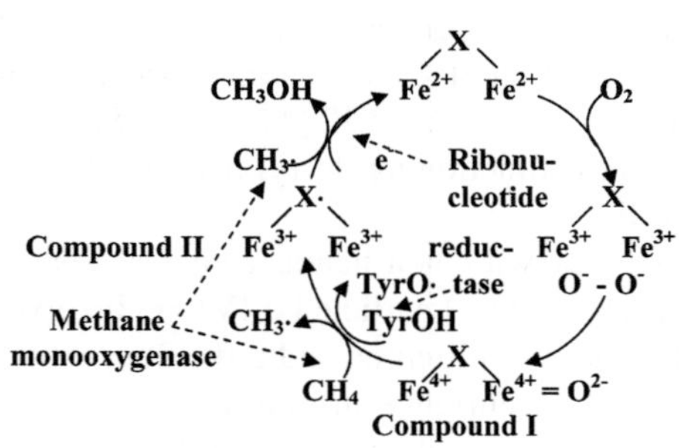

Fig. 3.13 Mechanism of enzymes interaction with non-hemic iron

Mononuclear non-hemic active centers of some enzyme also involve ferryl iron into the process mechanism. Thus, oxygen involvement does not proceed simultaneously at some dioxygenases, which indicates the possible presence of ferryl particles as intermediates [182].

3.2 Enzymes and Their Modeling

3.2.1 Mechanisms of Some Enzymes' Functioning and Their Modeling: Modeling of Monooxidases in Alkanes, Alkenes, and Other Substrates Hydroxylation and Epoxidation

In the first chapter, O–O bond activation in dioxygen, hydrogen peroxide, and hydroperoxides was mostly considered in the presence of transition metal ions and coordination compounds as catalysts.

In this chapter, biochemical catalytic processes are described, the most important role in which is played by oxenoids considered in the first chapter (O_2, H_2O_2, ROOH, etc.) as oxidants. Thus, only the enzymes are considered (briefly) and their models (in more detail) with regard to the catalyzing oxygen atom or molecule (monoxygenase or dioxygenase) entrance into substrate, hydrogen peroxide (superoxide dismutase) formation from superoxide anion radicals, decomposition of the latter (catalase), various substrates (peroxidase) oxidation with hydrogen peroxide, and four-electron reduction of the oxygen molecule to water (oxidase).

Metal enzymes, transition metal ions, and coordination compounds catalyze biochemical processes most often. The structure, properties, and details of metal enzymes' work will not be considered here, as extensive and well-generalized literature is devoted to these problems, namely, to monooxygenases [183–190]. Most attention will be given to generalizations on some of the most important monooxygenases modeling (e.g., cytochrome P-450), catalytic properties of such mimetic systems, their actitity, selectivity, stability, and comparison of metal porphyrinates work mechanisms with those of their models. Creation of such imitating (mimetic) catalytic systems has at least twofunctions: (1) to use the results obtained on such systems study for a deeper understanding of the appropriate metal enzymes action and (2) to use the results obtained, sooner or later, for creating sufficiently active, selective, and stable systems for industrial production of certain chemical products from the main alkanes sources—oil and natural gas. This is especially important because from the approximately 1 billion tonnes of oil obtained per year, several percent are subject to chemical processing, and direct alkanes functionalizationrepresents a small percentage of this.

To achieve this, the most important requirement is to create such systems which would activate C–H and C–C bonds in various hydrocarbons with sufficient rate and selectivity, close to that of the appropriate metal enzymes (or exceeding them). Activation of C–H bonds of alkanes under rather strict conditions (high temperature, strong oxidants, superacids, free radicals, carbenes, etc.) has long been known. Discovery of catalytic systems based on Pt(II) complexes able to activate the C–H bond in methane at 120°C is reflected in the remarkable pioneering work of Shylov and colleagues [191–194]. Reactions of H–D exchange in alkanes and their oxidation in systems containing platinum complexes have been carried out later by many researchers [195–198].

Numerous reactions of alkanes with metal hydride complexes and metal organic compounds were also discovered; the appropriate review is given in [199]. However, in all these reactions of alkane activation, oxenoids (oxide and peroxide systems), being the most important participants of C–H bonds activation in biological systems, do not take part. Therefore, these systems will not be considered here.

Modeling of some oxygenases, dioxygenases, and oxidases will be considered further. According to [26], difference between these factors are as follows. *Oxygenases* activate molecular oxygen catalytically and introduce one (monooxygenases) or two (dioxygenases) oxygen atoms into the substrate with electron donor participation. *Oxidases* catalytically activate hydrogen, but not dioxygen, and realize a variety of substrate oxidation by oxygen reduction with one, two, or four electrons (products: $HO_2^{\bullet}$, H_2O_2, H_2O).

There is a number of conjugated reactions in vivo, successively catalyzed with different enzymes: phenylamine hydroxylase catalyzing the reaction of phenylamine hydroxylation into tyrosine; tyrosine hydroxylase—tyrosine hydroxylation to 3,4-dioxyphenylalanine; dopamine-β-hydroxylase—truncated 3,4-dioxyphenylalanine (without COOH group) hydroxylation—dopamine to noradrenaline:

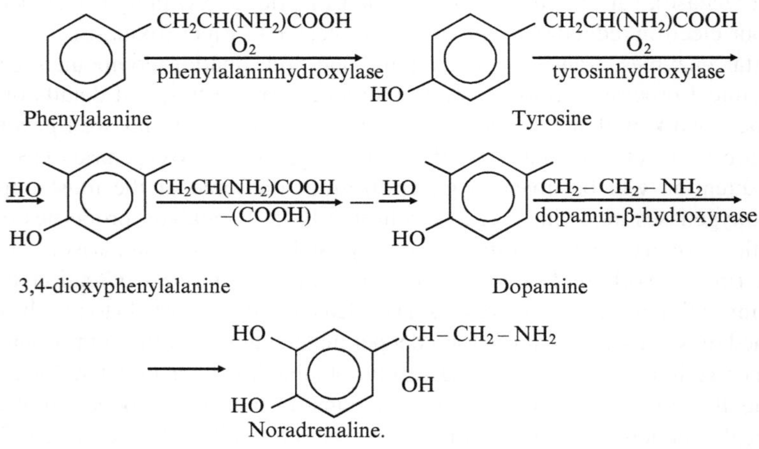

It is evident from this chain of catalytic reactions that noradrenaline biosynthesis proceeds (this substance is a mediator of nervous pulses in the synapses of sympathetic nervous system). Change in noradrenaline contents in cerebrum under the influence of some phsychothropic drugs, in particular its excess accumulation, results in inhibition of dopamine hydroxylation. Imitation (mimetics) of some of these enzymes (tyrosine, dopamine-β-hydroxylase) will be considered below.

Selective activation of C–H bonds under relatively mild conditions (temperature close to room temperature, usual oxidants of oxenoids type) became possible only recently. Such a possibility is caused by a number of reasons, the main one being the use of suitable catalytic systems with metal complexes, the interaction of which with oxenoids results in active particles formation, capable of introducing oxygen atom into the C–H bond of alkanes. Examples of such systems are enzymatic systems of cytochrome P-450 type, MMO, etc. Such enzyme modeling is the concrete embodiment of the IVth conceptual system of "evolution catalysis" [200].

One of the most significant differences between enzymes functioning in usual catalytic chemical reactions and those with metal complexes' participation is their exclusive selectivity caused by the influence of protein surroundings and rigid enough organization of their active centers. Such organization provides both selectivity and stability of enzyme activity. Another peculiarity of enzymes consists in their ability to work according to the conjugation principle, i.e., one active compound can be involved in two reactions. Optimization of two conjugated reactions can be reached on condition that active particles are formed in primary reaction, initiating the secondary reaction in which they are consumed. After this, one of the products enters into the reaction; as a result, active particles involved in the final products of primary reaction formation will be regenerated [201].

One of the possible conjugation examples is mitochondrial processes of oxidative phosphorylation [202]:

respiration process

$$\left.\begin{array}{l} CH_3COOH + 2o_2 \longrightarrow 2CO_2 + 2H_2O \\ H_2O \overset{ATP-ase,}{\leftrightarrows} H^+ + OH^- \end{array}\right\} \text{primary reaction}$$

oxidative phosphorylation process

$$ATP^{3-} + P^{3-} + 2H^+ = ATP^{4+} + H_2O\} \text{ secondary reaction,}$$

where $P^{3-} = PO_4^{3-}$.

Primary and secondary reactions are conjugated. In the primary reaction, a highly reactive intermediate compound will be formed, H^+ ion as oxidant causing the reaction of ATP oxidation. H^+ ion in its turn participates in both reactions' product formation (water molecules). Water molecule formed at once will be dissociated into H^+ and OH^- ions on mitochondrion membrane. Thus, OH^- anion will be desorbed into cytoplasm, and H^+ ion into matrix, where it takes part in water formation in the final stage of the respiration process. Conjunction between respiration and oxidation is realized by ATP-ase whose role consists in H^+ ion generation (from cytoplasm) with subsequent reaction of ATP synthesis [201].

This example illustrates, on the one hand, the principles of conjugation action on enzyme work, and points out and, on the other, possible ways of enzymatic activity modeling with more simple systems.

3.2.1.1 Mechanism of Alkanes Hydroxylation with Cytochrome P-450

It is known that monooxygenases (in particular, cytochromes P-450 representing the whole enzyme class) realize one oxygen atom entrance (oxene) from dioxygen into C–H bonds of alkanes (hydroxylation) or C = C bonds of olefins (epoxidation). In the first case, it corresponds to the reaction

$$RH + O_2 + 2e^- + 2H^+ \longrightarrow ROH + H_2O$$

in which source of electrons is conjugated oxidation of NADH or NADPH. These electrons, reducing metal ions of enzyme's active center, provide the possibility of its joining to the molecular oxygen and, thus, its reactions with alkane activation. Activation of O_2, resulting in metal oxo- and peroxo complexes formation, is one of the main functions of monooxygenases [201].

In almost all living organisms (from bacteria to mammals), hydroxylation of C–H bonds in alkanes is catalyzed by cytochrome P-450-monooxygenases (about 500 of their representatives are known), containing iron protoporphyrin in an active center, in which iron is bound to protein surroundings by means of a cysteine residue and a hydroxyl group of a certain part of the protein (S^- and OH^- in Fig. 3.14, where RH

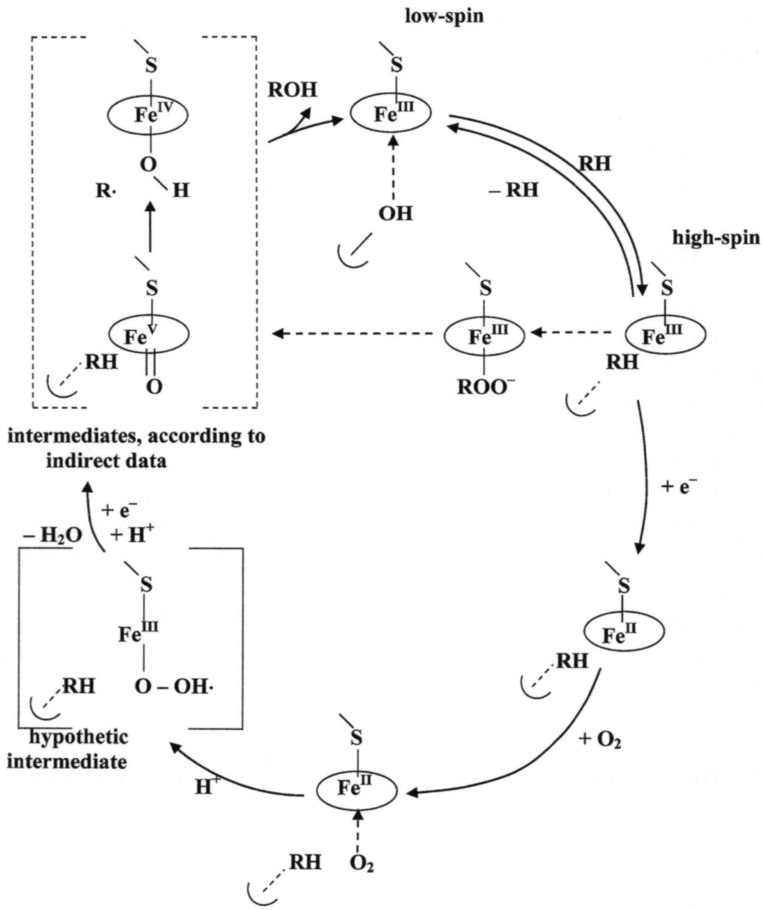

Fig. 3.14 Mechanism of alkane RH hydroxylation with cytochrome P-450 active center

is localized alkane on hydrophobic interaction with the protein part of the enzyme and hydrogen bonds formation near the active iron–porphyrin center;–porphyrin. It seems that the reason for porphyrin molecule choice as the main ligand of iron(III) in cytochrome P-450 active center is its stability and high conjugation of double bonds, which makes it possible to pass easily the part of electron density on iron or vice versa, and thus to create suitable redox potential. However, some monooxygenases able to hydroxylate C–H bonds of alkanes contain non-hemic iron–protein active centers. Such monooxygenases involve MMO, etc., [203–207].

Cytochromes P-450 fulfill various functions in steroids, aliphatic acids, and prostagladins biodegradation, drug metabolism, and removal of a number of toxic compounds from the organism. Equally with R–H bonds hydroxylation, they are able to transfer oxene on multiple bonds, aromatic rings, and heteroatoms of various substrates [208–211].

Despite the different origin of cytochrome P-450-monooxygenases, the mechanism of alkanes hydroxylation by them is the same.

In the initial state, two Fe(III) porphyrin complexes are in equilibrium: six-coordination (low spin) with cysteine (S–), and ligands containing hydroxyl groups (–OH), and five-coordination (high spin) with one axial ligand [212]. Alkane RH fixation on the protein hydrophobic part near heme results in an equilibrium shift in the direction of the five-coordination high-spin complex. The latter will be reduced by one-electron action with NADPH or NADH (by means of the electron transport chain). Its interaction with O_2 yields a new six-coordination low-spin complex, but this time an Fe(II) one. These first relatively stable complexes were synthesized as model intermediates and have been studied experimentally [213–215].

On further protonation, occurence of a hypothetical rather short-living intermediate (within dotted frame) is supposed, which cannot be detected by spectral methods and which is transformed, joining the electron, into short-living high-spin iron (IV) complex, oxygenating alkanes (within stroke frame). Its existence is supposed to proceed from indirect data only. Such highly reactive particles, due to their low concentrations, could have been studied in the tests, in particular, by the substrate selectivity in forming products (stereo- and radioselectivity) and the isotope effect [216]. This oxo-complex is formally $PFe^{V} = O$ (a similar model complex can be obtained and studied). Its properties correspond to the π-cation radical of iron porphyrin complex ($P^{\bullet+}Fe^{IV} = O$), as soon as the electron of the porphyrin ring passes to iron(V) (these two identical notions are often used in the literature). Thus, dioxygen, but not substrate, will be activated by particle $P^{\bullet+}Fe^{IV} = O$ formation.

The mechanism of alkanes hydroxylation with $PFe^{V} = O$ complex (proceeding from numerous experimental data) implies, first, hydrogen atom detachment with this complex with pair [PFe^{IV}–OH, $R^{\bullet}$] generation within the enzyme's cell, and then rapid recombination of $R^{\bullet}$ with PFe^{IV}–OH, yielding the reaction products PFe(III) and R–OH without $R^{\bullet}$ radical coming out into the bulk.

However, details of the recombination mechanism and product formation are still not quite clear, and there are various assumptions regarding this case [212].

Recent research on reaction radioselectivity of methyl cubane interaction with *tert*-butoxy radical enzymes, such as cytochrome P-450 and MMO, has confirmed that the reactionary oxygen atom in the active centers of enzymes is arranged close to the substrate's carbon atom at the moment of introduction. Thus, the ferryl particle $Fe^{V} = O$ under the influence of substrate will be turned, apparently, into the radical particle Fe^{IV}–$O^{\bullet}$, which interacts with substrate RH. Non-synchronic concept mechanism of oxygen atom detachment is supposed [217, 218]:

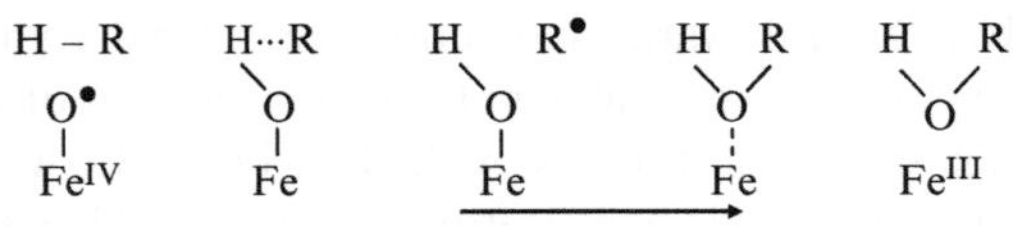

Thus, hypersensitive methods for radical detection were used, and it was found that the lifetime of $R^{\bullet}$ radical (shown with arrow) $\approx 1.5 \times 10^{-10}$ s (70 fs), and activation energy of oxygen atom detachment activation is equal to zero.

As the direct oxidant of substrate in cytochrome P-450, oxene complex of iron (V) $P^{\bullet+}Fe^{IV} = O$ (P-450 $Fe^{V} = O$) is being preferentially considered, similar to the active intermediate compounds, occuring on peroxidase functioning, although there are data available which do not comply with this assumption. This situation has been considered in detail using a lot of sources [219], although the nature of monooxygenating particles in cytochromes P-450 still remains disputable.

Under anaerobic conditions, cytochrome P-450 reacts with other donors of oxygen atoms (phenyl hypoiodite C_6H_5IO, hydrogen peroxide H_2O_2, hydroperoxides ROOH, and other oxenoids). Thus, the alkane hydroxylation process does not proceed by the long route (Fig. 3.14), but much shorter—through intermediate formation, resulting in the same high-valency iron oxoporphyrin ($PFe^{V}O$)–ferryl intermediate. This way is known as the "peroxide shunt" (in Fig. 3.14 this way is shown by dotted arrows).

Thus, for example, hydroperoxides ROOH can substitute the usual oxidant ($1e^- + O_2 + 1e^-$) with the intermediate. In such a shunting way, formation of I–$PFe^{V} = O$ ($P^{\bullet+}$ $Fe^{IV} = O$) compound is supposed:

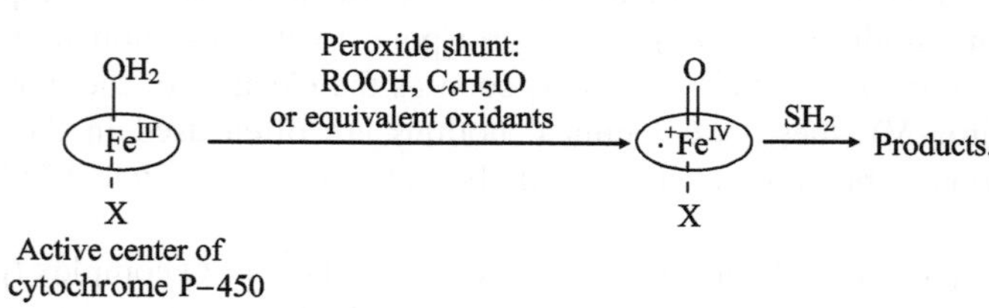

where SH_2 represents Oxidized substances.

According to some data [210], the structure of active intermediate compounds forming on PhIO interaction with cytochrome P-450 resembles that of peroxo particles (Fe–O–O), rather than iron oxene (Fe = O), and proceeding from this, it is supposed sometimes to use iron peroxide or hydroperoxide as intermediate in shunting systems. The possible mechanism of O–O bond breakage in ROOH (homolytic or heterolytic) in this way in the case of cytochrome P-450, as well as for catalases and peroxidases, will be discussed below.

It is to be noted that some of the reactions which are realized with Cyt P-450 can be realized with other enzymes as well. Such reactions involve, for example, oxidation dealkylation of *N*-alkylamines (horseradish peroxidase + H_2O_2; catalase + H_2O_2) [220, 221], or aniline oxidation (cytochrome C + cumile hydroperoxide) [222]. However, there are also differences in cytochrome P-450 and aforesaid enzymes, i.e., not all the reactions catalyzed with cytochrome P-450 are catalyzed with other enzymes (such are reactions of ether peroxide, O-dealkylation [223], and steroids oxidation [224]). The main difference seems to be in the hydroxylation reactions mechanism. The mechanism of P-450 action does not involve the stages with free radicals participation, passing out into the bulk. Mechanisms of other hemoproteins in this regard are more complex, and it is possible that in some cases [220, 221] such stages may be involved [26].

Besides possessing knowledge of alkanes hydroxylation mechanism with cytochromes P-450, to create efficient model systems it is important to consider their selectivity (regioselectivity and stereoselectivity), hydroxylation rate, products yield of this reaction, and catalytic system's stability.

Considering all experimental data on alkanes hydroxylation with cytochromes P-450, the model system created on the basis of iron porphyrins must correspond to the following requirements [211]:

1. With high valency iron oxo-compound ($Fe^V = O$) generation in model system, internal isotope effects must be high ($\kappa_H/\kappa_D \sim 10$), and hydroxylation of tertiary bonds C–H (as the least strong) must be predominating.
2. Hydroxylation rate must reach from several up to 100 cycles per minute (in cytochrome P-450, 6–70 cycles per minute) [225], yields calculated per oxidant consumed are 50–100%.
3. Having changed the nature and spacial arrangement for substitutes in porphyrin, one can reach certain arrangement of alkanes and subsequent limitation of intermediate radical $R^\bullet$ movement, which must provide good regio- and stereoselectivity. Besides these requirements, the most important for model system is its stability, oxidant cheapness, working capacity in many solvents, and the ability to introduce oxene in various substrates.

3.2.1.2 Modeling of Monooxidases Belonging to Cytochromes P-450 Class

All model systems of alkanes' hydroxylation and alkenes' epoxidation can be subdivided into two groups, namely, hematic and non-hematic systems. In cytochrome P-450 models containing porphyrin complexes of iron(III), the complexes can be the nearest to the structure of their active center. Iron(III), manganese(III), and chromium(III) complexes are the most efficient in alkanes hydroxylation among porphyrin systems. Their activity diminishes in the following sequence: PFe(III)

> PMn(III) > PCr(III). Therefore, further model systems will be considered, containing mostly iron(III) and manganese(III) porphyrin complexes.

Hematic Model Systems

*Model systems of M(*III)–P–RH–O_2 *type*. In these systems, M(III) is Fe(III), Mn(III), Cr(III), and some other metal ion, and P is more often tetraphenylporphyrin (TPP), tetra(*o*-fluorine-phenyl)porphyrin (TFPP), tetra(penta-fluorine-phenyl)-β -octabromporphyrin (TPPF$_{20}$-β-Br$_8$), tetra(3-chlorine-2,4,6-trimethylphenyl)-β-Octachlorporphyrin (Cl$_{12}$TMP), tetramezytilporphyrin (TMP), tetra(2,6-dichlor-phenyl)porphyrin (TDCPP), tetra(*p*-tolyl)porphyrin (TTP), tetra(penta-fluorine-phenyl)porphyrin (TF$_5$PP), tetramezytil-β-octabromporphyrin (Br$_8$TMP), tetra(*o*-pyvaloil-amidophenyl)porphyrin (TpivPP), tetra(*p*-hexadecyloxophenyl)porphyrin (THDOPP), tetra(2,4,6-trimethyloxyphenyl)porphyrin (TTMPP), tetra(2,4,6-triphenylphenyl)porphyrin (TTPPP), tetra(3-chlorine-5-sulfonato-2,4,6-trimeth-ylphenyl)-β-octachlorporphyrin (Cl$_{12}$TSMPP), tetra(4-pyridyl-phenyl)porphyrin (TpyPP), tetra(4-*N*-methylpyridyn)porphyrin (T4-MPyP), or mezo-tetra-*p*-metoxyphenylporphyrin (TMPhP).

Such a variety of substituted porphyrins used as ligands is due to the fact that iron(III) complexes with porphyrins or its simple substitutes such as iron(III) tetraaryl porphyrins represent the subjects to oxidative destruction in the reactions of substrates' hydroxylation. The use of halogen-substituted aryl groups introduced into porphyrin has sharply decreased catalyst destruction (increased its stability) and has resulted in a significant increase in rate, products yield, and selectivity [226, 227].

As reducers, borohydrides (such as $NaBH_4$ or (n-Bu$_4$N)(BH$_4$)), sodium ascorbate, hydrogen H_2 (adsorbed on colloidal Pt), zinc amalgam Zn (Hg), and others were used, and as oxidant, O_2 was used. Thus, as catalysts Mn(TPP)Cl, Fe(TpivPP)Cl complexes were used with some additives. All these and similar systems hydroxylated or epoxidized various substrates (cyclohexane, cyclohexene, and 1,2-dimethylcyclohexane [228]) with the formation of appropriate alcohols, ketones, and epoxides as products [212, 229–234]. For cyclohexane hydroxylation, other iron porphyrins were used as well in acetonitrilic solutions in the presence of carbonic acids (proton donor) and zinc powder (as reducer) in the presence and absence of viologen, making easier electron transfer from the reducer to TpivPPFe(O_2). Without methylviologen, dioxygen activation in these systems proceeds almost like in cytochrome P-450 [235].

For oxene formation from dioxygen and its transfer to iron porphyrin complex, it is important to bind the residue of water (O^{2-}) with some acceptors. A suitable acceptor is acetic anhydride. This anhydride interacts with O_2 and iron porphyrin forming under −70°C complex PFe–O–O–Ac, which under -50°C will be transformed into $P^{\bullet +}Fe^{IV} = O$ [236]. Participation of this particle in the interaction with alkanes and olefins was proved by numerous research (especially spectral) [236–242].

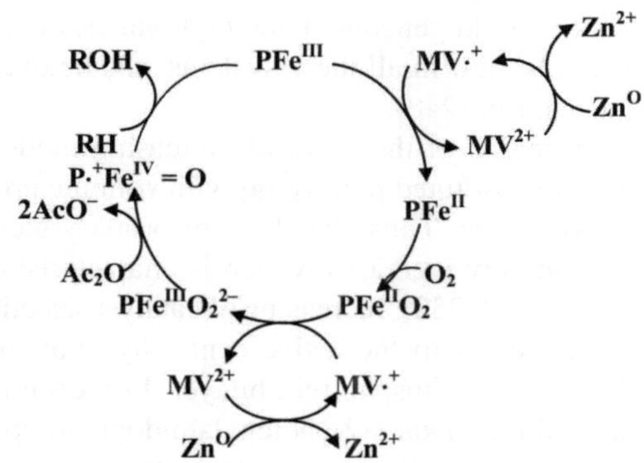

Fig. 3.15 Model system $PFe^{3+}–O_2–RH–Ac_2O–MV^{2+}–Zn^o$ functioning

One of the possible examples of such modeling catalytic systems is the system involving one or another iron porphyrin complex (PFe(III)), electrons carrier (methylviologen MV^{2+}), acetic anhydride (Ac_2O), O_2, substrate (RH), and zinc powder as reducer, functioning as shown in Fig. 3.15 [190].

Alkane hydroxylation in the presence of O_2 and electrons source proceeds under normal temperature and pressure. Reaction selectivity and rate are dependent upon the nature of the catalyst and electron and proton donor [216]. For example, by catalyst Fe(TPPF$_{20}$-β-Br$_8$)Cl using, isobutane hydroxylation (and that of alkanes containing tertiary C–H bonds), it proceeds under room temperature with the selectivity of 92% and about three cycles per minute, and catalytical activity is preserved within about 3 days [243]. The efficiency is much lower for secondary C–H bonds and disappears at all in the case of primary C–H bonds [219].

These model systems (especially in organic solvents), in general, qualitatively imitate the main reactions of cytochrome P-450. However, as a rule, products yield with regard to the reducer is low, compared to cytochrome P-450, because active oxene complex of iron porphyrin forming on hydroxylation will readily react with both substrate and reducer, which is in excess in the system. Thus, reactions of metal oxo-complex (Fe, Mn) run quicker with the reducer than with substrate [244, 245]. Usually, this yield calculated with regard to the reducer falls within the limits 0.01–0.5% for hydroxylation, and 0.1–0.5% for epoxidation in borohydric or H_2(Pt) systems [246]. Catalytic activity, i.e., number of cycles per minute, is also low. For example, in the system Mn(TTP)Cl–cyclooctene-O_2 with acylating agent $(PhCO)_2O$, the rate of entering electrons is too low, and its activity in alkanes epoxidation reaches only 0.03 cycles per minute [247]. Even the best model systems of iron(III) and manganese(III) in the presence of dioxygen are only approaching the appropriate indexes for cytochrome P-450 (for instance, by kinetic isotope effects); by other indexes, they are below the enzyme [212].

The main difference of model systems of the appropriate enzymes is low stability and selectivity of model catalysts, which is caused to a significant degree by radical chain reactions predominating in them, having low selectivity. The model systems of the basis of molecular oxygen and in the absence of reducers are still unknown, which would generate metal high-valency active oxo-compounds, reacting directly with alkanes. In all these systems, alkane oxidation proceeds by free radical chain mechanism [248].

A search of the ways of increasing model systems' selectivity resulted in the use of substituted porphyrins with volumic groups, causing steric obstacles near the active center. Thus, the share of primary alcohols and ketones in oxidation products sharply increases, which is characteristic for hydroxylation with cytochrome P-450 [249–252]. Increasing of catalyst selectivity and stability can also be achieved on transition to the active center hydrophobic surroundings by metaloporphyrin synthesis in phospholipid bilayer. In the same way, it became possible to synthesize Mn–tetrakis (*O*-cholenylamidophenyl)porphyrin, incorporated into the lipid membrane [253]. This incorporated complex catalyzes the reactions of alkanes and cholesteryne hydroxylation with dioxygen in the presence of ascorbic acid. This model system is stable, though rather complex. However, only little is known about model monoxygenase systems on the basis of manganese and iron porphyrin complexes, in which substrates would be biologically active natural compounds or medicine drugs. It was shown recently that model systems of PM–sitosterol type (or N,O-diacylsolysidine)–O_2–$NaBH_4$ (where P represents various derivatives of tetraphenylporphyrin, and M represents Mn or Fe) hydroxylate with high stereo- and regioselectivity the aforesaid steroid olefins (product yield regarding the initial substrate reaches 40–100%). Thus, the product of sitosterol hydroxylation is its 5-α-hydroxyderivative:

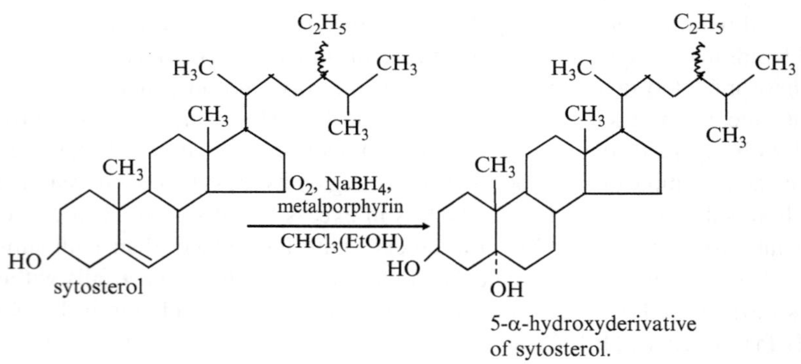

It is supposed that hydroxylation mechanism in these model systems is similar to the appropriate mechanism of cytochrome P-450 functioning, namely, reduction of manganese(III) porphyrin complex to manganese(II), coordination of O_2 to it, and formation of manganese(IV) high-valency porphyrin oxo-complex, which turns out to be active in oxene transfer on substrate.

Under the same conditions as in the system TMPhPMn(III)–O_2–$NaBH_4$, oxidation of polynuclear aromatic hydrocarbons proceeds in the mixture $CHCl_3$/EtOH (3:2), in particular, 2,6-dimethylnaphthalene with 61% conversion of initial substance. Thus, formation of seven main products proceeds [254]:

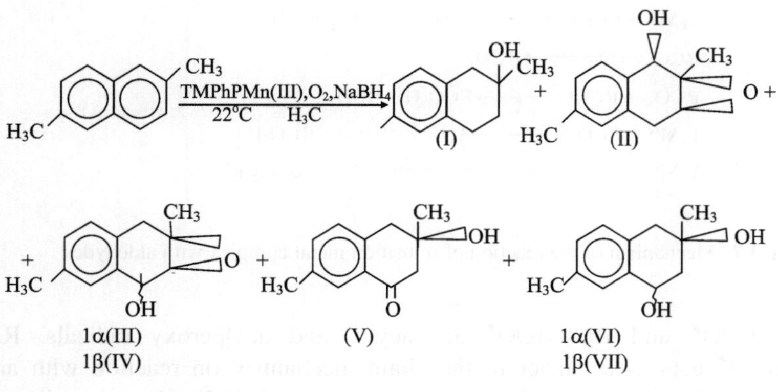

In this system, it is possible to obtain polyoxidized tetramine derivatives; by changing experimental conditions, the reaction can be directed mainly by way of epoxyalcohols or 1,3-diols formation. Stereodirection of the process can also be regulated. Stereoschematically individual epoalcohol (III) and 1,3-diol (VII) can be obtained with yields of ~60%. Similar oxidation of substituted naphthalenes can be used in the synthesis of bicyclic sesquiterpenoids and other cyclic natural compounds [254].

The rate of reaction, product yield, and the number of turnovers in the case of model systems with Mn(III) complexes participation are much higher than for the same Fe(III) complexes. It is caused by the difference between metalporphyrin orbital structures of Mn(IV)(TPP) and Fe(IV)(TPP) high-spin states. The results of models research reveal the possibility of catalytic metalporphyrin systems usage for steroid olefin modification and various hormonal drugs production. However, in many cases, such as, for example, in the reactions of olefins epoxidation, using dioxygen as oxidant is not always expedient. In this case, other oxidants are used such as hydrogen peroxide (H_2O_2), organic hydroperoxides (ROOH), or phenyl hypoiodite (PhIO), as on using O_2 as oxidant, these reactions either do not proceed, or side products are obtained. However, if aldehyde is added to the reaction model system containing transition metal complexes, dioxygen epoxidizes olefins with high yields, with simultaneous aldehyde oxidation to carbonic acids [255–259]:

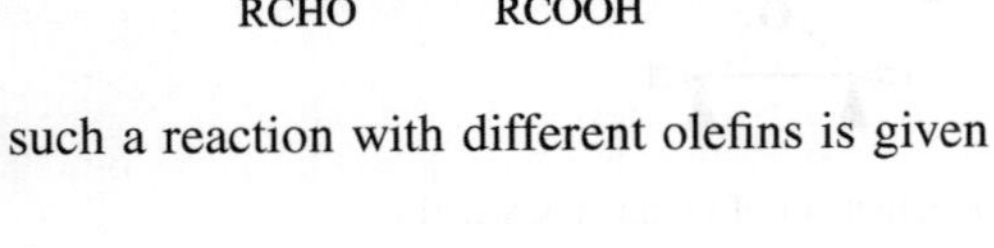

Mechanism of such a reaction with different olefins is given in [260, 261] (see Scheme 3.7):

$$\left.\begin{array}{l} L_nM^{z+} + RCHO \longrightarrow L_nM^{(z-1)+} + RCO\cdot + H^+ \\ RCO\cdot + O_2 \longrightarrow RCO_3\cdot \\ RCO_3\cdot + RCHO \longrightarrow RCO_3H + RCO\cdot \\ L_nM^{z+} + RCO_3H \longrightarrow L_nM^{(z+2)+}=O + RCO_2H \\ L_nM^{(z+2)+}=O + \rangle\!=\!\langle \longrightarrow L_nM^{z+} + \text{epoxyde} \end{array}\right\} \text{(route A)}$$

Scheme 3.7 Mechanism of the reaction of transition metal complex with aldehydes

where $RCO^{\bullet}$ and $RC(O)OO^{\bullet}$ are acylic and acylperoxy radicals. Radical $RC(O)OO^{\bullet}$ acts as a carrier in the chain mechanism on reaction with another aldehyde molecule with the formation of peroxyacid (RCO_3H) and acylic radical. Formation of oxidation product (epoxide) proceeds through the high-valency metal-oxo intermediate.

However, this mechanism scheme (route A) is not the only one possible. On olefins' oxidation (cyclohexane, cys-stybenem norbornene) with dioxygen and aldehyde in the presence of metal complexes [Cr(TPP)Cl, Mn(TPP)Cl, Fe(TPP)Cl, Co(TPP), Ni(TPP), Cu(TPP), Zn(TPP), $Mn(cyclam)^{2+}$, $Fe(cyclam)^{2+}$, $Co(cyclam)^{2+}$, $Ni(cyclam)^{2+}$, $Cu(cyclam)^{2+}$, or $Zn(cyclam)^{2+}$] in $Cl–CH_2–CH_2–Cl$ solution for metal porphyrinates and in CH_3CN solution for metal cyclam complexes, the mechanism scheme can be different (route B). The general mechanism of olefin epoxidation with dioxygen and aldehyde is supposedly as Scheme 3.8 [261].

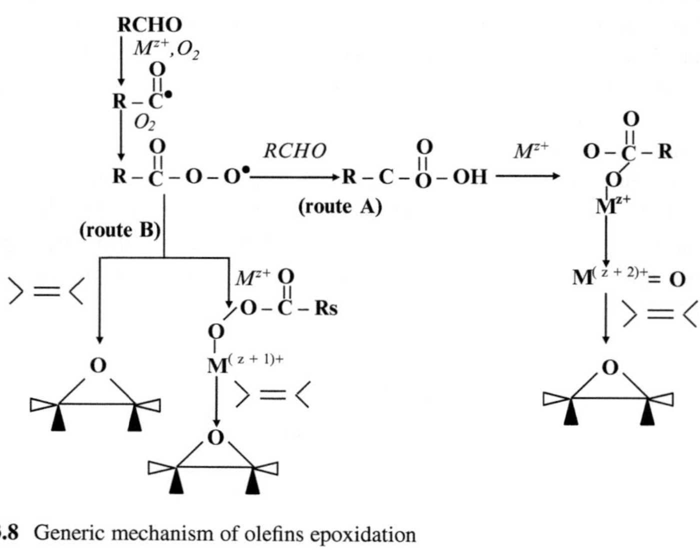

Scheme 3.8 Generic mechanism of olefins epoxidation

In this mechanism scheme, route B is usually more favorable than route A in the experimental conditions [261]. Route B is realized through acyl peroxy radical

$$(R - \overset{\displaystyle O}{\overset{\|}{C}} - O - O^{\bullet})$$

interaction either with olefin with the direct epoxide formation, or with metal complex with the formation of the first metal peroxide rather than that of epoxide. Metal peroxide (by route B), having interacted with olefin, is directly transformed into epoxide, or (by route A) forms metal-oxo particle (through the O–O bond of peroxide breakage) which, having reacted with substrate, yields epoxide. Thus, the competition is observed between the reaction's routes A and B so that epoxide formation is not the direct proof of high-valency metal-oxo complexes' presence, as metal alkyl peroxides [262] and metal peroxy acids [263] can transform olefins into epoxides.

An example of route B realization (Scheme 3.8) is oleic acid epoxidation with oxygen, with good yield in the presence of benzaldehyde and catalyst Co(TPP). The result of this process can be used in commercialapplicatins, as the resulting epoxide is a good stabilizer of polyvinylchloride and other polymers [264].

As has already been mentioned, hydroxylation and epoxidation with model systems M(III)–P–RH–O_2 are not very efficient for many such systems. Photocatalytic activation is a new method of alkenes activation. Really, interaction of various alkenes with dioxygen in benzene or toluene (but with admixture of some water) is photocatalyzed with iron(III) complexes with 5,10,15,20-tetraarylporphyrins, namely, iron(III) chloro (tetraphenylporphyrinate) (PFeCl) [265]. In the general case of such photocatalytic ($\lambda = 350$–440 nm) substrate (SH) oxygenation in the presence of free radicals $Cl^{\bullet}$ trap, the series of reactions proceed as follows:

$$PFe^{III}Cl \xrightarrow{h\nu} PFe^{II} + Cl^{\bullet}$$

$$2PFe^{II} + O_2 \longrightarrow PFe^{III} - O - O - Fe(III)P$$

$$PFe^{III} - O - O - Fe(III)P \longrightarrow 2PFe^{IV} = O$$

$$PFe^{IV} = O + PFe^{II} \rightleftarrows PFe^{III} - O - Fe(III)P$$

$$PFe^{IV} = O + RH \xrightarrow{h\nu} PFe^{III}OH + R^{\bullet}$$

Further oxygenation can proceed in various ways depending on alkenes used and reaction conditions [265]:

1. By alkyl hydrogen removal from the substrate and subsequent recombination of hydroxyl with alkyl radical within the cell of porphyrin complex with $R^{\bullet}$:

$$PFe^{IV} = O + RH \longrightarrow [PFe^{III} - OH \ldots R^{\bullet}] \longrightarrow PFe^{II} + ROH$$

with enantioselective formation of allyl alcohols.

2. By diffusion of alkyl radicals $R^{\bullet}$ from $[PFe(III)^{-}OH\ldots R^{\bullet}]$ and their reaction with O_2 yielding various products. Epoxides were formed from intermediate hydroperoxide radicals (or hydroperoxides) by means of oxygen atom transfer either from peroxyl radicals or from porphyrin complexes.
3. Direct epoxidation of deformed alkenes with oxoiron(IV) porphyrin complexes proceeds similarly with the reaction $P^{\bullet+}\ Fe^{IV}=O$, although with less reactivity in the sense of oxygen atom transfer compared to $P^{\bullet+}\ Fe^{IV}=O$.

Unlike less reactive $P^{\bullet+}Fe^{IV}=O$, which mostly epoxidizes alkenes, its $PFe^{IV}=O$ complex reduced in the one-electron way realizes a wide spectrum of oxygenation reaction routes, like microsomal cytochrome P-450, i.e., it occupies an intermediate position in the reactions of cytochrome P-450 [265].

Synthetic iron complexes with porphyrins are widely used as models of catalysts of olefins oxidation, with the scope to understand in detail the mechanisms of enzymatic oxidation processes [150,266,267]. Such catalysts are model compounds of iron(III) porphyrins, interacting with peracids at 78 K [268]:

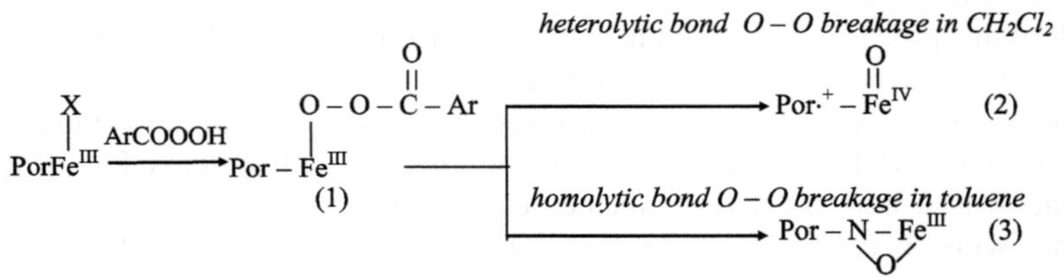

where Por = a: TMP = 5,10,15,20-tetramesitylporphyrin;

B: TDCPP = 5,10,15,20-tetrakis(2,6-dichlorphenyl)porphyrin;

$X = OH^{-}$ or $RCOO^{-}$; ArCOOOH = t-$BuCH_2COOOH$, etc.

Thus, porphyrin complex of acylperoxoiron(III) will be formed [(TMP)Fe(III)(RCO_3)] (1a), which on O–O bond heterolytic breakage in CH_2Cl_2 will be transformed into oxo-ferryl porphyrin π-cation radical [$(TMP^{\bullet+})\ Fe^{IV}=O$] (2a), and in toluene, (TMP)Fe(III)N-oxide will be obtained (3a) [269]. Intermediate complexes of $Por^{\bullet+}\ Fe^{IV}=O$ type (2) even at –78°C will oxidize olefins into epoxides, while complex (3) does not realize this reaction. However, it was proved that iron porphyrin complex (1) can catalyze epoxidation, not involving in the process the highly oxidized iron intermediates of type (2), as can be seen from Scheme 3.9

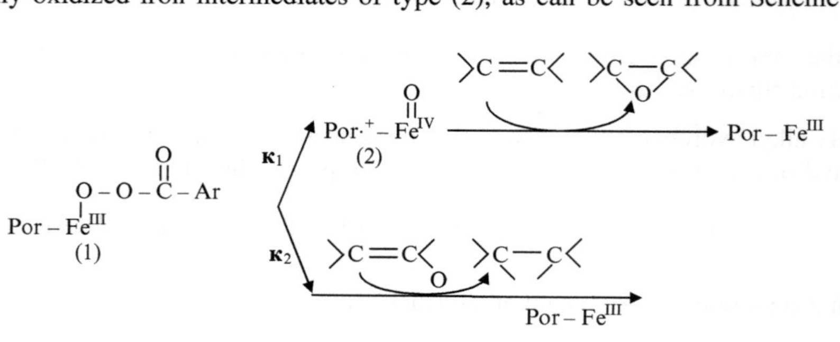

Scheme 3.9 Catalyzing of epoxidation with iron–porphyrin complex

[268]. The process route through the formation of (2) or directly from (1) on higher temperatures will be determined by the appropriate rate constants k_1 and k_2.

Model systems of Mn(III)–P–RH–H_2O_2(ROOH) type. Hydrogen peroxide is a stronger oxidant than molecular oxygen. It is easily accessible and cheap, and it was expected that model systems with its participation would be efficient enough. However, in a system such as Fe(TDCPP)Cl–cyclooctane–H_2O_2, only rapid H_2O_2 decomposition proceeds, hydroxylation almost does not proceed, and epoxidation of initial substrates proceeds only to a low degree (imidazole addition to this system increases the epoxide yield only slightly in relation to the oxidant) [212]. Substitution of Fe(TDCPP)Cl by Mn(TDCPP)Cl in this system in the presence of imidazole results in considerable increase in epoxide yield, and slightly increases alcohol (cyclooctanol) and ketone (cyclooctanone) yields. Model system Mn(TDCPP)Cl–H_2O_2–imidazole was able to oxidize various alkanes (cyclohexane, cyclooctane, ethylbenzene, tetramine, etc.). On separate cyclooctene and cyclohexane oxidation (substrates were taken in excess), products yield reached 95 and 60%,respectively. Imidazole, apparently, contributes to the rapid breakage of the O–O bond in H_2O_2 (due to the formation of manganese–imidazole porphyrin complex with subsequent transformation of the latter into the appropriate Mn^V-oxo intermediate) and decrease of H_2O_2 decomposition to O_2 and H_2O.

The above-mentioned results show that iron complexes with various porphyrins are of low efficiency as model systems. They act somewhat better in alkenes epoxidation with hydrogen peroxide. Thus, epoxidation can proceed by the various reaction routes [270]: homolytic with $OH^{\bullet}$ generation, heterolytic with oxoferryl particles formation, and catalase with O_2 and H_2O formation [270, 271].

Besides, model systems with H_2O_2 are of low selectivity. Thus, in the system Mn(TDCPP)Cl–imidazole–H_2O_2, cyclohexane is oxidized to alcohol and ketonein the ratio 3:1 [272], and in the system Fe(TPP)Cl–cumylhydroperoxide in the ratio 2:1 [273]. However, the mechanism of alkanes hydroxylation is different to that in cytochrome P-450 (active oxidation particle is not joined to metal—it appears to be radical cumyl–$O^{\bullet}$ or cumyl–$OO^{\bullet}$, obtained from hydroperoxide) [274]. Thus, iron(V) oxoporphyrin will be formed as a preliminary, but it will react sooner with hydroperoxide than with alkane [275]:

$$\mathrm{PFe(III) + ROOH \longrightarrow PFe(IV) - OH + RO^{\bullet}}$$

$$\mathrm{PFe(III) + ROOH \longrightarrow PFe(V) = O}$$

$$\mathrm{PFe(V) = O + ROOH \xrightarrow{-ROH} PFe(IV) - OH + ROO^{\bullet}}$$

This results in many products, and reduces the selectivity.

Addition of imidazole (Im) in catalytic amounts significantly improves alkenes epoxidation (in particular, epoxide yield) in both systems of Fe(TPP)Cl–RH–cumylhydroperoxide type [276], and mostly in the systems of Mn(TPP)Cl–RH–cumylhydroperoxide (or *tert*-butyl-hydroperoxide) type [276, 277]. The possible explanations of imidazole influence can be that the more probable becomes heterolytic breakage of O–O bond in ROOH (it will contribute to high yield of

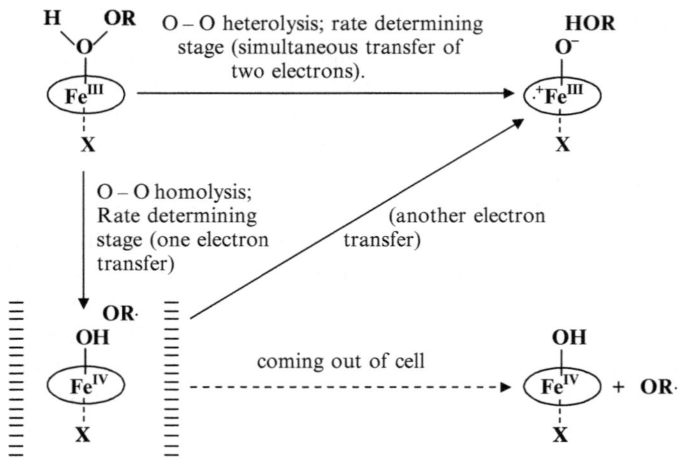

Fig. 3.16 Mechanism of bond O–O heterolysis or homolysis in ROOH

PFe(V) = O compared to $RO^{\bullet}$), formed intermediate ImPFe(V) = O is less reactive, compared to ClPFe(V) = O, in relation to ROOH rather than to alkene, and manganese complex catalyzes oxygen atom transfer from ROOH on imidazole, the product forming can transferring oxene on alkene [211].

It was also recently discovered [278] that porphyrin modification, namely, thyol anion (–S–CH_2–O–C_6H_4–OCH_2–) joining to iron porphyrin, results in considerable increase in product yield (from 11% with Fe(TPP)Cl to 88% with joined thyol anion), and also in the selectivity of primary alcohol formation by adamantane oxidation with μ-chlorperoxybenzioc acid.

H_2O_2 substitution with hydroperoxides results in their deterioration with iron(III) porphyrinate complexes on epoxidation as a result of homolytic decomposition of ROOH with the formation of iron(V) oxo-porphyrinate complex and alkoxy radical $RO^{\bullet}$[269, 279]. The problem of revealing the epoxidation mechanism is complex. Numerous works have been devoted to this problem; its critical review is given in [280]. As model water systems [280], the following were studied: P(1)Fe(III)$(X)_2$–ROOH, P(2)Fe(III)$(X)_2$–ROOH, and P(1)$Mn^{III}(X)_2$, P(2)$Mn^{III}(X)_2$, where P(1) is mezo-tetrakis (2,6-dimethyl-3-sulfonatophenyl)porphyrin, X is H_2O or OH^- axial ligands, and ROOH is $(Ph)_2(CH_3OCO)COOH$, $Ph(CH_3)_2COOH$, or *t*-BuOOH.

Hydroperoxide interaction with catalysts Fe(III) and Mn(III) is possible by means of homolytic or heterolytic breakage of RO–OH bond (Fig. 3.16) [281].

The yield of reaction products of ROOH interaction with calalysts Fe(III) and Mn(III) testifies that the rate-determining stage is homolytic breakage, although heterolysis of this bond can also proceed in the case of acyl peroxides and peroxides in which the RO^- group is easily detached [281, 282].

Certain arguments have been presented in favor of heterolytic breakage of O–O bonds on H_2O_2 or *t*-BuOOH interaction with iron(III) tetrakis (pentafluorophenyl)-chloride and other electronegative substituted porphyrins in dichloromethane/methanol mixtures (1:3) in cyclooctene, and norbornene epoxidation with high epoxide yield of 60–100% [270].

Proof of O–O bond heterolysis is the fact that epoxynobornane yield is obtained with the same ratio of eco/endo-isomers as on using pentafluoorphenyl hypoiodite as oxidant, which forms intermediate $[P^{\bullet +}Fe^{IV} = O]^+$ characteristic of heterolysis. Thus, these and other proofs are given of hydrogen peroxide and hydroperoxides heterolysis in proton solvents.

Provided that iron(III) tetraphenylporphyrin chloride is the epoxidation catalyst, the process proceeds with low epoxide yield and low stereospecificity. Introduction of electronegative substituents into porphyrin in the case of iron(III)–tetrakis(pentafluorophenyl)-chloride complex promotes epoxides yield and increases their stereospecificity, due to the fact that the oxene particle $[P^{\bullet +}Fe^{IV} = O]^+$ formed will preferrably interact with alkene, but not with hydroperoxide ROOH.

Intermediate formation in model systems proceeds by means of both heterolytic disintegration of oxidants (H_2O_2, ROOH, etc.)

$$>Fe^{3+}X + ROOH \rightleftharpoons\ >Fe^{3+}OOR + HR$$

$$>Fe^{3+}OOR + HX \longrightarrow -\overset{\bullet+}{}Fe^{4+}{=}O + ROH\ (\text{or } RO^-) + X^-$$

and homolytic processes [200]

$$>Fe^{3+}OOR \longrightarrow Fe^{4+} = O + RO^{\bullet}$$

If $Fe^{3+}OOR$ is formed at low temperatures in protonic solvents in the presence of strong bases, intermediates formation ($-^{\bullet+}Fe^{4+} = O$) proceeds more often in homolytic way [283]. In hydroxyl solvents, oxidants are usually disintegrated in heterolytic way [284, 285].

Thus, the problem of O–O bond heterolysis or homolysis in ROOH is still in dispute, and it appears that in various cases and under different conditions (various solvents), the mechanism can be different.

In the system P(3)Fe(III)$(H_2O)_2$–t-BuOOH–water, where P(3) is synthesized model ligand mezo-5,10,15,20-tetrakis (2,4,6-trimethyl-3,5-bis($\alpha - N$, N, N-trimethyl-ammoniummethyl) phenyl) porphyrin, on pH change from 0.9 to 12, various intermediates can be formed: P(3)Fe(III)$(H_2O)(t - \text{BuOOH})$, P(3)Fe(III)(H_2O) $(t - \text{BuOO}^-)$, and P(3)Fe(III)$(OH^-)(t - \text{BuOO}^-)$.

Proceeding from products yield and other data on all possible mechanisms of O–O bond breakage in t-BuOOH (heterolysis, radical chain, P(3)Fe^{II} and P(3)Fe(III) states formation according to Fenton's scheme, and homolysis), homolytic mechanism only fully explains all the experimental data obtained [281]. Its detailed scheme in the case of (3) Fe(III)$(X)_2$ interaction with $(CH_3)_3COOH$ looks as follows (Scheme 3.10 is taken from [281]):

In general, $PFe^{II}(X)_2$ (where P is P(1) or P(2)) interaction with $(CH_3)_3COOH$ looks as follows (Scheme 3.11) [280, 286–292].

Association:

$$P(3)Fe(X)_2 + (CH_3)COOH \longrightarrow \| P(3)Fe^{IV}(X)(OH) + (CH_3)_3CO\cdot\|;$$

Reactions in cell:

$$\{ \| P(3)Fe^{IV}(X)(OH) + (CH_3)_3CO\cdot\| \longrightarrow$$
$$\longrightarrow \| P(3)\cdot{}^{+}Fe^{IV}(X)(O) + (CH_3)_3COH \| \},$$
$$\| P(3)Fe^{IV}(X)(OH) + (CH_3)_3CO\cdot\| \longrightarrow$$
$$\longrightarrow \| P(3)Fe^{IV}(X)(OH) + (CH_3)_2CO + CH_3\cdot\|,$$
$$\| P(3)Fe^{IV}(X)(OH) + CH_3\cdot\| + H_2O \longrightarrow$$
$$\longrightarrow \| P(3)Fe^{III}(X)(H_2O) + CH_3OH \|;$$

Reactions outside the cell:

$$\{ P(3)\cdot{}^{+}Fe^{IV}(X)(O) + P(3)Fe^{III}(X)_2 \longrightarrow 2P(3)Fe^{IV}(X)(OH) \},$$
$$(CH_3)_3CO\cdot \longrightarrow (CH_3)_2CO + CH_3\cdot,$$
$$P(3)Fe^{IV}(X)(OH) + CH_3\cdot + H_2O \longrightarrow$$
$$\longrightarrow P(3)Fe^{III}(X)(H_2O) + CH_3OH;$$

Reactions of continuation:

$$P(3)Fe^{III}(X)(OH) + CH_3\cdot + H_2O \longrightarrow$$
$$\longrightarrow P(3)Fe^{II}(X)(H_2O) + CH_3OH,$$
$$2P(3)Fe^{II}(X)(H_2O) + O_2 \longrightarrow P(3)(X)Fe^{III}O - ODe^{III}(X)P(3),$$
$$P(3)(X)Fe^{III}O - ODe^{III}(X)P(3) \longrightarrow 2P(3)Fe^{IV}(X)(O),$$
$$P(3)Fe^{IV}(X)(O) + P(3)Fe^{II}(X)(H_2O) \longrightarrow 2P(3)Fe^{III}(X)(OH);$$

Hydroperoxide oxidation:

$$P(3)Fe^{IV}(X)(O) + (CH_3)_3COOH \longrightarrow$$
$$\longrightarrow P(3)Fe^{III}(X)(OH) + (CH_3)_3COO\cdot,$$
$$(CH_3)_3COO\cdot \longrightarrow 1/2O_2 + (CH_3)_3CO\cdot \longrightarrow (CH_3)_2CO + CH_3\cdot,$$
$$\{(CH_3)_3COO\cdot + CH_3\cdot \longrightarrow (CH_3)_3CO - O - O - CH_3\}.$$

Scheme 3.10 Mechanism of $P(3)Fe^{III}(X)_2$ with $(CH)_3COOH$

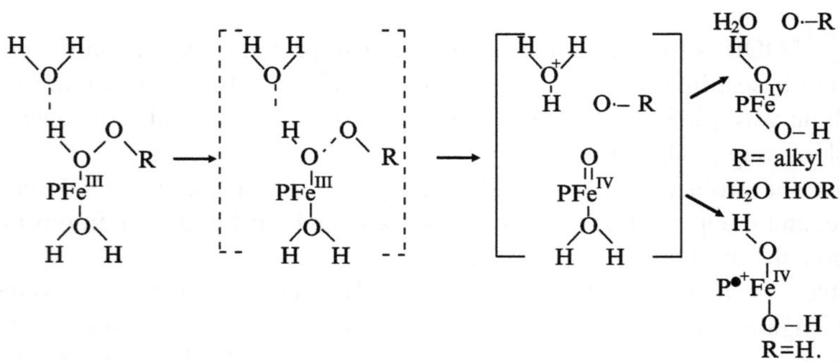

Scheme 3.11 Interaction of $PFe^{II}(X)_2$ with $(CH)_3COOH$

The mechanism of active epoxidizing particle interaction with olefins was also broadly discussed. It was supposed that the following intermediates are formed on such interaction, namely, methoxyethane (I) [293], a derivative of π radical cation alkene (II) [286, 289, 294], carbocation (III) [286, 289–292], carbonic radical (IV) [295–299], and a coordinated entrance of oxene into alkene double bond (V) [300–303].

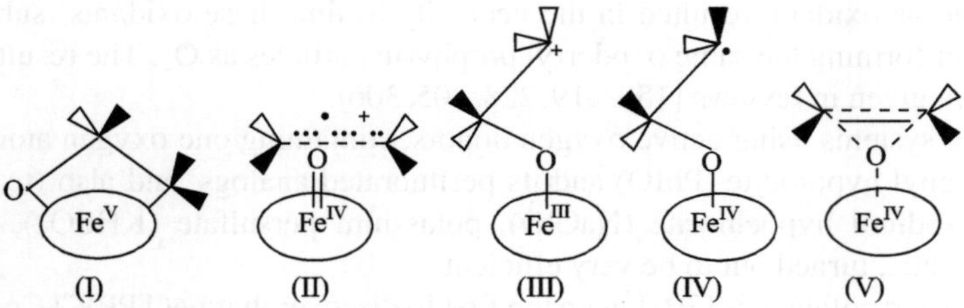

These intermediates' occurence was supposed proceeding from data on epoxidation products' stereochemistry, although stereochemistry relates only to the mechanism of epoxides formation, but not of other products. Critical assessment [280] of epoxidation probability through these intermediates has shown that the rate-determining stage is not the formation of intermediates I–IV, but that of the complex (V) with charge transfer (CTC) with further rapid change of oxygen spin state in $P^{\bullet+}Fe^{IV} = O$ from triplet to singlet, due to the mixing of P_x and P_y Oxygen orbitals with d_{xz} and d_{yz} iron orbitals [304]. Subsequent transformation will be determined by the relative rate of oxygen atom penetration from $P^{\bullet+}Fe^{IV} = O$ into alkene, and by the rate of alkene electrophyl attack by the same oxygen with the formation of carbocation as intermediate. Formation of the latter favors the interlapping of alkene p-orbitals with nitrogen atoms of porphyrin. Possible processes are shown in Scheme 3.12 [280].

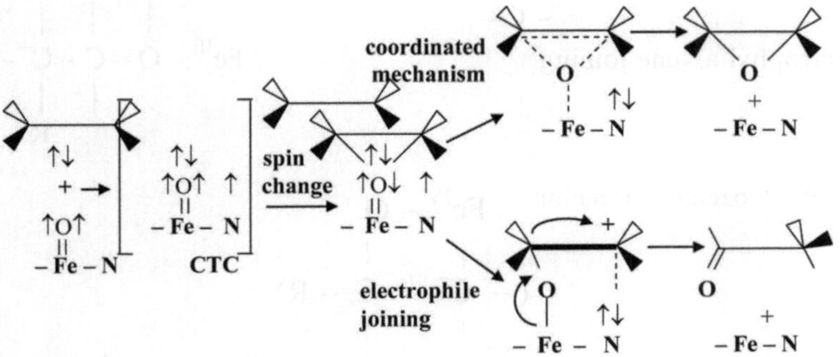

Scheme 3.12 Mechanism of active epoxidizing particle interaction with olefin

In general, processes following the rate-determining stage are caused by a number of factors, namely oxidation potentials of alkenes and active oxidant ($P^{\bullet+}Fe^{IV} = O$), reagents' steric and electron structures, and capacity of various substrates to be subject to re-construction [280].

Model systems of M(III)–P–RH–ArIO type (other oxidants with one oxygen atom). A low enough efficiency of model systems (by their oxidation rate, products yield, and especially, by regio- and stereoselectivity) in which O_2, H_2O_2, or ROOH were used as oxidant, resulted in the necessity to find these oxidants' substitutes, capable of forming the same oxoderryl porphyrin particles as O_2. The results of this search are given in reviews [150, 219, 228, 305, 306].

Model systems using active oxygen donors, containing one oxygen atom, especially phenyl hypoiodite (PhIO) and its perfluorated analogs, and also (to a lesser degree) sodium hypochlorite (NaClO), potassium persulfate ($KHSO_5$), amines, N-Oxide, etc., turned out to be very efficient.

Grows and colleagues [307] were the first to discover that Fe(TPP)Cl Complexes are able to catalyze oxene transfer from alkanes to PhIO in a rather similar way to cytochrome P-450 (high isotope effects $\kappa_H/\kappa_D \approx 13$ [245, 306, 308, 309]), bond C–H configuration is partly preserved [307, 310], and tertiary bonds C–H react first of all [308]. Thus, selectivity of Fe(TPP)Cl–PhIO system in alkanes oxidation is comparable with that of the enzymatic one. The process proceeds in an accelerated (shunted) way, as shown with dotted arrow in Fig. 3.14.

Olefin epoxidation in this system is stereospecific: from *cis*-olefins only *cis*-epoxides are obtained. In this system, oxoiron(IV) porphyrin π-cation radical $P^{\bullet+}Fe^{IV} = O$ is obtained, which is a substrate oxidant (like in cytochrome P-450) and which can in different ways interact with alkenes, generating intermediates due to direct oxene transfer

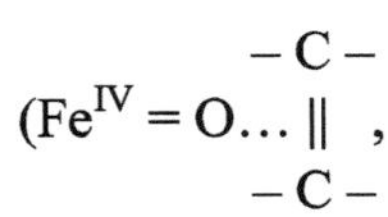, by free-radical 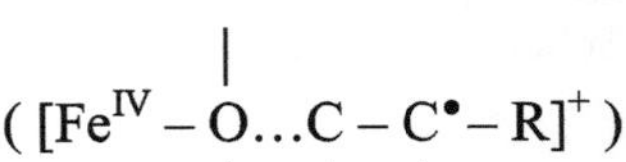

or electrophyl alkene joining 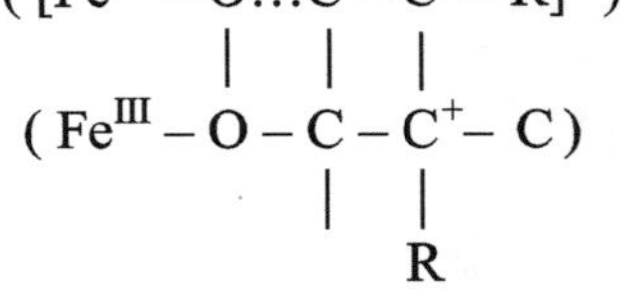

through Fe^{IV}-oxetane formation 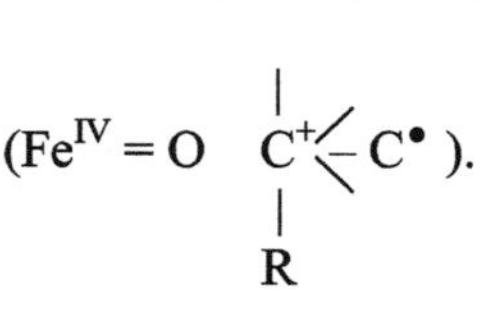

or through electron transfer with the formation of the alkene cation radical [150, 242]

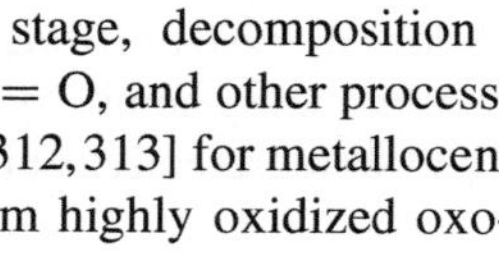

As the rate-determining stage, decomposition of Fe^{IV}-Oxanate to epoxide, electron transfer to $P^{\bullet+}Fe^{IV} = O$, and other processes were considered [150, 311]. Scheme 3.13 was proposed [312, 313] for metallocene intermediate formation [314] on oxygen atom transfer from highly oxidized oxo-iron porphyrinates to alkenes [306, 312, 313, 315, 316].

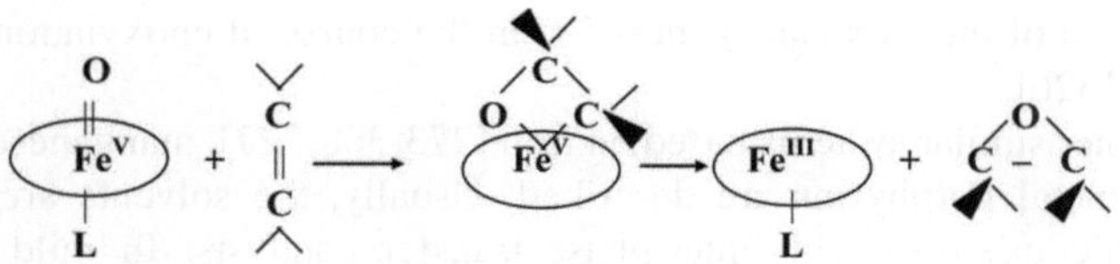

Scheme 3.13 Intermediate formation of metallocene

Formation of an iron intermediate complex was not confirmed by spectral methods (EPR, NMR) [315]. Distribution of reaction products served as proof of its formation.

However, simple iron tetraarylporphyrins (such as Fe (TPP)Cl) deteriorate on alkanes hydroxylation (oxidation destruction), i.e., catalyst stability is low. Introduction of halogen-substituted aryl groups (pentachlorophenyl [317], 2, 6-dichlorophenyl [318], pentafluorophenyl, etc.) into such porphyrins makes the catalysts with their participation much more efficient and stable. For instance, in the system Fe(TDCPP)Cl–C_6F_5IO–alkene, the initial rate of epoxidation reaches more than 300 cycles per second [319], and about 100,000 moles of epoxide are formed per mole of catalyst without its destruction [212]. Use of steric complicated catalysts, such as Mn (III)(TTPPP)Cl, results in the enhancing of regioselectivity and assymetry of alkanes hydroxylation. The highest selectivity is characteristic for the systems with PhIO and similar iron(III) porphyrinates. Formation of Fe^{IV}-Oxetane in the process of *cis*-cyclooctene epoxidation within the system $(F_{20}TPP)F(III)Cl–RH–C_6F_5IO$ is given in Scheme 3.14 [320, 321].

$$(F_{20}TPP)Fe^{III} + C_6H_5IO \rightleftharpoons (F_{20}TPP)Fe^{III} - OIC_6H_5,$$
$$(F_{20}TPP)Fe^{III} - OIC_6H_5 \longrightarrow (F_{20}\cdot^{+}TPP)Fe^{IV}O + C_6H_5I,$$
$$(F_{20}\cdot^{+}TPP)Fe^{IV}O + \text{alkene} \rightleftharpoons Fe^{IV}\text{-oxetane},$$
$$Fe^{IV}\text{-oxetane} \longrightarrow (F_{20}TPP)Fe^{III} + \text{epoxide}.$$

Scheme 3.14 Fe^{IV}-oxetane formation on epoxidation of cyclooctene

The last stage is rate-determining [321].

However, on a *trans*-cyclooctane and other alkenes oxidation study in similar systems, epoxidation was considered not as the result of iron-oxetane intermediate formation but as a result of electron transfer from alkene to $P^{\bullet +}Fe^{IV}O$ [242, 322–324]. In other case, when ethylene played the role of substrate for epoxidation, formation of an earlier biradical intermediate than iron-oxetane was supposed [325]. Therefore, the possibility of epoxidation reaction proceeding in different ways should be considered, especially as the proofs of metaloxetane formation are incomplete. This situation somewhat resembles the consideration of the oxyferryl ion occurrence probability [30]. If ironoxethane possesses high reactivity (hence, it is a short-living compound), it is hard or impossible to detect it experimentally.

Thus, the idea of metaloxetane generation in the course of epoxyiation remains to be debated [326].

Many other similar systems based on iron [273, 308, 327], manganese [328, 329], and other metal porphyrins are described. Usually, the solvents are organic or water-organic mixtures with inter-phase transfer catalysts. In mild conditions, oxidation of alkanes to the appropriate alcohols, ketones, and, sometimes, other substances proceeds with a rate close to the cytochrome P-450 oxidation rate and products yield of about 60–90% [216].

One of the prospective ways of olefin selective epoxidation with iodobenzene (PhIO), on using transition metal porphyrin complexes as model catalysts, is the creation of special, sterically complicated porphyrins—so-called porphyrins in the form of a "picnic basket" (PBP) [330]. PBP complexes have different sized rigid space and the appropriate "hand" (–R–) on the surface of the porphyrin ring. Such rigid conformation contributes to the formation of an active oxygenating (epoxidizing) center inside porphyrin superstructure, and its interaction with olefin results in selective epoxidation. Thus, large-sized cyclooctene will be epoxidized into acetonitrile with the rate constant much lower than that of small-sized acyclic olefins, like *cis*-2-octene (Fig. 3.17) [330].

Selectivity was also achieved by choosing a large enough axial ligand (L)—3, 5-di-*tert*-butyl phenoxide (OAr), which blocked access to the PBP open surface (to the "bottom" of the picnic basket from the outer side) (Fig. 3.18) [330].

Comparison of epoxidation selectivity of many pairs of olefins using various "picnic-basket" manganese(III) porphyrinates with the protective ligands OAr and usual manganese porphyrinates (with the ligands of tetraphenylporphyrin or tetramesitylporphyrin type) has revealed much higher selectivity (sometimes 1,000 times higher) of "picnic-basket" complexes of Mn(C_6PBP)(OAr) and

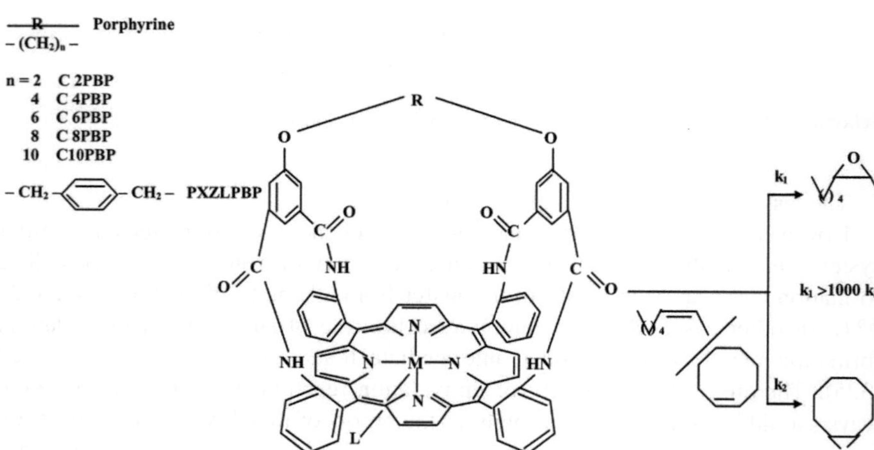

Fig. 3.17 Selective competing epoxidation of the pair of acyclic *cis*-2-octene and cyclooctene with iodosylbenzene in the presence of "picnic-basket" complex Mn(PXYLPBP)(OAr) [330]

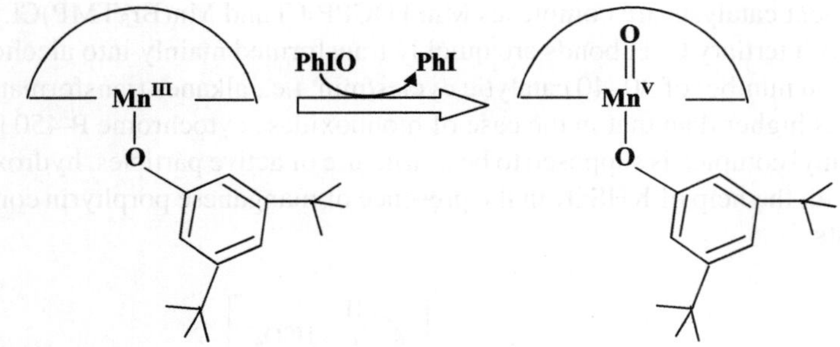

Fig. 3.18 Transformation of "picnic-basket" catalyst Mn^{III} with iodosylbenzene into $Mn^V = O$ in the presence of OAr [330]

Mn(PXYLPBP) (OAr) types. This reveals the possibility to predict both hydroxylation and epoxidation for a number of such synthetic model catalysts. Perhaps it will soon become possible to realize highly selective epoxidation of polyenes such as steroids and terpenoids using the collateral "picnic-basket" porphyrinates as ligand [310].

With phenyl hypoiodite as oxygen atom donor, another strong oxidant can be used in hydroxylation and epoxidation reactions—potassium monopersulfate ($KHSO_5$):

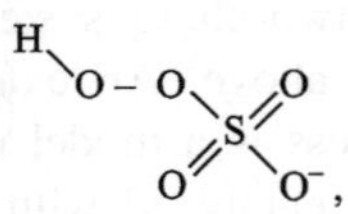

in which the O–O bond is asymmetric, and its length is 1.46 Å. On heterolytic breakage of the O–O bond, detachment of sulfate group (SO_4^{2-}) proceeds.

Olefin epoxidation proceeds in the two-phase system water/dichloromethane [329] as follows:

$$\text{Olefin} + KHSO_5 \xrightarrow[\text{agent of phase transfer, buffer } H_2O/CH_2Cl_2]{M^{III}(\text{porphyrin})Cl/4\text{-}R\text{-pyridin}} \text{Epoxide.}$$

Here, M represents Mn or Fe; TPP, TMP, TDCPP, Br_8TMP, F_{28}TPP, TPPS, TMP_yP mean dianions of meso-tetraphenylporphyrin, meso-tetramesitylporphyrin, meso-tetrakis (2,6-dichlorophenyl)porphyrin, meso-tetramesityl-β-octabromoporphyrinate, meso-tetrakis (pentafluorophenyl)-β-octafluorOporphyrin, meso-tetra(n-sulfonatophenyl)porphyrin, and meso-tetrakis (4-N-methylpyridinehumil) porphyrin, respectively; phase transfer agent—tetraalkylammonium salt; 4-R-pyridine—proximal ligand (usually pyridine derivative, R = Me or 4-*tert*-butyl), significantly increasing catalyst efficiency. Olefin epoxidation proceeds with high catalytic activity. Thus, cyclohexene epoxidation in the presence of Mn(TPP)(OAc)/4-*tert*-butyl-pyridine compound reaches 80% within 5 min with a circulation of 13 cycles/min. In these model systems, hydroxylation of tertiary, and to a lesser degree secondary C–H bonds of alkanes, also readily proceeds. The most efficient catalysts are complexes Mn(TDCPP)Cl and Mn(Br_8TMP)Cl. Hydrocarbons with tertiary C–H bonds are quickly transformed mainly into alcohols with a circulation number of 30–40 catalytic cycles/min, i.e., alkanes transformation rate is ten times higher than that in the case of monooxidase cytochrome P-450 [329].

Manganyl complex is supposed to be a structure of active particles, hydroxylating alkanes with the help of $KHSO_5$ in the presence of manganese porphyrin complexes as catalysts:

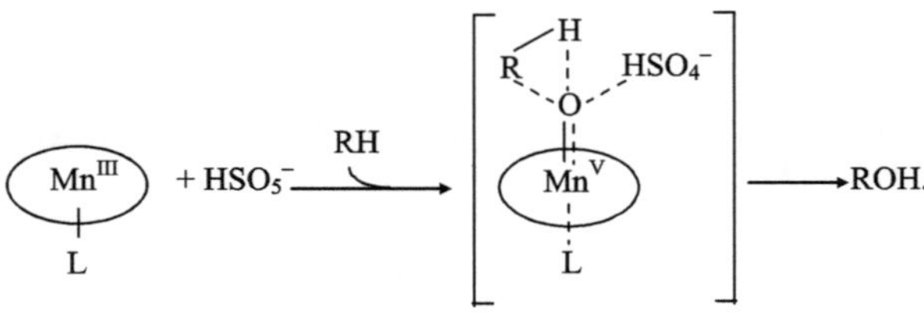

However, benzene and methane hydroxylation in these systems is still not very efficient. It appears that using such water-soluble asymmetric derivatives of peroxide (such as $KHSO_3$) could help to overcome the activation stage needed to generate highly oxidized metal-oxo complexes, efficiently realizing the processes of alkane and alkene hydroxylation and epoxidation [329].

Influence of catalyst stability, metal complex with porphyrin on catalytic cycle circulation number, as well as two-phase system influence on conversion and selectivity, has been mentioned above. An example of the influence of these parameters on the catalytic process is a model reaction of olefin epoxidation in two-phase system H_2O/CH_2Cl_2, catalyzed with manganese(III) complexes with various porphyrin derivatives. Oxygen-containing hypochloric acid (HOCl) was taken as oxidant. The most stable to oxidation, efficient, and selective turned out to be manganese(III) complexes (I–III) with various electron-attaining substituents in the porphyrin ring, providing steric protection of the catalyst's active center and its stability [331]:

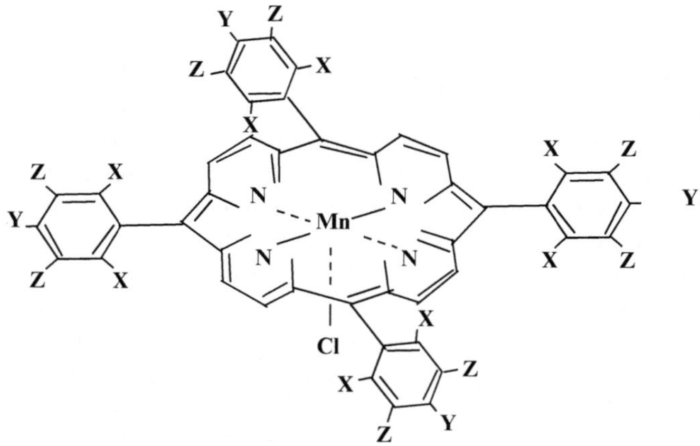

In complexes I–III, such substituents were, respectively:

	X	Y	Z
I	Cl	H	H
II	Cl	Cl	CH_3
III	Br	Br	CH_3

The epoxidation process is given in Fig. 3.19 [331].

At pH 9.5–10.5, maximal rates of alkenes rich in electrons (namely, cyclooctene) epoxidation were observed. Initial ratio L:P = 1:1, where L is *tert*-butyl pyridin or *N*-hexylimidazole. Stability of complex I was high, and on epoxidation, circulation numbers reached 60,000–100,000, and the process was proceeding with very high selectivity.

The totality of the above-mentioned material concerning oxygen atom transfer with metal porphyrin complexes onto substrate C–H bonds in model systems, mimeting cytochrome P-450, shows its complexity. Equally, with the molecular and hidden radical detachment–recombination mechanism (likewise for cytochrome P-450), in model systems the usual radical process can proceed by hydroperoxide formation and its transformations. Oxidation reactions with the involvement of non-selective $OH^{\bullet}$ radical are characterized by low kinetic isotope effects—*K* IE ($\kappa_H/\kappa_D \approx$1–2). Higher KIE values indicate more selective intermediate involvement as oxidant (namely, metallocene) [219]. Model systems of substrate C–H bond hydroxylation (like cytochrome P-450) are often characterized by high KIE value ($\kappa_H/\kappa_D > 9$), which apparently is caused by proton tunneling contribution to the mechanism, and its increase is accompanied by the enhanced stereoselectivity [332]. Many of the model metal porphyrin systems considered are imitating monooxygenases not only in a functional way but also by the intermediate active particles and mechanism of substrates oxidation [219].

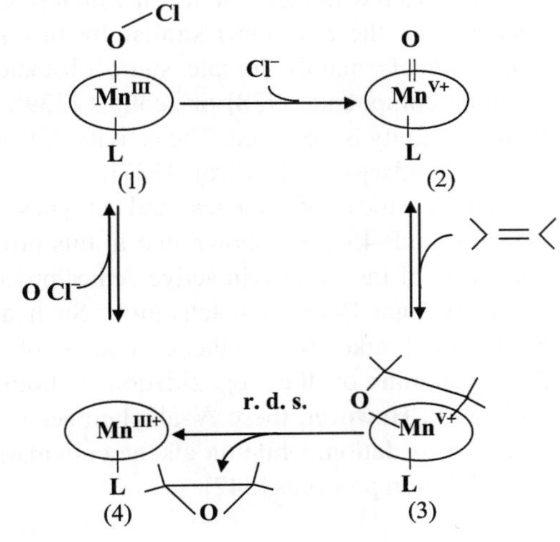

Fig. 3.19 Supposed catalytic cycle

Non-Hematic Model Systems

Non-hematic monooxidases represent a rather large group of enzymes, among which the most interesting are ω-hydroxylase, tytosine hydroxylase and MMO. All of them oxidize unactivated C–H bonds, introducing oxygen atoms in them, but the active center of the first two hydroxylases involve mononuclear iron(III) complex, and that of third hydroxylase–binuclear iron(III) complex. Modeling of certain dioxygenases is also interesting, realizing the introduction of one oxygen atom O_2 on hydroxylation and, conjugated with this, another oxygen atom introduction into a suitable substrate such as dioxygenase ketoglutarate.

$\omega-$Hydroxylase and its model systems obtained from various bacterial sources possess some functional peculiarities. They catalyze terminal hydroxylation of methyl groups in alkanes and aliphatic acids, and oxidize n-alkanes (from C_5 to C_{10}) into n-alcohols and then into n-acids. Separated ω-hydroxylases not only hydroxylate, but also epoxidate sulfides and demethylize ethers in a stereospecific way [333] and with high KIE ($\kappa_H/\kappa_D = 6$–8) [334].

It is supposed that in the reactions of ω-hydroxylase, a high-valency iron compound of ferryl type ($Fe^V = O$) is a possible oxidative intermediate particle [249, 308], detaching hydrogen atom from alkane methyl group with subsequent recombination of alkyl radical formed with iron hydroxo-complex (the situation is very similar to that realized in alkanes hydroxylation process with cytochrome P-450) (Fig. 3.14). Thus, the alcohol will be obtained:

$$R - CH_3 + O = Fe^{IV} \longrightarrow [Fe^{IV} - OH, CH_2^{\bullet} - R] \longrightarrow Fe(III) + R - CH_2 - OH^{\bullet}$$

Many systems containing non-porphyrin iron complexes are able to hydroxylate alkanes to alcohols, although with usual regioselectivity [188, 198, 228, 335, 336], i.e., first the weakest tertiary bonds C–H will be hydroxylated, then secondary ones, and only then the strongest primary bonds.

There are few model systems in which selectivity in alkanes oxidation changes, thus making these systems similar by this parameter to ω-hydroxylases [337]. These models mainly imitate steric obstacles with the help of solid matrix—clathrate compounds [338] or zeolites [339]. Thus, change in both substrate and regioselectivity is observed. The activity of primary C–H bonds is higher compared to the secondary ones in alkane [337].

Careful study of alkenes and alkynes oxidation process, catalyzed with cytochrome P-450, has shown that in this process, as with substrates epoxidation, alkylation of the porphyrin active center proceeds near pyrrole nitrogen, resulting in cytochrome P-450 autoactivation. Such autoactivation is caused more often by terminal alkenes. Synthetic models of cytochrome P-450 will also form N-alkylhemins on their epoxidation of both terminal and non-terminal alkenes [340–343]. However, these N-alkylamines in model systems are able to catalyze alkene epoxidation, while on alkene epoxidation with cytochrome P-450, enzyme autoinhibition proceeds [342].

With high circulation and moderate concentrations of oxidant (pentafluoroiodobenzene (PFIB)) in model system PFe(III)–PFIB, epoxidizing alkenes, being a model of cytochrome P-450, hemine PFe(III) will be rapidly and completely alkylated with formation of a mixture of primary and secondary N-alkylhemines [344]. They are catalysts of alkenes epoxidation with PFIB, although they are very unstable and are disintegrated during the course of the reaction. As system PFe(III)–PFIB mimics cytochrome P-450 well, it apparently will not be decomposed during the process. Such behavior of the model is similar to the reversible inhibition of cytochrome P-450 with ligands joined to iron of the active center.

Methane Monooxygenase and Its Models

There exists a large group of MMO, obtained from various bacterial sources, which realize efficient methane transformation into methanol [345–348]. They rather significantly differ by their activity in the process of methane hydroxylation. The general reaction of inert methane hydroxylation with oxygen can be given in the following form:

$$CH_4 + NADH + H^+ + O_2 \xrightarrow{\text{methane monooxygenase}} CH_3OH + H_2O + NAD^+.$$

This reaction proceeds in many steps at low temperature with high activity and selectivity [147, 349–353]. Besides C–H bond of methane oxidation, they can also oxidize edge methyl and adjacent to it methylene groups in n-alkanes (C_2–C_8), as well as epoxidize olefins and hydroxylate aromatic compounds. In subsequent reactions, methanol formed from methane will be oxidized through formaldehyde to formic acid and then to CO_2. Due to this process, energy for bacterial vital activity is obtained.

MMO Consists of proteins A, B, and C, the active center being in the first of them. The active center includes binuclear iron(III) μ-oxo (μ-hydroxo) complex (Fe_2-hydroxylase MMOH, 245 kDa) [354]. The most probable hydroxo complex is considered to combine both iron atoms. Besides, these iron atoms are linked together with glutamine carboxylate as bridge. Three other glutation carboxylates are monodentate ligands. Two histidine imidazoles are also ligands.

Each iron atom has coordination surroundings close to octaherdal. The active center

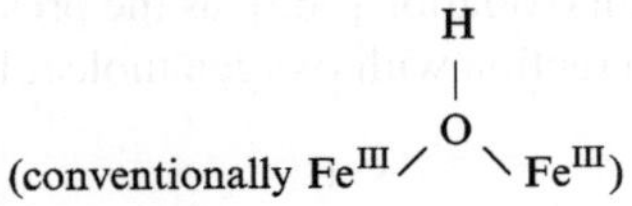

is within the hydrophobic space, in which CH_4 binding proceeds [355]. Using EXAFS method [356, 357], it was found that each of the two iron atoms in MMO is surrounded by four to six O- or N-ligands. Length Fe...Fe (it can be slightly changed) is equal to 5.1 Å, and they are linked between each other in oxidized or semi-oxidized form with OH bridge

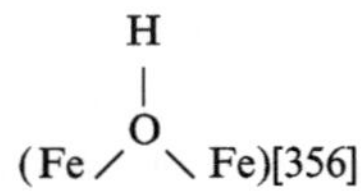

or oxygen from water [358].

In the reduced form in the active center, each of the iron atoms is five-coordinated. With each iron(III) atom, five oxygen atoms (belonging to the carboxylate groups of glutamic and asparagic acids) and one nitrogen atom (from the histidine imidazole group) are coordinated [359]. Two iron atoms in MMO active center are bound to each other with μ-oxo-μ-carboxylate bridges [360–363].

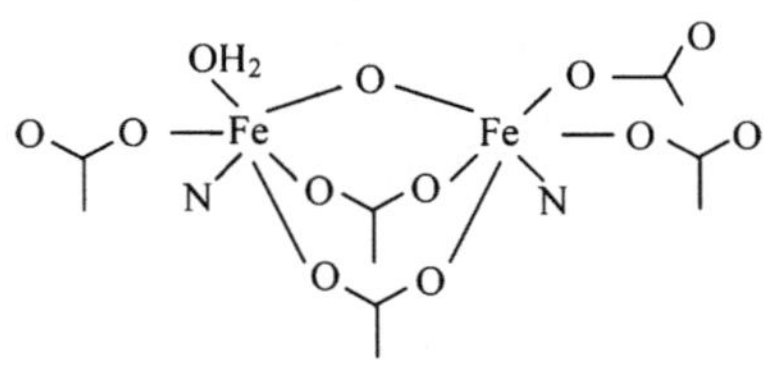

There are two Fe_2O centers in hydroxylase component A (protein A MMOH), situated symmetrically at 45 Å distance from each other. Each iron atom is close to octaherdal coordination surroundings. It is still unclear why the most inert hydrocarbon–methane—is the most efficiently bound with MMO compared to other alkanes. Active center redox properties are strongly influenced by its joining to protein B. Functions of regulatory protein B (MMOB, 15.5 kDa), containing no metal or cofactors, are rather important; as they regulate electron transfer in catalase from reductase to hydroxylase, this two-electron process proceeding only in the presence of substrates [364]. It also increases substrate oxidation rate and changes the regioselectivity of oxidation. This protein is situated in the Fe_2O center, apparently, conformationally changing its coordination sphere, which results in change in spectroscopic, redox, and catalytic properties.

Protein C is reductase (Fe_2S_2-flavin reductase MMOR, 40 kDa), and realizes electrons transfer from NADH to hydroxylase. This transfer is controlled and regulated by protein B.

All three proteins A, B, and C are closely connected with each other and are needed for substrate efficient oxidation [365], as the presence of protein B increases the rate of active center interaction with oxygen molecule

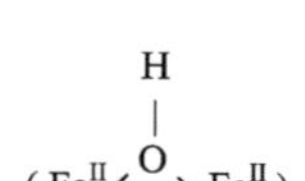

and the formation of binuclear ferryl particle $[Fe^{IV}–O–Fe^{IV}] = O$ [366]. Protein B also shifts the redox potential of the iron active binuclear center in the negative direction by 132 mV [367]. The formation of MMOH–MMOB complex changes the conformation of

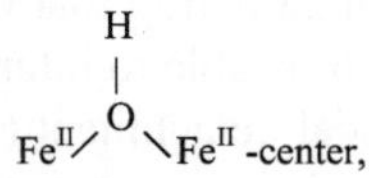

so that the substrate can join this center more strongly than simple MMOH. Thus, the space of the ligand surroundings of center $Fe^{II} \ldots Fe^{II}$ will also be changed.

The mechanism of methane oxidation in the presence of MMO, probable to a great degree, although still not certain (similar to cytochrome P-450), is given in Fig. 3.20 [368, 369].

In isolated MMO enzyme, two iron ions are in oxidation degree 3+, but can be reduced to degree 2+ [370], and in this form the active center becomes capable of joining oxygen molecules. On partial reduction, the active center can be in the mixed-oxidized state

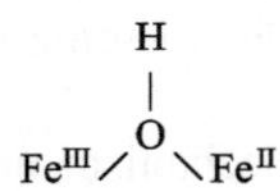

with the distance between the iron ions equal to 3.1 Å [370].

Processes proceeding in this cycle can be described in the following way [337]. The active center of protein B (1) joins methane (2) and NADH (3), after which, with the involvement of proteins B and C, rapid two-electron transformation of this triple complex (3) into μ-hydroxo–Fe_2^{II} complex (4) proceeds. The latter, interacting with O_2 by means of two-electron transfer, will form hydroperoxide intermediate (5). Then heterolytic breakage of O–O bond proceeds with the formation of water and

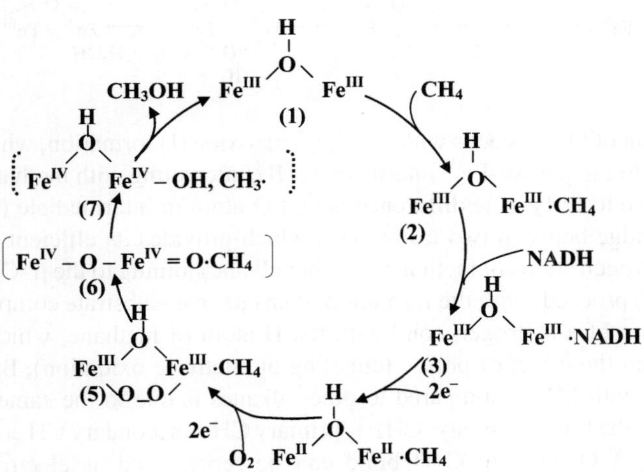

Fig. 3.20 Supposed mechanism of methane oxidation in the presence of methane monooxygenase

relatively long-living $[Fe^{IV}Fe^{IV}] = O$ intermediate of diferryl or iron (IV) bis-μ-oxo complex type (6) [337, 371, 372].

It was shown by the method of EPR-trap that $OH^{\bullet}$ radical is absent in catalytic MMO cycle [373]. Intermediate (6) is able to interact with alkane coordinated with it by the mechanism of $CH_3^{\bullet}$ radical detachment (7) and its further recombination. Intermediate disintegration (7) limits the process rate, and its structure is represented as diferryl or bis-μ-oxo complex [346, 374]. Recombination can proceed slowly; therefore, complete preservation of configuration does not proceed on stereoisomers of some alkane oxidation, namely, dimethylcyclohexane. This made it possible to detect $CH_3^{\bullet}$ radicals with EPR-traps. However, intermediate formation (7) and its decomposition by the detachment–recombination mechanism are still disputable.

Direct introduction of oxygen atoms from diferryl complex (6) into C–H bonds of methane with ferryl particle is also possible. Oxygen atom introduction into C–H bonds of methane directly from peroxide intermediate (5) can also occur. Generation of intermediates in this cycle remains hypothetical, but has lately received confirmation [375–377]. In this research, consecutive formation of three intermediates was confirmed, similar to some degree with intermediates (5), (6), and (7).

However, the validity of the mechanism needs to be assessed, especially of its last stages (stages 6, 7, and partially 5), as it explains neither preferrence for methane as substrate compared to other alkanes, nor regioselectivity. Doubt is also expressed regarding the possibility of ferryl particles formation under biological conditions in the case of non-hematic systems [238, 263, 378–380], especially as oxygen atom can be joined to two iron atoms at the same time in the form of Fe^{IV} bis-μ-oxo complex (I), which is oxenoid agent for both peroxide and acid, and is the specific alternative to the ferryl particle [359]. On this basis, a new mechanism of O_2 activation for binuclear comlexes was proposed [359, 381].

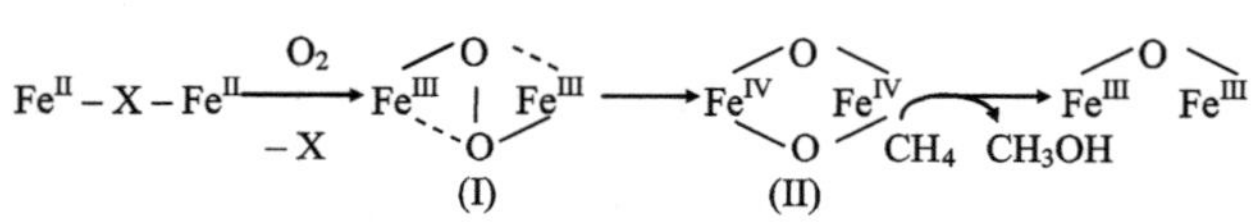

Activation of O_2 Proceeds with μ–η^2, η^2-peroxide (I) formation, which will then be turned into bis-μ-Oxo-Fe_2^{IV} intermediate (II), interacting with methane and being an alternative to ferryl (the difference is that O atom in intermediate (II) plays the role of a bridge between two iron atoms, which provides its efficient stabilization [219]). Hydrogen atoms of methane (or other alkane) joining to the μ-Oxo-bridge of complex (II) proceeds with the formation of an enzyme–substrate complex (through the formation of a hydrogen bond with the H atom of methane, which in its turn could explain the effect of proton tunneling on methane oxidation). Better joining of methane with MMO compared to other alkanes is due to the same reasons, as methane has the highest acidity: CH_4 > primary CH > secondary CH > tertiary CH.

Transfer of O atom to C–H bond can be represented as electron attack of substrate on C–H bond, realized through many-centers' cyclic state [382]. Thus, the mechanism of substrate (CH_4) transformation in the final stage of product formation can be given as follows [359]:

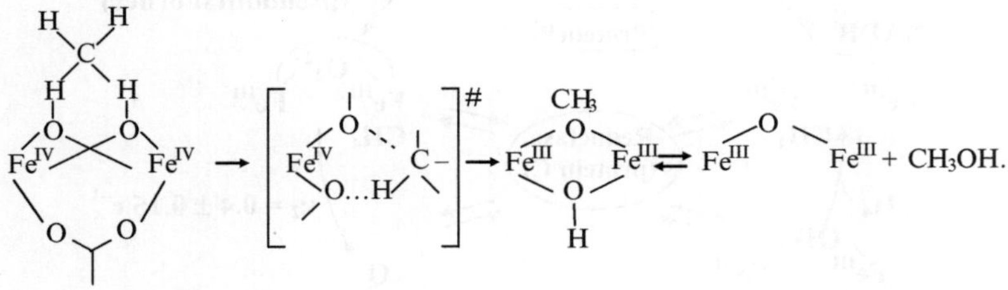

Such a mechanism of O atom transfer is suitable only for CH_4 and terminal CH_3 group (primary bond C–H). For secondary and tertiary C–H bonds, this mechanism is less probable because of steric obstacles, and it becomes clearer why CH_4 hydroxylation proceeds more efficiently than for other alkanes, and this new hypothesis needs further experimental confirmation. The need for binuclear iron complex in the active MMO center is apparently caused by the following circumstances [219]:

1. Possibility of two electrons' simultaneous transfer onto O_2 molecule

$$Fe^{II} - O - Fe^{II} \xrightarrow{O_2} Fe(III) - O - Fe(III)^{\bullet}O_2^{2-}$$

2. Capacity to accept O^{2-} (after heterolytic disintegration of O–O bond) with two Fe(III) ions without preliminary protonation
3. Necessity to avoid unstable superoxidized state of iron(V) in conditions of non-hematic surroundings

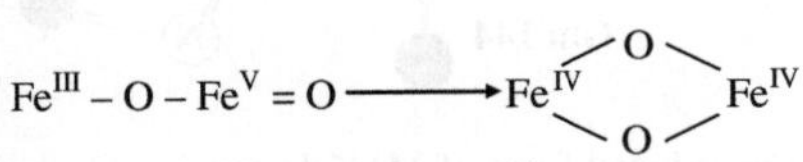

The mechanism of methane hydroxylation with soluble MMO from the bacterial source *Methylococcus capsulatus* (Bath), being the result of experimental data available for the mid-1990s, is given (with certain simplifications) in Fig. 3.21 [383].

On protein A (active center of hydroxylase) the joining of substrate (CH_4) and electron transfer from reductase occurs, with the formation of the reduced form of methamemomooxidase (MMOH) active center

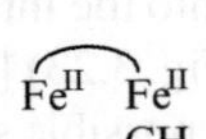

of methane monooxidase, shown in Fig. 3.22 [384, 385].

Two iron(II) ions in the reduced MMOH are joined to each other with two carboxylate groups of glutamate (Glu 144 and Glu 243). Coordination sphere of

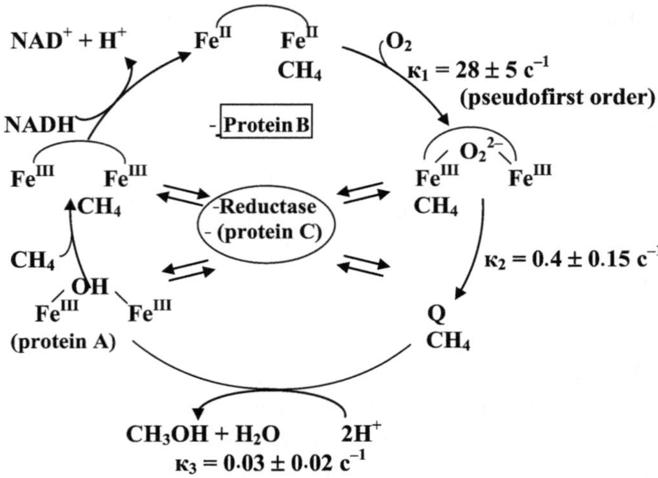

Fig. 3.21 Catalytic hydroxylating cycle of soluble MMO from *Methylococcus capsulatus* (Bath)

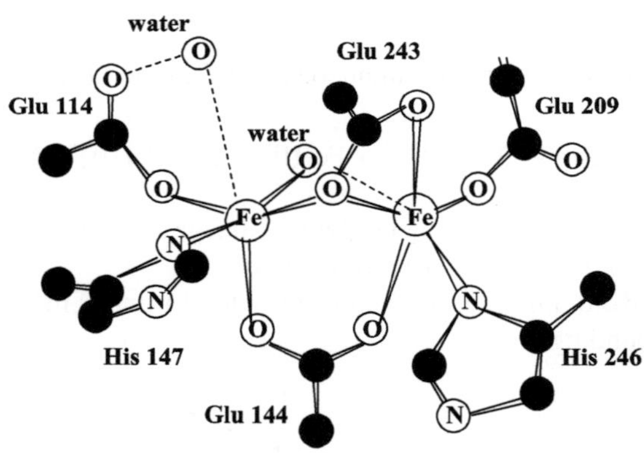

Fig. 3.22 Active center of the reduced form of MMOH causing O_2 activation

the active center also involves two monodentate glutamine residues (Glu 144 and Glu 209), two weakly bound water molecules, and hystidine residues (His 147 and His 246).

Oxygen molecule reacts with this reduced form MMOH with rate constant k_1 and forms the peroxo-form (several possible structures are proposed). The latter with rate constant k_2 is transformed into the intermediate Q with uncertain structure (the possible structures are given in Fig. 3.23) [383].

It can be seen from this figure that possible structures can include FeIV = O and FeIII–O$^{\bullet}$–particles formed on homolytic breakage of O–O bond in peroxo-form. There is also another possibility: when the oxygen atom in supposed Fe(IV)-oxo or Fe(III)-oxyl radical particles are joined symmetrically between two iron atoms. This can take place on bond O–O protonation and heterolytic breakage, resulting in Fe(IV) = O or Fe(III)–O$^{\bullet}$–particles formation. It seems that O–O bond breakage

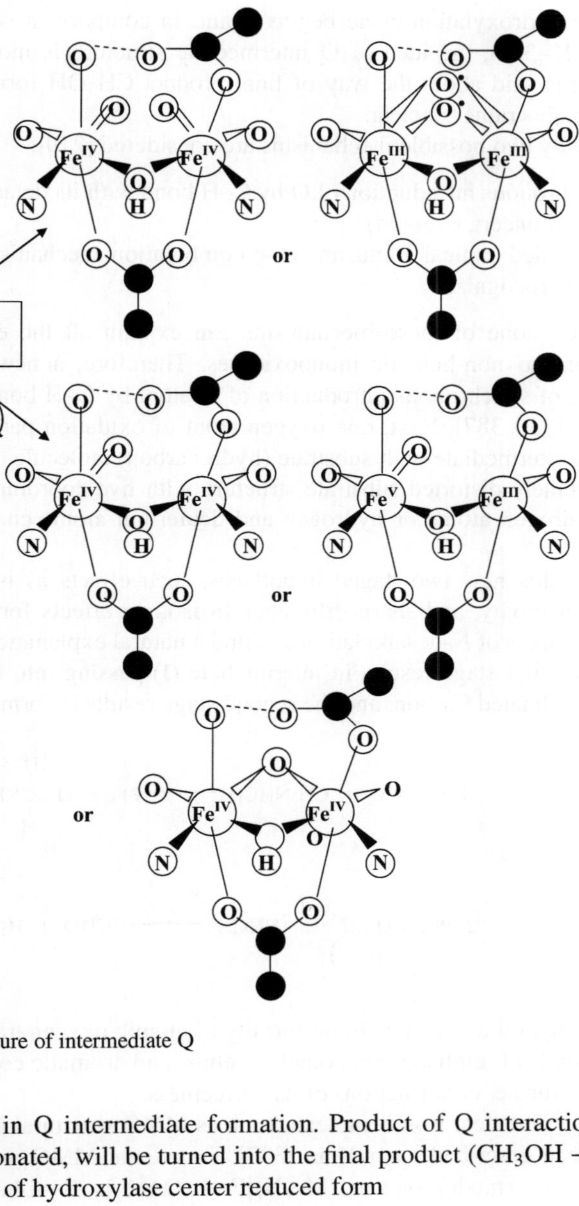

Fig. 3.23 Possible structure of intermediate Q

is the limiting stage in Q intermediate formation. Product of Q interaction with substrate, being protonated, will be turned into the final product ($CH_3OH + H_2O$) with the regeneration of hydroxylase center reduced form

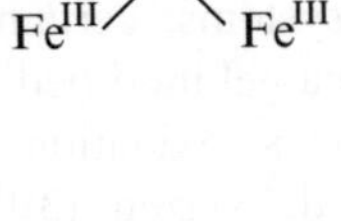

and the hydroxylation cycle begins again. In comparison with the data given in Figs. 3.21–3.23, the idea of Q intermediate structure is more detailed, although nothing is said about the way of final product CH_3OH formation. This problem still remains rather unclear.

Usually two possible mechanisms are considered [219]:

1. Synchronous introduction of O by C–H bond with its detachment from methane atom (concert, oxenoid)
2. Concealed radical (detachment–recombination mechanism), which is mostly finally recognized

However, none of these mechanisms can explain all the experimental data for hematic and non-hematic monooxidases. Therefore, a new hypothesis was put forward of synchronous introduction of O atom by C–H bond, which is briefly as follows [386, 387]. First, one oxygen atom of oxidation pariticle forms an active oxygen intermediate with substrate (hydrocarbon) molecule. This intermediate has an extremely distorted substrate structure with five-coordinated carbon, in which two hydrogen atoms or hydrogen and deuterium atoms change places relatively readily.

With this new two-staged hypothesis, such effects as isomerization, unusual stereoselectivity, and sharp difference in isotope effects for similar substrates or in the process of N-desalkylation can find a natural explanation [219]. For the latter case, the first stage results in intermediate (I) passing into intermediate (II) with five-coordinated Carbon, and the second stage results in formaldehyde detachment:

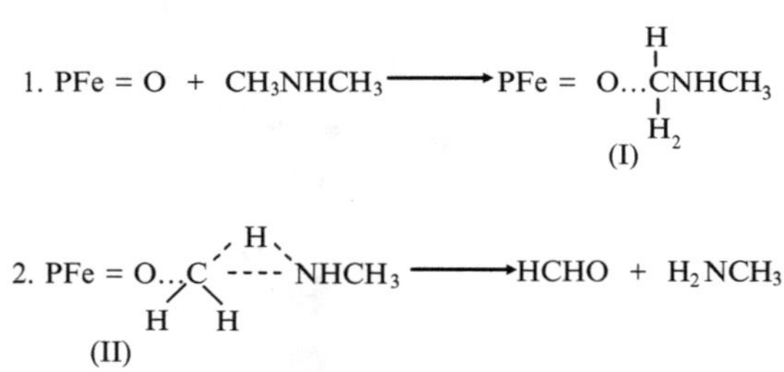

This hypothesis results in uniformity of atomic oxygen transfer mechanisms on C–H bonds of aliphatic compounds, olefins, and aromatic compounds; however, it requires further confirmations of its correctness.

Some binuclear iron complexes such as MMO structural and functional models have been described recently in [337]. Functional MMO models able to epoxyiate alkenes (model system $(Me_4N)[Fe_2L(OAc)_2]$–*cis*-stilbene–H_2O_2, where L– is tripodal heptadentate ligand) [388] or to hydroxylate alkanes (model system $Fe_2O(OAc)_2Cl_2(dipy)_2$–alkane-*tert*-BuOOH in acetonitrile) [389], etc., [337] turned out to be rather unstable.

In order to stabilize model systems, certain binuclear iron complexes were synthesized on the surface of silica gel modified with imidazole [390], which were active in methane and other alkanes' oxidation, but the active particle turned out to be $OH^{\bullet}$ radical. In another model system [391], $(Fe_2OCl_4(DTBC)_2$–alkane–O_2

in the presence of hydrazobenzene as reducer, DTBC (di-*tert*-butyl catechol)—the binuclear catalyst—was stable, and binuclear oxoferryl complex was supposed to be the reducer. In general, reactions of the reduced form of MMO with O_2 proceeds over many steps [349, 350, 352, 353, 392–397].

As has been mentioned above, the initial stage is two-electron oxidation of di-iron(II) center, with unstable peroxo-particle formation further decomposed on intermediates, which have not been studied well enough (these could be di-ferryl or other kinds of particles). In spite of the fact that a number of model compounds were synthesized, none of them reproduces the catalytic activity of MMO [398]. For example, three relative compounds were studied, each of them containing (μ-alkoxo) (μ-carboxylato) di-iron(II) coordinated with N, N, N', N'-tetrasubstituted 1,3-diamino-2-hydroxypropane ligand. These model compounds interact with O_2, forming unstable μ-peroxo intermediates at low temperatures, studied by the stopped flow method. At ordinary temperatures, these intermediates are decomposed with the formation of (μ-oxo) poly-iron(III) products. Decomposition of all three peroxide intermediates resulted in tetranuclear particles' formation in transition state. All these peroxide particles are still unsuitable for the particles that are convenient for oxygen transfer into MMO. A variety of O_2 joining ways to produce binuclear MMO center or its models, in spite of numerous experimental results, resulted in the attempt to resolve this problem on the basis of quantum-chemical calculations using the enlarged semi-empirical Huckel method [399]. As the initial experimental data, the results used were obtained by the EXAFS method according to the bonds Fe–O and Fe–N lengths in active MMO center [352, 400]. As theoretic models for quantum chemical activity, those resulting from the numerous research projects were chosen:

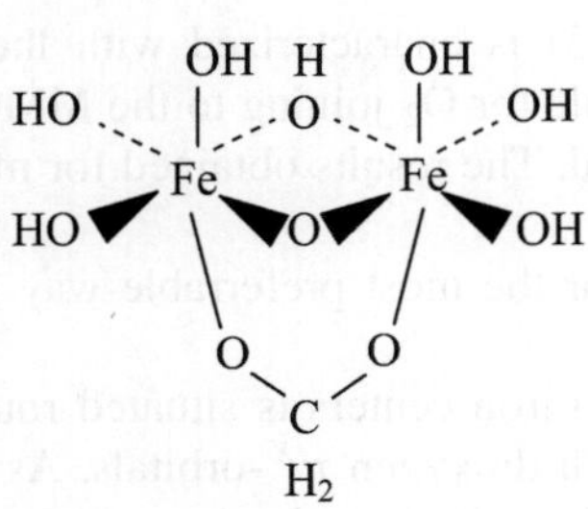

(1)

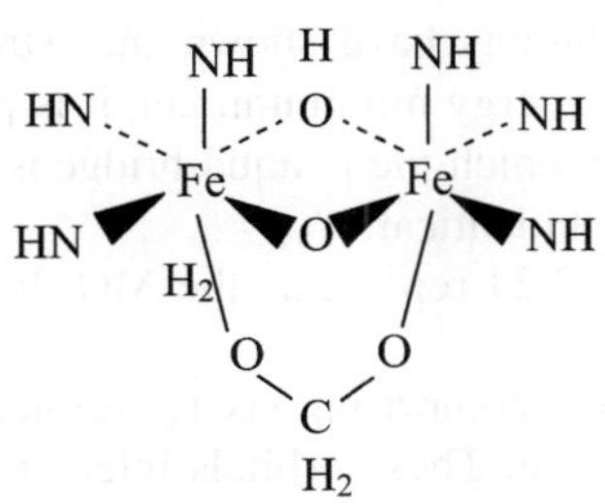

(2)

These calculation models contain hydroxo-, carboxylato-, and aqua-bridges, joining two iron atoms. Thus, OH^- and NH^{2-} ligands were used for the remaining part of coordination sphere modeling. Each iron atom is in surroundings close to octaherdal. Calculated d-Orbital disintegration of these molecules is characteristic for two iron atoms octaedrically coordinated and weakly interacting. In the reduced form of MMO center ($Fe^{II} \ldots Fe^{II}$) which interacts with O_2, the block of low-arranged d-orbitals is completely occupied, and only highly situated orbitals are vacant. To overcome the spin obstacle between triplet dioxygen $^3O_2$ and singlet organic substrate, oxygen molecule at the first stage reacts with the reduced form of enzyme, forming, apparently, di-iron(III) peroxide intermediate (III) [397]. By this time, it was experimentally proven that there are several ways of O_2 joining to the active MMO center [401]. For quantum chemical calculations, the following important geometric arrangemnents were considered:

$$(3) – (\mu – \eta^1 : \eta^1 – O_2^{2-}), (4) – (\mu – O_2^{2-}), и (5) – (\mu – \eta^2 : \eta^2 – O_2^{2-}):$$

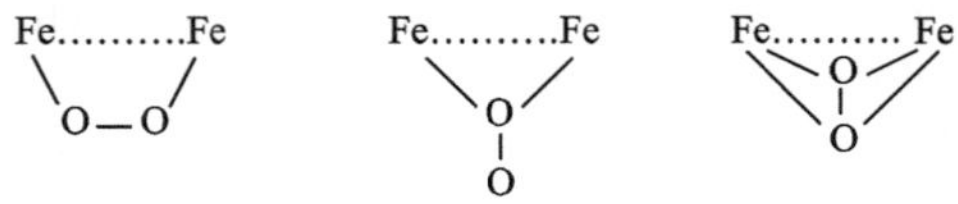

(3) (4) (5)

It is difficult to determine, proceeding from the spectral data, which of these structures is the most probable on O_2 interaction with the MMO active center. Therefore, potential energy curves were calculated for these three binding structures (3, 4, and 5) of the hydroxide model (1):

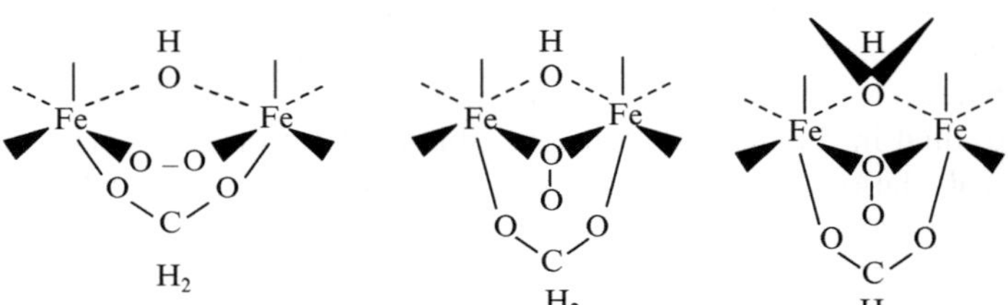

(3) (4) (5)

Calculations have shown that structure (3) is characterized with the lowest potential energy minimum, i.e., it is preferrable for O_2 joining to the MMO active center, in which the μ-aqua-bridge is removed. The results obtained for model (2) are almost identical.

Figure 3.24 represents the MO diagram for the most preferrable way of binding (3).

A block comprising six t_{2g}-orbitals of two iron centers is situated roughly on -12 eV value. These orbitals interact well with dioxygen π^X-orbitals. As double-degenerated π^X-orbitals in unbound O_2 are filled with two electrons, it appears that on the early stages of reaction two electrons are transferred from energetic higher arranged Fe, Fe block of t_{2g}-Orbitals to O_2, reducing it to O_2^{2-}. Thus, this way of dioxygen binding (3) $(\mu–\eta^1: \eta^1–O_2^{2-})$ is preferrable for O_2 joining to the MMO active center.

The number of model systems is described with iron binuclear complexes as catalysts, and with peroxide oxidants (H_2O_2, *tert*-BuOOH) in organic solvents which were able to oxidize methane and other alkanes [402–405]. Thus, involvement of ferryl particles in hydroxylation process was supposed.

The influence of the substituents' nature was also studied in model binuclear iron complexes of $[Fe_2OL_4(H_2O)_2](ClO_4)_2$и $[Fe_2OL_2(PhCOO)_2(H_2O)_2](ClO_4)_4$ (L = α, $\alpha\prime$-dipyridyl, 1,10-phenanthroline and their various substitutes) type on alkanes

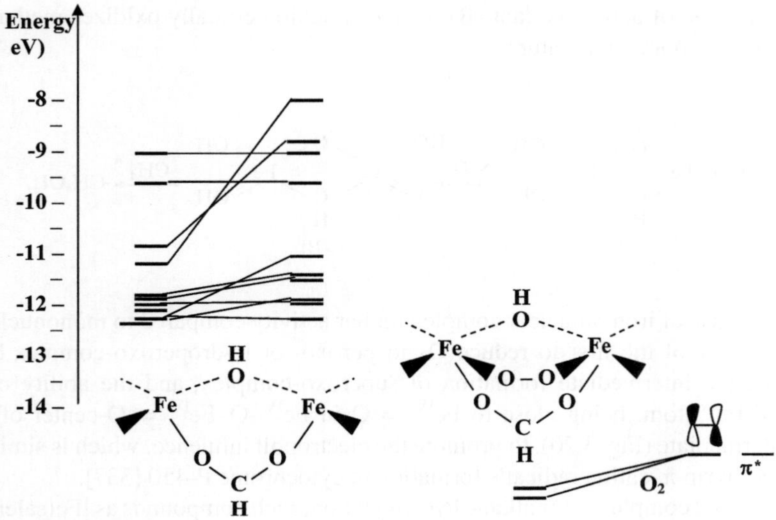

Fig. 3.24 Diagram of the most preferrable binding interaction (3), of dioxygen with $Fe_2(\mu\text{–}OH)(\mu\text{–}O_2CH)(OH)_6$

oxidation (methane, ethane, hexane, and cyclohexane) with hydrogen peroxide or *tert*-butyl hydroperoxide in acetonitrile [406]. However, a relatively low number of circulations for alkanes oxidation was discovered, which, supposingly, can be explained by competing processes of alkane (RH) oxidation and peroxides (R′OOH) decomposition:

$$\begin{cases} Fe(III)_2O + R'OOH \longrightarrow Fe_2^{IV}O(O) + R'OH, \\ Fe_2^{IV}O(O) + RH \longrightarrow Fe(III)_2O + ROH, \\ Fe_2^{IV}O(O) + R'OOH \longrightarrow Fe(III)_2O + R'OH + O_2, \end{cases}$$

where $R' = H, {}^tBu$.

In the case of methane oxidation, the catalytic activity of model complexes increases on substituent electronegativity increase in the ligand. On cyclohexane oxidation, the inverse dependence is observed. Ethane oxidation takes intermediate position. In general, substituent electrophilicity plays an important role in C–H bonds in alkane activation with similar model binuclear iron complexes.

The weak part of these systems is binuclear complex *instability* on alkane oxidation. It is also to be noted that ferryl complex participation is disputable. One of the best MMO models which is stable enough is the system iron-containing zeolite ZSM-5–methane–N_2O, able to oxidize methane, benzene, and some other compounds [407]. α-Centers of iron-containing zeolites ZSM-5 are surface hydroxide clusters of iron (A). Oxygen atom detachment from N_2O proceeds on them with the formation of active oxidant (B), which stoichiometrically oxidizes methane to methanol at room temperature:

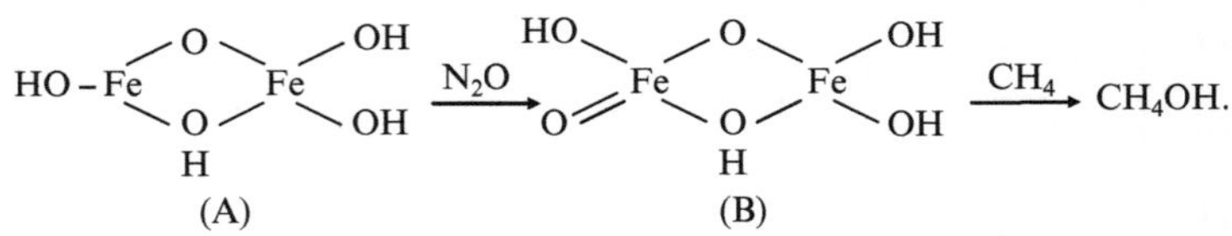

The cause of iron binuclear complex higher activity compared to mononuclear is the capacity of the first to reduce O_2 to peroxo- or hydroperoxo-complex better without the intermediate formation of superoxo-complex, and the ability of the second Fe^{IV}atom, being close to $Fe^{IV} = O$ in Fe^{IV}–O–Fe^{IV} = O-center of type (6) intermediate (Fig. 3.20), to promote the electrophil influence, which is similar to the porphyrin π-cation radical's formation in cytochrome P-450 [337].

As model complexes for alkane hydroxylation, such compounds as $[Fe(salen)_2O]$ [408], $[Fe_2O(HB(pz)_3)_2(OAc)_2]$ [409, 410], $[Fe_3O(OAc)_6(Py_3)]$ [411, 412], $[Fe_2OL_4(H_2O)_n]$ [413–415], $[Fe_2O(OAc)_2(bpy)_2Cl_2]$ [416, 417], $[Fe_2O(H_2O)_2(tmima)_2](ClO_4)4$ [417], and $[Fe(PA)_2]$ [418] were also described. Here, salen = $N,N\prime$-ethylenebis (salicylideneimine), $HB(pz)_3$ = hydrotris (pyrazolyl) borate, OAC = acetate, PA = picolinic acid, Py = pyridine, bpy = bipyridine, phen = 1, 10-phenanthroline, tmima = tris ((1-methylimidazole-2-yl) methyl) amine, and L = bpy, substituted bpy, phen.

These catalysts in the presence of oxygen donors or peroxides will hydroxylate alkanes. An efficient catalyst of cyclohexane oxidation (with cyclohexanol, cyclohexanone, and other products) formation in the presence of ^{t}BuOOH is (μ-oxo) di-iron(III) complex $[Fe_2(TPA)_2O(OAc)](ClO_4)_3$ (where TPA = tris (2-pyridylmethyl) amine) [419, 420]. However, the nature of these iron-containing catalysts and the way of their action are still not quite clear.

To reveal the protein influence on MMO active center functioning, a model complex of this center was synthesized $[Fe_2(BIPhMe)_2(O_2CH)_4]$ (1a) (where BIPhMe = 2, 2′-bis-(1-methylimidazolyl) phenylmethoxymethane), and a kinetic study of this complex interaction with O_2 was carried out using the stopped flow method. This reaction in $CHCl_3$ is a first-order reaction by $[O_2]$ and a second-order one by [1a].

The main peculiarity of such a supposed reaction mechanism is the bimolecular way, involving the formation in transition state of four-nuclear (μ^4-peroxo) di-iron(II) di-iron(III) particle (Scheme 3.15) [421].

Model complex (1a) reacts with O_2 in a different way than in enzymatic systems. The initial stage of oxygen molecule binding is similar to peroxo-complex formation in the active center

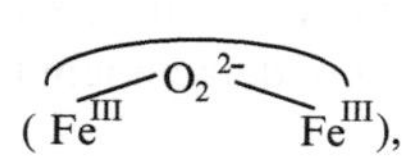

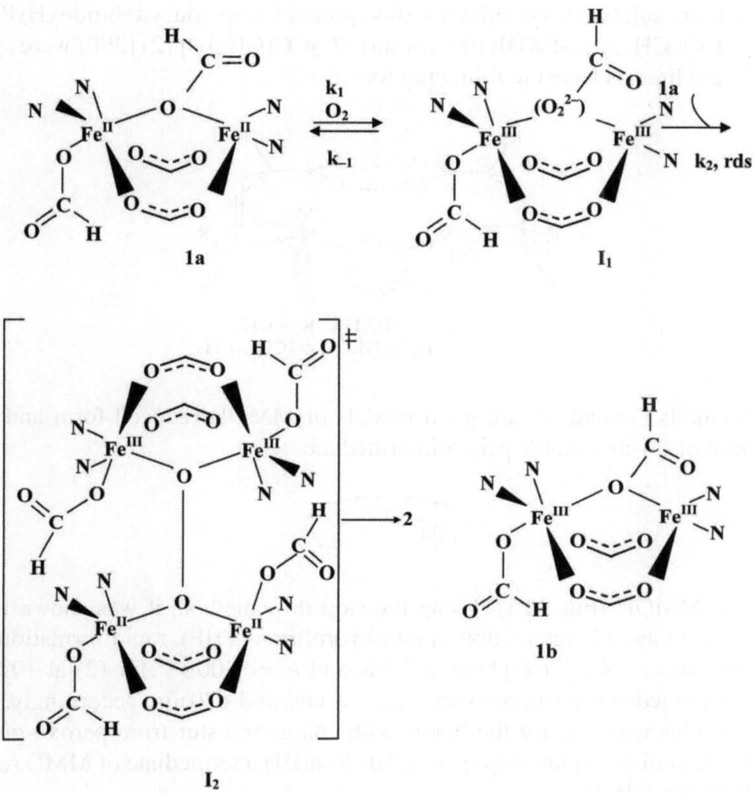

Scheme 3.15 Interaction of Fe(II) binuclear model complex with oxygen

(Fig. 3.21) of soluble MMO, but then in the case of model complex (1a), a number of rapid reactions proceed, resulting in iron polynuclear particles formation. Protein MMO protects the active center from such reactions by active di-iron center surrounding with polypeptide. Under these conditions, the reactions, resulting in substrate activation, become predominant [421]. Perhaps, in the systems modeling MMO active center, ligands with more steric difficulties should be used, or a model complex should be fixed on a suitable carrier, in order to make the formation of iron polynuclear particles more difficult [421].

In reactions $1a \longrightarrow I_1 \longrightarrow I_2 - 1\varepsilon$ (Scheme 3.15), moving of the carboxylate group proceeds, this process also taking place in the appropriate enzymatic systems, resulting in O_2 reactivity change. Carboxylate moving, creating pre-conditions for oxygen joining to the active center, can be increased on protein B or reductase (protein C) binding with MMOH active center, which to some degree explains the regulating role of these proteins. To modelize the active center of MMOH reduced form, binuclear complexes of iron(II) with dicarboxylate ligands μ-xelendiamine bis (Camptriacyd) imide (H_2XDK) $[Fe_2(\mu\text{–}XDK)(\mu\text{–}O_2CPh)(ImH)_2(O_2CPh)(MeOH)]$ (1)

and with more soluble μ-xylendiamine bis (propyl Camp-triacyd) imide (H_2PXDK) $[Fe_2(\mu\text{–}O_2CCCH_3)_3(\mu\text{–}PXDK)(N\text{–}MeIm)_2(O_2CC(CH_3)_3)]$ (2) [390] were synthesized. These ligands have the following form:

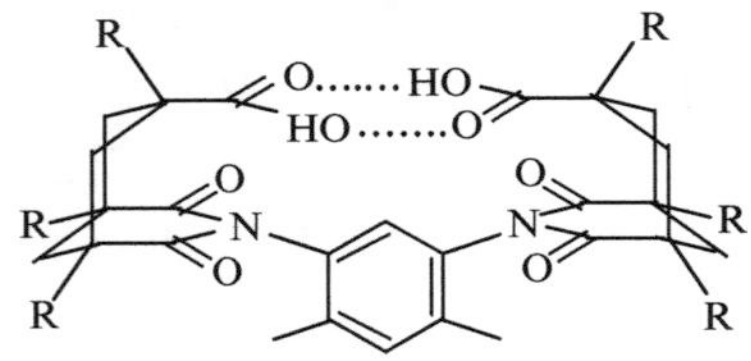

H_2XDK, R = CH_3
H_2PXDK, R = $(CH_2)_2CH_3$

Compounds (1) and (2) are good models of MMOH reduced form and under interaction of O_2-metastable peroxointermediate

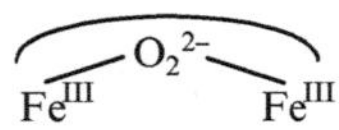

react with MMOH (Fig. 3.21). Using the stop flow method, it was shown that on complexes (1) and (2) interaction in tetrahydrofuran (THF), rapid formation (with rate constants $\approx 74\,s^{-1}$ for (1) at 202.5 K and $k \approx 300\,s^{-1}$ for (2) at 197 K) of colored intermediates proceeds with $\lambda_{max} \approx 660$ and 670 nm, accordingly. These values are characteristic for the bonds with charge transfer from peroxo-group to iron and are similar to that of (μ-peroxo) di-iron(III) intermediate of MMO reaction cycle MMO (ри. II.21).

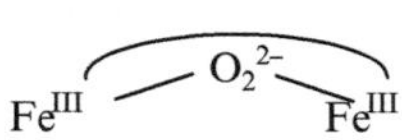

Thus, the problem of MMO modeling is far from being resolved: the nature of binuclear complexes and their surroundings as catalysts are not always clear; little is known about the nature of possible intermediates, and the stability of the complexes needs to be assessed.

Model Systems of Tyrosine Hydroxylase

Another non-hematic monooxygenase, the center of which represents iron(II) mononuclear complex, is tyrosine hydroxylase (TH). Tyrosine hydroxylase catalyzes hydroxylation of L-tyrosine (HTyr)

$$HTyr + O_2 + DH_2 \xrightarrow{\text{enzyme}} HOtyr + D + H_2O$$

to L-3,4-dihydroxyphenylamine (HOtyr), which is the initial and rate-determining stage in catecholamine neurotransmitters synthesis. However, the role of iron compounds in O_2 catalytic activation and tyrosine hydroxylation has not been revealed [422].

Model systems [423] imitating tyrosine hydroxylase were proposed: $Fe^{II}(DPAH)_2$–O_2–PhNHNHPh–RH, $Fe^{II}(PA)_2$–O_2–PhNHNHPh–RH, and $Fe^{II}(dipy)_2$–O_2–PhNHNHPh–RH, where DPAH is 2,6-dicarboxylpyridine, PAH is picolinic acid, dipy is 2, 2′- dipyridyl, PhNHNHPh is the reducer (DH_2), and RH is 4–t–BuC_6H_4OH. The same systems but with other substrates RH [benzene (PhH), phenol (PhOH), and substituted phenols 4–ClC_6H_4OH, 4–$MeOC_6H_4OH$] successfully realize hydroxylation reaction. Thus, the following transformations proceed: PhH $\longrightarrow$ PhOH, PhOH $\longrightarrow$ o,p–$C_6H_4(OH)_2$, 4–ClC_6H_4OH $\longrightarrow$ 4–$ClC_6H_3(OH)_2$, 4–$MeOC_6H_4OH$ $\longrightarrow$ 4–$MeOC_6H_3(OH)_2$, 4–t–BuC_6H_4OH $\longrightarrow$ 4–t–$BuC_6H_3(OH)_2$.

It was recently shown [418, 424] that the interaction of iron complex compounds with hydrogen peroxide (Fenton's reagent) in the presence of the base (B) yields nucleophilic adduct sooner than free $OH^\bullet$ radical:

$$Fe^{II}L_x + HOOH \overset{B}{\rightleftarrows} [L_x^{-}Fe^{II}OOH(BH^+)]$$

On aromatic compound hydroxylation with Fenton's reagent, substrates reactivity and distribution of hydroxylation products differ from what could have been, provided the oxidizing particle was $OH^\bullet$ radical.

Hydroxylation depends on transition metal, ligand, and solvent. It is supposed that by $OH^\bullet$-particle transfer from (1) on substrate, carbonic radical (2a) will be formed first by means of Fe–C bond, and then the hydroxylation product will be obtained from it:

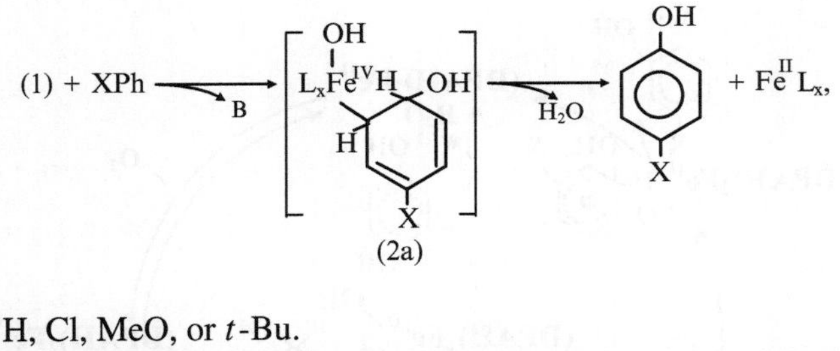

where X is H, Cl, MeO, or t-Bu.

In the case where the substrate is phenol (PhOH), the ortho-hydroxylated product (catechol) will be the main product:

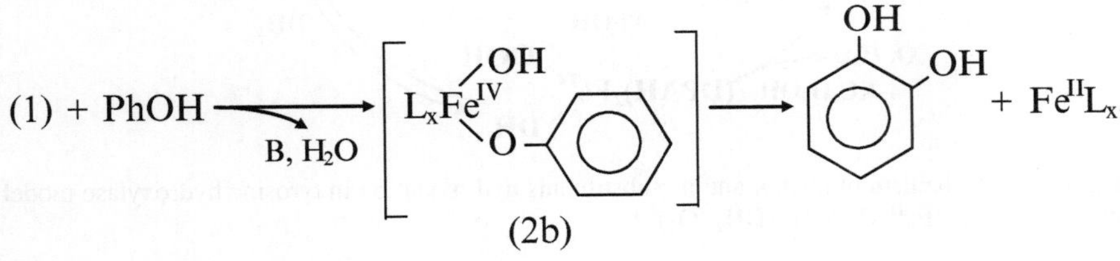

If instead of H_2O_2, O_2, reducer (DH_2), and saturated hydrocarbon substrates (RH) are introduced into the system, the reactive intermediate will appear during the process (1c), the reactivity of which is close to that of intermediate (1), and hydroxylation will proceed as follows:

$$Fe^{II}L_x + DH_2 + O_2 \rightleftharpoons \underset{(1c)}{\left[L_xFe^{IV} \begin{matrix} OOH \\ DH \end{matrix} \right]} \xrightarrow[D,\ H_2O]{} ROH + Fe^{II}L_x,$$

where $Fe^{II}L_x = Fe^{II}(DPAH)_2$, $Fe^{II}(PA)_2$, $Fe^{II}(Dipy)_2^{2+}$, and $DH_2 = PhNHNHPh$.

The scope of such systems study was to establish their correspondence as tyrosine hydroxylase models [423]. In the model system $Fe^{II}(DPAH)_2$–PhNHNHPh–O_2, phenol hydroxylation reaches about 100% efficiency:

$$PhOH + DH_2(PhNHNHPh) + O_2 \xrightarrow{Fe^{II}(DPAH)_2} o, p - C_6H_4(OH)_2 + PhN = NPh.$$

Hydroxylation of substituted phenols yields 4-substituted catechols [$4–XC_6H_3(OH)_2$]. Substrates reactivity corresponds with their electron-removing capacity: Cl > H > *t* − Bu > MeO.

Proceeding from the research of the aforesaid model systems' results, the mechanism of phenol hydroxylation reaction and its substituents with the system $Fe^{II}(DPAH)_2$–PhNHNHPh–O_2 in acetonitrile:pyridine (3:1) solvent [404] was proposed (Fig. 3.25).

Complex $(DPAH)_2Fe^{II}$ reversibly joins O_2 to DH_2 and is transformed into an intermediate able to interact with substrate. Substrate intermediate formed is similar

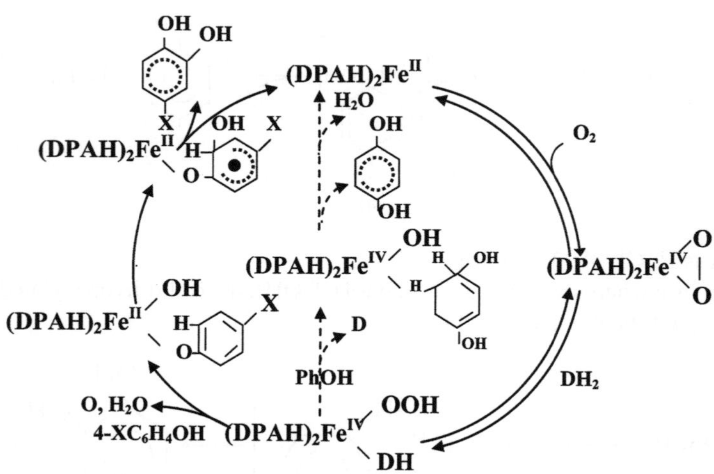

Fig. 3.25 Mechanism of phenol and its substituents hydroxylation in tyrosine hydroxylase model systems (system $Fe^{II}(DPAH)_2$–DH_2–O_2)

to intermediate (2a) of Fenton system. Then, transfer of OH$^{\bullet}$ radical from iron to substrate's carbon proceeds, arranged in the ortho-position. Finally, H atom transfer follows from hydroxylated carbon of substrate, with the formation of product (catechol) and initial catalyst regeneration.

The proposed model system with the involvement of $(DPAH)_2Fe^{II}$ as catalyst, and substrate 4–*t*–BuC_6H_4OH, is apparently a suitable model, imitating the processes proceeding on substituted aromatic compound hydroxylation with tyrosine hydroxylase. This mechanism of phenol and its substitutes' hydroxylation reflects schematically the probable route for the enzymatic system as well. It follows from the aforesaid that the role of the reduced iron form of enzyme consists in O_2 binding, phenoxy radical intermediate stabilization, and making easier hydrogen atom with catechol formation [423].

3.2.2 *Dopamine β-Hydroxylase and Its Models*

Along with the above-mentioned monooxygenases and their model systems, the active centers of which contained iron mono- or binuclear complexes, there exist monogenases with copper compounds as active centers. To such enzymes belong dopamine-β-monooxygenase (DβH), catalyzing benzene hydroxylation of phenylethylamines, such as dopamine (DPa) (and phenylethylamine, thyramine, and epyrine), turning it into noradrenaline (NAD) in the presence of ascorbic acid (AA) and O_2:

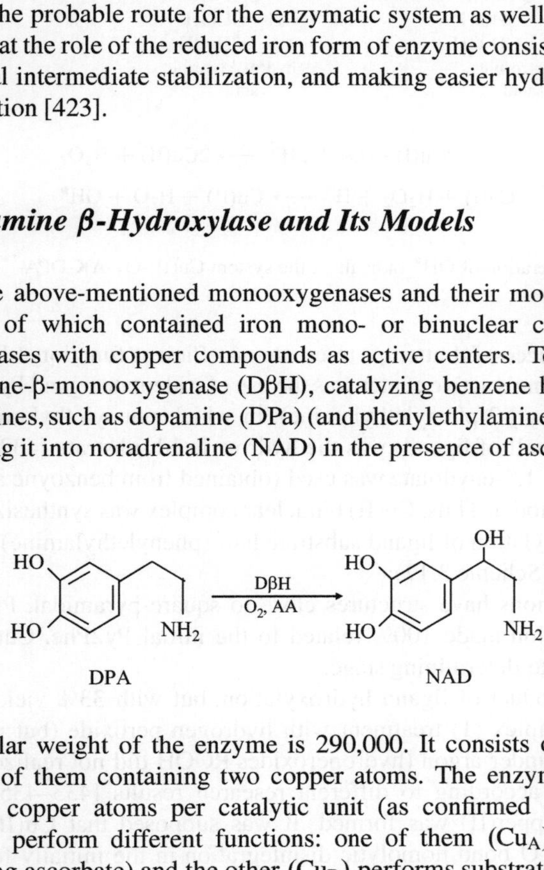

The molecular weight of the enzyme is 290,000. It consists of four identical subunits, each of them containing two copper atoms. The enzyme contains two non-equivalent copper atoms per catalytic unit (as confirmed on model compounds) which perform different functions: one of them (Cu_A) is an electron acceptor (joining ascorbate) and the other (Cu_B) performs substrate and O_2 joining [425–429].

Two electrons were delivered to Cu_B of the active center, from ascorbate through the Cu_A center in the process of hydroxylation, and reduced dioxygen joining Cu_B to peroxide complex Cu(II)–O–OH. On homolytic disintegration of this complex, radicals Cu(II)–O$^{\bullet}$ were formed, being active particles in hydroxylation reactions. It was shown by the EPR method that, during this enzymatic process, the oxidation state of copper was changed between Cu(II) and Cu(I) [430].

It was established later that DPA can be oxidized to NAD in the presence of O_2 and Cu(I) ion, which was formed on Cu(II) ion reducing with ascorbic acid, although product yield made only ≈1% of DPA [431]. In the system Cu(II)–O_2–AK–DPA, $OH^\bullet$ radicals were also generated as active particles (Scheme 3.16):

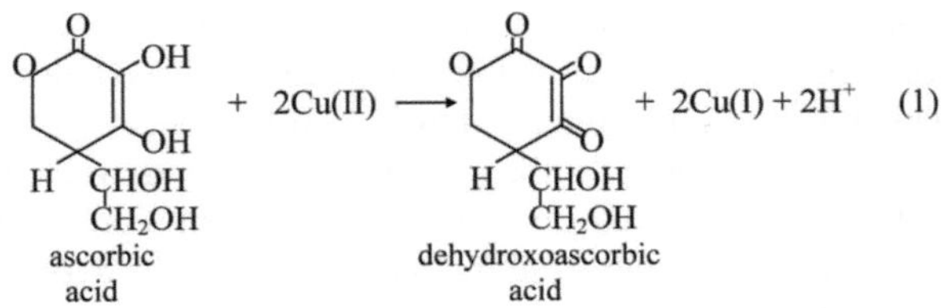

$$2Cu(I) + O_2 + 2H^+ \longrightarrow 2Cu(II) + H_2O_2 \quad (2)$$

$$Cu(I) + H_2O_2 + H^+ \longrightarrow Cu(II) + H_2O + OH^\bullet \quad (3)$$

Scheme 3.16 Generation of $OH^\bullet$ radicals in the system Cu(II)-O_2-AK-DPA

A much more successful attempt to create an efficient functional DβH model and imitate this enzyme's active center was the use of tridentate ligand Py2Phe (*N*,*N*-bis [2-(2-pyridyl)ethyl]-2-phenylethylamine), in which phenylethylamine (introduced into the second half of ligand) was the hydroxylated substrate [432]. Here, instead of ascorbic acid, 1,2-enydolate was used (obtained from benzoyne and base) as the electron donor model. Thus, Cu(II) binuclear complex was synthesized (2) in which selective hydroxylation of ligand substrate half (phenylethylamine) was realized in benzyl position (Scheme 3.17):

Both copper ions have structures close to square-pyramidal. Product yield on such hydroxylation made 100% related to the initial Py2Phe. Cu(II) reducing to Cu(I) was the rate-determining stage.

The same product of ligand hydroxylation, but with 33% yield, was obtained after Cu(II) complex (1) treatment with hydrogen peroxide (but not with O_2) in methyl alcohol under argon (hydroperoxides ROOH did not realize hydroxylation process). Thus, according to different research results [433–435], the complex (μ-peroxo) dicopper(II) was formed. It was supposed that Cu(II)–O$^\bullet$ particles, generated by O–O bond homolytic disintegration in the initially formed complex (μ-peroxo) dicopper(II), can be responsible for ligand Py2Phe Py2Phe hydroxylation in this case. Cu(II)–O$^\bullet$ particles could remove hydrogen atom in the benzyl position, generating Cu(II)–OH and carbon-centralized radical, which attacked other Cu(II)–*l*O$^\bullet$ particles with the formation of the final product. Thus formed, Cu(II)–OH was reduced to Cu(I) state, which was once again able to take part in hydroxylation reactions. As has already been mentioned, Cu(II)–O$^\bullet$ particles were also generated in the enzymatic system, which were active in hydroxylation reaction.

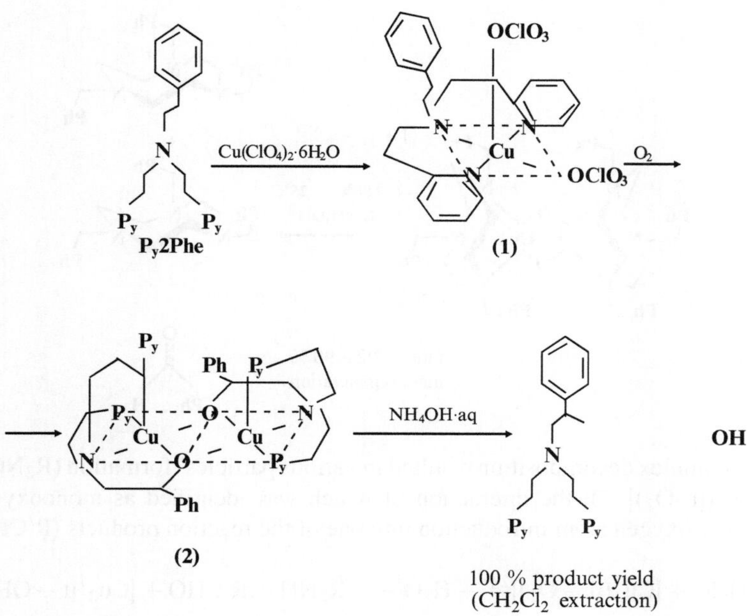

Scheme 3.17 Modeling of phenylethylamine hydroxylation, catalyzed by DβH

Therefore, the above-mentioned system can be considered as a model, imitating (mimeting) dopamine-β-hydroxylase.

Another interesting model monooxygenating system imitating the influence of dopamine-β-hydroxylase and secondary aminooxygenase was proposed, resulting from the synthesis of bis(μ-oxo) copper complex in high oxidation degree $[(L^{Bn3}Cu^{III})_2(\mu{-}O_2)]X_2$, under −80 °C in the presence of O_2 from initial compound $[L^{Bn3}Cu^{I}(CH_3CN)]X(L^{Bn3}$ is 1;4,7-tribenzyl-1,4,7-triazacyclononane, X is various amines) 9435]:

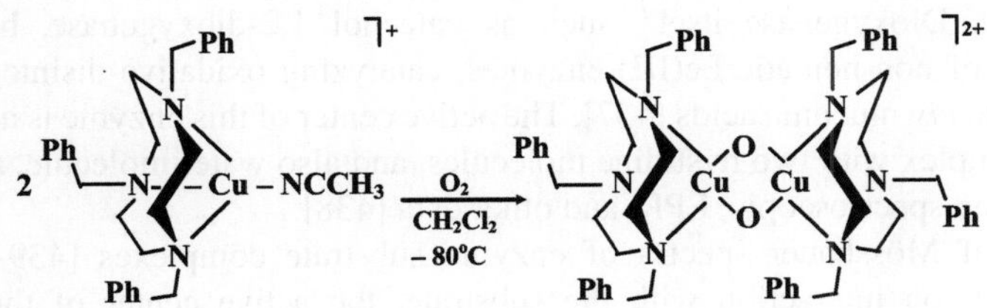

This complex with the nuclei $[Cu_2(\mu{-}O_2)]^{2+}$ is thermally unstable, and will be decomposed on heating to +25 °C.

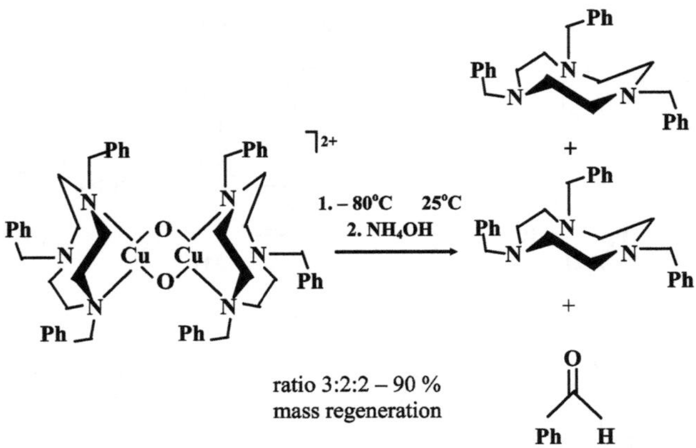

This complex decomposition resulted in various particles' formation (R_2NCH_2R' and $[Cu_2(\mu–O_2)]^{2+}$), the interaction of which was identified as monooxygenase reaction of oxygen atom introduction into one of the reaction products (R′CHO):

$$R_2NCH_2R' + [Cu_2(\mu - O_2)]^{2+} + H_2O \longrightarrow R_2NH + R'CHO + [Cu_2(\mu - OH_2)]^{2+}.$$

The mechanism of bis(μ-oxo) dicopper complex decomposition is in accordance with the inner-molecular cutting of α–C–H bond with electrophilic nuclei $[Cu_2(\mu–O_2)]^{2+}$), and is shown in Scheme 3.18 [436].

3.2.3 *Model Systems of Dioxygenases*

3.2.3.1 Models of Iron-containing Dioxygenases

Unlike numerous model systems of monooxygenases, considered above, on the basis of iron coordination compounds, appropriate model systems of dioxygenases are scanty. Dioxygenase itself, such as catechol 1,2-dioxygenase, belongs to the group of non-hematic Fe(III) enzymes, catalyzing oxidative disintegration of catechols to *cis*-muconic acids [437]. The active center of this enzyme is a high-spin Fe(III) complex with two hystidine molecules, and also water molecule, as follows from Raman spectroscopic, EPR, and other data [438].

Study of Mössbauer spectra of enzyme–substrate complexes [439–441] has shown that, on interaction with the substrate, the active center of the enzyme represents itself as Fe(III) complex, and UV and visible spectra obtained by stopped flow method [442, 443] testify to the charge transfer from tyrosine to iron(III). It was impossible to discover participation of iron(II) in the catalytic cycle.

The active iron-containing enzyme center with two coordination molecules of tyrosine is characterized by low potential, and catechol's appropriate coordination to this center results in its further decreasing, so iron(III) reduction becomes

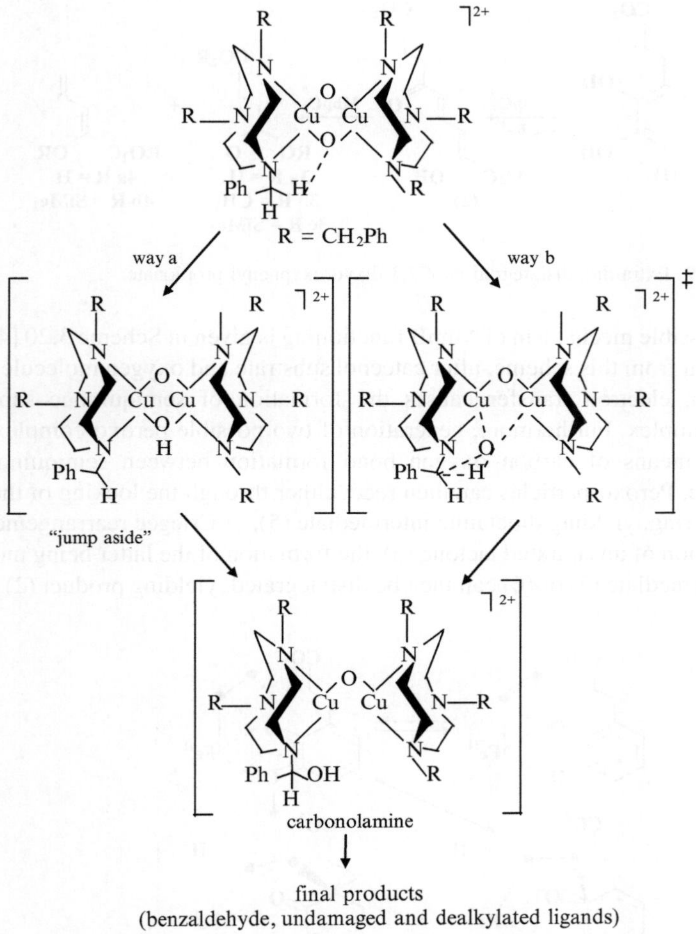

Scheme 3.18 Mechanism of bis(μ-oxo)dicopper complex disintegration

impossible [444]. Proceeding from this, the proposed scheme of the substrate (S) activation mechanism, contained in the center of the enzyme, looks as follows [445]:

$$Fe(III) + S \longrightarrow Fe(III)^{\bullet}S + O_2 \longrightarrow Fe(III)^{\bullet}OOR \longrightarrow Fe(III) + \text{products}.$$

Catechol 1,2-dioxygenase belongs to the class of the so-called inner-diol dioxygenases, the functioning mechanism of which involves anhydride intermediate formation and Criegee rearrangement [446].

The representative of another class of the so-called out-diol dioxygenases is iron(II)-dependent 2,3-dihydroxyphenyl-propyonate 1,2-dioxygenase (MhpB) from *Escherichia coli*. They catalyze extra-diol disintegration of 2,3-dihydroxyphenyl propionate (1) with product (2) formation [447], which is the substrate for hydroxylase (MhpC) (Scheme 3.19).

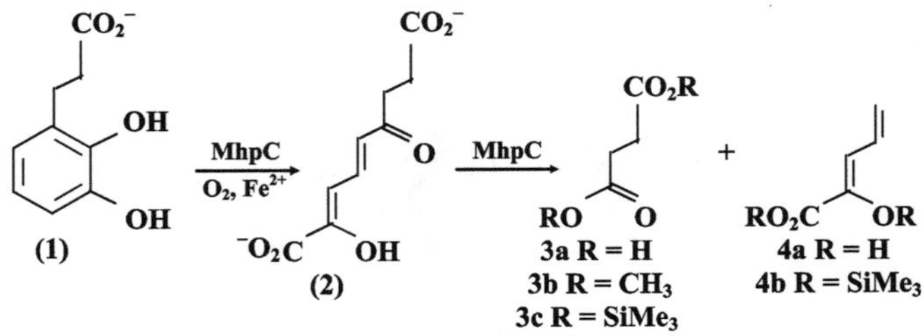

Scheme 3.19 Extra-diol disintegration of 2,3-dixydroxyphenyl propionate

The possible mechanism of MhpB functioning is given in Scheme 3.20 [448]. As can be seen from this scheme, after catechol substrate and oxygen molecule joining to iron(II), electron transfer causes the formation of semiquinone–superoxide iron(II) complex. Furthermore, generation of two possible peroxo-complexes proceeds by means of carbon–oxygen bond formation between semiquinone and superoxide. Peroxo-particles can then react either through the locking of the inner-molecular ring, yielding dioctanite intermediate (5), or Criegee rearrangement with the formation of unsaturated lactone (6), the formation of the latter being more possible. Intermediate (5) or (6) can then be disintegrated, yielding product (2) [448].

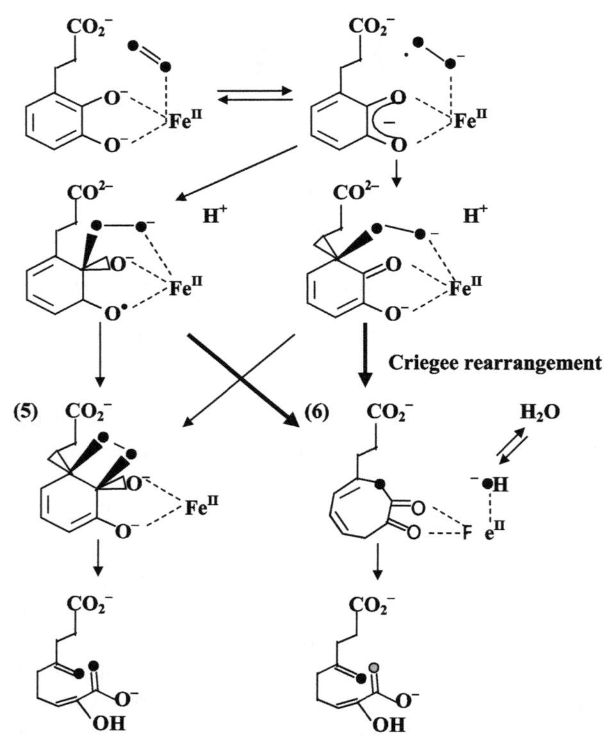

Scheme 3.20 Possible mechanism of MhpB functioning

Another non-hematic iron-containing enzyme is ketoglutarate-dependent dioxygenase [198, 236]. It realizes the catalytic process of C–H bond hydroxylation in various organic compounds (substrates RH) [449]. Co-substrates are α-ketoglutaric acid (or in general, α-ketoacids) and O_2. The oxidation rate of various substrates with this dioxygenase falls within the limits 3–50 cycles/s. Mechanism of catalytic process, realized with ketoglutarate dioxygenase, can be presented as follows [199]:

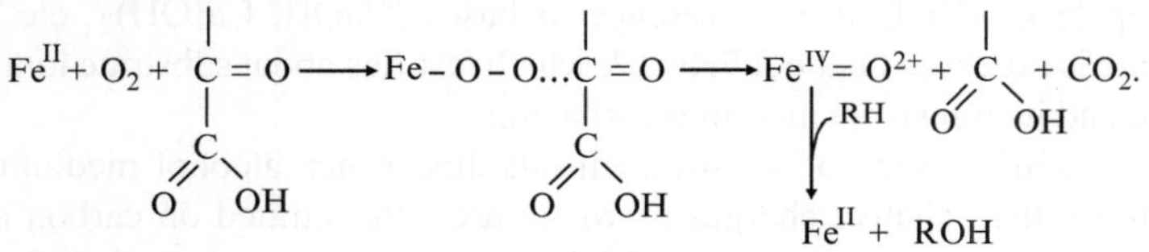

The homolytic breakage of the O–O bond and formation of succinic acid and CO_2 proceed, thus making the formation of ferryl particle $Fe^{IV}{=}O^{2+}$ thermodynamically easier, this particle being an efficient oxidant of hydrocarbon RH. In this chain of reactions, one oxygen atom is introduced into the organic substrate, and another into succinic acid. However, the activation mechanism of O_2 and the C–H bond of the substrate is still not quite clear, and an active oxidant of substrate formation in the form of ferryl particle is presupposed.

A model system of this dioxygenase, working at room temperature, is $Fe^{II}(ClO_4)_2$–O_2^{2-}–α-ketoacid–Zn(Hg)–methyl viologen–acetonitrile solution. Thus, zinc amalgam is a reducer, and methyl viologen makes electron transfer easier. Substrate cyclohexane is oxidized into cyclohexanol and cyclohexanone. Thus, generation of ferryl particle $Fe^{IV}{=}O^{2+}$ is supposed due to electrophilic contribution of ketoacid α-carbonyl group, which apparently is an active oxidant of substrate $RH^{\bullet}$ $OH^{\bullet}$ radical formation is also possible [450].

Another group of model compounds involves Fe(II) complexes with α-ketoacid of $[Fe^{II}L(PhCOCOO)]ClO_4$ type, where L represents tridentate ligands—tris (2-pyridylmethyl)amine and its 6-methyl-2-pyridyl analog [451]. When these complexes interact with O_2 and with substrate RH, decarboxylation of α-ketoacid proceeds, with quantitative formation of benzoic acid. Thus, one oxygen atom is introduced into substrate RH, and another into benzoic acid (like enzyme). In such a way, interaction of α-ketoacid with active dioxygenase center containing Fe^{2+} is modeled.

One more non-hematic iron-containing dioxygenase is lipoxydioxygenase which catalyzes dioxygenation of polyunsaturated aliphatic acids [452]. The interest in it was caused by the discovery that it is involved in leukotrienes biosynthesis, which plays an important role in supersensitivity and inflammation processes [453]. The active form of native monooxygenase-1 form soya beans contains iron(III), which in the course of the dioxygenation process is able to form intermediate of organoiron. One of the lipoxydioxygenase models is the system $[FeCl_2(cyclam)]ClO_4$–O_2–linoleic acid (cyclam-1,4,8,11-tetraazacyclotetradecane) [454]. This system possesses a high capacity to linoleic acid peroxidation and simultaneous formation of $^1O_2$ ($^1\Delta_g$). The supposed mechanism of dioxygenation in general comes in the transformation of triplet O_2 ($^3\sum_g$) weakly joined to iron(III) complex into singlet $^1O_2$ and linoleic acid dioxygenation by the latter.

3.2.3.2 Models of Copper-Containing Dioxygenases

One of the best models of copper-containing dioxygenases is the relatively simple catalytic system CuL_2^{2+}–O_2–substrate–OH^-–solvent, where L is o-phenanthroline or bipyridyl [455]. It is very active in ketone oxidation (cyclohexanone, α-tetralone, acetyne, heptanone-3, methylnaphthylketone, etc.) in water–alcohol media at temperatures up to 40–50 °C in the presence of bases (NaOH, $Ca(OH)_2$, etc.). The maximal rate is observed at about PH=12, which testifies about substrate ionization importance and its transfromation into carbanion.

Therefore, active forms of substrates in alkaline water–alcohol medium (1:1) are carbanions, the negative charges of which are concentrated on carbon atoms. The extreme activity of carbanions in joining reactions was mentioned earlier [456]. They also play an important role in processes catalyzed with dioxygenases [455, 457]. Copper(II) complex with *o*-phenanthroline serves as an active form of catalyst. Its composition depends on the ratio between ligand and Cu^{2+}ion. At ratio 1:1 и 2:1 and at PH 8–11.5 it exists in the forms $[Cu(phen)(OH^-)]_2^{2+}$ and $[Cu(phen)_2(OH^-)]^+$, accordingly. At PH>12, significantly less catalytic active dihydroxy complexes $[Cu(phen)(OH^-)_2]$ are formed. The main products of ketone oxidation are appropriate acids. Rate and selectivity of ketones oxidation are strongly dependent on oxygen partial pressure.

Ketone oxidation (anion form S^-) proceeds within the triple complex $[O_2\ldots L_nCu^{2+}\ldots S^-]$ [455]; likewise this proceeds in pyrocatechol dioxygenases [458], and also in the model systems of catechol-1,2-dioxygenase (bipyridyl Fe^{3+} Complexes with O_2 in 3,5-di-*tert*-butylcatechol oxidation) [459]:

$$L_nCu^{2+} + S^- \longrightarrow [L_nCu^{2+}\ldots S^-] \xrightarrow{O_2} [O_2\ldots L_nCu^{2+}\ldots S^-] \longrightarrow \text{products}$$

In this triple complex, substrate oxidation proceeds by a two-electron concerted mechanism, but not through the formation of free or coordinated radicals. Thus, appropriate peroxo-complex formation is supposed as an active intermediate:

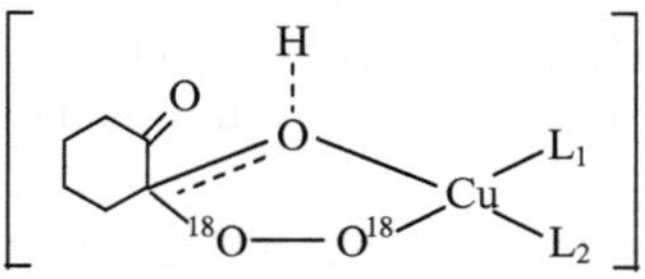

The same model system (Cu^{2+}–phen–O_2–S–OH^- solvent) is rather efficient in primary (methanol, *n*-propanol, benzyl alcohol, etc.), secondary (isopropanol, fluorobutanol, cyclohexanol) and many-atomic alcohols oxidation (sorbitol, mannitol, etc.) to acids and carbonyl compounds [455]. Many aspects here are similar to the aforementioned ketones' oxidation, namely, need of bases, alcohol transformation into anionic form, *o*-phenanthroline, bipyridine, or their derivatives use as ligand, use of the same central metal ion, similar (more similar than on ketone oxidation) dependence of rate and oxidation selectivity of number of aliphatic alcohols on oxygen partial pressure, significant influence of solvent on rate change, and selectivity of alcohols oxidation. Like in the case of ketones, on alcohol oxidation, triple complex formation is supposed $[Cu^{2+} \ldots S^- \ldots O_2]$($S^-$–alcoholate-ion), the activity of which depends to a great degree on the nature of the solvent.

This system with Cu(II) complexes, efficiently catalyzing low-temperature dioxygenation of primary alcohols and polyoles to acids, represents a chemical model of dioxygenases. Thus, catalase efficiency in the model system (biomimetic catalysts) is comparable (close) to dioxygenase efficiency.

In the process of alcohol dioxygenation, aldehydes are not final product predecessors—acids; this fact indicates the non radical oxidation character. In water solutions alcohol oxidation, proceeding much more slowly than for ketones, is apparently caused by more difficult formation of carbanion (alcoholates) which is a limiting stage.

Change in alcohol nature (one-atomic alcohols $\longrightarrow$ polyols) will change (tens and hundreds of times increases) not only the process rate but its mechanism as well, and that is why, for example, mannitol and sorbitol oxidation in water solutions is closer to ketones oxidation regularities and oxidation rate than those of primary alcohols. This is explained by the fact that carbanions are formed much easier from polyols than from one-atomic alcohols. Therefore, the mechanism of polyols dioxygenation is close to that of ketones [455].

On passing from water to non-water media, alcohol oxidation rate is often increased hundreds of times, and values of circulation numbers reaches 1–2 s^{-1} and higher, which is close to enzymatic processes. Thus, alcohols oxidation in alkaline aprotonic media proceeds, in one stage, to acids. Dependence of oxidation rate and selectivity on oxygen pressure apparently testifies to the fact that it takes part in the limiting stage of the process and two-electron reducing of oxygen is realized, proceeding by concerted mechanism without free radicals formation [455]:

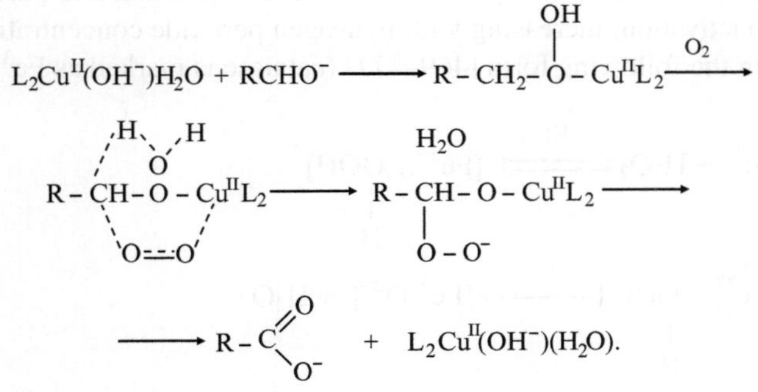

Thus, the catalyst remains active in practise, and the substrate half-transformation time reaches no more than 5–30 min under mild oxidation conditions (10–60 °C), the number of catalytic cycles per one copper ion reaches 2–5 s^{-1}, which makes this model catalytic system very efficient, selective, and rather similar to dioxygenase by its functional influence.

3.2.4 Model Systems of Catalase

Catalases, in great amounts, present in the animal, vegetable, and bacterial world, realize an extremely important reaction of rather toxic hydrogen peroxide excess removal from the organism:

$$2H_2O_2 \longrightarrow 2H_2O + O_2.$$

Each catalase has a molecular weight of about ~250,000 and contains four equal sub-units, which involve iron(III) and protoporphyrin IX. Three spacious enzymes structures were established from mushrooms and beef liver [460–462]. Axial ligand of reactive center is tyrosine; its hydroxyl group joins the fifth coordination place of iron(III) porphyrin complex. The imidazole part of histidine neighbors the reaction center. The role of the latter is apparently in being catalyst-base. The reagent is undissociated H_2O_2 at rather low PH [463,464].

High-valency oxo-iron particles were detected in catalases (like in peroxidases), which were marked as compounds I and II [463,465,466]. Compounds I and II are iron(IV)-oxo porphyrin π-cation radical [$P^{\bullet+}$(TyrO)Fe^{IV}O] (here iron is formally in Fe^{V} state) and iron (IV)-oxo porphyrin [P(TyrO)Fe^{IV}O] (where P is porphyrin and TyrO is tyrosine) [467,468]. Compound I is formed according to the following reaction:

$$\mathrm{P(TyrO)Fe(III)} + H_2O_2 \longrightarrow \mathrm{P^{\bullet+}(TyrO)Fe^{IV}O} + H_2O.$$

Interaction of compound I with the second H_2O_2 molecule results in the initial iron(III) complex regeneration and H_2O_2 reducing to O_2:

$$\mathrm{P^{\bullet+}(TyrO)Fe^{IV}O} + H_2O_2 \longrightarrow \mathrm{P(TyrO)Fe(III)} + H_2O + O_2.$$

The simplified mechanism of H_2O_2 enzymatic decomposition and parallel process of catalase inactivation, increasing with hydrogen peroxide concentration rising, is represented in the following form [469–471] (catalase is marked as Fe^{3+}):

$$Fe^{3+} + H_2O_2 \underset{}{\overset{K_1}{\rightleftharpoons}} [Fe^{3+}...\underset{\substack{|\\ H}}{O}OH] \quad (3.4)$$

$$[Fe^{3+}...\underset{\substack{|\\ H}}{O}OH] \longrightarrow [Fe^{5+}O^{2-}] + H_2O \quad (3.5)$$

$$[Fe^{5+}O^{2-}] + H_2O_2 \xrightarrow{K_2} Fe^{3+}O_2 + H_2O \quad (3.6)$$

$$[Fe^{3+}...\underset{\substack{|\\ H}}{O}OH] + H_2O_2 \xrightarrow{K_{in}} Fe^{3+} + H_2O + OH^{\bullet} + HO_2^{\bullet} \quad (3.7)$$

$$Fe^{5+}O^{2-} + 2H_2O_2 \xrightarrow{K'_{in}} Fe^{3+} + H_2O + 2OH^{\bullet} + HO_2^{\bullet} \quad (3.8)$$

Reaction (3.4) proceeds very quickly ($\kappa_1 = 0.6–1.1 \times 10^7\,M^{-1}\,s^{-1}$) [472] and does not limit the rate of the catalase process. Catalase peroxo-complex is turned into compound I (reaction (3.10)), which, having reacted in a two-electron way with the second molecule of hydrogen peroxide by reaction (3.5), results in products O_2 and H_2O generation ($\kappa_2 = (2–4)^{\bullet}10^7\,M^{-1}\,s^{-1}$) [472]. Simultaneously with compound 1 heterolytic decomposition (reaction (3.5)), homolysis of catalase peroxide complex also proceeds with active radicals $HO_2^{\bullet}$, and then $OH^{\bullet}$ formation (reaction (3.7)) [473]. Free radicals formed attack porphyrin active center of catalase, resulting in its inactivation [474, 475]. Free radical formation rates are increased with H_2O_2 concentration enhance, which is reflected in brutto-process (3.7). Thus, as can be seen from the scheme, the process of H_2O_2 decomposition to O_2 and H_2O is characterized most completely by the rate constant k_2. Inactivation rate constant (k_{in}) determines the rate of catalase inactivation due to the unproductive H_2O_2 decomposition, influencing enzyme deterioration. These rate constants k_2 and k_{in} can be determined simultaneously on kinetic curves linearization [471].

On H_2O_2 excess, catalase compound I can be transformed into compound II [471], and one more way of catalase destruction becomes possible (reaction (3.8)); thus the process of enzyme inactivation becomes more complicated. It was found in the comparative study of catalase from bull liver and yeasts *Candida boidinii* that enzyme stability is 35 times higher in the second case. Besides, the advantage of yeasts catalase is cheapness of its formation source, and simultaneous obtaining of alcohol oxidase, which opens up the possibility to apply such bi-enzymatic preparation for methanol-polluted waste water treatment [471].

Numerous model tests were carried out in order to reveal in detail the mechanism of H_2O_2 decomposition with catalase. However, many of these research works, carried out earlier, were made on the level of phenomenologic description, or were incomplete. One of the model systems, which imitates rather successfully the influence of catalase active center, is the system based on iron(III) complex with [5,10,15,20-tetrakis(2,6-dimethyl-3-sulfonatophenyl)porphyrinate] iron(III) hydrate [P′Fe(III)(H_2O)] instead of porphyrin complex with iron(III) as in catalase [476]:

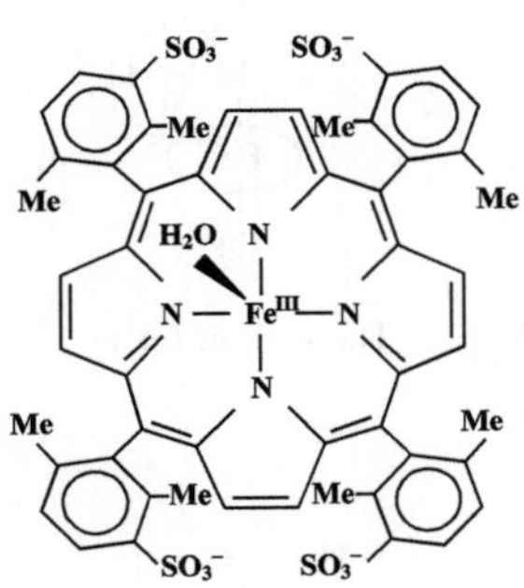

This complex is water soluble in the absence of protein, does not aggregate, and formation of μ-oxodimer is sterically aggravated. It forms high-oxidation oxo-complex of Fe(V) with H_2O_2 (it can be detected by the appropriate agent trap), which corresponds to the requirements of catalase active center modeling. As agent trap for iron(IV) oxo-complex detection, forming in the reaction with H_2O_2, 2,2′-asino-di-(3-ethyl-benzthazoline)-6-sulfonate (ABTS) was used.

The rate-determining stage is oxygen transfer from H_2O_2 or HO_2^- to [P′Fe(III)(H_2O)] Complex with the formation of iron(IV) oxo-complex, which will then rapidly oxidize ABTS tO $ABTS^{\bullet+}$ (peroxidase reaction type), and hydrogen peroxide to O_2 and H_2O (catalase reaction type). The rate constant of second order (k) of such transfer on limiting stage was determined spectrophotometrically by radical $ABTS^{\bullet+}$ ($\lambda = 660$ nm) accumulation. At PH 1–4, 7–9, and 11–12, $[P'Fe(H_2O)(H_2O_2)]^+$ (I), [P′Fe(OH)(H_2O_2)] (II), and $[P'Fe(OH)(HO_2)]^-$ (III) will be formed,respectively.

Proceeding from the dependence log k—pH, it became possible to identify three transition states (on low, intermediate, and high pH), which yielded [476]:

Low PH (I):

1) 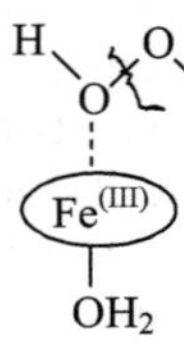 or 2) 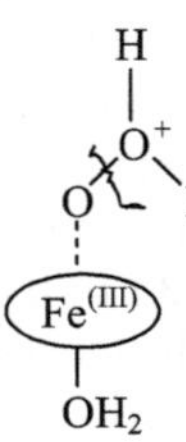

Homolytic and partially heterolytic breakage of O - O bond, which can result in iron (IV) complexes.

Heterolytic breakage of O-O bond on iron(IV)-oxoporphyrin π-cation formation and water molecule detachment.

Intermediate PH (II):

3) 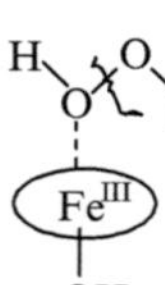 or 4) 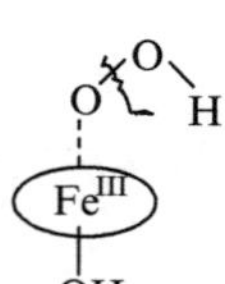 or 5) 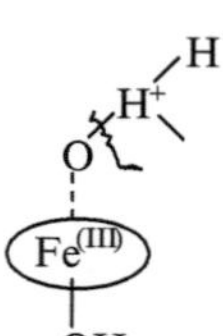

Thee same as for 1) The same as for 1) The same as for 2)

High PH (III):

6) 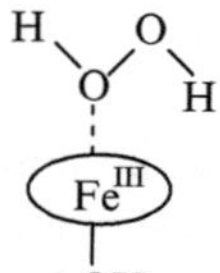

The same as for 1).

Products of O–O bond heterolytic and homolytic breakage for the cases I–III are Fe^{IV} oxo-complexes and π-cation radical complexes $[P^{\bullet+}Fe^{IV}O(OH)]$. The latter complex will react with H_2O_2 by a mechanism similar to the influence of catalase. Addition of nitrogenous bases (namely, 2,4,6-collydine, etc.) at low and mean pH value catalyzes oxygen transfer from hydrogen peroxide to $[P'Fe(III)(H_2O_2)]$. The mechanism of nitrogenous bases influence implies not only heterolytic breakage of O–O bond, but catalytic proton transfer to oxygen going away as water.

The model system was also studied, decomposing H_2O_2 to O_2 and H_2O on the basis of manganese(III) complex with the same ligand, but in neutral (PH 7.6) and highly alkaline (PH 12.1) media [477]. Thus, in dependence on pH, complexes $[P'Mn^{III}(OH)(H_2O_2)]$ (Ia), $[P'Mn^{III}(H_2O)(H_2O_2)]$ (Ib), $[P'Mn^{III}(OH)(HO_2)]$ (II), and $[P'Mn^{III}(OH)_2(HO_2)]$ (III) will be formed. Products of oxygen transfer from hydrogen peroxide ($X = H_2O_2$, $HO_2^{\bullet -}$) in complexes Ia and Ib are manganese(IV)–oxo porphyrin π-cation radical $[P^{\bullet+}Mn^{IV}O(Y)](Y = OH^-$ or $(OH^-)_2)$, and in complexes (II) and (III), manganese(V)–oxo porphyrins $[P'Mn^{V}O(OH)$ и $P'Mn^{V}O(OH)_2$, accordingly.

The following mechanism of hydrogen peroxide peroxidase and catalase decomposition is proposed:

$$\text{catalyst} + X \rightarrow \text{catalyst}\cdot X$$

$$\text{catalyst}\cdot X \xrightarrow{\text{rate-determining stage}} [\text{catalyst}(Mn^V) = O]^+ + H_2O(\text{or } OH^-)$$

$$[\text{catalyst}(Mn^V) = O]^+ + \text{ABTS} \xrightarrow{\text{rapidly}} [\text{catalyst}] + \text{ABTS}^{\bullet+} + H^+$$

$$[\text{catalyst}(Mn^{IV}){=}O] + \text{ABTS} \xrightarrow{\text{rapidly}} [\text{catalyst}(Mn^{IV}){=}O] + \text{ABTS}^{\bullet+} + H_2O$$

$$[\text{catalyst}(Mn^V) = O]^+ + X \xrightarrow{\text{rapidly}} [\text{catalyst}] + O_2 + H_2O.$$

Manganese(IV)–oxo porphyrin π-cation radical will better oxidize H_2O_2, while manganese(V)–oxo porphyrin, forming at PH>11.5, will preferrably oxidize ABTS, compared to $H_2O_2^{\bullet}$ Mn(V)-oxo-porphyrin realizes two one-electron reductions of ABTS (two $\text{ABTS}^{\bullet+}$ cation radicals will be formed), while iron(IV)–oxo porphyrin π-cation radical will more easily be subject to simple two-electron reduction. Manganese(IV)–oxo porphyrin formation at ~PH 11.5 is explained with the interaction of manganese(V) with unreacted manganese(III) porphyrin:

$$P'Mn^{III}(OH)_2 + P'Mn^{V}(OH)_2 \longrightarrow 2P'Mn^{IV}(OH)_2$$

Thus, ABTS oxidation (peroxidase reaction) is more favorable at high pH, and hydrogen peroxide oxidation at lower pH. This system is a model of manganese-containing non-hematic pseudocatalase that has been obtained from some microorganisms and other bacteria, which are unable to synthesize heme [478]. It represents six subunits, associated in a non-covalent way (molecular weight ~172,000), Containing one manganese atom per subunit. The active center of the enzyme is manganese binuclear complex, able to exist in four oxidation states: Mn(II), Mn(II); Mn(II), Mn(III); Mn(III), Mn(III); and Mn(III), Mn(IV) [478–481].

Other models of manganese-containing pseudocatalase from *Lactobacillus plantarum* were also proposed, such as Mn^{2+}–ethylendiamine-*N*,*N*′-diacetate (edda)–H_2O_2 systems. Reduction of H_2O_2 to H_2O was initiated by Mn^{II}(edda) reaction with active binuclear μ-oxo-bridge complex forming in this system

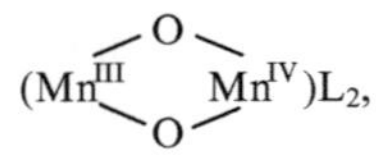

which was characterized with the same EPR signal as the active center of Lactobacillus [482], although its optic properties were different [478, 483].

The mechanism of hydrogen peroxide reduction (or *tert*-butylperoxide) in this model system is a two-electron one (without $OH^{\bullet}$ radical formation). It is similar to Lactobacillus and other manganese-containing bacterial catalases [484] (Scheme 3.21)

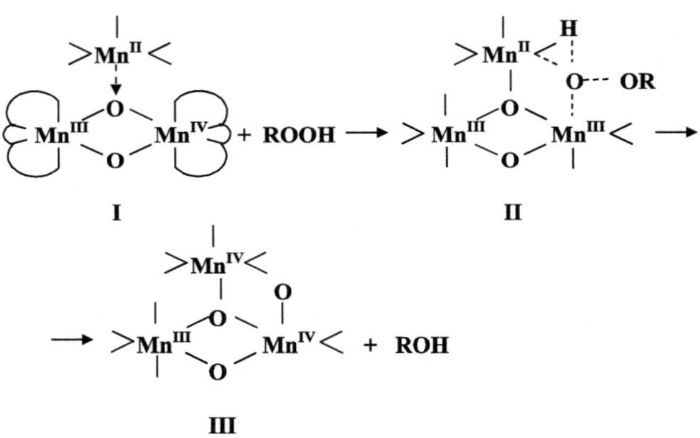

Scheme 3.21 Mechanism of peroxide reducing with Mn-containing catalase

where R = H or $(CH_3)_3C$.

In this proposed mechanism scheme, Mn^{II}(edda) complex reacts with the bridge of mixed-valent complex, forming a three-nuclear structure, which interacts with hydroperoxides (I⟶II). Breakage of the O–O bond in hydroperoxides is heterolytic and becomes easier by oxygen introduction into the μ-oxo-bridge between Mn^{IV} atoms in structure III. However, there are no direct proofs that such structures are really formed. Besides, it is unclear whether such structures occur in biological catalases.

The influence of ligand nature, bound to the binuclear center of manganese model complex, and also that of solvent, is apparently rather high and can result in different mechanisms of hydrogen peroxide dismutations. Thus, in the case of $[LMn_2^{II,II}(\mu - CH_3COO^-)](ClO_4)_2$ (where LH–*N*, *N*, *N*′, *N*′-tetrakis (2-methylbenzimidazole)-1,3-diaminopropane-2-ol) in methanol or acetonitrile (under similar conditions)

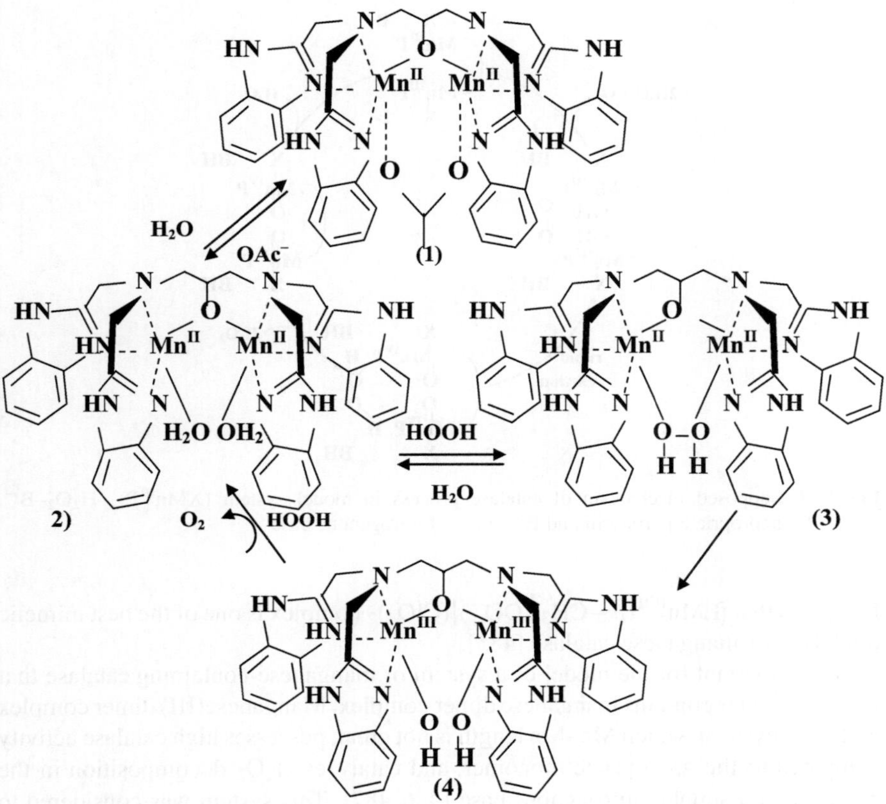

Fig. 3.26 Supposed mechanism of complex $[LMn_2^{II,II}(\mu-CH_3COO)](ClO_4)_2$ catalase activity

using the complex as a model of catalase, the mechanism of hydrogen peroxide dismutation is realized by means of two consecutive two-electron stages:

$$H_2O_2 + 2e^- + 2H^+ \longrightarrow 2H_2O$$

$$H_2O_2 \longrightarrow O_2 + 2e^- + 2H^+$$

which are caused by oxidation degree change of model catalyst active center $Mn_2^{II,II} \leftrightarrow Mn_2^{III}$. The scheme of the supposed mechanism of model complex $[LMn_2(CH_3COO)](ClO_4)_2$ catalase activity is given in Fig. 3.26 [485].

Compound $[LMn_2^{II,II}(\mu-CH_3COO)]^{2+}$ (form (1)) is catalase inactive. Hydrolysis (form (2)) transforms it into an active form, which, after interaction with H_2O_2, forms the appropriate peroxo-complex (form (3)). Subsequent transformation (3) results in HO–OH bond breakage on account of its reduction with $Mn^{II}Mn^{II}$ complex formation. Within the area of system existence, neither $Mn_2^{II,III}$, nor $Mn_2^{III,IV}$ formation was discovered in conditions of hydrogen peroxide catalase dismutation.

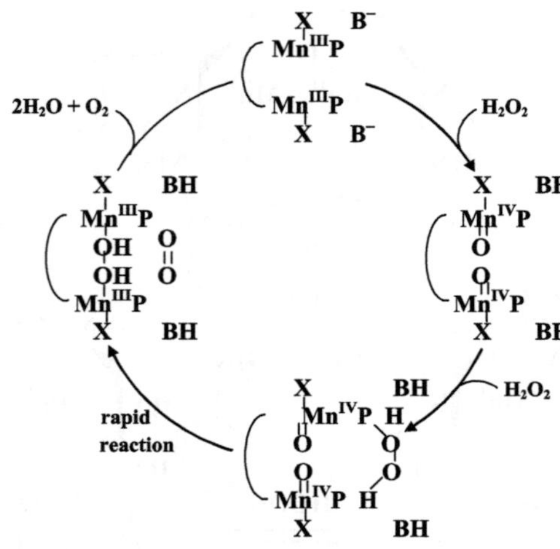

Fig. 3.27 Supposed mechanism of catalase process in model system $(XMn^{III}P)_2$–H_2O_2–B^-, where P is appropriate porphyrin and B^- anion of nitrogenous base

It appears that $[LMn_2^{II,II}(\mu-CH_3COO^-)](ClO_4)_2$ complex is one of the best mimetic models of di-manganese catalase [485].

It is important for the model of a system of manganese-containing catalase that its active center contains manganese dimer complex. Manganese(III) dimer complex with porphyrin, in which Mn–Mn length is not great, possesses high catalase activity compared to the appropriate monomer, and catalyzes H_2O_2 decomposition in the presence of a suitable nitrogenous base [486, 487]. This system was considered to be a functional part of manganese-containing catalase, which also contains in its active center two manganese ions. This model reaction proceeds via the formation of an appropriate $Mn(IV)_2$ complex involving homolytic breakage of the O–O bond intermediate. $Mn(IV)_2$ complex formation is the rate-determining stage.

The mechanism of this catalase process is given in Fig. 3.27 [488]:

Taking into account that for the model system a close approach of two manganese(III) ions isrequired, dimanganese(III) complex was synthesized with 1,8-bis [5-(10,15,20-trimesityl)porphyrinyl] anthracene (1) [488], which can be oxidized with hypochlorite into the appropriate $Mn(IV)_2$ complex (2):

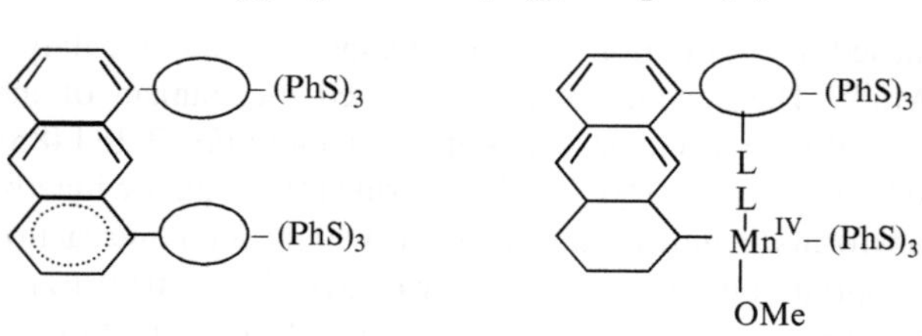

where Mn–$(PhS)_3$ is thrice-substituted triphenyl porphyrin, L = MeO or OH.

On hydrogen peroxide influence on complex (2) (λ_{max} = 422 nm) in CH_3CN at a temperature of −40 to −20 °C, the catalase process of H_2O_2 decomposition to O_2 and H_2O proceeds, and complex (1) is reduced (λ_{max} = 468 nm). The rate-determining stage is $Mn(IV)_2$ dimer formation.

Another mimetic approach to pseudocatalase modeling is binuclear manganese complex $[Mn_2L^{4,4}(O_2CMe)_3]$ (a) and $[Mn_2L^{3,3}(O_2CMe)_3]$ (b), where L is a macrocyclic ligand Containing two 2,6-bis (iminomethyl-4-methyl) phenolate parts, linked to each other (through nitrogen atoms of iminogroup) with bridges from $(CH_2)_m$ and $(CH_2)_n$ groups (m, n = 2,3;2,4;3,3;4,4—the amount of CH_2 groups in the bridge):

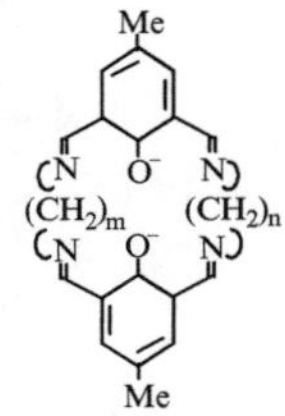

Two acetic groups O_2CMe are arranged in axial position with regard to manganese, and two phenol oxygens in equatorial position. Each surrounding manganese ion is pseudooctaedric, and Mn...Mn distance is 2,978 Å. It was thus established that complexes (a), and especially (b), possess catalase activity, realized by means of internal transformation between $Mn^{II}Mn^{III}$ and $Mn^{II}Mn^{IV}(=O)$ states of binuclear center [489]. Evidence of the manganyl-like state $Mn^{II}Mn^{IV}(=O)$ occurence in the reaction course of H_2O_2 decomposition is the appearance of a charge transfer band at λ = 530 nm [489] and fixing of bond Mn = O oscillation frequency at 0 °C, equal to 730 cm^{-1} [490].

3.2.5 *Peroxidases and Their Model Systems*

3.2.5.1 Functioning of Peroxidases and Their Models

Peroxidases represent a large group of enzymes, catalase relatives, widely distributed in animals, vegetation, and aerobic organisms, using hydrogen peroxide and hydroperoxides [491] or peracids [492] for substrates oxidation (alcohols, phenols, aromatic amines, ferrocyanide, and other compounds).

The active center of enzyme represents hematic iron(II or III) (ferro- or ferriperoxidase, respectively) bound to histidine (N_{His}). Within the hemtic iron's interaction with molecular oxygen, intermediates I, II, and III will be formed. The formula of all three formations and their marks are as follows [493]:

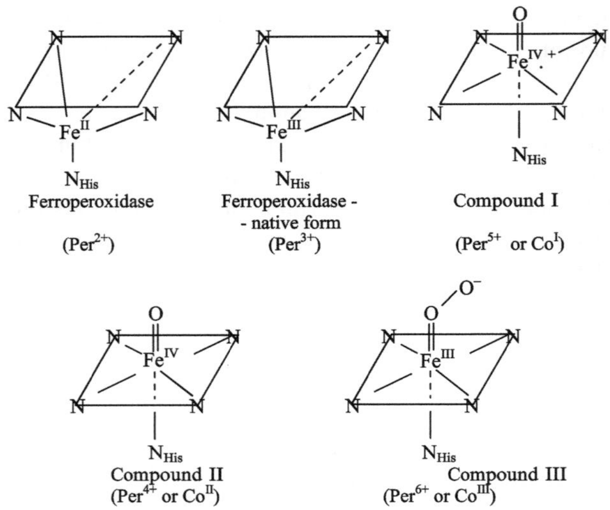

In compound I, the formal oxidation degree of iron corresponds to 5+, and in compound II, 4+. In compound III, molecular oxygen in the reduced form is bound to iron as axial ligand. For research, horseradish peroxidase is used more often (HRP). In this case, compounds I and II are often marked as HRP-I and HRP-II,respectively.

Compounds I and II rapidly oxidize, for example, phenols and aromatic amines ($\kappa \approx 10^7\,\mathrm{M}^{-1}\cdot\mathrm{s}^{-1}$) [494] according to the following scheme:

$$\mathrm{HRP} + \mathrm{H_2O_2} \longrightarrow \mathrm{HRP} - \mathrm{I}$$

$$\mathrm{HRP} - \mathrm{I} + \mathrm{RH} \longrightarrow \mathrm{HRP} - \mathrm{II} + \mathrm{R}.$$

$$\mathrm{HRP} - \mathrm{II} + \mathrm{RH} \longrightarrow \mathrm{HRP} + \mathrm{R}.$$

The supposed mechanism of compound I formation was based on the well-known cytochrome C peroxidase (CCP) structure, and is shown in Fig. 3.28 [495, 496].

Histidine and arginine play an important role in compound I formation. It was thus found that hydrogen peroxide forms hydroxide bonds with histidine's N_ε. Using a quantum chemical method (INDO/S–ROHF/Cl) [497], the electron structures of two possible intermediates—one with HOOH and another with COOH ligand—were calculated and compared with known spectral data. The possibility was shown in these calculations of both peroxide and peroxoanionic intermediate formation [495].

It appears that during the evolution, peroxidases appeared earlier than catalases, and this is revealed via their composition. Thus, horseradish peroxidase has a molecular weight of about 44,000 and contain one iron atom joined to protoporphyrin IX and histidine, while catalase is made in a more complex way (molecular weight is $\approx$ 250000). The latter enzyme involves four subunits, each of them having one

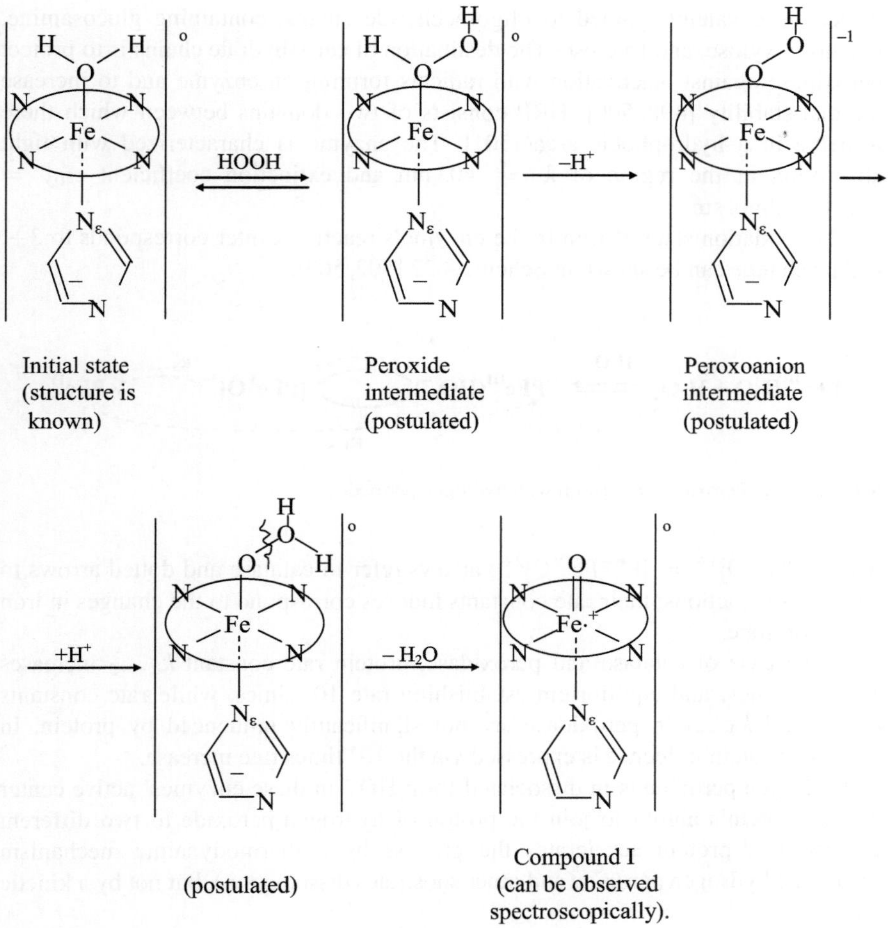

Fig. 3.28 Supposed stages of compound I formation

iron atom. In peroxidase the fifth place in the iron complex coordination sphere is occupied by histidine imidazole, but in catalase by carboxylate group. There are also other differences. A short comparison of these two enzyme groups will begiven later.

Peroxidases (like catalases), unlike other enzymes, are much less specific regarding substrate, although the reasons for such low substrate specificity are not clear. They belong to a type of most active enzymes. Besides, some peroxidases (namely, chloroperoxidase) manifest catalase activity with histidine ligand, and catalase dissociated on monomeric subunits possesses peroxidase activity.

In the most often studied peroxidase, HRP, iron(III) complex with protoporphyrin IX (1,3,5,8-tetramethyl-2,4-divinylporphyrin-6,7-dipropionic acid) (protoheme PFe(III)) and with axial histidine ligand is non-equivalently joined to protein.

Protein is covalently joined to oligosaccharide chains, containing glucosamine, mannose, xylose, and fructose. The destination of carbohydrate chains is to protect peroxidase against inactivation with radicals forming on enzyme and to increase thermal stability [498, 500]. HRP consists of two domains between which there is heme in a hydrophobic area [501]. The enzyme is characterized with light absorption in the region of $\lambda = 403\,\text{nm}$ and extinction coefficient $\varepsilon_{403} = 102000\,\text{L/mol sm}$.

The oxidation state of iron in the enzyme's reaction center corresponds to 3+, and its change can be shown in Scheme 3.22 [502, 503]:

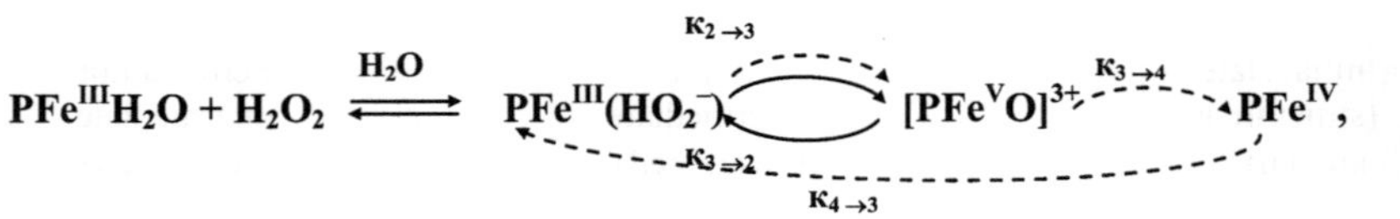

Scheme 3.22 Peroxidase oxidation with hydrogen peroxide

where $[PFe^{V}O]^{3+} = [P^{\bullet+}Fe^{IV}O]^{2+}$; arrows refer to catalase and dotted arrows to peroxidase reactions; their rate constants indexes correspond to the changes in iron oxidation state.

In the case of catalase and peroxidase, protein rate constant $k_{2\longrightarrow 3}$ increases 10^3–10^5 times, and equilibrium establishing rate 10^7 times, while rate constants $k_{3\longrightarrow 4}$ and $k_{4\longrightarrow 3}$ in peroxidase are not significantly influenced by protein. In catalase, protein influence is expressed via the 10^5 times rate increase.

Hydrogen peroxide is in dissociated form HO_2^- in these enzymes' active center due to protein's ability to join the proton of hydrogen peroxide to two different centers, i.e., protein accelerates the process by a thermodynamic mechanism (changing hydrogen peroxide and other substrates dissociation), but not by a kinetic one.

Iron porphyrin in solution is rapidly degraded with hydrogen peroxide. One of protein's functions is, apparently, its protecting of deterioration by active hydrophobic encircling of the iron porphyrin center [502, 503]. Due to the protein, enzyme is able to limit and control small alkyl hydroperoxides (ROOH) and H_2O_2 access to the hematic active center. Enzyme is able to promote the heterolysis of the O–O peroxide bond and generation of oxoiron(IV) porphyrin π radical cation—compound I, by acid–base catalysis and active center encircling. The result of peroxidase's active center encircling by protein is that subsequent one-electron oxidation of substrate proceeds near the active center, without its direct involvement in the oxoiron(IV) complex; it proceeds likewise in the case of cytochrome P-450 monooxygenase.

The general sequence of substrates (SH) oxidation reactions, catalysed with peroxidase from horse, is the following [504]:

$$HRP + ROOH \longrightarrow HRP - I + ROH,$$

$$HRP - I + SH \longrightarrow HRP - II + S$$

$$HRP - II + SH \longrightarrow HRP + S,$$

$$S^{\bullet} + S^{\bullet} \longrightarrow \text{Products.}$$

where R = H or alkyl group.

Reduction of HRP-II (Compound II) to HRP is usually the rate-limiting stage of the process so that absorption spectrum of compound II is usually observed [505]. Reactions rates of peracids, alkyl hydroperoxides, and hydrogen peroxide with HRP are similar, while in the case of peracid derivatives, the O–O bond is broken much easier than with hydroperoxide derivatives. The difference in peroxidase and cytochrome P-450 influence on substrate consists of the fact that peroxidases remove electrons from their substrates at the same time with the accompanying ferryl oxygen reducing to water, while monooxygenases transfer ferryl oxygen atom onto substrate.

This confirms the earlier idea on breakage in the place of substrate joining in monooxygenase (cytochrome P-450) and in peroxidase. Monooxygenase function of transfer from ferryl particle to substrate is realized, provided that the substrate is able to interact directly with ferryl oxygen, and peroxidase influence will predominate if the protein structure prevents this interaction [506,507]. Thus, it was confirmed that substrates transfer the electron, being close to heme, without direct contact with ferryl oxygen. Therefore, substrate radical formed on electron transfer is able to be removed more rapidly than if it had been bound to the radical-capable atom of ferryl oxygen [508–511].

The reason for peroxidase model systems' study is the need for a more detailed revealing of high-valency (intermediate) compounds, their structure, and mechanism of their interaction with substrates. To reveal the mechanism of peracids, hydroperoxides, and hydrogen peroxide interaction with peroxidases, model systems, resembling the HRP active center, were elaborated and studied in organic (CH_3OH) and water organic solvents (with the aim of preventing porphyrin complexes aggregation) [466].

The peroxidase model system involved various model derivatives of hemin (catalysts), μ-chlorperbenzoic acid (oxidant), and tri-*tert*-butylphenol (substrate). Reaction of hemin derivatives with peracid RCOOOH or ROOH was proceeding by heterolytic disintegration, with the formation of acid and two-electron oxidized hemin, corresponding to the compound HRP-I. The limiting stage was oxidized intermediate formation

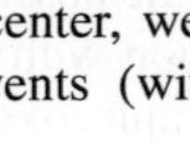
.

Base B (imidazole), bound to hemin complex, raises the possibility of electron transfer from it to ROOH with subsequent formation of ferryl particle at the rate-limiting stage:

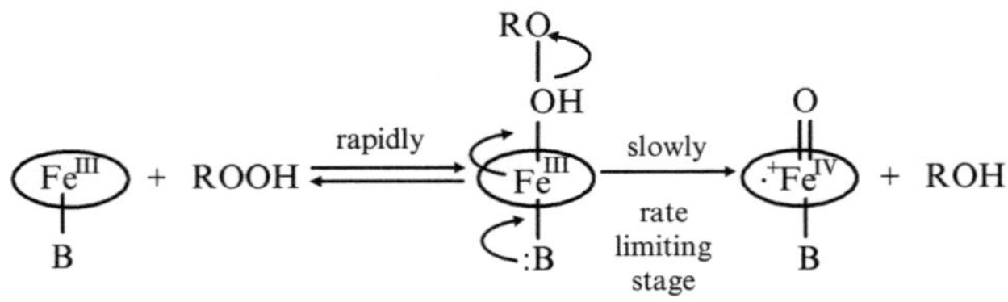

Rate of peracid (or ROOH) transformation by second-order reaction corresponded to

$$\frac{-\mathrm{d[RCOOOH]}}{\mathrm{dt}} = k + 2[\mathrm{hemin}][\mathrm{RCOOOH}].$$

Reaction rate of hemin derivatives with RCOOOH or ROOH increases with the capacity of R increasing to detach the electron.

However, the rate constant of hydrogen peroxide interaction with hemin derivatives in methanol (k_2), compared to interaction with HRP, in the case of the model systems turns out to be much smaller, than for peroxidase ($\kappa_2 = 12\,\mathrm{M}^{-1} \cdot \mathrm{s}^{-1}$ and $1.8 \cdot 10^7\mathrm{M}^{-1} \cdot \mathrm{s}^{-1}$, accordingly). One of the reasons can be a much lower possible concentration of the base B (~0.2 M) in the model systems compared to its local concentration in protein (>20 M). In spite of this, the model systems demonstrate functional similarity with HRP[447].

The most often studied model systems of peroxidase are the systems of $PFe^{III}L–H_2O_2–S$ type (where P is protoporphyrin IX or its substituted derivatives, L is imidazole and other ligands, and S is substrate: phenols, aromatic amines, etc.), in which oxo-iron(IV) porphyrin complexes are formed imitating compounds I and II of peroxidase. Because of their low stability, they are usually used on low temperatures and non-water media.

However, it was shown [512–514] that ionic oxo-iron(IV) porphyrins are stable in alkaline water solutions. Therefore, the possibility to compare the process of synthetic oxo-iron(IV) porphyrins oxidation with their enzymatic analogs has arisen. Such oxo-iron(IV) complex is, for example, oxo-iron(IV) tetra (2-*N*-methylpyridyl) porphyrin (O = Fe(IV)T2MPyP), which is rather stable, especially in water media. Oxidation mechanism study of various phenols with this oxo-complex leads to the conclusion that the latter, having no steric limitations for the reaction with ferryl oxygen, will oxidize phenols by hydrogen atom removal from these substrates.

Phenols oxidation with the compound II HRP, due to the steric obstacles, created by protein surroundings, will be realized only near (on the edge of) the hematic ring [504]. Thus, the study of compound II HRP helps one to understand the peculiarities of peroxidase active center interaction with substrate.

Various oxidants containing oxygen atom, such as nadacids [515–518], imidazole [311, 518–520], hypochlorite [521], and also OBr^-, ClO_2^-, BrO_3^-, IO_4^-, μ-nitrobenzoic acid, etc. [522], transfer it onto the iron(III) complexes with porphyrin, forming oxo-iron(IV) porphyrin π-cation radical particles. Thus, intermediate compound I of horseradish peroxidase and catalase is imitated, which is generated by these enzymes on interaction with hydrogen peroxide. Rate constant of peroxidase interaction with these oxidants fluctuates within the range 10^8–10^6L/mol s depending on oxidant structure (it was studied by the stop flow method [522]).

Compound I is the direct oxidant of many peroxidase substrates, and the reaction proceeds more often by one-electron mechanism with the formation of compound II, and the latter substance will regenerate the initial peroxidase by one-electron mechanism as well.

As was already stated, iron is present as ferryl complex in compounds I and II [523], and additional oxidizing equivalent in compound I is localized either on porphyrin macrocycle, or on one of the protein functional groups [524]. Thus, the properties of compounds I and II do not depend on the substrate nature. Substrates can realize direct contact with heme (ferrocyanide, sulfite, nitrite, thyocyanate ions), or do not realize it (aromatic phenols and amines, various organic compounds) [525]. Substrates of the first type are electron donors and those of the second type hydrogen atom donors. The nature of the first type substrates of peroxidase with enzyme is not quite clear. With second type substrates (organic substrates), peroxidic oxidation peripheric mechanisms of electrons transfer will be realized with functional groups including the enzyme's active center. These groups realize transfer from substrates (adjoining the proteic globulae) on the hematic center [500].

It was shown in model peroxidase systems that electron transfer can proceed via the heme periphery, even if the iron atom of heme is quite accessible [526]. Taking this into consideration, it is supposed that oxidation with free radical formation proceeds only in the case when outer-sphere mechanisms of electron transfer are realized [500].

The study of model peroxidase reaction $PFe^{III}L$–H_2O_2–S–B, (where PFe^{III} is protohemin, L is histidin, S is substrate, B is various bases: imidazole, 1-methylimidazole, 2-methylimidazole, 4-methylimidazole, or 1,2-dimethylimidazole) have revealed the role of bases in this system activation [527]. In the absence of bases, hemin manifests weak activity in potassium ferrocyanide (substrate) to its ferricyanide in basic solutions (up to PH 10); in acid solutions (PH 5 and lower), it is completely absent. It was shown that hydroxyl ion joining to iron(III) hemin causes peroxidasic activity. Adding bases into the system has contributed to the reaction rate increasing on basic pH values. It is caused by one base joining to a highly oxidized iron state in intermediate porphyrin complex. Joined base, being a good donor, induces electron transfer to the oxo-ferryl center, thus activating its oxygen.

Thus, analogy can be traced with the action of peroxidase compound II. This, being bound to histidin's imidazole group, will also strongly activate peroxidase reaction. On higher excess of the base, i.e., on two ions of base joining by fifth and sixth coordination places, peroxidase reaction will be inhibited, which is caused in this case by the complication of H_2O_2 access to the active center. It was mentioned above that peroxidases are characterized with much lesser specificity in relation to substrates than other enzymes. One of peroxidase types, namely, chloroperoxidase, is able to catalyze halogenation of substrate's C–H activated bond. The active center of this enzyme represents iron complex with protoporphyrin IX, and histidin of fifth coordination place is substituted with thiolate ligand. This enzymatic active center is joined to protein by means of bridge hydrogens of two propionate side chains and thiol ligand, coordinated with iron and bound through hydrogen with two amide groups of peptide.

Chloroperoxidase, which has first been extracted from fungus [528], not only catalyzes substrates halogenation, but also realizes many other reactions characteristic for peroxidases, catalases, and cytochromes P-450 [529]. Unlike these enzymes, the mechanism of halogenation with chloroperoxidase is still not quite clear. Halogenation reaction with chloroperoxidase proceeds in the presence of H_2O_2 and Cl^- at PH 3. Substrates can, for example, be such diketones as monochlordymedin (1), which on interaction with enzyme are turned into dichlordymedon (2) (Scheme 3.23). Chlorination with the enzyme's active center proceeds in the following way [530]:

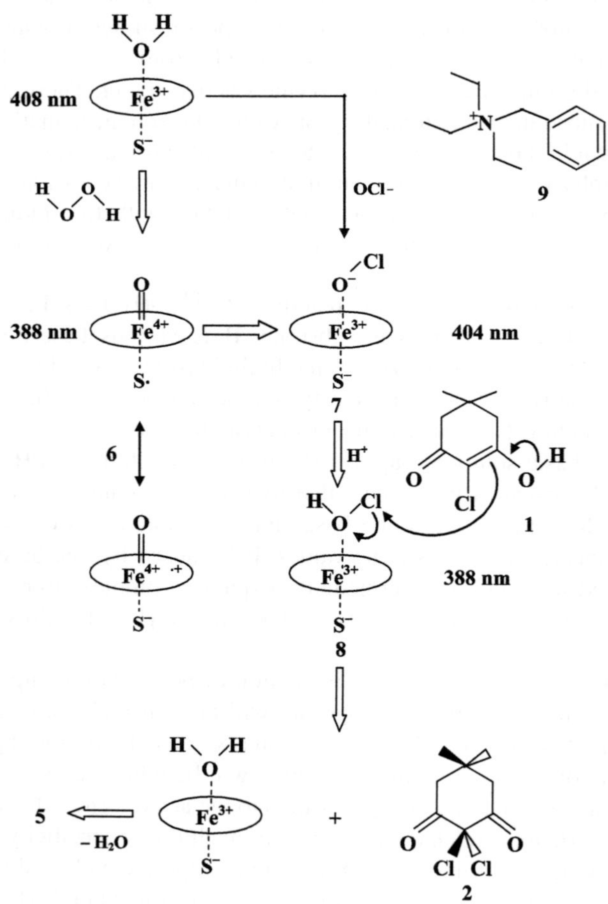

Scheme 3.23 Mechanism of chloroperoxidase catalytic cycle

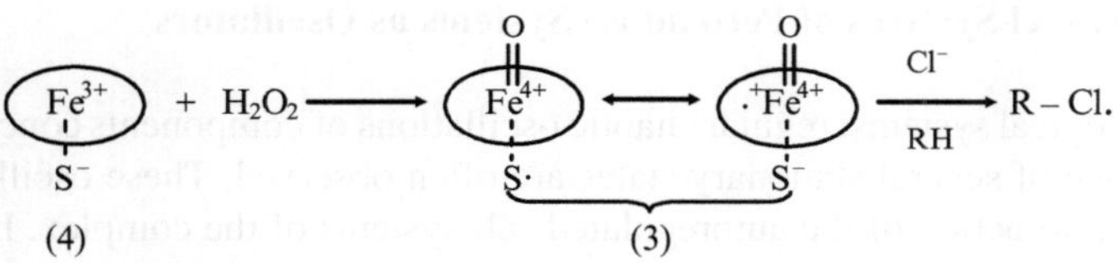

via intermediate (3), which is formed by the initial active center of chloroperoxidase (4) interaction with hydrogen peroxide, and its further reaction with Cl^- and RH.

As identification of chloroperoxidase active states in chlorination reaction proceeding is still difficult, analogs of chloroperoxidase active center were synthesized (model systems of heme-thiolate type), namely, compound (5) [531]. Complex (5) is a high-spin iron(II) porphyrinate, the waterless form of which absorbs at $\lambda_{max} = 400$ nm, which is identical with the initial active center of chloroperoxidase ($\lambda_{max} = 399$ nm).

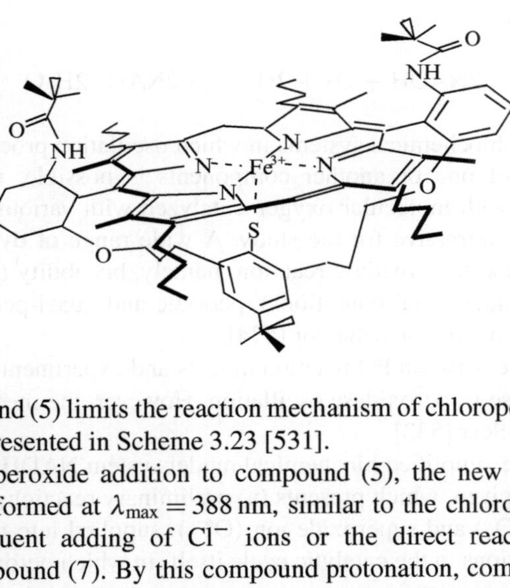

Model compound (5) limits the reaction mechanism of chloroperoxidase reaction center, which is presented in Scheme 3.23 [531].

On hydrogen peroxide addition to compound (5), the new complex (6) will be quantitatively formed at $\lambda_{max} = 388$ nm, similar to the chloroperoxidase active state (3). Subsequent adding of Cl^- ions or the direct reaction of (5) with OCl^- yields compound (7). By this compound protonation, compound (8) will be obtained, which, interacting with substrate (1), will realize chlorination reaction. Final products of this process include the initial compound (5) and chlorinated substrate (2). Therefore, intermediate compounds (7) and (8) were identified, and it was shown that compound (5) is a model of heme-thiolate proteins [531]. This is especially important, since the compounds of (7) and (8) type have apparently not been identified in an enzymatic chlorination process because of their short lifetimes. Introduction of benzyltriethylammonium salt (9) into the system promotes the formation of sufficient OCl^- in CH_2Cl_2 so as to transfrom (5) into (7) quantitatively.

3.2.5.2 Model Systems of Peroxidase Systems as Oscillators

In real biological systems, regular chaotic oscillations of components concentrations and presence of several stationary states are often observed. These oscillations are caused by interaction of the autoregulated sub-systems of the complex. In the case of oscillating systems with O_2 and H_2O_2 participation, the presence of oxygen oscillator can be spoken of, the main peculiarity of which is oxygen inexhaustibility in the gaseous phase and the unlimited possibility of its passing into solution [532]. In the presence of reducers (or enzymatically in biological systems), hydrogen peroxide may be formed, often playing the role of one of the components of an oxygen oscillator. A review of similar oscillating biochemical systems is given in [532]. A number of works [533, 534] are devoted to the study and review of peroxidase–oxidase (PO) oscillator model in vitro (systems in which H_2O_2 can be substituted with O_2 on such substrates oxidation as NADH).

In the presence of NADH as hydrogen donor, the general stoichiometry of such PO reaction will be

$$2NADH + O_2 + 2H^+ \longrightarrow 2NAD^+2H_2O$$

Among numerous biochemical systems in which oscillation processes the contents, where a change of one or another components is possible, model reaction of NADH oxidation with molecular oxygen catalyzed with various catalases in vitro became especially attractive for the study. A wide range of dynamic behavior is manifested in peroxidase–oxidase reaction, namely, bistability (occurrence of two stable positions under equal conditions), periodic and quazi-periodic oscillations, and various forms of chaotic behavior [534].

Correspondence between PO reaction models and experiment is being gradually revealed in the case of peroxidase oscillation. However, the mechanism of the PO reaction is still unclear [533].

In Fig. 3.29, the simplified biochemical model system NADH–O_2–Mn^{2+}–Per^{3+} of PO reaction is given, which presents (a) preliminary reactions producing hydrogen peroxide (H_2O_2) and superoxide ion ($O_2^{\bullet -}$), supplied into the catalytic cycle, and (b) main reactions in the catalytic cycle itself, in which peroxidase (Per^{3+}) is a catalyst.

In this model system, H_2O_2 is needed to initiate NADH oxidation with peroxidase in the catalytic cycle. The main source of H_2O_2 is NADH reaction with O_2, which proceeds spontaneously, but very slowly [535]. Methylene blue (MBH) catalyzing this reaction is introduced additionally into the system.

Another way of H_2O_2 production is $O_2^{\bullet -}$ generation and NADH oxidation with it:

$$O_2^{\bullet -} + NADH + H^+ \longrightarrow NA^{\bullet} + H_2O_2$$

which is catalyzsed with Mn^{2+}[536] ions:

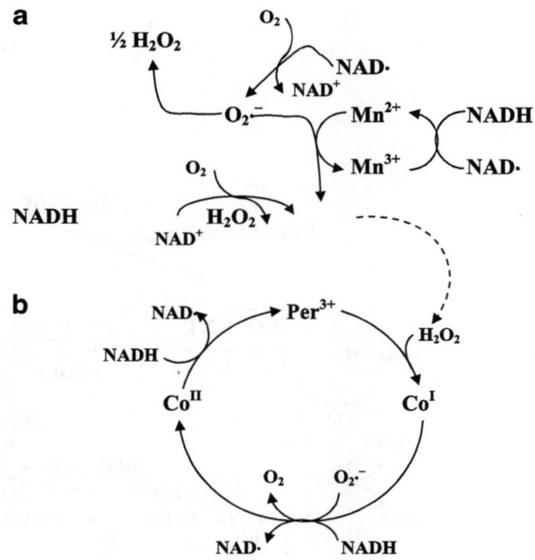

Fig. 3.29 Simplified biochemical model system showing (**a**) preliminary reactions and (**b**) main PO reactions [533]

$$Mn^{2+} + O_2^{\bullet} \xrightarrow{2H^+} Mn^{3+} + H_2O_2$$
$$Mn^{3+} + NADH \longrightarrow Mn^{2+} + NAD^{\bullet} + H^+$$

Mn^{2+} acts here like superoxide dismutase, which agrees with Mn^{2+} influence on peroxidase–oxidase oscillator [84].

H_2O_2 formed one way or another will be involved in the catalytic cycle b, in which it starts the main reactions of this cycle:

$$Per^{3+} + H_2O_2 \longrightarrow co\,I + H_2O$$
$$co\,I + NADH \longrightarrow co\,II + NAD^{\bullet}$$
$$co\,II + NADH \longrightarrow Per^{3+} + NAD^{\bullet} + H_2O$$

where CO I and CO II represent compounds I and II, accordingly.

Reactions of H_2O_2 interaction with co I and co II turn out not to be involved in this cycle:

$$co\,I + H_2O_2 \longrightarrow Per^{3+} + O_2$$
$$co\,II + H_2O_2 \longrightarrow Per^{3+} + HO_2$$
$$co\,I + H_2O_2 \longrightarrow co\,II + HO_2^{\bullet}$$

and also many other reactions of co III, Per^{2+}, and other particles formation and consumption, described in detail in [533]. Reactions of co III render a regulating influence on NADH oxidation by Per^{3+} inactivation with oxygen.

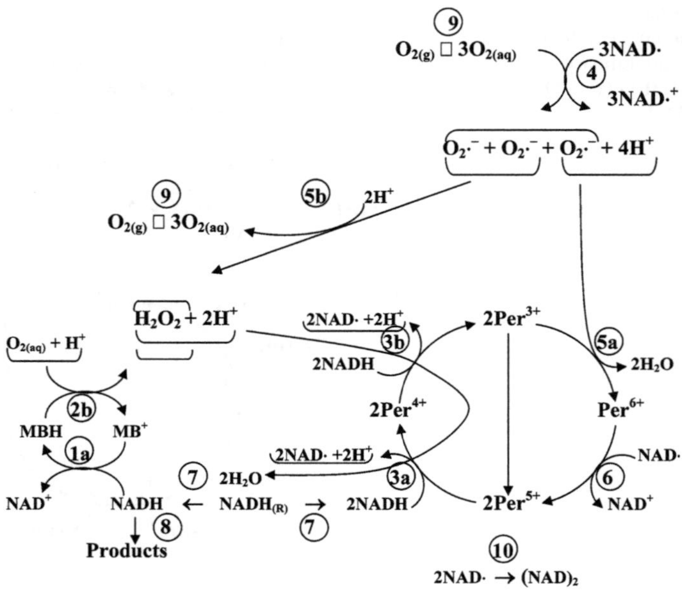

Fig. 3.30 Model of oscillator peroxidase-NADH. Numbers 1–6 represent the necessary sequence of events, which will form the cycle on their interaction

Recently, new, so-called "detailed models" of PO reaction were created [528, 537, 538]. Among them, the so-called "Urbanolator" model is to be mentioned, described in detail in [533], and also the model proposed in [528], in which not only the influence of peroxidase itself is imitated, but also the totality of reactions, comprizing the system, in which peroxidase is only one of the ingredients. It is important to note that, under certain conditions, this model works as an oscillator, as soon as periodic oscillations of the system ingredients contents are observed, namely, increasing and decreasing of dissolved O_2 concentration, initial HRP, compound III (co III), $O_2^{\bullet -}$, and H_2O_2. Initial ingredients of such a model system were HRP, O_2 (gas), O_2 (solution), NADH, and MBH. The scheme of the model peroxidase–NADH of oscillator functioning is given in Fig. 3.30 (some of the model work's simplified scheme is taken from [531]). In this model, encircled numeration of reactions corresponds to the sequence of their running, which creates the oscillation. Numbers of reactions with indexes "a" and "b" are conceptual bound. The brackets below and above the particles correspond to the resulting stoichiometry. HRP is marked here as Per^{3+}, peroxidase COmpounds I, II, III—Per^{5+}, Per^{4+}, Per^{6+}, respectively, gaseous O_2–O_2(g), dissolved in water O_2–O_2(aq).

Model construction begins from reaction 5a, in which initial peroxidase Per^{3+}, interacting with $O_2^{\bullet -}$ and H^+, will be slowly turned into Per^{6+}, and via forming short-living intermediates Per^{5+} and Per^{4+} which will be rapidly reduced back into Per^{3+} (reactions 6, 3a, and 3b) [539].

Hydrogen peroxide H_2O_2 can at once oxidize Per^{3+} to Per^{5+} (reaction 2); and Per^{5+} will then react with NADH with the formation of Per^{4+} and $NAD^{\bullet}$ (reaction 3a). Per^{4+}, having interacted in its turn with NADH, will be transformed into Per^{3+} and $NAD^{\bullet}$, and therefore, the oscillation cycle of Per^{3+} contents in model system is over. In this oscillation process, catalyzed with peroxidase, Per^{6+} is the regulating intermediate in NADH oxidation with oxygen and, thus, peroxidase starts to obtain the capacity of self-regulation, by means of oscillation. Cyclic renewal of Per^{6+} is needed for periodic work of system. Reactions 5a and 5b function in common on the basis of superoxide $O_2^{\bullet-}$ as switch, realizing the control over decreasing and increasing of oxygen contents, which depends on $O_2^{\bullet-}$ reactivity.

Per^{3+} oxidation to Per^{6+} by reaction with $O_2^{\bullet-}$ proceeds until all Per^{3+} is consumed. The rate of $NAD^{\bullet}$ interaction with $O_2(aq)$ (reaction 4) is rapidly decreased during the process, and the result is that $NAD^{\bullet}$ starts the cascade of reactions in order to reduce Per^{6+} to Per^{3+}. Regenerated Per^{3+}, interacting with $O_2^{\bullet-}$, starts the oscillation cycle anew.

Creation of schematical switches on the basis of enzyme intermediates was used for finding suitable targets for new drugs [540,541], and also for switching behavior study in biochemical processes [528]. In general, complex oscillation is apparently a widespread phenomenon in biology. The peroxidase system is still the most studied oscillating enzymatic system, although indications have appeared that catalase can also manifest oscillator behavior under certain conditions [152].

3.2.6 Superoxide Dismutase and Its Models

During the biological reactions of O_2 reduction, free radicals can occur, in particular superoxide radicals $O_2^{\bullet-}$ (as cellular respiration products), which represent a potential threat to the respiratory cells' integrity. They take part as mediators in tissue inflammation, ischemic damage and aging, and cellular degenerative processes which are being promoted by AIDS [542–549]. Thus, $O_2^{\bullet-}$ occurence (in excessive amounts) in living organisms represents a great threat to their normal existence.

This threat is removed by the special enzymes—superoxide dismutases (COD), which realize the following reaction:

$$O_2^{\bullet-} + 2H^+ \longrightarrow H_2O_2 + O_2$$

The rate of this reaction rate remains constant within the pH range 5–9.5 [550].

Native superoxiddismutase contains in its active center Cu and Zn, or Fe, or Mn. In the first case, the catalytic mechanism involves copper reduction and oxidation with superoxide ion radical [550]:

$$O_2^{\bullet-}Cu^{II}Zn^{II}SOD \longrightarrow O_2 + Cu^{I}Zn^{II}SOD$$
$$O_2 + 2H^+Cu^{I}Zn^{II}SOD \longrightarrow H_2O_2 + Cu^{II}Zn^{II}SOD$$

Rate constants of these reactions, and also general rate constant (κ_{gen}), are close to each other, and reach $k_1 \approx k_2 \approx k_{gen} = 2.0–3.7 \times 10^9$ L/mol s [551]. Copper-zinc superoxide dismutase (CuZnSOD) is a dimer protein (molecular weight = 31,000) with two identical subunits (each of them containing one Cu^{2+} ion and one Zn^{2+} ion), realizing efficient dismutation of $O_2^{\bullet -}$ [552]. The high rate of $O_2^{\bullet -}$ dismutation reaction, close to diffusion-controlled, points out that superoxide ion interaction with enzyme is an electrostatically ordered process. It is realized by $O_2^{\bullet -}$ anion brought to the Cu^{II} active center through the narrow positively charged canal [553, 554]. Independence of dismutation rate on pH within a certain range is provided, apparently, by an imidazole bridge between Cu^{2+} and Zn^{2+} in SOD and influence of Zn^{2+}. Removing of the imidazole bridge results in a pH-dependent rate of $O_2^{\bullet -}$ dismutation with Cu–SOD enzyme (without Zn).

It is supposed that the zinc-imidazole part of Cu^{II}–imidazole–Zn^{II}SOD enzyme plays an important role in the catalytic mechanism of enzyme functioning, thus providing, by the respective protonation, rapid release of hydrogen peroxide (formed as a result of $O_2^{\bullet -}$ dismutation) from the Cu^{II} coordination sphere [555].

The role of zinc does not lie in enzyme activation but in the structural function fulfilled by it. It contributes to active center stability, being bound to copper via an imidazole bridge. It is also interesting that dimer SOD form (Cu_2Zn_2SOD) can also be formed, possessing the same activity as the native enzyme. Copper and zinc substitution on cobalt substantially decreases enzyme's activity (rate constant decreases by about three orders), although the mechanism is similar to that characterizing (Cu, Zn)–SOD.

Fe(III)–SOD COnsists of two identical subunits. Catalytic mechanism is described like (Cu, Zn)–COD [556]:

$$\text{Fe(III)} + O_2^{\bullet -} \longrightarrow \text{Fe(II)} + O_2$$

$$\text{Fe(II)} + O_2^{\bullet -} \xrightarrow{2H^+} \text{Fe(III)} + H_2O_2$$

With the general rate constant equal to 5.5×10^8 L/mol s, it is constant within the pH range 6.2–9.0 [551]. Mn–SOD activity is about one order lower than that of (Cu, Zn)–SOD[537].

However, superoxide dismutase activity is not the prerogative of the appropriate enzymes only. Dismutase reaction is also realized by metal porphyrins and other transition metal complexes. Thus, aqua-ions $Cu(H_2O)_6^{2+}$ possess the appropriate activity, twice exceeding that of native (Cu, Zn)–enzyme, while other copper complexes, namely, Cu(II)–EDTa, do not possess activity at all, and Fe(III)–EDTa possesses weak activity only [551]. As model systems, copper complexes with histidine ([Cuhis$_2$H]) were also studied, $k_{gen} = 3.4 \times 10^8$ L/mol s, $2 < \text{PH} < 7$ [557].

Iron aqua-ions do not possess activity compared to copper aqua-ions. However, various copper(III) complexes (and also of CO(III)) are active in superoxide dismutation reaction [558–560]. Thus, iron(III) tetrakis (4-N-methylpyridyl) porphyrin complex [Fe(III)TMPyP] at PH $\approx$ 8 will form an adduct with $O_2^{\bullet -}$, which reacts with the second $O_2^{\bullet-}$ [560]:

$$O_2^{\bullet-} + [Fe(III)TMPyP] \xrightarrow{K_2} [Fe(III)TMPyP]\cdot O_2^{\bullet-}$$

$$[Fe(III)TMPyP]\cdot O_2^{\bullet-} + O_2^{\bullet-} \xrightarrow[k_4]{2H} [Fe(III)TMPyP] + H_2O_2 + O_2.$$

where $k_3 = 2\times 10^9$ L/mol s and $k_4 = 2.3\times 10^9$ L/mol s, i.e., this complex had the same activity as (Cu, Zn)–enzyme, and in the reaction course, an intermediate compound was formed—iron(III) peroxo-complex with porphyrin.

Transition metal porphyrin complexes can be arranged in the following line by their activity [561]:

Fe(III)TMP*y*P>>Mn(III)TMP*y*P>Co(III)TMP*y*P $\cong$ Mn(III)TAP>Fe(III)TPPS.

where TAP is tetra (4- *N*,*N*,*N*-trimethylanilinium) porphyrin and TPPS is tetra (4-sulfonatophenyl) porphyrin.

Although many other iron and manganese complexes were also studied, which act out the roles of (Cu, Zn)–SOD [562, 563], Fe–SOD [564–566], and Mn–SOD [567, 568], there is a number of problems, connected with the indirect methods of superoxide quantitative determination, participating in reaction [569, 570]. Because of difficulties in intermediate compounds determination, the catalytic mechanism of this process has still not been revealed.

In many cases of the aforesaid diseases, caused by the occurence of significant $O_2^{\bullet-}$ amounts in an organism, addition of SOD enzymes produces a favorable effect. However, for different reasons, use of these enzymes does not seem to be possible (cost, stability, immunogenity, etc.,) as pharmaceutical remedies. Therefore, attempts are being made to find suitable imitating substances for these applications.

One of the most active SOD imitators turned out to be iron(II) complex with three leg-like ligand TPAA (TPAA = tris [2-[*N*-(2-pyridylmethyl)amino]ethyl]amine) [566, 570, 571].

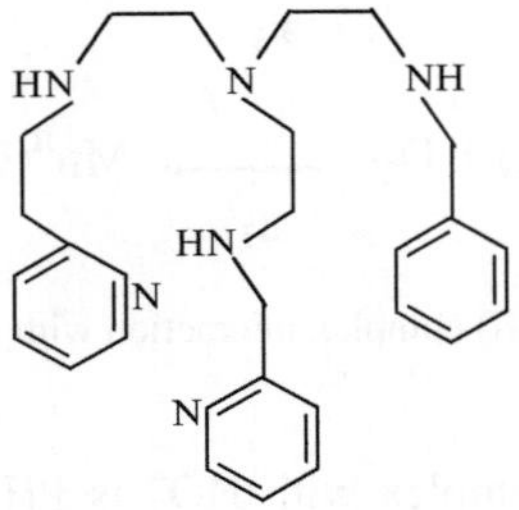

With the same pharmacologicapplication, Mn(II) Complex with TPAA was synthesized and studied by spectroscopy, electrochemistry, and γ- and pulse-radiolysis), as manganese is less toxic for the organism than iron, copper, and other metals [571, 572]. This complex is seven-coordinated (all seven atoms of the ligand's nitrogen are involved) and has the composition [Mn–TPAA] $(PF_6)_2$, where

$(PF_6)_2$ = bis – hexafluorophosphate. This complex interacts well with $O_2^{\bullet-}$ (in a stoichiometric but not catalytic way) with rate constant $\sim 1 \times 10^7$ L/mol s. The process proceeds via the intermediate $[(Mn\text{–}TPAA)^{2+} \cdot O_2^{\bullet-}]^+$ formation. However, the kinetic scheme of this complex superoxide activity is in the stage of elaboration now.

With the same application as possible pharmaceutical reagents, a number of manganese(II) complex compounds with macrocyclic ligands L were synthesized and studied. The most active of them turned out to be Mn(II) Complex with 1,4,7,10,13-pentaazacyclopentadecane $[Mn([15]aneN_5)\,Cl_2]$ (general rate constant $\kappa = 4.13 \times 10^7$ L/mol s, under pH physiological value of 7.4). Here, $(Mn[15]aneN_5)$ is macrocyclic ligand:

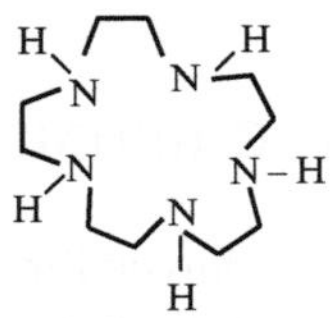

[15]ane N5

The general process mechanism can be given by Scheme 3.24 [573].

$$Mn^{II}(L) + HO_2\cdot \xrightarrow[\kappa_1]{} Mn^{III}(L) + HOO^-$$

$$Mn^{II}(L) + O_2\cdot^- \xrightarrow[\kappa_2]{} Mn^{III}(L)(OO^{2-})$$

$$Mn^{III}(L)(OO^{2-}) + H^+ \xrightarrow[\text{rapidly}]{} Mn^{III}(L)(O_2H)$$

$$Mn^{III}(L)OOH + O_2\cdot^- \xrightarrow[\kappa_3]{H^+} Mn^{II}(L) + O_2 + H_2O_2$$

$$Mn^{III}(L) + O_2\cdot^- \xrightarrow[\kappa_3']{} Mn^{II}(L) + O_2.$$

Scheme 3.24 Mechanism of Mn(II) complex interaction with $HO_2^{\bullet}$

Interaction of $Mn^{II}(L)$ Complex with $HO_2^{\bullet}$ is PH dependent and proceeds via outer-sphere electron transfer with the intermediate $Mn^{III}(L)$ formation. The next pH-dependent stage leads to the formation of $Mn^{III}(L)(OO^{2-})$ peroxo-complexes, and then $Mn^{III}(L)OOH$. Reactions of $Mn^{III}(L)OOH$ and $Mn^{III}(L)$ with $O_2^{\bullet-}$ yield final products O_2 and H_2O_2. Thus, $[Mn([15]aneN_5)\,Cl_2]$ complex, excellent imitating (mimicking) SOD, is stable to dismutation and hydrolysis at physiological pH value. This complex behavior as model was studied in vivo in the case of miocardium heart attack. It was found that it is an efficient remedy for damage restoration, connected with superoxide during the ischemia [574].

With the scope to recieve even more stable and efficient SOD imitators, carbon-substituted derivatives of $[Mn([15]aneN_5)Cl_2]$ type complex were synthesized [575]:

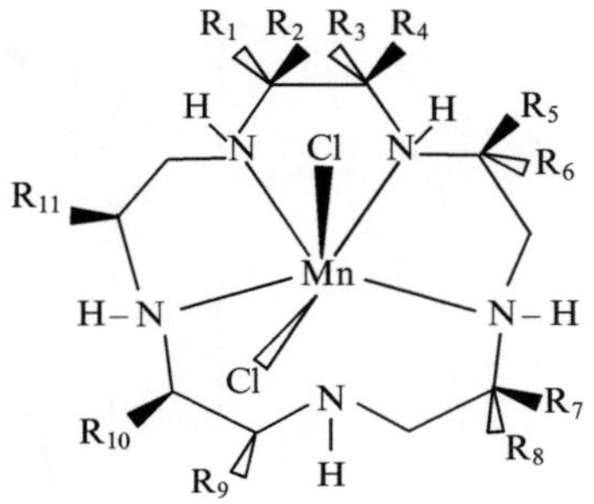

where R_1=H, Me, $-CH_2CH_2CH_2CH_2-$, $-CH_2CH_2CH_2-$, $-CH_2CH_2CH_2CH_2CH_2-$, and $-CH_2CH_2(t-Bu)CH_2CH_2-$ in various combinations. The most active in $O_2^{\bullet -}$ dismutation turned out to be the complex in which $R_1 = R_4 = -CH_2CH_2CH_2CH_2-$ (other R_1 = H). Its thermodynamic stability ($\log k_{stab} = 11.65$) is about one order higher than that of non-substituted manganese complex ($\log \kappa_{stab} = 10.85$), and the efficiently was more than twice as high (general rate constants $k = 9.09 \times 10^7$ and $\kappa = 4.13 \times 10^7$ L/mol s, respectively).

As has been already mentioned, enzyme's (Cu(im)Zn)–SOD (where im is imidazole bridge) activity is the highest ($\sim 3 \times 10^9$ L/mol s) among other metal-substituted SOD. Hence, numerous attempts to create the appropriate model complexes imitating (Cu(im)Zn)–SOD became clear. Thus, imidazole-bridge binuclear copper(II) complexes and imidazole-bridge hetero-binuclear copper(II)–zinc(II) complexes with macrobicyclic ligand L (L = 1,4,12,15,18,26,31,39-Octaazapentacyclo[13.13.13.1.1.1.] tetratetracontan-6,8,10,20,22,24,33,35,37-nonaene) were synthesized and studied by the methods of Roentgen structural analysis, electrochemistry, magnetic susceptibility, EPR, and electronic spectroscopy (Fig. 3.31) [576].

In Fig. 3.32, $[(Cu(im)Zn)L]^{3+}$ complex is shown schematically [576].

Here, im is dissociated imidazole form (imH). The EPR spectrum is very close to that of $[Cu_2Zn_2SOD]$. $[(Cu(im)Zn)L]^{3+}$ also had a similar structure. In both model complexes, copper and zinc were penta-coordinated with ligand. Bond metal-N length falls within the limits 1.91–2.33 Å, and Cu–Cu and Cu–Zn lengths are equal to 5.95 and 5.93 Å, respectively (that is only slightly shorter than for the appropriate bonds in (Cu–Cu)SOD and (Cu–Zn)SOD). Both complexes are stable within a wide pH range: for $[(Cu(im)Cu)L]^{3+}$ within the pH range 4.5–12, and for $[(Cu(im)Zn)L]^{3+}$, within the pH range 6–10.5.

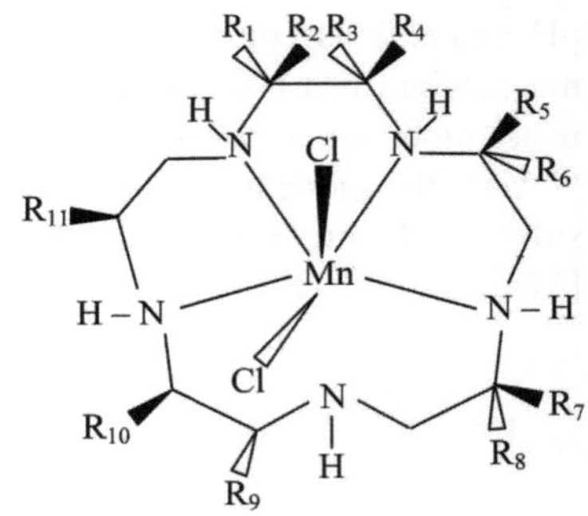

Fig. 3.31 Structural formula of macrobiocyclic ligand L

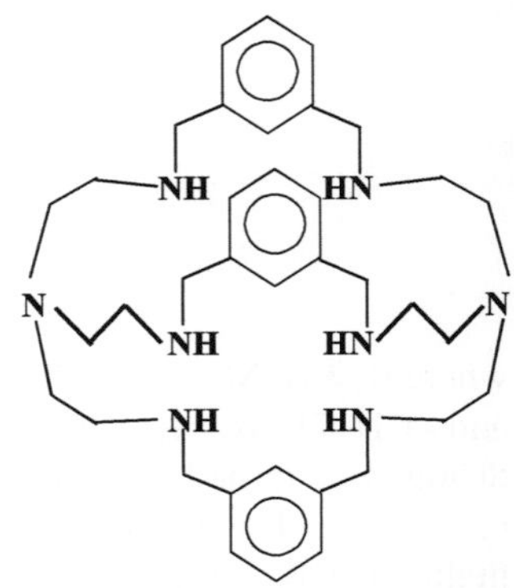

Fig. 3.32 Schematic representation of $[(Cu(im)Zn)L]^{3+}$ complex

Both model complexes, being rather stable compounds, catalyse superoxide dismutation on biological PH, and activity thus is kept in the presence of protein, which usually extracts copper from almost all model compounds, and thus, the activity of such systems diminishes. $[(Cu(im)Cu)L](ClO_4)_3$ and $[(Cu(im)Zn)L](ClO_4)_3$ complexes are good model catalysts of dismutation, and the mechanism of their action is under study now.

3.2.7 Models of Oxidases (Laccase, Cytochrome C, and Galactose Oxidase)

Almost all the above-mentioned enzymes (monooxygenases, cytochrome P-450, methane monooxygenases, catalase, peroxidase, etc.) activate oxygen (O_2), hydrogen peroxide (H_2O_2), hydroperoxides (ROOH), or oxygen-containing compounds of the iodobenzene type. All of them use iron porphyrinates as active centers. Meanwhile, there exist many enzymes whose active centers involve copper ions. They are oxidases, which catalyze reactions of various substrates' oxidation due to the direct reduction of dioxygen.

The number of copper ions, making parts of these enzymes, and the enzyme molecular weights reach 1–8 and 160,000–290,000,respectively. A characteristic feature of these biocatalysts ("blue" copper-containing oxidases) is their capacity to reduce O_2 to H_2O in a four-electron and synchronous way, without intermediate hydrogen peroxide formation. Such biocatalysts are laccase, cytochrome oxidase C, ascorbate oxidase, and ceruloplasmin. We shall consider only the first two oxidases and their models.

3.2.7.1 Laccase and Its Models

Laccases from different sources (they are widely distributed in plants and microorganisms) have different molecular weight of 110,000–140,000 Da, contain four to six copper atoms and are able to oxidize phenols (in general, aromatic alcohols), aromatic diamines, and a number of inorganic ions (namely, ferricyanide ion) with molecular oxygen. Laccase catalyzes oxidation of substrates by the following scheme:

$$2SH_2 + O_2 \longrightarrow 2S + 2H_2O$$

The active center contains four copper ions with different properties. The first and second copper ions are well detected by EPR method, though they differ from each other. They both absorb light in the region of ~600 nm; however, the extinction coefficient of the first copper ion ("blue" ion) is much higher than that of the second. The third and fourth copper ions form EPR-undetectable dimer Cu^{2+}–Cu^{2+} with unpaired electrons spins. This makes it possible to get or give away two electrons. It is supposed that the enzyme's active center has the following composition:

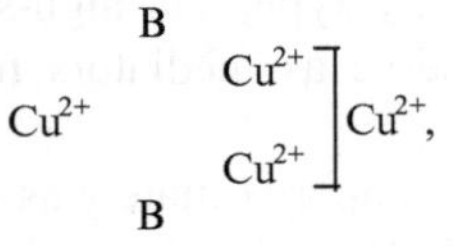

where B represents bases (imidazole groups of histidine, aminogroups of lisine, etc.). Scheme 3.25 of functioning mechanism of active center is given in [577].

Stage I here is O_2 entrance into the active center; stage II one-electron reduction of active center with the scope of better O_2 retention in it; and stages III, IV, VI, and VII are stages of bases B protonation. The protonated form of two bases makes it easier to transfer the electrons from donor on the two-electron acceptor, and is the donor of two protons at the stage of water formation. Stage V is intermediate enzyme particle formation, registered by an EPR method. Stage VIII is structural change of oxygen $\bar{O}$ two-electron atom. It seems that at this stage Cu^+ ion forms a common electron system with dimer Cu^{2+}–Cu^{2+}, taking part in the stages of interaction with $\bar{O}$. The active center of laccase, consisting of four copper atoms, apparently serves for mechanism switching from one-electron to two-electron working (stages II and V).

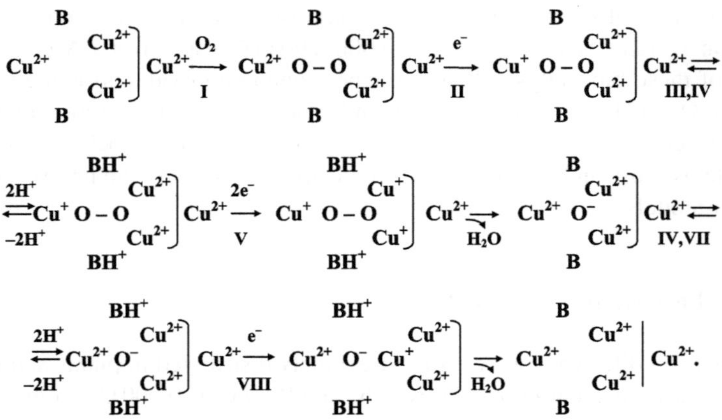

Scheme 3.25 Mechanism of laccase active center functioning

3.2.7.2 Cytochrome C Oxidase and Its Models

Cytochrome C oxidases represent metal enzymes bound with membrane, which are final products of the respiratory chain in mitochondrium and in many aerobic bacteria [578]. Enzymes of this chain catalyze four-electron reduction of O_2 to water:

$$O_2 + 4H^+ + 4e^-(\text{from cytochrome C}) \longrightarrow 2H_2O$$

The functional active center serves the multi-nuclear system, which involves copper complex of Cu_A^{2+} type in the initial oxidized state, low-spin porphyrin complex Fe_a^{3+}, copper complex of CuB2+ type, and high-spin porphyrin complex Fea_3^{3+}. Cu_A^{2+} and Fe_a^{3+} Complexes serve as mediators in the electrons transfer to the binuclear center.

Structure of Cu_A^{2+} complex is approximately as given in Fig. 3.33 [579–581]. S of methionine (Met 227) and C=O of glutathione (Glu 218) are weakly coordinated with copper atoms. A dotted line means possible bond Cu···Cu. Each copper atom is bound with two sulfur atoms of cysteine (Cys 216 and Cys 220) and nitrogen of imidazole in histidine (His 224 and His 181) ($r \approx 2.25$ and 1.95 Å, respectively). Cu···Cu length reaches ≈2.55 Å. Angle Cu S Cu reaches ≈ 70° [579].

Such thermodynamically advantageous structures make the complete delocalization of unpaired electron between two copper atoms easier (Cu(1, 5)···Cu(1, 5)) even at 10 K, which causes insignificant reorganization energy for electron transfer. Perturbation of end-ligands at increasing pH or their substitution with others (namely, substitution of His 224 or Met 224) causes non-equivalency of two copper atoms with the formation of state Cu(I)···Cu(II). Complex Cu_A^{2+} (forming part of sub-unit II of protein) obtains electrons from cytochrome C and transfers them to low-spin complex Fe_a^{3+}:

Fig. 3.33 structure of cytochrome oxidase–Cu_A^{2+} complex Approximate

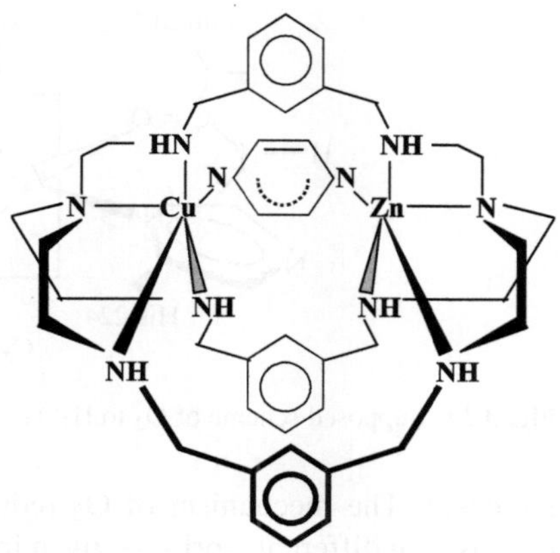

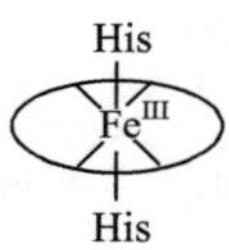

The latter, in its turn, transfers electrons to high-spin porphyrin complex Fe_A^{3+}:

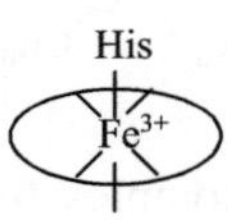

making part of subunit I jointly with complex Cu_B^{2+}:

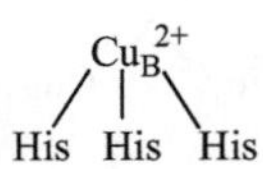

Complexes Cu_B^{2+} and Fe_a^{3+} are in close to each other at a distance of 4.5 Å [581] and form paired center Fe(III)–X–CuII [582], where X represents oxo (O^{2-}), hydroxo, chloro, sulfide, cysteinate, or imidazole group (sometimes X is completely absent)[563]. O_2 reducing to water proceeds on this binuclear center $Cu_B^{2+}\ldots Fea_3^{3+}$ [583] by accepting electrons from cytochrome C (through Cu_B^{2+} and Fe_a^{3+}). Proton transfers limit the reaction in cytochrome C oxidase. Electron transfer is constantly retarded in the course of reaction between O_2 reducing and electron transfer. Thus, the concentration of oxygen intermediates (bound with iron oxy, peroxy, ferryl, and hydroxyl particles) reaches such a level at which they can be detected spectrophotometrically [578, 584–587].

The binuclear center in the reduced state acquires the capacity for O_2 binding. Thus, the rate of electron transfer from Cu_A^{2+} and Fe_a^{3+} to $Fea_3^{3+}\ldots Cu_B^{2+}$ sharply

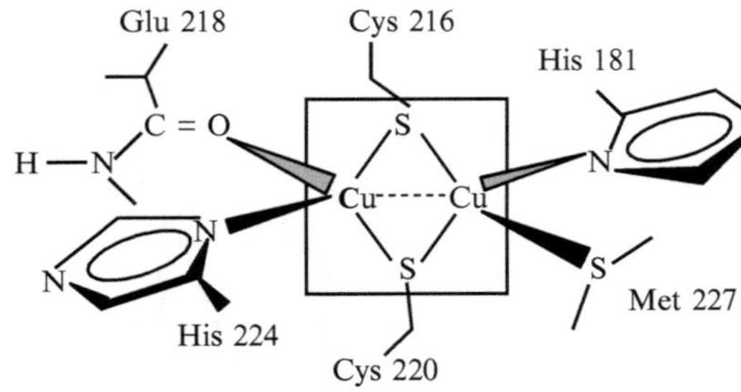

Fig. 3.34 Supposed scheme of O_2 to H_2O reducing mechanism with cytochrome C oxidase

increases. The mechanism of O_2 reducing to water with cytochrome C oxidase, proposed in different works, is given in Fig. 3.34 [583, 588–590].

By two-electron transfer from cytochrome C, via Cu_A^{2+}– and Fe_a^{3+}– intermediate, the initially unoxidized unactive center

$$Fe_{a_3}^{3+}\diagup X \diagdown Cu_B^{2+}$$

will be transformed into the active one

$$Fe_{a_3}^{2+}\diagup X \diagdown Cu_B^{+},$$

with which O_2 will interact. Oxo-complex formation proceeds rapidly with rate constant $k = 3.5 \times 10^8\,M^{-1}\bullet s^{-1}$ [591] within first $10\,\mu s$ of reaction [592]. The formed complexes,

$$Fe_{a_3}^{2+}\diagup X \diagdown Cu_B^{+},$$

the maximal accumulation of which proceeds within approximately $30\,\mu s$ [593], will be rearranged, and O_2 is found near Fea_3^{2+}. The process of O_2 position rearrangement proceeds with the rate constant $\kappa = 9\times 10^4\,s^{-1}$. Then inner transfer of two electrons follows on coordinated O_2, from Fea_3^{2+} and Cu_B^{+} with the formation of peroxo-iron(III) intermediate ($\kappa = 3 \times 10^4\,s^{-1}$ [594, 595]) which, after protonation and third electron transfer from Fe_a^{3+}, will be decomposed with oxo-ferryl particle Fea_3^{IV} generation:

$$\left[Fe_{a_3}^{4+}{=}O \diagup X \diagdown Cu_B^{+}\ldots H_2O\right].$$

On fourth electron transfer (also from Fe_a^{3+}) and protonation, two water molecules will be formed, and the initial oxidized state of active center will be restored.

$$Fe_{a_3}^{3+} \diagup X \diagdown Cu_B^{2+}.$$

Thus, the mechanism of cytochrome C oxidase functioning (reduction of O_2 to H_2O), with the energy evolved used for ATN synthesis, is one of the variants (additional involvement of Cu_A^{2+} and Cu_B^{2+}) of hematic systems used by nature for various reactions realization. Thus, on the basis of enzymatic systems choice realizing various reactions, the principle of most economical use of construction material available and creationof the maximal variety of processes is realized. In this sense, this process is similar to that of such enzymes as peroxidase and cytochrome P-450.

Modeling of the cytochrome C oxidase binuclear center could have helped to reveal the mechanism of O_2 reduction to water in more detail. With this application in mnd, attempts with such mixed iron(III)–copper(II) complexes modeling were undertaken. These were connected with each other by various bridge ligands (oxo-, imidazolo-, etc.) and a study made of their interaction with O_2. Thus, oxo- and hydrooxo-bridge compounds $[F_8\text{–TPPFe}^{III}\text{–}(O^{2-})\text{–Cu}^{II}(TMPA)]^+$ (1) and $[(F_8\text{–TPP})Fe(III)\text{–}(OH^-)\text{–Cu}^{II}(TMPA)]^{2+}$ (2) [588, 596–598] (where Fe_8TPP = tetrakis (2,6-difluorophenyl)porphyrinate (2-), TMPa = tris (2-pyridylmethyl)amine), were synthesized and studied by various methods, which for many parameters (especially substance (2)) were similar to cytochrome C oxidase active center:

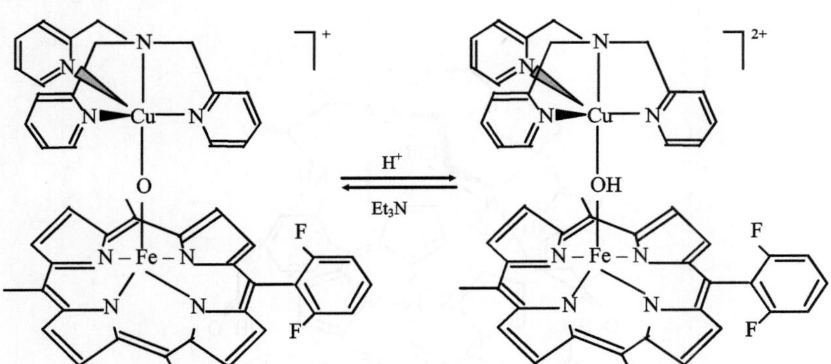

(1) (2)

Thus, hydroxo-complex (2) can be generated by protonation of (1), and vice versa, by base triethylamine (Et_3N) addition to (2), complex (1) will be obtained [597].

The value of deprotonation pK_a (2) reaches 8 ± 2.5. Such acid–base transformation is important, as the center heme-copper is the site of electron transfer. Thus, oxo-ligands, but not nitrogen atoms of pyridyl or porphyrin ligands, are sites of protonation [590]. Structural, magnetic, and other properties of compound (2) make it a possible candidate for active center imitation

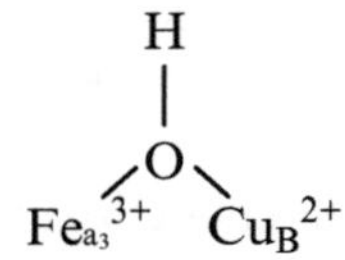

of cytochrome C oxidase. However, there is a difference:

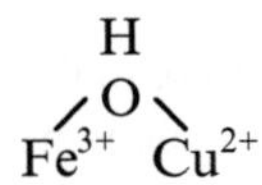

The angle of the group in model compound (2) is close to linear (~157°), while in cytochrome C active center

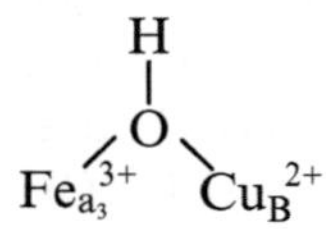

It is non-linear. There are other differences as well.

The structurally similar model compound (A) was also synthesized with the center of native cytochrome C oxidase, which is supposed to be joining dioxygen [599]:

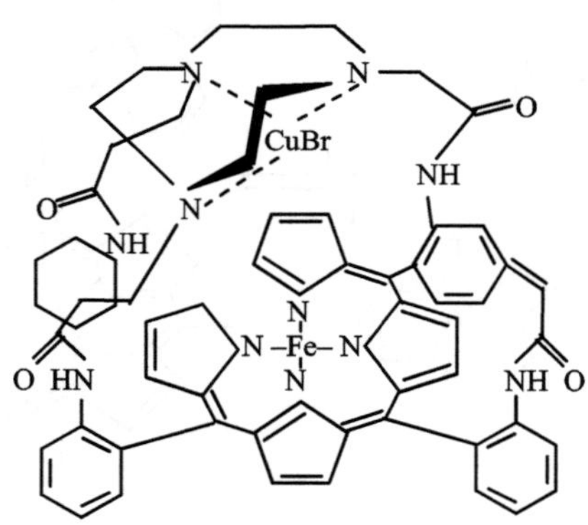

(A)

Like the native enzyme, this model compound contains porphyrin iron(II) complex and copper(I) situated near it. Copper(I) is bound to three tertiary amines in a coordinated way.

Dissolution of this compound in toluene and treatment of this solution with dioxygen leads to the formation of irreversible adduct $A^{\bullet}O_2$ in the ratio 1:1. Copper(I) plays the determining role in O_2 binding. On this adduct $A^{\bullet}O_2$ treatment with cobaltocene (electrons source), it turned out that four cobaltocene equivalents are needed ($4e^-$) for complete oxygen reduction in adduct $A^{\bullet}O_2$. Oxygen structure in this adduct is represented as a bridge peroxo-particle. Thus, A compound turned out to be one of model complexes of cytochrome C oxidase.

3.2.7.3 Galactose Oxidase and Its Models

Unlike mono- and dioxygenases, activating O_2 and introducing one or two oxygen atoms into substrate, oxidases activate hydrogen, which accepts O_2, catalysing free radical $HO_2^{\bullet}$, hydrogen peroxide, or water formation. Copper, iron, or molybdenum can be an active center of oxidases. One of these oxidases is copper-containing galactose oxidase. This enzyme, consuming oxygen, will oxidize galactose and some primary alcohols, thus generating hydrogen peroxide:

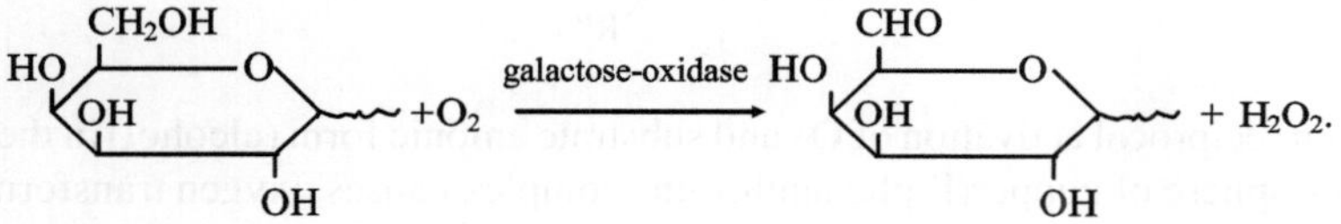

The copper of the enzyme's active center is coordinated with two nitrogen atoms from histidine imidazole ring with carboxylate or phenolate ligands, arranged within the square plane. Water seems to be in an equatorial position [600]. There were also earlier assumptions regarding the structure of the enzyme's active center [601].

The mechanism of galactose oxidases' influence is not clear now, although there are two kinds of assumptions: either the Cu(II) form is inactive and redox cycle Cu(I)–Cu(III) [602] is important, or the Cu(II) form is active and it simply makes easier sequential one-electron transfers from alcohol to O_2 [603].

Various model systems of galactose oxidase were considered. The most efficient and at the same time relatively simple is Cu^+–o–phen–O_2–R–OH–OH^- solvent [455]. In the case of primary alcohol oxidation at low degrees of conversion, and also benzyl alcohol oxidation, selectivity of aldehydes and ketones formation in this system is high and close to 100%. Secondary alcohols oxidation at primary stages of the process yields high-selective formation of ketones, which are rapidly oxidized to acids, especially on oxidation of alcohols with unsubstituted α-CH_2-groups, namely, cyclohexanol, butanol-2, etc.). Isopropionic alcohol in the presence of $[Cu(phen)_2]^+$ complex in basic medium is oxidized particularly rapidly and selectively (selectivity is close to 100%). The number of catalytic cycles for this complex reaches 1–2 s^{-1}, which exceeds by thousands of times the activity of known oxidase models, and is close to the activity of galactose oxidase [455].

In oxidation of both primary and secondary alcohols, catalytically active particles are complexes $[Cu(phen)_2]^+$. It is supposed that alcohol oxidation, namely of isopropanol, proceeds by the formation of intermediate triple complex $[(phen)_2Cu^I\ldots(CH_3)_2CH(O^-)\ldots O_2]$, and the limiting stage of the process is hydride ion H^- transfer from alkoxy anion on O_2 molecule with acetone and hydrogen peroxide formation. This model system is similar to galactose oxidase by its kinetic behavior, in which the active center also contains copper(I) compound [455, 604]. High stability of *o*-phenanthroline in active complex $[Cu(phen)_2]^+$ on alcohols catalytic oxidation (it is not destroyed after several thousands of catalytic cycles) makes it possible to deny both the participation of Cu^{III} in the reaction (which was established in [605]), and that of $O_2^{\bullet-}$ (which was supposed in [605]), as the latter efficiently destroys *o*-phenanthroline.

The most probable supposed mechanism is as follows [455]:

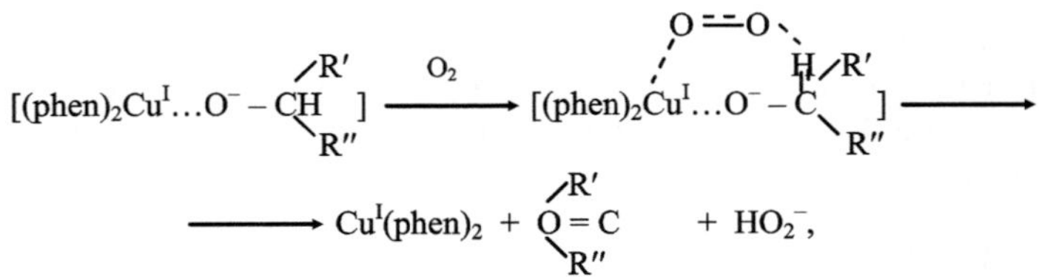

in which reciprocal activation of O_2 and substrate anionic form (alcohol) in the coordination sphere of copper(I) phenanthroline complex causes oxygen transformation into one of the strongest oxidants in such reactions.

References

1. G. Rehner, Ernährungs–Umschau **30**(2), 35 (1983)
2. L.A. Macmillan-Crow, D.L. Cruthirds, Free Radic. Res. **34**(4), 325–336 (2001)
3. I.A. Chernavina, *Physiology and Biochemistry of Microelements* (Vysshaia Shkola, Moscow, 1970) p. 77 [Russian]
4. Ya.V. Peive, *Guidance on Microfertilizers Application* (Selikhozgiz, Moscow, 1963) [Russian]
5. V.K. Rita, Manganese influence on some biochemical processes in grapes, Ph.D. thesis, Odessa, 1967, [Russian]
6. E.J. Unterwood, *Trace Elements in Human and Animal Nutrition*, 4th edn (Academic, New York, 1977) p. 170
7. L. Wilson, *Nutritional Balancing and Hair Mineral Analysis* (L.D. Wilson Consultants, Inc., Prescott, AZ, 2005), p. 2010
8. K. Szentmihályi, P. Vinkler, J. Fodor, J. Balla, B. Lakatos, Orv. Hetil. **150**(15), 681–687 (2009) [Hungarian]
9. J.M. Wrigglesworth, H. Baum, in *Iron in Biochemistry and Medicine*, vol. 2, ed. by A. Jacobs, M. Worwood (Academic, London, 1980), pt. 2, p. 29
10. A.S. Manoguerra, A.R. Erdman, L.L. Booze, G. Christianson, P.M. Wax, E.J. Scharman, et al. Clin. Toxicol. (Phila). **43**(6), 553–570 (2005)
11. K. Raja, R. Simpson, T. Peters, Biochim. Biophys. Acta. **1135**, 141 (1992)
12. M.A. Barrand, R.C. Hider, B.A. Callingham, J. Pharm. Pharmacol. **42**, 279 (1990)
13. A.E. Thumser, A.A. Rashed, P.A. Sharp, J.K. Lodge, Food Chem. **123**(2), 281–285 (2010)
14. P. Aisen, in *Iron in Biochemistry and Medicine*, vol. 3, ed. by A. Jacobs, M. Worwood (Academic, London, 1980), pt. 4, p. 84
15. E.H. Morgan, Mol. Aspects Med. **4**, 1 (1981)

16. S. Singh, R.C. Hider, in *Free Radical Damage and Its Control*, ed. by C.A. Rice-Evans, R.H. Burdon, (Elsevier Science B.V., Amsterdam, 1994), pp. 189–216
17. J.J. Bullen, E. Griffiths, *Iron and Infection* (Wiley, New York, 1987)
18. E.D. Weinberg, Drug Metab. Rev. **22**, 531 (1990)
19. H. Munro, M.C. Linder, Physiol. Rev. **58**, 317 (1978)
20. G.G. Ford, P.M. Harrison, D.W. Rice, J.M.A. Smith, A. Treffry, J.L. White, J. Yariv. Philos. Trans. R. Soc. London Ser. B **304**, 551 (1984)
21. I. Fridovich, in: *Free Radicals in Biology*, vol. 1., ed. by W. Pryor (Mir, Moscow, 1979), pp. 272–314. [Russian]
22. D.I. Metelitsa, G.S. Arapova, A.N. Eremin, Biokhimya [Biochemistry] **62**, 460 (1997) [Russian]
23. J.H. Ward, J.P. Kushner, J. Kaplan, in *Current Hematology and Oncology,* ed. by V.F. Fairbanks (Medicina, Moscow, 1987), pp. 11–57 [Russian]
24. B.J. Bolann, R.J. Ulvik, Eur. J. Biochem. **193**, 899 (1990)
25. Yu.A. Vladimirov, O.A. Azizova, A.I. Deev, A.V. Kozlov, A.N. Osipov, *Itogi nauki i tekhniki. Biofizika* [Sci. and Tech. Ser. Biophysics], vol. 29 (VINITI, Moscow, 1977), p. 51 [Russian]
26. D.I. Metelitsa, *Oxygen Activation with Enzymatic Systems* (Nauka, Moscow, 1982), p. 34 [Russian]
27. P. Hochstein, Isr. J. Chem. **21**, 52 (1981)
28. R.A. Floyd, C.A. Levis, Biochem. J. **22**, 2645 (1983)
29. D.C.H. McBrien, T.E. Slater (eds.) *Free Radicals, Lipid Peroxidation and Cancer* (Academic, London, 1982)
30. C.A. Rice-Evans, in *Free Radical Damage and Its Control*, ed. by C.A. Rice-Evans, R.H. Burdon (Elsevier Science B.V., Amsterdam, 1994), p. 131
31. A. Benedetti, M. Comporti, Bioelectrochem. Bioenerg., **18**, 187 (1988)
32. K. Fukuzawa, T. Fujii, T. Todokoro, K. Mitsuta, J.M. Gebicki, in O. Hayaishi, E. Niki, M. Kondo, T. Yoshikawa, eds. *Medical, Biochemical and Chemical Aspects of Free Radicals*: Proc. of the 4th Biennial General Meeting of the Soc. for Free Radical Research, Kyoto, Japan, 9–13 April 1988. Amsderdam: Elsevier Science B.V., 1989, p. 71
33. H. Kappus, H. Muliawan, M.E. Scheulen, in W. Bors, M. Saran, D. Tait, eds. *Oxygen Radicals in Chemistry and Biology*: Proc. of the 3rd Int. Conf., Neuherberg, Federal Republic of Germany, 10–15 July 1983. Berlin; New York: Walter der Gruyter, 1984, p. 359
34. Y. Hayashi, Y. Ueda, A. Nakajima, Y. Mitsuyama, Brain Res. **1025**(1–2), 29–34 (2004)
35. M. Tien, B.A. Svingen, S.D. Aust, in *Oxygen and Oxy-Radicals in Chemistry and Biology,* ed. by M.A.J. Rodgers, E.L. Powers (Academic, New York, 1981), p. 147
36. M.G. Repetto, F.N. Ferrarotti, A. Boveris, Arch. Toxicol. **84**(4), 255–262 (2010)
37. I.B. Afanasiev, N.S. Kuprianova, A.V. Letuchaia, in W. Bors, M. Saran, D. Tait, eds. *Oxygen Radicals in Chemistry and Biology*: Proc. of the 3rd Int. Conf., Neuherberg, Federal Republic of Germany, 10–15 July 1983. Berlin; New York: Walter der Gruyter, 1984, p. 17
38. Aust, S.D., J.R. Bucher, M. Tien, in W. Bors, M. Saran, D. Tait, eds. *Oxygen Radicals in Chemistry and Biology*: Proc. of the 3rd Int. Conf., Neuherberg, Federal Republic of Germany, 10–15 July 1983. Berlin; New York: Walter der Gruyter, 1984, p. 147
39. E. Niki Free Radic. Biol. Med. **47**(5), 469–484 (2009)
40. J.F. Mead, in *Free Radicals in Biology*, vol. 1,ed. by W.A. Pryor (Academic, New York, 1976)
41. B. Halliwell, J.M.C. Gutteridge, in O. Hayaishi, E. Niki, M. Kondo, T. Yoshikawa, eds. *Medical, Biochemical and Chemical Aspects of Free Radicals*: Proc. of the 4th Biennial General Meeting of the Soc. for Free Radical Research, Kyoto, Japan, 9–13 April 1988. Amsterdam: Elsevier Science B.V., 1989, pp. 21–31
42. J.M. Branghler, J.F. Pregenzer, R.L. Chase, L.A. Duncan, E.J. Jacobson, J.M. McCall, J. Biol. Chem. **262**, 10438 (1987)
43. J.P. Kehrer, Toxicology **149**, 43–50 (2000)
44. Mladenka, P., T. Simůnek, M. Hübl, R. Hrdina. The role of reactive oxygen and nitrogen species in cellular iron metabolism. Free Radic. Res. **40**(3), 263–272 (2006)
45. J.M.C. Gutteridge, FEBS Lett. **201**, 251 (1986)

46. M.J. O'Connell, R.J. Lee, G.E. Cartwright, J. Clin. Invest. **229**, 135 (1985)
47. P. Biemond, H.G. van Eijk, A.J.G. Swaak, J.F. Koster, J. Clin. Invest. **73**, 1576 (1984)
48. J.F. Koster, R.G. Slee, FEBS Lett. **199**, 85 (1986)
49. C.E. Thomas, L.A. Morehouse, S.D. Aust, J. Biol. Chem. **260**, 3275 (1985)
50. P. Biemond, A.J.G. Swaak, C.M. Beindorff, J.F. Koster, Biochem. J. **239**, 169 (1986)
51. G.R. Buettner, T.P. Doherty, L.K. Patterson, FEBS Lett. **158**, 143 (1983)
52. J. Butler, B. Halliwell, Arch. Biochem. Biophys. **218**, 174 (1982)
53. M.E. Aliaga, C. Carrasco-Pozo, C. López-Alarcón, C. Olea-Azar, H. Speisky, Bioorg. Med. Chem. **19**(1), 534–554 (2011)
54. G.R. Buettner, Bioelectrochem. Bioenerg. **18**, 29 (1987)
55. J.M.C. Gutteridge, D.A. Rowley, E. Griffiths, B. Halliwell, Clin. Sci. **68**, 463 (1985)
56. B. Halliwell, J.M.C. Gutteridge, Mol. Aspect. Med. **8**, 89 (1985)
57. J. Kanner, S. Harel, Free Radic. Res. Commun. **3**, 309 (1987)
58. S. Muraoka, T. Miura, Life Sci. **74**(13), 1691–1700 (2004)
59. E. Graf, J. Am. Oil Chem. Soc. **60**, 1861 (1983)
60. E. Graf, J.W. Eaton, Cancer **56**, 717 (1985)
61. E. Doria, L. Galleschi, L. Calucci, C. Pinzino, R. Pilu, E. Cassani, E. Nielsen, J. Exp. Bot. **60**(3), 967–978 (2009)
62. H. Jenzer, H. Kohler, C. Broger, Arch. Biochem. Biophys. **258**, 381 (1987)
63. O.I. Aroma, B. Halliwell, Biochem. J. **241**, 273 (1987)
64. B. Halliwell, J.M.C. Gutteridge, D. Blake, Philos. Trans. R. Soc. London, Ser. B. **311**, 659 (1985)
65. M. D'Hardemare Adu, S. Torelli, G. Serratrice, J.L. Pierre, Biometals **19**(4), 349–366 (2006)
66. P. Baret, C.G. Beguin, H. Boukhalfa, C. Caris, J.-P. Laulhere, J.-L. Pierre, G. Serratrice, J. Am. Chem. Soc. **117**, 9760 (1995)
67. T.U. Schraufstater, D.B. Hinshaw, P.A. Hislop, R.G. Spragg, C.G. Cochrane, J. Clin. Invest. **77**, 1312 (1986)
68. M.E. Hoffman, A.C. Mello-Filho, R. Meneghini, Biochim. Biophys. Acta. **781**, 234 (1984)
69. H.C. Sutton, C.F. Vile, C.C. Winterbourn, Arch. Biochem. Biophys. **256**, 462 (1987)
70. C.C. Winterbourn, Free Radic. Biol. Med. **3**, 33 (1987)
71. T.N. Novodarova, E.M. Kolosov, L.L. Kyseleva, M.Yu. Tuvin, M.E. Volpyn, in *Problems of Moderm Bioinorganic Chemistry*: Mater. of field session, 26–28 March 1984. Novosibirsk, 1984, p. 67. [Russian]
72. NV. Gist-Brocades, I. Wateringseweg, Delft, Holland. Patent 2002746 (UK). 1978–06–16. Chem. Abstr., 1979, **91**, 39515
73. Patent 498201 (Australia). Chem. Abstr., 1979, 91, 33055
74. A. Sultan, G. Cade, L. Dumble, G.M. Laycock, Arzneim-Forsch **22**, 154 (1972)
75. B. Coyle, K. Kavanagh, M. McCann, M. Devereux, M. Geraghty, Biometals **16**(2), 321–329 (2003)
76. D.R. Winge, R.K. Mehra, Int. Rev. Exp. Pathol. **31**, 47 (1990)
77. A. Samuele, A. Margagalli, M.T. Armentero, et al., Biochimica et Biophysica Acta (BBA) **1741**(3), 325–330 (2005)
78. E.M. Messa, C. Giulivi, Free Radic. Biol. Med. **14**, 559 (1993)
79. A.C. Fernandes, P.M. Filipe, C.F. Mensa, Eur. J. Farmacol. **220**, 211 (1992)
80. A. Mori, M. Niramatsu, R. Adamatsu, M. Kohno, in O. Hayaishi, E. Niki, M. Kondo, T. Yoshikawa, eds. *Medical, Biochemical and Chemical Aspects of Free Radicals*: Proc. of the 4th Biennial General Meeting of the Soc. for Free Radical Research, Kyoto, Japan, 9–13 April 1988. Amsterdam: Elsevier Science B.V., 1989, p. 1249
81. M.J. Davies, J.Clin.Biochem. Nutr. **48**(1), 8–9 (2011)
82. G. Calvin, K.E. Arfors, J. Free Radic. Biol. Med. **1**, 437 (1985)
83. J. Bylund, K.L. Brown, C. Movitz, C. Dahlgren, A. Karlsson, Biol. Med. **49**(12), 1834–1845 (2010)
84. D.K. Das, R.M. Engelman, J.A. Rousou, R.H. Breyer, H. Otani, S. Lemeshow, Basic Res. Cardiol. **81**, 155 (1986)

85. D.K. Das, D. Bagghi, R.M. Engelman, R. Subramanian, M.R. Prasad, R. Jones, G. Cordis, H. Otani, in O. Hayaishi, E. Niki, M. Kondo, and T. Yoshikawa, eds. *Medical, Biochemical and Chemical Aspects of Free Radicals*: Proc. of the 4th Biennial General Meeting of the Soc. for Free Radical Research, Kyoto, Japan, 9–13 April 1988 (Elsevier Science B.V., Amsterdam, 1989), p. 1355
86. T. Tanigawa, T. YoshikawA, Y. Naito, S. Ueda, T. Takemura, Y. Morita, K. Tainaka, N. Yoshida, S. Sugino, M. Kondo, in O. Hayaishi, E. Niki, M. Kondo, T. Yoshikawa, eds. *Medical, Biochemical and Chemical Aspects of Free Radicals*: Proc. of the 4th Biennial General Meeting of the Soc. for Free Radical Research, Kyoto, Japan, 9–13 April 1988. Amsterdam: Elsevier Science B.V., 1989, p. 873
87. Fujii, H., K. Kakinuma, in O. Hayaishi, E. Niki, M. Kondo, and T. Yoshikawa, eds. *Medical, Biochemical and Chemical Aspects of Free Radicals*: Proc. of the 4th Biennial General Meeting of the Soc. for Free Radical Research, Kyoto, Japan, 9–13 April 1988. Amsterdam: Elsevier Science B.V., 1989, p. 861
88. J.W. Heineche, Free Radic. Biol. Med. **3**, 65 (1987)
89. M. Rotheneder, G. Striegl, H. Esterbaner, in *Free Radicals Lipoproteins and Membrane Lipids*, ed. by A. Crastes de Paulet, L. Douste-Blazy, R. Paoletti (Plenum Press, New York, 1990), p. 203
90. Hempel, S.L., G.R. Buettner, D.A. Wessels, G.M. Galvan, Y.Q. O'Malley. Arch. Biochem. Biophys. 1996, **330**, 401
91. D. Yin, H. Lingnert, B. Ekstrand, U.T. Brunk, Free Radic. Biol. Med. **13**, 543 (1992)
92. W.H. Koppenol, Bioelectrochem. Bioenerg. **18**, 3 (1987)
93. A.Ya. Sychev, V.G. Isak, Uspekhi khimii [Russian Chem. Rev.] **64**, 1183 (1995) [Russian]
94. M.R. Prasad, R.M. Engelman, D.K. Das, in O. Hayaishi, E. Niki, M. Kondo, T. Yoshikawa, eds. *Medical, Biochemical and Chemical Aspects of Free Radicals*: Proc. of the 4th Biennial General Meeting of the Soc. for Free Radical Research, Kyoto, Japan, 9–13 April 1988. Amsterdam: Elsevier Science B.V., 1989, p. 997
95. D. Bonnefont-Rousselot, Glucose and reactive oxygen species. Curr. Opin. Clin. Nutr. Metab. Care. **5**(5), 561–568 (2002)
96. A.G. Maxina, B.A. Dayniak, T.V. SebcO, Yu.A. Knyazev, Biofizika [Biophysics], **42**, 671 (1997)
97. S.K. Jain, Free Radic. Med. Suppl. 1, 163 (1990)
98. D.S. Sergeev, V.F. Zarytova, Uspekhi khimii [Russian Chem. Rev.], **65**, 977 (1996) [Russian]
99. S.H. Hecht, in *Bleomycin Chemical, Biochemical and Biological Aspects,* ed. by S.M. Hecht (Springer, New York, 1979), p. 1
100. F. Hutchinson, L.F. Povirk, H. Umezawa, Cancer. Treat. Rep. **68**, 137 (1984)
101. M. Watanabe, Y. Takabe, T. Katsumata, J. Antibiot. **26**, 417 (1973)
102. D.S. Sergheev, T.S. Gorodnikova, V.F. Zarytova, Bioorg. Khim. [Russ. J. Bioorg. Chem.], **21**(9), 695–702 (1995) [Russian]
103. J. Chen, M.K. Ghorai, G. Kenney, J. Stubbe, Nucleic. Acids Res. **36**(11), 3781–3790 (2008)
104. P.C. Dedon, J.H. Goldberg, Chem. Res. Toxicol. **5**, 311 (1992)
105. D.H. Petering, R.W. Byrnes, W.E. Antholine, Chem.-Biol. Interact **73**, 133 (1990)
106. J. Stubbe, J.W. Kozarich, Chem. Rev. **87**, 1107 (1987)
107. A. Kittaka, J. Sugano, M. Otsuka, S. Kabayashi, M. Ohno, J. Sugiura, H. Umerawa, Tetrahedron Lett **27**, 3631 (1986)
108. P. Fulmer, D.H. Petering, Biochemistry **33**, 5319 (1994)
109. M.A.J. Akkerman, E.W.J.F. Nejjman, S.S. Wijmenga, C.W. Hibbers, W. BERMEL, J. Am. Chem. Soc. **112**, 7462 (1990)
110. N.S. Katrich, in *Fiziologya rastitelinyh organizmov i roli metallov* [Physiology of vegetal organisms and role of metals]. ed. by N.M. Chernavskaya (MGU, Moscow, 1989) p. 136. [Russian]
111. T. Matsuura in W. Ando, Y. Morooka, eds. *The Role of Oxygen in Chemistry and Biochemistry*: Proc. of an Int. Symp. on Activation of Dioxygen and Homogeneous Catalytic Oxidations, Tsukuba, Japan, 12–16 July 1987. Amsterdam: Elsevier Science B.V., 1988, vol. 33. p. 353

112. Y. Sugiura, J. Am. Chem. Soc. **102**, 5208 (1980)
113. W.J. Caspary, D.A. Lanzo, C. Niziak, Biochemistry **21**, 334 (1982)
114. Y. Sugiura, T. Kikushi, J. Antibiot. **31**, 1310 (1978)
115. L.W. Oberley, G.R. Buettner, FEBS Lett. **97**, 47 (1979)
116. N. Murngesan, S.M. Hecht, J. Am. Chem. Soc. 107, 493 (1985)
117. R.M. Burger, J. Peisach, S.B. Horwitz, J. Biol. Chem. **256**, 11636 (1981)
118. J.W. Sam, X.-J. Tang, J. Peisach, J. Am. Chem. Soc. **116**, 5250 (1994)
119. K.E. Loeb, J.M. Zaleski, T.E. Westre, R.J. Guajardo, P.K. Mascharak, B. Hedman, K.O. Hodgson, E.I. Solomon, J. Am. Chem. Soc. **117**, 1309 (1995)
120. S.H. Hecht, Acc. Chem. Res. **19**, 383 (1986)
121. D.G. Knorre, O.S. Fedorova, E.I. Frolova, Uspekhi khimii [Russian Chem. Rev.] **62**, 70 [Russian] (1993)
122. M.S. Chow, L.V. Liu, E.I. Solomon, Proc. Natl. Acad. Sci. USA. **105**(36), 13241–13245 (2008)
123. C.E. Holmes, B.J. Carter, C.M. Hecht, Biochemistry **32**, 4293 (1993)
124. G.H. McGall, L.E. Rabow, G.W. Ashley, S.H. Wu, J.W. Kazarich, J. Stubbe, J. Am. Chem. Soc. **114**, 4958 (1992)
125. C.-C. Cheng, J.G. Goll, C.A. Neyhart, Th.W. Welch, Ph. Singh, H.H. Thorp, J. Am. Chem. Soc. **117**, 2970 (1995)
126. B.N. Ames, R.L. Saul, in *Theories of Carcinogenesis,* ed. by O.H. Iversen (Hemisphere Publ. Corporation, Cambridge, 1988), p. 203
127. G. Harris, Immunol. Today, **4**, 109 (1983)
128. S. Blount, H.R. Griffiths, J. Lunec, FEBS Lett. **245**, 100 (1988)
129. P.G. Winyard, D. Perrett, D.R. Blake, G. Harris, J.K. Chipman, Anal. Proc. **27**, 224 (1990)
130. P. Merry, P.G. Winyard, G.J. Morris, M. Grootveld, D.R. Blake, Annu. Rheum. Dis. **48**, 864 (1989)
131. W.K. Pogozelski, T.J. McNeese, T.D. Tullius, J. Am. Chem. Soc. 1995 **117**, 6428
132. J. Emerit, M. Keck, A. Levy, J. Feingold, A.M. Michelson, Mutat. Res. **103**, 165 (1982)
133. P.D. Lawley, R. Topper, A.M. Denman, W. Hylton, I.D. Hill, G. Harris, Annu. Rheum. Dis. **47**, 445 (1988)
134. C.-H. Chang, J.L. Dalas, C.F. Meares, Biochem. Biophys. Res. Commun. **110**, 959 (1983)
135. J.P. Albertini, A. Garmer-Suillerot, Biochemistry **21**, 6777 (1982)
136. C.-H. Chang, C.F. Meares, Biochemistry **23**, 2268 (1984)
137. K.D. Goodwin, M.A. Lewis, E.C. Long, M.M. Georgiadis, Proc. Natl. Acad. Sci. USA **105**(13), 5052–5056 (2008)
138. R. Subramanian, C.F. Meares, J. Am. Chem. Soc. **108**, 6427 (1986)
139. E.T. Farinas, J.D. Tan, P.K. Mascharak, Inorg. Chem. **35**, 2637 (1996)
140. J.D. Tan, S.E. Hudson, S.J. Brown, M.M. Olmstead, P.K. Mascharak, J. Am. Chem. Soc. **114**, 384 (1992)
141. R.J. Guajardo, P.K. Mascharak, Inorg. Chem. **34**(4), 802 (1995)
142. R.B. Patel, S.R. Kotha, S.I. Sherwani, S.M. Sliman, T.O. Gurney, B. Loar, S.O. Butler, A.J. Morris, C.B. Marsh, N.L. Parinandi, Int. J. Toxicol. **30**(1), 69–90 (2011)
143. T. Tetsuka, Proc. Jpn. Cancer. Ass. **37**, 195 (1978) [Japanese]
144. H. Kikuchi, T. Tetsuka, in O. Hayaishi, E. Niki, M. Kondo, T. Yoshikawa, eds. *Medical, Biochemical and Chemical Aspects of Free Radicals*: Proc. of the 4th Biennial General Meeting of the Soc. for Free Radical Research, Kyoto, Japan, 9–13 April 1988. Amsderdam: Elsevier Science B.V., 1989, p. 1101
145. J.R. Kanafsky, J. Biol. Chem. **261**, 13546 (1986)
146. T.E. Lehmann, J. Bio. Inorg. Chem. **7**(3), 305–312 (2002)
147. A.L. Feig, S.I. Lippard, Chem. Rev. **94**, 759 (1994)
148. Que, L. Jr. in *Bioinorganic Catalysis*, ed. by J. Reedijk (Marcel Dekker, , 1993), p. 347
149. M. Momenteau, C.A. Reed, Chem. Rev. **94**, 659 (1994)
150. B. Meuner, Coord. Chem. Rev. **92**, 1411 (1992)

151. T.J. McMurry, J.T. Groves, in *Cytochrone P-450: Structure, Mechanisms and Biochemistry*, ed. by P.R. Ortiz de Montellano (Plenum, New York, 1986), p. 3
152. R.J. Guajardo, F. Chavez, E.T. Farinas, P.K. Mascharak, J. Am. Chem. Soc. **117**, 3883 (1995)
153. R.J. Guajardo, S.E. Hudson, S.J. Brown, P.K. Mascharak, J. Am. Chem. Soc. **115**, 7971 (1993)
154. C. Nguyer, R.J. Guajardo, P.K. Mascharak, Inorg. Chem. **35**, 6273 (1996)
155. D.S. Sigman, Acc. Chem. Res. **19**, 180 (1986)
156. T.D. Tullius, W.A. Dombroski, Science **230**, 679 (1985)
157. T.D. Tullius, W.A. Dombroski, Proc. Natl. Acad. Sci. U S A **83**, 5469 (1986)
158. L.B. Chapnick, L.A. Chasin, A.L. Raphall, J.K. Barton, Mutat. Res. **201**, 17 (1988)
159. G.S. Silver, W.C. Trogler, J. Am. Chem. Soc. **117**, 3983 (1995)
160. S. Sarkar, S. Chaudhuri, T. Basu, Curr. Sci. 2002, **83**(11), 1376–1380
161. R.P. Herlzberg, P.B. Dervan, Biochemistry **23**, 3934 (1984)
162. M.W. Von Dyke, R.P. Herlzberg, P.B. Dervan, Proc. Natl. Acad. Sci. USA **79**, 5470 (1982)
163. L.F. Povirk, Mutat. Res. **257**, 127 (1991)
164. A. Natrajan, S.M. Hecht. in *Molecular Aspects of Anticancer Drug. DNA Interactions*, vol. 2, ed. by S. Neidle, M. Naring (CRS Press, Boca Raton, FL, 1994), p. 197
165. Ch.A. Ditmer III F.V. Pamatong, J.R. Bocarsly, Inorg. Chem. **35**, 6292 (1996)
166. F.V. Pomatong, Ch.A. Ditmer III, J.R. Bocarsly, Inorg. Chem, **35**, 5339 (1996)
167. F.V. Pomatong, Ch.A. Detmer III, J.R. Bocarsly, J. Am. Chem. Soc. **118**, 5339 (1996)
168. K.L. Malisza, B.B. Hasinoff, Arch. Biochem. Biophys. **321**(1), 51 (1995)
169. V. Melatesta, *Free Radicals in Synthesis and Biology*, ed. by F. Minisei (Kluwer Acad., Dortrecht, 1989), p. 437
170. X. Xu, H.L. Persson, D.R. Richardson, Mol. Pharmacol. **68**, 261–270 (2005)
171. N.S. Enikolopian, K.A. Bogdanova, K.A. Askarov, Uspekhi khimii [Russian Chem. Rev.] **52**, 20 (1983) [Russian]
172. J. Smegal, C. Hill, J. Am. Chem. Soc. **105**, 3515 (1983)
173. T. Mlodnicka, J. Mol. Catal. **36**, 205 (1986)
174. J.T. Groves, W.J. Kruper, R.C. Haushalter, J. Am. Chem. Soc. **102**, 6375 (1980)
175. C.E. Cooper, in *Free Radical Damage and Its Control*, ed. by C.A. Rice-Evans, R.H. Burdon (Elsevier Science B.V., Amsterdam, 1994), pp. 67–111
176. R. Dietz, W. Nastainczyk, H.H. Ruf, Eur. J. Biochem. **171**, 321 (1988)
177. R. Karthein, R. Dietz, W. Nastainczyk, H.H. Ruf, Eur. J. Biochem. **171**, 313 (1988)
178. T.L. Poulos, J. Kraut, J. Biol. Chem. **255**, 10322 (1980)
179. R. Rang, S.A. Martinis, S.G. Sligar, T.L. Poulos, Biochemistry, **30**, 11420 (1991)
180. G.T. Babcock, M. Wikström, Nature **356**, 301 (1992)
181. T.A. Dix, S.J. Benkovic, Acc. Chem. Res. **21**, 101 (1988)
182. A.I. Alayash, R.P. Patel, R.E. Cashon, Antioxid. Redox Signal., **3**(2), 313–327 (2001)
183. H. Fretz, W.D. Woggon, R. Voges, Helv. Chem. Acta. **72**, 391 (1989)
184. H. Kusunore, A. Sawamura, H. Kawashima, J. Biochem. 106, 194 (1989)
185. W.M. Atkins, S.G. Sligar, J. Am. Chem. Soc. **111**, 2715 (1989)
186. D.J. Schuller, W. Zhu, S. Stojiljkovic, A. Wilks, T.L. Poulos, Biochemistry, **40**, 11552–11558 (2001)
187. M. Sugishima, H. Sakamoto, et al., J. Biol. Chem. **277**, 45086–45090 (2002)
188. I. Sevriukova, H. Li, T.L. Poulos, J. Mol. Biol. **236**, 889–902 (2004)
189. A.M. Khenkin, A.E. Shilov, React. Kinet. Catal. Lett. **33**, 125 (1987)
190. R. Davydov, T. Matsui, H. Fujii, et al., J. Am. Chem. Soc. **125**(52), 16208–16209 (2003)
191. N.F. Goldshleger, N.B. Tyabin, A.E. Shilov, A.A. Shteinman, Zh. Fiz. Khim. [J. Phys. Chem.]. **43**, 2174. [Russian] (1969)
192. N.F. Goldshleger, V.V. Yeskova, A.E. Shilov, A.A. Shteinman, Zh. Fiz. Khim. [J. Phys. Chem.] **46**, 1358. [Russian] (1972)
193. V.V. Yeskova, A.E. Shilov, A.A. Shteinman, Kinetika i kataliz [Kinetics and catalysis], **13**, 534. [Russian] (1972)
194. G.B. Shulpin, Comptes rendu, Chimie **6**(2), 163–178 (2003)
195. D.E. Webster, Adv. Organometal. Chem **15**, 147 (1977)

196. V.P. Tretyakov, E.S. Rudakov, A.V. Bogdanov, G.P. Zymtsova, L.I. Kozhevina. Izv. AN SSSR, Ser. Khim. [Proc. Acad. Sci. USSR. Chem. Ser.] **249**, 878. [Russian] (1979)
197. L.S. Shulpina, A.R. Kudinov, G. Suss-Fink, et al., Petrol. Chem. **45**(5), 309–311 (2005)
198. A.E. Shilov, *Activation of Saturated Hydrocarbons by Transition Metal Complexes* (D. Reidel, Dordrecht, 1984)
199. A.E. Shilov, in *Activation and Catalytic Reactions of Alkanes*, ed. by K. Hill (Mir, Moscow, 1992), p. 21. [Russian]
200. V.I. Kuznetsov, *General Chemistry: Development Trends* (Vysshaya shkola, Moscow, 1989) [Russian]
201. Chauban, N., S.I. Thackray, S.A. Rafice, J. Am. Chem. Soc. **131**(2), 4186–4187 (2009)
202. R. Bernhardt, Rev. Physiol. Biochem. Pharmacol. **127**, 137 (1995)
203. R.E. White, M.J. Coon, Annu. Rev. Biochem. **49**, 315 (1980)
204. F.P. Guengerich, T.L. MacDonald, Acc. Chem. Res. **17**, 9 (1984)
205. J.P. Collman, T.N. Sorrgell, J. Am. Chem. Soc., Symp. Ser. **444**, 27 (1977)
206. V. Ullrich, W. Duppel, in *The Enzymes*, vol. 12, ed. by P.D. Boyer (N.Y. Academy, New York, 1975), p. 253
207. T.L. Poulos, B.C. Finzel, A.J. Howard, J. Mol. Biol. **195**, 687 (1987)
208. N. Ahmad, H. Mukhtar, J. Onvest. Dermatol. **123**, 417–425 (2004)
209. P.R. Oritz de Montellano, in *Cytochrome P-450: Structure, Mecanism and Biochemistry,* ed. by P.R. Oritz de Montellano (Plenum Press, New York, London, 1986), p. 217
210. D.F.V. Lewis, *Cytochromes P-450: Structure, Function and Mechanism* (Taylor and Francis, London, 1996)
211. Hamdane, D., H. Zhang, P. Hollenberg, Photosynth. Res. **98**(1–3), 657–666 (2008)
212. Mansuy, D., P. Battioni, in *Activation and Functionalization of Alkanes*, ed. by G.L. Hill (A. Wiley-Interscience, New York, 1989)
213. J.P. Collman, T.N. Sorrell, in *Drug Metabolism Concepts*, ed. by D.M. Jerina, (American Chemical Society, Washington, DC, 1977), p. 27
214. D. Mansuy, in *The Coordination Chemistry of Metalloenzymes*, ed. by I. Bertini, R.S. Drago, C. Luchinat (D. Reidel, Dordrecht, 1983), p. 343
215. M. Schappacher, R. Weiss, R. Montiel-Montoya, A. Trautwein, A. Tabard, J. Am. Chem. Soc. **107**, 3736 (1985)
216. E.I. Karasevich, Ros. Khim. Zh. [Russian Chem. J.] **39**, 31. [Russian] (1995)
217. M. Newcomb, M.-H. Le Tadic-Biadatti, D.L. Chestney, E.S. Roberts, P.F. Hollenberg, J. Am. Chem. Soc. **117**, 12085 (1995)
218. M. Newcomb, M.-H. Le Tadic-Biadatti, in *Free Radicals in Biology and Environment*, ed. by F. Minisei (Kluwer Academic, Dordrecht, 1997), p. 91
219. Karasevich, E.I., V.S. Kulikova, A.E. Shilov, A.A. Shteinman, Uspekhi khimii [Russian Chem. Rev.] **67**, 376 (1998) [Russian]
220. C.W. Criffin, FEBS Lett. **74**, 139 (1977)
221. Achrem, A.A., S.M. Belskyi, D.M. Metelitsa, Izv. AN BSSR. Ser. Khim. [Proc. Acad. Sci. BSSR. Chem. Ser.]. **1**, 87. [Russian] (1978)
222. D.I. Metelitsa, P.A. Kyselev, S.N. Kyseleva, Kinetika i kataliz [Kinetics and catalysis] **21**, 1436 (1980) [Russian]
223. N.P. Niraula, B.K. Kanth, et al. Enzyme Microb.Technol. **48**(2), 181–186 (2010)
224. E.G. Hrycay, J.-A. Gustafsson, M. Ingelman-Sundberg, L. Ernster, Biochem. Biophys. Res. Commun. **66**, 209 (1975)
225. McMurry, T.J. and J.T. Groves, in *Cytochrome P-450, Structure, Mechanism and Biochemistry*, ed. by P.R. Oritz de Montellano (Plenum Press, New York, 1986), p. 1
226. J.E. Lyones, P.E. Ellis, V.A. Durante, Stud. Surf. Sci. Catal. **67**, 99 (1991)
227. P.E. Ellis, J.E. Lyones, Catal. Lett. **8**, 45 (1991)
228. A.M. Khenkin, A.E. Shilov, New J. Chem. **13**, 659 (1989)
229. J.P. Collman, R.R. Cagne, C.A. Reed, T.R. Bert, G. Lang, W.T. Robinson, J. Am. Chem. Soc. **97**, 1427 (1975)

230. M. Schappacher, R. Weiss, R. Monteil-Montoya, A. Trautwein, A. Tabard, J. Am. Chem. Soc. **107**, 3736 (1985)
231. W.I. Kim, T.I. Kim, W.S. Ahm, et al. Catal. Lett. **91**(1–2), 123–127 (2003)
232. A.M. Khenkin, A.E. Shilov, Stud. Org. Chem. (Amsterdam) **33**, 563 (1988)
233. I. Tabuski, Coord.Chem. Rev. **86**, 1 (1988)
234. P. Battioni, J.F. Bartoli, P. Leduc, M. Fontecave, D. Mansuy, J. Chem. Soc., Chem. Commun. 791 (1987)
235. E.I. Karasevich, Yu.K. Karasevich, Kinet. Catal. **41**(4), 485–491 (2000)
236. J.T. Groves, T.E. Nemo, J. Am. Chem. Soc. **105**, 5786 (1983)
237. A.M. Khenkin, A.A. Shteinman, Oxid. Commun. **4**, 433 (1983)
238. A.A. Shteinman, Izv. AN Rossii, Ser. Khim. [Proc. Acad. Sci. Rus., Chem. Ser.] 272 (1993) [Russian]
239. A.M. Khenkin, A.A. Shteinman, J. Chem. Soc., Chem. Commun. 219 (1984)
240. A.B. Sorokin, A.M. Khenkin, C.F. Marakushev, A.E. Shilov, A.A. Shteinman, Dokl. AN SSSR [Proc. USSR Acad. Sci.] **279**, 939 (1984) [Russian].
241. B.R. Cook, T.J. Reinert, K.S. Suslik, J. Am. Chem. Soc. **108**, 7281 (1986)
242. T.G. Traylor, T. Nakano, B.E. Dunlap, P.S. Traylor, D. Dolphin, J. Am. Chem. Soc. **108**, 2782 (1986)
243. J.E. Lyons, P.E. Ellis, V.A. Durante, Stud. Surf. Sci. Catal. **67**, 99 (1991)
244. M. Fontecave, D. Mansuy, Tetrahedron **40**, 4297 (1984)
245. Tabushi, I., M. Kodera, M. Yokoyama, J. Am. Chem. Soc. **107**, 4466 (1985)
246. D. Mansuy, Pure Appl. Chem. **59**, 759 (1987)
247. S.E. Creager, S.A. Raybuck, R.W. Murray, J. Am. Chem. Soc. **108**, 4225 (1986)
248. A.E. Gekhman, I.P. Stolyarov, N.I. Moiseeva, I.I. Moiseev, C. R. Chim. **7**(8–9), 833–844 (2004)
249. K.S. Suslick, in *Activation and Functionalization of Alkanes*, ed. by C.L. Hill (Wiley-Interscience, New York, 1989), p. 219
250. B.L. Hay Kraus, D.J. Greenblatt, et al. Xenobiotica **30**(6), 575–588 (2000)
251. E.I. Karasevich, A.M. Khenkin, Biochemistry **51**, 1454 (1986)
252. K. Morohashi, H. Sadano, J. Okada, T. Omura, J. Biochem. **93**, 413 (1983)
253. J.T. Groves, R. Neumann. J. Am. Chem. Soc. 1989, **111**, 2900.
254. L.L. Vasilieva, E.A. Lukasheva, A.B. Soloviova, K.K. Nyvnitskii, Izv. AN SSSR, Ser. Khim. [Proc. Acad. Sci. USSR. Chem. Ser.] **6**, 1378. [Russian] (1990)
255. T. Nagata, K. Imagawa, T. Yamada, T. Mukaiyama, Chem. Lett. 1259 (1994)
256. K. Imagawa, T. Nagata, T. Yamada, T. Mukaiyama, Chem. Lett. 527 (1994)
257. T. Mukaiyama, T. Nagata, T. Yamada, K. Imagawa, Chem. Lett. 327 (1993)
258. Murahashi, S.-I., Y. Oda, T. Naota, J. Am. Chem. Soc. **114**, 7913 (1992)
259. R. Iwanejko, P. Leduc, T. Mlodnica, J. Poltowicz, in *Dioxygen Activation and Homogeneous Catalytic Oxidation*, ed. by L.I. Simandi (Elsevier Science, Amsterdam, 1991), p. 113
260. L.I. Simandi, in *Catalytic Activation of Dioxygen by Metal Complexes*, ed. by L.I. Simandi (Kluwer Academic, Dordrecht, 1992), pp. 318–331
261. W. Nam, H.J. Kim, S.H. Kim, R.Y.N. Ho, J.S. Valentine, Inorg. Chem. **35**, 1045 (1996)
262. H. Mimoun, M. Mignard, P. Brechot, L. Saussine, J. Am. Chem. Soc. **108**, 3711 (1986)
263. H. Orita, Inorg. Chem. **30**, 258 (1991)
264. T.-Ch. Chou, Sh.-Van Lee, Eng. Chem. Res. **36**, 1485 (1997)
265. L. Weber, R. Hommel, J. Behling, G. Haufe, H. Hennig, J. Am. Chem. Soc. **116**, 2400 (1994)
266. B. Minaev, Chem. and Chem.Technol. **4**(1), 1–16 (2010)
267. B. Ensing, F. Buda, M. Gribnau, E. Baerends, J. Am. Chem. Soc. **126**, 4355 (2004)
268. K. Machii, Y. Watanabe, I. Morishima, J. Am. Chem. Soc. **117**, 6691 (1995)
269. H. Tsurumaki, Y. Watanabe, I. Morishima, J. Am. Chem. Soc. **115**, 11784 (1993)
270. B. Meunier, Chem. Rev. **92**, 1411 (1992)
271. T.G. Traylor, Sh. Tsuchiya, Y.-S. Byun, Ch. Kim, J. Am. Chem. Soc. **115**, 2775 (1993)
272. Mansuy, D. Pure Appl. Chem. 1990, **62**, 741
273. Mansuy, D., J.F. Bartoli, M. Momenteau, Tetrahedron Lett. **23**, 2781 (1982)

274. B.R. Cook, T.J. Reinert, K.S. Suslick, J. Am. Chem. Soc. **108**, 7281 (1986)
275. T.G. Traylor, F. Xu, J. Am. Chem. Soc. **109**, 6202 (1987)
276. D. Mansuy, P. Battioni, J.P. Renaud, J. Chem. Soc. Chem. Commun. 1255 (1984)
277. P.N. Balasubramaman, A. Sinha, T.C. Bruice, J. Am. Chem. Soc. **109**, 1456 (1987)
278. T. Higuchi, K. Shimada, N. Hirobe, J. Am. Chem. Soc. **115**, 7551 (1993)
279. J.R. Lindsay-Smith, R.J. Lower, J. Chem. Soc. Perkin Trans. **2**, 31 (1991)
280. R.T. Arasasingham, Th. C. Bruice, in *The Activation of Dioxygen and Homogeneous Catalytic Oxidation*, ed. by D.H.R. Barton, A.E. Martell, D.T. Sawyer (Plenum Press, New York, 1993) pp. 147–169
281. Ö. Almarsson, Th. C. Bruice, J. Am. Chem. Soc. **117**, 4533 (1995)
282. Ö. Almarsson, H. Adalsteinsson, Th. C. Bruice, J. Am. Chem. Soc. **117**, 4524 (1995)
283. J.T. Groves, Y. Watanabe, J. Am. Chem. Soc **110**, 8443 (1988)
284. Yu.N. Sazanov, A.V. Gribanov, Russ. J. Appl. Chem. **83**(2), 177–197 (2010)
285. T.G. Traylor, F. Xu, J. Am. Chem. Soc. **112**, 1178 (1990)
286. T.G. Traylor, T. Nakano, B.E. Dunlap, P.S. Traylor, D. Dolphin, J. Am. Chem. Soc. 1986, **108**, 2782;
287. T.G. Traylor, A.R. Mikzstal, J. Am. Chem. Soc. **109**, 2770 (1987)
288. T.G. Traylor, T. Nakano, A.R. Mikzstal, B.E. Dunlap, J. Am. Chem. Soc. **109**, 3625 (1987)
289. O. Bartolini, B. Meunier, J. Chem. Soc. Perkin Trans. **2**, 1967 (1984)
290. J.R. Lindsay-Smith, P.R. Sleath, J. Chem. Soc. Perkin Trans. **2**, 1009 (1982)
291. A. Castellino, T.C. Bruice, J. Am. Chem. Soc. **110**, 158 (1988)
292. J.P. Collman, T. Kodadek, J.I. Brauman, J. Am. Chem. Soc. **108**, 2588 (1986)
293. J.P. Collman, T. Kodadek, S.A. Raybuck, J.I. Brauman, L.M. Papazian, J. Am. Chem. Soc. **107**, 4343 (1985)
294. G.-X. He, R.D. Arasasingham, G. Zhang, T.C. Bruice, J. Am. Chem. Soc. **113**, 9828 (1991)
295. F.P. Guengerich, T.L. MacDonald, Acc. Chem. Res. **17**, 9 (1984)
296. J.R. Groves, Y. Watanabe, J. Am. Chem. Soc. **108**, 507 (1986)
297. M. Fontecave, D. Mansuy, J. Chem. Soc., Chem. Commun. 879 (1984)
298. A. Castellino, T.C. Bruice, J. Am. Chem. Soc. **110**, 1313 (1988)
299. A. Castellino, T.C. Bruice, J. Am. Chem. Soc. **110**, 7512 (1988)
300. T. Watanabe, K. Akamatsu, Biochem. Pharmacol. **23**, 1079 (1974)
301. T. Watanabe, Y. Ueno, J. Imazumi, Biochem. Pharmacol. **20**, 912 (1971)
302. P.R. Oritz de Montellano, B.L.K. Mangold, C. Wheeler, K.L. Kunze, N.O. Reich, J. Biol. Chem. **258**, 4208 (1983)
303. R.P. Hanzlik, G.O. Shearer, Biochem. Pharmacol. **27**, 1441 (1978)
304. G.H. Loew, C.J. Kert, L.M. Hjemeland, R.F. Kirchner, J. Am. Chem. Soc. **99**, 3534 (1977)
305. D. Mansuy, Pure Appl. Chem. **59**, 759 (1988)
306. B. Meunier, Bull. Soc. Chem. Fr. **4**, 578 (1986)
307. Groves, J.T., T.E. Nemo, R.S. Myers, J. Am. Chem. Soc. **101**, 1032 (1979)
308. J.T. Groves, T.E. Nemo, J. Am. Chem. Soc. **105**, 6243 (1983)
309. T.J. MeMurry, J.T. Groves,in *Cytochrome P-450: Structure, Mechanism and Biochemistry*, ed. by P.R. Oritz de Montellano (Plenum Press, New York, 1986), p. 1
310. J.R. Lindsay-Smith, P.R. Sleath, J. Chem. Soc., Perkin Trans. **2**, 1165 (1983)
311. D. Ostovic, T.C. Bruice, J. Am. Chem. Soc. **111**, 6511 (1989)
312. J.P. Collman, J.I. Brauman, B. Meunier, S.A. Raybuck, T. Kodadek, Proc. Natl. Acad. Sci. U S A **81**, 3245 (1984)
313. J.P. Collman, J.I. Brauman, B. Meunier, T. Hayashi, T. Kodadek, S.A. Raybuck, J. Am. Chem. Soc. **107**, 2000 (1985)
314. K.B. Sharpless, A.J. Teranishi, J.-E. Backvell, J. Am. Chem. Soc. **99**, 3120 (1977)
315. J.P. Collman, T. Kodadek, J.I. Brauman, J. Am. Chem. Soc. **108**, 2588 (1986)
316. B. Meunier, Gazz. Chim. Ital. **118**, 485 (1988)
317. C.K. Chang, F.J. Ebina, J. Chem. Soc., Chem. Commun. 778 (1981)
318. P.S. Traylor, D. Dolphin, T.G. Traylor, J. Chem. Soc., Chem. Commun. 279 (1984)
319. T.G. Traylor, J.C. Marsters, T. Nakano, B.E. Dunlap, J. Am. Chem. Soc. **107**, 5537 (1985)

320. J.P. Collman, T. Kodadek, S.A. Raybuck, J.I. Brauman, L.P. Papazian, J. Am. Chem. Soc. **107**, 4343 (1985)
321. R.W. Lee, P.C. Nakagaki, P.N. Balasubramanian, T.C. Bruice, Proc. Natl. Acad. Sci. USA **85**, 641 (1988)
322. S. Funyu, T. Isobe, S. Takagi, D. Tryk, H. Inoue, J. Am. Chem. Soc. **125**, 5734–5740 (2003)
323. H. Shi, , Z. Zhang, Y. Wang, J. Mol. Catal. A: Chem. **238**(1–2), 13–25 (2005)
324. T.G. Traylor, A.R. Miksztal, J. Am. Chem. Soc. **111**, 7443 (1987)
325. K. Yamaguchi, Y. Tekahara, T. Fueno, in *Applied Quantum Chemistry*, ed. by V.H. Smith, H.F. Schaefer, K. Morokuma (D. Reidel, Dordrecht, 1986), p. 155
326. K.A. Jorgensen, , B. Schiott, Chem. Rev. **90**, 1483 (1990)
327. C.A. Tolman, J.D. Druliner, M.J. Nappa, N. Herron, in *Activation and Functionalization of Alkanes*, ed. by C.L. Hill, (Wiley, New York, 1989), p. 303
328. B. De Poorter, M. Ricci, O. Bortolini, B. Meunier, J. Mol. Catal. **31**, 221 (1985)
329. B. Meunier New J. Chem. **16**, 203 (1992)
330. J.P. Collman, X. Zhang, V. Lee, R.T. Hembre, J.I. Brauman. in *Homogeneous Tranition Metal Catalyzed Reactions*, ed. by W.B. Moser, D.W. Slocum (Washington, DC: Am. Chem. Soc., 1992), pp. 153–162
331. S. Banfi, A. Maiocchi, F. Montanari, S. Quici, La Chimica e L'Industria **72**(4), 304 (1990)
332. A.E. Shilov, E.I. Karasevici, in *Metalloporphyrin Catalyzed Oxidations*, ed. by F. Montanari, L. Cosella, (Kluwer Academic, Dordrecht, 1994), p. 87
333. A.G. Katopodis, H.A. Smith, S.W. May, J. Am. Chem. Soc. **110**, 897 (1988)
334. H. Fu, M. Newcomb, C.-H. Word, J. Am. Chem. Soc. **113**, 5878 (1991)
335. W. Nam, J. Ryu, I. Kim, C. Kim, Tetrahedron Lett. **43**(31), 5487–5490 (2002)
336. D.I. Metelitsa, *Modeling of Redox Enzymes* (Nauka i Tekhnika, Minsk, , 1984) [Russian]
337. A.M. Khenkin, A.A. Shteinman, Ros. Khim. Zh. [Russian Chem. J.] **39**, 41. [Russian] (1995)
338. A.M. Khenkin, A.A. Shteinman, in *Vth Int. Symp. on Connection between Homogeneous and Heterogeneous Catalysis*. (Novosibirsk, 1986), p. 270 [Russian]
339. N. Herron, New J. Chem. **13**, 761 (1989)
340. D. Mansuy, ,New J. Chem. **13**, 581 (1989)
341. T. Nakano, T.G. Traylor, D. Dolphin, Can J. Chem. **68**, 1504 (1990)
342. T.G. Traylor, T. Nakano, A.R. Miksztal, J. Am. Chem. Soc. **109**, 3625 (1987)
343. J.P. Collman, P.D. Hampton, J.I. Brauman, J. Am. Chem. Soc. **112**, 2977 (1990)
344. Z.-Q. Tian, J.L. Richards, T.G. Traylor, J. Am. Chem. Soc. **117**, 21 (1995)
345. V.J. DeRose, K.E. Liu, S.J. Lippard, B.M. Hoffman, J. Am. Chem. Soc. **118**, 121 (1996)
346. S.K. Lee, J.C. Nesheim, J.D. Lipsomb, J. Am. Chem. Soc. **268**, 21569 (1993)
347. S.K. Lee, B.G. Fox, W.A. Froland, J.D. Lipsomb, E. Münck, J. Am. Chem. Soc. **115**, 6450 (1993)
348. K.E. Liu, D. Wang, B.H. Hayng, D.E. Edmondson, A. Salifoglou, S.J. Lippard, J. Am. Chem. Soc. **116**, 7465 (1994)
349. P. Barbaro, F. Liguori (eds.), *Heterogenized Homogeneous Catalysis for Fine Chemicals Production*. (Springer, Dordrecht, 2010). p. 400
350. S.J. Lippard, J.M. Berg, *Principles of Bioinorganic Chemistry*. (Univ. Sci. Books, Mill Valley, CA, 1994)
351. C. Tinberg, S. Lippard, Biochemistry **49**, 7902–7912 (2010)
352. R.E. Stenkampf, Chem. Rev. **94**, 715 (1994)
353. E.I. Solomon, F. Tuczek, D.E. Root, C.A. Braun, Chem. Rev. **94**, 827 (1994)
354. B.G. Fox, M.P. Hendrich, K.K. Surezus, K.K. Andersson, W.A. Froland, J.D. Lipscomb, E. Münck, J. Am. Chem. Soc. **115**, 3688 (1993)
355. A.C. Rosenzweig, S.A. Frederick, S.J. Lippard, P. Nordlund, Nature **366**, 537 (1993)
356. R.C. Prince, G.N. George, J.C. Savas, S.P. Cramer, R.N. Patel, Biochim. Biophys. Acta. **952**, 220 (1988)
357. J.G. Dewitt, J.G. Bentsen, A.C. Rosenzweig, B. Hedman, J. Green, S. Pilkington, G.C. Papaefthymion, H. Dalton, K.O. Hodgson, S.J. Lippard, J. Am. Chem. Soc. **113**, 9219 (1991)

358. A.C. Rosenzweig, P. Nordlund, P.M. Takehara, C.A. Frederick, S.J. Lippard, Chem. Biol. **2**, 409 (1995)
359. A.A. Srainman, Izv. AN SSSR, Ser. Khim. [Proc. Acad. Sci. USSR. Chem. Ser.] 1011 [Russian] (1995)
360. R.C. Prince, G.N. George, J.C. Savas, S.P. Cramer, R.N. Patel, Biochim. Biophys. Acta. **952**, 220 (1988)
361. A. Ericson, B. Hedman, K.O. Hodson, J. Green, H. Dalton, J.-G. Bentsen, R.H. Beer, S.J. Lippard, J. Am. Chem. Soc., **110**, 2330 (1988)
362. M.R. Woodlaud, D.S. Pastil, R. Cammack, H. Dalton Biochim. Biophys. Acta. **873**, 237 (1986)
363. B.G. Fox, K.K. Surerus, E. Munck, J.D. Lipscomb, J. Biol. Chem. **263**, 10553 (1988)
364. S.C. Pulver, W.A. Froland, J.D. Lipscomb, E.I. Solomon, J. Am. Chem. Soc. **119**, 387 (1997)
365. B.G. Fox, W.A. Froland, J.E. Dege, J.D. Lipscomb, J. Biol. Chem. **268**, 10023 (1993)
366. B.G. Fox, Y. Liu, J.E. Dege, J.D. Lipscomb, J. Biol. Chem. **266**, 540 (1991)
367. K.E. Paulsen, Y. Liu, B.G. Fox, J.D. Lipsomb, E. Münck, M.T. Stankovich, Biochemistry **33**, 713 (1994)
368. J. Green, H. Dalton, J. Biol. Chem. **264**, 17698 (1989)
369. H. Dalton, P.C. Wilkins, Y. Jiang, Biochem. Soc. Trans. **21**, 749 (1993)
370. V.J. DeRose, K.E. Liu, S.J. Lippard, B.M. Hoffman, J. Am. Chem. Soc. **118**, 121 (1996)
371. K.E. Liu, D. Wang, B.H. Huynh, D.E. Edmondson, A. Salifoglou, S.J. Lippard, J. Am. Chem. Soc. **116**, 7465 (1994)
372. J.D. Lipsomb, Annu. Rev. Microbiol. **48**, 371 (1994)
373. N.D. Priestley, H.G. Floss, W.A. Froland, J.D. Lippsomb, P.G. Williams, H. Morimoto, J. Am. Chem. Soc. **114**, 7561 (1992)
374. M.A. Molmes, , R.E. Stenkamp, J. Mol. Biol. **220**, 723 (1991)
375. S.K. Lee, J.C. Nesheim, J.D. Lipsomb, J. Biol. Chem. **268**, 21569 (1993)
376. S.K. Lee, B.G. Fox, W.A. Florand, J.D. Lipsomb, E. Münck, J. Am. Chem. Soc. **115**, 6450 (1993)
377. K.E. Liu, D. Wang, B.H. Huynh, D.E. Edmondson, A. Salifoglou, S.J. Lippard, J. Am. Chem. Soc. **116**, 7465 (1994)
378. A.A. Shteinman, Izv. AN Rossii. Ser. Khim. [Proc. Acad. Sci. Rus. Chem. Ser.] **324**, 617 (1992) [Russian]
379. A.A. Shteinman, Izv. AN Rossii. Ser. Khim. [Proc. Acad. Sci. Rus. Chem. Ser.] 272 (1993) [Russian]
380. A.A. Shteinman, FEBS Lett. **362**, 5 (1995)
381. A.A. Shteinman, Izv. AN Rossii. Ser. Khim. [Proc. Acad. Sci. Rus. Chem. Ser.] 1676 (1997) [Russian]
382. E.S. Rudakov, *Reactions of Alkanes with Oxidants, Metal Complexes and Radicals in Solutions* (Naukova Dumka, Kiev, 1985), p. 56 [Russian]
383. K.E. Liu, A.M. Valentine, D. Wang, B.H. Huynh, D.E. Edmondson, A. Salifoglou, S.J. Lippard, J. Am. Chem. Soc. **117**, 10174 (1995)
384. S. Herold, S.J. Lippard, J. Am. Chem. Soc. **119**, 145 (1997)
385. A.C. Rosenzweig, P. Nordlund, P.M. Takahara, C.A. Frederick, S.J. Lippard, Chem. Biol. **2**, 409 (1995)
386. A.F. Shestakov, A.E. Shilov, Zh. obshchei khim. [Rus. J. Gen. Chem.] **65**, 622. [Russian] (1995)
387. A.F. Shestakov, A.E. Shilov, J. Mol. Catal., A. **105**, 1 (1996)
388. B.P. Murch, F.C. Brodley, L. Que, Jr, J. Am. Chem. Soc. **108**, 5027 (1986)
389. J.B. Vincent, H.C. Huffman, G. Christou, Q. Lee, M.A. Nanny, D.N. Hendrickson, R.H. Fong, R.H. Fish, J. Am. Chem. Soc. **110**, 6898 (1988)
390. V.S. Belova, I.M. Gimanova, M.L. Stepanova, A.M. Khenkin, A.E. Shilov, Dokl. AN SSSR [Proc. of USSR Acad. of Sci.] **316**, 653 (1991) [Russian]
391. A.M. Khenkin, , M.L. Stepanova, in Mendeleev. Commun. 57 (1992) [Russian]
392. M. Torrent, D. Musaev, H. Basch, K. Morokuma, J. Compt. Chem. **23**(1), 59–76 (2009)

393. S.-K. Lee, J.C. Nesheim, J.D. Lipsomb, J. Biol. Chem. **268**, 21569 (1993)
394. S.-K. Lee, B.G. Fox, W.A. Froland, J.D. Lipsomb, E. Münck, J. Am. Chem. Soc. **115**, 6450 (1993)
395. J.M. Bollinger, Ir., W.H. Tong, N. Ravi, B.H. Huynh, D.E. Edmondson, J. Stubbe, J. Am. Chem. Soc. **116**, 8024 (1994)
396. B.F. Gherman, B.D. Dunietz, D.A. Whittington, S.J. Lippard, R.A. Friesnev, J. Am. Chem. Soc. **123**, 3836–3837 (2001)
397. K.E. Liu, A.M. Valentine, D. Qiu, D.E. Edmondson, E.H. Appelman, T.G. Spiro, S.J. Lippard, J. Am. Chem. Soc. **117**, 4997 (1995)
398. A.L. Feig, M. Becker, S. Schindler, R. Van Fldik, S.J. Lippard, Inorg. Chem. **95**, 2590 (1996)
399. Yochizawa, K.R. Hoffman, Inorg. Chem. **35**, 2409 (1996)
400. J.G. DeWitt, A.C. Rosenzweig, A. Salifoglou, B. Hedman, S.J. Lippard, K.O. Hodgson, Inorg. Chem. **34**, 2505 (1995)
401. K.E. Liu, C.C. Johnson, M. Newcomb, S.J. Lippard, J. Am. Chem. Soc. **115**, 939 (1993)
402. G.B. Shulpin, J. Mol. Catal. A: Chem. **189**, 39 (2002)
403. R.A. Leising, J. Kim, M.A. Perez, L. Que, Jr, J. Am. Chem. Soc. **115**, 9524 (1993)
404. S. Menage, J.M. Vincent, C. Lambeaux, G. Chottard, A. Grand, M. Fontecave, Inorg. Chem. **32**, 4766 (1993)
405. S. Menage, J.M. Vincent, C. Lambeaux, M. Fontecave, J. Chem. Soc., Dalton. 2081 (1994)
406. O.N. Gritsenko, G.N. Nesterenko, A.A. Shteinman,Izv. AN Rossii. Ser. Khim. [Proc. Acad. Sci. Rus. Chem. Ser.] , **12**, 2518 (1995) [Russian]
407. M.J. Filatov, A.G. Pelmenschikov, G.M. Zhidomirov, J. Mol. Catal. **80**, 243 (1993)
408. I. Tabushi, T. Nakajima, K. Seto, Tetrahedron Lett. **21**, 2565 (1980)
409. N. Kitajima , M. Ito, H. Fukui, Y. Morooka, J. Chem. Soc., Chem. Commun. 102 (1991)
410. N. Kitajima, H. Fukui, Y. Morooka, J. Chem. Soc., Chem. Commun. 485 (1988)
411. W. Nam, J.S. Valentine, New J. Chem. **13**, 677 (1989)
412. D.H.R. Barton, J. Boivin, M. Gastiger, J. Morzycki, R.S. Hay-Motherwell, W.B. Motherwell, N. Ozbalik, K.M. Schwartzentruber, J. Chem. Soc., Perkin Trans. 947 (1986)
413. S. Menage, E.C. Wilsinson, L. Que Jr., M. Fontecave, Angew. Chem., Int. Ed. Eng. **34**, 203 (1995)
414. S. Menage, J.M. Vincent, C. Lambeaux, M. Fontecave, J. Chem. Soc., Dalton Trans. 2081 (1994)
415. J.M. Vincent, S. Menage, C. Lambeaux, M. Fontecave, Tetrahedron Lett **35**, 6287 (1994)
416. R.H. Fish, M.S. Konougers, K.J. Oberhausen, R.H. Fong, W.M. Yu, G.C. Christon, J.B. Vincent, D.K. Coggin, R.M. Buchanan, Inorg. Chem. **30**, 3002 (1991)
417. R.M. Buchanan, S. Chen, J.F. Richardson, M. Bressan, L. Forti, A. Morvillo, R.H. Fish, Inorg. Chem. **33**, 3208 (1994)
418. D.T. Sawyer, C. Kang, A. Liobet, C. Redman, J. Am. Chem. Soc. **115**, 5817 (1993)
419. J. Kim, R.G. Harrison, S. Kim, L. Que Jr., J. Am. Chem. Soc. **118**, 4373 (1996)
420. J. Kim, E. Larka, E.C. Wilkinson, L. Que Jr., Angew. Chem., Int. Ed. Eng. **34**, 2048 (1995)
421. A.L. Feig, A. Masschellin, A. Bakac, S.J. Lippard, J. Am. Chem. Soc. **119**, 334 (1997)
422. P.E. Fitzpatrick, Biochem. Biophys. Res. Commun. **6**, 1 (1989)
423. J.P. Hage, D.T. Sawyer, J. Am. Chem. Soc. **7**, 17 (1995)
424. Kang, C. Redman, D.T. Sawyer, Bioorg. Med. Chem. **1**, 25 (1993)
425. M.C. Brenner, J.P. Klinman, Biochemistry **28**, 4664 (1989)
426. T.M. Pettingill, R.W. Strange, N.J. Blackburn, J. Biol. Chem. **266**, 16996 (1991)
427. N.J. Blackburn, S.S. Hasnain, T.M. Pettingill, R.W. Strange, J. Biol. Chem. **266**, 23120 (1991)
428. B.J. Reedy, N.J. Blackburn, J. Am. Chem. Soc. **116**, 1924 (1994)
429. B.J. Reedy, N.N. Murthy, K.D. Karlin, N.J. Blackburn, J. Am. Chem. Soc. **117**, 9826 (1995)
430. T. Scotland, L. Petterson, D. Backstrom, T. Ljones, T. Flatmark, A. Ehrenberg, Eur. J. Biochem. **103**, 5 (1980)
431. S. Ito, M. Takimura, K. Sasaki, Chem. Lett. 461 (1993)
432. Sh. Itoh, T. Kondo, H. Komatsu, Y. Ohshiro, Ch. Li, N. Kanehisa, Y. Kai, Sh. Fukuzumi, J. Am. Chem. Soc. **117**, 4714 (1995)

433. K.D. Karlin, Y. Gultneh, J.C. Hayes, J. Zubieta, Inorg. Chem. **23**, 521 (1984)
434. N. Kitajima, T. Koda, Y. Iwata, Y. Morooka, J. Am. Chem. Soc. **112**, 8833 (1990)
435. I. Sanyal, M. Mahroof-Tahir, M.S. Nasir, P. Ghosh, B.I. Cohen, Y. Gultneh, R.W. Cruse, A. Farooq, K.D. Karlin, S. Liu, J. Zubieta, Inorg. Chem. **31**, 4322 (1992)
436. W.B. Tolman, Acc. Chem. Res., **30**(6), 227 (1997)
437. L. Que, Jr, Adv. Inorg. Biochem. **5**, 167 (1983)
438. L. Que Jr., R.M. Epstein, Biochemistry **20**, 2545 (1981)
439. L. Que Jr., J.D. Lipsomb, R. Zimmermann, E. Münck, W.H. Orme-Johnson, N.R. Orme-Johnson, Biochim. Biophys. Acta. **452**, 320 (1976)
440. J.W. Whittaker, J.D. Lipscomb, T.A. Kent, E. Münck, J. Biol. Chem. **259**, 4466 (1984)
441. T.A. Kent, E. Münck, J.W. Pyrz, J. Widom, L. Que Jr., Inorg. Chem. **26**, 1402 (1987)
442. M. Pau, M. Davis, A. Orville, et al., J. Am. Chem. Soc. 27 (2007)
443. T.A. Walsh, D.P. Ballon, R.J. Mayer, L. Que Jr., J. Biol. Chem. **258**, 14422 (1983)
444. J.W. Pyrz, A.L. Roe, L.J. Stem, L. Que Jr., J. Am. Chem. Soc. **107**, 614 (1985)
445. L. Que Jr., D.D. Cox, R.C. Kolanczyk, in W. Ando, Y. Morooka, eds. *The Role of Oxygen in Chemistry and Biochemistry*: Proc. of an Int. Symp. on Activation of Dioxygen and Homogeneons Catalytic Oxidations, Tsukuba, Japan, 12–16 July 1987. (Elsevier Science B.V., Amsterdam, 1988), vol. 33, p. 399
446. L. White, , P. Nilson, L. Pignolet, L. Que Jr., J. Am. Chem. Soc. **106**, 8312 (1984)
447. W.W.Y. Lam, T.D.H. Bugg, J. Chem. Soc., Chem. Commun. 1163 (1994)
448. J. Sanvoisin, G.J. Langley, T.D.H. Bugg, J. Am. Chem. Soc. **117**, 7836 (1995)
449. H.M. Hanayske-Abel, V. Gunsler, J. Theor. Biol. **94**, 421 (1982)
450. V.S. Kulikova, A.M. Khenkin, A.E. Shilov, Kinetika i kataliz [Kinetics and catalysis] **29**, 1278 (1988) [Russian]
451. Y.-M. Chiou, L. Que Jr., J. Am. Chem. Soc. **114**, 7567 (1992)
452. G.A. Veldink, J.F.G. Vliigenhart, Adv. Inorg. Biochem. **6**, 139 (1984)
453. B. Samuelsson, Science **220**, 568 (1983)
454. Y. Nishida, N. Tanaka, Sh. Takahashi, Chem. Lett. 411 (1993)
455. I.P. Skibida, A.M. Sakharov, Ros. khim. zh. [Rus. Chem. J.] **319**(1), 14. [Russian] (1995)
456. P. Sikes, in *Reactions Mechanisms in Organic Chemistry,* (Khimya, Moscow, 1977), p. 319 [Russian]
457. J.W. Whittaker, J.D. Lipscomb, J. Biol. Chem. **259**, 4487 (1984)
458. L. Que Jr., J.D. Lipscomb, E. Münck, J.M. Wood, Biochim. Biophys. Acta. **485**, 60 (1977)
459. T. Funabiki, I. Yoneda, M. Ishikawa, M. Ujiie, Y. Nagai, S. Yoshida, J. Chem. Soc., Chem. Commun. 1453 (1994)
460. M.R.N. Marthy, T.J. Reid, A. Sicigano, N. Tanaka, M.G. Rossmann, J. Mol. Catal. **15**, 465 (1981)
461. I. Fita, M.G. Rossmann, J. Mol. Biol. **188**, 49 (1986)
462. I. Fita, M.G. Rossmann, J. Mol. Biol. **185**, 21 (1985)
463. G.R. Schonbaum, B. Chance, in *The Enzymes*, 3rd edn., vol. 13,ed. by P.D. Boyer (Academic, New York, 1976), p. 363
464. H.M. Palcic, H.B. Dunford, J. Biol. Chem. **255**, 6128 (1980)
465. J.E. Frew, P. Jones, in *Advances in Inorganic and Bioinorganic Mechanisms*, vol. 3, ed. by A.G. Sykes (Academic, New York, 1984), p. 175
466. T.G. Traylor, W.A. Lee, D.V. Stynes, J. Am. Chem. Soc. **106**, 755 (1984)
467. L.K. Hansen, C.K. Chang, M.S. Davis, J. Fajer, J. Am. Chem. Soc. **103**, 663 (1981)
468. S. Hashimoto, Y. Tatsuno, T. Kitagawa, Proc. Natl. Acad. Sci. USA **83**, 2417 (1986)
469. A.N. eremin, A.V. Litvinchuk, D.I. Metelitsa, Biokhimya [Biochemistry] **61**, 664. [Russian] (1996)
470. A.N. Eremin, D.I. Metelitsa, Biokhimya [Biochemistry] **61**, 1672 (1996) [Russian]
471. D.I. Metelitsa, A.N. Eremin, I.M. Artsukevich, I.P. Chernikevich, Biokhimya [Biochemistry] **62**, 444 (1997) [Russian]
472. J. Pratt, in *Methods and Advances of Bioinorganic Chemistry*, ed. by K. Mcauliff (Mir, Moscow, 1978) pp. 194–228 [Russian]

473. D.I. Metelitsa, in *Modeling of Redox Enzymes* (Nauka i Tekhnika, Minsk, 1984) p. 163 [Russian]
474. A.V. Litvinchuk, A.A. Morozova, A.N. Eremin, D.I. Metelitsa, Zh. prikladnoi khim. [Russ. J. Appl. Chem.]. **68**, 940 (1995) [Russian]
475. A.V. Litvinchuk, A.A. Morozova, A.N. Eremin, D.I. Metelitsa, Biokhimya [Biochemistry] **61**, 664 (1996) [Russian]
476. T.C. Bruice, M.F. Zipplies, W.A. Lee, Proc. Natl. Acad. Sci. USA **83**, 4646 (1986)
477. P.N. Balasubramanian, E.S. Smidt, T.C. Bruice, J. Am. Chem. Soc. **109**, 7865 (1987)
478. Y. Kono, I. Fridovich, J. Biol. Chem. **258**, 6015 (1983)
479. G.S. Allgood, J.J. Perry, J. Bacteriol. **168**, 563 (1986)
480. C.V. Khangulov, V.V. Barynin, S.V. Antonyuk-Barynina, Biochem. Biophys. Acta. **1020**, 25 (1990)
481. J.E. Penner-Harn, in *Manganese Redox Enzymes*, ed. by V.L. Pecoraro (Wiley-VCH, New York, 1992), p. 29
482. R.M. Fronko, J.E. Penner, C.J. Bender, J. Am. Chem. Soc. **110**, 7554 (1988)
483. J.E. Sheats, R.S. Czernuszewicz, G.C. Dismukes, A.L. Rheingold, V. Petronleas, J. Stubbe, W.H. Armstrong, R.H. Beer, S.J. Lippard, J. Am. Chem. Soc. **109**, 1435 (1987)
484. J.D. Rush, Z. Maskos, Inorg. Chem. **29**, 897 (1990)
485. P.J. Pessiki, G.C. Dismukes, J. Am. Chem. Soc. **116**, 898 (1994)
486. Y. Naruta, K. Maruyama, J. Am. Chem. Soc. **113**, 3593 (1991)
487. Y. Naruta, M. Sasayama, K. Maruyama, Chem. Lett. 1267 (1992)
488. M. Sasayama, Y. Naruta, Chem. Lett. 1, 63 (1995)
489. T. Aono, H. Wada, M. Yonemura, M. Ohba, H. Ōkawa, D.E. Fenton, J. Chem. Soc. Dalton Trans. 1527 (1997)
490. C. Higuchi, H. Sakiyama, H. Ōkawa, R. Isobe, D.E. Fenton, J. Am. Chem. Soc. Dalton Trans. 1097 (1994)
491. B. Chance, Arch. Biochem. Biophys. **37**, 235 (1952)
492. M. Morrison, G.R. Schonbaum, Annu. Rev. Biochem. **45**, 861 (1976)
493. J.H. Dawson, Science **240**, 433 (1988)
494. D. Jol, H.B. Dunford, Eur. J. Biochem. **66**, 607 (1976)
495. D. Harris, G.H. Loew, J. Am. Chem. Soc. **118**, 10588 (1996)
496. J.R. Collins, P. Du, G.H. Loew, Biochemistry **31**, 11166 (1992)
497. W.D. Edwards, M.C. Zerner, Theor. Chim. Acta. **72**, 347 (1987)
498. B.B. Kim, V.V. Pisarev, A.M. Egorov, Anal. Biochem. **1**, 129 (1991)
499. J.W. Tams, K.G. Welinder, J. Biochem. **108**, 111 (1991)
500. O.V. Lebedeva, N.N. Ugarova, Izv. AN Rossii, Ser. Khim. [Proc. Acad. Sci. Rus. Chem. Ser.] **1**, 25 (1996) [Russian]
501. A.P. Savitskii, N.N. Ugarova, I.V. Berezin, Bioorg. khimya [Bioorganic Chem.] **5**, 1210 (1979) [Russian]
502. J.M. Pratt, in *Techniques and Topics in Bioinorganic Chemistry,* ed. by C.A. Mcauliffe ,Univ. of Manchester Inst. of Science and Technology, (Mir, Moscow, 1978), p. 133
503. J.M. Pratt, in *Techniques and Topics in Bioinorganic Chemistry*, ed. by C.A. Mcauliffe Univ. of Manchester Inst. of Science and Technology, (Macmillan Press Ltd., London, 1975), p. 342
504. N. Colclough, J.R. Lindsay-Smith, in *The Activation of Dioxygen and Homogeneous Catalytic Oxidation*, ed. by D.H.R. Barton, A.E. Martell, D.T. Sawyer (Plenum Press, New York, 1993) p. 171
505. H.B. Dunford, in *Peroxidases in Chemistry and Biology*, vol. 2, ed. by J. Everse, K.E. Everse, M.B. Grisham (CRS Press, Boca Raton FL, 1991) p. 1
506. P.R. Ortiz de Montellano, Annu. Rev. Pharmacol. Toxicol. **32**, 89 (1992)
507. P.R. Ortiz de Montellano, Acc. Chem. Res. 1987, **20**, 289
508. M.A. Ator, S.K. David, P.R. Ortiz de Montellano, J. Biol. Chem. **262**, 14954 (1987)
509. J.S. Wiseman, J.S. Nichols, M.X. Kolpak, J. Biol. Chem. **257**, 6328 (1982)
510. D.J.T. Porter, H.J. Bright, J. Biol. Chem. **258**, 9913 (1983)
511. J. Sakurada, S. Takahashi, T. Hosoya, J. Biol. Chem. **261**, 9657 (1986)

512. S.J. Bell, P.R. Cooke, P. Inchley, D.R. Leonard, J.R. Lindsay-Smith, A. Robbins, J. Chem. Soc. Perkin Trans. **2**, 549 (1991)
513. S.-M. Chen, Y.D. Su, A. Robbins, J. Chem. Soc., Chem. Commun. 491 (1990)
514. K.R. Rodgers, R.A. Reed, I.O. Su,T.G. Spiro, Inorg. Chem. **31**, 2688 (1992)
515. T. Groves, Y. Watanabe, J. Am. Chem. Soc. **110**, 8443 (1988)
516. Y. Watanabe, K. Yamaguchi, I. Morishima, K. Takehira, M. Shimizu, T. Hayakawa, H. Orita, Inorg. Chem. **30**, 2581 (1991)
517. T.G. Traylor, F. Xu, J. Am. Chem. Soc. **112**, 178 (1990)
518. W.A. Lee, L.-C. Yuan, T.C. Bruice, J. Am. Chem. Soc. **110**, 4277 (1988)
519. L. Marnett, J. Am. Chem. Soc. **111**, 6621 (1989)
520. T.G. Traylor, W.-P. Fann, D. Bandyopadhyay, J. Am. Chem. Soc. **111**, 8009 (1989)
521. B. Meunier, E. Guilmet, M.-E. de Carvalho, R. Poilblanc, J. Am. Chem. Soc. **106**, 6668 (1984)
522. T.Araiso, R. Rutter, L.P. Hager, H.B. Dunford, J. Biol. Chem. **259**, 233 (1981)
523. H. Tsurumaki, Y. Watanabe, I. Morishima, J. Am. Chem. Soc. **115**, 11784 (1993)
524. I. Morishima, S. Ogawa, Biochem. Biophys. Res. Commun. **83**, 946 (1978)
525. S. Modi, D.U. Behere, S. Mitra, Biochim. Biophys. Acta. **1204**, 14 (1994)
526. T.G. Traylor, W.A. Lee, D.V. Stynes, Tetrahedron **40**, 553 (1984)
527. T. Uno, A. Takeda, S. Shimabayashi, Inorg. Chem. **34**, 1599 (1985)
528. D.R. Morris, L.P. Hader, J. Biol. Chem. **241**, 1763 (1966)
529. P. Ortiz de Montellano, in *Cytochrome P-450. Structure. Mechanism and Biochemistry*, 2nd ed, ed. by P. Ortiz de Montellano (Plenum Press, New York, , 1995) p. 153
530. R.D.Libby, A.L. Shedd, A. Kathryn Phipps, T.M. Beachy, S.M. Gerstberger, J. Biol. Chem. **267**, 1769 (1992)
531. H.-A. Waganknecht, W.-D. Woggon, Angew. Chem. Int. Ed. Eng **36**, 390 (1997)
532. P.E. Strizak, K.B. Yatsimirskii, E.B. Ribak-Akimova, Ros. khim. zh. [Rus. Chem. J.] **39**, 89 (1995) [Russian]
533. G.A. Scheeline, D.L. Olson, E.P. Williksen, G.A. Horras, Chem. Rev. **97**, 739 (1997)
534. Hauser, M.J.B., L.F. Olsen, T.V. Bronnikova, W.M. Schaffer, J. Phys. Chem. B. **101**, 5075 (1997)
535. C. Bernofski, S.-Y. Wanda, J. Biol. Chem., **257**, 6809 (1982)
536. B. Halliwell, 1978, **140**, 81 (Planta)
537. T.V. Bronnikova, V.R. Fed'kina, W.M. Schaeffer, L.F. Olsen, J. Phys. Chem. **99**, 9309 (1995)
538. Olson, D.L., E.P. Williksen, A. Scheelline, J. Am. Chem. Soc. **117**, 2 (1995)
539. D.L. Olson, A. Scheelline, Anal. Chim. Acta. **283**, 703 (1993)
540. R.C. Jackson, J. Natl. Cancer Inst. **85**, 539 (1993)
541. P.E. Morrisson, R.L. Dedrick, J. Natl. Cancer Inst. **85**, 518 (1993)
542. A.J. Davison, A.J. Kettle, D.J. Futur, J. Biol. Chem. **261**, 1193 (1986)
543. B. Halliwell, J.M.C. Gutteridge, Free radicals, ageing and disease, in *Free Radicals in Biology and Medicine*, 2nd. ed. (Clarendon Press, Oxford, 1989)
544. B.Halliwell, J.M.C. Gutteridge, C.E. Cross, J. Lab. Clin. Med. **119**, 598 (1992)
545. R.Schreck, P. Richer, P.A. Baewerle, EMBO J. **10**, 2247 (1991)
546. J.V. Bannister, W.H. Bannister, G. Rotilio, Crit. Rev. Biochem. Mol. Biol. **22**, 111 (1987)
547. J.K. Hurst, W.C. BARRETE. Jr. Crit. Rev. Biochem. Mol. Biol. 1989, **24**, 271.
548. J.M. McCord, Free Radic. Biol. Med. **4**, 9 (1988)
549. A.Petkau, Cancer Treat Rev. **13**, 17 (1986)
550. D. Klug-Roth, I. Fridovich, J. Rabani, J. Am. Chem. Soc. **95**, 2786 (1973)
551. G.V. Buxton, in *Advances in Inorganic and Bioinorganic Mechanisms*, vol. 3, ed. by A.G. Sykes (Academic, London, 1984), , p. 131
552. I. Fridovich, Annu. Rev. Biochem. **64**, 97 (1995)
553. F. Polticelli, G. Bottaro, A. Battistoni, M.T. Carri, K. Djinovic-Carugo, M. Bolognesi, P. O'neill, G. Rotilio, A. Desideri, Biochemistry **34**, 6043 (1995)
554. E.D. Getzoff, D.E. Cabelli, C.L. Fisher, H.E. Parge, M.S. Viezzoli, L. Banci, R.A. Halliwell, Nature **358**, 347 (1992)
555. L.M.Ellerby, D.E. Cabelli, J.A. Graden, J.S. Valentine, J. Am. Chem. Soc. **118**, 6556 (1996)

556. M.A. McAdam, R.A. Fox, F. Lavelle, E.M. Fielden, Biochem. J. **165**, 71 (1977)
557. J. Weinstein, B.H.J. Bielski, J. Am. Chem. Soc. **102**, 4916 (1980)
558. R.F. Pasternack, W.R. Skowronek, J. Inorg. Biochem. **11**, 261 (1979)
559. Y. Ilan, J. Rabani, I. Fridovich, R.F. Pasternack, Inorg. Nucl. Chem. Lett. **17**, 93 (1981)
560. D. Solomon, P.P. Peretz, M. Faraggi, J. Phys. Chem. **86**, 1842 (1982)
561. P. Peretz, D. Solomon, D. Weinraub, M. Faraggi, Int. J. Radic. Biol. **42**, 449 (1982)
562. E. Kimura, T. Koike, Y. Shimizu, M. Kodama, Inorg. Chem. **25**, 2242 (1986)
563. J.L. Pierre, P. Chautemps, S.M. Refaif, C.G. Beguin, A. El-Mazzouki, G. Serratrice, E. Saint-Aman, P. Rey, J. Am. Chem. Soc. **117**, 1965 (1995)
564. L. Iuliano, J.Z. Pedersen, A. Ghisielli, D. Pratico, G. Rotilio, F. Violi, Arch. Biochem. Biophys. **293**, 153 (1992)
565. M. Faraggi, P. Peretz, D. Weinraub, Int. J. Radic. Biol. **49**, 951 (1986)
566. T. Nagano, T. Hirano, M. Hirobe, Free Radic. Res. Commun. **12**, 21 (1991)
567. N. Kitajima, M. Osawa, N. Tamura, Y. Morooka, T. Hirano, M. Hirobe, T. Nagano, Inorg. Chem. **32**, 1879 (1993)
568. K.M. Faulkner, R.D. Stevens, I. Fridovich, Arch. Biochem. Biophys. **310**, 341 (1994)
569. B.H.J. Bielski, D.E. Cabelli, Int. J. Radiat. Biol. **59**, 291 (1991)
570. R.H. Weiss, A.G. Flickinger, W.J. Rivers, M.M. Hardy, K.W. Aston, U.S. Ryaor, D.P. Riley, J. Biol. Chem. **31**, 23049 (1993)
571. T. Nagano, T. Hirano, M. Hirobe, J. Biol. Chem. **264**, 9243 (1989)
572. A.E.G. Cass, in *Metalloproteins*. Part 1, ed. by P. Harrison (Verlag Chemie,Weinheim, 1985)
573. D.P. Riley, R.H. Weiss, J. Am. Chem. Soc. **116**, 387 (1994)
574. G.S. Friedrichs, K.S. Kilgore, L. Chi, U.S. Ryan, B.R. Lucchesi, FASEB J. **7**, 2457 (1993)
575. D.P. Riley, S.L. Henke, P.J. Lennon, R.H. Weiss, W.C. Neumann, W.J. Rivers Jr, K.W. Aston, K.R. Sample, H. Rahman, L. Chaur-Sun, J.-J. Shein, D.H. Busch, W. Szulbinski, Inorg. Chem. **35**, 5213 (1996)
576. J.-L. Pierre, P. Chautemps, S. Refaif, C. Beguin, A. Elmazzouki, G. Serratrice, E. Saint-Aman, P. Rey, J. Am. Chem. Soc. **117**, 1965 (1995)
577. S.D. Varfolomeev, in *Chemical and Biological Kinetics,* ed. by N.M. Emanuel, I.V. Berezina, S.D. Varfolomeev (MGU, Moscow, 1983), p. 114. [Russian]
578. G.T. Babcock, M. Wikström, Nature **356**, 301 (1992)
579. C.R. Andrew, R. Fraczkiewicz, R.C. Czernuszewicz, P. Lappalainen, M. Saraste, J. Sanders-Loehz, J. Am. Chem. Soc. **118**, 10436 (1996)
580. S. Iwata, C. Ostermeier, B. Ludwig, H. Michel, Nature **376**, 660 (1995)
581. T. Tsukihara, H. Aoyama, E. Yamashita, T. Tomizaki, H. Yamaguchi, K. Shinzawa-Itoh, R. Nakashima, R. Yaono, S. Yoshikawa, Science **269**, 1069 (1995)
582. E.P. Day, J. Peterson, M.S. Sendova, J. Schoonover, G. Palmer, Biochemistry **32**, 7855 (1993)
583. A. Nanthakakumar, S. Fox, S.M. Nasir, N. Ravi, B.H. Huynh, R.D. Orosz, E.P. Day, K.S. Hagen, K.D. Karlin. in *The Activation of Dioxygen and Homogeneous Catalytic Oxidation*, ed. by D.H.R. Barton, A.E. Martell, D.T. Sawyer (Plenum Press, New York, , 1993), p. 381
584. M. Wirkström, Chem. Scr. **278**, 53 (1987)
585. M. Wirkström, FEBS Lett. **231**, 247 (1988)
586. M. Wirkström, Nature **338**, 776 (1989)
587. M. Wirkström, G.T. Babcock, Nature **348**, 16 (1990)
588. K.D. Karlin, A. Nanthakumar, S. Fox, N.N. Murthy, N. Ravi, B.M. Hugnh, R.D. Orosz, E.P. Day, J. Am. Chem. Soc. **116**, 4753 (1994)
589. C.A. Varotsis G.T. Babcock, J. Am. Chem. Soc. **117**, 11260 (1995)
590. S. Fox, A. Nanthakumar, M. Wirkström, K.D. Karlin, N.J. Blackburn, J. Am. Chem. Soc. **118**, 24 (1996)
591. R.S. Blackmore, C. Grunwood, Q.H. Gibson, J. Biol. Chem. **266**, 19245 (1991)
592. G.T. Babcock, J.M. Jean, L.N. Johnston, G. Palmer, W.H. Woodruff, J. Am. Chem. Soc. **106**, 8305 (1984)

593. C.A. Varotsis, Y. Zhang, E.H. Appelman, G.T. Babcock, Proc. Natl. Acad. Sci. USA **90**, 237 (1993)
594. C.A. Varotsis, G.T. Babcock, Biochemistry **29**, 7357 (1990)
595. S. Han, Y.-C. Ching, D.L. Roussean, Proc. Natl. Acad. Sci. USA **87**, 2491 (1990)
596. K.D. Karlin, S. Fox, A. Nanthakumar, N.N. Murthy, N. Wei, H.V. Obias, C.F. Martens, Pure Appl. Chem. **67**, 217 (1995)
597. A. Nanthakumar, S. Fox, N.N. Murthy, K.D. Karlin, N. Ravi, B.H. Hugnh, R.D. Orosz, E.P. Day, K.S. Hagen, N.J. Blackburn, J. Am. Chem. Soc. **115**, 8513 (1993)
598. A. Nanthakumar, S. Fox, K.D. Karlin, J. Chem. Soc. Chem. Commun 499 (1995)
599. J.P. Collman, P.C. Herrman, B. Boitrel, X. Zhang, T.A. Eberspacher, L. Fu, J. Am. Chem. Soc. **116**, 9783 (1994)
600. D.J. Kosman, J. Peisach, W.B. Mims, Biochemistry **19**, 1304 (1980)
601. C. Cleveland, R.E. Coffman, P. Coon, L. Davis, Biochemistry **14**, 1108 (1975)
602. G.A. Hamilton, in *Copper Proteins*, ed. by T.G. Spiro (Wiley-Interscience, New York, , 1981), p. 193
603. M.J. Ettinger, D.J. Kosman, in *Copper Proteins,* ed. by T.G. Spiro (Wiley-Interscience, New York, 1981) p. 219
604. G.A. Hamilton, P.K. Adolf, J. de Jersey, G.C. DuBois, D. Libby, J. Am. Chem. Soc. **100**, 1899 (1978)
605. J.T. Groves, P. Kuk-Tae, in W. Ando, Y. Morooka, eds. *The Role of Oxygen in Chemistry and Biochemistry*: Proc. of an Int. Symp. on Activation of Dioxygen and Homogeneous Catalytic Oxidations, Tsukuba, Japan, 12–16 July 1987. (Elsevier Science B.V., Amsterdam, 1988), vol. 33, p. 541

Chapter 4
Catalytic Processes and Human Diseases

4.1 Catalytic Free Radical Processes In Vivo and Toxicity of Oxygen Reduction Products

Biological oxidation is rather complex and, therefore, a very economical process with efficiency ≈50–90%. Organic compound oxidation with oxygen practically does not proceed spontaneously because of the great amount of energy needed for bond O–O breakage in dioxygen (∼490 kJ/mol) and for organic substrate activation (∼300–400 kJ/mol) [1], as well as the spin restriction (main condition of O_2-triplet $^3\sum_g$, and that of organic substances—singlet), which has been mentioned in Chap. 2.

One possible way for this energetically advantageous process to proceed, i.e., organic substrates oxidation with oxygen, is O_2 activation on account of its complex formation with metal ions or compounds (mainly with Fe, Cu, Mn, Co, Ni, Zn, Se, and some others). The result of such complex formation is electron density re-distribution of interacting components and oxy radicals ($^1O_2$, $^{\bullet}HO_2$, $^{\bullet}O_2^-$, $^{\bullet}OH$) and intermediate particles (H_2O_2, $L_nFe^{IV} = O^{2+}$, $L_nMn^{IV} = O^{2+}$) formation.

Superoxide-ion radical $^{\bullet}O_2^-$, not a strong oxidant, formed by the Haber–Weiss reaction, is able to generate the strongest oxidant—hydroxyl radical $^{\bullet}OH$. Such reactions also proceed with the organic hydroperoxides ROOH, which, with catalysis by transition metal ions, results in alkyl ($R^{\bullet}$)-, alcoxyl ($RO^{\bullet}$)-, and alkylperoxide ($ROO^{\bullet}$) radicals. Thus, in this way of O_2, H_2O_2, and ROOH activation, the most important is these processes undergoing catalysis with transition metals ions or complexes.

The future destiny of these active reactive catalysis products in physiological conditions depends both on rate constants of their formation (their concentration in vivo), mainly on elementary rate constants of their interaction with biological targets, and other factors (pH of medium, concentrations of interacting components, their ratio, etc.).

The redox potential at low pH of the $Fe^{3+}{}_{aq}/Fe^{2+}{}_{aq}$ couple is rather high ($E^{0'} =$ 0.77 V in relation to the normal hydrogen electrode), which causes Fe^{2+} ion stability

G. Duca, *Homogeneous Catalysis with Metal Complexes*, Springer Series in Chemical Physics 102, DOI 10.1007/978-3-642-24629-6_4,

to oxidation. At pH > 2 the hydrolysis of Fe^{3+} ion proceeds, and as a result $E^{0\prime}$ will be reduced at pH 7 up to the value 0.48 V. This can proceed in very diluted solutions only, and at about pH 3 the precipitate $Fe(III)(OH)_3$ appears [2].

Ligand introduction into the system $M^{z+}{}_{aq} + O_2$ can substantially change the value $E^{0\prime}$ either to lower (on σ bond M^{z+}–L formation) or higher value (on π-bond M^{z+}–L formation) from the value 0.77 V. This is connected with the change in metal ion hydrolytic stability, its electrochemical properties, change of metal coordination sphere lability, and activation of O_2 mechanisms—which has been discussed in detail earlier (see Chap. 2).

The mechanism of iron(II) interaction with dioxygen at low pH involves the formation of an intermediate compound:

$$Fe^{2+} + O_2 \rightarrow Fe^{2+}O_2.$$

Interacting with the second iron(II) ion, it will be turned into iron(III):

$$Fe^{2+}O_2 + Fe^{2+} + 2H^+ \rightarrow 2Fe^{3+} + H_2O_2.$$

Oxidation of iron(II) with dioxygen is strongly dependent on pH value. Thus, at pH 7.03 the half-life of this process reaches 2,700 s, and at pH 7.45—about ten times higher [3]. To maintain the medium homogenity at such pH values, one has to introduce ligands into the system, which change half-life significantly ($\tau_{1/2}$). For instance, if EDTA is the ligand, $\tau_{1/2}$ reaches about 10 s at neutral pH. In these conditions intermediate oxygenated complex $EDTAFe^{II}{\cdot}O_2$ [4] will be formed, which then turns into hydrogen peroxide and an appropriate iron(III) complex. Reverse reaction of iron(III) complex reduction with superoxide proceeds with rate constant $1.9 \times 10^6\,M^{-1}\,s^{-1}$ [5]. In its turn, superoxide can also form the adducts with iron(II) complexes [6].

Dioxygen reduction with copper(I) complexes proceeds more rapidly than in the case of iron(II) complexes. For example, rate constants of this process are equal to 5×10^4 and $4 \times 10^4\,M^{-1}\,s^{-1}$ for complexes $Cu^I(phen)_2$ [7] and Cu^I $(histidine)_2$ [8], respectively. Copper(II) catalyzes superoxide dismutation with a rate close to the diffusion one ($8 \times 10^9\,M^{-1}\,s^{-1}$) [9].

Endothermic type of the reaction of O_2 interaction with organic compounds RH in vivo

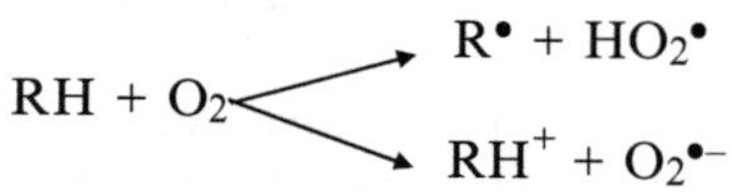

with $O_2^{\bullet-}$ ($HO_2^{\bullet}$) formation will strongly complicate this process, and it proceeds rather slowly. However, if the organic substance is a strong enough reducer, such as flavin (FH_2) in biological systems, perhydroxyl radical $HO_2^{\bullet}$ can be formed in a non-catalytic way [10].

$HO_2^\bullet$ radicals are formed mainly in vivo in a catalytic way with participation of various enzymes, such as cytochrome P-450 [11], xanthine–xanthinoxidase system, NADPH-oxidase, etc. [10, 12, 13]. Another way of free radicals generation is by using the Haber–Weiss catalytic reaction:

$$Fe^{3+} + O_2^{\bullet -} \rightarrow Fe^{2+} + O_2$$

$$Fe^{2+} + H_2O_2 \rightarrow Fe^{3+} + OH^- + OH^\bullet$$

$$O_2^{\bullet -} + H_2O_2 \overset{Fe(III)}{\rightarrow} O_2OH^- + OH^\bullet$$

Such a reaction of $OH^\bullet$ generation, having iron(III) and suitable ligand in vivo, can result in intermediate active particle formation of ferryl complex type:

$$LFe^{3+} + OH^\bullet \rightarrow (LFe^{IV} = O)^{2+} + OH^-.$$

Such types of active particles are formed in many enzymes functioning, such as peroxidases, cytochrome P-450, etc.

Free radicals, formed one way or another, will be involved in the usual biological processes in vivo. Thus, they take part in anti-microbial activity of phagocyte cells [14]. Dioxygen will be used by neutrophils and macrophages due to the influence of NADPH-oxidase complex on it, linked with the plasma membrane. The electrons will be released, consumed for dioxygen oxidation to superoxide radical. However, thermodynamic calculations have shown that superoxide $O_2^{\bullet -}$ cannot detach the hydrogen atom from R–H bond if this bond energy is higher than 276 kJ/mol [15]. Therefore, because of its rather low reactivity, $O_2^{\bullet -}$ radical is hardly responsible for the direct ruin of pathogenic microorganisms. More often, it will only start some reactions leading to such a result. For example, some kinds of bacteria will be ruined under the influence of hydrogen peroxide, formed by superoxides dismutation. It is also possible here that H_2O_2 itself is not acting as the killer, but the product of its reduction in the presence of iron(II) compound, namely, hydroxyl radical $OH^\bullet$. Another way of microorganism destruction involves the whole range of reactions yielding hypochlorous acid, which is able to oxidize many biological molecules, especially the reduced thiol group and the remains of bacteria methionine [16]:

$$O_2 + NADPH + H^+ \xrightarrow{\text{NADPH-oxidase}} O_2^{\bullet -} + NADP^+,$$

$$2O_2^{\bullet -} + 2H^+ \xrightarrow{\text{SOD}} H_2O_2 + O_2,$$

$$H_2O_2 + Cl^- \xrightarrow{\text{myeloperoxidase}} HOCl + OH^-,$$

$$OCl^- + RNH_2 \xrightarrow{\text{respiratory explosion}} RNHCl + OH^-.$$

This process, carried out by phagocytic cells, is called "respiratory explosion," resulting in hypochlorous acid discharge, oxidizing thiol groups of microorganisms.

Another example of such superoxide indirect influence is its rapid interaction with nitrogen oxide NO (formed by phagocytes and endothelium cells), resulting in peroxynitrite anion formation:

$$O_2^{\bullet-} + NO \rightarrow ONOO^-,$$

which, being protonated ($ONOO^- + H^+ \rightarrow ONOOH$), will be turned into the strong oxidant, able to destroy tumor cells and bacteria [17] when its decomposition is accompanied with generation of even stronger oxidant—$OH^\bullet$ radical [18, 19]:

$$ONOOH \rightarrow NO_2 + OH.$$

In this way, $O_2^{\bullet-}$, being a kind of physiological antagonist to $NO^\bullet$, can increase its cellular toxicity.

In acid localizations of vacuoles' phagocytes type, superoxide can be protonated, transforming into perhydroxyl radical $HO_2^\bullet$, which is more reactive than superoxide, thus having a more toxic influence on cells.

However, in some cases, $O_2^{\bullet-}$ can directly attack the enzymes of some bacteria [20–22].

Excessive outer-cellular formation of superoxide in vivo can result in a peculiar impact on mammals' in the form of an inflammatory process, thus initiating the process of neutrophils' activation [23].

Processes resulting in toxicity from dioxygen reduced forms ($O_2^{\bullet-}$, H_2O_2) are still subjects of debate. Superoxide radicals interaction with biological targets (Biol) has been represented by crypto-particles formation [24]:

$$\begin{aligned} &\text{Biol} - M^{n+} + O_2^{\bullet-} \rightarrow \text{Biol} - M^{(n-1)} + O_2, \\ &\text{Biol} - M^{(n-1)+} + H_2O_2 \rightarrow \underset{(\text{crypto-OH}^\bullet)}{\text{Biol} - M^{n+} \ldots OH^\bullet} + OH^-. \end{aligned}$$

Only this fact made it possible to explain the limited number of reactions with biological molecules (targets) in the presence of hydrogen peroxide [15].

Another hypothesis supposes the formation of intermediate iron(II) peroxocomplex $Fe(OOH)^+$, which will decompose into two reactive particles, namely FeO^+ and $OH^\bullet$, which therefore has to be a strong radical initiator [25].

However, there are still no experimental data confirming this hypothesis.

It should be considered that $O_2^{\bullet-}$ is a one-electron oxidant and that iron complexes can change their oxidation degree, forming the couples LFe(III)/LFe(II) or LFe(IV)/LFe(III), and likewise in the case of active centers of peroxidase. Provided the ligands L are good σ-donors, a redox-potential of LFe(IV)/LFe(III) is to be reduced, which will make it easier to stabilize in a more oxidized state, i.e., Fe(IV). Therefore, it was supposed [24, 26] that the unique superoxide's toxicity is caused by its capacity to induce local-specific and self-destroying generation of superoxidized metal-particles.

Superoxidized particles Fe(IV) are somewhat less active than $OH^{\bullet}$ radicals, so their half-life and oxidative specificity is higher than that of $OH^{\bullet}$ radicals. The main difference between these two particles consists in the fact that $OH^{\bullet}$ radicals are electrophilic, while ferryl particles are nucleophilic agents [15].

Another function of free radicals is their ability to regulate biochemical processes. For instance, lymphocytes (non-grain leucocytes) and fibroblasts (fiber-type cells layer) constantly generate small amount of superoxide radicals regulating the growth [27, 28]. Other non-phagocytic cells of endothelium type (layer of plain-shaped cells, going beyond the outside cells of blood and lymphatic vessels) and arterial smooth muscular cells can be stimulated, resulting in superoxide generation [16].

Cellular toxicity of organic peroxides ROOH is caused by their capacity to be decomposed into free radicals, stimulating the oxidation of organic substrates, and taking away hydrogen atoms from them [15]:

i. $RH + OH^{\bullet} \rightarrow R^{\bullet} + H_2O \qquad (\kappa_{RH+\acute{O}H} \approx 10^8\text{–}10^9 M^{-1} s^{-1}),$

1. $R^{\bullet} + O_2 \rightarrow ROO^{\bullet} \qquad (\kappa_{R'+O2} \approx 10^8\text{–}10^9 M^{-1} s^{-1}),$

2. $ROO^{\bullet} + R'H \rightarrow ROOH + R^{\bullet\prime} \qquad (\kappa_{RO\acute{O}+R'H} \approx 10^1\text{–}5 \times 10^2 M^{-1} s^{-1}),$

0. $ROO^{\bullet} + R'OO^{\bullet} \rightarrow ROOR' + O_2 \qquad (\kappa_{RO\acute{O}+R'O\acute{O}} \approx 10^6\text{–}10^8 M^{-1} s^{-1}).$

Within the concentrations range $\approx 10^{-9}$ to 10^{-7} M some lipid hydroperoxides LipOOH can represent the important physiological regulators [29]. On LipOOH concentration increase above this range ($\geq 10^{-6}$ M), irreversible structural changes will appear both in membranes, and the number of enzymes. LipOOH can also induce the breakage of DNA thread [30].

Alcoxy radical $RO^{\bullet}$, appearing in the presence of iron(II), will often be turned into cell-toxic aldehydes and epoxides [31], which can inactivate many enzymes. These are unsaturated alkenals, namely, 4-hydroxynonenal and malonic aldehyde [32].

Products of dioxygen reduction ($O_2^{\bullet -}$, H_2O_2, $OH^{\bullet}$, ROOH, $ROO^{\bullet}$, $RO^{\bullet}$) are within the cells and inter-cellular liquids in strictly determined, limited concentrations, the level of which is controlled by strong enzymatic and other regulatory systems. Various influences of the outer and inner media on living organisms will deteriorate such control, which can induce oxidation stress resulting in cell ruin.

Hydrogen peroxide forming on oxidation stress can provoke the activation of human immunodeficiency virus (HIV) within the infected cells, it seems, by oxidation of nuclear regulating protein NF_kB [33–35]. Easily penetrating through the cellular membranes, hydrogen peroxide can directly attack the number of cell targets (namely, number of enzymes) [36], though, because of limited reactivity, both superoxide radical and hydrogen peroxide will affect the cell targets not directly, but indirectly by consequent formation of more reactive radicals ($OH^{\bullet}$, $RO^{\bullet}$) and active intermediate particles ($LFe^{IV} = O^{2+}$), interacting with transition metal compounds.

Iron(III) compounds in vivo, not connected with protein, having interacted with $O_2^{\bullet -}$ and H_2O_2, can realize a Haber–Weiss reaction, resulting in $OH^{\bullet}$ radical formation. In this reaction various ligands can influence $O_2^{\bullet -}$ capacity differently to reduce Fe(III). For example, specific ligands deferoxamine and bathophenanthroline will inhibit this reaction (Fe(III) will not be reduced to Fe^{II}), and ethylenediaminetetraacetic acid, adenosine-triphosphate, and picolinic acid will promote it [37].

Specific ligands using, binding iron(III) and preventing its reduction, can readily suppress oxidation deteriorations on therapeutic treatment of heart, kidney, skin, and other tissue damage, as well as decrease post-traumatic tissue degenerations in cerebrum and spinal cord [38].

Hydroxyl radical $OH^{\bullet}$, generated by Haber–Weiss reaction, will act in a more destructive way and more specifically (directly) than the same radical formed by radiolysis. This is due to the fact that in the first case $OH^{\bullet}$ radical will interact with the nearest biological target at a distance no greater than 5–10 Å, and in the second case the interaction proceeds in the large bulk of solution [39].

There exist various ways of biological targets protection of deterioration, for instance by introducing into the system such substances as thiourea or mannitol [40,41]. The whole system of enzymes (superoxide dismutase, catalase, glutathione peroxidase, cytochrome-C-peroxidase) play the same role of organism protection [42–44]. Thermodynamic conditions of Haber–Weiss reaction occurrence can also be used with the same purpose. For this reaction to run, a low enough reduction potential of iron complex is needed ($\varphi = 0.32$ V under standard conditions).

Iron complexes, characterized by higher potential ($\varphi = 0.32 - 94$ V) will catalyze the dismutation, but not the Haber–Weiss reaction. Therefore, using various ligands, one can change the superoxide dependence. Superoxide toxicity suppressing or enhancing mainly depends on the change by the ligand of iron complex redox potential. Depending on this, either dismutation ($O_2^{\bullet -}$ toxicity is suppressed) or Haber–Weiss reaction ($O_2^{\bullet -}$ toxicity is increased) will be observed.

In spite of reactions generalized by the scheme of oxygen toxicity (see Fig. 2.3), there exists the large class of reactions (often called the reactions of metabolic activation), in which free radicals cause cell damage or deterioration (such as liver toxic damage). Toxic substances undergoing such metabolic activation by their catalysis with enzymatic systems are CCl_4, aromatic nitrocompounds, aromatic amines, nitrosoamines, hydrazines, quinones, etc. Due to the metabolic activation, these substances obtain electrons in vivo from flavoproteins or cytochrome P-450, being reduced to free radicals [45]:

$$CCl_4 + e^- \rightarrow [CCl_4]^{\bullet -},$$

$$RNO_2(\text{namely, nitroimadazoles}) + e^- \rightarrow [RNO_2]^{\bullet -}.$$

Then, in secondary reactions, the following products will be formed:

$$[CCl_4]^{\bullet -} \rightarrow CCl_3^{\bullet} + Cl^-,$$

$$CCl_3^{\bullet} + RH \rightarrow CHCl_3 + R^{\bullet}(\text{reaction of substrate RH oxidation}),$$

$$[RNO_2]^{\bullet -} + O_2 \rightarrow RNO_2 + O_2^{\bullet -}.$$

It is known that $O_2^{\bullet -}$ dismutation can proceed with H_2O_2 formation, which, in the presence of Fe^{2+} ions, will yield ȮH radical formation and nearby biological targets damage (oxidation).

Free radical formation in cells and tissues can also be provoked with radiation (ionizing, ultraviolet, visible, thermic) or transition metal or enzymes catalysts.

Free radicals formed one way or another can be extremely reactive (hydroxide radical $OH^{\bullet}$), of mediam reactivity (trichloromethyl peroxyl radical $CCl_3OO^{\bullet}$), weakly reactive (superoxide-, $O_2^{\bullet -}$) or rather stable (diphenylpicrylhydrazyl, DPPH [45]). Hydroxyl radical will interact with surrounding substances usually with high rate constants, within the limits 10^9–10^{11} M^{-1} s^{-1}, which diminishes its diffusion radius up to the value of about 10 Å. Other radicals of mean and weak reactivity are characterized by essentially lower rate constants of interaction with substrates-targets. Thus, their half-transformation time in case of interaction with unsaturated aliphatic acids reaches approximately 0.7×10^{-9} with ($OH^{\bullet}$), 0.7×10^{-6} with ($CCl_3OO^{\bullet}$) and 0.07 with ($O_2^{\bullet -}$) [45], and diffusion radius will accordingly be increased in this series.

Therefore, reactive free radicals can damage DNA in vivo, destroy nucleotide co-enzymes, damage SH-dependable enzymes and membranes, and carry out lipids peroxidation (LPO) and other catalytic and non-catalytic reactions, accompanied by cells damage, resulting in many diseases occurring (such as stress-related diseases, inflammatory diseases, atherosclerosis, miocardium sciatica, and age-related diseases).

4.2 Stress and Catalytic Processes

Briefly mentioned earlier, "stress" (oxidation stress), first introduced by Selye [46], can be translated as "tension." The influence of any environmental factor on an organism provokes the tension of regulatory systems, mobilizing the internal reserves and adapting the organism to the new living conditions (general adaptation syndrome). Therefore, stress is the activation of non-specific response reactions of organism to the outside (thermic shock, light influence, poisoning, ionizing irradiation, etc.) or internal (psychoemotional condition, etc.) influences, aimed at preserving and maintaining homeostasis. The stress value caused by the strength of exogenous or endogenous factors' influence on an organism is characterized by the degree of such activation.

Stress theory, its physiological and other manifestations are described in detail in numerous papers [47–52] and will not be considered here. The main attention will be paid to catalytic processes taking place during the stress, connected with free radical reactions of oxygen reduction products and activation with them, and with transition metal compounds and lipids peroxidation (LPO) (see Sect. 3.1).

LPO activation is a comprehensive process, proceeding in the organism both in normal and pathologic conditions. In the normal condition this process will be sustained at a low stationary level due to the presence in the organism of a variety of protective antioxidant (AO) enzymes, inactivating LPO primary products: dismutase—$O_2^{\bullet -}$, catalase—H_2O_2, glutathione peroxidase—ROOH [53]. The only $OH^\bullet$ radical will not be inactivated with the enzymes because of its high reactivity. The change of isobar potential of its interaction with hydrogen atom (ΔG) with water formation ($H + OH^\bullet \rightarrow H_2O$) reaches 263 kJ/mol. Such a high negative value of ΔG makes the reaction of H detachment from any organic compounds extremely advantageous from the thermodynamic point of view. Besides, the strength of the O–H bond is high in the reaction product—water (499 kJ/mol). The lifetime of the $OH^\bullet$ radical is 7×10^{-10} s only, while that of linolic acid alcoxyradical $RO^\bullet$ reaaches 7×10^{-7} s, and its peroxyradical $ROO^\bullet$—7×10^{-3} s [54]. Therefore, $OH^\bullet$ radical will oxidize practically any organic substance (it is not specific as an oxidant), and $OH^\bullet$ radical inactivation in the cells is produced by α-tocopherol and other antioxidants [55,56].

However, the ease with which LPO proceeds in the organism, and autocatalytic accumulation of oxidation products, due to the presence of strictly definite amount of initiators (namely, $O_2^{\bullet -}$, $R^\bullet$, $RO^\bullet$, $ROO^\bullet$) and transition metal compounds (especially those of iron(II) and iron(III) as catalysts, make this process universal [57]. Only concentration limits of LPO products generation and their variety will be changed depending on the organism living conditions, in particular stress factors.

Under normal conditions of organism functioning, pro- and antioxidant equilibrium will be realized, maintaining the homeostasis. Lipid hydroperoxide concentration under these conditions will reach about 10^{-9} to 10^{-7} M, and within this concentration range they are important physiological regulators both outside, and inside the cell [58]. Any damage of organism and strains of its regulatory systems, provoked with the increasing of physiological activity (getting the food, struggle with enemies, escape of danger, etc.), is accompanied by LPO (with the occurrence of the variety of active oxygen forms) and shift to the left of pro- and antioxidant equilibrium. Thus, LPO activation is a universal phenomenon accompanying both normal and pathologic conditions of the organism. Because of its universal character, LPO correlates with the stress as its component and an extremely important part. LPO activation is the factor causing direct damage of cellular membrane structures, enhancing membranes' lysosomes permeability and various enzymes coming out of them, in particular enzymes of lipids oxidation. This process, with enough intensity, will provoke heart, vessels, gastrointestinal tract mucous membrane, cerebrum, and retina reaction and other pathologies.

The key role of LPO in pathogenesis lies in the fact that the products of its free radical oxidation ($O_2^{\bullet -}$, H_2O_2, $OH^\bullet$, $R^\bullet$, $RO^\bullet$, $ROO^\bullet$, ROOH, Lip, $LipO^\bullet$, $LipOO^\bullet$, LipOOH, dienic conjugates, etc.) plays the part of primary mediator between the influence of irritating stress factor and reticular formation centers, hypothalamus [49]. The existence of some primary mediator has been supposed by Selye, though it was unknown then in what way it represents itself.

One of the recent hypothetical schemes of stress-reaction proceeding was proposed by [49] (Fig. 4.1), considering an earlier scheme by [50]. As can be

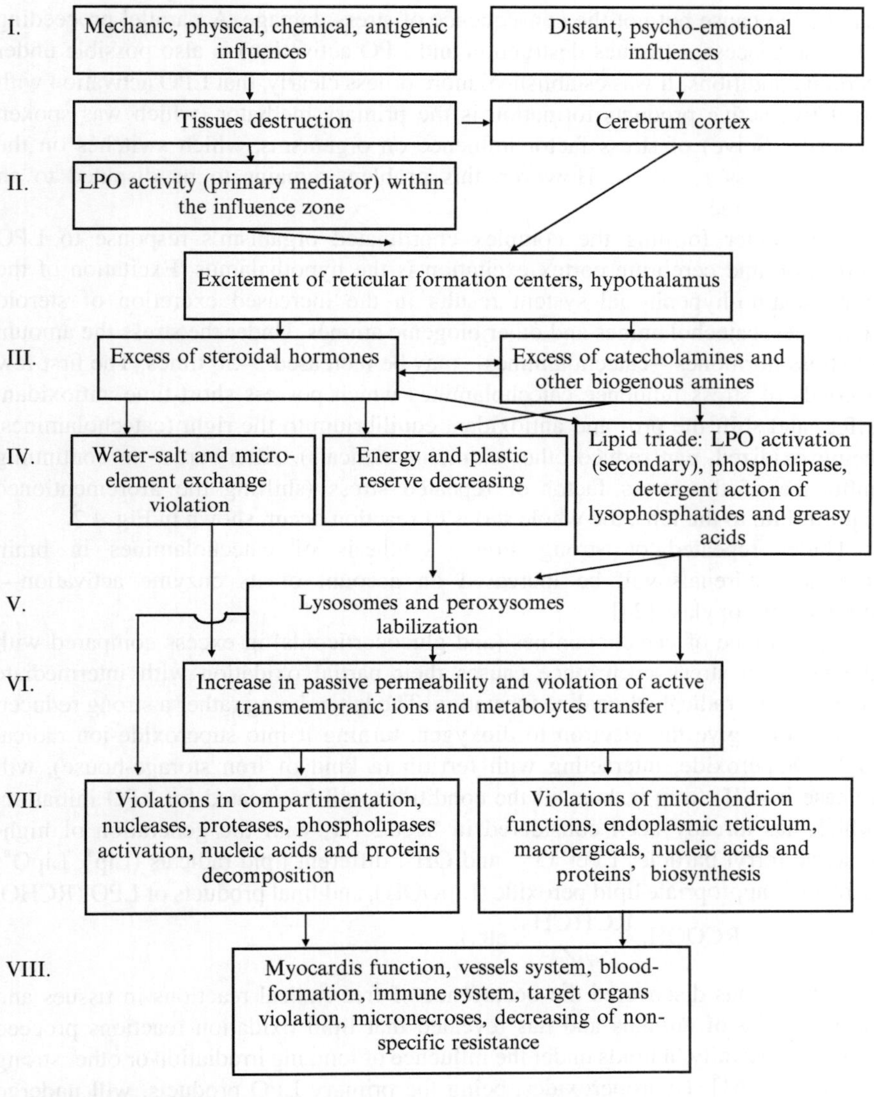

Fig. 4.1 Hypothetic scheme of stress-reaction proceeding (taken from [49])

seen from Fig. 4.1, exogenous and endogenous stress factors (such as thermic burn) provoke or do not provoke tissue destruction (psycho-emotional influence). Tissues destruction, on the one hand, causes the excitation of cerebrum cortex, and on the other, activates the LPO. It seems that LPO activation can occur as the result of tissues deterioration (Fig. 4.1) [36], and also without such primary deterioration (for example, on irradiation: high-reactionary oxygen radicals $OH^{\bullet}$ formation, etc., subsequent violation of organelles' functions and tissues deterioration) [49], i.e., it can be the cause but not the consequence of stress damage. A parallel proceeding of these processes (tissues destruction and LPO activation) is also possible under certain conditions. It was established, more or less clearly, that LPO activation with oxidative active products formation is the primary mediator (which was spoken about by Selye) of stress factor influence on organisms, which switches on the sequent stress reactions. However, this problem remains to be disputed to an essential degree.

The center forming the complex coordinated organism's response to LPO activation and cerebrum cortex excitation is the hypothalamus. Excitation of the hypothalamo-hypophysial system results in the increased excretion of steroid hormones, catecholamines and other biogenic amines. Under the stress the amount of stress hormones—catecholamines—may be increased 5–50 times. The first few seconds of stress influence catecholamines which possess short-time antioxidant effect and shift the pro- and antioxidant equilibrium to the right (catecholamines, being oxidized, can reduce other oxidizing radicals). Afterwards, on continuing influence of the stress factor of repeated stress (shifting the aforementioned equilibrium to the left), the whole series of reaction occur, shown in Fig. 4.2.

Under repeated or strong stress, synthesis of catecholamines in brain substance–adrenals–will be increased on account of its enzyme activation—tyrosinehydroxylase [24].

Appearance of catecholamines (and glucocorticoids) in excess compared with the norm in stress occurrence causes their partial oxidation with intermediate semiquinone radical adrenaline formation. The latter, being rather a strong reducer, will readily give the electron to dioxygen, turning it into superoxide-ion radical $O_2^{\bullet-}$. Superoxide, interacting with ferritin (a kind of iron storage-house), will release iron(II) from it. hus, all the conditions will be created for LPO initiation, which has already been considered in Sect. 3.1.1, with the generation of high-reactive ferryl particles $LFe^{IV}O^{2+}$ and $OH^{\bullet}$, different lipid radicals ($Lip^{\bullet}$, $LipO^{\bullet}$, $LipOO^{\bullet}$), appropriate lipid peroxide (LipOOH), and final products of LPO (RCHO, RCO, ROH, RCOOH, $\underset{\diagdown O \diagup}{RCHCH_2}$, etc.).

Tarusov has discovered the occurrence of free radical reactions in tissues and liquid media of animals and has revealed that lipid oxidation reactions proceed most successfully in lipids under the influence of ionizing irradiation or other strong factors [59–61]. Hydroperoxides, being the primary LPO products, will undergo further catalytic transformations (carried out with iron compounds via the formation of alcoxyradical $LipO^{\bullet}$) yielding the final products. As with the aforementioned products, the final product is malonic dialdehyde

$$(\underset{\underset{O}{||}}{CH}-CH_2-\underset{\underset{O}{||}}{CH}),$$

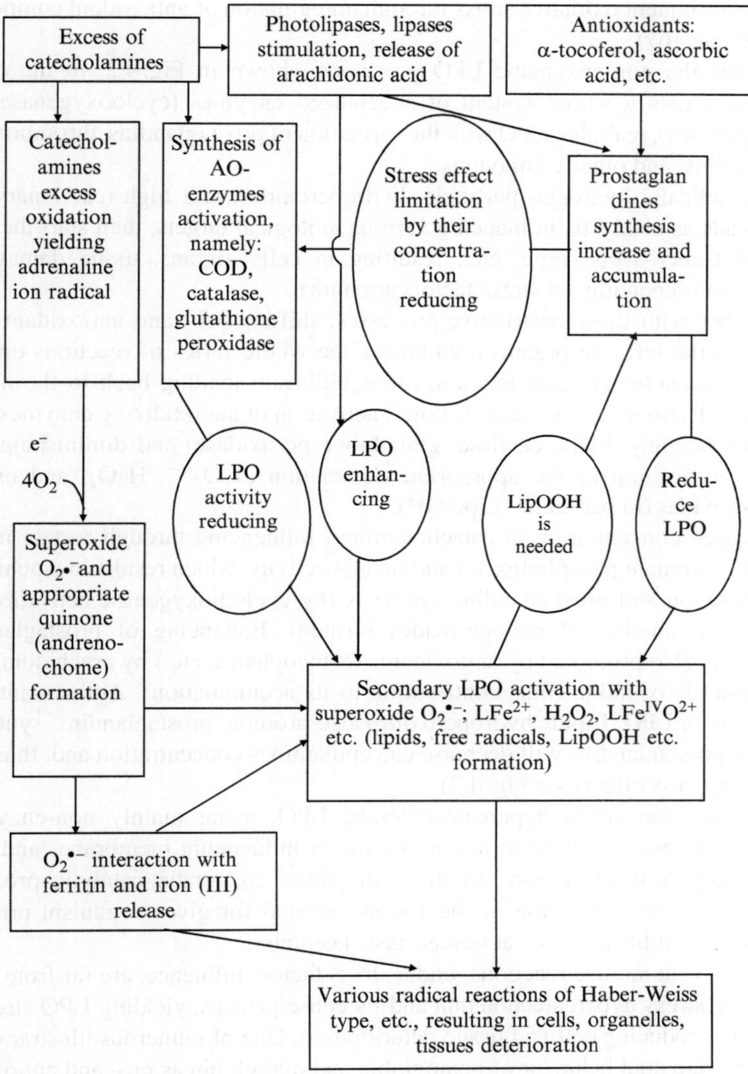

Fig. 4.2 Simplified scheme of reactions resulting in LPO secondary activation and catecholamines stress effect limitation

and products of lipid polycondensation–polymerization (resulting in fluorescent Schiff bases) and pigments (lipofuscin), appearing in time with the increasing amount in fat tissue [62–64]. Such LPO products as alkanals (malonic dialdehyde) [65], alkenals, and hydroxyalkenals (4-hydroxynonenal) [66] possess cytotoxic properties. Although many LPO products exert cytotoxicity, sublethal concentrations of LPO products induce cellular adaptive responses and enhance tolerance against subsequent oxidative stress through upregulation of antioxidant compounds and enzymes [67].

Besides the non-enzymatic LPO processes, shown in Fig. 4.2, in the organism there exists a whole system of specialized enzymes (cyclooxygenases and lypoxygenases), realizing LPO with the formation of prostaglandins, thromboxanes, prostacyclins, and other compounds.

Free radicals, hydrogen peroxide, hydroperoxides, and high-reactionary iron compounds as catalysts, influencing various biological targets, then start the reactions of Haber–Weiss type, etc., resulting in cells, organs, tissue damage, or destruction (depending on stress factors strength).

Together with these destructive processes, shifting pro- and antioxidant equilibrium to the left, the organism embraces the whole series of reactions directed at hypercatecholemia neutralization, the equilibrium tending back to the normal condition. These reactions are as follows: activation of antioxidizing enzymes (AO) synthesis (namely, SOD, catalase, glutathione peroxidase) and diminishing LPO activity on account of the appropriate destruction of $O_2^{\bullet-}$, H_2O_2, and organic hydroperoxides (in particular, LipOOH).

Increased concentration of catecholamines, influencing through certain mediators, will stimulate phospholipases and lipases activity, which results in arachidonic acid liberation and prostaglandins synthesis (by cyclodioxygenase activation and through the number of endoperoxides formed). Enhancing of prostaglandins' synthesis is also promoted by antioxidants (α-tocopherol, etc.) by arachidonic acid protection of oxidation (thus contributing to its accumulation). Antioxidants will also decrease LPO. Lipids hydroperoxides also promote prostaglandins' synthesis. Forming prostaglandins will decrease catecholamines concentration and, thus, will limit their stress effect (see Fig. 4.2).

Therefore, on stress hypercatecholemia, LPO, using mainly non-enzymatic catalytic processes, will be turned into a factor influencing membranes and other organic targets deterioration. At the same time, enzymatic catalytic processes are aimed at the restoration of the former, normal for given organism pro- and antioxidant equilibrium, i.e., at homeostasis keeping.

Naturally, numerous reactions, under stress factors influence, are far from being limited by stress hypercatecholemia and its consequences, yielding LPO strengthening and producing cell and tissue deterioration. One of numerous illustrations of this fact is the dual behavior of hemoglobin and myoglobin as pro- and antioxidant depending on the prevailing conditions [68].

As it was already pointed out, cell and tissue damage proceeds under stress situations. In these processes, superoxide ion-radical $O_2^{\bullet-}$ and hydrogen peroxide can play the role of mediator, the influence of which will be increased in the presence, for example, of hematic proteins, among them hemoglobin, myoglobin, cytochrome, or trace amounts of transition metals, mainly iron and copper compounds:

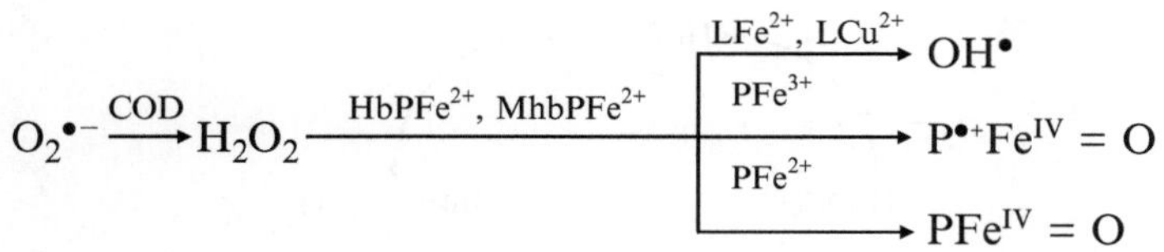

where $HbPFe^{2+}$ is deoxyhemoglobin (hemoglobin), $MHbPFe^{2+}$ is deoxymyoglobin (myoglobin), PFe^{2+} and PFe^{3+} are iron(II) and (III) porphyrine complexes, free of protein in pathologic situations, and $P^{+}Fe^{IV}O$ and $PFe^{IV} = O$ are ferry ion-radical and oxo-iron(IV) complex.

Under certain stress pathologic conditions and on damage, heme cells (hemoglobin, myoglobin) can also be degraded with the formation of iron porphyrin complexes (turning into high-reactive ferryl particles under the influence of hydrogen peroxide), or iron compounds with other (non-porphyrin) ligands (through which interaction with H_2O_2 can lead to $OH^{\bullet}$ radicals generation). The latter case is in accordance with the capacity of $O_2{}^{\bullet -}$ radicals, accumulated in the damaged sites (for instance, in case of rheumatic inflammation of joints), to extract Fe^{3+} ions from ferritin, which in the presence of H_2O_2 will provide the possibility of a Haber–Weiss catalytic reaction proceeding with $OH^{\bullet}$ radicals formation. Thus, iron release out of iron-containing proteins under many pathologic conditions (micro-hemorrages in eye, inflammatory processes, atherosclerotic damage, and many others) [16, 69] will turn it into the catalyst, generating the aforesaid high-reactionary particles and radicals affecting the cells and tissues. This takes place non-specifically ($OH^{\bullet}$ radicals action) or more specifically (action of $P^{+}Fe^{IV} = O$ and $PFe^{IV} = O$). Metal–proteins, namely hemoglobin and myoglobin, play the roles of pro-oxidants in these stress situations [16] as soon as ferrylheme protein radicals forming from myoglobin and hemoglobin degraded by stress can react with the membranes [70–72] and lipoproteins [73–75]. It is known that low density lipoproteins (LDL) within the artery walls take part in adipose layer formation and atherosclerotic follicles development fom it [76].

All kinds of cells in which LDL oxidation takes place (monocytes, macrophages, plain muscle cells, endothelium cells, etc.) will generate $O_2{}^{\bullet -}$ radicals [16], but their reactivity is not enough for LDL oxidation to proceed. Provided the destroyed erythrocytes are introduced here (hemoglobin, myoglobin, iron compounds free of globin PFe^{2+}, PFe^{3+}), LDL oxidation will be induced [74] due to $O_2{}^{\bullet -}$ radicals, ferryl ion-radicals, and oxo-iron(IV) complex formation. Therefore, hemoglobin and myoglobin will play the roles of LDL promotors.

Under normal physiological conditions of the organism, iron is mainly bound in the protein state, for example, in hemoglobin and methemoglobin of erythrocytes. A certain equilibrium is observed in erythrocytes between the contents of deoxyhemoglobin (hemoglobins-$HbPFe^{2+}$) and methemoglobin ($HbPFe^{3+}$) due to the redox cycle (modified Fig. 4.3 taken from [16]), provoked by the latter reduced with NADH-dependent methemoglobin-reductase (NADH-$HbPFe^{3+}$Red).

Oxidative particles $O_2{}^{\bullet -}$ forming, and others, seem to be passed toward hemoglobin, like through a wastewater pipe, and will deteriorate on it, getting

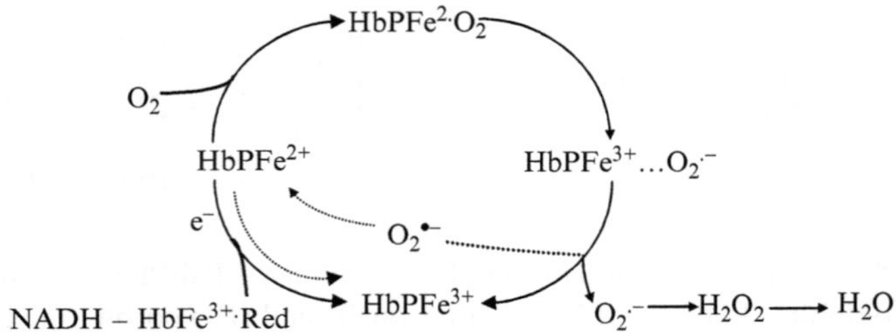

Fig. 4.3 Equilibrium hemoglobin ↔ methemoglobin

electrons from it (thus, hemoglobin will partially be turned into methemoglobin). These processes are shown in Fig. 4.3 with dotted arrows. Thus, while erythrocytes are undamaged, the hemoglobin in them, acting as antioxidant, will contribute to oxidants present in the normal quantity (oxygen concentration within the cell reaches $\approx 10^{-6}$ M, and that of active oxidation oxygen forms—10^{-13} M [49]). The capacity of hemoglobin to decompose catalytically hydroperoxylipids (LipOOH), like lipoxygenase [77], should also be mentioned.

As soon as it comes under the influence of stress, the well-regulated equilibrium system within erythrocytes will be broken, making use of hemoglobin capacity to become pro-oxidant and generating the initiating particle $O_2^{\bullet-}$ of lipids peroxidation (sheer arrow for $O_2^{\bullet-}$ generation in Fig. 4.3). Therefore, hemoglobin, depending on the condition of the organism (in particular erythrocytes), can act either as antioxidant, or pro-oxidant [16, 78].

In general, it becomes clear from the aforesaid that oxidation stress can enhance the contents of "active" iron as free radical reactions catalyst in accordance with Haber–Weiss reaction:

$$O_2^{\bullet-} + H_2O_2 \xrightarrow{LFe^{2+}} OH^{\bullet} + OH^{-} + O_2^{\bullet}.$$

This reaction can be inhibited by introducing the appropriate chelate-forming ligands, namely, desferoxamine, and thus provide the protection of oxidative deterioration. With the same purpose, various $OH^{\bullet}$-traps can be used (mannitol, salicylate, thiourea, deoxyribose), which do not interact as much with $OH^{\bullet}$, but play the roles of ligands, changing the redox potential and thus decreasing $OH^{\bullet}$ radicals generation by iron(II), which prevents much damage in biological systems [40,79]. The organism is able to manage the contents of "free" iron in cells and intercellular liquids [80, 81], using transport (transferrin) and reserve (ferritin) proteins, thus limiting its contents in "free" condition under normal physiological conditions. However, oxidation stress can enhance its concentration (for instance, from ferritin) through a mediator such as $O_2^{\bullet-}$ radical [82, 83], thus increasing the deterioration of tissues and organism as a whole.

Damage of tissue often causes hemorrhages, accompanied by erythrocyte deterioration and hemoglobin release out of them [36]. In the same way, hemoglobin is released under cardiac affections [84] and in other cases. Interaction of hemoglobin's erythrocyte released out of the degraded cell with various oxidants (namely, hydrogen peroxide) provokes a whole cascade of reactions. Thus, oxyhemoglobin ($HbPFe^{2+}{\cdot}O_2$), having interacted with hydrogen peroxide, can generate high-reactionary ferryl and amino acid radicals causing LPO initiation. Hydrogen peroxide excess can result in heme degradation, ion of iron(II) release out of it, $OH^{\bullet}$ radicals generation, and, again, in LPO initiation and other oxidation damage [36]. Therefore, oxyhemoglobin acts as pro-oxidant in these conditions.

As can be seen from Fig. 4.2, in LPO activation suppression, antioxidants play an important role. These antioxidants provide three-level protection, caused by the following:

1. Enzymes, controlling the formation and transformation of primary oxygen radicals and depending on trace amounts of Fe, Mn, Zn, and Se ions.
2. Influence of vitamins E, C, carotenoids, ureic acid, thus preventing secondary radical changes in chain LPO reactions.
3. Enzymes impeding secondary radicals formation [85].

Primary products of dioxygen reduction under stress situations are $O_2^{\bullet-}$ and H_2O_2. Their transformations result in violations which can provoke pathologic changes in tissues.

The first protection levels against cells and tissues deterioration with free radicals are anti-oxidant enzymes such as superoxide dismutases (SOD), catalases, and glutathione peroxidases. Cu/Zn-SOD are found in cytoplasm of the most eukaryotic cells (with nuclei), and Mn-SOD are present in mitochondria and realize one and the same reaction:

$$2O_2^{\bullet-} + 2H^+ \xrightarrow{\text{SOD}} H_2O_2 + O_2.$$

Thus, the rate of such enzymatic reaction is pH-independent within the range 5.3–9.5, unlike the non-enzymatic reaction. In the case of Cu/Zn-SOD the role of copper ions consists in superoxide reduction and oxidation:

$$\text{Fer} - Cu^{2+} + O_2^{\bullet-} \rightarrow \text{Fer} - Cu^{+} + O_2,$$
$$\text{Fer} - Cu^{+} + O_2^{\bullet-} + 2H^+ \rightarrow \text{Fer} - Cu^{2+} + H_2O_2,$$

the role of zinc being enzyme stabilization. The maturation and activation of Cu, Zn superoxide dismutase are highly regulated processes that require the presence of CCS and COMMD1 [86].

Mn-SOD are to a great degree prokaryotic enzymes, and were found in bacteria where they are often present along with Fe(III)-SOD. The mechanism of their action is much more difficult than for Cu/Zn-SOD, although its details are not known. It has been found recently that overexpression of SOD is insufficient to extend lifespan in mice [87]. Catalase (cat-Fe(III)) protects cytosol contents in the cells of mammals against their deterioration with hydrogen peroxide:

$$2H_2O_2 \rightarrow 2H_2O + O_2.$$

In human cells, catalase is contained in great concentrations in liver and erythrocytes, and in lesser amounts in cerebrum, heart, and skeleton muscles [88].

Hydroperoxide degradation in most tissues is realized with glutathione peroxidase containing selenium in the active center (Se–GPe) [89, 90], which turns hydrogen peroxide into water and organic peroxides into alcohols:

$$H_2O_2 + 2GSH \xrightarrow{Se-GP_x} 2H_2O + GSSG,$$

or

$$ROOH + 2GSH \xrightarrow{Se-GP_x} ROH + H_2O + GSSG.$$

The second reaction is characterized by a rate constant of about $0.5–6.0 \times 10^6\,M^{-1}\,s^{-1}$ [15].

It is important to add that glutathione peroxidase and catalase are inactivated with superoxide, and superoxide dismutase with the mixture of hydrogen peroxide and superoxide [91, 92].

Glutathione reductase (GRed) will catalyze GSSH reduction to glutathione (GSH) with the help of NADPH:

$$NADPH + H^+ + GSSG \xrightarrow{GRed} 2GSH^+ + NADP.$$

The actual concentration of GSSG, for instance, in liver cells, is close to $30\,\mu M$; however, GSSG concentration will be enhanced on oxidation stress, and glutathione reductase (GRed) will be activated.

Under oxidation stress, peroxide-degenerating enzymes rapid switching on, with the scope to protect the organism of H_2O_2, ROOH accumulation, will be limited by the copper and selenium ions availabel.

Antioxidants, namely, α-tocopherol, ascorbate, glutathione, will prevent the formation of free radicals and peroxides in excessive amounts. Under normal physiological conditions and stress situations the role of these substances is much more complicated and contradictory. Ascorbate under certain conditions can act (like hemoglobin and myoglobin) both as anti- and pro-oxidant. Protective systems can be divided into hydrophobic and hydrophilic [15]. To the first kind belong the systems with α-tocopherol α-TH (vitamin E), which, being the trap of hydrophobic peroxyl radicals $LipOO^{\bullet}$, will generate long-living radical α-T:

$$\alpha\text{-TH} + LipOO^{\bullet} \rightarrow \alpha\text{-}T^{\bullet} + LipOOH.$$

This reaction is characterized by the rate constant $\sim 10^6–10^7\,M^{-1}\,s^{-1}$ in organic solvents, and $\sim 10^5\,M^{-1}\,s^{-1}$ in lipid layers, winning the competition with other organic substances RH:

$$RH + LipOO^{\bullet} \rightarrow R^{\bullet} + LipOOH,$$

as soon as rate constants of such reactions are changed within the range 10^1–$5 \times 10^2\,M^{-1}\,s^{-1}$ [15].

Within the membranes the long-living α-T radical can interact with the second $LipOO^{\bullet}$ radical, turning into the final product α-monoferilquinone (α-TQ).

Ascorbate capacity to convert α-T radical to α-tocopherol is remarkable, thus playing the role of pro-oxidant. Therefore, depending on vitamin E presence or absence (in iron-managed LPO), vitamin C can be both pro- and antioxidant [93]. The synergetic effect of vitamins E and C was observed in vivo [94].

Ascorbic acid is co-factor of various iron- and copper-dependable hydroxylases (lysinhydroxylase, dopamine-β-hydroxylase, etc.), keeping metal ions in them (Fe^{2+}, Cu^{1+}) in the reduced state, which is necessary for hydroxylation. scorbate can trap such free radicals as $O_2^{\bullet -}$, $HO_2^{\bullet -}$, $OH^{\bullet}$, reduce thiol radicals $GS^{\bullet}$ (capable to initiate radical reactions), and deteriorate hypochlorous acid (HOCl) (which is important regarding the running of inflammatory processes) [85]:

$$\text{Ascorbate} + \text{HOCl} \rightarrow \text{semihydroascorbate} + H_2O + Cl^-.$$

Carotenoids and vitamin E also act in a synergetic way.

Carotenoids are very efficient extinguishers of singlet oxygen $^1O_2$. Protection of $O_2^{\bullet -}$ is extremely important in the case of oxidation stress, when superoxide, having interacted with thiols, will generate $^1O_2$.

To the hydrophilic protective systems belong a number of reducers, free radical traps, and certain enzymes. Such reducers and free radicals traps are glutathione (GSH) and ascorbate. Glutathione breaks the chains in chain catalytic aerobic reactions with the formation of free radicals ($GS^{\bullet}$ and $GSSG^{\bullet -}$), and then glutathione disulfide and superoxide ($O_2^{\bullet -}$) [95].

Here PS* and PS excited (for instance, with light as stress factor) and unexcited thiol containing proteins:

$$GSH + Lip^{\bullet} \rightarrow GS^{\bullet} + LipH,$$

$$GS^{\bullet} + GS^- \rightarrow GSSG^{\bullet -},$$

$$GSSG^{\bullet -} + O_2 \rightarrow GSSG + O_2^{\bullet}.$$

Thus, $O_2^{\bullet -}$ formation can play the role of specific passage for free radicals ($Lip^{\bullet}$ or, in general, $R^{\bullet}$). It is not excluded that electrons donor such as ascorbate (A^-), due to its low potential value ($E^{0'} = 0.282$ V), can interact with most of the hydrophilic free radicals, and thus forming $A^{\bullet}$ radical can be considered as oxidation stress index [15, 89].

The material considered makes it possible to conclude that dioxygen reducing activation on oxidation stress is the general source of free radical initiators. Protection of them in the stress conditions involves superoxide, hydrogen peroxide, hydroperoxide degradation, inhibition of continuation stage in autooxidation cycles by long-living peroxyl radicals trapping, and iron salts inactivation by complex

Fig. 4.4 O_2, H_2O_2, and ROOH activation and other active particles generation on stress

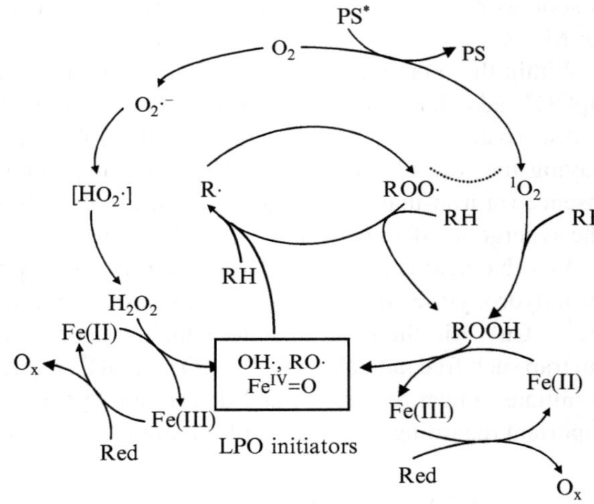

Fig. 4.5 Supposed mechanism of vitamins E, C, glutathione, and NADPH interaction during the protection of LPO. *Asc* ascorbate, *DHAsc* dihydroascorbate, *GSH* glutathione, *GSSG* glutathione disulfide

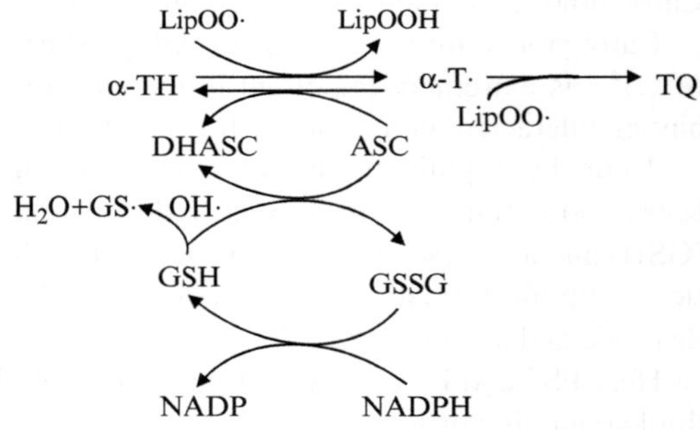

formation and their transfer into the iron(III) form, $^1O_2$ suppressing. Activation of O_2 and other active particles is shown in Fig. 4.4, taken from [15].

Under the conditions of oxidation stress, when an excessive amount of active particles of oxygen and iron compounds, released from storage proteins (hemoglobin, myoglobin, etc.), is formed, homolytic splitting of hydroperoxides and increasing cell damage occur. While vitamin E plays the role of chain terminator in hydrophobic medium (in membranes), the same role is played by ascorbate and glutathione in water surroundings [15]. Recently, melatonin enhancement of the antioxidant action of α-tocopherol and ascorbate against NADPH- and iron-dependent lipid peroxidation has been detected [96].

The proceeding interactions (synergetic, etc.) of α-tocopherol, ascorbate, glutathione, and NADPH, from a number of works [85, 97], can presumably be represented as in Fig. 4.5, although such an elegant interpretation is considered to be dubious [98].

Here:

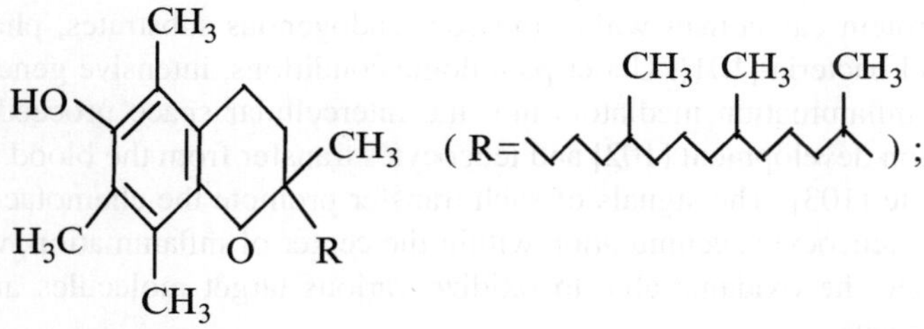

(α-TH)

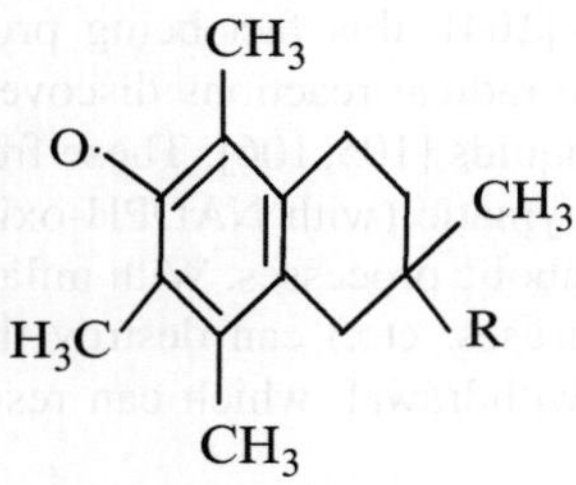

(α-T·)

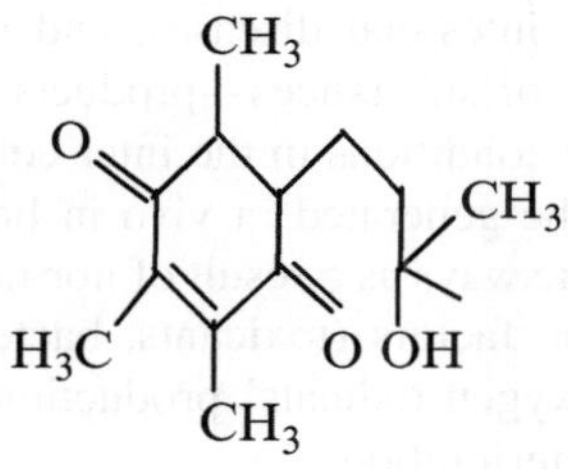

(α-TQ)

[α-Tocoferyl quinone]

Antioxidant protection can also be provided with uric acid, which is contained in plasma at a concentration of 0.25–0.45 mm. It is a perfect trapper of singlet oxygen $^1O_2$, hydroperoxides, and peroxyl and $OH^{\bullet}$ radicals [85].

In general, stress situations are accompanied by the most complex interlacing of both chemical (often catalytic) processes (responding to stress factors) and biological possibilities of these reactions being used by an organism.

Stress influence can be favorable for an organism, adapting it to changing living conditions. This situation was called "austress" by Selye. Provided the adaptation is embarrassed or impossible (for example, in case of ionizing irradiation), this is accompanied with certain destructive changes in organism, resulting in various types of pathology. In this case we speak of "distress" (by Selye). Distress, caused by both exogenous and psychoemotional factors, is one of the most important causes of many human diseases, namely cardiac-vascular, oncologic, inflammatory, endocrine, and neuropsychic [49]. Infact, oxidation processes take place in the majority of human diseases [36].

4.3 Inflammatory Processes and Role of Metal Compounds

Any damage of tissue from any cause results in inflammation. The main participants of inflammatory processes are neutrophils (polymorphous-nuclear leucocytes, PNL) and the products of their secretion which play the main role in organism protection from foreign cells and bacteria [99,100]. They are the sources of inflammation mediators. Oxidizing agents of neutrophils ($O_2^{\bullet-}$, H_2O_2 etc.) under normal conditions perform protein catabolism within the cell, endogenous substrates, phagocytated viruses, and bacteria [101]. Under pathologic conditions, intensive generation and release of inflammation mediators into the intercellular space proceeds, causing inflammation development [102] and leucocytes transfer from the blood toward the damaged site [103]. The signals of such transfer promote the chemotactic factors, resulting in leucocyte accumulation within the center of inflammation, where they will generate the oxidants able to oxidize various target molecules and foreign microorganisms.

Free radicals are involved in inflammatory processes, including rheumatic arthritis, atherosclerosis, lungs emphysema (increased contents of air in lungs), inflammatory intestinal diseases, and cancer [104], this fact being proved by a great number of substances—products of free radical reactions discovered under inflammatory conditions in the inter-cellular liquids [105, 106]. These free oxygen radicals can be generated in vivo in both enzymatic (with NADPH-oxidase) and non-enzymatic ways as a result of normal metabolic processes. With inflammations the pathologic factors (toxicants, bacteria, viruses, etc.) can destroy the normal balance of oxygen radicals' production and withdrawal, which can result in cell damage or deterioration.

On the initial inflammation stage, influenced by stimulators (bacteria, etc.), a so-called "respiratory explosion" proceeds. It was called this because it results in a sharp (20-fold) increase in oxygen demand and its simultaneous consumption for $O_2^{\bullet-}$, H_2O_2, $OH^{\bullet}$, and ClO^- generation.

Stimulators of this process can be bacteria and viruses, as well as some substances, such as cytosine, immunoglobulin, etc. The main oxygen consumption is connected with $O_2^{\bullet-}$ production.

This phenomenon is the basis of, for instance, pathogenic microbe destruction with the help of the mechanism involving the formation of oxidative oxygen particles by two separate enzymatic systems [107]. The first of them, a NADPH-dependent oxidase system, localized by the membranes and containing cytochrome, will generate $O_2^{\bullet-}$. The second, myeloperoxidase haemoprotein, is localized inside the neutrophillic granulae and will catalyze the formation of hypochlorite (hundreds of times stronger oxidant than H_2O_2):

$$Cl^- + H_2O_2 \xrightarrow{\text{myeloperoxydase}} ClO^- + H_2O.$$

It is supposed that hypochlorite is responsible for the bactericide activity of neutrophils [108, 109].

Polymorphonuclear leucocytes (neutrophils), leucocytes, and macrophages, containing membranous NADPH-oxidase, take part in inflammatory processes, playing the key role in them [104] by stimulating free radical oxygen particle formation, in particular, $O_2^{\bullet-}$.

In electron-transport respiratory chains, resulting mainly (95%) in four-electron reduction of oxygen molecule to water, approximately 5% of the electrons are spent on $O_2^{\bullet -}$ generation:

$$\text{Cytochrome C oxidase} + O_2 \begin{cases} \xrightarrow{4e^-,\ 4H^+} 2H_2O \\ \xrightarrow{e^-} O_2^{\bullet -}. \end{cases}$$

The source of electrons for superoxide formation is NADPH bound with the membrane:

$$2O_2 + \text{NADPH} \rightarrow 2O_2^{\bullet -} + \text{NADP}^+ + H^+.$$

Increased superoxide contents is characteristic of inflammatory processes and therefore it is supposed that phagocyte cells release $O_2^{\bullet -}$ into their surroundings. But this statement is still based on indirect data.

Under inflammatory conditions, xanthine dehydrogenase can become a source of $O_2^{\bullet -}$. Under the normal physiological conditions it exists in the form of dehydrogenase (in "D"-form), oxydizing xanthine to uric acid or its salts [110]:

$$\text{Xanthine} + \text{NAD}^+ + H_2O \xrightarrow{\text{Xanthine dehydrogenase}} \text{uricacid cationurate} + \text{NADH} + H^+.$$

The enzyme can be turned into oxidase ("O"-form), able to transfer electrons directly onto O_2. The "D"-forms are transformed into the "O"-form of enzyme, and with a sufficient supply of O_2 will contribute to superoxide production.

During the reperfusion phase, the oxygen molecule becomes accessible to the "O"-form of oxidase, and, as consequence, $O_2^{\bullet -}$ will be generated:

$$\text{Xanthine} + 2O_2 + H_2O \xrightarrow{\text{Xanthine oxigenase}} \text{uricacid cation} + 2O_2^{\bullet -} + 2H^+.$$

But there is no direct evidence for this assumption as well.

In temporary heart disease the concentration of O_2 will be reduced, which causes a diminishing of DNA to TNA phosphorylation in the respiratory chain. Thus, TNA concentration in the cell will be reduced, and the cell is no longer able to maintain the necessary gradient of ions transfer through its membranes. This results in an increase in cytosole concentration of Ca^{2+} ions, which will activate cellular proteases for the catalysing of xanthine dehydrogenase ("D"-form) into xanthine oxidase ("O"-form) transformation. Xanthinoxidase, as shown above, in its turn, will reduce O_2 to $O_2^{\bullet -}$.

$O_2^{\bullet -}$ radicals take part in $OH^{\bullet}$ radicals generation, which results in tissue damage at the sites of xanthine oxidase localization. It is supposed that the same processes and damage proceed in the case of local intestinal, myocardial, cerebral disease, as well as in the case of acute kidney disease.

It was shown that similar damage also occurs in inflammatory human joints. The main sources of joint inflammation are phagocytic cells and tissue ischemic reperfusion, due to the fact that, for instance, phagocytic cells are able to generate $O_2^{\bullet-}$ radicals [110].

It was first supposed that superoxide damages the tissues itself, but now it is clear that because of its relatively low oxidizing ability, it is only the source of a much stronger oxidant—$OH^{\bullet}$ radical, formed by catalytic reaction of Haber–Weiss type. The latter will efficiently perform hydrogen atom detachment from substrate [111].

There are plenty of $O_2^{\bullet-}$ radical generation sources in living organisms, which, by dismutation with LCD, will be turned into hydrogen peroxide. In further reactions of H_2O_2 transformation in $OH^{\bullet}$ radicals, iron and copper ions act as catalysts.

Iron ions in bound form are contained in the electron-transport chain (cytochrome C) in storage proteins such as ferritin and lactoferritin. Iron ions content in storage proteins depends on pH value. Low pH value in rheumatic-arthritic tissues favors Fe^{3+} ions release and $O_2^{\bullet-}$ turning into $OH^{\bullet}$ radical by the reaction of Haber–Weiss [112–115].

Equally, with Fe^{3+} ions in synovial liquid of patients with joints arthritis, Cu^{2+} ions were found [112], which can take part in many catalytic reactions, interacting with hydrogen superoxide:

$$Cu^{2+} + O_2{}^{\bullet}- \rightarrow Cu^{+} + O_2,$$

$$Cu^{+} + O_2{}^{\bullet}- \rightarrow Cu^{2+} + H_2O_2,$$

$$Cu^{+} + H_2O_2 \rightarrow Cu^{2+} + OH^{-} + OH^{\bullet},$$

$$2Cu^{2+} + H_2O_2 \rightarrow 2Cu^{+} + O_2 + 2H^{+}.$$

As a result of these reactions the strong oxidant, $OH^{\bullet}$ radical, will be formed, locally, deteriorating the tissues.

On inflammation connected with neutrophils, monocytes, and macrophages penetration into tissue, $O_2^{\bullet-}$, H_2O_2, and $OH^{\bullet}$ will be formed, and their influence causes fibrosis to appear in many organs [116]. Such correlation between H_2O_2, $O_2^{\bullet-}$-, and $OH^{\bullet}$ radicals formation, and pulmonary fibrosis was shown on bleomycin influence (anti-tumor remedy). The latter, forming complexes with iron and O_2, will generate $OH^{\bullet}$ radicals, which will destroy the lungs [117].

An important role of iron compounds as catalysts of reactionary oxygen particles appearance, promoting liver fibrosis, was observed in patients with an excessive iron content [118]. It should be added that fibrosis correlates with oxygen concentration in liver [119] and with LPO increase, also caused by the increase in reactionary oxygen particles content [120]. Such correlation between LPO and iron content is characteristic of fibrosis liver deterioration [121].

The inflammatory process, being an organism reaction to foreign invasion (that of toxins, microorganisms), is associated with the formation of free radical oxygen particles and LPO in the presence of iron and copper catalytic compounds. Thus correlation is observed in low density lipoproteins (o-LDL) oxidative change (modification). In the case of chronic inflammatory diseases, such as rheumatic arthritis or atherosclerosis, the reducing of o-LDL is observed, which is extremely sensitive to LPO. LDL can be oxidatively modified with neutrophils and macrophages by the release of reactionary oxygen radicals with the macrophages. $OH^\bullet$ radicals thus formed, can initiate the peroxidation of aliphatic acids long chains inside the LDL molecule with the formation of conjugate dienes and lipid hydroperoxyradicals ($LipOO^\bullet$) [104].

The process of LDL oxidative modification in many reactions has pro-inflammatory character, activating a number of stimulating factors and resulting in the development of atherosclerotic formations [122], though its mechanism is unclear. Copper compounds are potential catalysts of LDL oxidative modification in vitro [123], but in vivo most of the copper (>95%) in human serum is bound in ceruloplasmin enzyme. This enzyme is a potential LPO inhibitor. In inflammatory processes, inside capillary walls protease and oxygen radicals release can proceed with endothelium cells and macrophages that can result in copper ions release, which are able to induce LPO [104]. So, atherosclerotic progress is observed in persons with high copper concentration in serum [124]. Approximately the same picture is observed in patients with rheumatoid arthritis.

It is to be noted that superoxide has one more surprising capacity. It seems that superoxide in inflammatory processes is able to control neutrophil chemotaxis (i.e., directed neutrophils passing toward the sites of inflammation) by chemotaxic factors activation (factor of thrombocytes activation, anaphylatoxin C_{5a}, and other chemotaxis factors). This superoxide-dependent chemotaxic factor seems to be necessary for normal neutrophil-meditated inflammatory response realization. Prevention of this factor formation under the influence of SOD ($O_2^{\bullet-}$ radicals are withdrawn) makes clear the mechanism of this anti-inflammatory enzyme action [125].

Neutrophils, being the central participants of the inflammatory process, they ot only destroy foreign microorganisms during phagocytosis process (positive effect) but also will damage endothelium of living organism vessels (negative effect). This specific "cost" is paid by the organism for pathogenic microorganisms deterioration. The destructive influence of neutrophils on vessels is due to the fact that they generate various biologically active substances (proteolytic enzymes, oxygen radicals, and other highly toxic substances). Such direct influence, neutrophils also act indirectly by their influence on thrombocytes endothelium functions [88].

In the case of the inflammatory process in the living organism, there is a protective barrier against oxygen free radicals, first of all by a number of enzymes, such as SOD, catalase, glutathione peroxidase. These enzymes are localized mainly on the surface of endothelium cells, which testifies to their cellular-specific protective role [110]. It was shown that on some inflammations, introduction of Mn-SOD is accompanied by an anti-inflammatory effect [126]. On some inflammatory processes (liver inflammation [127]) additional synthesis of another protective enzyme, ceruloplasmin (copper-containing protein), will be activated.

Another non-enzymatic protection form against free oxygen radicals outside the cell on inflammations is strong chelato-binding of iron into "non-catalytic" state with transferrin [128].

Due to such binding, iron compounds concentration able to realize Haber–Weiss reaction in human serum is close to zero in conditions of normal organism functioning. Thus, the formation of $OH^{\bullet}$ radicals able to damage the tissues is prevented. Introduction of complex iron compounds with strong-binding ligands, such as dextrin, to rheumatoid patients, causes acute conditions, because iron appears able to catalyze oxygen radicals formation (in particular, $OH^{\bullet}$), causing destructive processes [129, 130].

Thiols, ascorbic acid, and uric acid are antioxidants, and therefore, in high concentrations they can inhibit LPO, clearing the system of $O_2^{\bullet-}$ and $OH^{\bullet}$ radicals. In contrast, in low concentrations of ascorbic acid in the presence of iron, LPO induction can occur. Vitamin E reacts with lipid peroxyl radicals ($LipOO^{\bullet}$) with the formation of tocopherol radical (α-$T^{\bullet}$) [98]. This vitamin prevents the stage of LPO chain continuation, withdrawing $LipOO^{\bullet}$ from the system [131]. Another antioxidant β-carotene will also trap $LipOO^{\bullet}$ in conditions of low values of pO_2 [132], which takes place in chronic inflammation (for example, in patients with vessels rheumatic arthritis) [133].

4.4 Role of Enzymes and Metal Compounds in Phagocytosis

In 1908 Russian scientist Ilya Mechnikov won the Nobel Prize for the discovery, as long ago as 1883, of the protective role of phagocyte cells against infections (phagocytosis). Such special cells are, in particular, neutrophils. Neutrophils, often called polymorphonuclear leucocytes or granulocytes (because their cytoplasm contains granulae limited with membrane), predominate among the white blood cells.

White blood cells (or leucocytes) make about 0.1% blood cells and take part mainly in infection microorganisms identification and destruction. Another group of leucocytes—monocytes—can develop further in the cells group known as macrophages. They were found in the tissues of lungs, liver, and other organs. In various organs, macrophages have different forms and functions. Neutrophils, monocytes, macrophages, and eosinophils, being stimulated by the invasion of foreign pathogenic microorganisms, will reduce oxygen to superoxide-anions $O_2^{\bullet-}$ (and also to H_2O_2), by their enzymatic systems. These free radicals (and hydrogen peroxide) will take part in subsequent catalytic processes of bactericide particles generation, directly destroying the microorganisms.

There are two granulae types. The first implies so-called primary granulae of neutrophils. They contain a lot of myeloperoxidase, the whole set of lysosome hydrolytic enzymes (taking part in digestion of nucleic acids proteins and foreign microorganisms enzymes), and also the group of bactericide cationic proteins. Another type of granulae involves specific (or secondary) ones, which also contain mesosomes rich in lactoferritin—iron-bound protein. The role of the latter consists

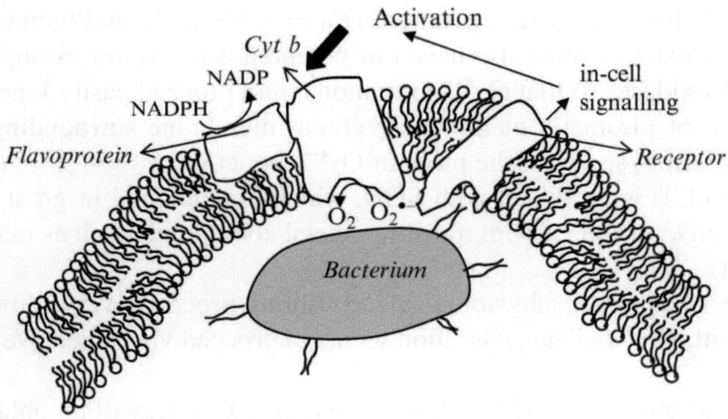

Fig. 4.6 Formation of phagosome

in Fe^{3+} ions release under certain conditions and their involvement into radical reactions of Haber–Weiss type.

Neutrophils and other granulocytes move along the blood capillaries in tissue, passing the joints of endothelium cells, to the infection places, and attract to themselves the molecules to which they serve as specific receptors. For neutrophils such chemo-attracting substances are some peptides, derivatives of leukotrienes from arachidonic acid, lymphokines, and other bacterial substances.

Binding of foreign microorganisms with the specific receptors on neutrophil surfaces initiates the complex sequence of events resulting in the formation of joint phagocyte vacuole (bubble) containing microbes (Fig. 4.6). Granulae passing through cytosole, join the membrane of phagocyte vacuole and interact with its contents. Within a few minutes the microbe will be killed and the destruction of its remains begins, which takes several hours. This overall process is called phagocytosis.

Microbes binding to the neutrophils receptors results in a high increase in oxygen demand by phagocytes. The mechanism of this process initiation is not quite clear, but it appears to include the activation of neutrophil inner-cellular phospholipase, which generates "secondary stimulators" activating the protein kinase, realizing catalytic phosphorylation of enzyme-oxidase neutrophil complex components (NADPH-oxidase) [134]. This oxidase complex, in its turn, initiates a so-called "respiratory explosion" proceeding for 15–20 min and characterizing the increased oxygen demand with $O_2^{\bullet-}$ and H_2O_2 generation.

Oxygen, having turned into its reduced compounds, becomes rather reactionary; this fact explains bactericide activity of neutrophils [135].

The initial product of increased respiratory activity of neutrophiles is $O_2^{\bullet-}$ [136], which initiates a complex chain of reactions:

$$O_2 + e^- \rightarrow O_2^{\bullet-}, \tag{4.1}$$

$$O_2^{\bullet-} + O_2^{\bullet-} + 2H^+ \rightarrow H_2O_2 + O_2, \tag{4.2}$$

$$O_2^{\bullet-} + H_2O_2 \rightarrow OH^{\bullet-} + OH^- + O_2. \tag{4.3}$$

Reaction (4.1) has a low redox potential ($E_0 = -330\,\mathrm{mV}$), and therefore electron donor for this reaction must also have low potential, such as, for example, NADPH (in NADPH-oxidase) so that $O_2^{\bullet-}$ formation could proceed easily kinetically. The outer surface of plasmatic membrane (cellular membrane surrounding the living cell's protoplasma) serves as the place of $O_2^{\bullet-}$ reduction.

Reaction (4.2) is catalyzed with SOD, which is contained in great amounts in neutrophil cell cytosole, or with transition metal compounds such as manganese(II) or copper(II).

Reaction (4.3) under physiological conditions proceeds very slowly, and its activation with $OH^{\bullet}$ radical generation seems to proceed via Haber–Weiss catalytic reaction.

Subsequent products, toxic for foreign microorganisms, will be obtained due to the following reaction catalyzed with myeloperoxidase:

$$H_2O_2 + Cl^- \xrightarrow{\text{myeloperoxidase}} OCl^- + H_2O.$$

The forming hypochlorite OCl^- is hundreds of times more bactericidal than hydrogen peroxide, and plays the significant role in phagocytosis. Thus, the overall system of catalytic reactions results in formation of such bactericide substances as $OH^{\bullet}$ radical and hypochlorite OCl^-.

Plasmatic membrane of neutrophil will be bent to the place of its joining to a microbe, covering it and including it into the so-called phagosome (Fig. 4.6 taken from [136]). Thus the breakage (melting of lysosome cellular membranes) proceeds and granulae contents comes into the phagosome. The contents of granulae are various, but myeloperoxidase and lactoferritin are especially important. The content of myeloperoxidase makes up 2–5% of the dry cell weight [137]. Myeloperoxidase coming from granulae into phagosome, is cationic protein, and the shell of the bacterial cell is charged negatively. Thus, their joining proceeds on account of electrostatic forces. It takes just a few minutes to identify the microbe and achieve its total isolation.

Catalytic oxidation reactions proceed in the process of phagocytosis, depending on O_2 and hemoprotein cytochrome b559 presence, being a component of complex NADPH-oxidase system. The absence of cytochrome b559, due to gene deficiency in some people,e causes granulomatous disease, characterized by neutrophils being able to bring about the respiratory explosion. Such people suffer periodic infections which resulted in early death before antibiotic therapy was available.

The respiratory explosion itself, in its turn, is caused by certain stimulators, which, being joined to outer-surface receptors of neutrophils, cause the release of granular lysosomes contents (cellular organelle serving as enzymes reservoir, digesting the proteins, nucleic acids, lipids).

Neutrophil's cytosole contains a number of enzymes (COD, catalase, glutathione peroxidase), which protect it from the products of respiratory explosion ($O_2^{\bullet-}$, H_2O_2), performing a number of reactions:

$$O_2^{\bullet -} + O_2^{\bullet -} + 2H^+ \xrightarrow{\text{COD}} H_2O_2 + O_2,$$

$$2H_2O_2 \xrightarrow{\text{catalase}} 2H_2O + O_2,$$

$$H_2O_2 + 2GSH \xrightarrow{\text{glutathione peroxidase}} 2H_2O + GSSG,$$

where GSH is glutathione, a GSSG-oxidized glutathione.

GSSG will be reduced to GSH with glutathione reductase, which uses NADPH as electron donor:

$$GSSH + NADPH + H^+ \xrightarrow{\text{glutathione reductase}} 2GSH + NADP^+.$$

It is also possible that vitamins E and C take part in phagocytes protection against the attacks of oxygen radicals.

The life of neutrophils (specific "biological self-killers") is very short in inflammation, and their remains, in common with those of bacterial victims, are the main components of pus. However, antioxidant protective systems of neutrophils make their survival longer, so that they are able to fulfil their main role.

In NADPH-oxidase systems, as shown above, there exists cytochrome b_{559}, characterized by low redox potential ($E = -245$ mV) [138], which can efficiently take part in $O_2^{\bullet -}$ production ($E = -160$ mV) [135]. Cytochrome b559 was found both in plasmatic membrane and in specific granulae. Purified cytochrome b559 is representative of glycoprotein with no definitely established molecular weight.

Circulating phagocytes will not form $O_2^{\bullet -}$ until they get the appropriate stimulus. This stimulus turns oxidase into the active form after a lag-period of 30 s. A characteristic stimulator is *N*-formyl peptide, which activates specific phosphodiesterase, realizing the hydrolysis of phospholipid phosphatidyl inositol. Diacylglycerol (DAG) and inositol *trans* phosphate (IP_3)-products of this hydrolysis are signal molecules [135]. DAG is an activator of Ca^{2+}-dependent protein kinase C, which can activate oxidase by its polypeptides phosphorylation. IP_3 appears to promote Ca^{2+} ions release out of the inner-cellular storage. In its turn, its increased concentration of Ca^{2+} ions is an activator of protein kinase. Other stimulators act in the same way [135].

As a result of NADPH-oxidase system activation the respiratory explosion proceeds, accompanied by a series of reactions (Fig. 4.7).

Under the influence of NADPH oxidase (1) and COD the formation of $O_2^{\bullet -}$ and H_2O_2 takes place. Their interaction yields the singlet oxygen $^1O_2$ (2) and hydroxyl radical (3) (by Haber–Weiss catalytic reaction). Hydrogen peroxide will react with Cl^- ions (4). This reaction is catalyzed by myleoperoxidase (MPO), which in neutral and weakly-acid solutions catalyzes two-electron oxidation with hydrogen peroxide of Cl^- ion physiological reducer, with the formation of bactericide hypochlorous acid HOCl. Under the influence of hydrogen peroxide, amines, and amino acids, it will be turned into various high-reactionary products, such as $^1O_2$, RNHCl, RCHO, etc. It was shown that the HOCl formed inhibits, for instance, DNA-synthase in *E. coli*, which is fatal for such microorganisms. This explains the universal character of

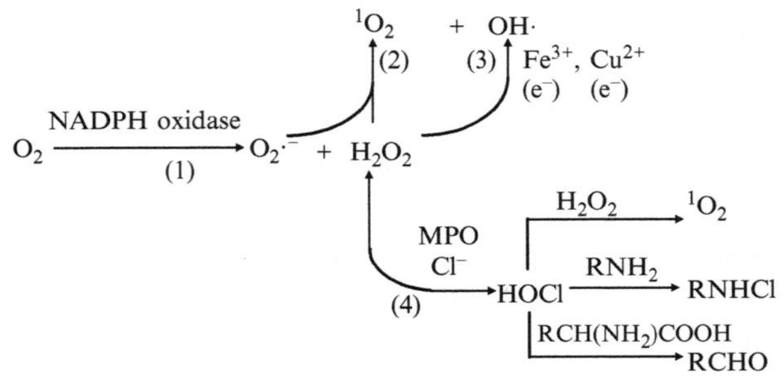

Fig. 4.7 Supposed ways of toxin generation by neutrophils (taken from [139])

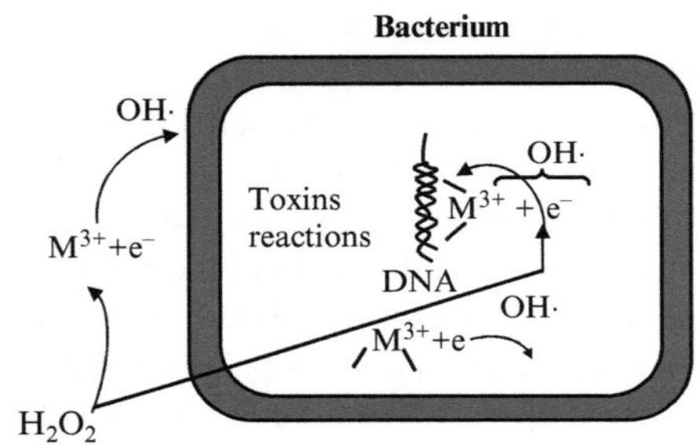

Fig. 4.8 Local-specific Haber–Weiss mechanism generating high-reactionary $OH^•$ radicals and their antimicrobial influence (taken from [139])

hypochloric acid toxicity, destroying energy-transferring cell elements by secondary reactions with involvement of metal-catalyzed reaction of Fenton type, resulting in $OH^•$ radical generation.

The role of superoxide-anion in oxygen toxicity is rather unclear, and, it seems, $O_2^{•-}$ will be involved in bactericide inner-phagosome reactions indirectly. Hydrogen peroxide is also ineffective in general as antimicrobial agent. However, the product of their interaction by Haber–Weiss reaction, $OH^•$ radical, is a strong bactericide agent. Though $OH^•$ radical is a non-selective oxidant, it was confirmed that in the cells its formation is local-specific [139] (Fig. 4.8).

Equally with myeloperoxidase use for HOCl generation, neutrophils possess other means to fight pathogenic microorganisms, namely, through the direct using of respiratory explosion products ($O_2^{•-}$ and H_2O_2), or $OH^•$ radicals will be generated in the places of metal (M^{3+}) compounds biological binding, i.e., system selectivity is provided due to the certain localization of catalyst [139].

Within the cell of a microorganism, metal ions can be bound with various biological important substances, particularly with DNA, which provide local-specific generation of $OH^•$ radicals and thereby destroying DNA and other physiologically important substances close to these radicals. Thus, viruses, microbes, and bacteria deterioration occurs to a great degree due to the variety of catalytic reactions proceeding in vivo.

4.5 Metal-Catalyzed Oxidation and Ageing

4.5.1 *Oxidative Damage of Proteins and Enzymes, and Ageing*

A great deal of literature is devoted to various aspects of the ageing problem and the influence of factors contributing to this process because of its importance (see the literature, for instance [140]). We shall consider here only the link between the catalytic processes realized with enzymes and transition metal compounds, yielding the appearance of highly reactive intermediate compounds, including free oxygen radicals, and some factors of ageing.

It has been known for a long time about free radicals involvement in the ageing process when initiated with certain catalysts [141]. Deterioration of cellular activity by free oxygen radicals is connected with all the diseases related to age—atherosclerosis, arthritis, cataract genesis, various neurologic diseases, and cancer.

First. several main causes are to be outlined, dealing with organism ageing. These include (1) oxidative damage of proteins in general and enzymes' proteins in particular in the places of their binding with metals (iron, copper), catalyzed with metals and resulting in reducing or total disappearance of enzymatic activity, (2) DNA deterioration with active particles, and (3) mitochondrial damage induced with free oxygen radicals and other causes. An important role in ageing retarding is played by the antioxidative protection with enzymes and antioxidants (vitamins E, C and other compounds).

The influence of these on organisms and their connection with ageing can be shown taking as an example loss of activity by glutamine synthetase (GS) under the influence of H_2O_2 and Fe^{2+} which are formed in an organism in many ways as considered above [142]. Oxidation of aminoacid residues of enzymes proceeds, often resulting in carbonyl derivatives formation (GS_{ox}–C$\cdots$R=O), which are very sensitive to proteolytic degradation and in their turn, under the influence of proteases, will be turned into peptides and amino acids. Thus, proteases will destroy oxidized proteins of enzymes more rapid than unoxidized enzymes. Therefore, reducTION of the total disappearance of enzyme activity takes place [143, 144].

Systems that can realize such proteins of certain enzymes oxidation catalyzed by metals are numerous. These are enzymatic (NAD(P)H-oxidases/NAD(P)H/Fe(III)/O_2, xanthine oxidase/hypoxanthyne/Fe(III)/O_2, cytochrome P-450 reductase/cytochrome P-450/NADPH/Fe(III)/O_2, etc.) and non-enzymatic systems (ascorbate/Fe(III)/O_2, mercaptane RSH/Fe(III)/O_2, Fe(II)/O_2, Fe(II)/H_2O_2, etc.) [145].

All these systems lead to metal-catalyzed oxidation (MCO) of certain enzyme's protein by metal joining. Such local-specific metal-ion catalyzed oxidation of enzyme protein is shown in Fig. 4.9.

Proofs of such local-specific mechanism of enzyme's protein oxidation can be as follows:

1. Enzymes, non-activated by MCO systems, are relatively insensitive to oxidation inhibition by free radical cleaners (formate, ethanol, and mannitol)

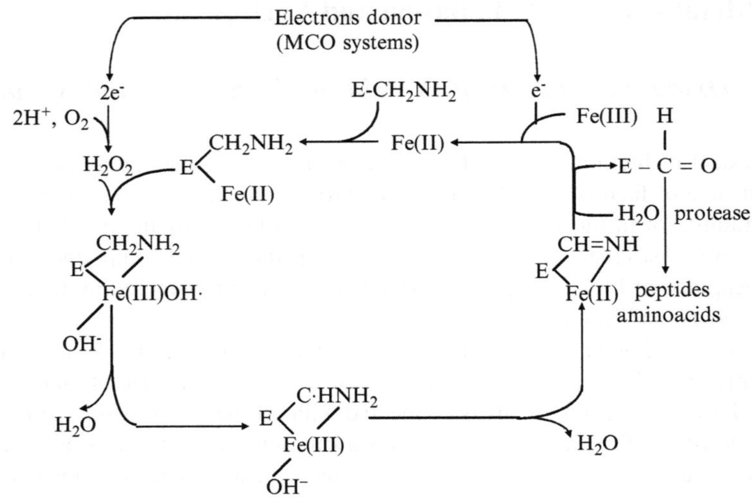

Fig. 4.9 Local-specific mechanism of enzyme's protein (E) oxidation with MCO systems (taken from [146]). Here (–CH_2NH_2)–ε-aminogroup of lysine is remainder

2. The amino acid residues in the protein, located close to the places of metal ion joining (proline, arginine, lysine), can be modified (changed) with MCO-systems, while on radiolysis all the amino acid residues of protein are subjected to free radical influence
3. All enzymes sensitive to their modification with MCO-systems need metal ions for catalytic activity and must contain local groups of atoms for their binding with metal ions [146]

The possibility of not only $OH^{\bullet}$ $(FeO)^{2+}$, $(Fe(OH)_2)^{2+}$, $(FeOH)^{3+}$, Fe(III), but Cu(II) involvement in such a mechanism of enzyme protein oxidation should also be taken into consideration. Catalase and SOD, withdrawing H_2O_2 and $O_2^{\bullet-}$ out of the system, will inhibit the modification (change) of enzymes-targets [146–149]. The same is realized with chelato-forming iron ligands (EDTA, *o*-phenanthroline).

Local-specific mechanism is realized via Fe^{2+} ion binding to certain site (L-chelato-forming local place of protein). Complex formation of this metal with lysine amino acid residue of enzyme's peptide and subsequent interaction of this complex with H_2O_2 by Fenton mechanism with $OH^{\bullet}$ radical or high-oxidized iron(IV) particles generation is as follows.

Hydrogen atom detachment of the ε-amino group of lysine residue with the formation of alkyl radical, electron transfer from this radical to Fe(III) with the formation of imine derivative (–CH=NH) and, finally, its hydrolysis with the formation of appropriate aldehyde derivative (formation of protein carbonyl group) are shown in Fig. 4.10, taken from [150]. Such a local-specific mechanism of protein carbonyl formation makes it clear why the cleaners of free radicals are unable to inhibit this process.

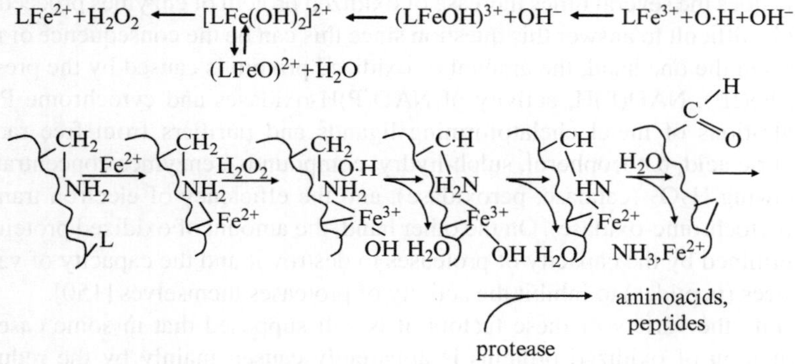

Fig. 4.10 Local-specific oxidation of enzymes

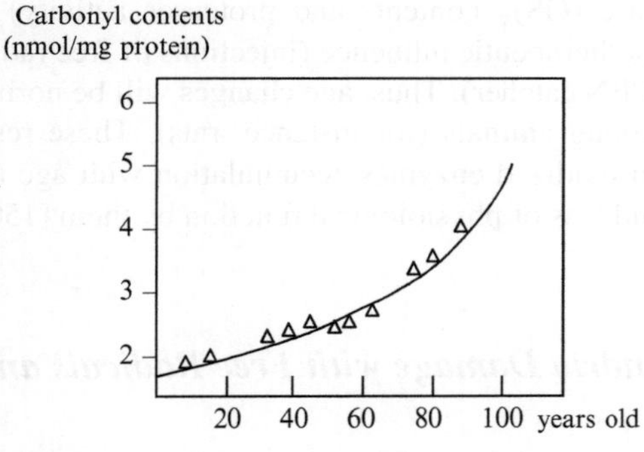

Fig. 4.11 Change in contents of protein carbonyl with age

Non-enzymatic MCO systems are also important, such as ascorbates and mercaptane being the electrons donors. Thus, the ascorbate–Fe(III)–O_2–system is a specific model of oxidase, which catalyzes oxygen-dependent hydroxylation of aromatic compounds [151], including lipids [152], nucleic acids [153], and proteins [149, 154, 155] oxidation.

During ageing, less-active or inactive heat-labile forms of some enzymes are accumulated in mammals [149]. It is supposed that the change (modification) of some enzymes is caused by the destructive action of oxygen free radicals [156, 157] and accumulation of damaged protein containing carbonyl group is a result of increasing the deterioration rate of some enzymes amino acid groups of protein with age [157, 158].

The contents of protein carbonyl in the cultures of human skin fibroblasts in young patients (less than 20 years) makes about 10% of total cell protein (2 nmol per milligram of protein) and increases in older age ($\approx$80 years) at least to 30%, though its actual contents apparently reaches 40–50% [159], increasing exponentially with age, as shown in Fig. 4.11.

Why does the several times increase of oxidized protein of enzymes proceed with age? It is difficult to answer this question since this can be the consequence of many factors. On the one hand, the amount of oxidized protein is caused by the presence of O_2, Fe(III), NAD(P)H, activity of NAD(P)H-oxidases and cytochrome P-450, concentrations of metal chelatoforming ligands and purifiers from free radicals (SOD, uric acid, α-tocopherol, sulph-hydryl compounds), enzymes concentrations, withdrawing H_2O_2 (catalase, peroxidase), and the efficiency of electron transport chain (cytochrome-oxidase). On the other hand, the amount of oxidized protein will be determined by the capacity of proteases to destroy it and the capacity of various substances (peptides) to inhibit the activity of proteases themselves [150].

Despite the variety of these factors, it is still supposed that in some cases the accumulation of oxidized proteins is apparently caused mainly by the reduction of protease activity with age [147–160], which is responsible for the deterioration of oxidized protein. Age changes (increase in protein carbonyl contents, decrease in glutaminsynthetase $(GS)_x$ contents and proteases activity) can be reversible under chronic chemotherapeutic influence (injections of free radicals *N*-*tert*-butyl-α-phenylnitrone—PBN catcher). Thus, age changes will be normalized to the level characteristic for young animals (for instance, rats). These results testify to the connection between oxidized enzymes accumulation with age (up to 50% of cell oxidized protein) and loss of physiological function by them [150].

4.5.2 Mitochondria Damage with Free Radicals and Ageing

General worsening of physiological functions of living organisms' organs, caused by the appropriate worsening of the same functions of cellular organelles (DNA, mitochondria, etc.), is usually associated with ageing. Thus excessive oxidation is the main source of cell ageing and degenerative diseases which accompany the ageing: cancer, cardio-vascular diseases, deterioration of immune system, brain dysfunctions, and cataracts. Food antioxidants (vitamins C, E, and carotenoids) protect against these degenerative diseases [161, 162].

Superoxide-anion, hydrogen peroxide, hydroxyl radicals, and, eventually, singlet oxygen ($^1O_2$), formed in catalytic enzymatic and non-enzymatic ways and being, as a rule, by-products of aerobic metabolism, will damage cellular macromolecules, including DNA [163], proteins [157], and lipids [164].

The accumulation of such damage has as a consequence organism ageing and degenerative diseases.

Mitochondria are the main physiological source of reactive oxygen particles during mitochondrial respiration. Superoxide radicals ($O_2^{\bullet -}$) will be formed in by-reactions of the mitochondrial electron transport chain or with NADH-independent enzyme, localized on the internal mitochondrial membranes. SOD presence in the cell results in hydrogen peroxide formation from $O_2^{\bullet -}$.

Later, by Haber–Weiss reaction, the formation of a strong oxidant, $OH^{\bullet}$ radical, proceeds. Mitochondria, apparently, are the greatest source of oxidants which is confirmed by the fact that mitochondrial electron-transport, one of the most important functions of mitochondria, is TNA producing, and therefore it is present in relatively large amounts in each cell of the body. Deficiency of cellular energy caused by the deterioration of this mitochondrial function weakens normal cellular activity and deteriorates cellular capacity to adaptation in the case of various physiological stresses. It is supposed that oxidative damage of mitochondria is the main factor in ageing [162, 165, 166].

Mitochondrial DNA (mtDNA) is the main subject of oxidative damage in the cell compared to nucleic DNA, and the rate of mtDNA mutation is much higher ($\sim$17 times) compared to the mutation of nucleic DNA [167, 168]. Such unequal level of different DNA damage is caused by mtDNA localization near the inner membrane where oxidizers are forming, the lack of protective proteins-histones, and the lack of activity in mtDNA reparation [162].

Oxidative damage of mtDNA are accumulated as a function of age in the muscles of the human diaphragm, human cerebrum, in rats liver, and other parts of the organism. The amount of mtDNA damage in old persons is $\sim$25 times higher than in young people. This damage result in mtDNA mutations, which are manifested in mitochondria dysfunctioning, and, in particular, in neurodegenerative disease occurence on cerebral mtDNA [169].

Accumulation of oxidative-degraded proteins in the cells, including mitochondrial protein (in particular with reactive carbonyl groups), can result in the reduction of energy production and an increase in oxidants formation.

With age the mobility of cellular membranes is reduced due to the oxidation of mitochondrial membrane lipid components (for instance, linolic acid). Besides, with age the proteins are accumulated, being connected with malondialdehyde—one of the products of lipids oxidation [162].

The electron-transport chain in which catalysis of DNA to TNA phosphorylation proceeds consists in general of five proteic complexes. Complexes I–IV are protein complexes with non-heme iron FeS(I, II, III, IV). They are involved in electron transfer from NADH by flavoproteins FP_{I-III} and cytochrome C_1, as well as the creation of an electrochemical gradient. Complex V is TNA-synthetase.

With age the increased deterioration or damage of mtDNA [161, 170–175], increase in the level of changed mtDNA forms (mutations), increase in the amount of oxidants produced [176, 177], and decrease in complexes I, II, IV activity [178] are observed, and also LPO is enhanced, increasing the formation of $O_2^{\bullet -}$ and H_2O_2. All this results in the loss or significant weakening of mitochondrial function.

A more detailed description of the mechanism leading to the deterioration of mitochondrial cells, consists of the following. The activity of protein complexes with non-heme iron FeS (complexes I and II) is apparently closely connected to each other. The lack in complex I caused by its deterioration or inhibiting with the appropriate substance (such as rotenone, etc.) results in the excessive using of complex II. Electrons passing in these conditions through the complex II produce four times more mitochondrial superoxide than electrons of complex I, which is accompanied with great generation of superoxide-anion.

The appropriate scheme of such excessive $O_2^{\bullet-}$ formation and its further involvement in $OH^{\bullet}$ radical formation, fatal for the cell, is shown in Scheme 4.1 [179].

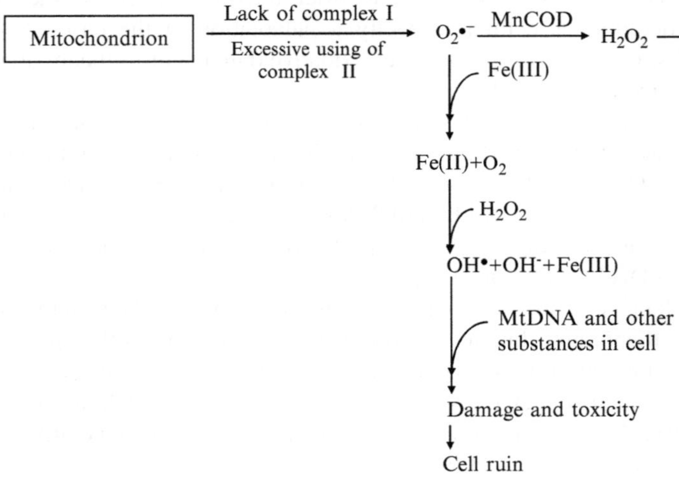

Scheme 4.1 Excessive formation of $O_2^{\bullet-}$ radical and its further transformations

It can be seen from the scheme that here, like in many other cases described above, catalytic reaction of Haber–Weiss proceeds—$O_2^{\bullet-}$ interacts with Fe(III) (usual oxidation degree of iron in the cell), turning it into Fe(II), and MnSOD generates H_2O_2. Further, the reaction of Fe(II) with H_2O_2 proceeds with the formation of $OH^{\bullet}$ radical, destroying mtDNA and other closely situated biological targets of the cell, resulting in its destruction. Thus, the most sensitive part of the electrons transfer chain in mitochondria is proteic complex I. Its damage or inhibition with any metabolism products, accumulating with age in animals or humans, must result in the reduction of mitochondrial function activity (for instance, the reduction of TNA synthesis, mutations, and other consequences). Therefore, the main cause of cells ageing is the instability of the mitochondrial genome because of the unbalance appearing with age between mitochondrial reparation (reducing) and mitochondria disorganization with free radicals of oxygen [180].

4.5.3 Mechanism of Lipofuscin Formation and Ageing

Lipofuscin (yellow-brown old age pigment) is accumulated with age in various cellular tissues of animals. Though lipofuscin accumulation is regarded as one of the ageing indexes, its importance and connection with the ageing process is not clear. Its chemical composition is as follows: 19–51% lipids, 30–50% proteins, 9–30% of acid residues of hydrolysis products, and also fluorescent substances. Lipofuscin is the final product of proteins and lipids peroxidation and polymerization, i.e., the product of cellular organ deterioration with oxygen free radicals, and is accumulated in lysosome [181]. Thus, the reactions of lipid peroxidation play an important role, which is shown, in particular, in the example of fluorescent substances of lipofuscin formation.

Peroxidation of membrane lipids catalyzed with iron compounds, resulting in a number of fluorescent products formation, is shown in Fig. 4.13 [181].

Acylic radicals ($Lip^{\bullet}$) formed from poly-unsaturated lipids (LipH) as a result of their initiation, will interact with O_2, generating $LipOO^{\bullet}$, and then by the following reaction:

$$LipOO^{\bullet} + LipH \rightarrow LipOOH + Lip^{\bullet}.$$

Lipid hydroperoxides (LipOOH), cyclic peroxides, and cyclic endoperoxides will be formed. Iron compounds will catalytically transform lipids hydroperoxides into radicals $LipO^{\bullet}$, which will then be turned into a number of products, including those of carbonyl character (malondialdehyde). Then, by reaction with proteins amino groups, free amino acids, aminophospholipids, nucleic acids, fluorescent Schiff's bases, and products of aldehyde polymerization will be formed.

There exist a number of hypotheses regarding lipofuscin genesis; one of them considers the influence of oxidative stress (i.e., the relation of pro-oxidants to antioxidants) as the main factor [182]. The confirmation of this hypothesis is iron ions and desferroxamine (complex forming agent with iron) influence on lipofuscin concentration in the miocytes of rats' hearts. Increase in iron ions content in culture has significantly enhanced the level of lipofuscin accumulation, while desferrioxamine has removed this effect. This becomes clear if we remember that iron(III) ions participation in oxidative stress by means of $O_2^{\bullet -}$ and other reducers (scorbic acid, glutathione, and also H_2O_2, by Haber–Weiss catalytic reaction, resulting in $OH^{\bullet}$ radical formation. $OH^{\bullet}$ radicals will initiate lipids peroxidation and thus cause an increase in lipofuscin concentration. Desferroxamine, being a specific ligand, makes it impossible for iron(III) to be turned into iron(II) and to realize a Haber–Weiss reaction. Thus, desferroxamine manifests the protective effect against lipofuscin accumulation [183].

Under normal conditions, lipofuscin localization in lysosome (cellular organoid serving as the enzymes reservoir taking part in protein, lipid, etc. digestion) can be expedient as secondary antioxidant protection; moreover, lipofuscin does not prevent cellular functions realization provided its content is insignificant [184]. Recently it has been shown that that lipofuscin is cytotoxic because of its ability to incorporate transition metals such as copper and iron, resulting in a redox-active surface, able to catalyze the Fenton reaction [185].

4.5.4 *The Role of Antioxidant Enzymes and Substances in Cellular Ageing*

Cells in normal conditions are strongly protected against free oxygen radicals by the whole series of antioxidant enzymes (SOD, catalase, and glutathione peroxidase), vitamins (E and C), and other reducing substances (quinones, glutathione, etc.).

Thus, Cu, Zn-SOD and Mn-SOD are contained in mitochondria, transforming $O_2^{\bullet-}$ to H_2O_2. Hydrogen peroxide is reduced by catalase to water. Glutathione peroxidase (GSP) is in the cytoplasm and mitochondrial matrix where it reduces hydroperoxides to alcohols.

Free oxygen radicals taking part in catalytic peroxidation of lipids, enzymes, proteins, and mtDNA damage, and in lipofuscin formation, have the major influence on ageing processes. In a young organism destructive processes are balanced with the action of the protective systems. With age, the first processes dominate the second.

Protective possibilities of the above-mentioned enzymes are different. Thus, it was shown in a number of experiments [186] that glutathione peroxidase is the strongest protective system, the activity of which was many times higher that of SOD and catalase. Inhibition of these three enzymes causes the enhanced cells sensitivity to free radicals influence.

It is to be noted that SOD activity and antioxidants concentration, such as urates and glutathione, turned out to be higher in long-living organisms that in short-living ones. Linear dependence was observed between the level of urates, SOD in liver, tissues, and maximal lifetime of various primates (including human). This correlation points out that long-living organisms possess higher relative contents of SOD than short-living ones [187].

However, withother antioxidants such correlation is not observed, and even inverse correlation is observed between antioxidant concentration and maximal lifetime. The reasons for this are not clear. Nevertheless, it was shown that long-living organisms possess a relatively higher level of many antioxidants, i.e., they are better protected. The rate of cerebrum substance was strongly correlated with malondialdehyde content.

The lowest level of malondialdehyde formation (organisms are protected better) is observed in long-living man, and a very high level in short-living species such as mice [188].

However, this is only one of possible factors influencing maximal lifetime. Other factors are as follows: better protection of DNA, excess of genes, detoxification processes, activity of guanylate cyclase, level of dehydroepiandrosterone, some vitamins, rate of protein transformation, lifetime of red blood cells, and other factors [189].

In many cases (model tests) the activity of such protective enzymes as catalase and glutathione peroxidase does not change with age (however, concentration of glutathione itself decreases with age). In a number of cultures the activity and SOD was unchanged, remaining the same in both young and old cells. All this points out that ageing cannot be explained by a lack of protection with antioxidant enzymes and that ageing in reality is a complex phenomenon [190, 191].

The increased sensitivity of old cells to the influence of free oxygen radicals cannot be also explained with the reducing of vitamins C and E contents in cells with age, since such reduction is low.

It is supposed that such increased cell sensitivity to free radicals is more probably caused by a general decrease of cellular metabolism and accumulation of changed cellular components by weakening of enzyme antioxidant protective system [187]. Old cells are more sensitive than young ones to stress influences and various cellular violations, which need additional energy for these violations removal. Apparently, a general decrease of cellular metabolism with age and possibilities of energy production by cells are the main factors of higher sensitivity of old cells to various kinds of stress and thus apparent activation of lipids peroxidation (as shown above), accompanied by the increase of free oxygen radicals level [187].

Thus, free radical influence on ageing is only one of the factors involved in this process. The synthesis of enzymes protecting against free radicals and maintenance of glutathione concentration need high energy consumption. With age or in stress situations the energy level produced by cells is decreased. As a consequence, the sensitivity to free radicals influence on cells increases and disorder of functions fulfilled by the cell rises, resulting at the end in ageing and death.

Another important reason for the change in normal cells condition and connected with this ageing is genetic instability. There are numerous data confirming this hypothesis from which it follows that mechanisms providing the stabilization of normal condition of the cells must play an important role in health maintenance and lifetime of various types of animals. Cancer diseases are also often connected with genetic instability, and good correlation is observed between the rate of cancer occurrence and ageing rate of various types of mammals. However, there is still not enough evidence that gene instability is the main cause of ageing. Therefore, it is supposed that ageing is not the result of genetic instability only; physiological ageing and old age diseases must have different reasons and control mechanisms [192].

Nevertheless, with age the changes in genes proceed, involving retroviruses or oncogenes and increasing with age of unusual kinds of cells—cellular metaplasia (transformation of one cell into another normal cell) [192].

Ageing is apparently a process depending on many factors such as free radicals influence, genetic instability, protection degree, etc. Long living species of mammals have a more stable genetic apparatus and better protection against reactive oxygen particles. Thus, almost linear dependence is observed between the ratio SOD/SRM (specific rate of metabolism) and potential of maximal lifetime of various mammals from mouse through ape to human. The highest value SOD-SRM (i.e., the highest protection) and the maximal lifetime ($\sim$110 years) is observed among these species in humans [192].

The connection between ageing and influence of various antioxidants and antioxidizing enzymes can be shown, to some degree, by the example of old age eye cataract occurrence. Cataract is eye crystalline lens dysfunction due to its partially or completely growing turbid.

During the 1980s about 50 million people in the world were blind because of cataracts, and more than 700 thousands of cataract removals were made annually, for which in the USA alone about 3.5 billion dollars were spent [193]. With age the number of people suffering from cataracts increases substantially. About 98% of crystalline lens solid mass is exclusively long-living protein which in the end gradually deteriorates. The reason for most of this damage is oxidative stress [194, 195].

Old proteins are accumulated, aggregated, and settled in the form of turbidity of eye crystalline lens, the source of oxidation influence being light and different highly-reactive free radicals and hydrogen peroxide [196].

The protective remedies for crystalline lens albumen are antioxidants (vitamins C, E, carotinoids, glutathione (GSH), and antioxidant enzymes, namely, glutathione peroxidase, superoxidedismutase, and catalase). The processes of cataract occurrence in eye crystalline lens and the photochemical and catalytic processes causing it are shown in Fig. 4.13.

The proteins of the eye crystalline lens under the influence of light and/or $OH^{\bullet}$ radicals generated according to the Haber–Weiss catalytic reaction are turned into oxidative-damaged proteins. Protection against these proteins' damage can be provided by antioxidant enzymes (SOD, catalase, glutathione-peroxidase) and antioxidants (vitamins C, E, carotenoids) which prevent $O_2^{\bullet-}$, H_2O_2, and Fe^{2+} accumulation, and thus prevent the formation of high-reactive $OH^{\bullet}$ radicals, and also weaken the influence of light (in Fig. 4.13 this is shown by dotted line).

Ascorbate and glutathione exist in the crystalline lens in millimolar concentrations. Thus, ascorbate level in crystalline lens is about 60 times higher than in human plasma [197], and the level of glutathione is several times higher than in blood.

Partly damaged proteins can still be removed with active proteases. However, with age active proteases are partly transformed into dysfunctioning proteases, and, therefore, the possibility of damaged proteins accumulation and aggregation appears, this fact resulting mainly in cataract occurrence with ageing.

There are a number of data showing that with age, ascorbate concentration decrease proceeds in the eye crystalline lens of animals [198] or at least their biosuitability changes on ageing. Glutathione (GSH) level also diminishes with age and in the case of cataract [199]. Also, a group of synthetic antioxidants were proved to prevent cataract.

Tocopherol and carotenoids concentration in crystalline lens is at the level of micromolar concentrations, but any change in this level with age was not proved. Some decrease of antioxidant enzymes activity with age is also observed. Thus, the weakening of protective antioxidants and antioxidizing enzymes functions proceeds with ageing.

The photooxidation process also plays a certain role in cataract occurrence.

Research has shown the possibility of some antioxidants level increase in eye crystalline lens with the help of dietary catering. The totality of research results (not always in agreement, more often contradictory) has still shown the possibility of regulation by diet of the level of ascorbate ($\geq$500 and $\geq$200 μg $\cdot$ per day for men and women, respectively), needed to reach its normal level in plasma, which is 80 μM [193].

4.6 Diseases of Joints and Tissues and Catalytic Free Radical Processes

Osteoarthritis and rheumatoid arthritis belong to the diseases connected with age. It is known that neutrophils are rapidly accumulated in the case of pathologic conditions in certain places, for instance, in inflamed joints. Thus, they will evolve into the environment high-reactionary oxygen particles, namely $O_2^{\bullet -}$, H_2O_2, HOCl, and $OH^{\bullet}$. The role of iron ions and compounds as catalysts of most of these particles formation and the interaction of the latter with joints tissue and inter-joint synovial liquid is generally recognized. It is also well known that in synovial liquid of patients with rheumatiod arthritis there are higher contents of iron compounds and ferritin from which iron is extracted than in the case of healthy people [200, 201].

There exists plenty of sources of oxygen free radical formation in vivo, as mentioned above (such as xanthinoxidase). Chondrocytes of rabbit, mouse, and bull are potential sources of free oxygen radicals. Human and bull immunoglobulin induce the formation of superoxide with the help of bull chondrocytes [202]. Normal chondrocytes from rabbit cartilage form more hydrogen peroxide than macrophages [203]. It was also shown that chondrocytes of man form $O_2^{\bullet -}$, H_2O_2, and $OH^{\bullet}$ and thus participate in cartilage destruction [204, 205]. Recent ex vivo studies have reported mitochondrial dysfunction in human osteoarthritis chondrocytes, and analyses of mitochondrial electron transport chain activity in these cells show decreased activity of complexes I, II, and III compared to normal chondrocytes [206].

In joint tissues study great importance is given to the components composition that form synovial liquid. Its predominating component is hyaluronic acid. This macromolecule determines the viscosity of synovial fluid. Its molecular weight normally reaches about 7 million. It is representative of a regular linear polymer with the structure A–B–A–B–...

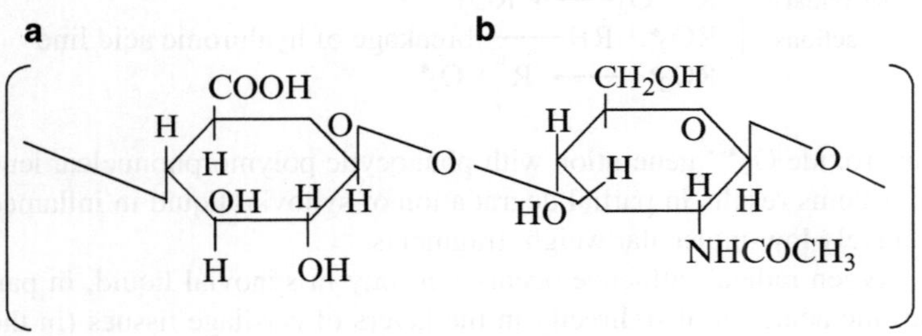

The biological role of hyaluronic acid consists in the maintenance of needed viscosity, which is created by the fact that it is able to retain a great amount of water. In rheumatoid synovial liquid the molecular weight of hyaluronic acid in average reduces to 4.8 million. Such deterioration of this polymer is realized with $OH^{\bullet}$ radicals and hypochlorous acid (HClO). The lowest value of separate fragments molecular weight after the destruction of hyaluronic acid reaches only 10,000. Therefore, the reason for a decrease in viscosity on degradation becomes clear.

The mechanism of strong destructive $OH^{\bullet}$ radicals formation is similar to that described above—this is a catalytic Haber–Weiss mechanism. The condition of the iron in rheumatoid synovial liquid is to a great degree in the form of complexes with tetradentate citrate as ligand, but is not bound with transferritin as normal, and it plays an important role in this mechanism realization. Thus, citrate and other iron(II) complexes in pathologic conditions are able to generate strong oxidizers—$OH^{\bullet}$ radicals, deteriorating hyaluronic acid.

Myeloperoxidase, released with stimulated neutrophils, in the presence of H_2O_2 will generate hypochlorous acid. The latter, having reacted with Fe(II), will generate dichloride radical ($Cl_2^{\bullet-}$):

$$\mathrm{Fe(II)} + \mathrm{HOCl} + \mathrm{HCl} \rightarrow \mathrm{Fe(III)} + \mathrm{Cl_2}^{\bullet-} + \mathrm{H_2O_2}.$$

At pH $>$ 4 dichloride ion-radicals will be rapidly retransformed into hydroxyl radicals [207]:

$$\mathrm{Cl_2}^{\bullet-} \rightarrow \mathrm{Cl}^{\bullet} + \mathrm{Cl}^{-},$$

$$\mathrm{Cl}^{\bullet} + \mathrm{H_2O} \rightarrow \mathrm{ClOH}^{\bullet-} + \mathrm{H}^{+},$$

$$\mathrm{ClOH}^{\bullet-} \rightarrow \mathrm{Cl}^{-} + \mathrm{OH}^{\bullet}.$$

At pH $\leq$ 4 ion-radicals $Cl_2^{\bullet-}$ are the main oxidative particles. These free radicals of chlorine can degrade hyaluronic acid more selectively than $OH^{\bullet}$ radicals [208].

Further fragmentation of hyaluronic acid (RH) is possible in accordance with the following (secondary) reactions [208]:

$$\mathrm{RH} + \mathrm{OH}^{\bullet} \longrightarrow \mathrm{R}^{\bullet} + \mathrm{H_2O}$$

$$\text{secondary reactions} \left\{ \begin{array}{l} \mathrm{R}^{\bullet} + \mathrm{RH} \longrightarrow \text{breakage of hyaluronic acid line,} \\ \mathrm{R}^{\bullet} + \mathrm{O_2} \longrightarrow \mathrm{RO_2}^{\bullet} \\ \mathrm{RO_2}^{\bullet} + \mathrm{RH} \longrightarrow \text{breakage of hyaluronic acid line} \\ \mathrm{RO_2}^{\bullet} \longrightarrow \mathrm{R}^{+} + \mathrm{O_2}^{\bullet-} \end{array} \right.$$

Thus, superoxide $O_2^{\bullet-}$ generation with phagocytic polymorphonuclear leucocytes or other systems results in partial degradation of synovial liquid in inflamed joints up to relatively low-molecular weight fragments.

Free oxygen radical influence occurs not only in synovial liquid, in particular, in hyaluronic acid, but also directly in the layers of cartilage tissues (in the fibers of proteoglycans and in collagen, or indirectly through the activation of latent collagenase) [209].

It was thus supposed that the destroying particle is the $OH^{\bullet}$ radical.

At the same time, hypochlorous acid generated with neutrophils will attack the joint in common with lysosomal enzymes [210, 211].

It is known that proteoglycans form aggregates with hyaluronic acid and this is important for the provision of a joint's mechanical properties. Under the influence of oxygen reactionary particles on proteoglycan, the capacity for interaction with hyaluronic acid is lost [212]. Collagen can also be degraded in solution under the influence of Haber–Weiss-type or Fenton-type reaction products (i.e., $OH^{\bullet}$ radicals [213]).

Collagen is the only protein sensitive to fragmentation with superoxide-anion. $OH^{\bullet}$ radicals in the presence of O_2 molecules will disintegrate the collagen in small peptides, and disintegration proceeds in a specific way to the remainders of proline or 4-hydroxyproline. It is interesting that hydroxyl radicals in the absence of oxygen or hypochlorous acid do not induce fragmentation of collagen molecules, but realize their polymerization by the formation of new-cross-binding bridges, though their nature is not clear [214].

In synovial liquid of patients with arthritis, increased contents of malondialdehyde are observed, this fact testifying to the activation of lipids peroxidation, primarily of unsaturated aliphatic acids such as arachidonic acid [215].

Chromosomal deteriorations (DNA deteriorations) were also observed in the case of patients with rheumatoid arthritis caused by the presence of high amounts of superoxide compares to the norm [203].

The fact that macromolecular substances of joint tissue (hyaluronic acid, collagen, elastin, proteoglycans, etc.) are unprotected against $OH^{\bullet}$ radicals (and also, possibly, against OCl^{-}, chloramines, and peroxides) causes the appropriate degradation of these substances.

The data on concentration change of protective antioxidant enzymes in the case of patients with rheumatoid arthritis are uncertain [203], although there are indications that their level in this case is decreased [216, 217].

The decreased level of antioxidant enzymes C and E and that of selenium (necessary element of the active center of antioxidant glutathione peroxidase) [203] is recognised. With regard to such patients, attempts at antioxidant therapy were undertaken. Sometimes such therapy with Cu, Zn-SOD solutions injection were successful, as in the case of knee joint osteoarthritis. In the case of rheumatoid arthritis it was less favorable. The imitation of SOD was also undertaken with the help of copper salicylate and D-penicillamine solutions in relation to patients with rheumatoid arthritis. However, these and other attempts at antioxidants introduction did not result in resolution of this arthritis problem and the many factors from the side effects [203].

4.7 Catalytic and Radical Processes on Atherosclerosis and Myocardial Ischemia

4.7.1 Catalytic and Radical Processes on Atherosclerosis

Numerous theories (endothelial and virus damage, thrombogenesis, lipid peroxidation, etc.) explaining atherosclerotic damage, do not provide explanations for their pathogenesis. There are many risk factors with atherosclerosis occurrence, and among these is the presence of high concentrations of cholesterol of low density lipoprotein (LDL).

Cardiovascular diseases (as well as the above considered stress and many others) are accompanied by the processes of lipids peroxidation activation (LPO) in blood. More and more data are accumulated, showing that LDL oxidation modification (change) can play an important (maybe determining) role in atherosclerosis [218, 219].

An understanding of initiation mechanism and damage progression becomes clear. From the large number of problems dealing with atherosclerosis, only a few will be considered (mainly LPO reactions) related to catalytic processes (enzymatic and non-enzymatic) resulting in atherosclerosis formation. In this problem statement many materials are used, but the main information and the order of consideration corresponds to the work of Fruchart and Duriez [220].

In the arteries, endothelial surface makes contact with highly oxygenated blood. Relatively high O_2 pressure in endothelium makes it a potential source of oxyradicals [221]. Small amounts of $O_2^{\bullet-}$ and H_2O_2 are products of normal mitochondrial activity [222].

The activity of xanthinoxidase, cyclooxygenase, lipoxygenase, and other oxy radical generating systems is high, like the rate of autooxidation reactions catalyzed by transition metal ions [223]. Reduction of oxygen pressure can also result in $O_2^{\bullet-}$ and H_2O_2 formation [224]. Reactions of oxyradicals, when their concentration becomes too high in relation to the protective structures (SOD, catalase, glutathione peroxidase) can lead to tissue deterioration [225, 226].

LDL resembles a large spherical particle with mean diameter of 22 nm and molecular weight of about 2.5 million Da, consisting of about 1,600 molecules of cholesterol ether, 600 molecules of free cholesterol, 700 molecules of phospholipids, and 170 molecules of triglycerides [227]. The number of various aliphatic acids making up the different LDL lipids amounts to about 3,000, about one half of them being polyunsaturated aliphatic acids (PUFAS), mainly linoleic acid and small amount of arachidonic and other aliphatic acids.

The following process of LDL oxidation is supposed [228]. LDL oxidation in the cells is in principle a free radical process. Thus, polyunsaturated aliphatic acids contained in LDL will be turned into lipid hydroperoxides and lipid radicals (LipOOH, Lip$^\bullet$, LipOO$^\bullet$), by means of LPO. Insignificant amounts of preliminary formed hydroperoxides (LipOH), which were discovered in LDL, are catalytically decomposed with trace amounts of transition metal compounds (Cu^{2+} or Fe^{2+}) to the reactive alcoxyradicals (LipO$^\bullet$) and peroxyradicals (LipOO$^\bullet$):

$$\mathrm{LipOOH} + \mathrm{Cu}^{2+} \rightarrow \mathrm{LipOO}^\bullet + \mathrm{Cu}^+ + \mathrm{H}^+$$

$$\underline{\mathrm{LipOOH} + \mathrm{Cu}^{+} \rightarrow \mathrm{LipO}^\bullet + \mathrm{Cu}^{2+} + \mathrm{OH}^-}$$

$$2\mathrm{LipOOH} \rightarrow \mathrm{LipOO}^\bullet + \mathrm{LipO}^\bullet + \mathrm{H_2O}.$$

These $LipO^{\bullet -}$ or $LipOO^{\bullet -}$radicals will initiate reactions of lipid peroxidation in which polyunsaturated aliphatic acids LDL are involved. In the presence of transition metal ions traces (mainly $Fe^{2+,3+}$ and Cu^{2+} ions) lipids hydroperoxides are constant sources of new alcoxy- and peroxy radicals, which initiate new chain reactions according to the new general scheme of lipids catalytic peroxidation considered above.

Lipid alcoxyradicals ($LipO^{\bullet}$) are decomposed into reactive aldehydes such as malondialdehyde, hydroxyalkenals, 2-alkenals, 2,4-alkenals, and alkanals. These oxyradicals also cause degradation of apolipoprotein B (apoB) into smaller protein fragments.

ApoB is a large protein with a molecular weight of about 500,000 Da. It contains in general 356 remainders of lysine, and some of them with positively charged ε-amino groups are responsible for apoB joining to LDL receptor. The modification of these amino groups of lysine (for instance, acylation or conjugation with MDA or other aldehydes) essentially increases their binding with macrophages' receptors [227].

Peroxidation products formed, aldehydes, are also bound in a covalent way with the side chains of apoB aminoacids. Thus, the negative charge of side amino acid chains of apoB is increased, and by this fact they can be recognized by the receptors of cleaning macrophages, that in the end will result in the formation of foamy cells (decisive stage in atherogenesis), greases destruction, and follicle formation (Fig. 4.12 taken from [227]).

On arterial walls activated macrophages can evolve $O_2^{\bullet -}$, H_2O_2, and hydrolytic enzymes [228]. These reactive species take part in LDL oxidation reactions and induce LPO [228]. Thus, LDL must be oxidized on the walls of arteries earlier than it can be caught with the receptors of monocytes/macrophages in order to generate "foamy" cells. Some forms of oxidized LDL cause macrophage transformation into "foamy" cells [229–231], which are component parts of atherosclerotic follicles on the walls of blood vessels [232]. Essential modification (oxidizing) of LDL in the cell is realized only in the presence of at least a small amount of copper or iron ions [233]. In higher concentrations they can serve as good catalysts of lipid oxidation in lipoproteins and in the absence of cells [230, 234], a fact which explains the often used modeling of oxidized lipoproteins obtained in vitro by copper-catalyzed oxidation in non-cellular medium [235].

Recent research [235] has shown that in various classes of lipoproteins, oxidation in LPO reactions take part only in the presence of complex-bound copper ions, bound with lipoprotein particles. This can be explained by the fact that only appropriate copper complexes possess a sufficient redox potential for radicals formation.

In LDL particles endogenous antioxidants are contained, mainly α-tocopherol and to a lesser degree γ-tocopherol and carotenoids, which protect polyunsaturated aliphatic acids from oxidation. The duration of lag-phase in LDL oxidation is linear and dependent on α-tocopherol content in LDL.

Insufficient LDL protection from free radicals results in LPO and a progressive loss in vitamin E and carotenoid within a short time. LPO products

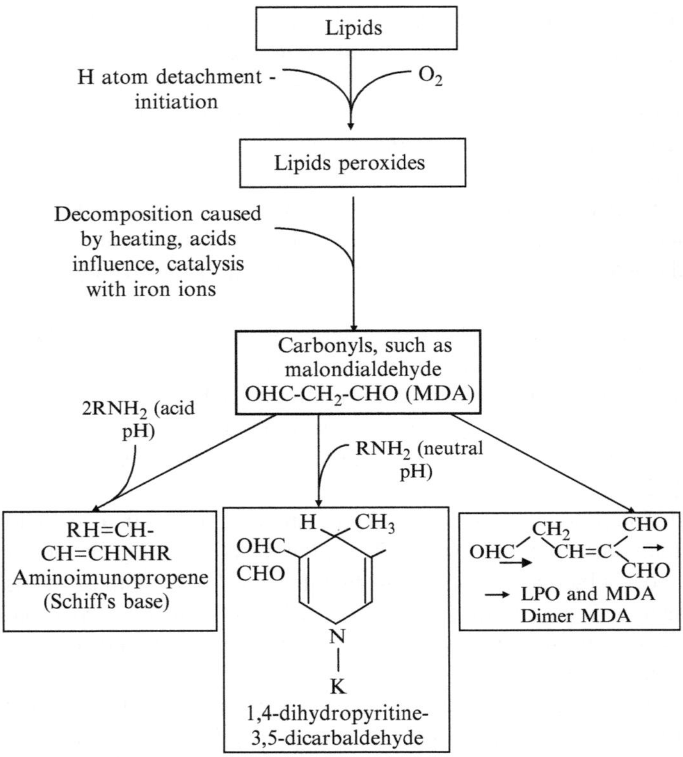

Fig. 4.12 Peroxidation of membrane lipids

are large amounts of aldehydes, namely 4-hydroxyhexanal, 4-hydroxyoctenal, 4-hydroxynonenal, propanal, butanal, pentanal, hexanal, 2,4-heptadienal, and malondialdehyde. LDL oxidation is initiated both with cells and transition metal ions, such as Cu^{2+}. Oxidized LDL (O_x-LDL) is toxic in relation to endothelial and plain-muscular cells, resulting in the deterioration of endothelial fibroblast layer of vascular system and fibroblasts. O_x-LDL, MDA and other LPO products take part in atherosclerotic damage. This confirms that LPO and aldehyde products play an important role in atherogenesis process' initiation.

Early manifestation of atherosclerosis is tight binding of blood monocytes with endothelial cells of arteries. Then, monocytes bound thus migrate in intima under the influence of chemotoxic factors. Monocytes inculcated into intima attract O_x-LDL. Inside the arterial wall monocyte is subject to phenotypic modification [236], turning into the macrophage's phenotype. Macrophage turning back into plasma is inhibited with oxidized LDL. Transformation of monocyte into macrophage can be made easier under the influence of minimal amounts of O_x-LDL [237].

LDL oxidation can take place only when LDL molecules are introduced into the wall of coronary vessels, and thus contact is provided with radical-generating

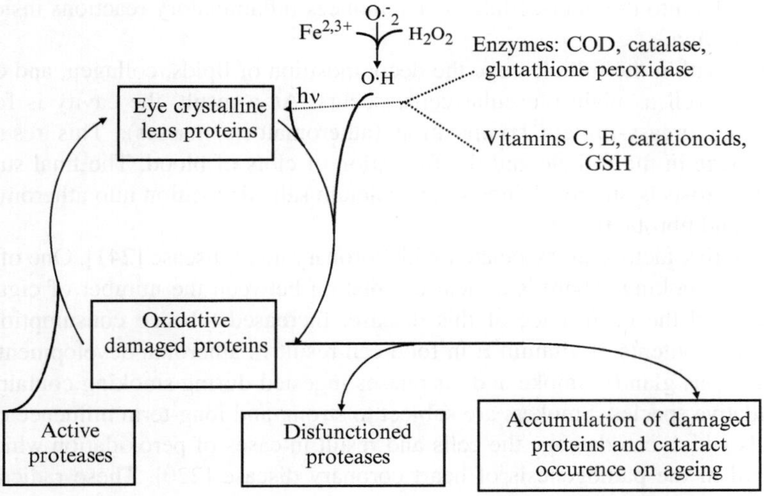

Fig. 4.13 Processes of proteins transformations of eye crystalline lens and cataract occurrence on ageing

cells (monocytes, macrophages). In the presence of copper or iron ions or on their addition, catalytic LPO processes begin. Thus, conjugated dienes are formed [236], which promote the occurrence of coronary atherosclerosis. LDL oxidation is prevented with antioxidants, such as vitamin E.

Monocytes transformation into macrophages occurs inside arterial intima [238]. These cells are characterized (in spite of evolving the reactive oxygen oxidants of protein) by high rates of phagocytosis, increased generation of superoxide anion, various enzymes excretion (collagenase, elastase, lipoprotein lipase), synthesis of a number of bioactive lipids [239], and involvement in foamy cells genesis. Foamy cells of macrophage origin are predominant in greasy strata, though not every greasy stratum will be turned into atherosclerotic follicles [65].

Thus, LDL is the main source of ligand infiltration into arterial walls, which precedes the formation of fibrotic follicles.

On progressive damage, cellular necrosis is realized, resulting in lipid release from foamy cells. Provided the supply of LDL particles and forming O_x-LDL exceeds the ability of macrophages receptors to remove O_x-LDL, O_x-LDL will be accumulated inside the arterial intima. As mentioned above, O_x-LDL is cytotoxic, inducing both damage and ruin of endothelial and plain-muscular cells, and also macrophages [240]. Ruin of fragile foamy cells of macrophage origin results in the release from them of lipids into the inter-cellular space of intima, giving rise to the growth of atherosclerotic follicle cores, rich in cholesterol ethers. Ruin and breakage of foamy cells also releases a number of lysosomal enzymes which can later damage the cells of surrounding intima. The release of a number of lipids from foamy cells into the inter-cellular space induces inflammatory reactions inside the arterial intima.

Thus, within the follicle zone the decomposition of lipids, collagen, and elastic fibers, as well as plain muscular cells, occurs. As a result, the cavity is formed containing greasy-protein composition (atheromatosis masses). This results in hemorrhage in the follicle and the formation of clots of blood. The final stage of atherosclerosis is atherocalcinosis, i.e., calcium salts deposition into atheromatosic masses and fibrotic tissue.

Many risk factors are associated with coronary heart disease [241]. One of these factors is smoking. There is a clear correlation between the number of cigarettes smoked and the occurrence of this disease. Increased tobacco consumption and decreased contents of vitamin E in food can result in atheroma development (cyst of sebaceous gland). Smoke and tar phases ingested during smoking contain a lot of oxidative species. Smokers are subject to strong and long-term influence of free radicals which can damage the cells and result in cases of peroxidation which are involved in the pathogenesis of heart coronary disease [220]. These radicals are subdivided into two different groups [220].

The first group represents long-living quinone–semiquinone radicals ($Q^{\bullet}$), which are generated on polycyclic aromatic hydrocarbons oxidation during the combustion process. The $Q^{\bullet}$ radical can reduce the oxygen molecule to superoxide ($O_2^{\bullet-}$) and hydrogen peroxide, and can also catalyze H_2O_2 transformation to the very destructive hydroxyl radical.

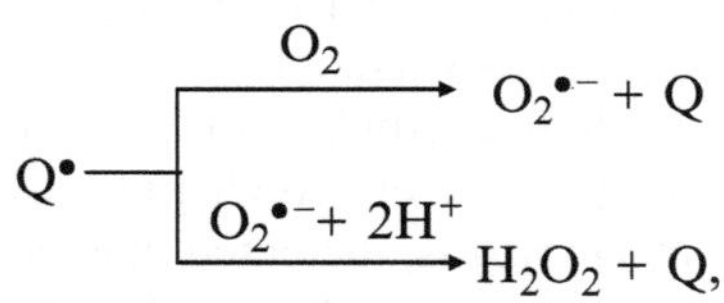

The second group represents short-living (less than 1 s) peroxy radical ($ROO^{\bullet}$) and other radicals. These short-living radicals can be formed during smoking by the slow oxidation of nitrogen monoxide (NO) to nitrogen dioxide (NO_2). The latter reacts with the components forming during smoking, such as aldehydes and olefins, with $ROO^{\bullet}$ formation. Moreover, NO and NO_2 can react with H_2O_2 forming the hydroxyl radical:

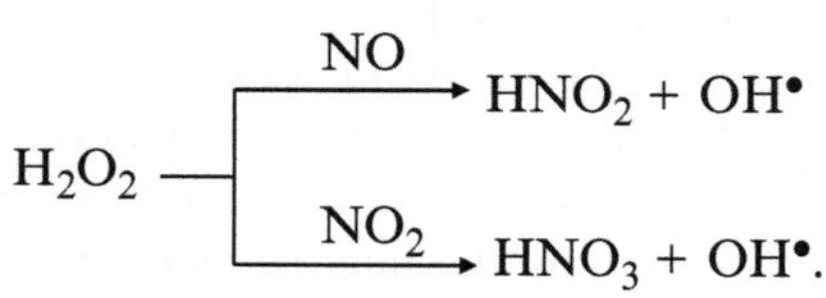

Oxidants production ($O_2^{\bullet-}$, H_2O_2) by phagocytes can be stimulated by nicotine. Cigarette smoke also contains Cu^{2+} and Fe^{2+} ions, which may catalyze $OH^{\bullet}$ formation by Fenton-type reaction. This smoke contains plenty of aldehydes, including formaldehyde, acrolein, acetaldehyde, propionaldehyde, and butyraldehyde. Their influence on LDL (along with the other factors considered above) causes its oxidation (O_x-LDL is formed with consequent results) [211]. LDL, extracted from the plasma of smokers, under the influence of macrophages' cultures, was degraded more rapidly in comparison with non-smokers, and was more sensitive to oxidation with plain-muscular cells cultures. The results of this research have confirmed the unfavorable effect of smoking on atherosclerosis occurrence.

As noted earlier, native LDL is very sensitive to metal-catalyzed oxidation or oxidation with reactive oxygen radicals evolved by neutrophils and macrophages [220]. This process is accompanied by the formation of O_x-LDL. In its turn, O_x-LDL induces release of endotheline (peptide consisting of 21 amino acids) from endothelial cells and macrophages, which makes vessel spasm occurrence more likely (including coronary) and the turning of a greasy layer into atheroma.

As to the mechanisms of LDL oxidative modification, it can be supposed that they correspond in general to the overall catalytic scheme of lipids peroxidation (which has been considered above), although there can be specific differences in some stages, which are to be established on the basis of advanced experiments.

It is also to be noted that in vivo initiator of LPO in lipoproteins can be released from erythrocytes hemoglobin which interacting with such oxidizers as hydrogen peroxide or hypochlorite ($HOCl/OCl^-$) and can result in the formation of stronger oxidants—methemoglobin (MetHb) and/or ferrylhemoglobin ($Hb\text{–}Fe^{IV} = O$) [242]. All these products (oxidants) induce LPO in low density lipoproteins (LDL).

4.7.2 *Myocardial Deteriorations and Catalytic Processes with Free Radicals Formation*

It is known that one of the most important causes of myocardial ischemia is insufficient supply of the latter with oxygen, which takes place with myocardial cell damage. In particular, it is observed in coronary atherosclerosis considered above. The damage of myocardial cells, as supposed, is caused by generation of toxic reactive oxygen species, namely superoxide anion radical, hydrogen peroxide, and hydroxyl radicals [243–247]. Such a statement is of great interest in myocardial surgery and heart transplantation [248, 249].

Reactive oxygen species can form both intracellular (mitochondria, xanthine oxidase) and intercellular sources (neutrophils, macrophages) [250–253]. If on normal aerobic metabolism myocardium generates most of its energy for cell needs and total catalytic reduction of O_2 to H_2O_2, then in the case of ischemia incomplete oxygen reduction proceeds in cellular respiration, and as a result oxy radical species are formed. $O_2^{\bullet -}$ appeared as a product of this process, and it can initiate a number of radical chain processes among both intercellular and intracellular species [37, 254–256], as shown in Fig. 4.14 (slightly modified and simplified figure taken from [257]).

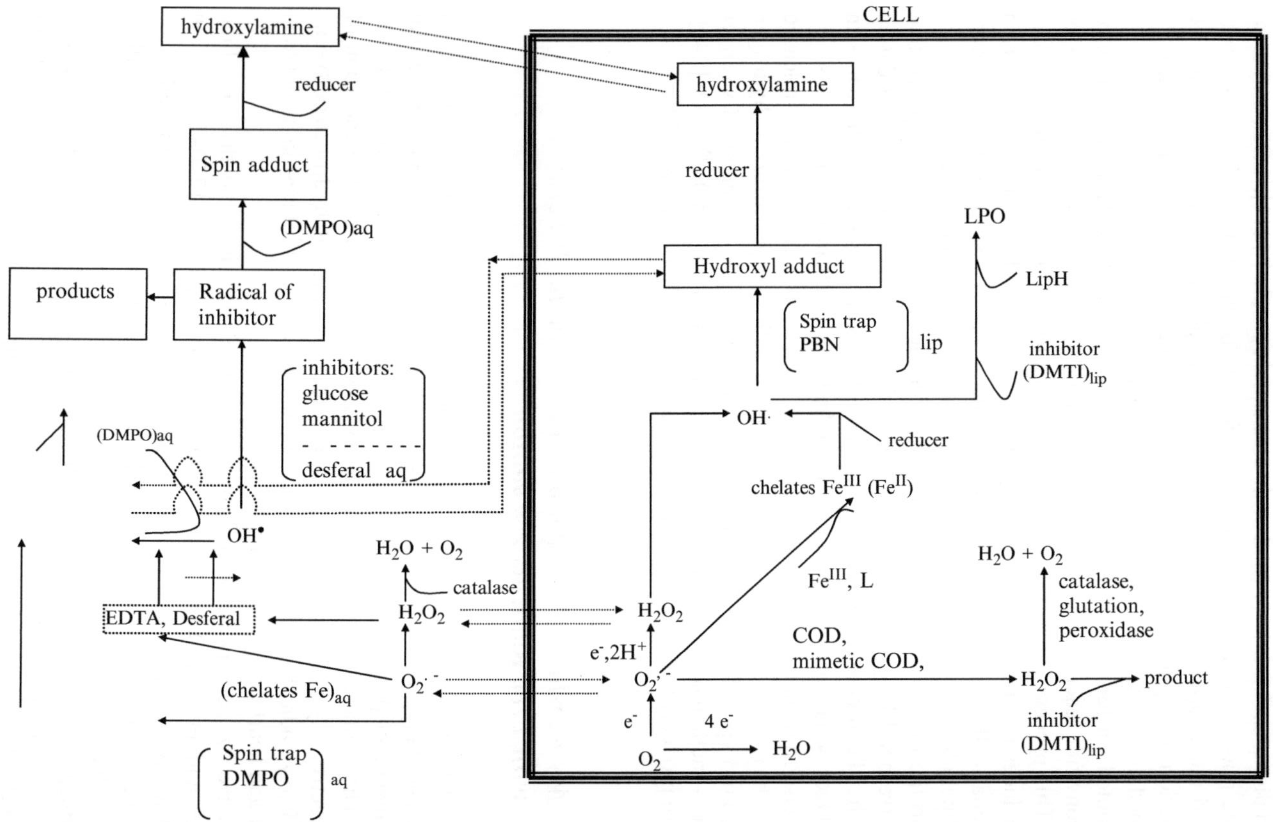

Fig. 4.14 Simplified model of catalytic and radical processes in myocardium tissues on ischemia, and also the influence of inhibitors, ligands, and spin traps on these processes

Reactive oxygen radicals appearing in intercellular space in the water phase are unable to react with the components of cell membrane. This can be seen from the fact that $OH^{\bullet}$ radicals formed catalytically in this space are inefficient in peroxidation initiation, while intracellular OH are very efficient in membranes lipid peroxidation initiation [257]. Thus, it becomes clear that cleaners of $OH^{\bullet}$ radicals such as mannitol or glucose can eliminate them without LPO inhibition. Cleaners of $OH^{\bullet}$ such as lipid-soluble dimethylthiourea and antioxidant butylated hydroxytoluene (BHT) can inhibit LPO (Fig. 4.13).

Small amounts of such species as superoxide anion and hydrogen peroxide can partly pass through the membrane in both directions (in Fig. 4.13 this is shown with dotted arrows). Catalase and glutathione peroxidase can deteriorate intracellular H_2O_2. Lipophilic dimethylthiourea (DMTU) can also remove intracellular H_2O_2 and $OH^{\bullet}$ [258].

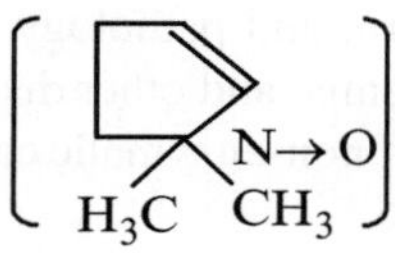

can catch $OH^{\bullet}$ radical only, formed in intercellular space (Fig. 4.13), while another lipophyl spin trap PBN

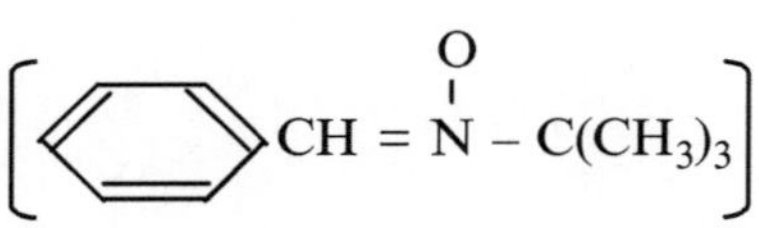

can catch $OH^{\bullet}$ radical formed inside the cell (Fig. 4.13). Spin adducts thus formed ($OH^{\bullet}$ with spin trap) can be subject to subsequent enzymatic and chemical reductions to hydroxylamines [259].

As repeatedly mentioned above, ligands, forming chelates with metal ions, play an important role in oxidant production by the Haber–Weiss catalytic reaction. Thus, under these conditions either $OH^{\bullet}$ radicals or ferryl complexes $L_n(Fe^{IV} = O)^{2+}$ can be formed depending on the nature of the chelate-forming agent [260, 261].

In myocardial ischemia favorable conditions are created for the formation of great amounts of these strong oxidants and for their influence on myocardial cells. While EDTA introduction can stimulate or inhibit the formation of intercellular $OH^{\bullet}$, introduction of desferal in low concentrations contributes to the inhibition of both inner- and intercellular $OH^{\bullet}$. At high concentrations desferal can act as a cleaner of intercellular $O_2^{\bullet-}$ and $OH^{\bullet}$ [262] (Fig. 4.14). Information obtained helps to understand at least generally the complexity of catalytic processes proceeding in myocardial ischemia, and on this basis can lead to improvements in the treatment of this disease [257].

The following is to be noted in conclusion:

1. The double role of the oxygen molecule: (1) as substance maintaining life by macroenergetic compound TNA reaction with mitochondria in the respiratory chain and its enzymatic system, realizing normal functioning of the organism as a whole, and its protection through the generation of small amounts of reactive oxygen species; (2) as a substance which can be catalytically activated with the formation of great amounts of free oxygen radicals and reactive species ($O_2^{\bullet-}$, $OH^{\bullet}$, H_2O_2, $L_n(M^{IV}=O)^{2+}$, $RO^{\bullet}$, $ROO^{\bullet}$, ROOH), exceeding the ability of the cells' and organism's protective systems and resulting in damage (diseases), ageing, and ruin of cells, organs, and the organism as a whole.
2. Uniformity of organism protective systems action (SOD, catalase, glutathione peroxidase, myeloperoxidase, vitamins C and E, carotenoids) both in norm and pathology.
3. Universal LPO process in norm and pathology (stress, inflammatory and joint diseases, atherosclerosis, ischemia, and other diseases).
4. Universal and the same type of non-enzymatic catalytic processes (Haber–Weiss and Fenton's reactions) in vivo.
5. Organism capacity to use relatively small set of means and substances for realization of numerous processes in vivo in norm and pathology, regulating these processes by changing the concentrations of these substances, their combination, ratios, sequences of their action, and other relatively simple ways.

References

1. Yu.A. Vladimirov, A.I. Archakov, *Peroxide Oxidation of Lipids in Biologic Membranes* (Nauka, Moscow, 1972), p. 252 [Russian]
2. W.H. Koppenol, in *Free Radical Damage and its Control*, ed. by C.A. Rice-Evans, R.H. Burdon (Elsevier, Amsterdam, 1994), p. 3
3. D.C. Harris, P. Aisen, Biochim. Biophys. Acta **329**, 156 (1973)
4. A.P. Purmal, Yu.I. Skurlatov, S.O. Travin, Izv. AN SSSR [Proc. Acad. Sci. URSS] **3**, 492 (1980)
5. G.R. Buettner, T.P. Doherty, L.K. Patterson, FEBS Lett. **158**, 143 (1983)
6. C. Bull, G.J. Mccline, J.A. Fee, J. Am. Chem. Soc. **105**, 5290 (1983)
7. S. Goldstein, G. Czapski, Inorg. Chem. **24**, 1087 (1985)
8. J. Weinstein, B.H.J. Bielski, J. Am. Chem. Soc. **102**, 4916 (1980)
9. J. Butler, W.H. Koppenol, E. Margoliash, J. Biol. Chem. **257**, 10747 (1982)
10. L.I. Simandi (ed.), *Advances in Catalytic Activation of Dioxygen by Metal Complexes* (Kluwer, Dordrecht, 2003), p. 159
11. T. Bugg, *Introduction to Enzyme and Coenzyme Chemistry* (Blackwell, Oxford, 2004), p. 23
12. M. Velayutham, C. Hemann, J.L. Zweier, Free Radic. Biol. Med. **51**(1), 160–170 (2011)
13. J. Bylund, K.L. Brown, C. Movitz, C. Dahlgren, A. Karlsson, Free Radic. Biol. Med. **49**(12), 1834–1845 (2010)
14. A. Cross, O.G. Jones, Biochim. Biophys. Acta **1057**, 281 (1991)
15. J. Chaudiere, in *Free Radical Damage and its Control*, ed. by C.A. Rice-Evans, R.H. Burdon (Elsevier, Amsterdam, 1994), pp. 25–65
16. C.A. Rice-Evans, in *Free Radical Damage and its Control*, ed. by C.A. Rice-Evans, R.H. Burdon (Elsevier, Amsterdam, 1994), pp. 131–153

17. M.A. Marletti, M.A. Tayeh, J.M. Hevel, Biofactors **2**, 219 (1990)
18. R. Radi, J.S. Beckman, K.M. Bush et al., J. Biol. Chem. **266**, 4244 (1990)
19. J.S. Beckman, T.W. Beckman, J. Chen et al., Proc. Natl. Acad. Sci. USA **87**, 1620 (1990)
20. I. Fridovich, J. Biol. Chem. **264**, 7761 (1989)
21. I.A. Implay, I. Fridovich, J. Biol. Chem. **266**, 6957 (1991)
22. P.R. Gardner, I. Fridovich, J. Biol. Chem. **266**, 19328 (1991)
23. W.F. Petrone, D.K. Englich, K. Wong, J. Mccord, Proc. Natl. Acad. Sci. USA **77**, 1159 (1980)
24. G. Czapski, S. Goldstein, D. Meyerstein, Free Radic. Res. Commun. **4**, 231 (1988)
25. I.B. Afanasev, *Superoxide Ion: Chemistry and Biological Implications*, vol. 1, (CRC, Boca Raton, 1990), p. 279
26. G. Czapski, S. Goldstein, Bioelectrochem. Bioenerg. **18**, 21 (1987)
27. G.A.C. Murrell, M.J.O. Francis, L. Bromley, Biochem. J. **265**, 659 (1989)
28. B. Meier, H.H. Radeke, S. Selle, M. Younes, H. Sies, K. Resch, G.G. Habermehl, Biochem. J. **263**, 539 (1989)
29. W.E. Lands, R.R. Keen, in *Biological Oxidation Systems*, vol. 2, ed. by C.C. Reddy, G.A. Hamilton, K.M. Madystha (Academic, London, 1990), pp. 657–665
30. A.B. Weitberg, D. Corvese, Carcinogenesis **10**, 1029 (1989)
31. C. Von Sonntag, *The Chemical Basis of Radiation Biology* (Taylor & Francis, London, 1987) p. 515
32. A. Benedetti, M. Comporti, H. Esterbauer, Biochim. Biophys. Acta **620**, 281 (1986)
33. R. Schreck, P. Rieber, A. Baeverle, EMBO J. **10**, 2247 (1991)
34. M. Roederer, P.A. Raju, F.J.T. Staal et al., AIDS Res. Hum. Retrovir. **7**, 563 (1991)
35. S. Legrand-Poels, D. Vaira, J. Pincemail et al., AIDS Res. Hum. Retrovir. **6**, 1389 (1990)
36. B. Halliwell, J.M.C. Gutterige, C.E. Cross, J. Lab. Clin. Med. **119**, 598 (1982)
37. B. Halliwell, J.M.C. Gutterige, Biochim. J. **219**, 1 (1984)
38. B. Halliwell, J.M.C. Gutterige, in *Medical, Biochemical and Chemical Aspects of Free Radicals*, ed. by O. Hayaishi, E. Niki, M. Condo, T. Yoshikawa (Elsevier, Amserdam, 1989), p. 21
39. A.Ya. Sychev, V.G. Isak, Uspekhi khimii [Russ. Chem. Rev.] **64**, 1183 (1995) [Russian]
40. J.M.C. Gutterige, Biochem. J. **224**, 761 (1984)
41. C.P. Moorhouse, B. Halliwell, M. Grootveld, J.M.C. Gutterige, Biochim. Biophys. Acta **843**, 261 (1985)
42. I. Dar, A. Barzilai, The DNA Damage Response: Implications on Cancer Formation and Treatment, in *Cellular Response to Oxidative Stress*, ed. by K.K. Khanna, Y. Shiloh (Springer, Dordrecht, 2009), pp. 65–79
43. A.R. Reddi, L.T. Jensen, A. Naranuntarat, L. Rosenfeld, E. Leung, R. Shah, V.C. Culotta, Free Radic. Biol. Med. **42**(2), 154–162 (2009)
44. G. Liu, S.I. Feinstein, Y. Wang, C. Dodia, D. Fisher, K. Yu, Y.-S. Ho, A.B. Fisher, Free Radic. Biol. Med. **49**(7), 1172–1181 (2010)
45. T.F. Slater, K.H. Cheeseman, in *Reactive Oxygen Species in Chemistry, Biology and Medicine*, ed. by A. Quintanilha (Plenum, London, 1988), pp. 1–14
46. H. Selye, *The Physiology and Pathology of Exposure to Stress* (Acta Inc. Medical Publishers, Montreal, 1950), p. 1025
47. Dipak K. Das (ed.), *Stress Adaptation, Prophylaxis and Treatment* (Kluwer, Dordrecht, 1999), p. 160
48. J.L. Holtzman, *Atherosclerosis and Oxidant Stress* (Springer, Berlin, 2007), p. 25
49. V.A. Baraboi, I.I. Brechman, V.G. Golovin, Yu.B. Kudreashev, *Peroxide Oxidation and Stress* (Nauka, Sankt-Petersburg, 1992) p. 148 [Russian]
50. P. Willmer, G. Stone, I. Johnston, *Environmental Physiology of Animals* (Blackwell, Malden, 2005), p. 6
51. Francis J. Turner, *Adult Psychopathology: A Social Work Perspective* (Simon and Schuster, New York, 1999) p. 696
52. B.E. Melnic, G. Tâbârnă, G.G. Duca, S. Gica, *Chimia, Stresul i Tumoarea* (Universul, Chiinu, 1997), p. 237

53. F. Shahidi, Y. Zhong, Chem. Soc. Rev. **39**, 4067–4079 (2010)
54. M.L. Katz, W.G. Kobson, E.A. Dratz, in *Free Radicals in Molecular Biology, Ageing and Disease*, ed. by D. Armstrong et al. (Raven, NY, 1984), pp. 163–180
55. K.N. Prasad, *Micronutrients in Health and Disease* (CRC/Taylor & Francis Group, Boca Raton, 2011), p. 391
56. L.R.C. Barclay, S.J. Locke, J.M. Mcneil et al., J. Am. Chem. Soc. **106**, 2479 (1984)
57. K. Jomova, M. Valko, Toxicology **283**(2–3), 65–87 (2011)
58. W.E. Lands, R.R. Keen, in *Biological Oxidation Systems*, vol. 2, ed. by C.C. Reedy, G.A. Hamilton, A. Madyastha (Academic, London, 1990), pp. 657–665
59. K.W. Lanks, Exp. Cell Res. **65**, 1 (1986)
60. Michael R. Barratt, Sam L. Pool, *Principles of Clinical Medicine for Space Flight* (Springer, New York, 2008), p. 596
61. B.N. Tarausov, A.I. Polivoda, A.I. Zuravlev, Biofizika [Biophysics] **6**, 490 (1981) [Russian]
62. E.B. Burlakova, N.G. Khrapova, Uspekhi khimii [Russ. Chem. Rev.] **54**, 1540 (1985) [Russian]
63. V.E. Kagan, O.N. Orlov, L.L. Prilipko, Itogi nauki i tehniki. Ser. Biofizika [Adv. Sci. Technol. Biophys. Ser.] **18**, 136 (1986) [Russian]
64. D. Armstrong, in *Free Radicals in Molecular Biology, Ageing and Disease*, ed. by D. Armstrong et al. (Raven, New York, 1984), pp. 163–180
65. A.L. Tappel, C.J. Dillard, Fed. Proc. **40**, 174 (1981)
66. H. Esterbauer, in *Free Radicals in Liver Injury*, ed. by G. Poli, K. Cheeseman, M.U. Dianzani, T. Slater (IRL, Oxford, 1985), pp. 29–47
67. E. Niki, Lipid peroxidation: physiological levels and dual biological effects. Free Radic. Biol. Med. **47**(5), 469–484 (2009)
68. U.B. Hendgen-Cotta, U. Flögel, M. Kelm, T. Rassaf, Unmasking the Janus face of myoglobin in health and disease. J. Exp. Biol. **213**, 2734–2740 (2010)
69. M. Doly, B. Bonhomme, J.C. Vennat, Ophthal. Res. **18**, 21 (1986)
70. C.A. Rice-Evans, G. Okunade, R. Khan, Free Radic. Res. Commun. **7**, 45 (1989)
71. J.J.O. Turner, C.A. Rice-Evans, M.J. Davies, E.S.R. Newman, Biochem. J. **377**, 833 (1991)
72. S. Harel, J. Kanner, Free Radic. Res. Commun. **3**, 309 (1987)
73. J. Dee, C.A. Rice-Evans, S. Obeyesekera, S. Meraji, M. Yacobs, K.R. Bruckdorfer, FEBS Lett. **294**, 381 (1991)
74. G. Paganga, C.A. Rice-Evans, R. Rule, D. Leake, FEBS Lett. **303**, 154 (1992)
75. C.A. Rice-Evans, E. Green, G. Paganga, C. Cooper, J. Wrigglesworth, FEBS Lett. **326**, 177 (1993)
76. D. Steinberg, S. Parthasarthy, T.E. Carew, J.C. Khoo, J.L. Witztum, New Engl. J. Med. **320**, 915 (1989)
77. T. Schewe, C. Hiebsch, P. Ludwig, S.M. Rapoport, Biomed. Biochim. Acta **815**, 426 (1983)
78. B.J. Reeder, Antioxid. Redox Signal. **13**(7), pp. 1087–1123 (2010)
79. B. Halliwell, Free Radic. Res. Commun. **9**, 1 (1990)
80. R.R. Crichton, M. Charlototeaux-Wauters, Eur. J. Biochem. **164**, 485 (1987)
81. E.D. Weinberg, Drug Metab. Rev. **22**, 531 (1990)
82. R. Biemond, H.G. Vaneijk, A.J.G. Swaak, J.F. Koster, J. Clin. Invest. **73**, 1576 (1984)
83. B.J. Bolann, M. Saran, W. Bors, Free Radic. Res. Commun. **2**, 369 (1987)
84. M. Odeh, New Engl. J. Med. **324**, 1417 (1991)
85. A.T. Diplock, in *Free Radical Damage and its Control*, ed. by A.C. Rice-Evans, R.H. Burdon, (Elsevier, Amsterdam, 1994), pp. 113–129
86. W.I.M. Vonk, C. Wijmenga, R. Berger, B. Van De Sluis, L.W.J.J. Klomp, Biol. Chem. **285**, 28991–29000 (2010)
87. V.I. Pérez, H.V. Remmen, A. Bokov, C.J. Epstein, J. Vijg, A. Richardson, Aging Cell **8**(1), 73–75 (2009)
88. S.L. Marklund, G. Beckman, T. Stigbrand, Cancer Res. **42**, 1955 (1982)
89. G.R. Buettner, B.A. Jurkiewicz, Free Radic. Biol. Med. **14**, 49 (1992)

90. L. Flohe, in *Glutathione: Chemical, Biochemical and Medical Aspects. Part A*, ed. by D. Dolphin, O. Avramovic, R. Poulson (Wiley, New York, 1980), pp. 643–731
91. J. Blum, I. Friedovich, Arch. Biochem. Biophys. **240**, 500 (1985)
92. R.M. Sinet, P. Garber, Arch. Biochem. Biophys. **212**, 411 (1981)
93. C.E. Thomas, L.R. Melean, R.A. Parker, D.F. Ohlweiler, Lipids **27**, 543 (1992)
94. D.E. Vance, J.E. Vance (eds.), *Biochemistry of Lipids, Lipoproteins and Membranes* (Elsevier, Amsterdam, 2008), p. 631
95. C.C. Winterbourn, Free Radic. Biol. Med. **14**, 85 (1993)
96. R. Milczarek, A. Hallmann, A. Sokołowska, C. Kaletha, J. Klimek, J. Pineal Res. **49**(2), 149–155 (2010)
97. A.L. Tappel, Vitam. Horm. **20**, 493 (1962)
98. G.W. Burton, U. Wronska, L. Stone, D.O. Foster, K.U. Ingold, Lipids **25**, 199 (1990)
99. R.R. Sandborg, J.E. Smolen, Lab. Invest. **59**, 300 (1988)
100. S. Weiss, J. New Engl. J. Med. **320**, 365 (1989)
101. L.A. Belova, Biokhimiya [Biochemistry] **62**, 659 (1997) [Russian]
102. Pamela C. Champe, Richard A. Harvey, Denise R. Ferrier, *Biochemistry* (Lippincott Williams & Wilkins, Philadelphia, PA, 2005) p. 534
103. J.P. Gaboury, D.C. Anderson, P. Kubes, Am. J. Physiol. **266**, H637 (1994)
104. P.G. Winyard, Ch.J. Morris, V.R. Winrow, M. Zaidi, D.R. Blake, in *Free Radical Damage and its Control*, ed. by C.A. Rice-Evans, R.H. Burdon (Elsevier, Amsterdam, 1994), pp. 361–383
105. B. Halliwell, S. Chirico, H. Kaur, O. Aruoma, M. Grootveld, D.R. Blake, in *Oxidative Damge and Repair; Chemical, Biological and Medical Aspects*, ed. by K.J.A. Davies (Pergamon, New York, 1993), pp. 846–855
106. P. Merry, P.G. Winyard, C.J. Morris, M. Grootveld, D.R. Blake, Ann. Rheum. Dis. **48**, 864 (1989)
107. H.J. Forman, M.J. Thomas, Annu. Rev. Physiol. **48**, 669 (1986)
108. P.R. Hansen, Circulation **91**, 1872 (1995)
109. N. Borregard, Eur. J. Haematol. **41**, 401 (1988)
110. S. Al-Assaf, G.O. Phillips, D.J. Deeble, B. Parsons, H. Starnes, C. Von Sonntag, Radiat. Phys. Chem. **46**, 207 (1995)
111. D.E. Deeble, G.O. Phillips, E. Bothe, H.-P. Schuchmann, C. Von Sonntag, Radiat. Phys. Chem. **37**, 115 (1991)
112. B. Halliwell, J.M. Gutteridge, D.R. Blake, Philos. Trans. R. Soc. Lond. B Biol. Sci. **311**, 659 (1985)
113. J.M. Gutteridge, Biochem. J. **245**, 415 (1987)
114. G.R. Buettner, W. Chamulitrat, Free Radic. Biol. Med. **8**, 55 (1990)
115. P. Biemond, A.J.G. Swank, H.G. Eijk, Arthritis Rheum. **29**, 1187 (1986)
116. B. Walaerd, P. Lassalle, F. Fortin, C. Aerts, F. Bart, E. Fournier, C. Voisin, Am. Rev. Respir. Dis. **141**, 129 (1990)
117. D.A. Chandlee, J.D. Fulmer, Am. Rev. Respir. Dis. **131**, 596 (1985)
118. C. Niederau, R. Fischer, A. Sennenberg, New Engl. J. Med. **313**, 1256 (1985)
119. L.M. Bassett, J.W. Halliday, L.W. Powell, Hepatology **6**, 24 (1986)
120. A.D. Heys, T.L. Dormandy, Clin. Sci. **60**, 295 (1981)
121. A.L. Cederbaunm, Ann. N. Y. Acad. Sci. **492**, 35 (1987)
122. M.L. Buja, J.T. Willerson, Hum. Pathol. **18**, 451 (1987)
123. M.J. Derelanko, M.A. Hollinger (eds.), *Handbook of Toxicology*, 2nd edn. (CRC, Boca Raton, 2002), p. 1414
124. J.T. Salonen, S. Yla-Herrtuale, R. Yamamoto, S. Butler, R. Salonon, K. Nyyssonen, W. Palinski, J.L. Wiztum, Lancet **339**, 883 (1992)
125. J.M. Mccord, R.S. Roy, Can. J. Physiol. Pharm. **60**, 1346 (1982)
126. E.J. Dowling, C. Chander, A.W. Claxson, C. Lillie, D.R. Blake, Free Radic. Res. Commun. **18**, 291 (1993)
127. P.G. Winyard, D.R. Blake, in *Free Radicals, Metal Ions and Biopolymers*, ed. by P.C. Beaument, D.J. Deeble, B.J. Parsons, C. Rice-Evans (Richelieu, London, 1989), pp. 183–187

128. C.W. Trenam, P.G. Winyard, C.J. Morris, D.R. Blake, in *Iron and Human Disease*, ed. by R.B. Lanffor (CRC, Boca Raton, 1992), pp. 395–417
129. P.G. Winyard, Ch.J. Morris, V.R. Winrow, M. Zaidi, D.R. Blake, in *Free Radical Damage and its Control*, ed. by C.A. Rice-Evans, R.H. Burdon (Elsevier, Amsterdam, 1994), pp. 361–383
130. P.G. Winyard, D.R. Blake, S. Chirico, J.M.C. Gutteridge, J. Lunec, Lancet **329**, 69 (1987)
131. H. Esterbauer, M. Dieber-Rotheneder, G. Waeg, H. Puhl, F. Tatzber, Biochem. Soc. Trans. **18**, 1059 (1990)
132. G.W. Burton, D.O. Foster, B. Perly, T.F. Slater, I.C.P. Smith, K.U. Ingold, Philos. Trans. R. Soc. Lond. B Biol. Sci. **311**, 565 (1985)
133. D.R. Blanke, P. Merry, J. Unsworth, B.L. Kidd, J.M. Quthwaite, R. Ballard, C.J. Morris, L. Gray, J. Lunec, Lancet **330**, 289 (1989)
134. J.B. Badwey, M.L. Karnovsky, Curr. Top. Cell Regul. **28**, 183 (1986)
135. O.T.G. Jones, in *Reactive Oxygen Species in Chemistry, Biology and Medicine*, ed. by A. Quintanilha (Plenum, London, 1988), pp. 143–156
136. B.M. Babior, Blood **64**, 959 (1984)
137. J. Schultz, K. Kaminker, Arch. Biochem. Biophys. **96**, 465 (1962)
138. A.R. Cross, O.T.G. Jones, A.M. Harper, A.W. Segal, Biochem. J. **194**, 599 (1981)
139. J.K. Hurst, in *Oxygen Complexes and Oxygen Activation by Transition Metals*, ed. by A.E. Martell, D.T. Sawyer (Plenum, New York, 1988), pp. 149–174
140. L.A. Gavrilov, N.S. Gavrilova, *Biology of Life-Time* (Nauka, Moscow, 1991), p. 280 [Russian]
141. D. Harmon, J. Gerontol. **11**, 298 (1956)
142. S.P. Fernandes, R. Dringen, A. Lawen, S.R. Robinson, Neurosci. Lett. **490**(1), 27–30 (2011)
143. R.J. Rivett, Arch. Biochem. Biophys. **243**, 624 (1985)
144. K.J.A. Davies, S.W. Lin, R.E. Pacifici, J. Biol. Chem. **262**, 9914 (1987)
145. C. Quiney, S. Finnegan, G. Groeger, T.G. Cotter, Protein Rev. **13**, 57–78 (2011)
146. E.R. Stadtman, C.N. Olivier, J. Biol. Chem. **266**, 2005 (1991)
147. R.L. Levine, C.N. Olive, R.M. Fuks, E.R. Stadtman, Proc. Natl. Acad. Sci. USA **78**, 2120 (1981)
148. L. Fucci, C.N. Oliver, R.M. Fuks, E.R. Stadtman, Proc. Natl. Acad. Sci. USA **80**, 1521 (1993)
149. R.L. Levine, J. Biol. Chem. **258**(1) 1828 (1983)
150. E.R. Stadtman, P.E. Starke-Reed, C.N. Olive, J.M. Carney, R.A.Floyd. in *Free Radicals and Aging*, ed. by I. Emerit, B. Chance, Eds., (Birkhaser Verlag, Basel, 1992), p. 64–71
151. H.S. Mason, Adv. Enzymol. **19**, 79 (1957)
152. V.M. Samokyszyn, C.D. Aust, in *Medical, Biochemical and Chemical Aspects of Free Radicals*, ed. by O. Hayaishi, E. Niki, M. Kondo, T. Yoshikawa (Elsevier, Amsterdam, 1989), pp. 41–48
153. M. Chevion, Free Radic. Biol. Med. **5**, 27 (1988)
154. A. Amici, R.L. Levine, L. Tsia, E.R. Stadtman, J. Biol. Chem. **264**, 3341 (1989)
155. V.V. Deshpande, J.G. Joshi, J. Biol. Chem. **260**, 757 (1985)
156. C.N. Oliver, R.L. Levine, E.R. Stadtman, in *Metabolic Interconversation of Enzymes*, ed. by H. Holzer (Springer, Berlin, 1981), pp. 259–268
157. E.R. Stadtman, Trends Biochem. Sci. **11**, 11 (1986)
158. G. Li, L. Liu, H. Hu, Q. Zhao, F. Xie, K. Chen, S. Liu, Y. Chen, W. Shi, D. Yin, Clin. Hemorheol. Microcirc. **46**(4), 305–311 (2010)
159. E.R. Stadtman, Science **257**, 1220 (1992)
160. J.M. Carney, P.E. Starke-Reed, C.N. Olive, R.W. Landum, M.S. Cheng, J.F. Wu, R.A. Floyd, Proc. Natl. Acad. Sci. USA **88**, 3633 (1991)
161. B.M. Ames, M.K. Shigenaga, T.M. Hagen, Proc. Natl. Acad. Sci. USA **90**, 7915 (1993)
162. B.N. Ames, M.K. Shigenaga, T.M. Hagen, Biochim. Biophys. Acta **1271**, 165 (1995)
163. C.G. Fraga, M.K. Shigenaga, J.-W. Park, P. Degan, B.N. Ames, Proc. Natl. Acad. Sci. USA **87**, 4533 (1990)
164. L.J. Marnett, H. Hurd, M.C. Hollstein, D.E. Esterbauer, B.N. Ames, Mutat. Res. **148**, 25 (1985)

165. A.D. Romano, G. Serviddio, A. De Matthaeis, F. Bellanti, G. Vendemiale, J. Nephrol. **23**(15), S29-S36 (2010)
166. D.C. Wallace, Environ. Mol. Mutagen **51**(5), 440–450 (2010)
167. D.C. Wallace, J.H. Ye, S.N. Neckelmann, G. Singh, K.A. Webster, B.D. Greenberg, Curr. Genet. **12**, 81 (1987)
168. C.B. Park, N.G. Larsson, J. Cell Biol. **193**(5), 809–818 (2011)
169. P. Mecocci, U. Mcgarvey, A.E. Kaufman, D. Koontz, J.M. Shoffner, D.C. Wallace, M.F. Beal, Ann. Neurol. **34**, 609 (1993)
170. G.A. Cortopassi, N. Arnheim, Nucleic Acids Res. **18**, 6927 (1990)
171. A.W. Linnane, A. Baumer, R.J. Maxwell, H. Preston, C.F. Zhang, S. Marzuki, Biochem. Int. **22**, 1067 (1990)
172. J.A. Free, Biochem. Sci. **7**, 84 (1982)
173. M. Hayakawa, K. Torii, S. Sugiyama, M. Tanaka, T. Ozawa, Biochem. Biophys. Res. Commun. **179**, 1023 (1991)
174. L. Piko, A.J. Hougham, K.J. Bulpitt, Mech. Aging Dev. **43**, 279 (1988)
175. K. Asano, S. Amagase, E.T. Matsuura, H. Yamagishi, Mech. Ageing Dev. **60**, 275 (1991)
176. R.S. Sohal, B.H. Sohal, U.T. Brunk, Mech. Ageing Dev. **53**, 217 (1990)
177. R.S. Sohal, U.T. Brunk, Mutat. Res. **275**, 295 (1992)
178. D. Boffoli, S.C. Scacco, R. Vergari, G. Solarino, G. Santacroce, S. Papa, Biochim. Biophys. Acta **1226**, 73 (1994)
179. G. Cortopapassi, E. Wang, Biochim. Biophys. Acta **1271**, 171 (1995)
180. J. Miquel, E. De Juan, J. Sevila, in *Free Radicals and Aging*, ed. by I. Emerit, B. Chance (Birkha¨user, Basel, 1992), pp. 47–57
181. M.R. Marzabadi, D. Yin, U.T. Brunk, in *Free Radicals and Aging*, ed. by I. Emerit, B. Chance (Birkha¨user, Basel, 1992), pp. 78–88
182. U.T. Brunk, R.S. Sohal, in *Membrane Lipid Oxidation*, vol. 2, ed. by C. Vigo-Pelfrey (CRC, Boca Raton, 1991), p. 191
183. J. Johnson, J. Kenealey, R.J. Hilton, D. Brosnahan, R.K. Watt, G.D. Watt, J. Inorg. Biochem. **105**(2), 202–207 (2011)
184. U.T. Brunk, E. Cadenas, APMIS **96**, 3 (1988)
185. A. Höhn, T. Jung, S. Grimm, T. Grune, Free Radic. Biol. Med. **48**(8), 1100–1108 (2010)
186. C. Michiels, J. Remacle, Biochim. Biophys. Acta **967**, 341 (1988)
187. J. Remacle, C. Michiels, M. Raes, in *Free Radicals and Aging*, ed. by I. Emerit, B. Chance (Birkha¨user, Basel, 1992), pp. 99–108
188. R.G. Gutler, Proc. Natl. Acad. Sci. USA **82**, 4798 (1985)
189. R.G. Gutler, in *Testing the Theories of Ageing*, ed. by R. Adelman, G. Roth (CRC Press, Boca Raton, 1982), pp. 25–114
190. G.B. De Quiroga, M. Lorez-Torres, R. Perez-Campo, in *Free Radicals and Aging*, ed. by I. Emerit, B. Chance (Birkha¨user, Basel, 1992), pp. 109–123
191. S.E. Espinoza, H. Guo, N. Fedarko, A. De Zern, L.P. Fried, Q.L. Xue, S. Leng, B. Beamer, J.D. Walston, J. Gerontol. A Biol. Sci. Med. Sci. **63**(5), 505–509 (2008)
192. R.G. Gutler, in *Free Radicals and Aging*, ed. by I. Emerit, B. Chance (Birkha¨user, Basel, 1992), pp. 31–47
193. A. Taylor, in *Free Radicals and Aging*, ed. by I. Emerit, B. Chance (Birkha¨user, Basel, 1992), pp. 266–279
194. M.C. Ho, Y.J. Peng, S.J. Chen, S.H. Chiou, J. Clin. Gerontol. Geriatr. **1**(1), 17–21 (2010)
195. H.J. Hoenders, H. Bloemendal, in *Molecular and Cellular Biology of the Eye Lens*, ed. by H. Bloemendal (Wiley, New York, 1992), pp. 327–366
196. V.M. Berthoud, E.C. Beyer, Antioxid. Redox Signal. **11**(2), 339–353 (2009)
197. A. Taylor, P.F. Jacques, D. Nadler, F. Morrow, S.I. Sulsky, D. Shepard, Curr. Eye Res. **10**, 751 (1991)
198. J. Berger, D. Shepard, F. Morrow, A. Taylor, J. Nutr. **119**(5), 734 (1989)
199. V.N. Reddy, Exp. Eye Res. **150**, 771 (1990)
200. D.R. Blake, P.J. Gallagher, A.R. Potter et al., Arthritis Rheum. **27**, 495 (1984)

201. P. Biemond, A.J. Swaak, H.G. Elijk et al., Arthritis Rheum. **29**, 1187 (1986)
202. D. Cooke, S. Saura, R. Uno, A. Scuddamore, J. Rheumatol. **18**(suppl. 27), 114 (1991)
203. Y. Henrotin, G. Deby-Dupont, C. Deby, P. Franchimont, I. Emerit, in *Free Radicals and Aging*, ed. by I. Emerit, B. Chance (Birkha¨user, Basel, 1992), pp. 308–322
204. W. Sutipornpalangkul, N.P. Morales, T. Harnroongroj, J. Med. Assoc. Thai. **92**(6), S268–S274 (2009)
205. K. Fukuda, Clin. Calcium **19**(11), 1602–1606 (2009)
206. F.J. Blanco, I. Rego, C. Ruiz-Romero, Nat. Rev. Rheumatol. **7**, 161–169 (2011)
207. G.G. Jayson et al., Int. J. Radiat. Phys. Chem. **7**, 363 (1975)
208. B.J. Parsons, in *Free Radical Damage and its Control*, ed. by C.A. Rice-Evans, R.H. Burdon (Elsevier, Amsterdam, 1994), pp. 281–300
209. H. Burkhardt, M. Schwingel, H. Meninger, H.W. Mccartney, H. Tscheshe, Arthritis Rheum. **29**, 379 (1986)
210. J.C. Kowanko, E.J. Bates, A. Ferrante, Immunol. Cell. Biol. **67**, 321 (1989)
211. M.S. Baker, in *Free Radical Damage and its Control*, ed. by C.A. Rice-Evans, R.H. Burdon (Elsevier, Amsterdam, 1994), pp. 301–317
212. C.R. Roberts, J.S. Mort, P.J. Roughley, Biochem. J. **247**, 349 (1987)
213. S.F. Curran, M.A. Amusoro, B.D. Goldstein et al., FEBS Lett. **176**, 155 (1984)
214. J.C. Monboisse, J.P. Borel, in *Free Radicals and Aging*, ed. by I. Emerit, B. Chance (Birkha¨user, Basel, 1992), pp. 323
215. J.S. Bomalski, F. Hirata, M. Clark, Biochem. Biophys. Res. Commun. **139**, 115 (1986)
216. A. Ymadaya, K. Terasawa, R. Tosa et al., J. Rheumatol. **15**, 1968 (1988)
217. J.L. Scott, C. Gabrielides, R.K. Davidson, T.E. Swingler, I.M. Clark, G.A. Wallis, R.P. Boot-Handford, T.B.L. Kirkwood, R.W. Talyor, D.A. Young, Ann. Rheum. Dis. **69**, 1502–1510 (2010)
218. F. Bonomini, S. Tengattini, A. Fabiano, R. Bianchi, R. Rezzani, Histol. Histopathol. **23**(3), 381–390 (2008)
219. S. De Rosa, P. Cirillo, A. Paglia, L. Sasso, V. Di Palma, M. Chiariello, Curr. Vasc. Pharmacol. **8**(2), 259–275 (2010)
220. J.C. Fruchart, P. Duriez, in *Free Radical Damage and its Control*, ed. by C.A. Rice-Evans, R.H. Burdon (Elsevier, Amsterdam, 1994), pp. 257–280
221. D.W. Crawford, D.H. Blankenhorn, Atherosclerosis **89**, 97 (1991)
222. H.J. Forman, A. Boveris, in *Free Radicals in Biology*, ed. by W.A. Pryor (Academic, New York, 1982), pp. 65–90
223. B.A. Freeman, J.D. Crapo, Lab. Invest. **47**, 412 (1982)
224. V.M. Victor, M. Rocha, E. Sola, C. Bañuls, K. Garcia Malpartida, A. Hernandez Mijares, Curr. Pharm. Design. **15**(26), 2988–3002 (2009)
225. R.C. Sharma, D.W. Crwford, D.M. Kramisch, A. Sevanien, Q. Jiao, Arterioscl. Throm. Vas. Biol. **12**, 403 (1992)
226. A. Mezzetti, D. Lapenna, A.M. Calfiore, G. Proietti-Franceschilli, E. Porreca, D. De Cezare, N. Neri, C. Di Ilio, F. Curcurullo, Arterioscl. Throm. Vas. Biol. **12**, 92 (1992)
227. H. Esterbauer, G. Waeg, H. Puhl, M. Dieber Rotheneder, F. Tatzber, in *Free Radicals and Aging*, ed. by I. Emerit, B. Chance (Birkha¨user, Boston, 1992), pp. 145–156
228. H. Esterbauer, M. Rotheneder, G. Striegl, G. Waeg, A. Ashy, W. Sattler, G. Jürgens, Fat Sci. Technol. **91**(8), 316 (1989)
229. U.P. Steinbrecher, M. Lougheed, W.C. Kwan, M. Dirks, J. Biol. Chem. **264**, 15216 (1989)
230. U.P. Steinbrecher, J.L. Witztum, S. Parthasarathy, D. Steinberg, Atherosclerosis **7**, 135 (1987)
231. S. Parthasarathy, D.J. Printz, D. Boyd, L. Joy, D. Steinberg, Atherosclerosis **6**, 505 (1986)
232. F. Lupu, J. Danaricu, N. Simionescu, Atherosclerosis **7**, 127 (1987)
233. J.W. Heinecke, H. Rosen, A. Chait, J. Clin. Invest. **74**, 1890 (1984)
234. M.L. Lenz, H. Hughes, J.R. Mitchell, D.P. Via, J.R. Guyton, A.A. Taylor, A.M. Gotto, C.V. Smith, J. Lipid Res. **31**, 1043 (1990)
235. T.V. Vakhrusheva, E.S. Dremina, V.S. Sharov, O.A. Azizova, Biofizika [Biophysics] **42**, 662 (1997) [Russian]

236. H. Esterbauer, O. Quehenberger, G. Jürgens, in *Free Radicals, Methodology and Concepts*, ed. by C.A. Rice-Evans, B. Halliwell (Richelieu, London, 1988), pp. 243–268
237. D. Steinberg, Circulation **84**(3), 1420 (1991)
238. D. Steinberg, S. Parthasarathy, T.E. Carew, J.C. Khoo, J.L. Witztum, New Engl. J. Med. **320**, 915 (1989)
239. C.J. Schwartz, A.J. Valente, E.A. Spraque, J.L. Kelley, C.A. Suennam, D.T. Graves, M.M. Rozek, E.H. Edwards, Semin. Thromb. Hemost. **12**, 79 (1986)
240. M.K. Cathart, D.W. Morel, G.M. Chisolm, J. Leucocyte Biol. **38**, 341 (1985)
241. B.Z. Simkhovich, M.T. Kleinman, R.A. Kloner, Curr. Opin. Cardiol. **24**(6), 604–609 (2009)
242. A.N. Osipov, E.V. Briukhanova, T.V. Vakhrusheva, O.M. Panasenko, Yu.A. Vladimirov, Biofizika [Biophysics] **42**, 400 [Russian] (1997)
243. H.D. Hoberman, R.J. Can George, Biochem. Toxicol. **3**, 105 (1988)
244. R.A. Kloner, K. Przyklenk, P. Whittaker, Circulation **80**, 115 (1989)
245. J. Kehrer, Free Radic. Res. Commun. **5**, 305 (1989)
246. P.J. Simpson, B.R. Lucchesi, J. Lab. Clin. Med. **110**, 13 (1987)
247. P. Thompson, M.L. Hess, Prog. Cardiovasc. Dis. 449 (1996)
248. T.J. Gardner, Free Radic. Biol. Med. **4**, 45 (1988)
249. P. Menasche, A. Piwnica, Ann. Thorac. Surg. **47**, 939 (1989)
250. V.M. Victor, N. Apostolova, R. Herance, A. Hernandez-Mijares, M. Rocha, Curr. Med. Chem. **16**(35), 4654–4667 (2009)
251. J.M. Downey, Annu. Rev. Physiol. **52**, 487 (1990)
252. B.R. Lucchesi, S.W. Werns, J.C. Fantone, J. Mol. Cell. Cardiol. **21**, 124 (1989)
253. K. Areimer, C.E. Murry, W.J. Richard, J. Mol. Cell. Cardiol. **21**, 1225 (1989)
254. J. Fridovich, Arch. Biochem. Biophys. **247**, 1 (1986)
255. S.D. Aust, L.A. Morehouse, C.E. Thomas, J. Free Radic. Biol. Med. **1**, 3 (1985)
256. H.S. Sutton, C.C. Winterbourn, Free Radic. Biol. Med. **6**, 53 (1989)
257. B. Kalyanaraman, E.A. Konorev, J. Joseph, J.E. Baker, in *Free Radical Damage and its Control*, ed. by C.A. Rice-Evans, R.H. Burdon (Elsevier, Amsterdam, 1994), pp. 333–359
258. N.B. Parker, E.M. Berger, W.E. Curtis, M.E. Muldrow, S.L. Linas, J.E. Repine, J. Free Radic. Biol. Med. **1**, 415 (1985)
259. G.M. Rosen, H.J. Halpern, L.A. Brunstig, D.P. Spencer, K.E. Strauss, M.K. Bowman, A.H. Wechsler, Proc. Natl. Acad. Sci. USA **85**, 7772 (1988)
260. C.C. Winterbourn, J. Free Radic. Biol. Med. **3**, 33 (1987)
261. W.H. Koppenol, J. Free Radic. Biol. Med. **1**, 281 (1985)
262. K.M. Morehouse, W.D. Flitter, R.P. Masson, FEBS Lett. **222**, 246 (1987)

Chapter 5
Catalytic Oxidation of Oxyacids and Natural Polyphenols

5.1 Metabolic Cycles of Tri- and Dicarbonic Acids

Redox processes with the involvement of oxygen or hydrogen, catalyzed by transition metal compounds (iron, copper, manganese, cobalt) or enzymes have been considered in Chaps. 2–4. Out of the numerous examples of such type, we should especially mention those which proceed not by being isolated from each other but by constituting cycles. The most important of these are the cycles of tricarbonic acids (Krebs cycle, glyoxalate cycle) and the cycle of dicarbonic acids (Baroud's cycle), which are connected with each other through glyoxalic acid (Fig. 5.1) [1].

With the help of these cycles, the decomposition of relatively simple compounds takes place in the organism in the process of aerobic breathing in mitochondria, proceeding through a number of intermediate stages catalyzed by the appropriate enzymes [2]. Krebs cycle will not be considered here since it is well known and described in numerous works. It is to be mentioned though that the most important function of this cycle is to supply the respiratory chain with electrons. These reactions of the respiratory chain and Krebs cycle proceed in the subcellular structures—mitochondria—and come to oxygen being reduced to water with the appropriate phosphorylation of ADN. Some bacteria (such as *Pseudomonas, Escherichia coli*) and mold fungus can realize intermediate carbon exchange with the help of a "shortened" Krebs cycle through glyoxalic acid (GC). This twice reduced variant of Krebs cycle is called the glyoxalic acid cycle and usually proceeds in critical cases of energy lack in a metabolism (Scheme 5.1).

Peculiar features of the GC cycle are reactions I and II, resulting from which isocitric acid turns into GC, and then into maleic acid. Reaction I is catalyzed by isocitrate lyase, and reaction II by malat sintase. Other reactions of the GC cycle are catalyzed by the same enzymes as the reactions of the complete Krebs cycle.

G. Duca, *Homogeneous Catalysis with Metal Complexes*, Springer Series in Chemical Physics 102, DOI 10.1007/978-3-642-24629-6_5,

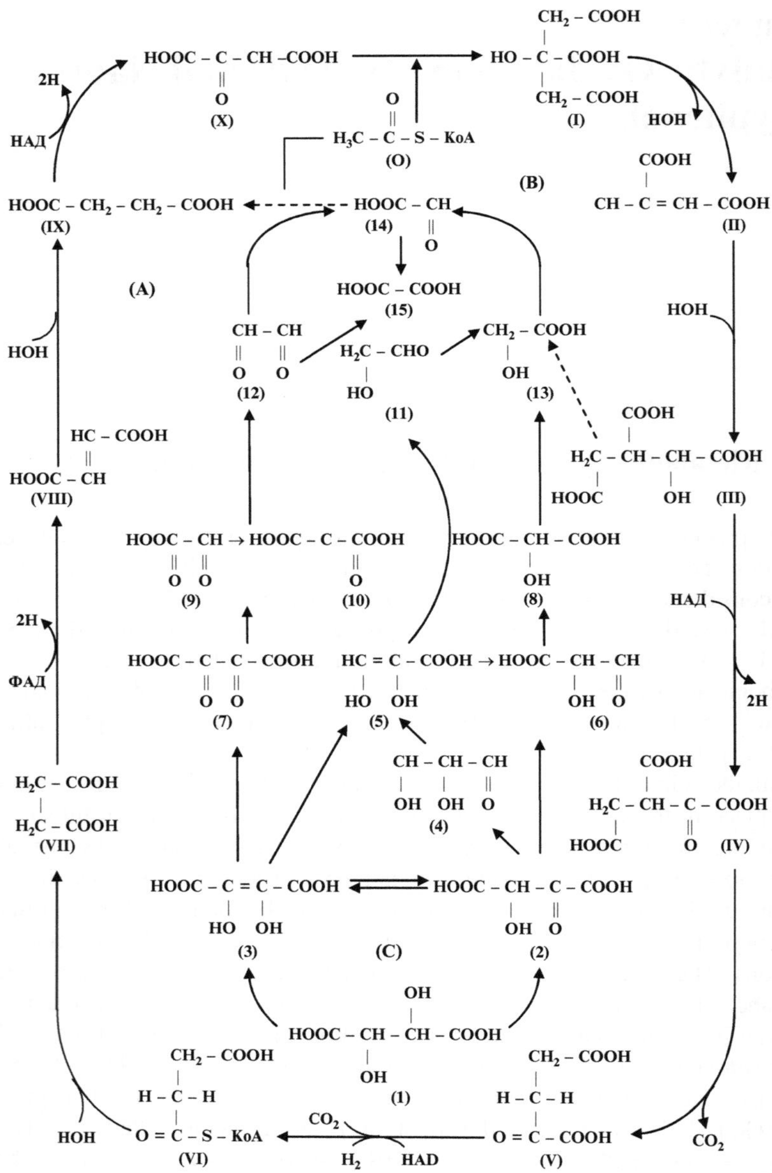

Fig. 5.1 Interrelation of metabolic cycles: (A) cycle of tricarbonic acids (Krebs cycle), (B) glyoxalate cycle, (C) cycle of dicarbonic acids (Baroud' cycle). I citric acid, II cys-acinitic acid, III isocitric acid, IV oxalic–succinic acid, V α–ketoglutalic acid, VI succinoxo-enzyme A, VII succinic acid, VIII fumaric acid, IX maleic acid, X oxalic–acetic acid. 0 acetyl coenzyme A; 1 tartaric acid, 2 oxaloglycolic, 3 dihydroxyfumaric acids; 4 reductone of triose, 5 dixydroxyacrilic acid, 6 semialdehyde of tartronic acid; 7 diketosuccinic acid, 8 tartronic, 9 glyoxylcarbonic, 10 mezoxalic acids; 11 glycolic aldehyde, 12 glyoxale; 13 glycolic acid, 14 glyoxalic acid, 15 oxalic acid

Scheme 5.1 Glyoxalic acid cycle

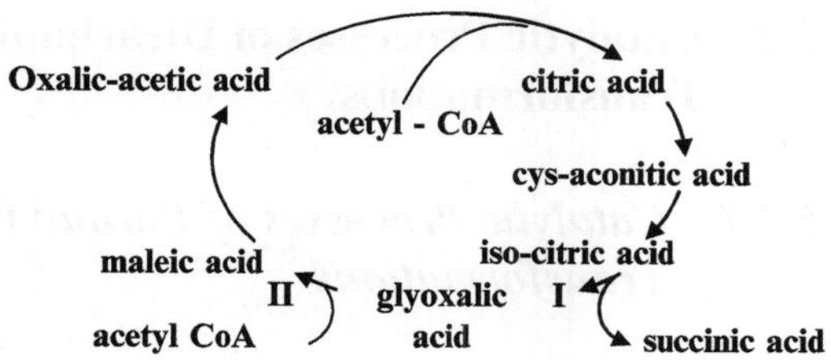

Reactions of the GC cycle are based on oil turning into carbohydrates on sprouting of seeds rich with oil. Reactions of glyoxalic cycle in this case proceed in cellular structures called glyoxysomes. In them, under the influence of isocitrate lyase, glyoximate will be transformed into maleic acid and later into oxalic-acetic acid, which will come into the cycle once again. Oxalic-acetic acid, apparently being outside the mitochondria, will be decarboxylized forming phosphoenolpyruvate, from which trioses and then hexose will be formed. These reactions are based on oils transformation into sucrose proceeding on oil-bearing seeds sprouting.

All the metabolic transformations of the Krebs cycle are catalytic processes realized with enzymes; in many cases, manganese or magnesium ions are needed for their activation [3].

As can be seen from Fig. 5.1, the third main metabolic cycle is the Baroud cycle of tartaric acid and its intermediate products transformation to oxalic acid. Oxidation of tartaric acid in foodstuffs and plants was first described and represented as the Baroud scheme [4]. Nowadays, the Baroud scheme (Fig. 5.1) should be considered as a simplified version of tartaric acid transformation in aerobic conditions. Unlike the Krebs cycle, information on the catalytic transformations in the Baroud cycle was, until recently, scanty and often uncertain. In this cycle, only the constants of dihydroxyfumaric (DHF_4) and oxaloglycolic (OXG) acids tautomerization were known, as well as details of the latter acid decarboxylation to dihydroxyacrylic acid (DHA). Taking into account the importance of these transformations in the food- and wine-making industry [5–7] and the fact that separate stages of these transformations are catalyzed by iron and copper ions, a detailed study was carried out [8–17], which made it possible to deepen the knowledge of the reactions mechanisms. This research contributes to the revealing of general regularities of oxyacids transformations, and its results can be used for complex chemical and biological processes modeling of the production of wine, juice, etc.. With this aim, the kinetics of Baroud cycle components transformation (tartaric, dihydroxyfumaric, and other acids) will be considered in some detail, as well as the influence of oxygen, hydrogen peroxide, and some transition metal ions on these processes.

5.2 Catalytic Processes of Dicarbonic Acids of Baroud Cycle Transformations

5.2.1 *Catalytic Processes of Baroud Cycle Oxyacids Transformations*

5.2.1.1 Catalytic Oxidation of Tartaric Acid

Tartaric acid is broadly used in the chemical, pharmaceutical, and perfume industries, oil production, radio engineering, food production, and other fields [8, 18–20]. Tartaric acid (H_4T) has three isomers (D-, L-, and mezo-):

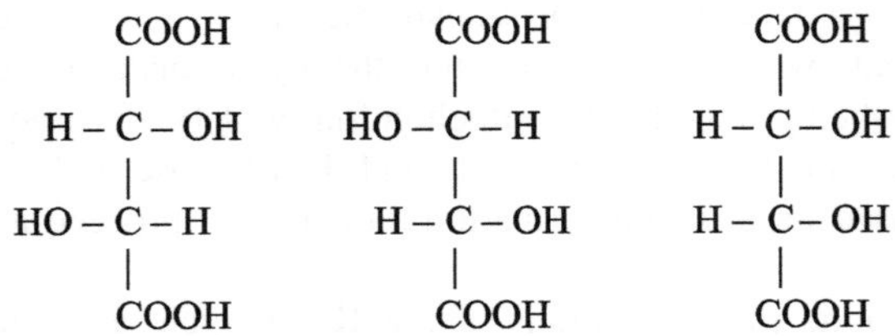

(D-) (L-) (Mezo-)

Usually, H_4T is contained mainly in the form of the D-isomer. It can be met in many kinds of fruit and berries (for instance, in ashberries) and especially in grapes (2–10 g/kg), which makes it possible to obtain it on an industrial scale from wine-making wastes [8]. Processes of tartaric acid transformation are strongly influenced by oxygen, hydrogen peroxide, and iron and copper compounds. Tartaric acid takes part in respiration processes, assimilation in plants, biosynthesis of 3-phosphoroglyceric acid, redox reactions proceeding in wine-making and other reactions. The product of its transformation—DFH_4—also plays an important role in wine-making, being a strong antioxidant.

System $Fe^{2\pm}$–H_4T–O_2

Interaction of Fe^{2+} ions with O_2 in acid medium proceeds slowly because of high potential $\varphi_{Fe3+\cdot aq/Fe2+\cdot aq} = 0.77\,V$. Adding of H_4T will reduce redox potential $\varphi_{Fe(H2T)+/Fe(H2T)} = 0.53\,V$ at pH 2.5 [21] and, hence, one can expect observable acceleration of Fe^{2+} oxidation reaction with oxygen and H_4T oxidation through the formation of tartaric acid free radical.

Kinetic curves of iron(II) are shown in Fig. 5.2, where it can be seen that the process is self-catalytic.

Reaction acceleration is connected with hydrogen peroxide forming in the system. At the initiation stage it seems to precede the oxidation of substrate coordination carboxyl group. It is supposed that the tartaric acid oxidation process proceeds in the following way [10]:

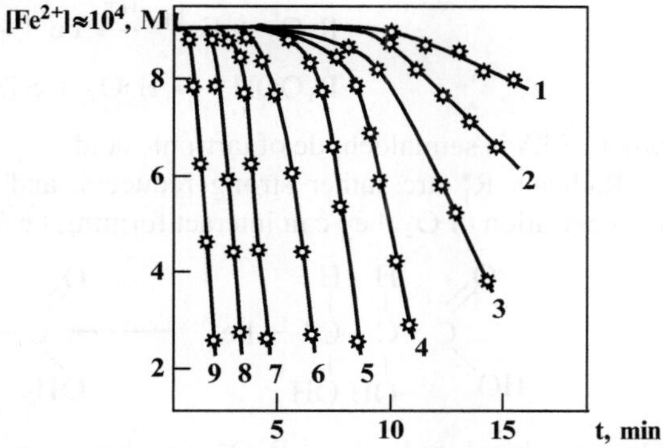

Fig. 5.2 Kinetic curves of iron(II) oxidation $[Fe^{2+}]_0 = 1 \times 10^{-3}$ M, pH 3.5, 25°C, $V_{O_2} = 100\,\text{mL min}^{-1}$; $[H_4T]_0$, M : 1–2×10^{-2}, 2–3×10^{-2}, 3–5×10^{-2}, 4–5×10^{-2}, 5–1×10^{-2}, 6–1.5×10^{-1}, 7–2×10^{-1}, 8–3×10^{-1}, 9–4×10^{-1}

$$Fe^{2+}H_2T^{2-} + O_2 \rightleftharpoons Fe^{2+}H_2T^{2-}O_2 \tag{5.1}$$

$$Fe^{2+}H_2T^{2-}O_2 \xrightarrow{2H^+} Fe^{4+}H_2T^{2-} + H_2O_2 \tag{5.2}$$

$$Fe^{4+}H_2T^{2-} \rightarrow Fe^{3+} + R^{\bullet} \tag{5.3}$$

where $R^{\bullet}$ is a carboxyl radical of the following type:

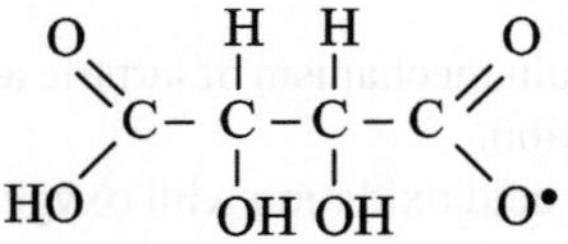

which will be decarboxylated with α-alcohol radical $R_1^{\bullet}$ formation:

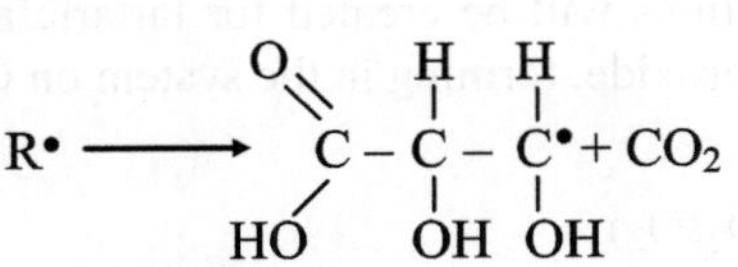
$(R_1^{\bullet})$ (5.4)

$R_1^{\bullet}$ will efficiently interact with O_2:

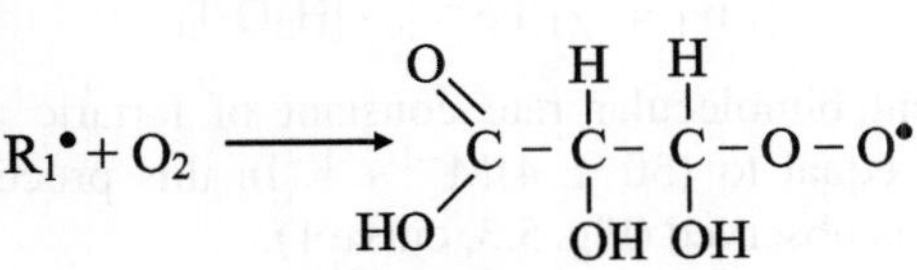
$(R_1O_2^{\bullet})$ (5.5)

Peroxide radical posesses oxidative properties:

$$R_1O_2^{\bullet} + Fe^{2+} \xrightarrow{H^+} Fe^{3+} + R_1OOH \tag{5.6}$$

$$R_1OOH \longrightarrow H_2O_2 + STA \tag{5.7}$$

where STA is semialdehyde of tartronic acid.

Radicals $R_1^{\bullet}$ are rather strong reducers, and in conditions of low stationary concentration of O_2 they can interact forming Fe^{3+} ions:

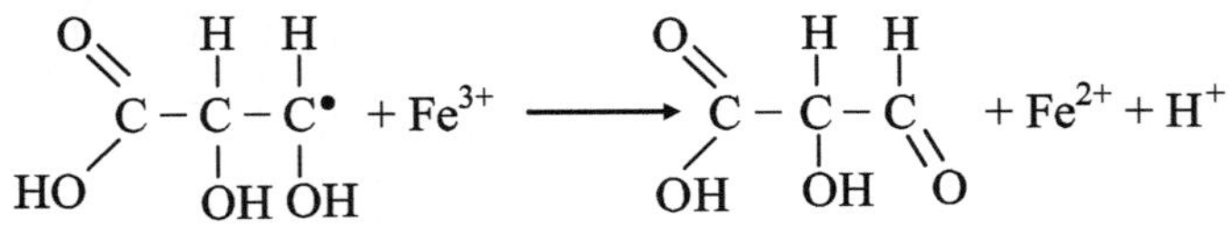

(5.8)

On other hand, radicals of $R_1O_2^{\bullet}$ type have some tendency to be decomposed to $HO_2^{\bullet}$:

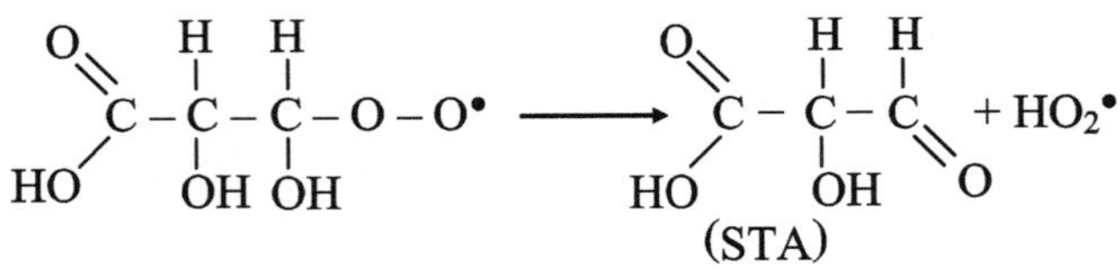

(5.9)

$$HO_2^{\bullet} + Fe^{2+} \longrightarrow Fe^{3+} + HO_2^- \tag{5.10}$$

$$HO_2^{\bullet} + Fe^{3+} \longrightarrow Fe^{2+} + O_2 + H^+ \tag{5.11}$$

Thus, on $[H_4T] > [Fe^{2+}]$, chain mechanism of tartaric acid oxidation will take place, conjugated with Fe^{2+} oxidation.

Brutto reaction of tartaric acid oxidation with oxygen at the initial process stages can be represented in the following form:

$$2Fe^{2+} + H_4T + 2O_2 \longrightarrow 2Fe^{3+} + 2H_2O_2 + CO_2 + PTA$$

For real systems, conditions will be created for tartaric acid joint oxidation with oxygen and hydrogen peroxide, forming in the system on O_2 reduction.

System $Fe^{2\pm}$–H_4T–$H_2O_2(O_2)$

Tartaric acid oxidation with Fenton reagent was studied under anaerobic conditions. The kinetic expression for oxidation rate is as follows:

$$W_1 = \chi_1[Fe^{2+}]_0 \cdot [H_2O_2]_0$$

where χ_1 is efficient bimolecular rate constant of tartaric acid oxidation with hydrogen peroxide, equal to $(50 \pm 4)M^{-1}\,s^{-1}$. In this process, stable oxidation degree of Fe^{2+} ions is observed (Fig. 5.3, curve 1).

The results of stoichiometric titration of Fe^{2+} ions with hydrogen peroxide in the presence of tartaric acid under anaerobic (Fig. 5.4, curve 1) and aerobic (Fig. 5.4, curve 2) conditions differ greatly. In the first case, Fe^{2+} concentration does not depend on H_2O_2 additives, in the second, two-thirds of the Fe^{2+} ions will be transformed into Fe^{3+} ions.

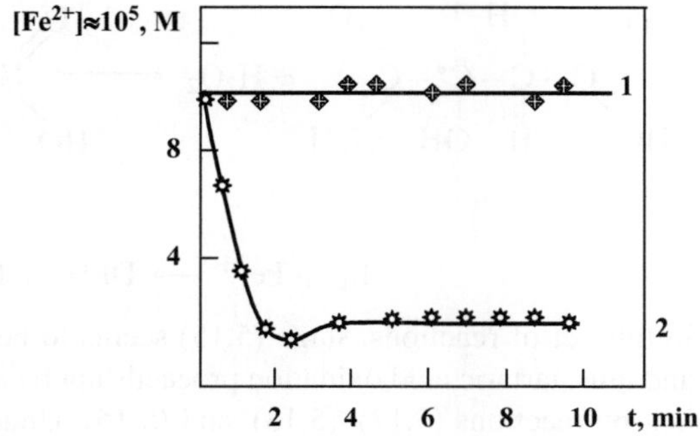

Fig. 5.3 Fe^{2+} concentration change within the systems Fe^{2+}—H_4T—H_2O_2 (1) and Fe^{2+}—H_4T—H_2O_2—O_2 (2); $[Fe^{2+}]_0 = 1 \times 10^{-4}$ M, $[H_4T]_0 = 1 \times 10^{-2}$ M, $[H_2O_2]_0 = 1 \times 10^{-2}$ M, pH 3.5

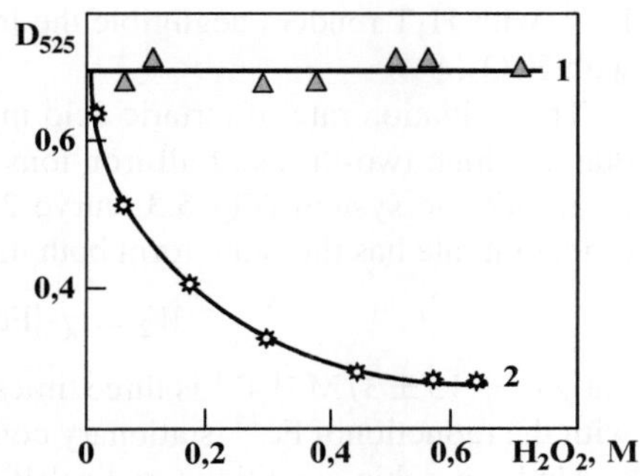

Fig. 5.4 Stoichiometric titration within the system Fe^{2+}–H_4T with hydrogen peroxide in the absence (1) and in the presence of O_2 (2). $[Fe]_0 = 7 \times 10^{-5}$ M, $[H_4T]_0 = 2 \times 10^{-2}$ M, pH 3.5

The rate constant of the reaction of interaction

$$Fe^{2+} + H_2O_2 \longrightarrow Fe^{3+} + OH^- + OH^\bullet \tag{5.12}$$

is $k_{12} = 50\,M^{-1}\,s^{-1}$ [22], which coincides with the rate constant of tartaric acid oxidation.

Stability of Fe^{2+} concentration in the course of the reaction (Figs. 5.3 and 5.4) is the result of Fe^{2+} ions regeneration, caused by substrate radicals R_α possessing reducing properties interactiing with Fe^{3+} ions:

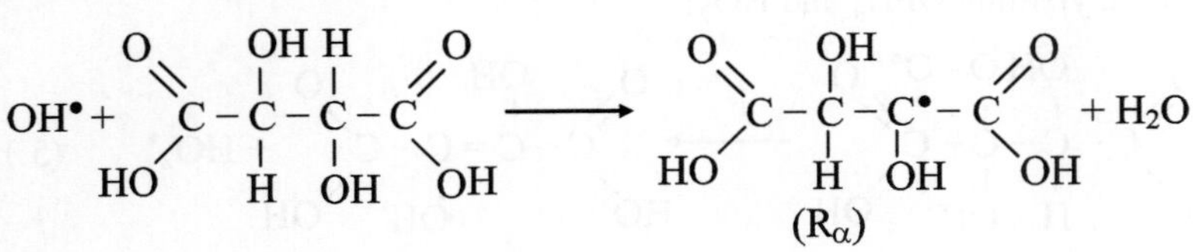

(5.13)

Under anaerobic conditions, R_α can interact either with H_2O_2 or with Fe^{3+} ions with the formation of dihydroxyfumaric acid (DFH_4); the presence of the latter has been proved experimentally [23]:

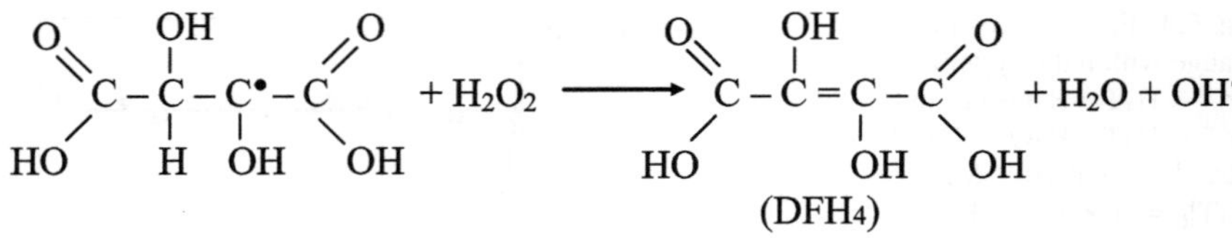

(5.14)

$$R_\alpha + Fe^{3+} \longrightarrow DFH_4 + Fe^{2+} + H^+ \quad (5.15)$$

In this set of reactions, stage (5.15) seems to be predominating over stage (5.14), and thus, tartaric acid oxidation proceeds not by a chain but by a cyclic mechanism, i.e., by reactions (5.12), (5.13), and (5.15). Under experimental conditions, it was possible to ignore the reactions of $OH^\bullet$ with H_2O_2 and Fe^{2+}. Complex formation of Fe^{2+} with H_4T renders negligible the influence on the rate constant of interaction with H_2O_2.

The oxidation rate of tartaric acid in Fenton's system is influenced by oxygen, due to which two-thirds of all iron ions will be transformed into the form of Fe^{3+} in an aerobic system (Fig. 5.3, curve 2). The kinetic expression for tartaric acid oxidation rate has the same form both under aerobic and anaerobic conditions:

$$W_2 = \chi_2[Fe^{2+}]_0 \cdot [H_2O_2]_0$$

but $\chi_2 = (15 \pm 3)\,M^{-1}\,s^{-1}$ is three times less than χ_1 constant. This is in agreement with the reductionof Fe^{2+} stationary concentration in the course of the process.

Under aerobic conditions, radical R_α will interact mainly not with Fe(III), but with O_2, resulting in peroxide radical ($RO_2^\bullet$) formation:

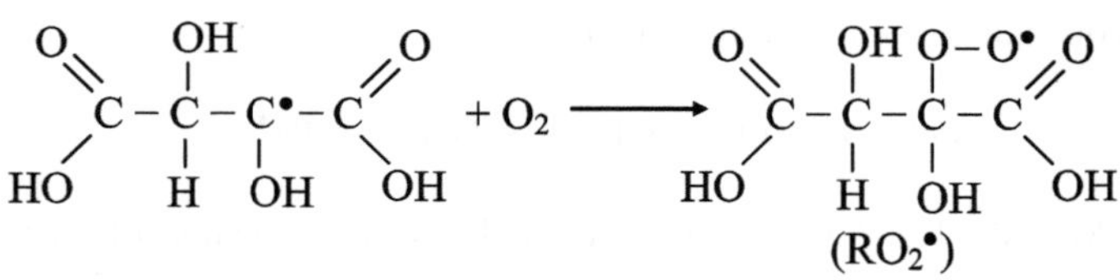

(5.16)

The elementary rate constant for similar reactions reaches $\sim 10^9\,M^{-1}\,s^{-1}$. Therefore, at a low enough concentration of Fe(III), reaction (5.16) will predominate over reaction (5.15). Radicals of $RO_2^\bullet$ type are rather unstable and will be readily decomposed, yielding DFH_4 and $HO_2^\bullet$:

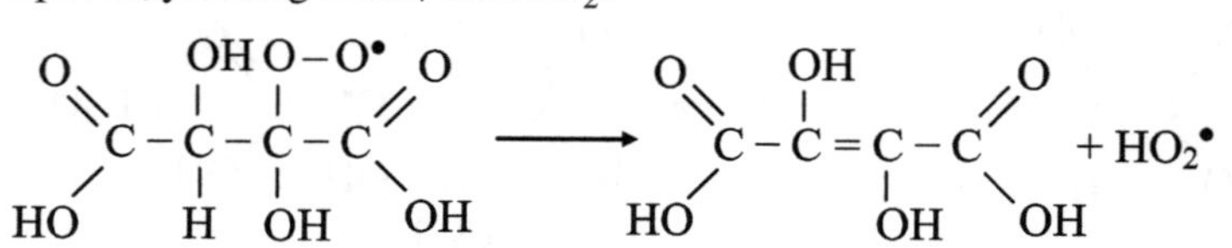

(5.17)

Radical $HO_2^\bullet$ will interact either with Fe^{2+} or with Fe^{3+} by reactions (5.10) and (5.11).

Radicals $RO_2^\bullet$, being rather strong oxidants, can also interact with the compounds of Fe(II):

$$RO_2^\bullet + Fe^{2+} \longrightarrow ROOH + Fe^{3+} \quad (5.18)$$

The hydroperoxide ROOH forming will either be decomposed to DFH_4 and H_2O_2:

$$ROOH \longrightarrow DFH_4 + H_2O_2 \tag{5.19}$$

or will interact with Fe^{2+} yielding alcoxyl radical $RO^\bullet$:

$$ROOH + Fe^{2+} \longrightarrow FeOH^{2+} + \quad (5.20)$$

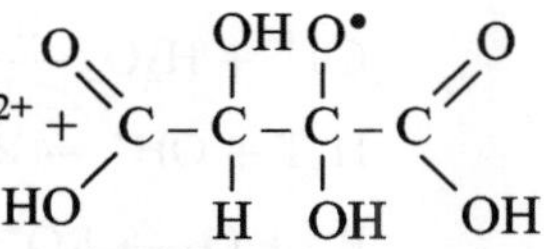

$(RO^\bullet)$

with further interaction with H_4T and repeated formation of R_α radical and DFH_4:

$$RO^\bullet + H_4T \longrightarrow R_\alpha + DFH_4 + H_2O \tag{5.21}$$

Detailed consideration demonstrates the complexity of the chemical transformation mechanism in, at first sight, a relatively simple system, Fe^{2+}–H_4T–$H_2O_2(O_2)$.

These data were obtained under conditions when H_2O_2 content in the initial system was high enough compared to O_2 or Fe^{2+} ions concentration. Provided H_2O_2 content in the system is low, oxidation rate of H_4T will be determined by the rate of $Fe^{2+}H_2T^{2-}$ with O_2 interaction. However, beginning at a certain moment, oxidation rate of H_4T on account of H_2O_2 formation will start increasing rapidly, attaining the stationary value. It seems, under the conditions in Figs. 5.3 and 5.4, the mechanism of tartaric acid oxidation in stationary reaction can be given by stages (5.12), (5.13), and (5.16)–(5.19).

Therefore, a kind of cyclic mechanism will be realized (although free radicals will be formed), in which H_2O_2 will be consumed on Fe^{2+} oxidation, while as a result of H_4T oxidation with oxygen, H_2O_2 will be formed again in the system. Such a scheme can be self-regulating enough, so that the process of tartaric acid oxidation, once started, can proceed to the end with constant rate.

Aerobic System $Cu^{2\pm}$–H_4T–H_2O_2

Copper ions, and likewise iron ions, are usually admixed components of natural water [24] and foodstuff such as wine and juice [7]. Often they determine catalytic processes proceeding in these substances in natural water (see Chap. 6) and in foodstuff when stored (see Chap. 7). To produce tartaric acid oxidation with oxygen, Cu^{2+} ions reducing to Cu^+ are needed; this process proceeds in the presence of hydrogen peroxide, which is an electron donor in its dissociated form (HO_2^-). Therefore, in the system Cu^{2+}–H_4T–H_2O_2, intensive oxidation of tartaric acid takes place. The kinetic expression for the process rate is as follows:

$$W_3 = \chi_3[Cu^{2+}]_0 \cdot [H_4T] \cdot [H_2O_2]_0/[H^+]$$

where $\chi_3 = (9 \pm 1) \times 10^{-2}\,M^{-1}\,s^{-1}$. The tartaric acid oxidation rate correlates with the decomposition rate of triple complex $H_2TCu(II)(HO_2^-)$, formed within this system. The mechanism of tartaric acid oxidation can be presented in the following form:

$$H_2TCu(II)(HO_2^-) \rightarrow H_2TCu(I) + HO_2^\bullet \quad (5.22)$$

$$HO_2^\bullet + Cu^{2+} \rightarrow Cu^+ + O_2 + H^+ \quad (5.23)$$

$$Cu^+ + H_2O_2 \xrightarrow{H^+} Cu^{2+} + OH^\bullet \quad (5.24)$$

$$H_4T + OH^\bullet \rightarrow R_\alpha \quad (5.25)$$

$$R_\alpha + O_2 \rightarrow RO_2^\bullet \quad (5.26)$$

$$RO_2^\bullet \rightarrow DFH_4 + HO_2^\bullet \quad (5.27)$$

$$RO_2^\bullet + Cu^+ \xrightarrow{H^+} ROOH + Cu^{2+} \quad (5.28)$$

$$R_\alpha + Cu^{2+} \rightarrow DFH_4 + Cu^+ \quad (5.29)$$

$$ROOH + Cu^+ \rightarrow Cu^{2+} + RO_2^\bullet + H^+ \text{etc.} \quad (5.30)$$

Thus, unlike the system Fe^{2+}–H_4T–H_2O_2, tartaric acid oxidation within the system Cu^{2+}–H_4T–H_2O_2 is a radical-chain process. This is testified to by the tests with the efficient acceptor of $HO_2^\bullet$ radicals and Cu^+ ions–tetranitromethane $C(NO_2)_4$ (TNM) introduced into the system. Thus, long-term induction periods in H_2O_2 decomposition are observed within the system (Fig. 5.5).

On TNM introduction at least some seconds after the reaction begins, induction periods will disappear. This is stipulated by tetranitromethane reduction due to the influence of DFH_4, formed during tartaric acid oxidation [25].

Experimentally determined rate value W_3 coincides qualitatively with the calculated value, found on the basis of the above mechanism scheme of tartaric acid oxidation in stationary conditions (W_{scheme}) on condition that the chain breakage is linear, and chain transporting particle $[Cu^{3+}(OH^-)_2]^+$ is involved [26]. The breakage of such a particle forming on the interaction of Cu^+ with $H_2O_2(Cu^+ + H_2O_2 \longrightarrow Cu(OH)_2^+)$, seems to be possible on its interaction with DFH_4.

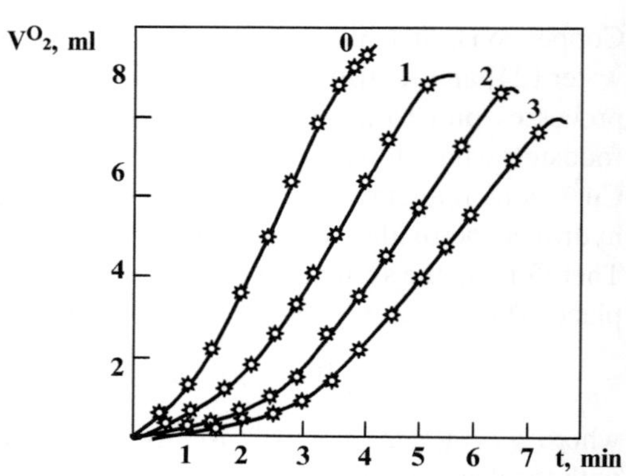

Fig. 5.5 Tetranitromethane influence on gas evolving (V^{O_2}) within the system Cu^{2+}–H_4T–H_2O_2.0 in the absence of TNM; 1 TNM = 2.5×10^{-7} M; 2–5×10^{-7} M; 3–8×10^{-7} M. $[Cu^{2+}] = 1 \times 10^{-2}$ M, $[H_4T] = 2 \times 10^{-2}$ M, $[H_2O_2] = 5 \times 10^{-2}$ M, pH 3.5

Tartaric acid and tartrate ions can also be oxidized by hydrogen peroxide with a cobalt chloride catalyst. Without the catalyst, the rate of reaction, as evidenced by the evolution of CO_2 gas (or lack of evolution of CO_2 gas), is very slow. Initially, the catalyst, $CoCl_2$, is pink, but when added to the reaction mixture, a green complex is formed and the reaction becomes vigorous. As the tartrate is oxidized, the catalyst will be regenerated and the solution will return to a pink color. By taking some of the spent reaction mixture and adding it to a fresh tartrate–peroxide mixture, the entire reaction sequence will be repeated, complete with the green activated complex. This shows that the original catalyst was regenerated at the end of the reaction and is still available to catalyze another reaction.

The Effect of Temperature

By direct comparison of the reaction at various temperatures (25, 45, 55, and 65°C), the effect of temperature on reaction rate can clearly be observed. If desired, the reaction can be timed. Start the timer when the catalyst is added and stop it when the pink color of the catalyst is regenerated.

The green-activated complex is a result of a superoxide linkage between two octahedral coordinated cobalt atoms.

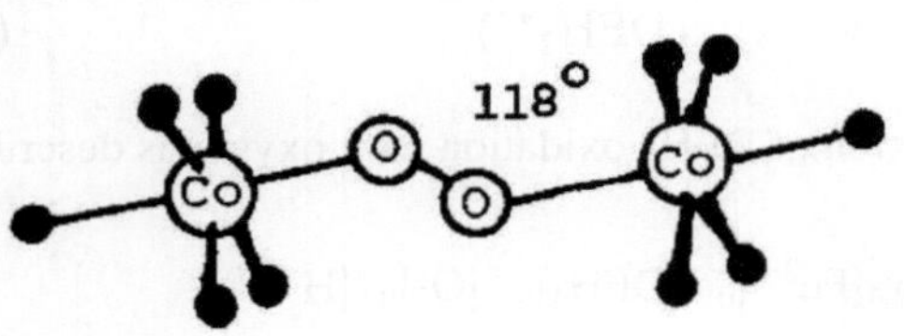

The O-O bond length is 131 pm

In the presence of the hydrogen peroxide, the Co atoms are linked by a peroxo linkage which oxidizes to a superoxo complex. (Note: peroxo complexes of cobalt are usually brown and the superoxo linkages are known to be green.) The unpaired electrons belonging to the O_2 reside in a molecular orbital of π symmetry relative to the planar Co–O–O–Co grouping and are delocalized over these four atoms. The Co atoms are formally described as Co(III) ions [27–29].

5.2.1.2 Catalytic Oxidation of Dihydroxyfumaric Acid

System $Fe^{2\pm}$–DFH_4–O_2

Dihydroxyfumaric acid exists in solution in tautomeric equilibrium with oxaloglycolic acid (OGA):

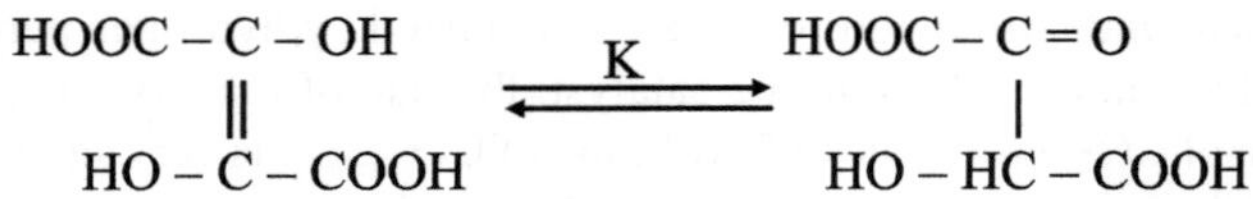

enolic form (DFH_4) ketoform (OGA)

IR-spectra, thermogravimetric, and mass spectrometric analyses and other research results have shown that the acid in solid state is in ketoform (OGA) [30]. In solution at low pH, it is in the same form, but, with pH increasing, the equilibrium will be shifted more and more to the enolic form (at pH 4, there are 80% OGA and 20% DFH_4) and at pH 12 and higher it is practically only in the form of DFH_4.

Depending on the oxidant nature, formation of three types of DFH_4 radicals is possible [31–33]:

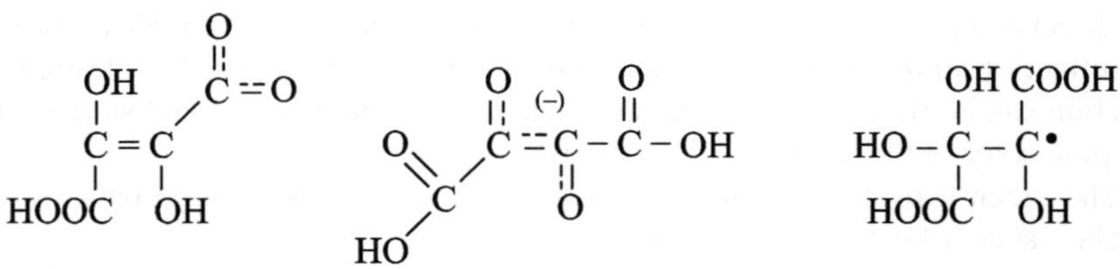

($RCOO^•$) ($DFH_2^{•-}$) (R_α)

The rate of catalytic reaction of DFH_4 oxidation with oxygen is described by the following [33]:

$$W_4 = \chi_4[Fe^{2+}]_0 \cdot [DFH_4]_0 \cdot [O_2]_0/[H^+]^{0.3}$$

where on pH 3.5 $\chi_4 = (1.5 \pm 0.5) \times 10^5\,M^{-2}\,s^{-1}$.

Mechanism of DFH_4 oxidation can be both cyclic, with oxidation degree of iron ions changing ($Fe^{2+} \longrightarrow Fe^{3+}$), and activation ($O_2$ is activated on account of coordination to iron compound).

In the cyclic mechanism, oxygen will oxidize Fe^{2+} dihydroxyfumaric complexes, and interaction of formed Fe^{3+} ions with DFH_3^- will yield again the regeneration of Fe^{2+} ions, and final product formation—diketosuccinic acid (DKH_2):

$$Fe^{3+} + DFH_3^- \xrightarrow{-H^+} Fe^{2+} + DFH_2{}^{•-} \quad (5.31)$$

$$FeDFH_3^+ + O_2 \xrightarrow{H^+} FeDFH_3^{2+} + HO^•{}_2 \quad (5.32)$$

$$FeDFH_2 + O_2 \xrightarrow{H^+} FeDFH_2^+ + HO^•{}_2 \quad (5.33)$$

$$HO_2^• + DFH_3^- \rightarrow DFH_2{}^{•-} + H_2O_2 \quad (5.34)$$

$$DFH_2{}^{•-} + Fe^{3+} \rightarrow Fe^{2+} + DKH_2 \quad (5.35)$$

In the activation mechanism, oxygen will be activated on account of the formation of triple metal–substrate–oxygen complex:

$$DFH_3^+ + O_2 \rightarrow DFH_3FeO_2^+ \tag{5.36}$$

$$DFH_3FeO_2^+ \xrightarrow{H^+} Fe^{3+} + RCOO^\bullet + H_2O_2 \tag{5.37}$$

or

$$DFH_3FeO_2^+ \xrightarrow{H^+} Fe^{3+} + DFH_2{}^{\bullet-} + H_2O_2 \tag{5.38}$$

with further DFH_4 radicals transformation into DKH_2 and into the products of DFH_4 decarboxylation—semialdehydes of tartronic and mesoxalic acids.

As soon as hydrogen peroxide is formed in the process of DFH_4 oxidation, its influence on the mechanism of DFH_4 catalytic oxidation should be considered.

System $Fe^{2\pm}$–DFH_4–H_2O_2

Hydrogen peroxide in the system Fe^{2+}–DFH_4–H_2O_2 stipulates the oxidation of dihydroxyfumaric acid. H_2O_2 decomposition to O_2 practically does not proceed. The reaction rate of DFH_4 oxidation is weakly dependent on $[H^+]_0$ and is in accordance with the following expression:

$$W_5 = \chi_5[Fe^{2+}]_0 \cdot [DFH_4]_0 \cdot [H_2O]_0$$

where at pH 3.5 $\chi_5 = 1.5 \times 10^4\,M^{-2}\,s^{-1}$. As in the oxygen system, first orders on the reaction components concentrations show that the process is nonchain. Weak dependence of oxidation rate on $[H^+]_0$ (efficiency order reaches 0.3) points to Fe(II) complexes with mono- and dianionic forms of dihydroxyfumaric acid participation in DFH_4 oxidation:

$$FeDFH_3^+ + H_2O_2 \xrightarrow{H^+} FeDFH_3^{2+} + OH^\bullet \tag{5.39}$$

$$FeDFH_2 + H_2O_2 \xrightarrow{H^+} FeDFH_2^+ + OH^\bullet \tag{5.40}$$

Radicals $OH^\bullet$ will interact with DFH_4, forming appropriate adhesion radicals by the double bond of DFH_4, which due to inner molecular transformation will turn into the more stabilized radical $DFH_2^{\bullet-}$. Interaction of $FeDFH_3^{2+}$ complexes with $R^\bullet$, $DFH_2^{\bullet-}$, and DFH_3^- yields in iron(II) regeneration, i.e., the typical cyclic process mechanism will be realized with alternating change of iron compounds oxidation degree and intermediate formation of free radicals. On such a process running, reactivity of $FeDFH_4$ Complexes with H_2O_2 is hundreds of times higher than that with O_2.

On H_2O_2 addition to the system Fe–DFH_4–O_2, a perceptible increase in the oxidation rate of dihydroxyfumaric acid to diketosuccinic one proceeds.

Thus, the cyclic mechanism of DFH_4 oxidation with molecular oxygen and hydrogen peroxide leads to the following reactions:

Table 5.1 Oxyacids (L) influence on iron redox potential, rate of DFH_4 oxidation and on interaction constants Fe(II) with H_2O_2 (κ_{41}) and Fe(III) with DFH_4 (κ_{42}). $[Fe^{2+}]_0 = 10^{-5}$ M, $[DFH_4] = 10^{-4}$ M, $[H_2O_2]_0 = 10^{-4}$ M, pH 3, $[L]_0 = 10^{-3}$ M

Ligand (L) (acids)	$\varphi_{LFe3+/LFe2+}$, V	$W, (M \cdot s^{-1})$	$\kappa_{41}, (M^{-1} \cdot s^{-1})$	$\kappa_{42}, (M^{-1} \cdot s^{-1})$
Dihydroxyfumaric	0.318	2.2×10^{-7}	5.2×10^{2}	1.2×10^{4}
Tartaric	0.295	3.0×10^{-6}	4.5×10^{3}	5.1×10^{3}
Tartronic	0.112	8.0×10^{-8}	2.0×10^{4}	2.7
Glyoxalic	0.035	1.6×10^{-8}	3.8×10^{4}	4.8×10^{-1}
Citric	0.227	2.6×10^{-6}	6.2×10^{3}	4.9×10^{3}
Glycolic	0.166	2.3×10^{-7}	8.9×10^{3}	3.6

$$Fe^{2+} + O_2(H_2O_2) \rightarrow FeOH^{2+} + HO_2^{\bullet}(OH^{\bullet}) \tag{5.41}$$

$$FeOH^{2+} + DFH_3^- \xrightarrow{H^+} Fe^{2+} + DFH_2^{\bullet-} \tag{5.42}$$

Resulting from these reactions, independent of oxidation degree in which iron ions are introduced into the system, its redistribution will occur between the states Fe(II) and Fe(III). Under stationary condition, the rates of iron ions oxidation and reduction are equal to each other:

$$k_{41} \cdot [Fe(II)][O_2(H_2O_2)] = k_{42} \cdot [Fe(OH)^{2+}][DFH_3^-].$$

The maximal rate of the catalytic process corresponds to the equality of rate constants k_{41} and k_{42}, and $[DFH_4]_0 = [H_2O_2]_0$ iron ion will be equally distributed between two oxidation states, as follows from the above equation [19].

Varying the nature of the ligand, one can change its potential within broad limits, and, accordingly, the rate constants of the elementary stages of interaction with the system components, as seen from the Table 5.1.

Systems $Cu^{2\pm}$–DFH_4–$(O_2)H_2O_2$

The rate of DFH_4 oxidation with oxygen, catalyzed by Cu^{2+} ions, is described by the following expression [34]:

$$W_6 = \chi_6 \frac{[Cu^{2+}] \cdot [DFH_4] \cdot [O_2]^{0.5}}{[H^+]^{1.5}}$$

where $\chi_6 = 0.15\,c^{-1}$. Fractional order by $[O_2]$ points to the chain process character. Initiation stages could have been the reactions resulting in Cu^{2+} ions formation. It was proved with the help of acetonitrile as inhibitor that Cu^+ ions are generated within this oxygen system [35], and DFH_4 oxidation mechanism represents the totality of the following stages:

$$Cu^+ + O_2 \longrightarrow CuO_2^+ \tag{5.43}$$

$$CuO_2^+ + DFH_3^- \rightleftharpoons DFH_3CuO_2 \tag{5.44}$$

$$DFH_3CuO_2 \longrightarrow Cu^{2+} + DFH_3^- + HO_{21}^\bullet$$
$$\longrightarrow Cu^{2+} + H_2O_2 + R^\bullet \qquad (5.45)$$
$$\longrightarrow Cu^+ + H_2O_2 + DKH_2$$

$$R^\bullet + Cu^{2+} \rightarrow Cu^+ + DKH_2 \qquad (5.46)$$

$$CuO_2^+ + Cu^+ \xrightarrow{2H^+} 2Cu^{2+} + H_2O_2 \qquad (5.47)$$

The triple complex in stage (5.45) can be decomposed by various kinetic equivalent directions. The main channel of dihydroxyfumaric radicals $R^\bullet$ transformation will be stage (5.46). Ruin of Cu^+ and CuO_2^+ ions proceeds at the breaking stage (5.47).

DFH_4 oxidation with hydrogen peroxide without catalyst (EDTA was added) proceeds slowly ($k = 7.3 \times 10^{-4}\,M^{-1}\,s^{-1}$) [36]. In the presence of Cu^{2+} ions in the system Cu^{2+}–DFH_4–H_2O, significant increase in DFH_4 oxidation rate took place, realized by the chain mechanism with quadrupole chains breakage. Initiation rate is determined by the rate of Cu^+ generation in this system. The primary product of Cu^+ ion interaction with H_2O_2 in acid medium is Cu^{3+} ion [37]:

$$Cu^+ + H_2O_2 \xrightarrow{2H^+} Cu^{3+} + 2H_2O \qquad (5.48)$$

In chain breakage the carrier participates, forming in weakly acid medium—cupryl-ion [38]:

$$Cu^{3+} \xrightarrow{H_2O} CuO^+ + 2H^+. \qquad (5.49)$$

The oxidant of DFH_4 is cupryl-ion:

$$CuO^+ + DFH_4 \longrightarrow Cu^+ + DKH_2 \qquad (5.50)$$

Depending on the conditions, in DFH_4 oxidation can take part directly with Cu^{3+} ion with other products formed—semialdehydes of mesoxalic and tartronic acids.

The kinetics of the hexacyanoferrate(III) oxidation of dihydroxyfumaric acid to hexacyanoferrate(II) and diketosuccinic acid was studied [39]. The kinetic effect of alkali metal ions, transition metal impurities, and substrate concentrations has also been analyzed. The observed inhibition effect brought about by addition of the reaction product, hexacyanoferrate(II), is a sign of a complex mechanism. Depending on the medium acidity, three mechanisms can be put forward, which involve different kinetically active forms. At low acidity, the rate-determining step involves a radical cation and both the neutral and the anion substrate forms are equally reactive ($k_1 = k_2 = 2.18 \pm 0.05\,M(-1)\,s(-1), k - 1 = 0.2 \pm 0.03$). When the medium acidity is boosted, the rate-determining step involves the neutral dihydroxyfumaric acid and two hexacyanoferrate(III) forms. In the intermediate region, the rate constant diminished with rising [H (+)] ($k'1 = 0.141 \pm 0.01$ and $k'2 = 6.80 \pm 0.05$). No specific catalytic effect by binding of alkali metal ions to oxidant has been observed. In all instances, it was assessed that the substrate decomposition is slow in comparison with the redox reaction [39].

5.2.1.3 Catalytic Oxidation of Tartronic, Glycolic, and Glyoxalic Acids, and Triose Reductone

Systems $Fe^{2\pm}$–(TA, GA, GxA, Red)–H_2O_2

Catalytic decarboxylation of oxaloglycolic acid (OGA) (tautomer of dihydroxyfumaric acid) turns it into the semialdehyde of tartronic acid [40]:

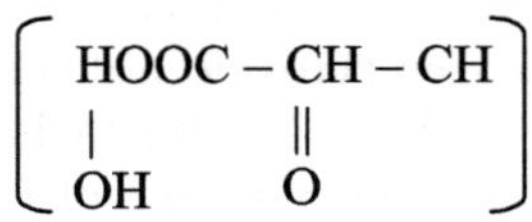

which, being oxidized (with O_2, H_2O_2), turns into tartronic acid (TA), and then, being decarboxylated, will be transformed into glycolic acid (GA), which further will be oxidized (O_2, H_2O_2) yielding glyoxalic acid (GxA):

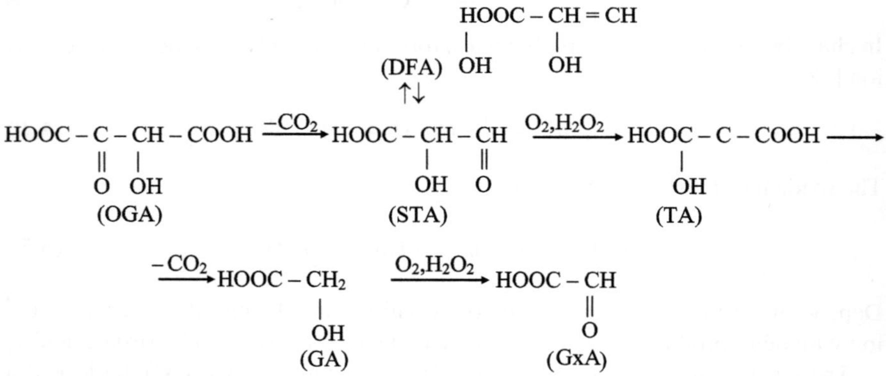

These transformations represent one of two branches of the Baroud cycle. Almost nothing was known about the catalytic transformation of TA, GA, and GxA, as well as one more component of this cycle—triose reductone:

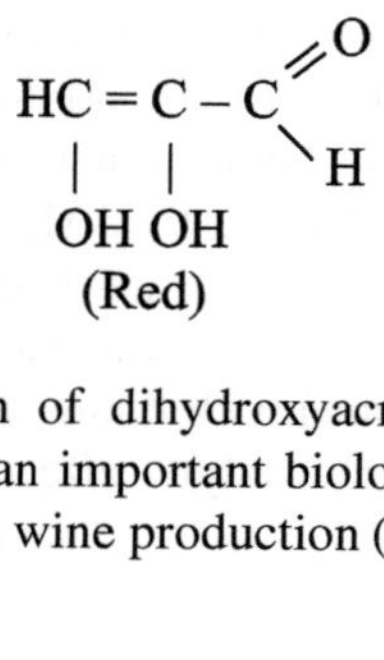
(Red)

(connected with transformation of dihydroxyacrylic acid—DHA) with Fenton reagent. Meanwhile, they play an important biological role in some plants respiration (TA), in foodstuffs, and in wine production (OGA, TA, GA, GxA, Red).

Systems H_2O_2–TA, GA, GxA are rather unstable. However, in the presence of Fe^{2+} ions they will be oxidized at a rate somewhat less than that of Fe^{3+} ions oxidation ($W_{(Fe2+\longrightarrow Fe3+)} = (56 \pm 2)\,M^{-1}\,s^{-1}$):

$$W_{TA} = \chi_1 \cdot [Fe^{2+}] \cdot [H_2O_2],\; W_{GA} = \chi_2 \cdot [Fe^{2+}]^{0.5} \cdot [GA]^{0.5} \cdot [H_2O_2],$$
$$W_{G\times A} = \chi_3 \cdot [Fe^{2+}] \cdot [H_2O_2]$$

where χ_1, χ_2, χ_3 reach (35 ± 2), (27 ± 1), (31 ± 2) $M^{-1}\,s^{-1}$, respectively. In these systems, $OH^\bullet$ radical will be formed taking part in GA, GxA, TA oxidation:

$$Fe^{2+} + H_2O_2 \longrightarrow FeOH^{2+} + OH^\bullet \tag{5.51}$$

$$OH^\bullet + GA \longrightarrow R_1 \tag{5.52}$$

$$OH^\bullet + G \times A \longrightarrow R_2 \tag{5.53}$$

$$OH^\bullet + TA \longrightarrow R_3 \tag{5.54}$$

Radicals R_2 and R_3 possess more pronounced reduction properties and can interact with Fe^{3+} ions:

$$Fe^{3+} + R_2 \longrightarrow Fe^{2+} + HOOC - COOH \tag{5.55}$$

$$Fe^{3+} + R_3 \longrightarrow Fe^{2+} + HOOC - \underset{\underset{O}{\|}}{C} - COOH \tag{5.56}$$

(MA – mesoxalic acid).

It seems, however, that the main reaction resulting in MA formation is interaction of R_3 with O_2.

The approximate (qualitative) scheme of one of these acids oxidation mechanism, namely, tartronic acid, can be presented in the following form:

$$Fe^{2+} + H_2O_2 \longrightarrow FeOH^{2+} + OH^\bullet \tag{5.57}$$

$$OH^\bullet + Fe^{2+} \longrightarrow FeOH^{2+} \tag{5.58}$$

$$OH^\bullet + H_2O_2 \longrightarrow HO_2^\bullet + H_2O \tag{5.59}$$

$$Fe^{2+} + HO_2^\bullet \longrightarrow Fe^{3+} + HO_2^- \tag{5.60}$$

$$OH^\bullet + HOOC - \underset{\underset{OH}{|}}{CH} - COO^- \longrightarrow HOOC - \underset{\underset{OH}{|}}{C^\bullet} - COO^- + H_2O \tag{5.61}$$

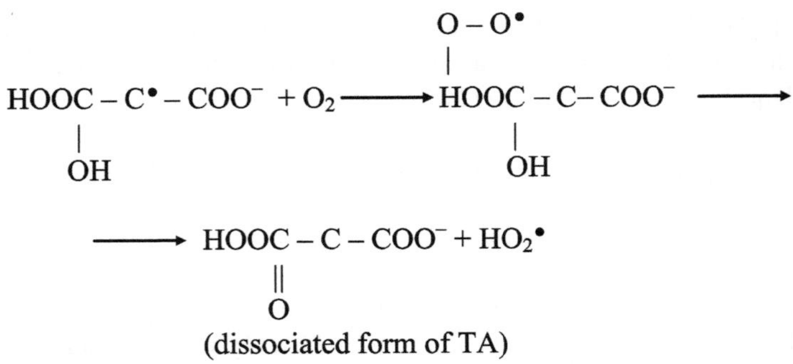

(5.62)

$$Fe^{3+} + HO_2^{\bullet} \longrightarrow Fe^{2+} + H^+ + O_2^{\bullet} \qquad (5.63)$$

Reactions of Fe^{3+} with dissociated form of TA are also possible.

Systems $Fe^{2\pm}$–Red–$(O_2)H_2O_2$

A very important compound of the Baroud cycle is triose reductone (Red) [41]:

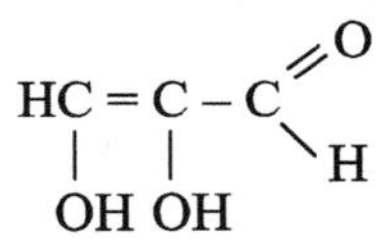

which in trace amounts is present in grapes and can be used in wine production [42]. It is obtained by glucose oxidative splitting in alkaline medium on heating to 90°C [43]. Apparently it is one of the electron donors in the photosynthesis process [44]. It was found among the metabolites from tartaric acid catalytic oxidation with Fenton reagent.

Water solutions of triose reductone are also rather stable in the presence of transition metal ions trace quantities under aerobic conditions (oxidation rate constant $<10^{-5}\,s^{-1}$) and on fourfold excess of H_2O_2 over reductone. Additives of EDTA (10^{-5} M), binding the trace amounts of iron ions, will completely inhibit Red oxidation, even on bubbling oxygen through the solution. Hence, neither oxygen itself nor hydrogen peroxide will oxidize reductone. Oxidizing particles will be formed only as a result of the initial molecules' activation with catalyst [45].

Triose reductone oxidation rate with oxygen in the presence of Fe^{2+} ions is described by the following kinetic expression:

$$W_{Red,1} = \chi_4 \cdot [Fe^{2+}]_0^{0.5} \cdot [Red] \cdot [O_2]$$

where $\chi_4 = (85 \pm 9)\,M^{-3/2}\,s^{-1}$. Within the pH range 1.5–3.5 reductone is mainly in nondissociated form (dissociation constants $\kappa_{a,1} = 6\times10^{-5}$ M, $\kappa_{a,2} = 1.4\times10^{-11}$ M [45]).

The mechanism of Red catalytic oxidation to dihydroxyacrylic acid (DHA) ($\varepsilon_{240} = 4{,}080\,M^{-1}\,cm^{-1}$) [46] is mainly reflected in the following reactions:

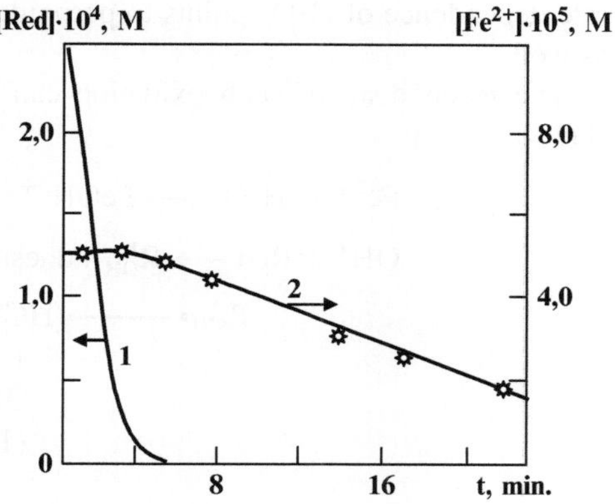

Fig. 5.6 Kinetic curves of triose reductone oxidation (1) and iron ions concentration change (2) on $[Red]_0 = 2.5 \times 10^{-4}$ M, $[Fe^{2+}]_0 = 5 \times 10^{-5}$ M, $[H_2O_2]_0 = 10^{-4}$ M, pH 2.15, $\mu = 0.01$, $T = 25°C$

$$Fe^{2+} + Red \rightleftharpoons RedFe^{2+} \tag{5.64}$$

$$RedFe^{2+} + O_2 \rightleftharpoons C_1 \qquad K_{57} \approx 10^{-10}\,M^{-1}[47] \tag{5.65}$$

$$C_1 + Fe^{2+} \rightleftharpoons Fe^{3+} + H_2O_2 + Red \qquad K_{58} \approx 10^{6}\,M^{-1}\,s^{-1}[47] \tag{5.66}$$

$$Fe^{3+} + Red \rightleftharpoons C_2 \qquad K_{59} \approx 10^{4}\,M^{-1}\,s^{-1} \tag{5.67}$$

$$C_2 \longrightarrow Fe^{2+} + R^{\bullet} \tag{5.68}$$

$$C_2 + Fe^{3+} \longrightarrow 2Fe^{2+} + DHA \tag{5.69}$$

During the catalytic oxidation of Red with hydrogen peroxide under aerobic conditions, the kinetic expression will have the following form:

$$W_{Red,2} = W_{Red,1} + \chi_5 \cdot [Fe^{2+}] \cdot [Red]^{0.5} \cdot [H_2O_2]^2$$

where $\chi_5 = (11 \pm 9) \times 10^7\,M^{-5/2}\,s^{-1}$ [48].

The rate of the Red consumption up to the deep conversion degree is subordinated to zero order. And only by the end of reaction ions of oxidized iron start accumulating in solution (Fig. 5.6). This fact may be explained by the high reduction properties of Red, which transforms Fe^{3+} into Fe^{2+}, thus creating the reversible cycle

$$Fe^{2+} \underset{Red}{\overset{H_2O_2}{\rightleftharpoons}} Fe^{3+}.$$

As follows from Fig. 5.6, with decreasing of [Red], rate of Fe^{3+} reduction becomes less than the rate of Fe^{2+} oxidation, resulting in Fe^{3+} accumulation. The rate $W_{Red,1}$ is directly proportional to catalyst concentration. Acid–basic equilibria do not have any influence within the studied pH range, which is testified by the absence of $W_{Red,1}$ dependence on pH.

On H_2O_2 concentration decreasing, the main channel of substrate consumption becomes its oxidation with oxygen (under the same initial conditions). Quadrupole rate dependence of H_2O_2 points to peroxide involvement in other oxidation stages as well.

The mechanism of such oxidation can be given as a combination of certain stages:

$$Fe^{2+} + H_2O_2 \longrightarrow FeOH^{2+} + OH^{\bullet} \quad (5.70)$$

$$OH^{\bullet} + Red \longrightarrow R^{\bullet}_{OH}(\text{adhesion radical by C=C bond}) \quad (5.71)$$

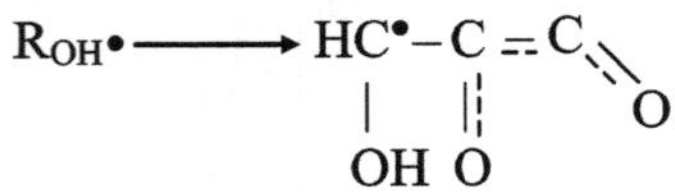

$(R^{\bullet})$

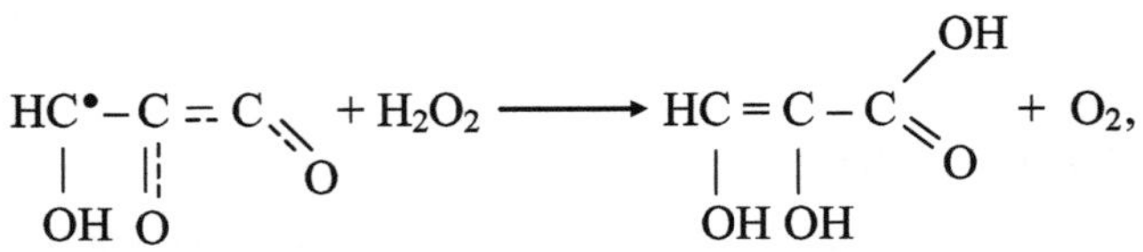

(DHA)

(5.72)

$$R^{\bullet} + Fe^{3+} \xrightarrow{H^+} Fe^{2+} + DHA \quad (5.73)$$

In this process, generation of $OH^{\bullet}$ radicals by the Fenton reaction proceeds, and consequently adhesion radicals $R^{\bullet}_{OH}$ will be formed. The latter will be transformed into $R^{\bullet}$ radicals, possessing stronger reduction properties than triose reductone. Radical $R^{\bullet}$ ($\lambda = 398\,\text{nm}, \varepsilon = 5.5 \times 10^4\,\text{M}^{-1}\,\text{sm}^{-1}$) will be formed by hydrogen atom detachment from aldehyde group, and unpaired electrons will be delocalized by dicarbonyl structure with π-electron density of C=O bonds redistribution [49].

Chain breakage is linear, resulting from $R^{\bullet}$ interaction with Fe^{3+}. The process mechanism is radical chain; this fact is testified to by quadrupole dependence on Red oxidation degree on H_2O_2.

5.2.2 *General Regularities of the Reduction–Oxidation Conversions of the Baroud Cycle Oxyacids*

Tartaric acid oxidation within the Baroud cycle (Fig. 5.1) results in the formation of many products [19]. Among them, DFH_4 and Red possess strong reducing properties. In addition, the following substances will be formed: oxaloglycolic, tartronic, diketosuccinic, mesoxalic, glycolic, glyoxalic, dihydroxyacrylic, reductonic, and oxalic (OA) acids, and semialdehydes of tartronic and mesoxalic acids. The product of O_2 reduction is H_2O_2, which makes it necessary to study tartaric acid and its metabolites oxidation with hydrogen peroxide under both anaerobic and aerobic conditions. Study of catalytic system $Fe^{2+}(Cu^{2+})$–H_4T–O_2 has shown that it will be decomposed into a variety of aerobic and anaerobic subsystems including H_2O_2 as well; Fe^{2+}–H_4T–H_2O_2, Fe^{2+}–H_4T–H_2O_2–O_2, Cu^{2+}–H_4T–$O_2(H_2O_2)$,

Fe^{2+}–DFH_4–$O_2(H_2O_2)$, Cu^{2+}–DFH_4–$O_2(H_2O_2)$ and other systems are considered in Sect. 5.2.1.

Oxidation of GA, TA, and GxA with molecular oxygen proceeds in accordance with the rate of Fe^{2+} ions interaction with O_2. In the presence of copper ions, only DFH_4 and Red will be oxidized by oxygen.

All the oxyacids considered above will be oxidized with hydrogen peroxide in the presence of iron ions, and in the presence of copper ions—only H_4T, DFH_4, GxA, and Red. OGA, $D\kappa H_2$, and MA will be subject to decarboxylation in the presence of iron and copper ions. Oxidation and decarboxylation processes proceed in living organisms in a conjugate way.

Model catalytic systems of oxyacids of Baroud cycle oxidation with oxygen and hydrogen peroxide can be conditionally subdivided into two types:

I. $Fe^{2+} - L - O_2(H_2O_2)$
II. $Fe^{3+} \cdot Cu^{2+} - L - O_2(H_2O_2)$

In systems of types I and II, metal ion is in the reduced and oxidized forms, respectively.

In systems of type I, three main mechanisms of reaction can be outlined:

(1) Metal ion does not change its valency, oxidant activation proceeds within the inner sphere, and it interacts with oxyacids in a two-electron way without intermediate radical formation:

$$LFe^{2+} + O_2(H_2O_2) \rightleftharpoons C_1(\text{intermediate compound})$$
$$C_1 + L \longrightarrow LFe^{2+} + H_2O_2(H_2O) + P_1$$

(2) Primary free radicals ($HO_2^\bullet, OH^\bullet$) will be formed in the systems, interacting with oxyacids, and secondary radicals ($R^\bullet$) possess reducing properties:

$$C_1 \longrightarrow Fe^{3+} + HO_2^\bullet(OH^\bullet)$$
$$HO_2^\bullet(OH^\bullet) + L \longrightarrow R^\bullet$$
$$R^\bullet + Fe^{3+} \longrightarrow Fe^{2+} + P_1$$

(3) Reaction of Fe^{2+} with O_2 (H_2O_2) will proceed in a radical process without metal ion involvement in the cycle continuation reactions (for example, in the case of glyoxalic acid oxidation):

$$Fe^{2+} + H_2O_2 \longrightarrow Fe^{3+} + OH^- + OH^\bullet$$
$$OH^\bullet + GxA \longrightarrow R^\bullet$$
$$R^\bullet + H_2O_2 \longrightarrow P + OH^\bullet$$

where P—here and later—will be reaction product.

Catalytic reactions of type II are subdivided into two kinds:

(1) Metal ion does not change its oxidation degree in the catalytic process, and thus active intermediate particles will not be formed (for example, in the case of H_4T oxidation in the presence of Cu^{2+} ions):

$$Cu^{2+} + H_4T \longrightarrow H_2TCu + 2H^+$$

$$H_2TCu + H_2O_2 \longrightarrow H_2TCuH_2O_2 \longrightarrow Cu^{2+} + DFH_4$$

(2) In the course of the catalytic process, active intermediate particles will be formed, participating in substrate oxidation (for example, in the case of DFH_4 oxidation in the presence of Cu^{2+} ions):

$$Cu^{2+} + DFH_4 \longrightarrow Cu^+ + R^\bullet$$

$$Cu^+ + H_2O_2 \longrightarrow Cu^{2+} + OH^\bullet$$

Many intermediate substances have not yet been detected in the process of such catalytic reactions, so a model test often helps one to obtain missing information. Proceeding from the results obtained in model systems of the tartaric acid cycle, a brutto-scheme can be proposed for tartaric acid and its metabolites oxidation mechanism (Scheme 5.2), where the products are marked, registered by a chromatographic method during the process of tartaric acid oxidation.

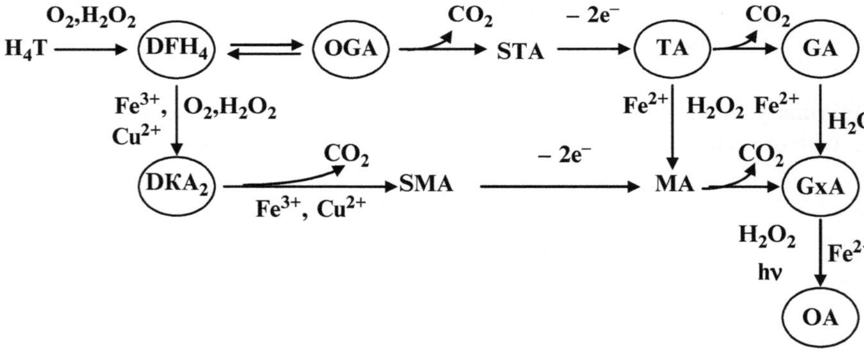

Scheme 5.2 Mechanism of tartaric acid and its metabolites oxidation

Each stage in these schemes represents the result of numerous reactions proceeding jointly with Fe^{2+} and Cu^{2+} ions involvement as catalysts, as well as that of intermediate compounds and radicals. Some stages are accelerated by light. Thus, the detailed mechanism of tartaric acid and its metabolites in real conditions is rather complicated. Nevertheless, it became possible to investigate more deeply and complete essentially the theoretical and experimental material considered above, dealing with the scheme of the dicarbonic acids transformations cycle.

5.3 Catalytic Transformations of Natural Polyphenols

5.3.1 *Role of Natural Polyphenols and Their Catalytic Oxidative Transformations*

Natural polyphenols are extremely widespread in the plant world, where they are formed in various plants' organs in the form of monomeric, oligomeric, and polymeric phenolic compounds. Nowadays the number of identified natural polyphenols approaches 3,000 [50–67]. The most widespread are flavonoids, whose main representatives are catechins, as well as antocyans. Hydroxyl groups of catechins in the ortho position play an important role in many redox transformations proceeding in both enzymatic and nonenzymatic character [50–64]. The relatively simple oxidizability of catechins stipulates their important biochemical properties such as: P—vitamin activity and spasmolytic effect. Antocyans give certain colorations (red, blue, and violet) to vegetable tissues, water foodstuff, and soil.

Natural phenols take part in respiration, photosynthesis, immunity formation, cold resistance, and other manifestations of plant growth and development. At the end of the respiratory process oxidized polyphenols–quinones serve as hydrogen acceptors. Thus, they will be reduced and repeatedly oxidized with enzyme (catecholoxidase) into quinones [65].

Redox transformations of natural polyphenols are of great importance, especially for foodstuffs production, since these transformations can be connected with foodstuffs deterioration and environmental pollution.

Numerous catechins, mainly (±)-catechins and (–)-epicatechins are encountered in the environment. Structural formula of catechins isomers are as follows [66]:

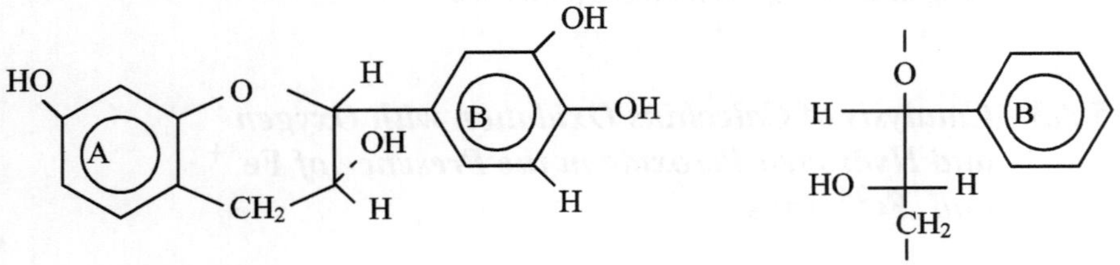

(+)-catechin

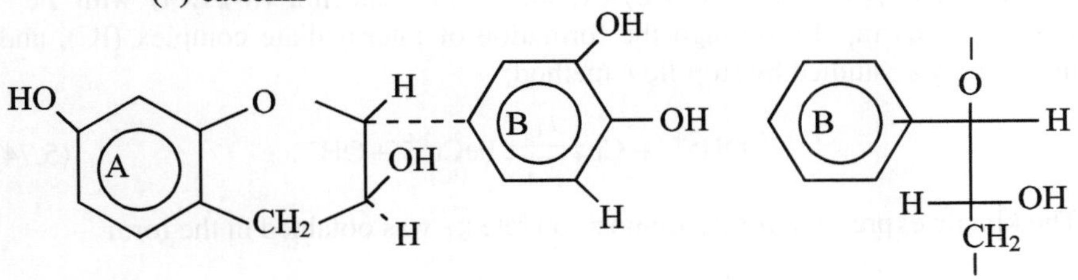

(–)-catechin

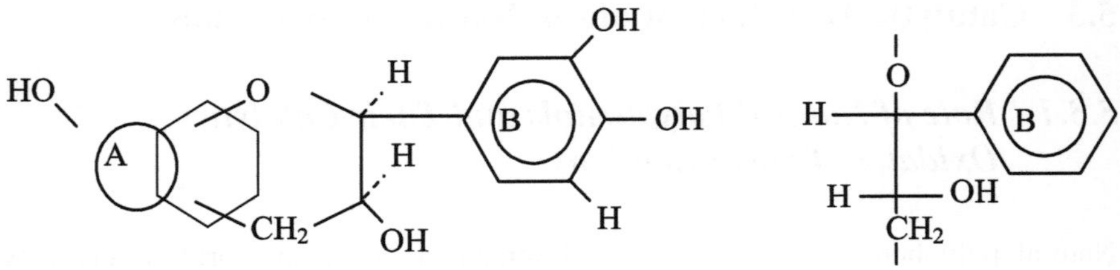

(–)-epicatechin

The main functional groups of catechins are phenolic hydroxyls. Antocyans are structurally related to catechins and capable of fixing the positive charge on C_2 atom:

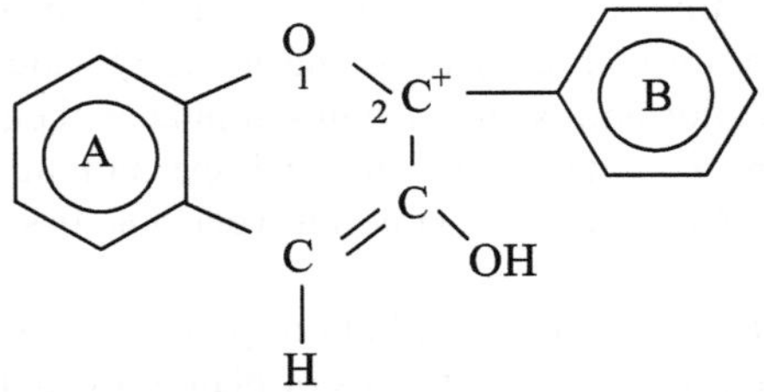

Antocyans
(hydroxyl groups are missing)

Polyphenols' condensation proceeds by two mechanisms, depending mainly on pH: acid catalysis (pH 2) and oxidative polymerisation (pH 4) (enzymatic and chemical oxidation). Until recently, all the condensation schemes presented have had a phenomenological character, without studying process kinetic peculiarities, transition metal ions influence, quantitative influence of O_2 and H_2O_2, and the influence of other substances (oxyacids). Below some research results are given, compensating to some degree for these problems.

5.3.2 *Catalysis of Catechins Oxidation with Oxygen and Hydrogen Peroxide in the Presence of* Fe^{3+} *and* Fe^{2+} *Ions*

Interaction of (±)-catechin ((±)-Ct) and (–)-epicatechin ((–)-ECt) with Fe^{3+} ions proceeds rapidly through the formation of intermediate complex (IC), and, therefore, was studied by stop-flow method:

$$FeOH^{2+} + Ct \underset{\kappa-1}{\overset{k_1}{\rightleftharpoons}} \underset{(IC)}{FeCt^{3+}} + OH^-. \tag{5.74}$$

The kinetic expression for decomposition rate κ_T was obtained in the form

$$W_7 = -\frac{d[Ct]}{dt} = \chi_3 \cdot \frac{[Fe^{3+}]_0 \cdot [Ct]_0}{[H^+]_0} \tag{5.75}$$

The intermediate complex formed will interact mainly with the second Fe^{3+} ion by a two-electron transfer mechanism:

$$FeCt^{3+} + Fe^{3+} \longrightarrow 2Fe^{2+} + P + CO_2, \tag{5.76}$$

i.e., (+)-catechin oxidation proceeds mainly with two electrons participation.

In the same way, interaction of Fe^{3+} ions with C (–)-epicatechin proceeds with the formation of intermediate complex C:

$$FeOH^{2+} + ECt \rightleftharpoons C \longrightarrow P \tag{5.77}$$

It was shown by EPR method that Fe^{3+} ion forms a complex with hydroxyl groups from the ring of B catechins. Apparently, the initiation stage is this complex decomposition, yielding epicatechin radical $(OH)RO^{\bullet}$ and Fe^{2+} ion with further radical-chain process proceeding:

$$C \longrightarrow (OH)RO^{\bullet} + Fe^{2+} \tag{5.78}$$

$$(OH)RO^{\bullet} + R(OH)_2 \longrightarrow (OH)RORO^{\bullet} + H_2O \tag{5.79}$$

where $R(OH)_2$ is an enodiolic fragment of ECt.

Stage (5.71), apparently, can be involved not only in chain continuation but also its breakage:

$$(OH)RO^{\bullet} + R(OH)_2 \longrightarrow \text{inactive radical} \tag{5.80}$$

Thus, catalytic polycondensation of ECt under anaerobic conditions proceeds. Research on (±)-catechin and (–)-epicatechin oxidation kinetics in aerobic conditions in the presence of Fe^{2+} ions did not reveal great differences; therefore, the discussion will deal with all catechin types.

Catechins oxidation with molecular oxygen proceeds very slowly. With catechin and Fe^{2+} ions, the low concentrations process rate has the following form:

$$W_8 = -\frac{d[\kappa_T]}{dt} = \chi_8 \cdot [Fe^{2+}] \cdot [Ct] \cdot [O_2]$$

where $\chi_8 = 70\,M^{-2}\,s^{-1}$ for (+)Ct, $58\,M^{-2}\,s^{-1}$ for (–)Ct, and $96\,M^{-2}\,s^{-1}$ for (–)ECt.

Interaction of $Fe^{2+}Ct$ Complex with O_2, taking into account the reducing properties of catechins, can proceed without $O_2^{\bullet-}$ radicals coming out into the bulk. The supposed reaction mechanism can include the following stages:

$$Fe^{2+} + Ct \rightleftharpoons Fe^{2+}Ct \tag{5.81}$$

$$Fe^{2+}Ct + O_2 \rightleftharpoons C \tag{5.82}$$

$$C \longrightarrow Fe^{2+} + \text{quinone} + H_2O_2 \tag{5.83}$$

$$C \longrightarrow Fe^{3+} + R^{\bullet} + H_2O_2 \tag{5.84}$$

However, the results obtained by massspectrometry and NMR C^{13} radiospectroscopy did not confirm the formation of quinones, i.e., in this case stage (5.84) of C complex decomposition, one-electron transfer products predominate. Then, interaction of Fe^{3+} ion with Ct proceeds with formation of further polymeric products:

$$Fe^{3+}Ct \rightleftharpoons C \longrightarrow Fe^{2+} + R^{\bullet} \tag{5.85}$$

$$R^{\bullet} + Fe^{2+} \xrightarrow{H^+} Fe^{3+} + Ct \tag{5.86}$$

$$R^{\bullet} + Fe^{3+} \xrightarrow{H^+(2H^+)} Fe^{2+} + P + CO_2 \tag{5.87}$$

$$R^{\bullet} + Ct(2Ct) \longrightarrow \text{dimer(trimer, etc.) of catechin} \tag{5.88}$$

In accordance with stage (5.84), the intermediate compound is hydrogen peroxide and therefore it was used for intensifying this process. In its presence, the kinetic expression for Ct oxidation rate in aerobic conditions (O_2 does not perceptibly influence the rate) can be given in the following form:

$$W_9 = -\frac{d[Ct]}{dt} = \chi_9 \cdot [Fe^{2+}] \cdot [H_2O_2] \cdot [Ct]$$

where $\chi_9 = (3 \pm 0.5) \cdot 10^4\, M^{-2}\, s^{-1}$ [57].

Stages sequence of catechin oxidation product formation in the presence of H_2O_2 looks as follows:

$$H_2O_2 + Fe^{2+} \rightleftharpoons (FeH_2O_2)^{2+} \tag{5.89}$$

$$(FeH_2O_2)^{2+} \longrightarrow FeOH^{2+} + OH\cdot \tag{5.90}$$

$$(FeH_2O_2)^{2+} + Ct \longrightarrow (CtFeH_2O_2)^{2+} \longrightarrow Fe^{2+} + P + H_2O \tag{5.91}$$

Rate independence of $[H^+]$ is explained by the fact that Fe^{2+} ion (within the pH range studied) does not undergo acid–basic dissociation.

In the case of Fe^{3+}–H_2O_2–ECt system, the most probable mechanism of the ECt oxidation process involves the stage of radicals initiation with participation of intermediate complex C, as well as the stage of chain breakage:

$$Fe^{3+} + ECt \rightleftharpoons C \tag{5.92}$$

$$C \longrightarrow R_1^{\bullet} + Fe^{2+} \tag{5.93}$$

$$R_1^{\bullet} + H_2O_2 \longrightarrow P_1 + OH^{\bullet} \tag{5.94}$$

$$OH^{\bullet} + ECt \longrightarrow R_2^{\bullet} \tag{5.95}$$

$$R_2^{\bullet} + ECt \longrightarrow \text{dimer} \tag{5.96}$$

$$R_1^{\bullet} + Fe^{3+} \longrightarrow \text{chain breakage} \tag{5.97}$$

The scheme proposed makes it possible to describe the kinetic regularities observed of ECt consumption in the presence of Fe^{3+} ions and H_2O_2.

Oxidation of antocyans—malvine (MV) within the aerobic system Fe^{3+}–H_2O_2–MV is accompanied by malvine destruction (condensation does not occur) by a radical-chain mechanism:

$$FeOH^{2+} + MV \rightleftharpoons C, \quad k_{90} = 1.4 \times 10^3\,M^{-1} \tag{5.98}$$

$$C \longrightarrow Fe^{2+} + R_3^{\bullet} \quad k_{91} = 5 \times 10^{-4}\,s^{-1} \tag{5.99}$$

$$Fe^{2+} + H_2O_2 \longrightarrow FeOH^{2+} + OH^{\bullet} \quad k_{92} = 50\,M^{-1}\,s^{-1} \tag{5.100}$$

$$R_3^{\bullet} + O_2 \longrightarrow P_3 + HO_2^{\bullet} \tag{5.101}$$

$$HO_2^{\bullet} + FeOH^{2+} \longrightarrow Fe^{2+} + O_2 + H_2O \quad k_{94} = 4 \times 10^8\,M^{-1}\,s^{-1} \tag{5.102}$$

$$OH^{\bullet} + MV \longrightarrow R_3^{\bullet} \quad k_{95} = 2.1 \times 10^9\,M^{-1}\,s^{-1} \tag{5.103}$$

$$Fe^{2+} + R_3^{\bullet} \xrightarrow{H^+} FeOH^{2+} + MV \tag{5.104}$$

where P_3 is the product of malvine oxidation.

Under the influence of UV-light Ct, ECt will condense, and MV will destruct. For instance, ECt molecule activated with light (by ring B) can interact with another ECt molecule (by ring A), leading to the formation of dimers:

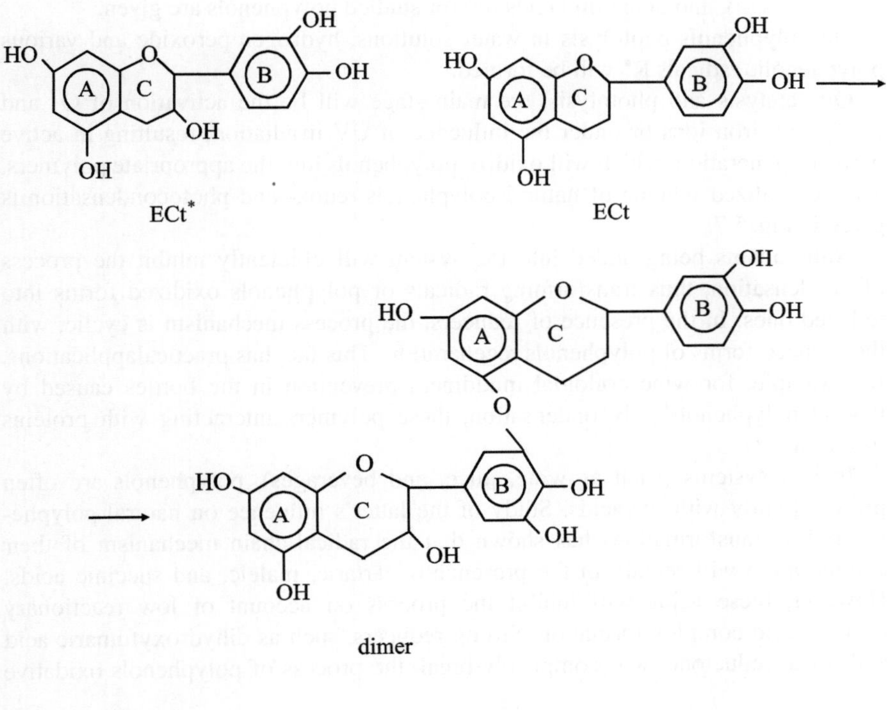

Table 5.2 Some photochemical parameters of polyphenols photolysis

Polypheno	(+)Ct	(−)Ct	(−)ECt	MV
$\kappa_d \cdot 10^4, s^{-1}$	2.0	2.7	6.6	4.6
$\varepsilon, M^{-1} sm^{-1}, \lambda = 280\,nm$	4×10^3	4×10^3	4×10^3	2.4×10^3
φ	0.1	0.1	0.22	0.37

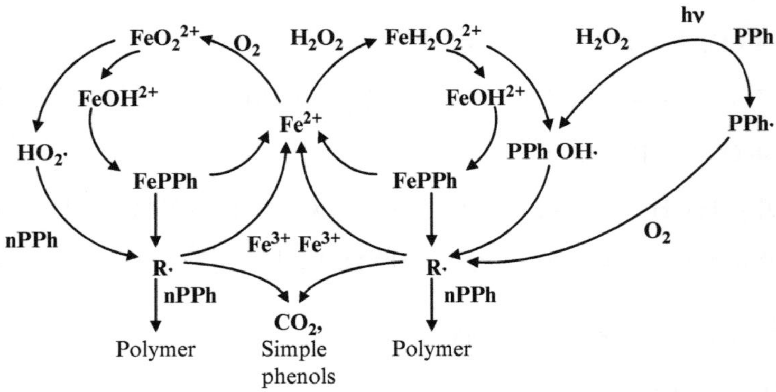

Fig. 5.7 Scheme of polyphenols redox and photolytic transformation

In Table 5.2, the values of direct photolysis rate constants (k_d), extinction coefficients (ε), and quantum yields (φ) for studied polyphenols are given.

On polyphenols photolysis in water solutions, hydrogen peroxide and various polyphenolic radicals $R^{\bullet}$ will be formed.

On catalysis and photolysis, the main stage will be the activation of O_2 and H_2O_2 with iron ions or under the influence of UV irradiation, resulting in active particles generation, which will oxidize polyphenols into the appropriate polymers. The generalized scheme of natural polyphenols redox- and photocondensation is given in Fig. 5.7.

Antioxidants being added into the system will efficiently inhibit the process of condensation, thus transforming radicals or polyphenols oxidized forms into reduced ones. In the presence of reducers, the process mechanism is cyclic, with the reduced forms of polyphenols regeneration. This fact has practicalapplications, for example, for wine colloidal muddiness prevention in the bottles caused by natural polyphenols polycondensation, these polymers interacting with proteins (see Chap. 7).

In real systems (such as wine, juice, and beverages), polyphenols are often present jointly with oxyacids. Study of the latter's influence on natural polyphenol redox transformations has shown that the radical-chain mechanism of their condensation will remain in the presence of tartaric, maleic, and succinic acids. However, these acids will inhibit the process on account of low reactionary iron oxyacid complex formation. Strong reducers, such as dihydroxyfumaric acid and triose reductone, will completely break the process of polyphenols oxidative condensation, apparently on account of interaction with intermediate radicals. The considered mechanisms of natural polyphenols transformation seem to have a more general character, modeling, for example, processes proceeding in natural water (see Chap. 6). In particular, consideration of oxidative and photochemical polyphenols condensation mechanisms can help reveal the origin of humus substances. Under conditions of solar irradiation, metal ions in various oxidation degrees, and hydrogen peroxide presence in water, monomer polyphenols oxidative radical condensation proceeding is possible, with formation of their macromolecular oxidative forms.

References

1. A.Ya. Sychev, G.G. Duca, Yu.V. Scutaru, V.V. Muntean, Catalytic oxidation of glycolic Fenton's acid, (MoldNIINTI, Chisinau, 1996), p. 21 [Russian]
2. D.L. Fridman, *Biokhimiya [Biochemistry]* (Vysshaya Shkola, Moscow, 1962), p. 56 [Russian]
3. Y. Margalit, *Concepts of Wine Chemistry* (The Wine Appreciation Guild, San Francisco, 2004), p. 476
4. I. Baroud, Ann. Chim. 1954, **9**, 535
5. A. w/f. Rodopulo, *Biochemistry of Wine-Making* (Food Industry Publications, Moscow, 1971), p. 48 [Russian]
6. J. Goode, *Science of Wine* (University of California Press, Berkeley, 2005), p. 166
7. R.B. Boulton, V.L. Singleton, L.F. Bisson, R.E. Kunkee *Principles and Practices of Winemaking* (Chapman and Hall New York 1998) p. 604
8. A.Ya. Sychev, S.O. Travin, G.G. Duca, Yu.I. Scurlatov, *Catalytic Reactions and Environmental Protection* (Stiinta, Chisinau, 1983), p. 143
9. A.Ya. Sychev, G.G. Duca, Zh. Fiz. Khimii [J. Phys. Chem.] **53**, 510–512 (1979) [Russian]
10. A.Ya. Sychev, G.G. Duca, Zh. Fiz. Khimii [J. Phys. Chem.] **59**, 1933 (1985) [Russian]
11. A.Ya. Sychev, G.G. Duca, Zh. Fiz. Khimii [J. Phys. Chem.] **60**, 78 (1986) [Russian]
12. A.Ya. Sychev, G.G. Duca, Proc. Acad. Sci. MSSR. Biol. Chem. Ser. **37** (1981) [Russian]
13. A.Ya. Sychev, G.G. Duca, Zh. Strukt. Khimii [J. Struct. Chem.] **27**, 142 (1986) [Russian]
14. G.G. Duca, Yu.V. Scutaru, A.Ya. Sychev, Zh. Fiz. Khimii [J. Phys. Chem.] **61**, 2266 (1987) [Russian]
15. A.Ya. Sychev, G.G. Duca, Zh. Fiz. Khimii [J. Phys. Chem.] **58**, 858 (1984) [Russian]
16. A.Ya. Sychev, G.G. Duca, Zh. Fiz. Khimii [J. Phys.Chem.] **53**, 513 (1979) [Russian]
17. A.Ya. Sychev, G.G. Duca, Zh. Fiz. Khimii [J. Phys. Chem.] **4**, 521 (1982) [Russian]
18. A.Ya. Sychev, G.G. Duca, *Catalytic oxidation of tartaric acid*, in *Fundamental Research in Homogeneous Catalysis*, ed. by Gordon and Breach, (New York, Sci. Publ., 1986), 3, p. 1235–1247
19. G.G. Duca, Mechanisms of ecochemical processes in water environment. Hab. Dr. Thesis. Odessa: Odessa State University, p. 46 [Russian], 1988
20. E.S. Severin, et al. *Biokhimiya [Biochemistry]* (Geotar-Media, Moscow, 2005), p. 780 [Russian]
21. A.Ya. Sychev, G.G. Duca, in *Chemistry of Coordination Organic and Inorganic Compounds* (Stiinta, Chisinau, 1978), p. 44
22. H. Anbar, P. Neta, Int. J. Appl. Rad. Isotopes **18**, 493 (1967)
23. A.Ya. Sychev, G.G. Duca, M.V. Gontsa, Patent USSR 4031715. 1987–03–28
24. W. Stumm, J.J. Morgan, *Aquatic Chemistry: An IntroduCtion Emphasizing Chemical Equilibria in Natural Waters* (Wiley, New York, 1981), p. 137

25. A.Ya. Sychev, G.G. Duca, Izv. AN MSSR. Ser. Biol. i Khim. Nauk [Proc. Acad. Sci. MSSR. Biol. Chem.Ser.] **6**, 37 [Russian] (1981)
26. E.V. Shtamm, Yu.I. Scurlatov, A.P. Purmal. Zh. Fiz. Khimii [J. Phys. Chem.] **51**, 3136 (1977)
27. D.A. Katz, Colorful catalysis: the oxidation of tartaric acid by hydrogen peroxide with a cobalt chloride catalyst, 2001. [online] Available from: http://www.chymist.com/colorful%20catalysis.pdf
28. F.A. Cotton, G. Wilkinson, *Advanced Inorganic Chemistry,* 5th edn. (Wiley, New York, 1988), pp. 468–470, 727–728, and 735–737
29. N.N. Greenwood, A. Earnshaw, *Chemistry of the Elements,* 1st edn. (Pergamon Press, New York, 1984), pp. 719–722, 1306–1307, 1311–1312
30. A.Ya. Sychev, G.G. Duca, Yu.V. Scutaru, Zh. Strukt. Khimii [J. Struct. Chem.] **27**, 142 (1986)
31. J. Jamazaki, L. Piette, Biochim. Biophys. Acta. **77**, 47 (1963)
32. B. Halliwal, S. Ahlumala, Biochem. J. **153**, 513 (1976)
33. G.H. Duca, Yu. Scurlatov, A. Sycev, *Redox Catalysis and Ecological Chemistry* (USM, Chisinau, 2002), p. 316
34. Gh. Duca, Iu. Scurlatov, *Ecological Chemistry* (USM, Chisinau, 2002), p. 289
35. A.Ya. Sychev, G.G. Duca, Zh. Fiz. Khimii [J. Phys. Chem.] **57**, 1938 (1983)
36. A.Ya. Sychev, G.G. Duca, Zh. Fiz. Khimii [J. Phys. Chem.] **53**, 510 (1979)
37. R.S. Shukla, R.P. Pant, J. Colloid Interface Sci. **268**, 168 (2003)
38. Yu.I. Scurlatov, A.P. Purmal, Zh. Fiz. Khimii [J. Phys. Chem.] **43**, 1580 (1969)
39. B. García, R. Ruiz, J.M. Leal, J. Phys. Chem. A **112**(22), 4921–4928 (2008)
40. A.Ya. Sychev, G.G. Duca, Yu.V. Skutaru. J. Struct. Chem. **27**(5), 802–804 (1986)
41. B. Fernández, M.A. Ríos J. Mol. Struct.-Theochem. **226**(3–4), 181–196 (1991)
42. T.A. Postolat II, Yu.V. Scutaru, A.Ya. Sychev, G.G. Duca, Patent USSR 1621504. 1990–09–15
43. H. Euler, B. Eistert, *Chemie und Biochemie der Reduktonen und Reduktonaten* (Ferdinant Enke Verl, Stuttgart, 1957)
44. R.E. Blankenship, *Molecular Mechanisms of Photosynthesis* (Blackwell Science, Oxford, 2002), p. 321
45. Yu.V. Scutaru, A.Ya. Sychev, Zh. Fiz. Khimii [J. Phys. Chem.] **65**, 1642 (1991)
46. A.Ya. Sychev, Yu.V. Scutaru, G.G. Duca. Zh. Fiz. Khimii [J. Phys. Chem.] **60**, 2045 (1986)
47. Yu.I Scurlatov, Kinetic studies of elementary activation mechanisms of oxygen and hydrogen peroxide. Ph.D. Thesis. Inst. Chem. Phys. Acad. Sci. USSR. Moscow, 1980
48. A.Ya. Sychev, Yu.V. Scutaru, Zh. Fiz. Khimii [J. Phys. Chem.] **65**, 1645 (1991)
49. H. Horri, Y. Abe, S. Taniguehi, Bull. Chem. Soc. **58**, 2751 (1985)
50. R. Nicholson, W. Vermeris, *Phenolic Compound Biochemistry* (Springer, Berlin, 2011), p. 276
51. F. Galeotti, E. Barile, P. Curir, M. Dolci, V. Lanzotti, Phytochem. Lett. **1**, 44 (2008)
52. C.A. Rice-Evans, N.J. Miller, G. Paganga, Free Radic. Biol. Med. **20**(7), 933–956 (1996)
53. M.G.L. Hertog, E.J.M. Feskenshollman, M.B. Katan, D. Kromhout, Lancet **342**, 1007–1011 (1993)
54. J.B Harborne, H. Baxter, *The Handbook of Natural Flavonoids,* vol 2. (Wiley, West Sussex, 1999)
55. J.B. Harborne, *The Flavonoids: Advances in Research since 1986* (Chapman and Hall, London, 1993), pp. 499–535
56. R.J. Williams, J.P. Spencer, C. Rice-Evans, Free Radic. Biol. Med. **36**, 838–849 (2004)
57. J. Harborg, *Biochemistry of Phenol Compounds* (Mir, Moscow, 1968) [Russian]
58. P. Ribereau-Gayon et al. *Handbook of Enology,* 2nd edn. (Wiley, Chichester, 2006), p. 456
59. J.W. Drynan, M.N. Clifford, J. Obuchowicz, N. Kuhnert, Nat. Prod. Rep. **27**, 417–462 (2010)
60. S. Hirotaka, A. Andria, O. Kazuyoshi, M. Shoji, S. Partomuan, Chem. Pharm. Bull. **53**(7), 866–867 (2005)
61. A.K. Broz, J.M. Vivanco, M.J. Schultz, L.G. Perry, M.W. Paschke, in *Plant Physiology*, 4th edn, ed. by L. Taiz, E. Zeiger (Sinauer, Sunderland, 2006), p. 137
62. M. Yamamoto, S. Nakatsuka, H. Otani, K. Kohmoto, S. Nishimura, Phytopathology **90**(6), 595–600 (2000)

63. Z. Chen, J. Liang, C. Zhang, C.J. Rodrigues Jr, Biotechnol. Lett. **28**(20), 1637–1640 (2006)
64. R. Bernini, E. Mincione, G. Provenzano, G. Fabrizi. Tetrahedron Lett. **46**(17), 2993–2996 (2005)
65. A.E. Rodopulo, *Biochemistry of Wine-Making* (Food Industry Publications, Moscow, 1971) [Russian]
66. J. Bakker, *Wine: Flavour Chemistry* (Wiley Blackwell, Ames, Iowa, 2011), p. 336
67. A.Ya. Sychev, V.G. Isak, G.G. Duca, M.V. Gonta, in *Catalytic Reactions in Liquid Phase*. Book of Proc. VI Conf. Alma-Ata, 1983, part II, p. 13

Chapter 6
Catalytic Processes in Ecological Chemistry

6.1 Ecological Chemistry and Catalysis with Metal Complexes

Certain catalytic processes occurring in the environment (in the atmosphere, and especially in natural water and wastewater) are considered in this chapter. To understand catalytic homogeneous redox reactions, which occur in the atmosphere and natural water with participation of natural and anthropogenic pollutants, as well as to elaborate optimal catalytic analytical methods (catalymetry), the mechanisms of such catalytic transformations are to be revealed. In natural processes ozone, oxygen, hydrogen peroxide, and reactionary particles (free radicals $OH^{\bullet}$, $OH_2^{\bullet}$), being the products of O_2 or H_2O_2 transformations, more often take part as oxidants. The catalysts turn out to be ions or complexes of metals such as Fe, Mn, Cu, Co, Ni, Cr, and Mo. More than 300 organic substances, which may play the role of ligands, were discovered in atmospheric aerosol particles. For this reason various elements in natural water can exist both in the form of metal ions and (more often) as their complexes. Complex formation of metal ions brings significant complications into the mechanisms of catalytic processes running within the environmental bodies. The involvement of pollutants, which more often play the role of substrates (S) or ligands (L_S) in catalytic reactions, in these processes makes the mechanisms even more complicated. Therefore, it is rather difficult to reveal the mechanisms of such processes even on a good qualitative level, and in many cases (due to the influence of a great number of environmental factors) it is absolutely impossible to do. For this reason, models of objects in the environment are used more often, with less or greater approximation degree, such as will be considered further, taking as an example the catalytic oxidation of SO_2 to H_2SO_4 (acid rain) in atmospheric moisture or aerosols. As soon as "natural" oxidants O_2, H_2O_2, O_3, and free radicals $OH^{\bullet}$ and $HO_2^{\bullet}$ obtained from them oxidize the pollutants, both in the atmosphere and in natural water, reactions of these particles with the catalysts (metal ions and complexes) and substrates (pollutants) are mostly important in revealing the process mechanism.

G. Duca, *Homogeneous Catalysis with Metal Complexes*, Springer Series in Chemical Physics 102, DOI 10.1007/978-3-642-24629-6_6,

During the occurrence of these processes, a variety of intermediate and then final products will be formed from the initial substances. Because of such system and process complexity, low concentrations of their participants, and other reasons, identification of certain compounds often represents a rather difficult problem. To resolve such problems, after which it will be possible to "compose," the mechanism of a running redox process like a mosaic, with more or less probability, one must use rather simple, suitable, sensitive, and selective experimental methods. Out of the all methods known, the spectroscopic ones, especially spectrophotometry in the ultraviolet and visible fields, luminescence, chemoluminescence, in some cases IR-spectroscopy, and EPR methods perfectly suit these purposes. All these methods are well described in numerous books and reviews. Therefore, we shall now only mention where they can be used on environmental bodies or their models' study and how they can be applied to calculate certain constants and parameters, characterizing the conditions of some compounds in the environment or their transformations.

Nowadays, in connection with the increase in production scale, the environmental protection problem becomes more and more topical. Study of the environmental problems is extremely difficult because of the fact that they are interlinked, involving the total system of living organisms and nonliving nature relations. It is natural that ecology, being the science which studies certain phenomena in biosphere, is closely related to the problems of biology, chemistry, chemical technology, agriculture, economy, medicine, and many other fields.

Physical and chemical processes are founded on biosphere composition change. To describe and manage the dynamic equilibrium in the biosphere, one has to know the chemical mechanisms of separate subsystem interaction. This branch of ecology has become the separate discipline *chemical ecology*, the notion of the latter being "the science of chemical interactions between living organisms and nonliving nature" [1]. The biological aspect is dominating in this definition. The tasks of *chemical ecology* include, in particular, determination of the degree of certain kinds of anthropogenic influence on living nature in order to establish the admissible load levels with regard to these, and also prediction of possible environmental consequences.

Meanwhile, another aspect of interrelations of chemistry and biology can easily be seen. We mean qualitative and quantitative composition of chemical anthropogenic pollutants of the biosphere, resulting from industrial and agricultural activities and the mechanisms of chemical transformations of substances in the environment. The chemical aspect is dominating in the resolution of these problems. This branch of science, which would be better called *ecological chemistry*, falls rather under the competence of chemical rather than biological specialists. Under the notion of *ecological chemistry* "the science of anthropogenic chemical pollutants and mechanisms of their transformations in biosphere" [2] is understood. The tasks of ecological chemistry involve the following: to reduce as much as possible the level of anthropogenic influence due to the elaboration of new or modification of already existing technological processes (more "ecological" technologies'), to create efficient treatment methods for production wastes, and to elaborate the methods of chemical pollution level prediction and regulation within the environmental bodies. Ecological chemistry involves first understanding how the uncontaminated environment works, which chemicals in what concentrations are present naturally, and with what effects. Without this it would be impossible to study accurately the effects humans have on the environment through the release of chemicals [3].

Among the other tasks we should mention are: recommendations on new methods and preparations development, to be used in agriculture and the household; recommendations on other processes prophylaxis, resulting in the environmental pollution; resolving of problems dealing with the deterioration of foodstuffs and destruction of construction materials, creating environmentally clean chemicals, assessment and prediction of pollutants' influence on the environment, etc. To resolve the problem of environmental protection the kind of research needed will need to answer the questionof the consequences brand new chemicals may have, coming into circulation in the biosphere as a result of human activities. In other words, a new qualitative approach is needed to make an assessment of human–environment interaction and its influence on anthropogenic and natural factors velocity and direction [3, 4]. Instead of simply ascertaining the changes in the environment, often negative, one should concentrate on environmental quality prediction and management.

Catalysis plays an important role in resolving of all these tasks, equally with the other branches of science. Depending on the task, the requirements of catalysis may be different. Thus, in creating new, more "ecological" technologies, the increase in catalytic reaction selectivity is the most important factor, being the way to reduce energy and raw material consumption, waste, and toxic products. Gas emission treatment is dealing with the selectivity and degree of pollutants' removal, and wastewater treatment—with the absence of selectivity of the most efficient processes of pollutant deep destruction. In the same way, to create ecologically pure substances, one must choose efficient catalysts and conditions for their use. Assessment and prediction of the environmental situation can be made, taking into account the rates and mechanisms of pollutant transformations as functions of environmental parameters.

Nowadays all the numerous research results have created the basis for understanding the catalytic activation (especially with metal complexes) and mechanisms of "environmentally friendly" oxidants action (such as ozone, oxygen, and hydrogen peroxide). Two latter oxidants (O_2 and H_2O_2) are not only used by living organisms but are also involved in the natural process of self-purification from the pollutants of anthropogenic and natural origin. This process, often catalytic, is especially characteristic for natural water self-purification, since transition metal ions and their complexes are widespread in water medium and atmospheric moisture. The sources of oxidant (H_2O_2) in natural water can be, among others, photochemical processes. The schemes of oxidation reactions mechanisms of some substrate (S), catalyzed by transition metal complexes (ML_i), are rather complicated.

If the elementary rate constant of at least one mechanism stage is absent, only a qualitative mechanism of reaction can be established. If the elementary rate constants are known for all the stages of the mechanism, one can, having compared the specific rate constant from the expression for W_{exp}, with the set of elementary rate constants from W_{scheme}, determine quantitatively the principal mechanism of catalytic reaction.

The most widely distributed, suitable, and informative methods, making it possible to determine many of the aforesaid parameters and constants, which characterize catalytic redox process, are numerous varieties of spectroscopy methods (electron, oscillation, EPR spectra, etc.). Having determined with the help of these methods (more often in combination with others) the mechanism of catalytic reaction, one can use it for reaction management (in analytical chemistry, chemical technology, or for environmental protection).

The main regularities of numerous redox catalytic reactions in environmental media and the atmosphere can be most easily established by the choice and study of an appropriate model. This model must be relatively simple, and at the same time universal and applicable for redox reactions running on various mechanisms.

Almost the only reaction complying with all these requirements is redox reaction of hydrogen peroxide catalytic decomposition (see Sect. 2.1).

This model is relatively simple. The reaction course can be easily followed by the initial peroxide disappearance (permanganatometric method) or oxygen evolution (gas-volumetric method):

$$2H_2O_2 \longrightarrow 2H_2O + O_2.$$

It follows from this reaction that one of the peroxide molecules serves as an oxidant and the other as a reducer, both having the same chemical nature. The procedure for H_2O_2 determination (spectrophotometric method) or that for the amount of O_2 evolved is simple and rapid; this has essential meaning, since thousands of kinetic tests are needed to reveal the complex dependence of redox process rate on many parameters. Besides, this reaction is catalyzed in living organisms with one of the most active enzymes—catalase.

Intermediate active and inactive, monomeric and dimeric peroxocomplexes, which appear in the process of hydrogen peroxide interaction with catalyst, can be identified and studied (for instance, their composition and stability constants can be determined), using spectrophotometric, potentiometric, and other methods.

Formation of free radical particles ($OH^{\bullet}$, $HO_2^{\bullet}$, or $O_2^{\bullet-}$) can also be identified and their concentration can be found (EPR, and especially inhibitor method).

Ease of H_2O_2 redox transformation running stipulates the possibility of the proceeding of these processes by various mechanisms, depending on the catalyst nature, its composition, and testing conditions. Thus, depending on certain circumstances, electron transfer can proceed by both inner-sphere and outer-sphere mechanisms. Besides, during the run of this reaction, one can manage to follow the change of process activity and mechanisms on catalyst change from the simplest aquated ion via the complex compound to the enzyme: $M_{aq}^{z+} \rightarrow ML_i^{z+} \rightarrow$ catalase. Most important within the context of the discussed problem is the fact that H_2O_2 is an oxidant widely distributed in the environment, often playing a key role in the oxidation processes in atmosphere and natural water.

Therefore, catalytic decomposition of H_2O_2 is, maybe, the only and simplest reaction where all the main mechanisms and peculiarities of redox transformations can be observed. Moreover, this is one of few reactions investigated today from the quantitative side.

For all these reasons, this reaction has been chosen as a model of redox transformations.

6.2 Catalytic Processes in the Atmosphere and Acid Rain Formation

Processes of intensive and far from rational industrial and agricultural development in many countries have resulted in the fact that during just one-generation lifetime the condition of both atmosphere and natural water has been significantly degraded. Catalytic processes in the atmosphere seem to be extremely important, because from the result, through the formation of so-called "acid rain" and other substances, they contribute to the pollution of not only the air medium but also the water surface and all the bodies on the Earth's surface.

Occasional pH readings in rain and fog water have been reported in industrialized areas. Industrial acid rain is a substantial problem in China and Russia [5] and areas downwind from them. The problem of acid rain has not only increased with population and industrial growth, but has also become more widespread. The use of tall smokestacks to reduce local pollution has contributed to the spread of acid rain by releasing gases into regional atmospheric circulation [6, 7]. Often deposition occurs a considerable distance downwind of the emissions, with mountainous regions tending to receive the greatest deposition (simply because of their higher rainfall).

A large number of substances exist in atmospheric pollution. The main ones are sulfur dioxide (and other compounds of sulfur), nitrogen oxides, and volatile organic compounds. They can be of either natural or anthropogenic origin. Natural sources of sulfur-containing pollutants result from the activities of microorganisms and plants. Dimethylsulfide, equally with sulfur dioxide, is among the most important pollutants. Natural sources of nitrogen oxides produce about 12% of all the atmospheric emissions and play a significantly more important role than natural sources of sulfur-containing compounds. They include, equally with the activities of soil microorganisms, water microflora of rivers, lakes, seas, and oceans, lightning discharges, volcanic activities, etc. In addition, to all these pollutants are added plants' and animals' vital activity products, natural gases, and biomass burning products (volatile organic substances, etc.), but all these pollutants are, to a great degree, the results of anthropogenic influence on the atmosphere. Thus, for example, the dynamics of the main pollutants SO_2 and NO emissions into the atmosphere during the years 1900–1990 is marked by their increase from ~10 and 3×10^6 tons/year in 1900 to ~20 and 19×10^6 tons/year in 1990. Since 1975, when a number of laws were adopted in the USA regarding pollution of the atmosphere, some decrease in emissions has been observed (especially SO_2). of the volatile organic compounds, 50% includes alkanes (propane, *n*-butane, etc.), 23% olefins (ethylene, propylene, etc.), and 8% aldehydes, and ketones (formaldehyde, acetone, etc.). About 65% of all volatile organic pollutants come to the atmosphere from natural sources resulting mainly industrial plant activity.

On March 10, 2005, EPA issued the Clean Air Interstate Rule (CAIR). This rule provides states with a solution to the problem of power plant pollution that drifts from one state to another. CAIR will permanently cap emissions of SO_2 and NO_x in the eastern United States. When fully implemented, CAIR will reduce SO_2 emissions in 28 eastern states and the District of Columbia by over 70% and NO_x emissions by over 60% from 2003 levels [8]. Since the 1990s, SO_2 emissions have dropped 40% and, according to the Pacific Research Institute, acid rain levels have dropped 65% since 1976 [9]. However, this was significantly less successful than conventional regulation in the European Union, which saw a decrease of over 70% in SO_2 emissions during the same time period [10]. In 2007, total SO_2 emissions were 8.9 million tons.

Out of the components of atmosphere, consisting mainly of nitrogen, oxygen, ozone, and water, due to the stream of solar energy, thunderstorm discharges, and other processes, numerous products of oxygen, nitrogen reduction, and water oxidation are formed. These products are reactive substances and free radicals: H_2O_2, ROOH (hydrogen peroxide and hydroperoxide), $OH^\bullet$, $HO_2^\bullet$ (hydroxyl radicals), superoxide (hydroperoxide) radicals, peroxide radicals $ROO^\bullet$, nitrogen oxides NO_x, atomic oxygen O, and hydrogen H. By influencing the components of atmosphere and the pollutants of all three types, they start running the most complex processes in the atmosphere, which result in the end in acid rain and other phenomena. These processes' complexity can be judged proceeding from the Scheme 6.1 [11] of generation and equilibria of $OH^\bullet$ and $HO_2^\bullet$ radicals in the atmosphere.

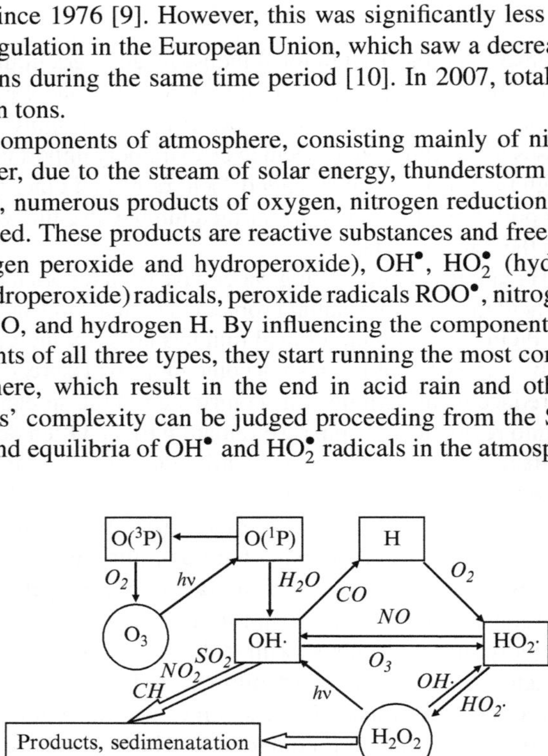

Scheme 6.1 Generation and equilibria of $OH^\bullet$ and $HO_2^{-\bullet}$ radicals in atmosphere

Rate constants of $OH^\bullet$ interaction with HSO_3^- and SO_3^{2-} reach 4.5×10^9 and $5.2 \times 10^9\,M^{-1}\,s^{-1}$, respectively [12]. The presence of transition metal compounds in the troposphere significantly complicates the radical processes, which result, in particular, in SO_2 oxidation to H_2SO_4.

These processes would not have proceeded such intensively, and maybe would have been proceeded in other way, if there were no water and transition metal ions in the atmosphere (Fe, Mn, Cu, Co, etc.), which themselves, or by complex formation with many inorganic and volatile organic substances, represent excellent catalysts for a variety of redox transformations. This takes place because water vapors, condensing into the drops and microdrops of different sizes, represent distinctive water reservoirs, catching up oxidants, transition metal compounds, and pollutants and providing favorable conditions for numerous catalytic and noncatalytic processes [13–15]. Such a process is catalytic transformation of SO_2 into H_2SO_4 in the atmosphere ("acid rain" formation [16–18]).

A comparative-kinetic review of the works in this field and the role of metal compounds is given in [19, 20].

An especially detailed review of atmospheric processes and mechanisms of catalytic oxidation processes of SO_2 into H_2SO_4 with transition metal compounds is given in [21]. Therefore, only brief consideration of catalytic processes' problems with participation of transition metal compounds and their mechanisms, taking as an example SO_2 oxidation to H_2SO_4, will be given here. As experimental data are available and their interpretation in both model tests and field conditions is contradictory in many relations [21], only the main problems of homogeneous catalytic processes' mechanisms with transition metal compounds will be considered here, especially as many atmospheric catalytic processes can proceed in parallel and be dependent on conditions.

In microwater reservoirs of the atmosphere, sulfite-ion HSO_3^- will be formed from sulfur dioxide which may be most easily oxidized with dioxygen to sulfate and sulfuric acid, but this noncatalytic process, at the pH characteristic of the drops, proceeds extremely slowly [22]. The mean rate constant of this process in the absence of transition metal ions (Fe, Mn, and Cu) (they are blockaded by complex-forming agents by adding EDTA or 1,10-phenanthroline) reaches about $1 \times 10^{-7}\ s^{-1}$, and in their presence (iron and manganese concentrations in rain water make approximately 3×10^{-6} and 3×10^{-7} M, respectively)—about $1 \times 10^{-3}\ s^{-1}$ being the difference. Even when metal ion concentrations are lower than 10^{-7} to 10^{-8} M (for example, for copper and manganese compounds), they can exert catalytic influence on many oxidation processes. Thus, transition metal ions exert strong catalytic influence on the process of SO_2 oxidation in the atmosphere [2, 23, 24]. The presence of ligands (EDTA, 1,10-phenanthroline, etc.) and antioxidants (phenol, hydroquinone, and pyrocatechol) is inhibiting S(IV) oxidation with Fe(II), Mn(II), and Cu(II) ions [25].

From the potential catalysts of S^{IV} transformation into S^{VI} the most meaningful are the compounds of iron (and also manganese [26]):

$$SO_2 + H_2O + \frac{1}{2}O_2 \xrightarrow{M^{z+}L_n} H_2SO_4 \rightleftarrows HSO_3^- + H^+.$$

Thus, for instance, Mn^{2+} ion catalyzes SO_2 oxidation in diluted water solutions, depending on concentrations, temperature, and pH. The highest oxidation rate is observed at high pH and with temperature increase. The addition of Fe^{2+} or Fe^{3+} ions to the solutions of Mn^{2+} increases the rate of SO_2 oxidation. The catalyst is $[Mn^{2+}(SO_3^{2-})_3]^{4-}$ complex. In acid solutions (pH $<$ 5) Cu^{2+} ions are neither efficient catalysts of SO_2 oxidation nor inhibitors of this process catalysis with

Mn^{2+} ions [27]. Under normal atmospheric conditions in cloud or rain water, the catalytic oxidation of sulfur dioxide plays a less important role in comparison with catalytic activation of O_2 and H_2O_2 with iron compounds in these reactions [28]. In fact, hydrogen peroxide forming in the atmosphere, more often in a catalytic or photocatalytic way, plays an important role in SO_2 oxidation to H_2SO_4.

What kinds of oxidants exist in the atmosphere? It was considered for a long time that O_3 is the main oxidant in the atmosphere. However, in the 1980s it was shown that hydrogen peroxide (equally with oxygen), formed in the atmosphere by photochemical means, is one more oxidant which can to a considerable degree influence air quality, especially by acid rain formation. H_2O_2 also deteriorates trees, bringing harm to the plant cells [13, 14, 29–32] and thus resulting in the green plantation decline [33, 34].

In which way is hydrogen peroxide forming in the atmosphere? The main ways resulting in hydrogen peroxide formation and decomposition (and also sedimentation) in the atmosphere are given in Scheme 6.2 [35].

Reactions in gaseous phase:

1. $NO_2 + h\nu \longrightarrow O(^3P) + NO,$
2. $O(^3P) + O_2 \xrightarrow{M} O_3,$
3. $O_3 + h\nu \longrightarrow O(^1D) + O_2,$
4. $O(^1D) + H_2O_2 \longrightarrow 2OH^\bullet,$
5. $OH^\bullet + CO \xrightarrow{O_2} 2OH^\bullet + CO_2,$
6. $HCHO + h\nu \xrightarrow{O_2} 2HO_2^\bullet + CO,$
7. $NO_3 + HCHO \xrightarrow{O_2} HNO_3 + HO_2^\bullet + CO,$
8. $HO_2^\bullet + HO_2^\bullet \longrightarrow H_2O_2 + O_2,$
9. $HO_2^\bullet\ H_2O + HO_2 \longrightarrow H_2O_2 + O_2 + H_2O,$
10. $HO_2^\bullet + RO_2 \longrightarrow ROOH + O_2,$
11. $HO_2^\bullet + NO \longrightarrow OH^\bullet + NO_2,$
12. $NO_2 + OH^\bullet \longrightarrow HNO_3,$
13. $O_3 + C_2H_4 \longrightarrow HCHO + 0{,}4CH_2O_2 + 0{,}4CO + 0{,}12HO_2^\bullet,$
14. $H_2O_2 + OH^\bullet \longrightarrow H_2O + HO_2^\bullet,$
15. $H_2O_2 \xrightarrow{h\nu} 2OH^\bullet,$

Reactions in water phase

16. $HO_2^\bullet(aq) + O_2^{\bullet-} \xrightarrow{H_2O} H_2O_2 + O_2 + OH^-,$
17. $HSO_3^- + H_2O_2(aq) + OH^\bullet(aq) + H^+ \longrightarrow SO_4^{2-} + 2H^+ + H_2O,$
18. $H_2O_2(aq) + OH^\bullet(aq) \longrightarrow H_2O + HO_2^\bullet(aq),$
19. $H_2O_2(aq) \xrightarrow{h\nu} 2OH^\bullet.$

Scheme 6.2 Main reactions resulting in hydrogen peroxide formation and disintegration in the atmosphere

It can be seen from Scheme 6.2 that H_2O_2 will be formed by photochemical reactions in the atmosphere. Thus, the reaction of two $HO_2^\bullet$ radicals with each other is predominating [reaction (8)]. Concentration of H_2O_2 in rain water and in clouds falls within the limits 10^{-6} to 10^{-2} M. The level of H_2O_2 is higher in summer and in the afternoon, and is lower in winter and night. In water phase H_2O_2 will be formed by reaction (16) [35].

Besides H_2O_2, organic hydroperoxides and peroxide radicals can be formed in gaseous phase by reaction (10)[2]. Total concentration of organic peroxides in rain and cloud water is far lower than that of hydrogen peroxide (according to some data [36] less than 5% of overall concentration of peroxides [H_2O_2 + ROOH]). Thus, concentrations of CH_3OOH, $HOCH_2OOH$ and 1-hydroxylethyl hydroperoxide [$CH_3CH(OH)OOH$] in rain water reach about 0.1–0.5, 0.2–0.8, and 0.1–0.3 μM, accordingly.

The results of model experiments (field tests made in Italy) have shown that the presence of atmospheric pollutants such as NO_x, volatile organic compounds, and CO, and the increase in solar irradiation, temperature, and water vapor condensation contribute to the formation of gaseous H_2O_2. High concentrations of NO_x, due to the catching of free radicals $HO_2^\bullet$, by reaction (11), and $OH^\bullet$ by reaction (12), are inhibiting H_2O_2 formation. In contrast, high concentrations of NO_x favor O_3 formation [reactions (1) and (2)] [35].

Higher concentrations of volatile organic compounds contribute to the formation of both H_2O_2 and O_3 due to larger photochemical formation of free radical particles ($OH^\bullet$, $RO^\bullet$, $HO_2^\bullet$, and $RO_2^\bullet$) from hydrocarbons and ketones. It was shown [37] that in polluted air, formaldehyde photolysis is the most important source of free radical formation, in particular, $HO_2^\bullet$ radicals [reaction (6)]. The clean atmosphere photolysis of O_3 is the main source of free radicals, in particular, $HO_2^\bullet$ radicals [reactions (3)–(5)].

The other sources of free radicals are reactions such as acetaldehyde and other aldehydes, and ketones' photolysis, reactions of O_3 with olefine [reaction (13)], and those of $NO_3^\bullet$ with organic compounds [reaction (7)] [38, 39]. High concentration favors the formation of H_2O_2, as can be seen from reaction (5). High concentrations of volatile organic compounds and CO and low concentrations of NO_x contribute to H_2O_2 formation, and vice versa.

High levels of solar radiation promote the formation of free radicals $OH^\bullet, HO_2^\bullet, RO^\bullet$, and $RO_2^\bullet$ due to formaldehyde and O_3 photolysis [38, 40]. Research has shown that the reduction of O_3 concentration by 10% results in H_2O_2 concentration redoubling.

Increasing air temperature, and also water vapor, significantly increases H_2O_2 concentration. In the latter case, this takes place because of the formation of hydrated hydroperoxide radicals ($HO_2 \cdot H_2O$), which actively interact with $HO_2^\bullet$ radicals with H_2O_2 formation [reaction (9)] [37]. It was also confirmed under field conditions that H_2O_2 level depends on the level of atmospheric pollutants such as SO_2.

The main way of H_2O_2 deterioration in the air is its interaction with $OH^\bullet$ radicals [reaction (14)] and its photolysis [reactions (15) and (19)] [28]. Hydrogen peroxide is the main oxidant of S(IV) at pH < 5. At pH > 5 its role will be decreased and the main oxidants will become $OH^\bullet$, O_3, and Fe(III)–O_2 [39–44].

In the gaseous phase various reactions of SO_2 with $OH^\bullet$ radicals take place, but the rate of their oxidation is much slower than the rate of SO_2 oxidation with hydrogen peroxide and O_3 in the water phase of the atmosphere, where the appropriate reactions run quickly.

As already mentioned, in water drops, metal ions (such as iron, copper, and manganese) are present, and their concentration is enough to provide significant acceleration of sulfite transformation to sulfate and sulfuric acid. It becomes much more rapid because of complex formation of different volatile organic substances (ligands) with transition metal ions, and, thus, formation of enough catalysts to efficiently activate oxygen molecules. The rate of this, now catalytic, process of sulfite oxidation will be dependent on a variety of factors, in particular, on metal ion and ligand ratio, pH of medium, etc. Therefore, observable rates of such an oxidation process at various sites and at different times can also be different. But, more often, in drops of mist, sulfur dioxide, in its significant part, will be very rapidly transformed into sulfuric acid. For example, in summer time 50% of air emissions SO_2 and NO_2 will be transformed into H_2SO_4 and HNO_3 in 180 and 14 h, respectively [14]. The detailed mechanism of these processes is still poorlyunderstood. Equally with oxygen activated with metal compounds and playing the role of oxidant for SO_2 transformation into H_2SO_4, an even more important role under certain conditions (up to pH < 5), as was mentioned above, is played by hydrogen peroxide in this oxidation process. One of its sources is the reaction of disproportioning of relatively low-reactive superoxide radicals $HO_2^\bullet$, present in the atmosphere [45]:

$$HO_2^\bullet + HO_2^\bullet \longrightarrow O_2 + H_2O_2.$$

Levels of $HO_2^\bullet$ in the air may reach 0.03 ppv, which corresponds to the rate of H_2O_2 formation (0.2 ppb/h in dry and up to 0.6 ppv/h in moist air in summer) [46]. An important source of $HO_2^\bullet$ is photolysis of formaldehyde [46]:

$$HCHO + h\nu \longrightarrow H + HCO^\bullet,$$

$$H + O_2 \longrightarrow HO_2^\bullet,$$

$$HCO^\bullet + O_2 \longrightarrow HO_2^\bullet + CO,$$

and reaction of formaldehyde with $OH^\bullet$ radicals:

$$HCHO + OH^\bullet \longrightarrow HCO^\bullet + H_2O^\bullet.$$

In the same way (catalytically or photochemically) iron (and manganese) complexes participate in S^{IV} oxidation in the water phase [36, 47]. Besides, these metals compounds may be concentrated in wet aerosol particles and, due to this, the oxidation rate of SO_2 will be significantly increased.

The concentration of hydrogen peroxide and iron compounds in drops of moisture reaches about 10^{-6} to 10^{-4} M [48]. Thus, typical catalytic redox systems will be formed: M^{z+} (Fe, Mn, Cu, etc.) –H_2O_2–SO_2, M^{z+}–L–H_2O_2–SO_2, M^{z+}–L–O_2–SO_2, and M^{z+}–L–O_3–SO_2, in which sulfur dioxide will be oxidized to sulfuric acid:

$$SO_2 + H_2O_2 \xrightarrow{Fe^{z+}L_n} H_2SO_4.$$

In drops of moisture the catalysts (Fe(III) compounds) on interaction with SO_2 will be reduced to Fe(II) compounds, which in turn will be oxidized with H_2O_2 to Fe(III), i.e., a reversible catalytic cycle will be realized, supporting SO_2 transformation into H_2SO_4 [48]. As was shown above, similar catalytic systems appear in the atmosphere in the phase of drops with O_3 and O_2.

In water solutions iron ions are in the form of hydrates or complex particles of various composition $Fe^{z+}(H_2O)_x(OH^-)_y$ or $Fe^{z+}L_n(H_2O)_x(OH^-)_y$ [49]. In activated metal ions or their complexes the value of pK_a for water dissociation is much less than that of free water (6–8 and 15.7, respectively [49]), i.e., on water molecule coordination to metal ion the bond O–H strength in water molecules will be significantly decreased. The solubility of aqua-hydroxocomplexes is strongly dependent on water phase pH, which under atmospheric conditions fluctuates within the limits ~3–9. Thus, solubility will be significantly decreased with pH increase due to the formation of metal hydroxide or oxide particles. The interaction of iron ions with various ligands usually yields an increase in the solubility of these complexes. Ligands, most often met in aerosols, in the order of their decreasing contents, are $SO_4^{2-}(HSO_4^-) > NO_3^- > Cl^- > NH_3 \approx CO_3^{2-}(HCO_3^-) \approx OH^- \approx SO_2 > H_2S$ (pH 5), and also different organic compounds. On transition metal complex compound formation, water molecules coordinated with metal ions are very labile (i.e., can be easily substituted with ligand or substrate molecule) in the case of V^{2+}, Ni^{2+}, Co^{2+}, Fe^{2+}, Mn^{2+}, Cr^{2+}, and Cu^{2+} ions, and less labile for Co^{3+} and Cr^{3+} ions, which is important for the creation of catalytically active complex of one metal or another.

In the atmosphere, besides carbonate equilibrium:

$$CO_2 + H_2O \rightleftarrows H_2CO_3 \overset{-H^+}{\rightleftarrows} HCO_3^- \overset{-H^+}{\rightleftarrows} CO_3^{2-},$$

similar sulfate ones exist [49]:

$$SO_2 + H_2O \rightleftarrows H_2SO_3 \overset{-H^+}{\rightleftarrows} HSO_3^- \overset{-H^+}{\rightleftarrows} SO_3^{2-}.$$

Appropriate distribution of various particles S(IV) as pH function is shown in Fig. 6.1 [50]. Particles of SO_2, HSO_3^-, and SO_3^{2-} are reactive, and in the presence of metal aqua-hydroxocomplexes at $3 < pH < 8$ can react rapidly with them, forming bound-via-oxygen sulfito-complexes. At the latter pH the transoformation into the complexes, bound via sulfur, will be far slower. In the process $S^{IV} \rightarrow S^{VI}$ various

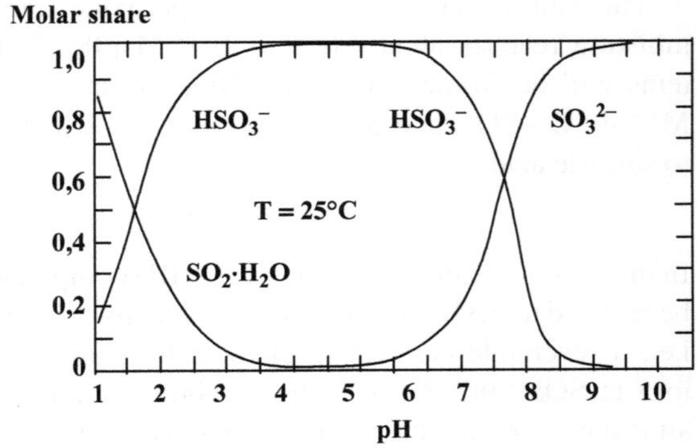

Fig. 6.1 Particles distribution S(IV) as pH function. $[S(IV)] = 1.5 \times 10^{-2}$ M, ionic force = 0.05 M

substrate radicals ($SO_3^{\bullet -}$, $SO_4^{\bullet -}$, $SO_5^{\bullet -}$) and intermediate iron compounds ($FeHSO_3^+$, $Fe(OH^-)(HSO_3^-)$, $Fe(O_2)(OH^-)^+$, etc.) are reacting, this fact being confirmed by EPR and spectrophotometric methods using the method of inhibitors. Final products are sulfates $SO_4^{2-}(H_2SO_4)$ and dithionates $S_2O_6^{2-}$. The most important oxidants are H_2O_2, O_3, $OH^\bullet$ radical, and O_2. The mechanisms of the reactions of S(IV) oxidation with hydrogen peroxide are still not clear. Numerous proposed mechanisms of sulfur catalytic oxidation into sulfuric acid in model systems Fe^{2+}–L–$O_2(H_2O_2)$–SO_2, studied under laboratory conditions, were analyzed and compared in a number of works [21, 51–56]. These mechanisms can in general be subdivided into (1) nonradical [57–60], (2) radical [61–69], and (3) mixed nonradical and radical ones [70].

Proofs of chain mechanism of the processes were obtained using substances fixing free radicals such as mannitol, tertiary butyl alcohol, thiosulfate, and hydroquinone, inhibiting the reaction.

Under anaerobic conditions the overall process of catalytic transformation S(IV) → S(VI) can be represented by the totality of reactions [21] (Scheme 6.3).

$$M^{3+} + HSO_3^- \rightleftarrows MSO_3^+ + H^+$$
$$MSO_3^+ \rightleftarrows M^{2+} + SO_3^{\bullet -}$$
$$M^{2+} + HSO_3^- \rightleftarrows MSO_3 + H^+$$
$$MSO_3 + M^{3+} \longrightarrow 2M^{2+} + SO_3^{\bullet -}$$
$$MSO_3 + SO_3^{\bullet -} \longrightarrow M^{3+} + 2SO_3^{2-}$$
$$M^{3+} + SO_3^{\bullet -} \xrightarrow{H_2O} M^{3+} + SO_4^{2-} + 2H^+$$
$$SO_3^{\bullet -} + SO_3^{\bullet -} \longrightarrow S_2O_6^{2-}$$

Scheme 6.3 Mechanism of catalytic transformation of S(IV) in S(VI)

The inhibiting stage is the formation of iron sulfite complex of various composition Fe(III):HSO_3^- = 1:1, 1:2, and 1:3 [60, 66, 71]. These complexes are decomposed spontaneously with the formation of reduced metal ion M^{2+} and sulfite radical $SO_3^{\bullet -}$ [72].

In the presence of oxygen, peroxomonosulfate radical $SO_5^{\bullet -}$ will appear, which is much more reactive than O_2, and can oxidize either sulfite ion HSO_3^- or reduced metal ion M^{2+}. Interaction of $SO_5^{\bullet -}$ with HSO_3^- results in the formation of sulfate radical $SO_4^{\bullet -}$, which in turn can interact with many participants of S(IV) into the S(VI) transformation process. Hydroperoxomonosulfite anion HSO_5^- will be also formed in this process, which equally with $SO_5^{\bullet -}$ is ionic oxidant and is able to realize many reactions. In general, the supposed mechanism, based on computer modeling of SO_2 catalytic oxidation with iron (III) compounds in the presence of oxygen, may be represented in the form of a sum of separate stages [68] (Scheme 6.4).

$$SO_3^{\bullet -} + O_2 \longrightarrow SO_5^{\bullet -}$$
$$SO_5^{\bullet -} + HSO_3^- \longrightarrow SO_3^{\bullet -} + HSO_5^-$$
$$SO_5^{\bullet -} + HSO_3^- \longrightarrow SO_4^{\bullet -} + HSO_4^-$$
$$SO_4^{\bullet -} + HSO_3^- \longrightarrow SO_3^{\bullet -} + HSO_4^-$$
$$HSO_5^- + HSO_3^- \longrightarrow 2SO_4^{2-} + 2H^+$$
$$SO_5^{\bullet -} + SO_5^{\bullet -} \longrightarrow 2SO_4^{\bullet -} + O_2$$
$$M^{2+} + SO_5^{\bullet -} \xrightarrow{H^+} M^{3+} + HSO_5^-$$
$$M^{2+} + HSO_5^- \longrightarrow M^{3+} + SO_4^{\bullet -} + OH^-$$
$$M^{2+} + HSO_5^- \longrightarrow M^{3+} + SO_4^{\bullet -} + OH^{\bullet}$$
$$M^{2+} + SO_4^{\bullet -} \longrightarrow M^{3+} + SO_4^{2-}$$
$$M^{2+} + OH^{\bullet} \longrightarrow M^{3+} + OH^-$$
$$HSO_3^- + OH^{\bullet} \longrightarrow SO_3^{\bullet -} + H_2O$$
$$HSO_5^- + OH^{\bullet} \longrightarrow SO_5^{\bullet -} + H_2O$$
$$HSO_5^- + HSO_5^- \longrightarrow SO_4^{\bullet -} + SO_4^{2-} + O_2^{\bullet -} + 2H^+$$
$$SO_3^{\bullet -} + SO_3^{\bullet -} \longrightarrow S_2O_6^{2-} + O_2$$

Scheme 6.4 Supposed mechanism of catalytic oxidation with iron (III) compounds

Some of these stages are not elementary and can proceed through the number of intermediate compounds' formation (for example, reaction $HSO_3^- + HSO_5^-$). Metal ion acts as initiator of S(IV) oxidation to S(V) (in the radical $SO_3^{\bullet -}$). It also brings SO_2 and oxidant O_2 together in the bridge-type intermediate compound $O_2^{\bullet}ML_n^{\bullet}OSO_2$, thus contributing to electron transfer from sulfur to oxidant. Very similar processes proceed on catalytic oxidation of NO and NO_2 to HNO_2 and HNO_3 through the formation of intermediate nitrito-complexes $L_nM^{\bullet}ONO$.

Equally with oxidants such as as $HO_2^{\bullet}$, H_2O_2, O_3, and free radicals, SO_2 can play an important role in metal ions oxidation, turning them into catalytically active super oxidized states [73]. Such redox cycles for iron were found both in the atmosphere [74–82] and in water solutions [82–84]. Similar redox cycles are supposed for cobalt(II), copper(II), and manganese(II) [63, 64, 74, 85–87] as well.

pH value exerts a strong influence on the mechanism of catalytic transformation of S(IV) into S(VI), but the data available are uncertain, and therefore it is impossible to formulate an expression for the rate of SO_2 catalytic oxidation over a wide range of pH [21]. In general, clearly expressed pH dependence is observed of process rate with the maximum within the pH range 2–4 [65, 88–92]. Distribution of S(IV) into S(VI) oxidation products, SO_4^{2-} and $S_2O_6^{2-}$, is also dependent on pH. Thus, in the case of SO_2 catalytic oxidation (catalyst—iron(III)) $S_2O_6^{2-}$ formation was observed at pH > 2, and at pH > 5.7 $S_2O_6^{2-}$ ions were not detected. Adding ligands such as EDTA to the system made it possible to discover $S_2O_6^{2-}$ within the pH range 5–8 as well [21].

As has been pointed out, the most efficient catalysts of SO_2 oxidation are iron and manganese compounds, but it is disputable which of them predominates in this process [21]. Under the usual atmospheric conditions ($4 \leq \text{pH} \leq 6$) a higher rate constant of overall process was observed for catalysis realized with manganese. Meanwhile, iron concentration in the atmosphere is about one order higher than that of manganese. Besides, the mechanism of oxidation process S(IV) → S(VI) is complicated and extremely sensitive to the changing of its realization conditions.

There is little known about the catalysis of SO_2 oxidation to H_2SO_4 with manganese ions. The data on kinetics and mechanism of this catalytic oxidation process with dioxygen on different pH values are contradictory even at the model level. Many mechanisms were proposed (chain and nonchain) of catalytic and noncatalytic (manganese(II) ions) processes. Thus, for instance, noncatalytic process at pH 4.5 is represented by all the stages of the chain process as shown in Scheme 6.5.

Peroxysulfate HSO_5^- and disulfate $S_2O_7^{2-}$ ions are intermediate particles. Introduction of manganese ion into the system results in the oxidation reaction S(IV) → S(VI) proceeding by catalytic chain mechanism. Thus, manganese ion will replace bisulfite ion HSO_3^- in reactions (6.2) and (6.3):

$$Mn^{2+} + SO_5^{\bullet -} \rightarrow \qquad (6.11)$$

In the presence of manganese(II) ions stages (6.2 and 6.3) will be replaced with stages (6.12 and 6.13):

$$Mn^{2+} + SO_5^{\bullet -} + S_2O \rightarrow HSO_5^- + Mn(OH)^{2+}, \qquad (6.12)$$

$$Mn(OH)^{2+} + HSO_3^- \rightarrow Mn^{2+} + SO_3^{\bullet -} + H_2O. \qquad (6.13)$$

Modification of stages (6.12 and 6.13) is also possible in the following form:

$$Mn^{2+} + SO_5^{\bullet -} \rightarrow Mn^{III}SO_5^+, \qquad (6.14)$$

$$Mn^{III}SO_5^+HSO_3^- \rightarrow Mn^{2+} + HSO_5^- + SO_3^{\bullet -}. \qquad (6.15)$$

These reactions in combination with reactions (6.1)–(6.10) represent the process mechanism, which leads to the rate expression (a) being in accordance with experimental data.

Chain continuation:

$$SO_3^{\bullet -} + O_2 \longrightarrow SO_5^{\bullet -}, \ (1)$$

$$SO_5^{\bullet -} + HSO_3^- \longrightarrow HSO_5^- + SO_3^{\bullet -}, \ (2)$$

$$SO_5^{\bullet -} + HSO_3^- \longrightarrow SO_4^{\bullet -} + SO_4^{2-} + H^+, \ (3)$$

$$SO_4^{\;-} + HSO_3^- \longrightarrow SO_4^{2-} + SO_3^{\bullet -} + H^+, \ (4)$$

$$SO_5^{\bullet -} + SO_5^{\bullet -} \longrightarrow O_2 + SO_4^{\bullet -} + SO_4^{\bullet -}, \ (5)$$

Initiating:

$$HSO_5^- + HSO_3^- \longrightarrow SO_4^{\bullet -} + SO_3^{\bullet -} + H_2O, \ (6)$$

Breakage:

$$SO_5^{\bullet -} + SO_5^{\bullet -} \longrightarrow O_2 + S_2O_8^{2-} \ (7)$$

Accompaning reactions:

$$HSO_5^- + HSO_3^- \longrightarrow S_2O_7^{2-} + H_2O \ (8)$$

$$HSO_5^- + HSO_3^- \longrightarrow SO_4^{2-} + SO_4^{2-} + 2H^+ \ (9)$$

$$S_2O_7^{2-} + H_2O \longrightarrow 2SO_4^{2-} + 2H^+ \ (10)$$

Scheme 6.5 Non-catalytic oxidation of sulfur-containing ions. In this Scheme I would ask to change with places the stages of Chain Initiation, that should be first, and Chain Continuation, that should follow the Initiation Stage

Rate expression, describing quantitatively catalysis with manganese ion, has the following form:

$$-\frac{d[O_2]}{dt} = \frac{k_6}{k_7(k_8 + k_9)}\{(k_2+k_3)[HSO_5^-]+k_{11}[Mn^{2+}]\}\{k_2[HSO_3^-]+fk_{11}[Mn^{2+}]\},$$

which involves the rate of both noncatalytic and catalytic processes (f is the fraction of reaction (6.11), which produces $SO_3^{\bullet -}$ and HSO_5^-). Thus, it is supposed that $S_2O_8^{2-}$ is initiator, and manganese ion is both initiator and multiplier of reactive particles, providing a peculiar synergism of both of these catalysts in chain reaction [93].

Summing up numerous pieces of research on catalytic oxidation mechanisms with transition metal ions (first of all, iron(III)), we should note the great importance of the kinetic modeling in reaction mechanisms established for complex kinetic systems [66, 67, 94–98].

However, the mechanisms of S(IV) oxidation into S(VI) in the atmosphere are much more sophisticated than in model systems considered above. Indeed, in all the mechanisms represented, which are often subjects of conflicting discussions, the influence of many processes taking place in the atmosphere is not considered.

One of the main components of the atmosphere, which ensures safe levels of UV-irradiation close to and on the Earth's surface for flora and fauna, and which also takes part in the maintenance of certain climate conditions on the Earth and controls the contents of some anthropogenic and natural admixtures in the atmosphere, is ozone. The amount of ozone in the atmosphere reaches 3.3 billion tons, and its maximal concentration is observed from a height of 20–25 km. Any decrease in its content will have unfavorable consequences, such as skin cancer in people, and will bring unpleasant consequences to all living creatures on the Earth's surface.

The ozone molecule is able to absorb light with a wavelength less than 290 nm, i.e., within the wavelength range where all living systems will be damaged. Besides, ozone absorbs IR-irradiation from the Earth with a wavelength of about 9–10 mkm, and thus increases the greenhouse effect caused by CO_2 by about 10%.

The bond energy of one oxygen atom in the three atoms in the ozone molecule is low, so it is highly reactive. When the ozone molecule encounters light (λ = 300–310 nm) excited unstable atomic oxygen will be formed, $O(^1D)$ having the energy of 1.97 eV, which will react with a number of substances (H_2O, H_2, and CH_4) in the atmosphere and will generate even more reactive hydroxyl radicals interacting with a variety of admixtures of anthropogenic and natural origin, thus determining their lifetime in the atmosphere. The hydroxyl radical is produced by the photodecomposition of ozone and is very highly reactive with any type of chemical compounds. It does not require a catalyst and it is approximately 108–109 times more abundant in the atmosphere than molecular oxygen [99].

All the aforesaid can be illustrated as in Scheme 6.6 [30].

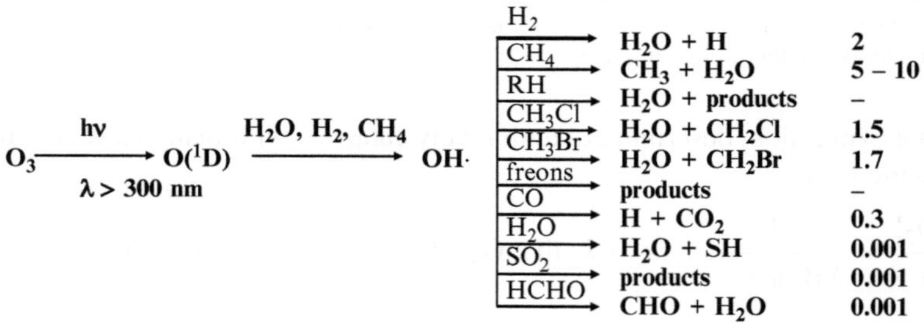

Scheme 6.6 Transformations of ozone under the influence of light ($\lambda = 300$–310 nm)

This is only the part of a possible mechanism of admixture removal from the troposphere (up to 11 km from the Earth surface), the most important of them being CO_2, CO, NO, NO_2, NH_3, CH_4, H_2, SO_2, CH_2O, and H_2S.[31].

The most dangerous regarding ozone layer anthropogenic factors are freons, which supply mainly atomic chlorine into the atmosphere. Under the action of UV-irradiation in the stratosphere, as well as on interaction with $OH^\bullet$ radicals, they dissociate evolving halogen atoms, which then deteriorate ozone molecules:

$$Cl + O_3 \rightarrow ClO + O_2,$$

$$ClO + O \rightarrow Cl + O_2.$$

As can be seen, in these pair reactions atmospheric chlorine will be preserved, which makes it possible for one of its atoms to come into the stratosphere to destroy up to 100,000 ozone molecules [30]. Thus freons are dangerous for the ozone layer.

In Scheme 6.6 the key role of a very strong oxidant, the $OH^{\bullet}$ radical, can be seen. Its determination in the atmosphere was done using the method of fluorescence (laser-excited) and also IR-spectroscopy [32]. Another strong oxidant in the atmosphere is the $HO_2^{\bullet}$ radical, which is determined by EPR method [32], and also by spectrophotometric methods using tetranitromethane as indicator.

Usually $OH^{\bullet}$ radicals are generated by pulse-radiolysis, and their accumulation is followed directly or indirectly using spectroscopic methods. The extinction coefficient of the $OH^{\bullet}$ radical in electronic spectra of neutral solution is low ($370\,M^{-1}\,SM^{-1}$ at $\lambda_{Max} = 200\,nm$), and to detect it, competitive reactions are used. Thus, in the case of $SO_2(HSO_3^-)$ radical oxidation with $OH^{\bullet}$ radicals, the rate will be determined in comparison with the reaction of $OH^{\bullet}$ with CO_3^{2-}:

$$OH^{\bullet} + CO_3^{2-} \rightarrow CO_3^{\bullet -} + OH^{-}.$$

The carbonate radical obtained is stongly absorbing at $\lambda = 600\,nm$, and the rate constant $k = 9.5 \times 10^9\,M^{-1}\,s^{-1}$.

Processes with reactive particle and free radical participation, on their interaction with atmospheric substances, may be followed by the totality of chemical reactions:

$$O_3 \xrightarrow{h\nu} O_2 + O.$$

Then, on atmospheric oxygen interaction with water vapor, carbon monoxide, organic and other substances, hydroxyl, superoxide, and peroxyl radicals, as well as hydrogen peroxide, an oxidant active in the liquid phase will be formed. Generally, these reactions may be presented as follows [33, 34]:

$$\begin{aligned}
&O + H_2O \rightarrow 2OH^{\bullet}, && OH^{\bullet} + CH_4 \rightarrow CH_3^{\bullet} + H_2O,\\
&OH^{\bullet} + CO \rightarrow CO_2 + H, && CH_3^{\bullet} + O_2 \rightarrow CH_3^{\bullet}O_2,\\
&H + O_2 \rightarrow HO_2^{\bullet}, && CH_3^{\bullet}O_2 + NO \rightarrow NO_2 + CH_3^{\bullet}O,\\
&RH + OH^{\bullet} \rightarrow ROH + H, && CH_3^{\bullet}O + O_2 \rightarrow HO_2^{\bullet} + CH_2^{\bullet}O,\\
&ROH + OH^{\bullet} \rightarrow ROH^{\bullet} + H_2, && NO_2 \xrightarrow{h\nu} NO + O,\\
&HO_2^{\bullet} + HO_2^{\bullet} \rightarrow H_2O_2 + O_2, && O + O_2 \rightarrow O_3,
\end{aligned}$$

and further, as in Scheme 6.6.

Therefore, a key particle providing transformation of other particles in the troposphere is the $OH^{\bullet}$ radical. In the presence of pollutants such as volatile organic substances (VOS) and NO in the atmosphere, the general scheme of interactions can be given as follows [14]:

$$(VOS) + 2OH^{\bullet} + 2NO \rightarrow 2HO_2^{\bullet} + (1-a)NO_2 + aRONO_2 + PAN + RCHO,$$

where RONO is organic nitrites, RCHO aldehydes, PAN peroxyacetylnitrate, and aa coefficient < 1.

Free radicals will also be formed in the nitrogen oxides photolysis, aldehyde photochemical oxidation, and PAN photodecomposition, in the reactions of O_3 with R–C = C–R′. In turn, aldehydes, ketones, and organic acids will be formed under the influence of light quantums from hydrocarbons in reactions proceeding in the gaseous phase. Gaseous phase reactions yield in the formation of main acid rain the products nitric and sulfuric acids. Pollutants of natural (H_2S, CS_2, $(CH_3)_2S$, COS, NH_3, and N_2O) and anthropogenic origin (sulfur and nitrogen oxides) are emitted into the atmosphere. It is known that the appearance of "acid rain" is connected with the increased emissions of NO_x and SO_2 in the troposphere. These oxidants, interacting in the gaseous phase with O_3, $HO_2^{\bullet}$, and $OH^{\bullet}$, are turned into molecules of nitric and sulfuric acids. At the same time, these oxides partly realize different chemical reactions, and partly remain unchanged (Scheme 6.7).

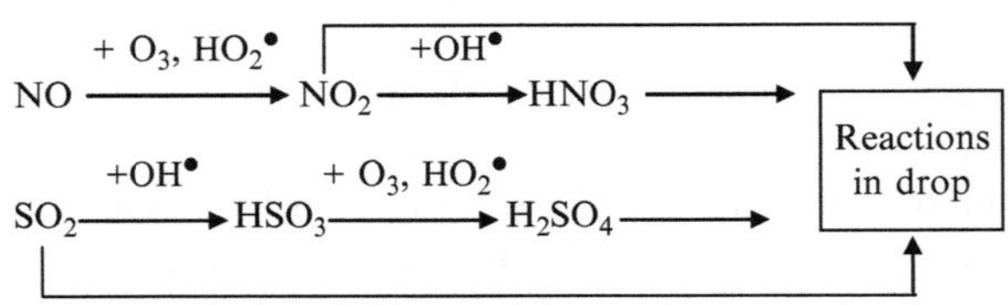

Scheme 6.7 Transformations of nitrogen and sulfur oxides in the atmosphere

Nitrogen oxide is the primary product in most of the anthropogenic pollution sources, which, reacting with hydroxyl, superoxide, and peroxide radicals, and also with ozone, will be transformed into nitrogen dioxide. The latter, interacting with hydroxyl radical, will be turned into nitric acid through the formation of nitrate ion and its further reaction with aldehydes and other compounds:

$$NO_2 + O_3 \rightarrow NO_3^- + O_2,$$

$$NO_3^- + RCHO \rightarrow HNO_3 + RCO^{\bullet}.$$

Thus, an important process is NO_x oxidation to nitric acid which often serves as the source of 20–40% acidity of cloud water [99] and which may also proceed catalytically [100–103].

Acid rain brings to the Earth's surface and into water reservoirs both nitric and sulfuric acid, and also nitrogen and sulfur oxides. An important role in these processes belongs to $OH^{\bullet}$ and $HO_2^{\bullet}$ radicals. If the aerosols' concentration is increased, on their surface ruin of $OH^{\bullet}$ and $HO_2^{\bullet}$ radicals willincrease, which may result in self-purification from tropospheric pollutants' deterioration.

Proceeding from such understanding of environmental processes, which proceed in the atmosphere with nitrogen oxides (NO and NO_2) and sulfur oxides (SO_2), various methods of reducing emissions into the atmosphere were proposed. Thus, an ammoniac method of purification from nitrogen oxides is based on similar reactions also proceeding in the atmosphere (Scheme 6.8).

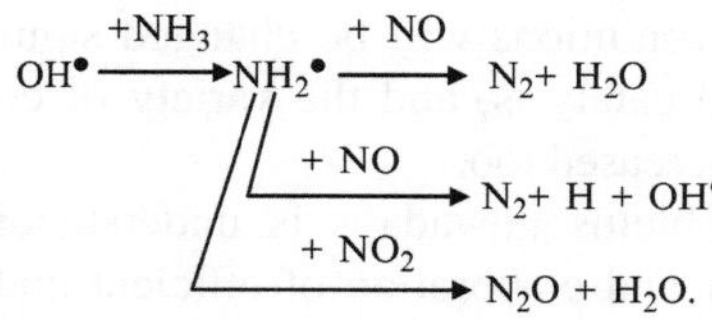

Scheme 6.8 Radical-chain process of nitrogen oxides transformation into the harmless products

This method is based on obtaining high enough concentrations of $OH^\bullet$ and $NH_2^\bullet$ radicals.

It becomes clear from this paragraph that the mechanisms of S^{IV} oxidation into S^{VI} in the atmosphere are much more complicated in reality than in the model systems discussed above. Indeed, in all the mechanisms presented, the influence of joint action of catalysts (iron, manganese, and other transition metals) was not taken into account, as well as the possible influence of complex formation of transition metal ions with organic and inorganic ligands from the atmosphere. All the various ways of free radicals and high-oxidated iron (manganese) intermediates are also not considered; photocatalytic processes are not always taken into consideration. For these and many other reasons, real atmospheric oxidation processes, including S^{IV} oxidation into S^{VI}, will significantly differ from model ones.

Acid rain damages not only the natural ecosystems but also man-made materials and structures. Marble, limestone, and sandstone can easily be dissolved by acid rain. Metals, paints, textiles, and ceramics can effortlessly be corroded. Acid rain can degrade leather and rubber. Man-made materials slowly deteriorate even when exposed to unpolluted rain, but acid rain helps to speed up the process. Acid rain causes carvings and monuments in stones to lose their features. In limestone, acidic water reacts with calcium to form calcium sulfate [99]:

$$CaCO_3 + H_2SO_4 \rightarrow CaSO_4 + H_2CO_3.$$

For iron, the acidic water produces an additional proton giving iron a positive charge:

$$4Fe(s) + 2O_2(g) + 8H(aq) \rightarrow 4Fe^{2+}(aq) + 4H_2O(l).$$

When iron reacts with more oxygen it forms iron oxide (rust):

$$4Fe^{2+}(aq) + O_2(g) + 4H_2O(l) \rightarrow 2Fe_2O_3(s) + 8H^+(aq).$$

6.3 Catalytic Redox Processes in Natural Water and Its Self-Purification

6.3.1 *Main Sources of* H_2O_2, $OH^{\bullet-}$ *and* $O_2^{\bullet-}$ $(HO_2^\bullet)$ *Radicals Formation in Natural Water*

On transition from catalytic redox processes proceeding in the atmospheric water microreservoirs to similar processes in the Earth's water reservoirs, the concentration and other conditions will be changed significantly, in particular in relation to substrates and catalysts, and the variety of components involved into these processes will be increased too.

One of the main problems nowadays is understanding the self-purification capacity of natural water and elaboration of efficient and environmental friendly methods of wastewater treatment. In all these processes and methods an important role is played by ions and complex compounds of iron, copper, and partly manganese as catalysts, and dioxygen and hydrogen peroxide (and also products of its decomposition) as oxidants. The hydrogen peroxide content in natural water is low (1×10^{-7} to 1×10^{-4} M). In sea- and freshwater it reaches about 1×10^{-6} to 1×10^{-5} M, respectively.

The main sources and appropriate reactions of hydrogen peroxide formation in natural water are shown in Scheme 6.9.

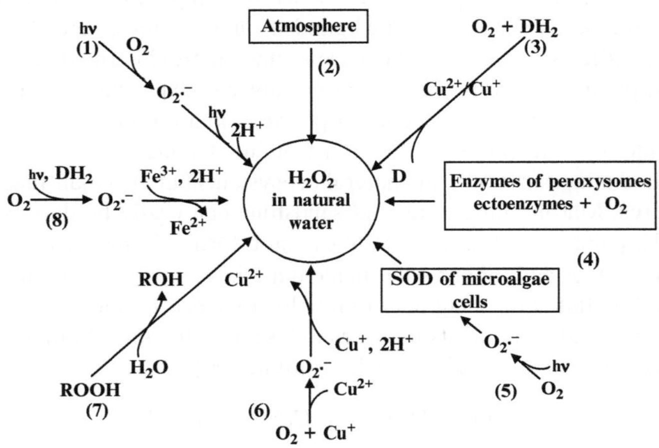

Scheme 6.9 Possible ways of hydrogen peroxide formation in water

Solar light plays a dominant role in H_2O_2 formation in natural water (1), as under its influence superoxide anion radical is soon formed as the predecessor of the main part of hydrogen peroxide generated in water. Some H_2O_2 comes from the atmosphere in the form of rain and other precipitates (2). A certain share of H_2O_2 is supplied by redox (DH_2-reducer) catalytic processes (3).

Large amounts of H_2O_2 are formed in peroxisomes of water microorganisms' cells, which to a great degree are capable of penetrating into the cytosole, and out of it, through the cell membrane, to reach natural water [104]. On the outer surface of the microorganisms of phytoplankton, so-called ectoenzymes (surface enzymes), such as deaminases, are capable of oxidizing nitrogen-containing organic substances with O_2, reducing to H_2O_2 and evolving the latter into water medium (4) [2].

In the presence of microalgae, the latter convert H_2O_2 into water in the process of their photosynthesis under the influence of the UV part of solar irradiation. $O_2^{\bullet-}$ formed in the cell, interacting with its superoxidedismutase (SOD), transforms into H_2O_2, and its excess will be excreted by the microorganism into water (5). O_2 interaction with Cu^+ also yields H_2O_2 (6) [105]. At last, hydroperoxides, the content of which in natural water is ten times less than that of H_2O_2, can be partly transformed into hydrogen peroxide (7).

In turn, hydroperoxide formation in natural water can be represented as a totality of a large number of catalytic and noncatalytic reactions [106–108]:

$$RH \xrightarrow{h\nu} R^\bullet + H$$

$$R^\bullet + O_2 \rightarrow RO_2^\bullet,$$

$$RO_2^\bullet + RH \rightarrow ROOH + R^\bullet,$$

$$RO_2^\bullet + DH^- \rightarrow ROOH + D^\bullet,$$

$$RO_2^\bullet + M^+ \xrightarrow{H^+} ROOH + M^{2+},$$

$$RO_2^\bullet + NO_2^- \xrightarrow{H^+} ROOH + NO_2,$$

$$RO_2^\bullet + HSO_3^- \rightarrow ROOH + SO_3^{2-},$$

$$M^+ + ROOH \rightarrow RO^\bullet + MOH^+.$$

The reactivity of $RO_2^\bullet$ radicals in these reactions is usually—three to five orders less than that of $RO^\bullet$ radicals. O_2 in the presence of light quanta or reducers DH_2, generated by microorganisms, will be reduced to $O_2^{\bullet-}$. $O_2^{\bullet-}$ in turn will reduce Fe^{3+} ions with H_2O_2 formation (8).

Hydrogen peroxide decomposition in natural water has been considered repeatedly [109–112] and it was shown that this process may proceed in both abiotic and biotic ways. Experimental and theoretical data on H_2O_2 decomposition catalysis with transition metal ions [113–117] may be used for the description of abiotic processes of H_2O_2 decomposition in natural water. In a water medium mostly copper ions, iron chelate complexes [118], and partly manganese ions [119] can be in homogeneous form. Catalytic decomposition of H_2O_2 (at pH $\approx$ 7) will proceed mainly on microcolloidal iron hydroxide (by nonradical mechanism) and on copper ions (through free radical formation) [120, 121]:

$$CuOH^+ + HO_2^- \rightarrow Cu^+ + O_2^\bullet$$

Further Cu^+ interaction with H_2O_2 yields CuO^+ particles (hydrolized ion Cu^{3+}), which will oxidize water with $OH^\bullet$ radical formation [122].

Possible ways of $OH^\bullet$ radicals formation in natural water are presented in Scheme 6.10.

In the presence of electron donor DH_2, hydrogen peroxide will be consumed for donor catalytic oxidation without $OH^\bullet$ free radical formation (1). If DH_2 possesses the properties of redox ligands, it can initiate $OH^\bullet$ radical formation (2). The latter can also be formed under hydrogen peroxide irradiation with light (especially UV) (3). The same takes place on hydroperoxide irradiation with light (4). The same picture is observed on their interaction with the reduced forms of metal

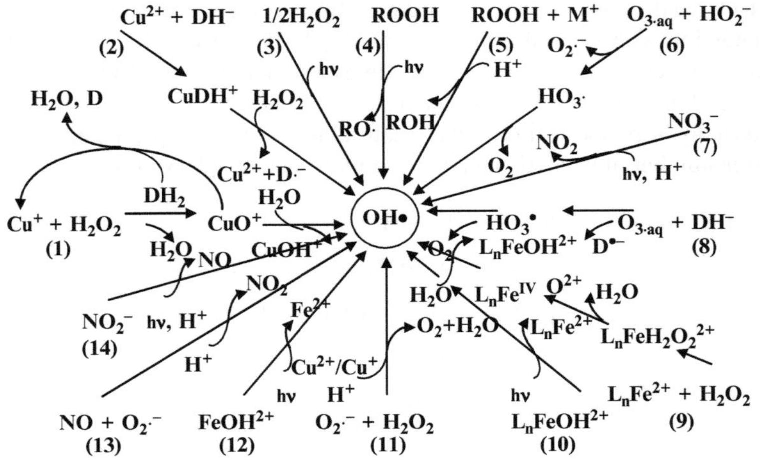

Scheme 6.10 Possible ways of OH radicals formation in water

M^+ (5). On ozone ($O_{3,aq}$) dissolving in water and interaction with dissociated form of peroxide its decomposition will proceed according to radical-chain mechanism with $O_2^{\bullet-}$ and $OH^\bullet$ radical formation (6). The last radical will also be formed on NO_3^- photochemical transformation (7). In the presence of reducers DH_2, ozone (its concentration in surface layers of water reaches 10^{-8} M) will interact with them and will also generate $OH^\bullet$ radicals (8).

Hydrogen peroxide interaction with iron(II) coordination compounds can result in the formation of superoxidized form of iron(IV) (provided the ligand is capable of stabilizing such form), and via this intermediate to form $OH^\bullet$ radicals (9). The latter will also be formed by irradiation with light of L_nFeOH^{2+}-type complexes (10). Superoxide anion radical interaction with hydrogen peroxide in water medium is catalyzed by metal compounds, in particular copper compounds, which results in the production of $OH^\bullet$ (11). Photochemical reactions of charge transfer with the same effect are also possible (12). There are nitrogen oxides, in particular NO, in natural water, which on reaction with superoxide makes the formation of $OH^\bullet$ radicals possible (13). The latter will be formed on photochemical transformation of NO_2^- as well (14).

In natural water there also exist other abiogenic and biogenic ways of free radicals' formation. Thus, for example, algae will induce DH_2 into water reducers, which in the presence of H_2O_2 and Cu^{2+} ions are capable of forming $OH^\bullet$ radicals in accordance with (2) in Scheme 6.10. Although qualitative assessment of the different ways of free radical initiation contribution is not very simple [123], their stationary concentration can be estimated both by direct measurement and by calculation proceeding from the rate of radical initiation and efficient rate constant of their ruin [110, 124].

Equally with H_2O_2 and $OH^{\bullet}$ radicals formation in natural water under the action of solar light, $O_2^{\bullet-}(HO_2^{\bullet})$ superoxide radicals will also be formed as primary ones. The main ways of superoxide radicals are presented in Scheme 6.11 [124–126].

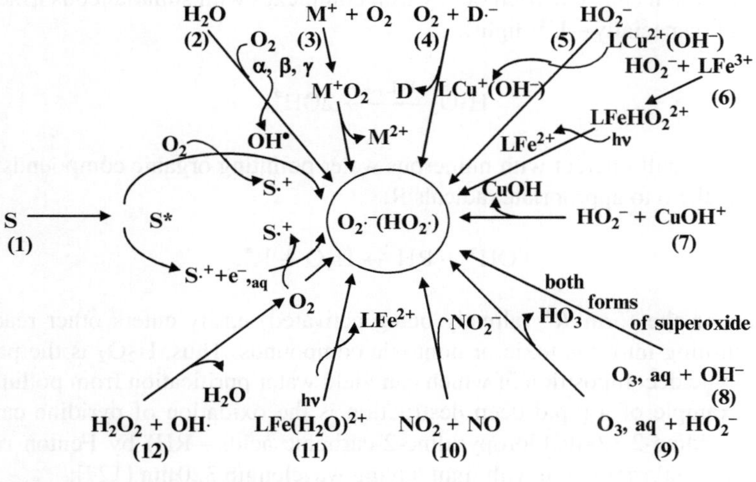

Scheme 6.11 Main ways of superoxide radicals generation

Organic substances (amino acids, primary alcohols, etc.) in natural water (S) under the influence of sunlight will be transformed into excited state (S*) and then, giving an electron to the molecule O_2, will generate $O_2^{\bullet-}$ (1). These radicals may be preceeded by hydrated electron ($e^-{}_{,aq}$) and other electron-excited particles. Such photoelectron transfer is widely distributed in natural water. Radioactive irradiation (α-, β-, and γ-rays) in the case of radiation pollution will result in $OH^{\bullet}$ radicals, hydrated electrons, and H atom formation, which, interacting with O_2, will form $O_2^{\bullet-}$ $(HO_2^{\bullet})$ superoxide radicals (2). As a result of secondary processes, reduced forms of metal M^+, reacting with O_2, will result in $O_2^{\bullet-}$ radicals (3). The transformation of O_2 into $O_2^{\bullet-}$ in natural water can result in radical-reducers $D^{\bullet-}$ (4). Superoxide anion radicals will also be formed on hydrogen peroxide reaction with copper ions and complexes (5). During the photolysis, transition metal peroxocomplexes will be decomposed in a reducing way with $O_2^{\bullet-}$ radical generation (6). Catalytic decomposition of hydrogen peroxide with $O_2^{\bullet-}$ radical formation will proceed in the presence of copper ions in a natural (close to neutral) water medium (7). Ozone dissolved in water ($O_{3,aq}$) as well can be a generator of both superoxide forms ($O_2^{\bullet-}$ and $HO_2^{\bullet}$) (8) or only one of its forms ($O_2^{\bullet-}$) (9). Nitrogen dioxide (NO_2) dissolved in water, which is capable of taking away an electron from donor (HO_2^-) (10), can take part in $HO_2^{\bullet}\cdot$ generation. Charge transfer photochemical reactions are also important (especially for iron complexes) in $HO_2^{\bullet}$ producing (11). Reactions of hydrogen peroxide with $OH^{\bullet}$ radicals in natural water will also be producing $HO_2^{\bullet}$ (12). Stationary concentration of $O_2^{\bullet-}$ radical in natural water reaches $\approx 10^{-8}$ to 10^{-9} M.

These are the main ways of the most important intermediates of the O_2 reduction process (hydrogen peroxide, $O_2^{\bullet-}$, and $OH^{\bullet}$ radical formation).

6.3.2 H_2O_2 and O_2 Activation in Natural Water and the Role of Metal Compounds

H_2O_2 activation can be carried out by iron complexes with simultaneous irradiation with light (especially—UV light):

$$H_2O_2 \xrightarrow{FeL_n, h\nu} 2OH^{\bullet},$$

OH radicals will interact with numerous water polluting organic compounds (RH), oxidizing them to appropriate radicals R:

$$OH^{\bullet} + RH \rightarrow H_2O + R^{\bullet}.$$

Pollutant in the form $R^{\bullet}$ (already being activated) easily enters other reactions, finally turning into less toxic or nontoxic compounds. Thus, H_2O_2 is the particle, the catalytic decomposition of which can yield water purification from pollutants.

An example of a rapid deep destruction is the oxidation of pyridine carbonic acids (pyridine-2,3,6-dichloropyridine-2-carbonic acids—RH) by Fenton reagent with additional irradiation with light having wavelength 320 nm [127]:

$$Fe^{2+}(L_nFe^{2+}) + H_2O_2 \rightarrow Fe^{3+}(L_nFe^{3+}) + OH^- + OH^{\bullet},$$
$$Fe^{3+}(L_nFe^{3+}) + h\nu \rightarrow Fe^{2+}(L_nFe^{2+}),$$
$$OHg^+RH \rightarrow Rg^+H_2O$$

Substrate $DH^{\bullet}$ radical formation in peroxidase reactions is also possible, provided there are substrate-donors of hydrogen atom in natural water:

$$H_2O_2 + Fe^{3+} \rightarrow FeO^{3+}H_2O,$$
$$FeO^{3+} + DH_2 \rightarrow FeOH^{3+} + DH^{\bullet}.$$
$$FeOH^{3+} + DH_2 \rightarrow Fe^{3+} + DH^{\bullet}.$$
$$DH^{\bullet} + DH^{\bullet} \rightarrow D + DH_2.$$
$$DH^{\bullet} \rightarrow \text{products of } DH_2 \text{ radical transformation.}$$

In the presence of Cu^{2+} ions in natural water, the following reactions with hydrogen peroxide can proceed in a homogeneous way:

$$Cu^{2+} + H_2O_2 \xleftrightarrow{-2H^+} CuO_2 \rightarrow Cu^+ + O_2^{\bullet -}.$$
$$Cu^{2+}O_2^{\bullet -} \rightarrow Cu^+ + O_2,$$
$$Cu^+ + H_2O_2 \rightarrow Cu(OH)_2 \rightarrow CuO^+ \rightarrow CuOH^+ + OH^{\bullet}.$$

If there are suitable ligands (bicarbonates, phosphates, etc.) in water medium, capable of transforming manganese(IV) oxide into manganese(IV) soluble complex compounds, the following one-electron reactions are possible [119]:

$$LMn(IV) \xrightarrow{e^-} LMn(III),$$

$$LMn(III) + H_2O_2 \xrightarrow{2H^+} LMn(II) + O_2^{\bullet -},$$

$$LMn(III) + H_2O_2 \xrightarrow{H^+} LMn(IV) + OH^{\bullet},$$

$$LMn(II) + OH^{\bullet} \longrightarrow LMn(III) + O_2.$$

Thus, during H_2O_2 interaction with transition metal compounds, both highly-reactive compounds of Cu(III), Fe(IV), Mn(III), and Mn(IV) and strong oxidant $OH^{\bullet}$ radicals are formed. Up to 50% of hydrogen peroxide can be transformed into free radicals, and stationary concentration of $OH^{\bullet}$ radicals in natural water reaches about 10^{-15} to 10^{-17} M [119]. Taking into account that $k_{OH^{\bullet}+RH} \approx 10^9$–$10^{10}\,M^{-1}\,s^{-1}$, the lifetime of many pollutants (RH) is only several days, which indicates the need for conditions of natural water optimal self-purification.

Catalytic redox processes in natural water have been considered in detail in [2, 104, 108, 110, 128]. Further development and amplification of these ideas are given in [129, 130]. In accordance with these concepts, natural water, from the chemical point of view, in environmental conditions represents itself as an open multicomponent, multiphase, photochemical, catalytic redox system of M^+–M^{2+}–O_2–H_2O_2–L–DH_2–S–In–$h\nu$ type, in which the processes of oxygen activation, hydrogen peroxide ($\approx 10^{-5}$ mol/L for 24 h), and pollutant decomposition proceed [2]. Thus, both stoichiometric reduction and oxidation of metal ions proceed with the appropriate activation of oxygen and hydrogen peroxide by them:

$$2M^{2+} + SH_2 \xrightarrow{-2H^+} 2M^+ + S$$

$$2M^+ + O_2 \xrightarrow{+2H^+} 2M^{2+} + H_2O_2,$$

$$2M^+ + H_2O_2 \xrightarrow{2H^+} 2M^{2+} + 2H_2O,$$

and catalytic processes of pollutant SH_2 oxidation:

$$SH_2 + O_2 \xrightarrow{M^{2+},M^+} S + H_2O,$$

$$SH_2 + H_2O_2 \xrightarrow{M^{2+},M^+} S + 2H_2O.$$

The concentration of iron, copper, and manganese ions in natural water reaches for the first ion $\approx 10^{-5}$ mol/L, and for the others: 10^{-6} to 10^{-8} mol/L. Ions and compounds of these transition metals are the most important in the realization of redox transformations of O_2 into H_2O_2. Copper ions are in natural water in homogeneous form in both oxidating and reducing states [118], together with iron ions, in oxidized form (in the form of microcolloidal hydroxide) and manganese ions, in the form of either free ions Mn^{2+} or in the form of microcolloidal oxide Mn(IV) [131]. For copper ions one- or two-electron reactions are characteristic with O_2 and H_2O_2 participation, while for iron and manganese ions, two-electron reactions of hydrogen peroxide catalytic decomposition [120, 132].

In natural water, redox reactions proceed and metal ions of transition valency play the role of process catalysts. Natural water pollution must be compensated with natural self-purification processes, in which a major role is played by O_2 and H_2O_2. It is most important that self-purification processes be considered from the positions of chemical kinetics and catalysis, as pollutants' transformations usually proceed in the reactions of catalysis, oxidation with free radicals, decarboxylation, hydrolysis, and other reactions, often catalyzed by transition metal ions and complexes.

In oxidized form iron ions are present in natural water (pH 5–8) in the form of either insoluble hydroxide or soluble complexes. In water medium at $10^{-6} \leq [Fe^{3+}] \leq 10^{-5}$ mol/L formation of iron polynuclear compound is insignificant. However, at pH 3.5 and $[Fe^{3+}] = 2 \times 10^{-5}$ mol/L the share of polynuclear iron form reaches 0.32. Hydroxide dissolving on account of Fe^{3+} ion binding in complex with ligands can take place only if the latter forms very strong complexes characterized by binding constants $\geq 10^{17}$ mol^{-1}, which is characteristic of polydentate ligands, forming chelate complexes. Oxy- and oxoacids and humic acids—components of the environment—play the role of such ligands.

Insoluble iron forms, settling in bottom deposits, find themselves under reductive conditions and can be reduced to Fe(II) compounds, possessing much higher solubility than Fe(III) in natural water:

$$Fe^{3+} - O - Fe^{3+} + DH_2 \xrightarrow{-H_2O} -Fe^{2+} \ldots Fe^{2+} - +D,$$

$$-Fe^{2+} \ldots Fe^{2+} - +O_2 \xrightarrow{H_2O} Fe^{3+} - O - Fe^{3+} + H_2O_2,$$

$$-Fe^{2+} \ldots Fe^{2+} - +H_2O_2 \rightarrow Fe^{3+} - O - Fe^{3+} + H_2O.$$

Diffusing from bottom deposits into water bulk, Fe^{2+} ions will once again be oxidized with oxygen and hydrogen peroxide dissolved in the water. In the case of their binding into complexes, the latter can take part in active intermediate particle generation, which will later oxidize pollutants. Thus, iron complexes play an important role in the self-purification of natural water.

O_2 will be activated within this process. the mechanism of O_2 activation with iron coordination compounds involves the formation of intermediate oxide, and then a peroxide complex (Scheme 6.12) [13, 111, 113].

The presence of oxygen and H_2O_2, jointly with iron complexes in natural water, results in redox processes proceeding with the formation of active free radicals.

Fe^{2+} ions or complexes will be oxidized with hydrogen peroxide, yielding ions or compounds of Fe^{3+} and hydroxyl radical $OH^{\bullet}$:

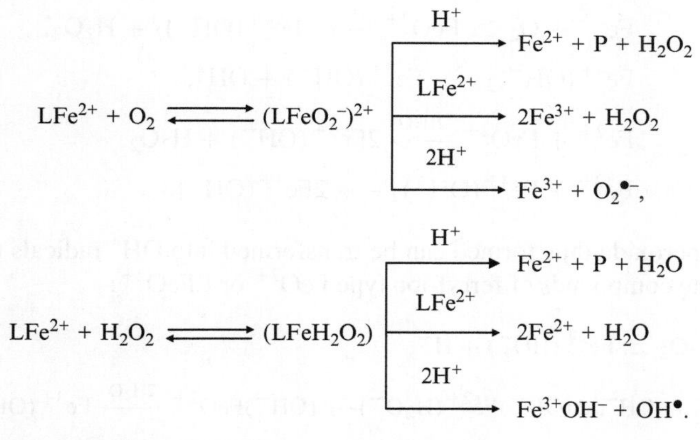

Scheme 6.12 Mechanism of oxygen activation by iron coordination compounds

$$Fe^{2+} + H_2O_2 \rightleftarrows FeHO_2^{+} \rightleftarrows (FeHO_2^{+})(OH^{-}) \xrightarrow{-H_2O} (OH^{-})FeO^{2+} \rightarrow Fe^{3+}(OH^{-})_2 + OH^{\bullet}.$$

The most efficient reactions of pollutant oxidation in water medium proceed by radical mechanism, hydroxyl radical being the most reactionary among all the radicals ($OH^{\bullet}$, $O_2^{\bullet-}$, $RO_2^{\bullet}$, etc.). It is initiated mostly in photochemical reactions (Scheme 6.10) with H_2O_2 participation:

$$H_2O_2 \xrightarrow{h\nu} 2OH^{\bullet},$$

$$FeOR^{2+} \rightarrow Fe^{2+} + RO^{\bullet},$$

$$Fe^{2+} + H_2O_2 \rightarrow FeOH^{2+} + OH^{\bullet},$$

$$LFeDH^{-} + H_2O_2 \rightarrow LFeOH^{2+} + D^{\bullet-} + OH^{\bullet}.$$

Unlike OH^{+}, superoxide radicals possess dual redox function. For this reason, these radicals in neutral and acid media are easily disproportioned to O_2 and H_2O_2. In basic medium they are stable. They play an important role in catalytic radical-chain processes of H_2O_2 decomposition:

$$2O_2^{\bullet-} + 2H^{+} \rightarrow H_2O_2\ (pH \leq 7),$$

$$HO_2^{\bullet} + Fe^{2+} \xrightarrow{H^{+}} HO_2^{-} + Fe^{3+},$$

$$O_2^{\bullet-}Fe^{3+} \rightarrow O_2 + Fe^{2+}$$

$$HO_2^{\bullet} + H_2O_2 \longrightarrow O_2 + 2OH^{\bullet} + H^{+}$$

The possible mechanism of oxygen activation with iron ions in natural water may be presented as two-electron transfer [2, 129]:

$$Fe^{2+} + O_2 \leftrightarrows FeO_2^{2+} \xrightarrow{2H_2O} Fe^{4+}(OH^-)_2 + H_2O_2.$$

$$Fe^{4+}(OH^-)_2 \longrightarrow Fe^{3+}(OH^-) + OH^\bullet,$$

$$Fe^{2+} + FeO_2^{2+} \xrightarrow{2H_2O} 2Fe^{3+}(OH^-) + H_2O_2,$$

$$Fe^{2+} + Fe^{4+}(OH^-)_2 \longrightarrow 2Fe^{3+}(OH^-).$$

Hydrogen peroxide thus formed can be transformed into $OH^\bullet$ radicals through the intermediate compounds of ferryl-ion type FeO^{2+} or $LFeO^{2+}$:

$$Fe^{2+} + H_2O_2 \rightleftarrows Fe^{2+}(HO_2^-) + H^+,$$

$$Fe^{2+}(HO_2^-)+OH^- \rightleftarrows (OH^-)Fe^{2+}(H_2O^-) \longrightarrow (OH^-)FeO^{2+} \xrightarrow{2H_2O} Fe^{3+}(OH^-)+OH^\bullet.$$

Unlike oxygen molecule activation with iron ions, its activation with copper ions proceeds by one-electron mechanism with $HO_2^\bullet$ superoxide radical, and then hydrogen peroxide formation [133]:

$$Cu^+ + O_2 \rightleftarrows CuO_2^+ \rightleftarrows Cu^{2+} + O_2^{\bullet-},$$

$$Cu^+ + O_2^{\bullet-} \xrightarrow{2H^+} Cu^{2+} + H_2O_2,$$

$$CuO_2^+ + Cu^+ \xrightarrow{2H^+} 2Cu^{2+} + H_2O_2.$$

In turn, hydrogen peroxide, through the formation of a number of intermediate compounds of "cupryl-ion" type (CuO^+), will be transformed into $OH^\bullet$ radical:

$$Cu^+ + H_2O_2 \rightarrow CuO^+,$$

$$CuO^+ + H_2O \rightarrow Cu^{2+} + OH^- + OH^\bullet.$$

$OH^\bullet$ radicals, as strong oxidants, will interact with pollutants, thus purifying natural water. Fe^{3+} or Cu^{2+} ions under the action of various reducers of water (for example, $O_2^{\bullet-}$, quantum of light) will be transformed into Fe^{2+} and Cu^+, and the cycle will be repeated.

To make an assessment of free radical concentration, one has to know the rate of their initiation (W_i) and efficient constant of their ruin (k_i), which, in the case of $OH^\bullet$ radicals, can be determined spectrophotometrically (by the method of competitive reactions) according to *p*-nitrosodimethylaniline coloration reducing (PNDMA—trap for $OH^\bullet$ radicals). On this basis the so-called "inhibiting capacity" parameter of water medium can be calculated $-\Sigma k_i[S_i]$, s^{-1} [110, 124]. In this case k_i is the rate constant of $OH^\bullet$ interaction with "trap," thevalue of which usually falls within the limits 10^8–$10^{10} M^{-1}s^{-1}$.

In natural water the share of H_2O_2, decomposing into $OH^\bullet$ radicals, is 10–50% [100]. Tentatively for $OH^\bullet$ initiation rate (W_i) the rate of photochemical formation of H_2O_2 in water medium can be taken (W_f), as soon as $W_i \approx W_f$. Proceeding from the value and parameter of $\Sigma k_i[S_i]$, stationary concentration of $OH^\bullet$ and efficient rate constant of any pollutant, radical oxidation (k_{RH}) can be estimated:

$$[OH^\bullet] = \frac{W_i}{\Sigma k_i[S_i]} \text{ and } k_{RH} = k_{OH^\bullet+RH}[OH^\bullet] = k_{OH+RH} W_i / \Sigma k_i[S_i]$$

where $k_{OH^\bullet+RH}$ is the rate constant of $OH^\bullet$ interaction with pollutant RH. Proceeding from these considerations, a dynamic chemical–biological redox model of natural water medium was proposed, based on the fact that biotic and abiotic processes of oxidative and reductive equivalents constantly proceed in it (Scheme 6.13 [124]).

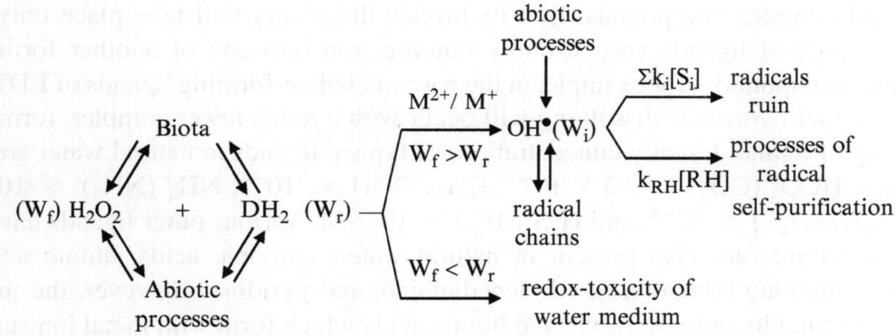

Scheme 6.13 Chemical–biological redox model of water environment

It can be seen from this scheme that biotic and abiotic processes of oxidative and reductive elements constantly proceed in natural water. The general result of their interaction is determined by the ratio of inner flows of active oxidative flows in water reservoirs (hydrogen peroxide W_f) (Scheme 6.9), and substance-reducers (DH_2), efficiently interacting with it (W_r). The sources of reducers in natural water are mainly blue-green algae, which thus adapted themselves to be used by the substance-reducers to neutralize toxic action of H_2O_2 (the latter even in concentration of 10^{-5} to 10^{-6} M will suppress these algae photosynthesis) [107, 109].

In the case of oxidative element predomination ($W_f > W_r$), the oxidative state of the medium will be realized. Reducers (DH_2) in water will be easily oxidized with hydrogen peroxide (more exactly, with radicals and their further ruin), or will initiate conjugate radical self-purification processes of water reservoirs containing pollutants (RH). In the inverse case (flow of reducers will exceed the flow of H_2O_2—i.e., $W_f < W_r$) quasi-reducing redox state of water environment will be observed. Thus in water various toxic effects appear, mainly connected with the rapid growth of blue-green algae in the reductive conditions and increased pH of medium sometimes up to ten and even higher. In these conditions pathogenic flora and intestinal pathogenic organisms (such as *Vibrio cholerae*) appear.

In all these processes, taking place in natural water, an important role is played by hydrated metal ions, especially iron and copper, metal hydroxides, and also their complex compounds with numerous ligands that catalyze proceeding in these conditions of variety of redox reactions. Iron(III) compounds are most often found in water medium. The solubility product (SP) for Fe(III) reaches the value 3.8×10^{-38}, at pH > 4 equilibrium concentration of iron aquatic form $[Fe^{3+}_{aq}]$ is low compared to hydroxide, at pH ≈ 5 it is in water solution in the form $Fe(OH)_2^+$, and at pH > 6 is is in the form $Fe(OH)_3$. Maximal concentration of $Fe(OH)_3$ in soluble form is 2×10^{-7} mol/L [107, 109].

Content of Cu^{2+} ions in surface water is approximately 3×10^{-7}M. Within the limits pH 7–9 mostly monohydroxocomplex $Cu(OH)^+$ will be in homogeneous form.

However, there are many ligands in natural water which form various kinds of metal complex compounds. Iron hydroxide dissolving will take place only in the presence of ligands very strongly binding iron into one of another form of complex compound—for example, in the case of chelate-forming ligands of EDTA-type. Copper hydroxide dissolving will occur with a much lower complex-forming capacity of ligand. Usually concentrations of typical ligands in natural water are as follows: $HCO_3^-(CO_3^{2-}) \approx 3 \times 10^{-3}$, $CH_3COOH \approx 10^{-4}$, $NH_4^+(NH_3) \approx 10^{-5}$, $H_2PO_4^-(HPO_4^{2-}) \approx 10^{-5}$, and $H_2S(SH^-) \approx 10^{-5}$ M. Various other ligands having organic natures are also present in natural water: carbonic acids, amino acids, sulfur-containing compounds, ethylenediamine, and pyridine. However, the main environmental ligands are fulvic and humic acids which form with metal ion stable compounds. Therefore, the presence of numerous ligands in natural water and also that of various pollutants (which can also be ligands), influence on pH value of medium, and changes in ratio of metal ion concentrations and ligands create conditions of different types of metal complex compound formation, characterized by extremely differing stability constants and redox potential values:

$$\varphi_{(\mathrm{Mz+Ln/M(z-1)+Ln})} = \varphi_{(\mathrm{M2+aq/M(z-1)+\bullet aq})} - RT\ \ln \frac{K_{\mathrm{Mz+Ln}}}{K_{\mathrm{M(z-1)+Ln}}}$$

where K_{Mz+Ln} and $K_{M(z-1)+Ln}$ are stability constants of metal complexes in oxidized and reduced forms, accordingly. In turn, the change in redox potential of metal ion on complex formation, and also the change in metal coordination sphere lability, its hydrolytic stability, its composition on account of multinuclear complexes formation, and the presence of numerous substrates and oxidants (O_2, H_2O_2, etc.) create possibilities for catalytic processes proceeding in various mechanisms, resulting in the self-purification of natural water. Of course, the dominating "motor force" of this process is solar light.

Thus, redox processes have an important influence on self-purification processes. In hydrogen peroxide formation within water surface layers the main role belongs to photochemical processes, and in bottom water layers to catalytic processes of oxidation with molecular oxygen of substances generated by water microorganisms. The rate of reducers DH_2 formation depends on the amount and specific diversity of water microorganisms. The most efficient catalysts of pollutant oxidation in natural water in homogeneous form are copper ions and complexes, and in heterogeneous–form Fe(III) and Mn(III) hydroxides.

The sources of oxidant (H_2O_2) formation in natural water can be photochemical and can take place within other processes (Scheme 6.9). On solar light influence ($\lambda \geq 300$ nm) on fulvic acids and other substances dissolved in water or microalgae, H_2O_2 will be formed. Radicals OH, HO_2, and others, which take part in dissolved pollutants oxidation, are, as a rule, the products of hydrogen peroxide decomposition processes.

6.3.3 Photocatalytic Transformation of Nitrogen Mineral Forms in Water

Among the most widespread pollutants of environmental water, compounds of nitrogen low-molecular forms, first of all nitrates, nitrites, and their derivatives, will be considered [134, 135].

The use of industrially manufactured nitrogen (N) fertilizers increased rapidly in the developed countries between 1960 and 1980. This facilitated a large increase in the production of feed and food grains (maize, wheat, and rice) per unit area of cultivated land, but in some regions it also contributed to enrichment of surface and groundwater with various forms of nitrogen [136]. Fertilizer, however, is not the only source of nitrogen that can cause contamination of surface waters. Biological nitrogen fixation, mineralization of soil organic nitrogen, and animal wastes can also contribute to nitrogen enrichment of water bodies. Additionally, under some conditions, nitrogen applied to the soil may be converted to gaseous or immobile forms of nitrogen that do not contribute to surface water contamination. Because of these various sources and transformations of nitrogen, the severity of surface water contamination by nitrogen fertilizer has been difficult to quantify precisely.

Existing research indicates that the amount of contamination from fertilizer varies depends on the amount of fertilizer applied, and characteristics of the soils, crops, climate, and the receiving water bodies. A large amount of nitrogen-containing compounds is discharged into water basins by cattle-breeding farms and municipal wastewater, and this is connected with nitrogen mineral compounds that often pass by transit through municipal biochemical treatment plants.

Complex environmental circulation of nitrogen in the biosphere (Fig. 6.2), providing vital activity of organisms and setting into motion the biological flow, is characterized by the capacity of self-regulation. Excessive and sometimes wrong use of nitrogen-containing fertilizers and factors of anthropogenic influence result in equilibrium violation in the cycle, and accumulation of one or another low-molecular compound of nitrogen, possessing toxic properties, such as nitrites or nitrosamines. The latter compounds are known for their carcinogenic properties. Thus, nitrite in acid media (pH 1.0–3.0), on coming in contact with foodstuffs or drinking water in the gastrointestinal tract, is a strong mutagen.

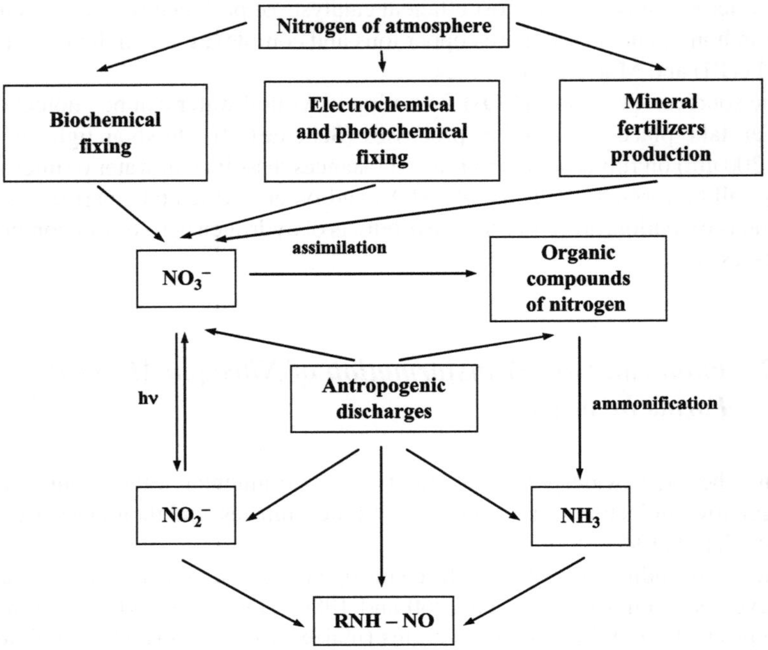

Fig. 6.2 Scheme of nitrogen-containing compounds transformation in ecochemical system

Being biological substances, such nitrogen compounds play an important role in the vital activity of hydrobionts [137, 138] and have a significant influence on both self-purification and self-pollution of water reservoirs.

The accumulation of certain compounds of bound nitrogen results not only in infringement of self-purification processes in water reservoirs, but also in an increase in the amount of toxic forms of nitrogen in agricultural crops.

The main sources of NO_3^- and NO_2^- ions penetrating into the human organism are foodstuffs and potable water.

It is to be noted that NO_3^- is often the source of NO_2^- formation. Under the influence of solar light in surface water during daytime efficient formation of NO_2^-, which results in the self-pollution of water reservoirs, takes place [139].

The toxicity of NO_2^- is manifested in that it oxidizes iron in the molecule of hemoglobin, transforming it from Fe^{2+} to Fe^{3+}. Methemoglobin thus formed is unable to realize reversible binding of O_2.

Nitrosamines will be formed, mainly, on NO_2^- interaction with amino acids or proteins in acid media. Nitrosamines mostly affect the liver, causing cirrhosis [140–148].

To reveal the main causes of water media pollution and to remove these pollutants, the kinetics and mechanisms of transformation of low-molecular forms of nitrogen should be known.

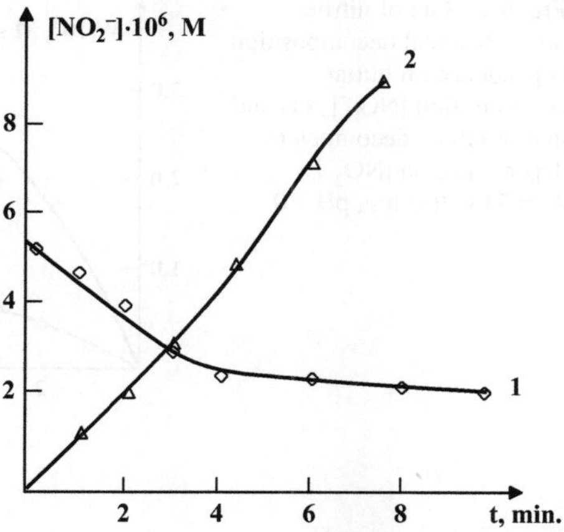

Fig. 6.3 Typical kinetic curves of nitrite (1) and nitrate (2) photolysis: $[NO_2^-]_o = 5 \times 10^{-6}$ M; $[NO_3^-]_o = 10^{-2}$ M; pH 7.0

Transformation of NO_3^- and NO_2^- can proceed under the influence of light ($\lambda \geq 300$ nm) in natural water, and also in model systems in the presence of hydrogen peroxide, alcohols, reducers, and iron and copper ions.

Typical kinetic curves of nitrate and nitrite photolysis are given in Fig. 6.3.

Due to the fact that most natural water components absorb within the nearby UV-region, this part of solar irradiation is the most efficient in water self-purification processes. The main ultraviolet part of solar irradiation is absorbed by the upper layers of atmosphere, and only the light of near UV and visible regions of the spectrum ($\lambda \geq 300$ nm) reaches the Earth's surface [2]. Within the wavelength limits 300–350 nm, solar light intensity increases abruptly. It is under this light influence that various photochemical transformations occur in natural water, in many relations determining self-purification capacity of water.

In distilled water photochemical transformations of NO_3^- and NO_2^- are possible only as a result of direct photolysis. Kinetics of NO_3^- and NO_2^- photolysis under the influence of filtered light of the DRS-250 lamp ($\lambda = 313$ and $\lambda = 360$ nm) is described by the equation of first-order rate (Fig. 6.4). Rate constants of direct photolysis of nitrate and nitrite, respectively, are $K_d' = (8 \pm 3) \times 10^{-7}\ s^{-1}$ and $K_d'' = (7 \pm 3) \times 10^{-5}\ s^{-1}$.

The photolysis of nitrates and nitrites under the influence of filtered and solar irradiation was studied [149, 150] in both distilled and various types of natural water. The photolysis of $NO_3^{-\bullet}$ and $NO_2^{-\bullet}$ is accompanied by free radical formation. Reaction products are NO_2 and OH radicals, as well as singlet oxygen:

$$1.\ NO_2^- \xrightarrow{h\nu\ (360\ nm)} (NO_2^-)^* \xrightarrow{H_2O} NO + OH^\bullet + OH^-$$

$$2.\ 2NO_3^- \xrightarrow{h\nu\ (313\ nm)} 2(NO_3^-)^* \begin{cases} \rightarrow 2NO_2^- + {}^1O_2 \\ \rightarrow 2NO_2^\bullet + OH^\bullet + OH^- \end{cases}$$

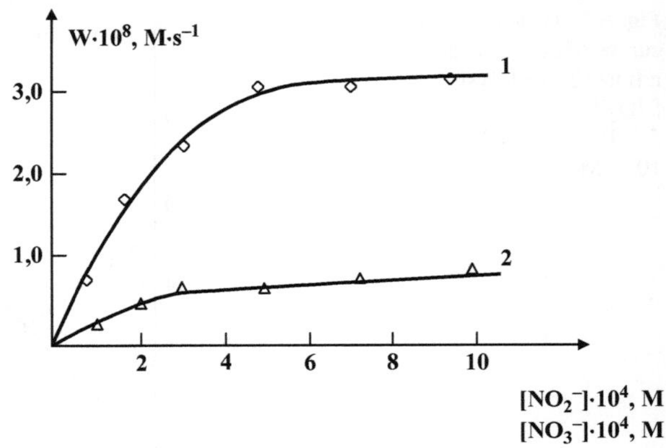

Fig. 6.4 Rate of nitrite photochemical decomposition dependence on initial concentration $[NO_2^-]_o$ (1) and that of nitrate decomposition dependence on $[NO_3^-]_o$; $\lambda = 313, 360$ nm, pH 7.0

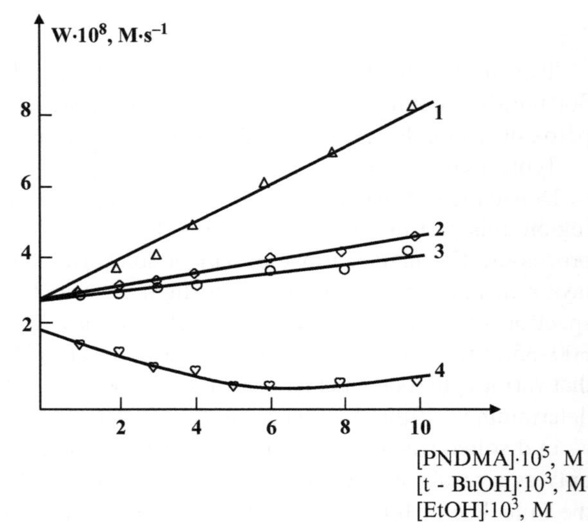

Fig. 6.5 Influence of PNDMA(1), *t*-BuOH(2) and EtOH(3) on nitrate photolysis, and PNDMA on nitrite photolysis (4): $[NO_3^-]_o = 10^{-2}$ M; $[NO_2^-]_o = 10^{-4}$ M; pH 7.0

As OH radical acceptors *tert*-butyl (*t*-BuOH) and ethyl (EtOH) alcohols were used, and also *p*-nitrosodimethylaniline (PNDMA), the additives of alcohols and PNDMA on nitrate photolysis resulting in an increase in the photolysis rate (Fig. 6.5). Proceeding from these data, rate expressions for NO_3^- photolysis rate were obtained in the presence of additives (S):

$$W = W_o(1 + \beta[S]),$$

where W_o is the rate of NO_3^- photolysis in the absence of additives, and W in the presence of PNDMA, *t*-BuOH and EtOH additives.

The increase in NO_3^- photolysis rate on adding the solutions of $OH^\bullet$ radicals' efficient acceptors indicates the $OH^\bullet$ radical interaction with nitrite-ions:

$$3.\ NO_2^- + OH^\bullet \rightarrow NO_2^\bullet + OH^-.$$

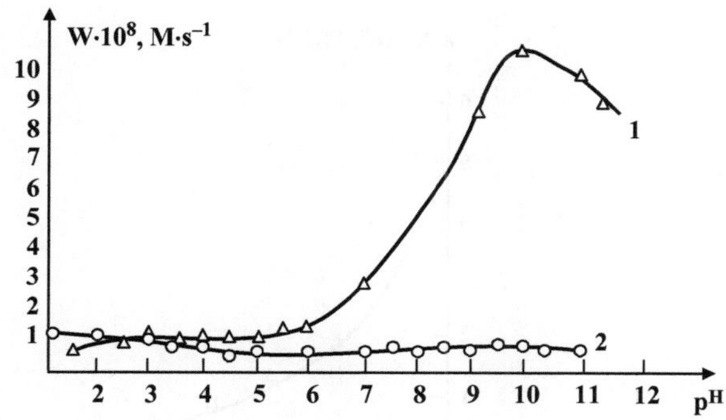

Fig. 6.6 Rate of nitrate (1) and nitrite (2) direct photolysis dependence on pH: $[NO_3^-] = 10^{-2}$ M, $[NO_2^-] = 5 \times 10^{-4}$ M

This fact is also proved by PNDMA influence on nitrite photolysis (Fig. 6.5, curve 4). Rate decreasing up to the lower limit ($0.4 \times 10^{-8}Ms^{-1}$) also points to the existence, besides reaction (1)—direct photolysis of nitrite, of reaction (3).

On comparing the PNDMA consumption rate ($0.6 \times 10^{-8}Ms^{-1}$) and rate of NO_2^- formation ($3 \times 10^{-8}Ms^{-1}$) during NO_3^- photolysis under the same conditions as in Fig. 6.5, one can see that $W_{PNDMA} < W_{NO_3^-}$. Thus, OH radical formation is not the only channel of nitrite photolysis. Excited anions of nitrate on their decomposition will also form nitrite anion [see reaction (2)].

The dependence of nitrate and nitrite photolysis on medium (Fig. 6.6) was studied. At pH > 7 the rate of nitrate photolysis will increase sharply and passes through a maximum at pH 10. This fact has an important meaning for surface water having high pH value. In these cases, during daytime an efficient photolysis of NO_3^- occurs, and nitrite concentration in water increases, which often has negative consequences. The rate constant of nitrite direct photolysis seems to be two orders higher than rate constant of nitrate direct photolysis, and, therefore, one should not fear NO_2^- accumulation due to nitrate present in natural water; moreover, mean pH value for natural water is ~7.6. However, a small increase in pH is enough to enhance the amount of nitrites. This is to be considered on hydrochemical index determination.

The presence of metal ions of transition valency and hydrogen peroxide in natural water can result in additional destruction of nitrites. Under the influence of light in the presence of microadditives of metals, free radical formation will take place [151]:

$$4.\ Fe^{3+}OH^- \xrightarrow{h\nu} Fe^{2+} + OH^{\bullet},$$

$$5.\ Cu^{2+}OH^- \xrightarrow{h\nu} Cu^+ + OH^{\bullet}.$$

During hydrogen peroxide irradiation its decomposition on hydroxyl radicals will occur:

$$6.\ H_2O_2 \xrightarrow{h\nu} 2OH^{\bullet}.$$

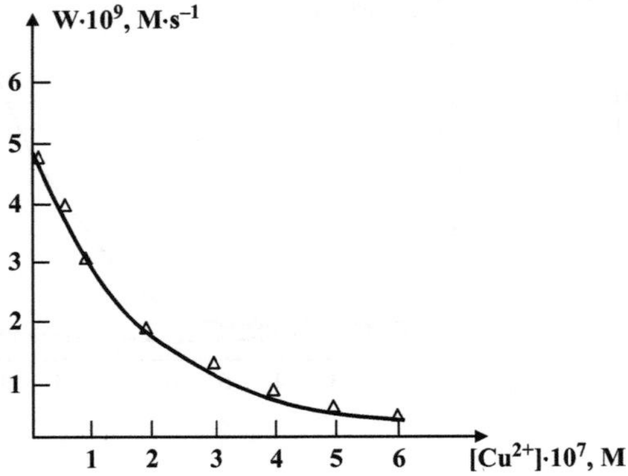

Fig. 6.7 Influence of Cu^{2+} ions on nitrites photolysis: $[NO_2^-] = 2.5 \times 10^{-5}$ M; pH 7.0

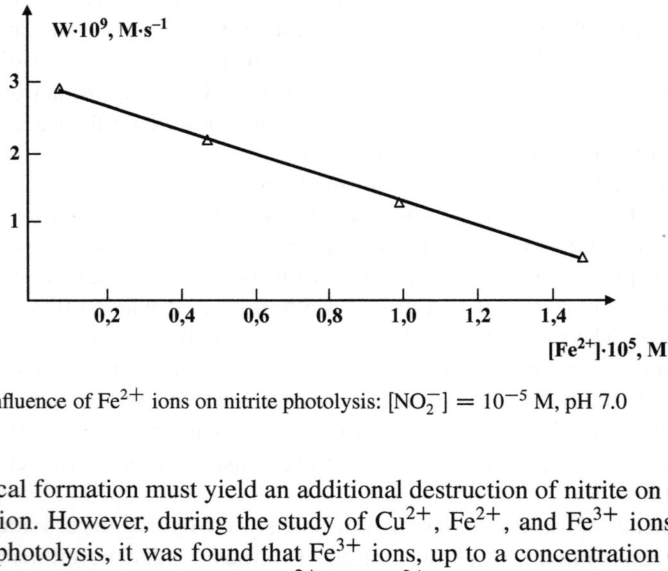

Fig. 6.8 Influence of Fe^{2+} ions on nitrite photolysis: $[NO_2^-] = 10^{-5}$ M, pH 7.0

$HO^\bullet$ radical formation must yield an additional destruction of nitrite on account of its oxidation. However, during the study of Cu^{2+}, Fe^{2+}, and Fe^{3+} ions influence on NO_2^- photolysis, it was found that Fe^{3+} ions, up to a concentration of 10^{-4}M, do not influence photolysis at all. Cu^{2+} and Fe^{2+} ions even inhibit photochemical transformation of NO_2^-; copper's influence can also be felt on $[Cu^{2+}] = 10^{-7}$ to 10^{-4}M (Figs. 6.7 and 6.8).

Copper ions seem to be necessary in the chain for NO_2^- anions regeneration. Mechanism of nitrate photolysis inhibiting with copper ions can be given by the following scheme:

1. $NO_2^- \xrightarrow{h\nu} NO + OH^\bullet + OH^-$,
7. $NO + CuOH^+ \longrightarrow Cu^+ + NO_2^- + H^+$,
8. $NO_2^- + OH^\bullet \longrightarrow No_2^\bullet + OH^-$,
9. $NO_2^\bullet + Cu^+ \longrightarrow No_2^- + Cu^{2+}$.

Table 6.1 Kinetics of nitrate (K_d') and nitrite (K_d'') transformation under the influence of solar light

No	Sample	$K_d' \cdot 10^7 (s^{-1})$	$K_d'' \cdot 10^5 (s^{-1})$
1.	Distilled water	8.0	7.0
2.	Dniester river water	7.2	0.3
3.	Water of Cuciurgani estuary	4.6	0.5
4.	Wastewater of the municipal treatment plant	5.5	3.4
5.	Water of Dniester estuary	0.9	2.1

The influence of iron(II) on NO_2^- photolysis can be observed on $[Fe^{2+}] = 10^{-5}$ M. Ions of iron(II) also will interact with $NO_2^\bullet$ radicals:

10. $Fe^{2+} + NO_2^\bullet \longrightarrow Fe^{3+} + NO_2^-$.

At the same time, Cu^{2+} and Fe^{2+} ions practically do not influence nitrate-anion photolysis.

Direct H_2O_2 interaction in darkness with nitrites or nitrates at pH 7 is low, and hydrogen peroxide plays the main role as the source of hydroxyl radicals on photolysis.

A large series of NO_2^- and NO_3^- photochemical transformation kinetic measurements was carried out under the influence of solar irradiation in the natural water of the Dniester river and some adjacent water reservoirs, and also in municipal wastewater. It was found that under different conditions, rate constants of NO_3^- and NO_2^- photolysis vary within the limits of one order (Table 6.1).

The comparison of K_d values, obtained in laboratory tests and under natural solar irradiation, shows that photochemical reactions induced by solar irradiation, can be successfully modeled using mercury lamps.

6.3.4 Photocatalytic Transformations of Anthropogenic Organic Pollutants in Water Compartments

There have been numerous studies of the photoinduced degradation with the Fe(III) aqueous complexes of organic compounds which do not absorb solar light ($\lambda_{abs} = 300$ nm). During the 1950s, the capacity of monomer Fe(III) species, and specifically, $Fe(OH)^{2+}$, to produce the high oxidative and nonselective $HO^\bullet$ radicals ($E_o = 2.80$ eV) was revealed by Bates and Uri (1953) [152] and Baxendale and Magee (1955) [153].

Since the 1980s, numerous research teams have focused on photochemical degradation studies in the systems containing Fe(III) aqueous complexes, including degradation under artificial UV-irradiation. Thus, degradation of numerous organic pollutants in water solutions under the artificial irradiation (UV–visible light), photoinduced by Fe(III), has been studied, including the following: 2-chlorophenol

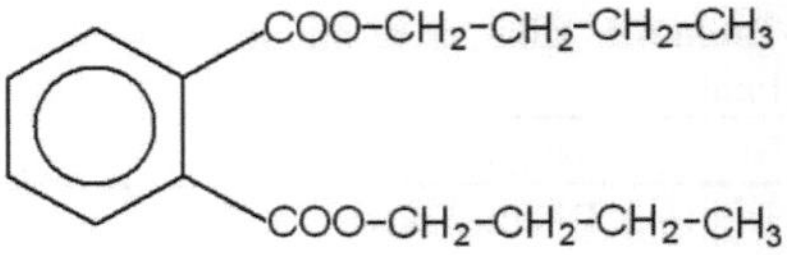

Fig. 6.9 Structure of dibutylphthalate (DBP)

[154], 3-chlorophenol [155], 4-chlorophenol [156], 2,6-dimethylphenol [157], herbicides of the phenylurea family, diuron (3-(3,4-dichlorophenyl)-1,1-dichloranilidyl) [158], asulam (4-amino-benzosulfonyl-methylcarbamate) [159] and monuron [160], benzoic acid [161], and dibutylphthalate [162].

It is to be noted that in all the cases cited, the organic molecules do not absorb or only slightly absorb solar light ($\lambda \geq 300$ nm) and are not subjected to direct photolysis. Moreover, these compounds do not interact with iron(III).

The degradation mechanism involves the only common stage in all the studied substances, specifically, the formation of hydroxyl radical catalyzed by the aqueous Fe(III) complex. On the other hand, the reaction of hydroxyl radical with the studied organics is strongly dependent on their chemical structure. Therefore, it can be concluded that the pathways of the pollutant degradation reactions during the interaction "Fe(III)–solar irradiation" may be very different. To illustrate this statement, we should consider some examples.

In the case of dibutylphthalate (DBP) [163], the authors have revealed that the hydroxyl radical attacks mainly on the aliphatic chain. As a result, two aliphatic chains in DBP are especially sensitive to this attack, as they possess nine H atoms susceptible to snatching by $HO^{\bullet}$, and, in addition, two hydrogen atoms are much more labile, being in the α-position with regard to the ester function (Fig. 6.9).

The degradation initiated by the hydroxyl radicals formed by the irradiation of Fe(III) aqueous complexes can be presented as follows:

- Radical $R^{\bullet}$ formed can react with oxygen, yielding an unstable $RO_2^{\bullet}$ radical, which can produce the hydroxyl product with Fe(III).
- Radical $RO_2^{\bullet}$, in turn, will result in formation of the same unstable hydroxyl product, which will be rapidly decomposed into carboxyl derivatives.

The DBP photoinduced degradation with iron salts in water solution is shown in Fig. 6.10. With much longer irradiation time, the complete mineralization of DBP occurs.

Hydroxyl radicals can also react directly with the aromatic nucleus. To illustrate this, we should consider the degradation of 3-chlorophenol (3-CP), photoinduced by Fe(III) [155]. Hydroxyl radicals resulting from the photolysis of $Fe(OH)^{2+}$ rapidly react with 3-CP, forming the intermediate radical compound with the aromatic ring of 3-CP—dihydroxychlorocyclohexadienyle [164]. In the presence of Fe(III) and oxygen this intermediate will form hydroxyl derivatives (chlorohydroquinone (I), 3-chlorocatechol (III), 4-chlorocatechol (IV), and chlorobenzoquinone (II) (Fig. 6.11).

This example shows the simultaneous intervention of hydroxyl radicals at the lateral chain and aromatic nucleus.

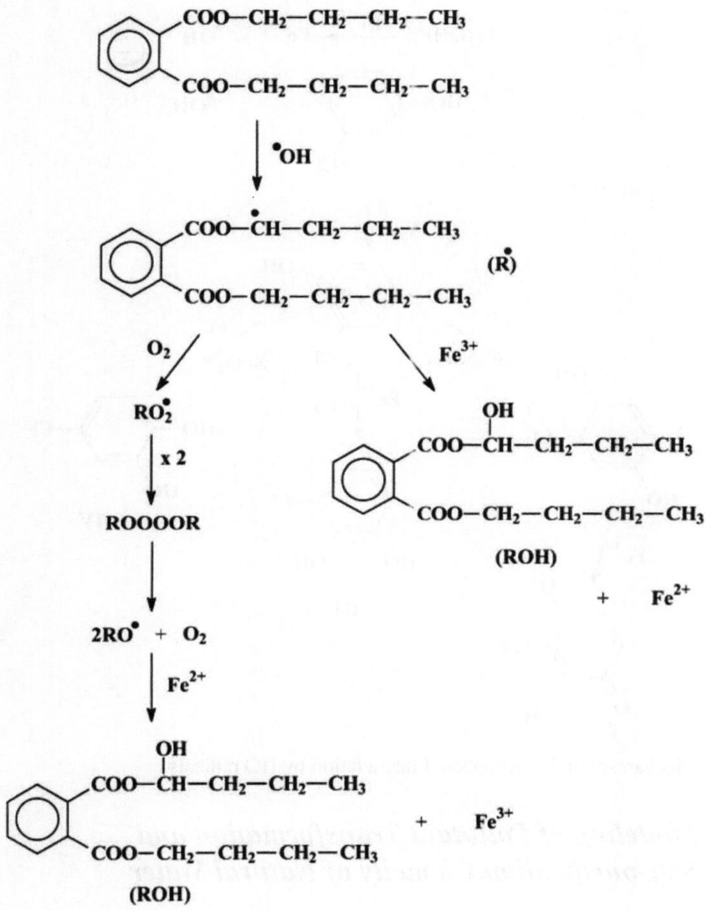

Fig. 6.10 Mechanism of dibutylphthalate (DBP) degradation with HO radicals

In the case of diuron [158] the identified products demonstrate that there are two preferential sites of attack by HO radicals in this molecule: methyl groups of urea functional group, and the aromatic ring (Fig. 6.12).

Due to the quantification of identified photoproducts, it can be stated that the attack of HO radicals mainly occurs through the detachment of a proton from one of the methyl groups of the urea functional group (Way A). Alkyl radical formed through the detachment of hydrogen can react with oxygen, forming ROO radical, and after the rearrangement two photoproducts (6 and 4) are formed. With time, the new attack of HO radicals on previous compounds (6 and 4) follows, which results in the formation of secondary photoproducts (5 and 3).

The second reaction way (Way B) involves the formation of radical intermediate with the aromatic cycle of diuron. This intermediate compound further forms the hydroxyl derivatives at the aromatic nucleus (1 and 2).

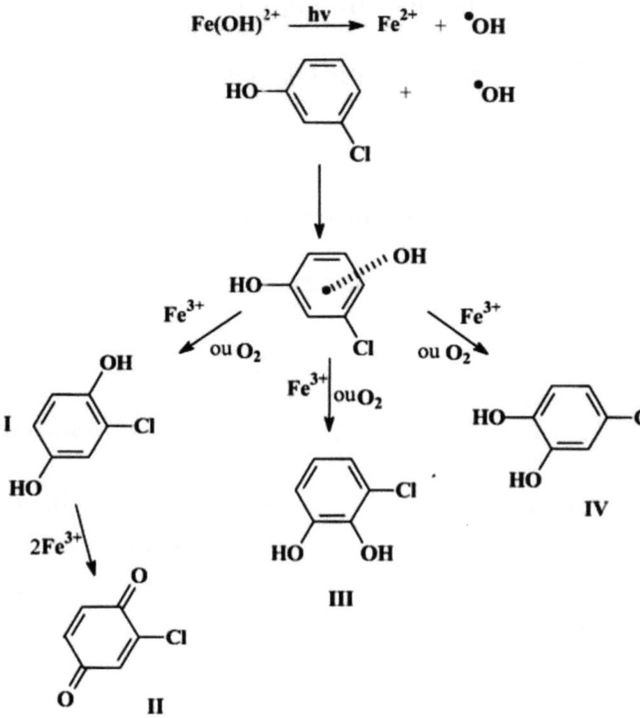

Fig. 6.11 Mechanism of 3-chlorophenol degradation by HO radicals

6.3.5 Modeling of Pollutant Transformation and Self-purification Capacity of Natural Water

The specificity of most processes proceeding in natural water is their low rate, the great amount of medium components, and the possibility of this medium state influencing with many physical–chemical factors. During the process modeling in natural water, the simplest system (distilled water + reagent-pollutant) is gradually complicated by various components introduced into it, which are present in natural water under environmental conditions—O_2, H_2O_2, M^{z+}, L, DH_2, etc. To obtain an overall picture of reagent chemical transformation (in an ideal case) one has to carry out kinetic research not only for the initial substance but also for various products of its transformation. It is extremely difficult and practically impossible, even using the method of high pressure liquid chromatography, which is widely used for these purposes.

As an example of some chemical transformation modeling, chemical transformation of 3,4-dichloroaniline (DCA) [165] can be considered, which is a metabolite of many widely used phenylamide herbicides which can enter the environment. The main processes of DCA oxidation in natural water are

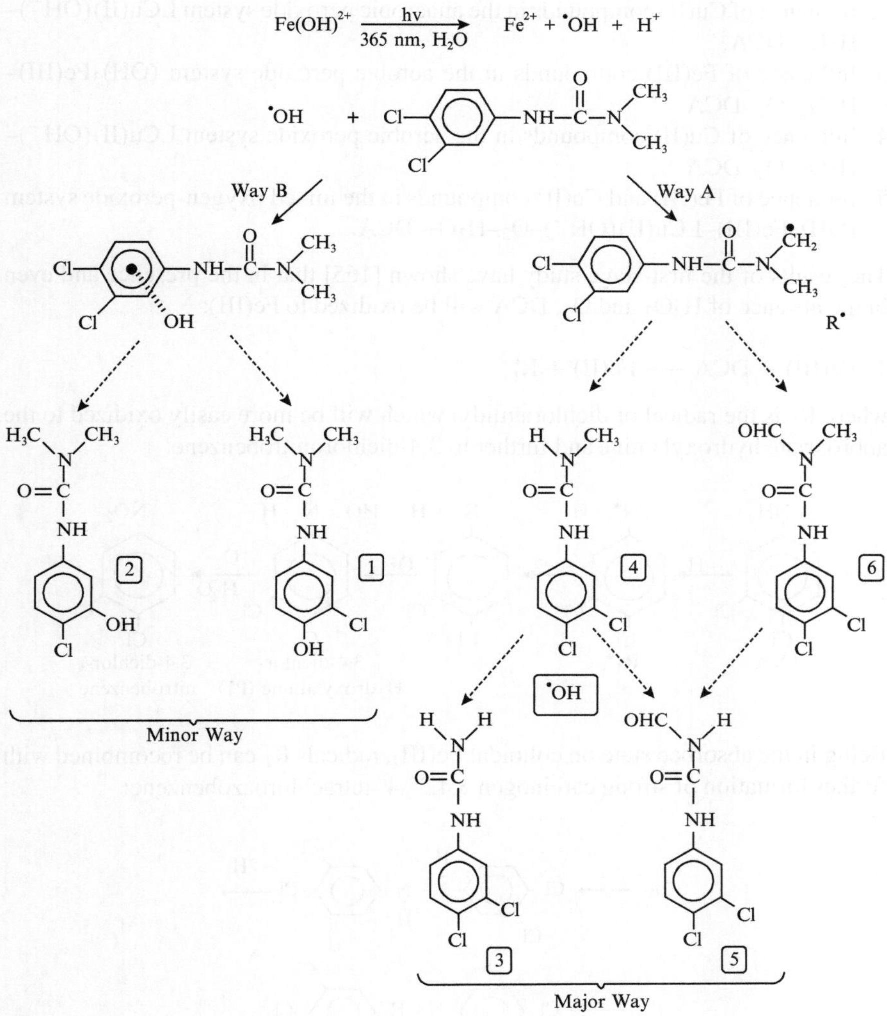

Fig. 6.12 Mechanism of diuron degradation by HO radicals

determined by mixed oxygen–peroxide systems catalyzed by the microcolloidal hydroxide of Fe(III) and hydrophosphate complex LCu(II), i.e., with the system $(OH)_3Fe(III)–LCu(II)(OH^-)–O_2–H_2O_2$–DCA within the pH range 4–9 (phosphate buffer).

The mechanism of the processes proceeding in such systems is diverse (it depends on many conditions) and complicated. Hence, the study involves several stages:

1. Influence of Fe(III) compounds in the anaerobic peroxide system $(OH)_3Fe(III)–H_2O_2$–DCA.

2. Influence of Cu(II) compounds in the anaerobic peroxide system LCu(II)(OH^-)–H_2O_2–DCA.
3. Influence of Fe(III) compounds in the aerobic peroxide system $(OH)_3$Fe(III)–H_2O_2–O_2–DCA.
4. Influence of Cu(II) compounds in the aerobic peroxide system LCu(II)(OH^-)–H_2O_2–O_2–DCA.
5. Influence of Fe(III) and Cu(II) compounds in the mixed oxygen-peroxide system $(OH)_3$Fe(III)–LCu(II)(OH^-)–O_2–H_2O_2–DCA.

The results of the first-stage study have shown [165] that in the presence and even in the absence of H_2O_2 and O_2, DCA will be oxidized to Fe(III):

1. $Fe(III) + DCA \longrightarrow Fe(II) + R_1^{\bullet}$,

where R_1 is the radical of dichloranilidyl which will be more easily oxidized to the appropriate hydroxylamine and further to 3,4-dichloronitrobenzene:

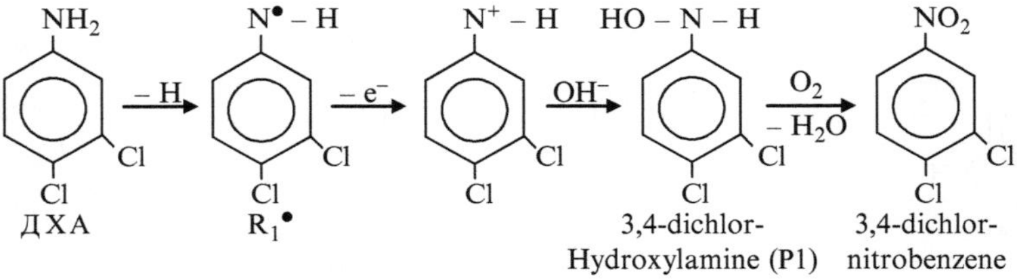

Being in the absorbed state on colloidal Fe(III), radicals R_1 can be recombined with further formation of strong carcinogen 3,4,3′, 4′-tetrachloroazobenzene:

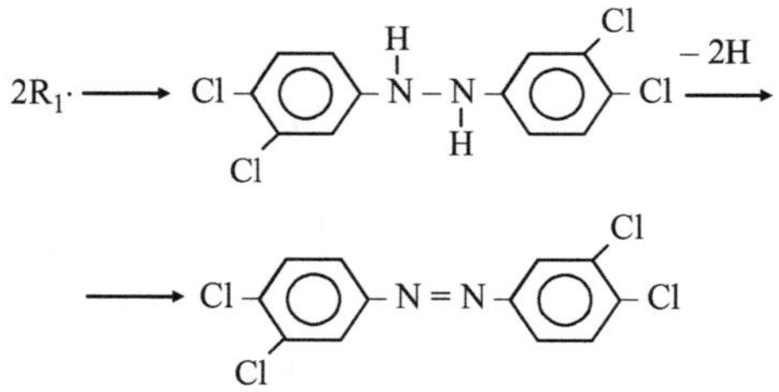

The results of the second-stage study have shown the possibility of the following reactions proceeding:

Stages of initiation:

2. $Cu(II) + H_2O_2 \longrightarrow Cu(I) + O_2^{\bullet -}$,
3. $O_2^{\bullet -} + Cu(II) \longrightarrow Cu(I) + O_2$.

Stages of continuation:

4. $Cu(I) + H_2O_2 \longrightarrow Cu(II) + OH^- + OH^\bullet$,
5. $OH^\bullet + DCA \longrightarrow R_2^\bullet$,
6. $R_2^\bullet + Cu(II) \longrightarrow Cu(I) + P_2$.

Linear breakage:

7. $OH^\bullet \xrightarrow{\text{phosphate oxidation}}$

where R_2 is the radical of $OH^\bullet$ adjoining to the benzene ring, and P_2 the product of DCA oxidation (3,4-dichloro-6-aminophenol):

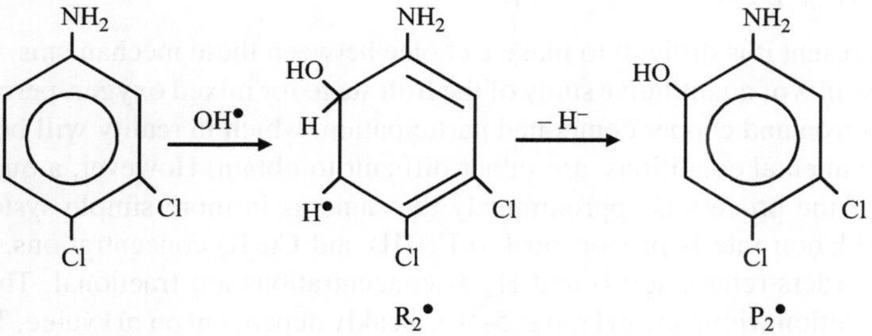

This mechanism is in agreement with experimentally found dependence:

$$W = \chi[Cu^{2+}][H_2O_2][DCA].$$

However, another alternative mechanism is also in agreement with the same dependence, in which predecessors of $OH^\bullet$ radicals in the reaction (4) are particles Cu^{3+} and $Cu(II)O^+$ (cupryl-ion and their hydroxocomplexes). Thus, DCA oxidation may proceed without O participation on account of DCA interaction with these particles.

The results of the third and fouth stages have shown that, like in peroxide systems (stages 1 and 2), the mechanism of DCA oxidation is radical chain. The stage of initiation is microcolloidal particle $Fe(OH)_3$ interaction with DCA [reaction (1)]. Then, on low concentrations of Cu^{2+} ions, $R_1^\bullet$ radicals formed at the stage of initiation will interact mostly with O_2 with the formation of $R_1O_2^\bullet$ peroxide radical, and then superoxide ion radical:

8. $R_1^\bullet + O_2 \longrightarrow R_1O_2^\bullet$,
9. $R_1O_2^\bullet \longrightarrow P_1O_2^{\bullet -}$.

Then, O_2^- will participate in Cu(II) ions changing in the system to Cu(I) ions. The latter, interacting with O_2, will form oxygen complex CuO_2^+, which possesses high reactivity in relation to hydrogen donors. The catalytic cycle will be closed provided CuO_2^+ reacts with DCA at the stage of chain continuation:

10. $O_2^{\bullet-} + Cu(II) \longrightarrow Cu(I) + O_2$,
11. $Cu(I) + O_2 \longrightarrow CuO_2^+$,
12. $CuO_2^+ + DCA \longrightarrow R_1^\bullet + Cu(II) + H_2O_2$.

and then reactions (8) and (9).

An alternative two-electron mechanism is also possible, when Cu(I) and O_2 will be formed in the initiation stage, and free radicals do not take part in the stage of chain continuation:

11. $Cu(I) + O_2 \longrightarrow CuO_2^+$,
12′. $CuO_2^+ + DCA \longrightarrow Cu^+ + P_1$.

For the present it is difficult to make a choice between these mechanisms.

The results of quantitative study of the fifth stage for mixed oxygen-peroxide system with iron and copper compound participation, which in reality will be realized in environmental conditions, are rather difficult to obtain. However, a quantitative picture of the process is approximately the same as in more simple systems, and DCA oxidation rate is proportional to Fe(III) and Cu(II) concentrations. Efficient reaction orders related to O_2 and H_2O_2 concentrations are fractional. The rate of DCA oxidation within the pH range 5–9 is weakly dependent on pH value. The chain processes of DCA oxidation with oxygen and hydrogen peroxide will be the same.

A detailed enough example of DCA catalytic oxidation [165] is a model of the same processes in natural water, though it is a simplified one. Thus, photochemical processes are not taken into account, proceeding in natural water and generating additional amounts of reactionary particles and free radicals, interacting both with each other and with the transition metal compounds.

The role of the free radicals considered above in the pollutants transformation in natural water is significant. Primary radicals appear [166] as a result of light ionizing radiation and the effect of other factors on saturated molecules:

$$R - R' \rightarrow R^\bullet + R^{\bullet\prime},$$

and also donors (DH^-) interaction with the acceptors of electrons (A):

$$DH^- + A \rightarrow DH^{\bullet-} + A^{\bullet-},$$

or electron donors with metal compounds in oxidized form (M^{2+}):

$$DH^- + M^{2+} \rightarrow DH^{\bullet-} + M^+,$$

There are great amounts of dissolved organic and inorganic substances in natural water, with which primary free radicals will interact forming secondary radicals. Besides, molecules of oxidizers O_2 and H_2O_2 are involved in radical processes.

Intermediates in electron transfer reactions serve the compounds of transition metals (iron, manganese, copper, etc.).

Under the influence of natural water components, solar beams, and radiation, $O_2^{\bullet -}$ and $OH^{\bullet}$ will be formed in natural water as primary free radicals:

$$(M^{2+}/M^{+}, O_2, H_2O_2) + DH^{-} \rightarrow O_2^{\bullet -} + OH^{\bullet}$$

natural water

$$H_2O_2 \xrightarrow{h\nu} 2OH^{\bullet}.$$

$O_2^{\bullet -}$ and $OH^{\bullet}$ radicals will also be formed as a result of numerous secondary processes:

$$M^{+} + O_2 \rightarrow M^{2+} + O_2^{\bullet -},$$
$$M^{+} + H_2O_2 \rightarrow M(OH)^{+} + OH^{\bullet}.$$

Free radicals formed ($O_2^{\bullet -}$) more often pass the electrons catalytically on H_2O_2, with $OH^{\bullet}$ radical formation:

$$O_2\bullet^{-} \xrightarrow{h\nu} H_2O_2 (\text{main reaction of } H_2O_2 \text{formation in natural water}),$$

$$O_2^{\bullet -} + H_2O_2 \xrightarrow{M^{2+}/M^{+}} O_2 + OH^{-} + OH^{\bullet}$$

Then, $OH^{\bullet}$ radicals will interact with the pollutant RH dissolved in the water by atom H isolation:

$$OH^{\bullet} + RH \rightarrow R^{\bullet} + H_2O,$$

$$R^{\bullet} + O_2 \rightarrow RO_2^{\bullet} (\text{stationary concentration in water} \approx 10^{-9}\text{M}),$$

$$RO_2^{\bullet} + RO_2^{\bullet} \rightarrow ROOH + O_2 + P(\text{product}),$$

$$ROOH \xrightarrow{h\nu} RO^{\bullet} + OH^{\bullet},$$

$$O_2^{\bullet -} + O_2^{\bullet -} \xrightarrow[\text{or } M^{2+}/M^{+}]{\text{COD}} O_2 + H_2O_2 (W \approx 10^{-6}\text{mol/l h}).$$

Reciprocal transformation of primary and secondary free radicals in natural water is shown in Fig. 6.13 (the picture is taken from [166]).

Therefore, natural water self-purification proceeds through the formation of H_2O_2, its further destruction to $OH^{\bullet}$, interaction of $OH^{\bullet}$ with pollutant RH, and its oxidation to the final product P. Self-purification capacity of natural water (SPCNW) can be estimated by the model test substance transformation rate, sensitive to water body state changes. Such a substance is dye para-nitrosodimethylaniline, the decolorizing rate of which, during the time of sample exposure (g l^{-1} h^{-1}), characterizes SPCNW [167]:

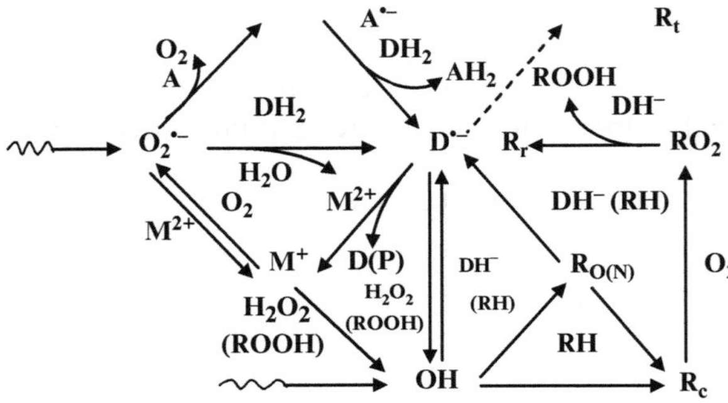

Fig. 6.13 Scheme of primary and secondary free radicals reciprocal transformations in natural water (without quadratic reactions of radicals ruin taken into consideration). M^+ is metal in reduced form, M^{2+} metal in oxidized form, DH_2 reducer-donor of H, A electron acceptor, R_c alkyl radical, R_r radical possessing reducing properties, $R_{O(N)}$ radical-oxidizers with free valency, localized on O, H atoms, $D^{\bullet-}$ radical-electron donor, R_t stabilized radical, and $D(P)$ products of $DH_2(RH)$ oxidation.

$$\mathrm{SPCNW} = \frac{M(A_b - A_e)}{\varepsilon \cdot \Delta\tau \cdot l}.$$

Here A_b, and A_e are the optical densities of water sample in the beginning and end of exposure, $M = 150$ the molecular weight of PNDMA dye, $\varepsilon = 3.42 \times 10^4$ the molar coefficient of PNDMA extinction ($\mathrm{l\,mol^{-1}\,sm^{-1}}$), $\Delta\tau$ the time of test exposure (h), and l the cuvette width (sm). SPNW of pure water is s $\approx 10^{-6}$ to 10^{-5} g L^{-1} h^{-1} and that of polluted water more than 5×10^{-5} g L^{-1} h^{-1}.

6.4 Wastewater Treatment and Catalytic Processes

6.4.1 Principles of Catalytic Wastewater Treatment

Natural water pollution is caused in may ways by industrial, municipal, and agricultural wastewater. At present more than 600 km^3 of wastewater (among them more than 300 km^3 of industrial origin) are discharged into the Earth's water reservoirs. About 70% of total river flow are used in industry and agriculture, and in the industrially developed regions water is repeatedly subjected to anthropogenic circulation.

The problem of wastewater treatment mainly comes to elaboration of new methods (including catalytic ones) of various organic substances' removal or decomposition. This refers to chlorinated anilines, phenols, benzene, aminocompounds, colorants, organic solvents, surfactants, and others, forming at different stages of technological processes as main or by-products.

An efficient means of chemically treating destructive wastewater (WW), which contains dissolved organic substances, is by means of oxygen, ozone, and hydrogen peroxide.

WW oxidation with air oxygen is usually performed at increased temperatures. The most efficient is the method of thermocatalytic oxidation at 260–540°. The degree of toxic substance neutralization then exceeds 95%. The catalyst is usually platinum (electroplated on nickel), as well as copper–chromium and manganese catalysts. With the latter in vapor-gaseous phase, practically complete WW purification can be reached, removing acetone, phenol, isopropyl alcohol, mesityl oxide, and picoline derivatives. The main drawbacks of this method are high energy consumption, high cost of equipment, its corrosion, scum formation, and catalysts poisoning.

Another oxidant of organic compounds is ozone catalyzed by transition metal ions and used for removal of chlorinated hydrocarbons, phenols, oil products, surfactants, etc. Direct and indirect ozone influence on dissolved organic substances yields OH and O_2^- radicals and their further interaction (Scheme 6.14).

$$O_3 + OH^- \longrightarrow O_3^{\bullet -} + OH^\bullet$$

$$O_3 + HO_2^- \longrightarrow O_3^{\bullet -} + HO_2^\bullet$$

$$O_3 + OH^\bullet \longrightarrow O_2 + HO_2^\bullet$$

$$O_2^{\bullet -} + HO_2^\bullet \longrightarrow O_2 + HO_2^-$$

$$HO_2^\bullet + HO_2^\bullet \longrightarrow O_2 + H_2O_2$$

Scheme 6.14 Formation of radicals on the ozone interaction with the dissolved organics in waste waters

Highly reactive particles, interacting with the substances dissolved in WW, will generate radical-oxidizers (R_{ox}), radical-reducers (R_{red}), radicals combining these properties, and low-reactionary and long-living radicals.

The third efficient source of $OH^\bullet$ radicals is hydrogen peroxide in the presence of transition metals. It possesses a number of technological advantages, such as high solubility in water, possibility of WW treatment within the wide range of pH and temperature values, ability to intensify the process in catalytic and photochemical way, and treatment of both diluted and concentrated wastes.

Hydrogen peroxide is often used jointly with other oxidants (such as O_3, Cl_2, and hypochlorites), in combination with biological treatment, for WW disinfection and solid wastes treatment.

H_2O_2 is most widely used for treating wastewater containing sulfur-containing compounds [168]:

$$H_2S + H_2O_2 \xrightarrow{pH \leq 8.5} S + 2H_2O,$$

$$H_2S + 4H_2O_2 \xrightarrow{pH \geq 8.5} H_2SO_4 + 4H_2O.$$

In relation to H_2S and sulfides, hydrogen peroxide is a selective oxidant. Iron(II) salts may be used as catalysts. Hydrogen peroxide is widely used in the presence of copper ions as catalysts for WW treatment containing cyanide admixtures and nitrogen compounds. H_2O_2 is used for WW deep purification from phenols and colorants with iron and copper ions' participation as catalysts. H_2O_2 used jointly with Fe^{3+} and Fe^{2+} ions at pH 3–5 makes it possible to oxidize 30–65% of organic admixtures in household wastewater.

At pH 6–8, characteristic for WW, copper ions are initiators of $OH^{\bullet}$. On WW acidification up to pH 4, iron ions will become more efficient at radical initiating.

Radicals $OH^{\bullet}$ will be formed on H_2O_2 interaction with metal reduced forms, and, as a rule, process rate will be limited by the rate of metal oxidized forms reducing. The general process of radicals' catalytic initiation with H_2O_2 participation can be given by the usual cyclic scheme (Scheme 6.15).

$$M^{2+} + DH^{-} \rightleftharpoons MDH^{+} \longrightarrow M^{+} + DH^{\bullet}$$

$$DH^{\bullet} \rightleftharpoons D^{\bullet -} + H^{+}$$

$$D^{\bullet -} + M_2^{+} \longrightarrow M^{+} + D$$

$$M^{+} + H_2O_2 \longrightarrow MOH^{+} + OH^{\bullet}$$

$$MDH^{+} + H_2O_2 \longrightarrow M^{2+} + D^{\bullet -} + OH^{\bullet}$$

$$2M^{+} + O_2 \xrightarrow{2H^{+}} 2M^{2+} + H_2O_2^{\bullet}$$

Scheme 6.15 Catalytic initiation of radicals with hydrogen peroxide involvement

The use of H_2O_2 in multicomponent wastewater treatment makes it possible to reduce significantly the chemical oxygen demand (COD) value on account of O_2 involvement in the processes of radical oxidation. H_2O_2 consumption in these processes may essentially be lower than the stoichiometric value. Thus, on WW treatment containing alcohols, H_2O_2 is a specific "priming-tube" initiating the formation of R_{α} radicals, which bring O_2 into circulation. Thus, H_2O_2 consumption is insignificant. The most efficient WW oxidation with hydrogen peroxide will proceed on process initiation with UV-irradiation in the presence of iron ion catalytic admixtures.

An example of the kinetic approach to the wastewater treatment problem is the research [163], carried out on strongly polluted wastewater from the vitamin-producing industry (Belgorod, Russia). Iron ion content in wastewater is 0.1–1 mg/L and they can participate in wastewater treatment. The observation was made over Fe(II) transformation in wastewater kinetics in conditions of darkness, under the influence of UV-irradiation, and on aeration. Also included were certain components of wastewater (such as oxy- and oxoacids, including citric, maleic, pyrotartaric, oxalic acids, etc., hydrocarbons, including glucose and sorbose) [163].

From Fig. 6.14 it follows that, on aerated wastewater being added to distilled water (DW), the rate of Fe(II) oxidation with oxygen will be increased proportionally.

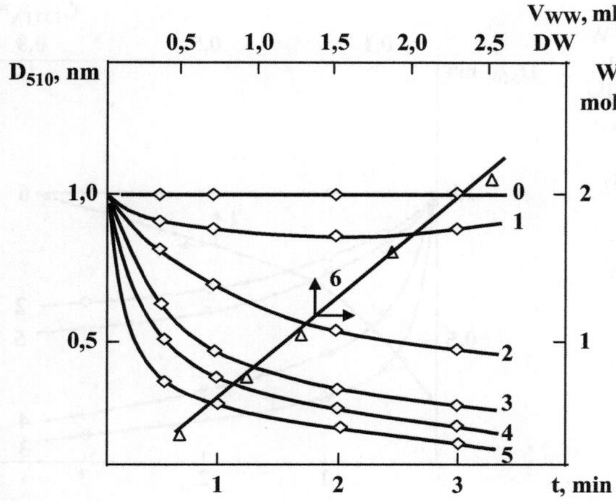

Fig. 6.14 Kinetic curves of Fe(II) oxidation in undiluted (0) and aerated wastewater, diluted 20- (1), 10- (2), 6,6- (3), 5- (4), and 4-times (5), and also process rate dependence (6) on dilution degree at $[Fe^{2+}] = 1 \times 10^{-4}$ mol/L and pH 7

All the data available made it possible to represent the mechanism of iron(II) complex oxidation in aerated wastewater by Scheme 6.16, where B^- represents anion components of WW, playing the roles of ligands.

$$1.\ Fe^{2+} + L \rightleftharpoons FeL^{2+}$$
$$2.\ FeL^{2+} + O_2 \rightleftharpoons L(FeO_2)^{2+}$$
$$3.\ L(FeO_2)^{2+} + B^- \xrightarrow{2H^+} LFe^{4+}B^- + H_2O_2$$
$$4.\ LFe^{4+}B^- + LFe^{2+} \longrightarrow 2LFe^{3+} + B^-$$
$$5.\ FeL^{2+} + H_2O_2 \longrightarrow FeL^{3+} + OH\cdot + OH^-$$

Scheme 6.16 Mechanism of Fe(II) oxidation in aerated waste water

As can be seen from Scheme 6.14, during the initial anaerobic wastewater aeration, in the process of $Fe^{2+}L$ oxidation in water medium radicals will appear, which may take part in various organic substances' oxidation. On reducers presence in wastewater, the cyclic process can proceed with alternating oxidation by air oxygen and Fe(II) reduction by substance-reducer.

Additives of various complex-forming substances in the wastewater (such as sodium-potassium tartrate, citric and glyoxalic acids, EDTA, reducing redox potential of Fe(III)/Fe(II) couple to the value of 0.117 B) result in multiple increases in Fe(II) oxidation rate (Fig. 6.15). Additives of glucose (reducer), possessing only weak ligand properties, do not practically influence Fe(II) oxidation rate. These data confirm the participation of chelate-forming ligands in complex formation with iron ions in mixed industrial–household WW. Wastewater irradiation with UV light is accompanied by H_2O_2 formation. The predecessor of H_2O_2 seems to be superoxide radical appearing on O_2 interaction with electron-excited components of

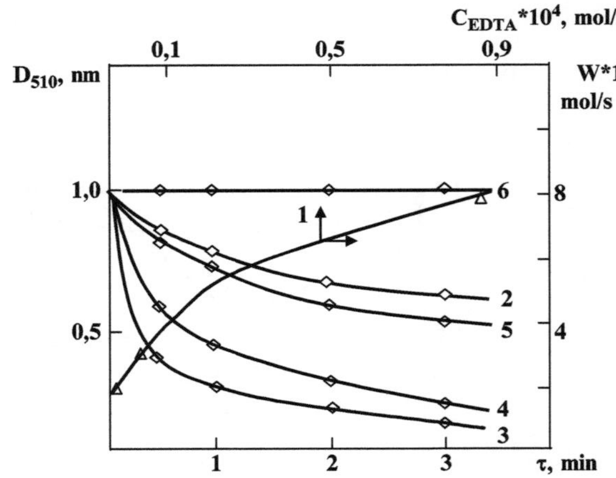

Fig. 6.15 Dependence of Fe(II) oxidation rate in WW on EDTA concentration (1), and kinetic curves of Fe(II) oxidation in EDTA absence (2) and presence (3), citric acid (4), glyoxalic acid (5), glucose (6) at $C_L = 5 \times 10^{-2}$, $C_{Fe(II)} = 10^{-4}$ mol/L; PH 7.7; WW was diluted ten times

wastewater. Superoxide ion radicals $O_2^{\bullet-}$ will reduce Fe(III) into Fe(II), while Fe(II), interacting with H_2O_2, results in $OH^{\bullet}$ radical formation (Haber-Weiss reaction), which later on will decompose pollutants in wastewater.

Thus, with the help of H_2O_2 one can intensify wastewater treatment in principle, especially from sulfur- and nitrogen-containing compounds of phenols, colorants, etc. In fact, the H_2O_2 molecule can be regarded as a "dimer" of radicals. This dimer will be decomposed on monomer radicals under the influence of UV light, or by joining an electron to O–O bond.

$OH^{\bullet}$ radical, the strongest oxidant known, will react with most organic substances with a rate constant of 10^9–$10^{10}M^{-1}s^{-1}$. The question is: from where can the reducing equivalents for $OH^{\bullet}$ formation be taken?

Municipal WW contains a number of substances possessing reducing properties, which will transform transition valency metals into reducing state. On their coming into aerotanks, metal ions will be oxidized with oxygen with H_2O_2, $OH^{\bullet}$ radical formation, which can oxidize barely degradable substances.

WW iron admixtures, forming efficient catalysts with many ligands, can serve as catalysts for conjugate oxidation processes of pollutants with oxygen and H_2O_2. In the presence of substance-reducers in wastewater, as with Fe(II) compounds oxidation with oxygen and hydrogen peroxide, Fe(III) reduction will take place; therefore cyclic processes' running is possible with alternating oxidation–reduction of LFe^{2+} to LFe^{3+}, which in turn can be used for efficiency determination of WW self-purification capacity.

In laboratory tests small concentrations of H_2O_2 can be used as initiators of WW radical treatment processes. However, under real conditions of large amounts of municipal WW, under an optimal regime, consumption of H_2O_2 for purification processes initiation is still great. Therefore, it is more expedient to use UV-irradiation for these purposes as well.

In this case under UV-irradiation influence, H_2O_2, $OH^{\bullet}$, $O_2^{\bullet-}$, and alkyl-free radicals will be formed in wastewater (Scheme 6.17) [169].

$$H_2O_2 \xrightarrow{h\nu} 2OH^\bullet$$
$$FeOH^{2+} \xrightarrow{h\nu} Fe^{2+} + OH^\bullet$$
$$FeOR \xrightarrow{h\nu} Fe^{2+} + RO\cdot$$
$$NO_2^- \xrightarrow{h\nu} NO + OH\cdot + OH^-$$
$$NO_3^- \xrightarrow{h\nu} NO_2 + OH\cdot + OH^-$$
$$>C=O \xrightarrow{h\nu} -\overset{|}{\underset{|}{C}}-O\cdot \xrightarrow{RH} R\cdot + \cdot\overset{|}{\underset{|}{C}}-OH\ (R_\alpha)$$
$$RH \xrightarrow{h\nu} (RH)^* + O_2 \longrightarrow R\cdot + HO_2^\bullet\ (O_2\cdot^- + H^+)$$
$$R\cdot\ (R_\alpha^\bullet) + O_2 \longrightarrow RO_2\cdot \longrightarrow P + O_2^{\bullet-}$$

Scheme 6.17 Radicals formation under the UV-irradiation in waste water

Thus, the cascade of radical reactions will be realized with participation of both hard- and easy-oxidative substances with the involvement of O_2 dissolving in water within this process. The presence of copper and iron ions as admixtures in WW will contribute to these reactions' proceeding.

Application of the kinetic approach to the analysis of catalytic redox and photochemical processes in natural and wastewater can be used to resolve many problems dealing with environmental protection.

6.4.2 Combined Redox-Photocatalytic Processes

Homogeneous catalytic processes are already applied for the treatment of water and air systems for the organic pollutant removal. There is no doubt that these processes will be perfected further, but their advantages are already clear, which determines their attractiveness and perspectives. These involve the simplicity, efficiency, and possibility to use solar light. The latter is very important for the future, as energy-saving technologies will undoubtedly soon be advantageous.

As in the air, in the water environment practically all organic compounds can be oxidized to CO_2 and H_2O. However, their complete oxidation time is long in many cases, to a certain degree due to the essentially slower diffusion of organic molecules in water than in air. Their typical diffusion coefficient in water is about $10^{-5}\ cm^2/s$, which is at least four orders below the same value in the air. That is why the stationary reactors are insufficiently effective.

However, this problem can be resolved by enhancing the intensity of mass-exchange and mass-transfer processes, or by combining homogeneous and heterogeneous catalysis, or combining with other chemical or microbiological technological processes.

Thus, pesticides used in agriculture are decomposed within several months. By adding small amounts of Fenton reagent, including iron salts and hydrogen peroxide, it is possible to reduce this time under solar irradiation. In this way, the pollutant oxidation rate can be increased by tens and even hundreds of times. Such systems are already used in practice for WW treatment. For example, the "photo-Fenton" system is commercialized by "Calgon Carbon Oxidation Technologies", USA.

Use of Goethite as a Nano-Catalyst

The development and using of nanomaterials, the structure of which is close to the colloidal or crystalline one, has lately become an incentive for science development throughout the world. Nanomaterials are widely used in various different fields of science and technology.

One of the prospective techniques is the obtaining of stabilized nanodisperse goethite [$\alpha - FeO(OH)$] particles as a result of the slow hydrolysis of iron(III) chloride ($FeCl_3$) [170]. Goethite particles at the moment of their formation are in disperse form. They can be stabilized in the presence of agar–agar or gelatin in low concentrations, having the surfactant properties. As a result, a rather stable gelatinous sediment is formed.

Nanodispersive goethite thus obtained can be used for organic substances' photocatalytic decomposition. The nanosize of goethite particles ensures their reactivity. Under UV-irradiation, photocatalytic reactions begins, the mechanism of which involves the formation of hydroxyl radical $OH^{\bullet}$ and $^{\bullet}O^{2-}$ radical. These radicals are formed as a result of the photodissociation of FeO(OH), from the hydroxyl group in composition of the goethite molecule.

Combined Photocatalytic Aerobic Reactor for the Destructive Elimination of Refractory Organics from the Wastewater

An integrated reactor (Fig. 6.16), combining homogeneous and heterogeneous catalytic processes, was proposed [171]. At the first stage, photocatalytic processes, under dynamic flowthrough conditions with the UV-irradiation, run in the upper part of the reactor. This results in the partial destruction of refractory pollutants, which cannot be decomposed by the microbiological way alone. At the second stage (bottom part of the reactor) the microbiological treatment proceeds.

The tubular reactors are arrangedappropriately, which extends the time and way of UV-treatment. To provide continuous operation of the reactor, there is a special mechanism of back-and-forth motion of brushes inside the quartz U-tubes. This permits one to retain the transparency of tubes during the treatment of heavily polluted liquids, and also to improve the mass exchange. In this way, the efficiency of photocatalytic destruction of the refractory organic molecules is improved, and makes the subsequent biochemical mineralization of treated solution easier under the aerobic conditions.

At the second treatment stage in the lower part of the reactor, the microorganisms are introduced. To improve the efficiency of microbiological destruction of organic substances, this part of the reactor is filled with the packing for microflora fixation and prevention of its being carried away from the reactor. The packing is a macroporous floating material. The aerated air is enriched with oxygen.

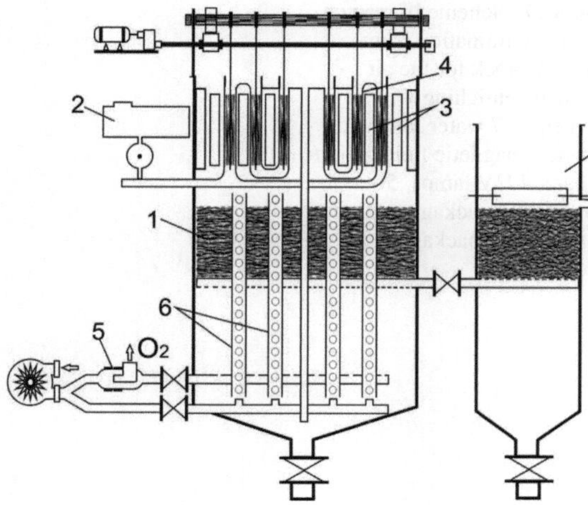

Fig. 6.16 Combined photocatalytic aerobic reactor: 1 bioreactor case with conic bottom, 2 dosing device, 3 sources of UV-irradiation, 4 U-shaped tubes with the system of their internal cleaning, 5 block of air enriching with oxygen, 6 air-lift, 7 filter

A very interesting principle was proposed regarding enriching the air with oxygen, connected with its magnetic susceptibility in the constant magnetic field. This can be explained by the paramagnetic properties of oxygen molecules, due to the unpaired electrons in the triplet state [172]. Therefore, only molecular oxygen, of all the gases making up air, has a positive magnetic susceptibility. This is manifested in the deviation of the molecular beam of oxygen in the magnetic field. It makes it possible to retract oxygen selectively from the air flow in the magnetic field and direct it into the treated water. This property of oxygen ensures at the same time the magnetic activation of its molecules and improvement of the bactericide effect on the treated water. In this way, it is possible to enhance the oxygen content in the aeration air from 18–18.5% to 23–25%, which improves the oxidation of pollutants during the water treatment.

Photocatalytic Removal of Hydrogen Sulfide from Water

A reactor (Fig. 6.17) has been developed for hydrogen sulfide photocatalytic removal from water [174].

The photocatalytic process (wavelength 180–380 nm and intensity 20–30 J/cm^2 min) occurs in the reactor in the presence of iron compounds. These compounds are generated due to the galvanochemical treatment in the field of short-circuit galvanic element appearing during the contact of iron and charcoal particles. The galvanic pair is thus formed because of the difference of electrochemical potentials of iron and charcoal. Therefore, iron dissolves, forming the iron ions.

Under the UV-irradiation, active radicals and peroxide compounds are formed in the treated solution, having a high free negative energy. They manifest the oxidation properties with regard to the hydrogen sulfide forms. Their effect on S^{2-}, HS^- ions, S^0, and the organic compounds in the treated water bears a multi-aspect character.

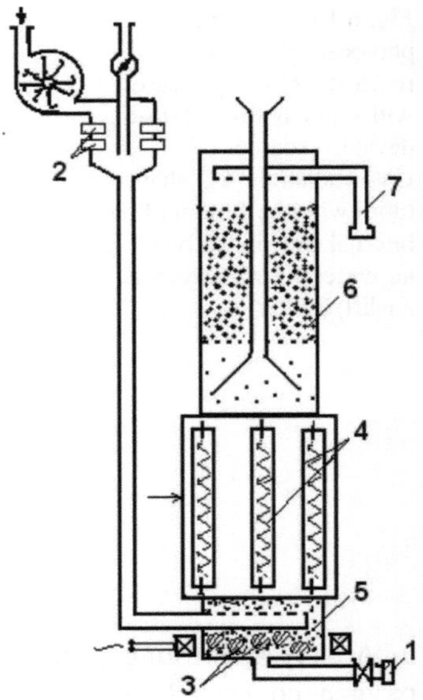

Fig. 6.17 Scheme of reactor for H_2S elimination from water: 1 block for the air magnetic enriching with oxygen, 2, 7 water inlet and outlet, 3 magnetic fluidization system, 4 UV-lamps, 5 iron-carbon package, 6 filter with floating package [173]

As a result, a step-by-step oxidation of sulfur compounds occurs to SO_3^{2-}, and then to nontoxic ions SO_4^{2-}. It is to be noted that the limited-admissible concentration of SO_4^{2-} ions in drinking water is 500 mg/L, whereas for H_2S and its compounds it is less then 1 mg/L.

Photocatalytic oxidation–reduction processes, which ensure elimination of various forms of hydrogen sulfide from water, can be summarized in the following way:

$$HS^-_{(ads)} + O_2^{\bullet -} \rightarrow Cat(S^*) + H_2O,$$

$$Kat(S^*) + O_2 \rightarrow SO_{2(ads)} + Cat,$$

$$SO_{2(ads)} + \frac{1}{2}O_2 \rightarrow SO_{3(ads)},$$

$$SO_{3(ads)} + H_2O_{(ads)} \rightarrow H_2SO_{4(ads)}.$$

Following similar mechanisms, transformation of hydrogen sulfide compounds occurs with the involvement of other active radicals. The primary processes are active radical formation under hard UV-irradiation, and the secondary processes are homogeneous and heterogeneous redox processes. The catalysts thus play a dual role: on the one hand they promote the formation of free radicals under photo-irradiation, and on the other, running of the redox processes in the bulk of treated water. At the same time, UV-irradiation has a ruinous effect on the microorganisms, pathogenic microflora, and thiobacteria. So, at the same time the bactericide treatment of water is ensured.

Enriching of aeration air with oxygen in the magnetic field improves its oxidation properties during the redox photocatalytic elimination of hydrogen sulfide, sulfides of organic components, and thiobacteria in the treated water. Consumption of oxygen, through its content in aeration air, reaches 5–10 mL O_2 per 1 g of hydrogen peroxide.

Oxidation is intensified under the conditions of efficient mass transfer and mass exchange during magnetic fluidization. Magnetic fluidization is provided due to the movement of spherical particles (barium hexaferrite) in the polygradient magnetic field.

Photoreactor Using the Solar Energy

To improve the efficiency of photocatalytic degradation of the refractory polyphenolic compounds in water compartments under UV-irradiation, a flowthrough reactor was developed (Fig. 6.18) with maximum use of solar energy. A rotating spiral brush cleans up the inner walls, plays the role of a screw pump, and ensures an improved mass exchange and mass transfer within the quartz tubes.

The system of automatic movement of a horizontal frame, following the sun during the day, makes it possible to switch over to artificial irradiation with UV-lamps in the case of low solar activity or at night.

The system combines homogeneous–heterogeneous photocatalysis, due to simultaneous introduction into the treated water of Fenton reagent and nanodisperse photocatalytically active particles (such as TiO_2). The synergetic effect that appears increases the efficiency of organic pollutants decomposition.

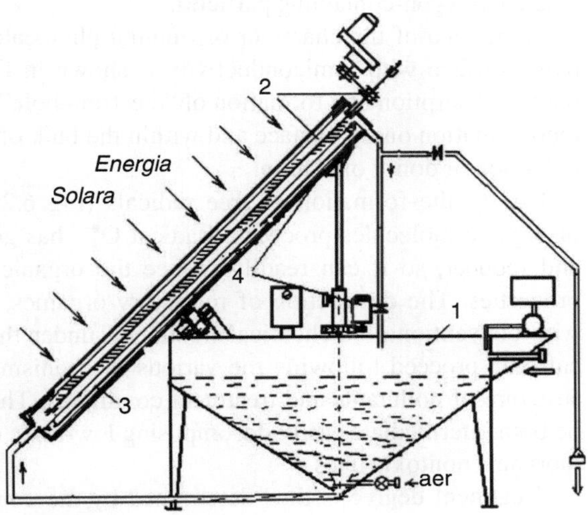

Fig. 6.18 Scheme of integrated photocatalytic reactor with solar irradiation: 1 accumulation–recirculation reservoir, 2 quartz flowthrough tubes with the system for hydromechanical cleaning, 3 UV-lamps and reflector, 4 inclined frame with programing device and photoelectrical sensor to ensure its automatic movement following the sun [175]

Catalytic Decomposition of Organic Pollutants with TiO_2

Due to its strong photocatalytical properties, TiO_2 is used for the decomposition of organic compounds [176]. With the decrease in the size of its particles, their specific surface, and accordingly, photoactivityincreases.

Electron orbitals in the Ti atom fill the levels $4s^2$ and $3d^2$, and in the O atom levels $2s^2$ and $2p^4$. In the molecule TiO_2 titanium ions are arranged within the distorted octahedron and have the electron configuration Ti^{4+} $(3d^0)$. The valency zone of TiO_2 mainly consists of 2p-orbitals of oxygen, hydrolyzed with 3d-state of Ti [177].

There exist three types of TiO_2 structure: (1) anatase, (2) rutile, and (3) brookite. The gap size between the valency zones in anatase is 3.2 eV, in rutile—3.02 eV, and in brookite—2.96 eV. Anatase possesses the best photocatalytic properties. When TiO_2 is subjected to UV-irradiation, the electrons from the valency zone (e^-) are excited into the conductance zone, forming the holes (h^+), i.e., $TiO_2 + h\nu \rightarrow TiO_2(e^- + h^+)$.

In this way, light with the wavelength $\lambda < 385$ nm excites the electrons from the valence zone to the conductance zone. Therefore, during UV-irradiation, simultaneous absorption of two reagents (H_2O/O_2) on TiO_2 may occur, and a free electron and an electron vacancy (hole) appear. This results in water molecule oxidation with positive holes and splitting into $OH^\bullet$ and H^+. Since oxygen is an easily reduced substance, its reduction with photoelectrons within the conductance zone results in the generation of superoxide radical anions ($^\bullet O_2^-$). In turn, these anions react with H^+, forming the dioxide hydrogen radical ($^\bullet HO_2$, hydroperoxyl). During further collisions with electrons, the hydrogen oxide radical is formed HO_2^-, and then, possibly, a hydrogen ion and H_2O_2 molecule are formed. The chain of reactions described above and other reactions result in the formation of highly reactive oxygen-containing particles.

Some idea of the character of running photocatalytical processes on the metal oxide surface with semiconductivity is shown in Fig. 6.19. At the first stage, the photon absorption and formation of "electron–hole" pairs occurs, followed by their recombination on the surface and within the bulk of solution, with further acceptor reduction of donor oxidation.

Due to the formation of free radicals (Fig. 6.20), an efficient decomposition of organic molecules proceeds. Radical $O_2^{\bullet -}$ has good reactivity as both oxidizer and reducer, so it can readily reduce the organic substances with the acceptor properties. The destruction of refractory organics, which cannot be decomposed with conventional biochemical treatment, under the action of forming free radicals can proceed following the various mechanisms, depending on the molecular structure of pollutants and treatment conditions. The results of such treatment can be both intermediate easily decomposing low-toxic organic compounds and simple inorganic nontoxic ones.

Treatment degree is thus determined by the concentrations of oxidant (H_2O_2) and TiO_2 introduced, and also by the UV-irradiation time. The role is played by the crystalline structure of TiO_2; anatase structure is preferable.

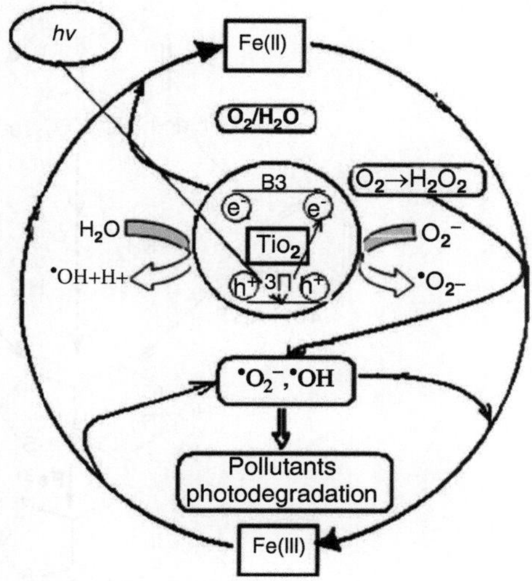

Fig. 6.19 Scheme of the main processes during the photocatalytic redox cycle in the combined system FeNTA–TiO_2–UV

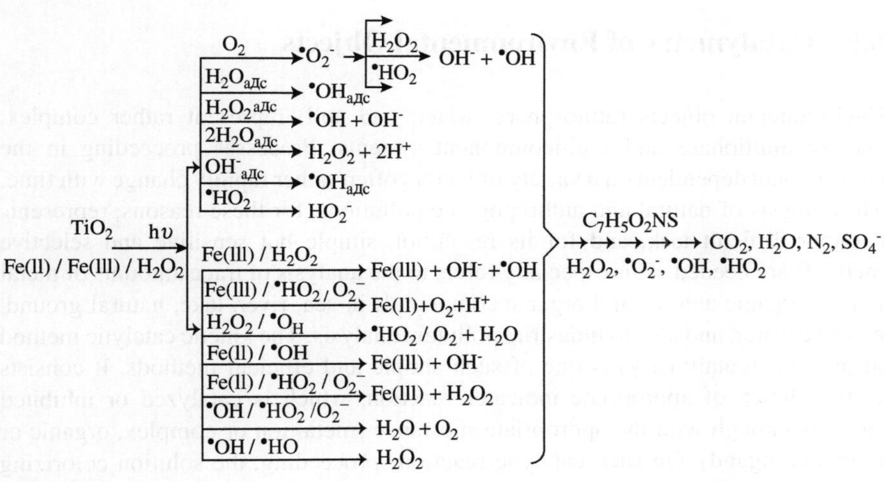

Fig. 6.20 A series of reactions with the involvement of reactive oxygen-containing particles

According to research [178], the intermediate compounds are formed during the ABT (amino-benzothiazole) transformation, up to the complete mineralization, following the scheme [Fig. 6.21].

Therefore, the use of a combined homogeneous–heterogeneous photocatalytic treatment of the solutions containing the refractory organic pollutants makes it possible to reach their efficient decomposition, in some cases up to complete mineralization.

Fig. 6.21 The scheme of ABT photocatalytic degradation

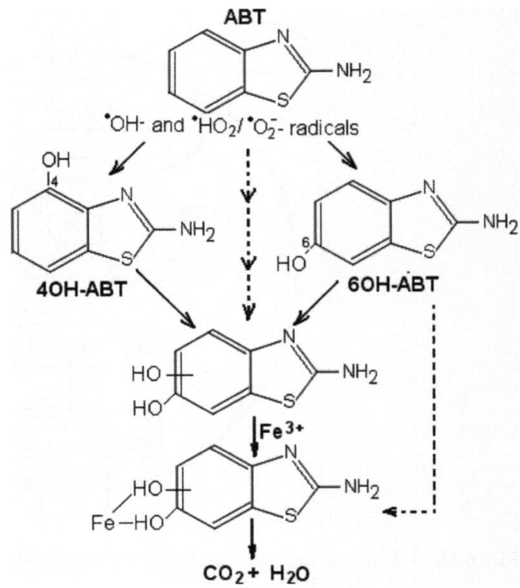

6.5 Catalymetry of Environmental Objects

Environmental objects (atmosphere, water, and soil) represent rather complex, usually multiphase and multicomponent systems. Processes proceeding in the environment dependent on a variety of factors often rather rapidly change with time. The analysis of natural and anthropogenic pollutions, for these reasons, represents a rather difficult task, and for its resolution simple but sensitive and selective methods are needed. This especially refers to the analysis of trace amounts of metal ions, inorganic anions, and organic compounds in sea, river, lake, natural ground, pipeline water, and also to industrial pollutant analysis. The kinetic catalytic method of analysis (catalimetry) is one of such simple and efficient methods. It consists in the choice of appropriate indicator reaction, which is catalyzed or inhibited strongly enough with the appropriate substance (metal ion or complex, organic or inorganic ligand). On such catalytic reactions proceeding, the solution colorizing may appear or disappear. The rate of colorizing change, as a rule, is proportional to the concentration of the solution sought. Catalymetric methods have been described in many papers and books, among them the works of more general [157, 179, 180] and more specific [181, 182] character are to be mentioned. Besides, reviews on the kinetic aspects of analytical chemistry appear [183].

Spectral methods in catalymetry will be considered here only in application to the analysis of water, atmosphere, and soil. From the kinetic methods only the most important will be considered further, based on homogeneous catalysis of redox reactions.

Peculiarities of catalymetry, with regard to environmental objects (water, atmosphere, soil), are as follows:

- Necessity of trace amounts of elements, inorganic anions, and organic substances' analysis, meaning about 10^{-4} to 10^{-5} mkg/mL or even less.
- Possibility of various forms of metal compounds and other substances' existence in such complicated systems (both regarding oxidation degree of central metal ion and the composition of complex compounds formed).
- Change of natural objects composition in time (within 24 h, season, etc.).
- Changing conditions of the processes proceeding (pH, presence of oxidants such as O_2, H_2O_2, O_3, etc.) and changing of their concentration over time, appearance of reactionary intermediate particles—free radicals $OH^{\bullet}$, $HO_2^{\bullet}$).
- Possibility of homogeneous–heterogeneous catalysis usage (in bottom deposits and suspended particles, on wet surfaces of aerosol particles, in water–soil conditions).

It is usually rather difficult to take all these factors and possible mechanisms of catalytic processes into consideration on catalymetry application to environmental objects. That is why it is so important to reveal the mechanism of these processes at least on model systems, in order to choose optimal conditions of analysis using such systems, and then to bring corrections on this basis, passing to the analysis of real environmental objects.

Catalytic reactions in catalymetry can be subdivided into several groups in terms of oxidizer character [182]: reactions of oxidation with hydrogen peroxide and dioxygen, anions of oxygen-containing acids (ClO_3^-, BrO_3^-, IO_3^-, IO_4^-, etc.), metal cations [184], enzymes [185], etc.

To determine element microquantities, homogeneous catalytic redox reactions are used more often, and the determination element is the catalyst of some suitable indicator reaction. For example, oxidation of some colorant or other organic substance (indicators) with hydrogen peroxide in the presence of metal-ion sought is a widely spread type of catalytic reaction.

Kinetic methods used for natural water analysis can be subdivided into the following reactions:

1. Reactions yielding the formation or deterioration of light absorbing substances within visible or ultraviolet region (reaction rate will be determined by optical density of solution change over time, using spectrophotometry or photoelectrocolorimetry).
2. Reactions accompanied by light irradiation emission (for its measurment, chemiluminescent photometers, or photographic methods are used).
3. Luminescent reactions (fluorimetric methods).

The rate of catalytic reaction,

$$\frac{dx}{dt} = \chi C_K P_K \quad \text{or} \quad \frac{\Delta x}{\Delta t} = \chi C_K P_K,$$

where x and Δx are current concentration of indicator substance-indicator or its transformation product, and change in this substance concentration during the time Δt, χ is specific rate constant, C_K catalyst concentration (element under determination), and P_K concentration of all reacting substances product.

Hence,

$$C_k = \frac{\Delta x}{\Delta t \cdot \chi P_k}.$$

The sensitivity is often determined by measuring Δx, as well as background reaction of indicator substance. In natural water analysis sensitivity usually amounts to 10^{-4} to 10^{-5} mkg/mL, and to increase it one has to reduce the rate of background reaction more often. The background reaction can also be catalyzed by some admixtures in the specimen; therefore special attention is to be paid to the reagents and glassware purity and thoroughness in analyzing.

Reaction specificity can be increased by choosing analysis conditions (pH, etc.) so that catalyst in solution (metal ion or complex compound) can be in certain form of ligand or composition. The latter is especially important, since the value of catalyst redox potential (φ_{cat}) must be between the values φ_{oxidant} and φ_{reducer} of the indicator reaction. Such a situation can be reached, changing both the nature of ligand and the composition of metal complexes formed (catalyst).

If concentrations of all participating reaction components (except of indicator substance concentration) remain practically unchanged, catalytic reaction rate can be presented as follows:

$$\frac{dx}{dt} = k(a - x)C_k,$$

where a and $(a - x)$ are initial and current concentration of substance-indicator, and x is current concentration of transformation product by substance-indicator. In the initial period $x = 0$, and

$$\frac{dx}{dt} = k \cdot a \cdot Ck.$$

After the last equation integration:

$$x = k \cdot a \cdot Ck \cdot t,$$

from which it follows that during the initial period of process the direct dependence is observed between the time of reaction and initial (indicator) substance concentration.

Variants of kinetic methods on the basis of the last equation are called differential ones [182]. If initial concentration change (more often indicator substance) cannot be neglected, the equation

$$\frac{dx}{dt} = k(a - x)C_k,$$

is to be integrated:

$$\lg\frac{a}{a - x} = k' \cdot C_k \cdot t, \quad \text{where } k' \frac{k}{2.303}$$

From the last equation C_k can also be determined, and such variants of kinetic methods are called integral ones. In the general case, differential methods of analysis are used when the concentration of indicator reaction products can be measured with high sensitivity during its initial period. When the initial indicator substance concentration change can be measured with good enough sensitivity, it is better using integral kinetic analytic methods [182].

To determine microquantities of elements (manganese, iron, etc.), the differential method of tangents is often used in kinetic methods, in accordance with which the tangent of the angle slope is determined from kinetic curves, describing substance–indicator concentration change over time in different catalyst's concentrations. These data are then used to build the graph in coordinates $tg \cdot \alpha / C_k$. Catalytic reaction rate can be determined from the initial segment of the kinetic curve:

$$W = \frac{\Delta D}{\Delta t \cdot \varepsilon \cdot I} - W_o(\mathrm{a}),$$

where ΔD is change of the solution optic density during the time $\Delta t(C)$, ε extinction coefficient, l thickness of absorbing layer, and W_o rate of background (noncatalytic) reaction. Thus, calibrating curve is built in coordinates W/C_k, and to determine unknown concentration of substance-catalyst, experimentally found values of $tg \cdot \alpha$ or W are used.

The sensitivity characterizes the lowest limit of determining element (substance) detection on the basis of indicator reaction chosen. Elements (substances) are determined on the basis of the graduating graph. While theoretical calculation shows that up to 10^{-16} mol/L of catalyst [157, 179] can be detected with the catalymetric method, practically the sensitivity does not exceed 10^{-9} to 10^{-10} g – ion/mL because of background reaction influence.

To increase the sensitivity of the catalytic method, activators [179, 180, 186] are often used, which are usually ligands for metal-catalyst. For example, use of certain activators can be considered, as well as that of a number of substance-indicators, for manganese ions microquantities' determination [23].

Bicarbonate-ion can be chosen as activator of Mn^{2+} ion. Activating influence of HCO_3^- on catalytic oxidation processes of lumomagneson (LM), alizarin (AL), alizarin C (ALC), as well as indigo carmine (IC), indigo monosulphonate (IMS), murexide (MU), and carmine (CA) with hydrogen peroxide in the presence of Mn^{2+} ions was found earlier [187–189]. Such action of bicarbonate-ion is connected with the formation of catalytically active complex Mn(II) with HCO_3^- within the reaction medium, which will efficiently interact with H_2O_2 [189, 190]. In the course of this process, $OH^\bullet$ and $Mn^{IV}(HCO_3^-)_2$ will be generated, due to which the efficient oxidation of substance indicators proceeds. Substrate (S) oxidation, having no complex-forming ability, will be done by $OH^\bullet$ radicals ($OH^\bullet + S \rightarrow$ products [187]), and substrate–ligands (S_L), forming complexes with manganese ions, will be oxidized within the coordination sphere of catalyst [188]:

Table 6.2 Optimal concentration conditions of Mn^{2+} ions microamounts' determination by catalymetric method in the systems Mn^{2+}–HCO_3^-–H_2O_2–S (substance indicator) and lower limit of Mn^{2+} ($C_{Mn^{2+}}$) ions detection

S	$[HCO_3^-]_o$, M	$[H_2O_2]_o$, M	$[S]_o$, M	pH	$C_{Mn^{2+}}$	
					g-ion/L	Mkm/mL
IMS	0.3	0.1	2×10^{-4}	7.5	5×10^{-9}	2.7×10^{-4}
IC	0.5	0.4	2×10^{-4}	7.3–7.4	2×10^{-8}	1.1×10^{-3}
MU	0.4	0.4	5×10^{-4}	8.0	1×10^{-8}	5.5×10^{-4}
CA	0.5	0.1	4×10^{-4}	8.0–8.2	6×10^{-8}	3.3×10^{-3}
LM	0.5	0.4	2.5×10^{-4}	8.0–8.2	8×10^{-9}	4.4×10^{-4}
AL	0.15	0.05	3×10^{-4}	8.0–8.2	5×10^{-9}	2.7×10^{-4}
ALC	0.175	0.1	3×10^{-4}	7.6	7.5×10^{-9}	4.1×10^{-4}

$$Mn(II)(HCO_3^-)_2S_L + H_2O_2 \leftrightarrows Mn(IV)(HCO_3^-)_2S_L + 2OH^-,$$

$$Mn(IV)(HCO_3^-)_2S_L \rightarrow Mn(II)(HCO_3^-)_2 + P.$$

It follows from the aforesaid that the activating role of HCO_3^- is dealing with the coordination compound $Mn(II)(HCO_3^-)_2$ formation within the reaction solution. Proceeding from the results of kinetic study of the above-mentioned substance indicator oxidation reactions, optimal concentrations of reagents were found, used for manganese ions microquantities' determination (Table 6.2).

As an example of Mn(II) microquantities' determination methods, reaction of lumomagneson oxidation with hydrogen peroxide can be taken, in the presence of bicarbonate-ion, which comes in a number of procedures. First, initial solutions are prepared and used: $C_{Na_2CO_3} = 1.0$ M; $C_{LM} = 1 \times 10^{-3}$M; $C_{H2O2} = 9$ M, and $HClO_4$ concentrated solution (concentration of manganese ion, which is under determination in the sample) must fall within the required limits (4–82) $\times 10^{-4}$ mkg/mL Mn(II).

Reagents are introduced into the cell (25°C), permitting one to control the change during the reaction course, in the following order: several drops of $HClO_4$ are added to 12.5 mL of Na_2CO_3 up to pH = 7.5–7.8, then a definite volume (up to 3 mL) with unknown concentration of Mn^{2+} ions is introduced (but within the aforementioned limits), and 4 mL of LM and 1.0–1.1 mL of H_2O_2 are added.

Total mixture volume must reach 25 mL (for this purpose, preliminary calculated volume of bidistilled water is introduced in the very beginning).

The moment H_2O_2 is introduced reaction begins. In some time (~1–2 min) the sample is taken from the cell, and is brought into the 1-sm cuvette and optic density is measured (D) at $\lambda = 508$ nm. Proceeding from the kinetic curve D/t, the difference of optic density ΔD is determined during the time Δt (considering the background reaction) and, using the aforementioned formula () and graduation graph, appropriate concentration of Mn^{2+} ions will be found.

In many works ethylenediamine (En) was used as an activator. The research on the number of individual substances (CA, MU, IC, LM, AL, and ALS) oxidation with hydrogen peroxide in the presence of Mn(II) coordination compounds with En [191] make the following conclusions possible:

– $MnEn_2^{2+}$ complex is the process catalyst.
– The oxidation substrates (S), which are not good complex-forming ones (CA, MU, and IC) proceed in accordance with the following scheme:

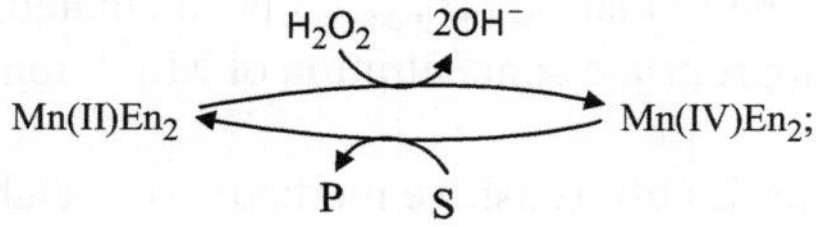

– Oxidation of complex-forming substrates S_L (LM, AC, and ALC) proceeds inside the coordination sphere of complex catalyst:

$$Mn(II)En_2S_L + H_2O_2 \leftrightarrows Mn(IV)En_2S_L + 2OH^-,$$
$$Mn(IV)En_2S_L \rightarrow Mn(II)En_2 + P.$$

The same process mechanism applies in the case of many other complex-forming colorants as well.

Hence, the role of activator (En) comes to formation of $MnEn_2^{2+}$ coordination compound, which will efficiently interact with H_2O_2. As a result, reactionary particles $Mn(IV)En_2$ will be generated in the reaction medium, oxidizing the substance indicator. H_2O_2 plays the role of oxidant, and substrate-reducer. In such systems a redox cycle will be formed, and central catalyst ion oxidation and reduction will proceed alternatively. Efficient proceeding of these processes will result in the fact that even the lowest amount of the catalyst provides the oxidation of significant amounts of substance indicator. Because of this, aforesaid methods make it possible to detect trace amounts of manganese.

The method of manganese microquantities' determination on the basis of lumomagneson oxidation reaction with hydrogen peroxide in the presence of En is similar to one of the systems $Mn(II)–HCO_3^-–H_2O_2$–LM described above. In the present case the optimal conditions are $[LM] = 5 \times 10^{-3}$ M, $[H_2O_2] = 5 \times 10^{-3}$ M, $[En] = 1 \times 10^{-2}$ M, and pH 8.5 (borate buffer solution). Under these conditions the lower limit of Mn^{2+} detection is 5×10^{-9} g–ion/L.

Histidine, triethyleneamine, 1,10-phenanthroline, and α, α'-dipyridyl are also activators of manganese. They will form with manganese catalytically active coordinative compounds, efficiently interacting with H_2O_2. Oxidation with substance indicators is done with $OH^\bullet$ radicals.

Another example of manganese microquantities' determination is based on the reaction of indigo carmine oxidation with hydrogen peroxide in the presence of histidine (His) in borate buffer solutions. To the solution studied (50 mL), containing $(0.6–2) \times 10^{-5}$ g–ion/L of Mn^{2+}, 0.3–0.6 g of His is added, mixed well to provide total dissolving, and is introduced quantitatively into the measuring 200-mL flask with the help of standard buffer solution with pH 7.3–7.5 (mixture of 0.2 M H_3BO_3 and 0.05 M $Na_2B_4O_7$), completed till the mark with the same buffer solution.

Then 25 mL of the solution obtained is brought into the cell (25°C), 0.5 mL of 5×10^{-3} M indigo carmine is added, and then 0.3 mL of 8M H_2O_2. From the reacting mixture samples are taken, are brought into the cuvette, and optic density (D) of the solution is measured at $\lambda = 615$ nm. Proceeding from the kinetic curve D/t, the rate of indigo carmine oxidation (W_{indcarm}) is calculated, and, with the help of the graduation graph, appropriate concentration of Mn^{2+} ions in studied solution is found.

In certain cases rather highly-sensitive methods were elaborated for manganese microquantities' determination in the absence of activators as well. The substance indicator itself can play the role of activator due to the formation of reactionary (in relation to H_2O_2) manganese(II) coordinative compound. Thus, lumomagneson will be efficiently oxidized by hydrogen peroxide in the presence of small amounts of manganese [192]. Lumomagneson will form with manganese(II) a rather strong coordinative compound $MnHLM^+$ (p$K = 6.02 \pm 0.28$ [193]), which will efficiently interact with H_2O_2 [194]:

$$\mathrm{Mn(II) + LM \rightleftarrows Mn(II)LM},$$

$$\mathrm{Mn(II)LM + H_2O_2 \rightarrow Mn(IV)LM + 2OH^-},$$

$$\mathrm{Mn(IV)LM \rightarrow Mn(II) + P}.$$

Hydrogen peroxide was used as Mn(II) compound oxidant. Substances indicators will be oxidized with the products of H_2O_2 decomposition—OH radicals and Mn(IV) compounds.

At present some kinetic methods of manganese microquantities' determination are developed, in which periodate is used as oxidant. A number of substances indicators, among them *o*-dianisidine, *p*-phenetidine, diethylaniline, iodide-ion, and other substrates, are effectively oxidized with iodate in the presence of Mn compounds as catalysts [23]. It can be supposed thatactivators such as α, α'-dipyridyl and 1,10-phenanthroline serve as ligand and form with manganese ion a catalytically active complex. The mchanism of many of these reactions also involves elementary stages of alternating oxidation of Mn(II) into Mn(IV) and reduction with the help of substance indicator into Mn(II). For instance, the scheme of arylamine (Ar) oxidation is as follows [195, 196]:

$$\mathrm{Mn(II) + IO_4^- \rightarrow Mn(IV) + IO_3^-},$$

$$\mathrm{Mn(II) + Mn(IV) \rightarrow 2Mn(III)},$$

$$\mathrm{2Mn(III) + IO_4^- \rightarrow 2Mn(IV) + IO_3^-},$$

$$\mathrm{Mn(IV) + Ar \rightarrow Mn(III) + P},$$

$$\mathrm{Mn(III) + Ar \rightarrow Mn(II) + P}.$$

The kinetic method of Mn(II) microquantities' determination by its oxidation in Mn(VII) by periodate was also elaborated [197].

Other catalytic methods of manganese microquantities' determination are also known, based on fluorescence reducing of berillium morine complex [198], the reaction of $KMnO_4$ with oxalate-ions [191, 199], 1,5-bis-(2-hydroxy-5-chlorophenyl)-3-cyanoformazane oxidation with air oxygen [200], and succinimide dioximine self-oxidation [201] or that of *o*-hydroxyphenylthiourea [202].

The drawback of almost all elaborated catalytic methods of elements microquantities' determination is the absence of enough selectivity. Elements' determination is often impeded by admixing metal ions that exert catalytic activity in the same reactions. These are, first of all, transition metal ions (iron, cobalt, copper, nickel, etc.) [23]. Inhibiting influence is exerted by the ligands forming with metal ions catalytically inactive compounds—$P_2O_7^{2-}$, EDTA, $C_2O_4^{2-}$, $H_2PO_4^-$, etc., as well as oxidants and reducers interacting with substances indicators and metal compounds (SO_3^-, $S_2O_3^{2-}$, NO_2^-, MnO_4^-, $Cr_2O_7^{2-}$, etc.). For example, as mentioned above, one of the highly selective methods of manganese microquantities' determination is based on substances indicators oxidation reaction (lumimagneson, lizarin, lizarin, indigo carmine, etc.) with hydrogen peroxide in the presence of bicarbonate-ions [23]. Mn^{2+} determination (in the amount of 1×10^{-8} g-ion/L) by this method is not prevented by 1,000-fold excess of Zn^{2+}, Ca^{2+}, Ba^{2+}, Fe^{3+}, Fe^{2+}, $Cr_2O_7^{2-}$, Co^{2+}, Cu^{2+}, Ni^{2+}, Cd^{2+}, Pb^{2+}, $Fe(CN)_6^{3-}$ ions, 10^4-fold excess of Mg^{2+}, NO_2^-, SCN^- ions, and 10^5-fold excess of Na^+, SiO_3^{2-}, MnO_4^-, SO_4^{2-}, HPO_4^{2-}, Li^+, Cl^-, CH_3COO^-, NO_3^-, Br^-, K^+, NH_4^+ ions. The high selectivity of this method is stipulated by the fact that bicarbonate complexes of metal ion admixtures (first of all Fe^{2+}, Fe^{3+}, Co^{2+}, Ni^{2+}, Cu^{2+}, etc., often preventing Mn^{2+} determination) do not possess catalytic activity in the reaction of H_2O_2 decomposition, or their activity is much weaker than that of Mn(II) bicarbonate complexes (as in the case of Co^{2+} [203]).

Various ways of increasing the selectivity of catalytic indicator reactions on metal ion determination were given in [157, 179, 180]. For example, the selectivity of Mn^{2+} ions' catalytic determination can often be improved by varying the reagent concentration and pH of solution within a broad range.

Impeding influence of admixing ions can be reduced by their masking. Thus, on Mn^{2+} ion determination, iron ions can be masked by citrate, or ethylenediaminetetraacetic acid, cobalt, nickel, copper, chromium, and silver ions by diethanolamine [180].

An example of catalytic analytical methods using (catalymetry) for the concrete environmental object study can serve for the determination of the same Mn^{2+} ions in soil samples or in grapes and vine plants using the systems HCO_3^-–H_2O_2–IC [23].

References

1. Th. Eisner, J. Meinwald, *Chemical Ecology: The Chemistry of Biotic Interaction* (National Academy, Washington, DC, 1995), p. 221
2. A.Ya. Sychev, S.O. Travin, G.G. Duca, Yu.I. Scurlatov, *Catalytic Reactions and the Protection of Environment* (Stiinta, Kishinev, 1983), p. 272 [Russian]

3. I. Williams, *Environmental Chemistry – A Modular Approach* (Wiley, Chichester, 2001) p. 388.
4. G.W. van Loon, S.J. Duffy, *Environmental Chemistry: A Global Perspective* (Oxford University Press, Oxford, 2000) p. 7.
5. J.N. Galloway, Z. Dianwu, X. Jiling, G.E. Likens, Science **236**, 1559–1562 (1987)
6. G.E. Likens, R.F. Wright, J.N. Galloway, T.J. Butler, Sci. Am. **241**(4), 43–51 (1979)
7. M. Satoh, *Atmospheric Circulation Dynamics and General Circulation Models* (Springer, Berlin, 2004) p. 643
8. US EPA. A Brief History of Acid Rain. Epa. gov., 2010
9. 'Cap-and-trade' model eyed for cutting greenhouse gases. San Francisco Chronicle, 3 December, 2007
10. T. Gilberston, O. Reyes, *Carbon Trading: How It Works and Why It Fails* (Dag Hammarskjöld Foundation, Sweden, 2009) p. 22
11. H. Güsten, in *Free Radicals in Biology and Environment*, ed. by F. Minisci (Kluwer, Dordrecht, 1997) pp. 87–408
12. A.P. Purmal, in *Free Radicals in Biology and Environment*, ed. by F. Minisci (Kluwer, Dordrecht, 1997) pp. 423–435
13. A.Ya. Sychev, V.G. Isak, *Homogeneous Catalysis with Iron-Containing Compounds* (Stiinta, Kishinev, 1988) p. 216 [Russian]
14. G.E. Zaikov, S.A. slov, V.L. Rubailo, *Acid Rains and Environment* (Khimya, Moscow, 1991) [Russian]
15. F.E. Brinckman, J.G. Olson, W.P. Jverson, in *Atmospheric Chemistry*, ed. by F.D. Goldberg (Springer, Berlin, 1982) p. 231
16. J.G. Calvert, in *SO_2, NO and NO_2 Oxidation Mechanism: Atmospheric Considerations*, ed. by J.G. Calvon (Butterworths, London, 1984) p. 3
17. R.G. Zika, J. Geophys. Res. **87**, 5015 (1982)
18. R.W. Johnson, G.E. Gordon (eds.) *The Chemistry of Acid Rain* (American Chemical Society, Washington, 1987) (ACS Symp. Ser., vol. 349).
19. P. Farmer, *Acid Rain and the Environment 1980–1984* (Technical Communication, Letchworth, 1984)
20. L. Grayson, *Acid Rain and the Environment 1984–1988* (Technical Communication, Letchworth and British Library Science Reference and Information Service, 1989) p. 1991
21. Ch. Brandt, R. van Eldik, Chem. Rev. **95**, 119–190 (1995)
22. L. Khprovat, *Acid Rains* (Stroiizdat, scow, 1990) [Russian]
23. A.Ya. Sychev, V.G. Isac, *Manganese Coordination Compounds in Catalysis* (Stiinta, Kishinev, 1990) p. 321 [Russian]
24. M.R. Hoffman, D.J. Jasov, in *SO_2, NO_2 and NO Oxidation Mechanisms: Atmospheric Considerations*, ed. by G. Calbert (Ann Arbor Science, Boston, 1984) p. 101
25. P.K. Lim, A. Huss, Jr., C.A. Eckert, J. Phys. Chem. **86**, 4233 (1982)
26. A.Ya. Sychev, V.G. Isak, Uspekhi khimii [Russ. Chem. Rev.] **62**, 303 (1993) [Russian]
27. L.A. Barie, H.W. Georgh, Atmos. Environ. **10**, 743 (1976)
28. W. Jaeschke, in *Chemistry of Multiphase Atmospheric Systems*, ed. by W. Jaeschke (Springer, Berlin, 1986) p. 3
29. L. Khprovat, *Acid Rains* (Stroiizdat, Mascow, 1990) [Russian]
30. V.L. Talrose, I.K. Larin, Uspekhi khimii [Russ. Chem. Rev.] **60**, 507 (1991) [Russian]
31. M. Capitelli, *Plasma Kinetics in Atmospheric Gases* (Springer, Berlin, 2000) p. 300
32. D.H. Stedman, J.V. Walega, C.A. Cantrel, J.P. Burrows, G. Tynall, in *Chemistry of Multiphase Atmospheric Systems*, ed. by W. Jaeschke (Springer, Berlin, 1986) p. 351
33. G.I. Skubnevskaya, N.. Bajin, Meteorol. Hydrol. **9**, 113 (1982) [Russian]
34. N.M. Bajin, V.V. Penenco, A.E. loian, G.I. Scubnevskaia, Uspekhi khimii [Russ. Chem. Rev.] **60**, 514 (1991)
35. H. Sakugawa, I.R. Kaplan, W. Tsai, J. Cohen, Environ. Sci. Technol. **24**, 1452 (1990)
36. M.R. Hoffman, D.J. Jacob, in *SO_2, NO and NO_2 Oxidation Mechanism: Atmospheric Considerations*, ed. by J.G. Galvert (Butterworths, London, 1984) p. 101

37. J.G. Calvert, W.R. Stockwell, Environ. Sci. Technol. **17**, 428 (1983)
38. L.I. Kleinman, J. Geophys. Res. **91**, 10889 (1986)
39. A.J. Hough, J. Geophys. Res. **93**, 3789 (1988)
40. B.J. Finlayson-Pitts, J.N. Pitts, *Atmospheric Chemistry: Fundamentals and Experimental Techniques* (Wiley, New York, 1986)
41. L.R. Martin, in *SO_2, NO and NO_2 Oxidation Mechanism: Atmospheric Considerations*, ed. by J.G. Calvert (Butterworths, London, 1984) pp. 63–100
42. T.E. Graedel, K.I. Goldberg, J. Geophys. Res. **88**, 10865 (1983)
43. D.J.J. Jacob, Geophys. Res. **91**, 9802 (1986)
44. S.N. Pandis, J.H. Seinfeld, J. Geophys. Res. **94**, 1105 (1989)
45. B.J. Heikes et al., J. Geophys. Res. **87**, 3045 (1982)
46. D.W. Gunz, M.R. Hoffman, Atmos. Environ. **24A**, 1601 (1990)
47. H. Harrison, T.V. Larson, C.S. Monkman, Atmos. Environ. **16**, 1039 (1982)
48. C.J. Weschler, T.E. Graedel, in *Heterogeneous Atmospheric Chemistry*, ed. by D.R. Schrazer (AGU, Washington, 1982) p. 196
49. R. van Eldik, in *Chemistry of Multiphase Atmospheric Systems*, ed. by W. Jaeschke (Springer, Berlin, 1986) p. 541
50. A. Huss, Jr., P.K. Lim, C.A. Eskert, J. Phys. Chem. **86**, 4224 (1982)
51. L.R. Martin, M.W. Hill, A.F. Tai, T.W. Good, J. Geophys. Res. **96**, 3085 (1991)
52. C. Seigneuz, P. Saxena, Atmos. Environ. **18**, 2109 (1984).
53. S.N. Pandis, J.N. Seinfeld, J. Geophys. Res. **94**(D1), 1105 (1989)
54. D.J. Jacob, E.W. Gottlieb, M.J. Plather, J. Geophys. Res. **94**(D10), 12975 (1989)
55. J. Kraft, R. van Eldik, Atmos. Environ. **23**, 2709 (1989)
56. I.A. Huss, P.K. Lim, C.A. Eckert, J. Phys. Chem. **86**, 4229 (1982)
57. M.H. Conklin, M.R. Hoffmann, Environ. Sci. Technol. **22**, 899 (1988)
58. M.H. Conklin, M.R. Hoffmann, Environ. Sci. Technol. **22**, 891 (1988)
59. D.J. Jacob, E.W. Gottlieb, M.J. Prather, J. Geophys. Res. **94**(D10), 12975 (1989)
60. D.E. Linn, Jr., S.D. Rumage, J.L. Grutsch, Jr., Int. J. Chem. Kinet. **25**, 489 (1993)
61. J.M. Anast, D.W. Margerum, Inorg. Chem. **20**, 2319 (1981)
62. D.E. Linn, Jr., M.J. Dragan, D.E. Miller, Inorg. Chem. **29**, 4356 (1990)
63. N. Coichev, R. van Eldik, Inorg. Chem. Acta **185**, 69 (1991)
64. N. Coichev, R. van Eldik, Inorg. Chem. **30**, 2375 (1991)
65. J. Berglund, L.I. Elding, in *Laboratory Studies of the Aqueous Chemistry of Free Radicals, Transition Metals and Formation of Acidity in Clouds: Final Report of Contract Na STEP-0005-C(MB)*, 1990, pp. 27–44
66. J. Berglund, S. Fronaeus, L.I. Elding, Inorg. Chem. **32**, 4527 (1993)
67. Ch. Brandt, I. Fabian, R. van Eldik, Inorg. Chem. **33**, 687 (1994)
68. S. Mukhopadhyay, R. Banerjee, J. Chem. Soc. Dalton Trans. 933 (1993)
69. J. Ziajka, F. Beer, P. Warneck, Atmos. Environ. **28**, 2549 (1994)
70. L.R. Martin, M.W. Hill, A.F. Tai, T.W. Good, J. Geophys. Res. **96**(D2), 3085 (1991)
71. J. Kraft, R. van Eldik, Inorg. Chem. **28**, 2297 (1989)
72. J. Kraft, R. van Eldik, Inorg. Chem. **28**, 2306 (1989)
73. R. van Eldik, N. Coichev, K. Bal Reddy, A. Gerhard, Ber. Bunsenges. Phys. Chem. **96**, 478 (1992)
74. T.E. Graedel, M.L. Mandich, C.J. Weschler, J. Geophys. Res. **91**(D4), 5205 (1986)
75. Y. Zuo, J. Hoigne, Environ. Sci. Technol. **26**, 1014 (1992)
76. Y. Erel, S.O. Pehkonen, M.R. Hoffmann, J. Geophys. Res. **98**(D10), 18423 (1993)
77. K. Bal Reddy, R. van Eldik, Atmos. Environ. **26A**, 661 (1992)
78. K. Bal Reddy, N. Coichev, R. van Eldik, J. Chem. Soc. Chem. Commun. **7**, 481 (1991)
79. P. Behra, L. Sigg, Nature **344**(6265), 419 (1990)
80. G. Zhuang, Z. Yi, R.A. Duce, P.R. Brown, Nature **355**(6360), 537 (1992)
81. D.L. Sedlak, J. Hoigne, Atmos. Environ. **27A**, 2173 (1993)
82. B.C. Faust, R.G. Zepp, Environ. Sci. Technol. **27**, 2517 (1993)
83. C.J. Miles, P.L. Brezonik, Environ. Sci. Technol. **15**, 1089 (1981)

84. B. Sulzberger, in *Aquatic Chemical Kinetics*, ed. by W. Stumm (Wiley, New York, 1990) pp. 401–429
85. J.M. Anast, D.W. Margerum, Inorg. Chem. **20**, 2319 (1981)
86. D.B. Hobson, P.J. Richardson, P.J. Robinson, E.A. Hewitt, J. Smith, J. Chem. Soc. Faraday Trans. **82**, 869 (1986)
87. N. Coichev, R. van Eldik, D.A. Franz, J. Chem. Educ. **71**, 767 (1994)
88. T. Ibusuki, H.M. Barnes, Atmos. Environ. **18**, 145 (1984)
89. I. Grgie, V. Hudnik, M. Bizjak, J. Levec, Atmos. Environ. **25A**, 1591 (1991)
90. I. Grgie, V. Hudnik, M. Bizjak, J. Levec, Atmos. Environ. **26A**, 571 (1992)
91. T. Ibusuki, K. Takeuchi, Atmos. Environ. **21**, 1555 (1987)
92. R.E. Connick, Y.-X. Zhang, Int. J. Chem. Kinet. **27**, 829 (1995)
93. R.E. Connick, Y.-X. Zhang, Inorg. Chem. **35**, 4613 (1996)
94. M. Von Piechowski, T. Nauser, J. Hoigne, R.E. Bühler, Ber. Bunsen. Phys. Chem. **97**, 762 (1993)
95. A. Warshel, *Computer Modeling of Chemical Reactions in Enzymes and Solutions* (Wiley, New York, 1991)
96. K. Chelkowska, D. Grasso, I. Fabian, G. Gordon, Ozone Sci. Eng. **14**, 33 (1992)
97. I. Fabian, G. Gordon, Inorg. Chem. **31**, 2144 (1992)
98. F.W. Lurmann, J.R. Yong, G.M. Hidy, in *Chemistry of Multiphase Atmospheric Systems*, ed. by W. Jaeschke (Springer, Berlin, 1986) p. 773
99. P. Mehta, South Asian J. Tour. Heritage **3**(2) (2010)
100. N. Oshino, D. Jamieson, B. Chance, Biochem. J. **146**, 53 (1975)
101. A. Boveris, N. Oshino, B. Chanel, Biochem. J. **128**, 617 (1972)
102. B. Palenic, F.M.M. Morel, Limnol. Oceanogr. **33**, 1606 (1988)
103. Gh. Duca, in *Free Radicals in Biology and Environment*, ed. by F. Minisci (Kluwer, Dordrecht, 1997) pp. 475–489
104. Gh. Duca, L. Romanchuk, Iu. Scurlatov, *Self-Purification Processes in Natural Waters*, Chisinau, 1995, p. 5. [Russian]
105. Gh. Duca, Mechanisms of ecochemical processes in water environment, Hab.Dr. Thesis, Odessa State University, Odessa, 1988, p. 46 [Russian]
106. Yu. Scurlatov, Gh. Duca, L. Ernestova, Proc. Acad. Sci. Mold. Biol. Chem. Ser. **5**, 3–20 (1983) [Russian]
107. Gh. Duca, Yu. Scurlatov, *Ecological Chemistry* (CE USM, Chisinau, 2002) p. 289 [Russian]
108. Gh. Duca, Yu. Scurlatov, A. Sycev, *Redox Catalysis and Ecological Chemistry* (CE USM, Chisinau, 2002) p. 316 [Russian]
109. Gh. Duca, Yu. Scurlatov, A. Misiti, M. Macoveanu, M. Suprateanu, *Chimie ecologica [Ecological Chemistry]*, 2nd ed. (CE USM, Chisinau, 2003) p. 304 [Romanian]
110. Gh. Duca, C. Zanoaga, M. Duca, V. Gladchi, *Redox Processes in the Environment [Manual for Students]* (CE USM, Chisinau, 2001) p. 382 [Romanian]
111. Yu.I. Scurlatov, G.G. Duca, D.G. Batir, S.. Travin, Koord. Khim. [Coord. Chem.] **15**(3), 291–307 (1989)
112. Yu. Scurlatov, Gh. Duca, *Chemistry and Life of Water* (Cartea Moldoveneasca, Chisinau, 1989) p. 124 [Russian]
113. Gh. Duca, L. Chub, Yu. Scurlatov, A. Sychev, Zh. Fiz. Khim. [J. Phys. Chem.] **63**(9), 2366–2371 (1989) [Russian]
114. Gh. Duca, A. Sychev, Khim. i tehnol. vody [Chem. Technol. Water] **9**(5), 454–456 (1987) [Russian]
115. S. Travin, Gh. Duca, Yu. Scurlatov, A. Sychev, Zh. Fiz. Khim. [J. Phys. Chem.] **60**(6), 1394–1398 (1986) [Russian]
116. A.Ya. Sychev, Gh. Duca, Zh. Fiz. Khim. [J. Phys. Chem.] **60**(1), 78–82 (1986) [Russian].
117. Gh. Duca, Iu. Scurlatov, A. Misiti, M. Macoveanu, M. Surpateanu, *Chimie ecologic [Ecological Chemistry]* (Matrix Rom, Bucuresti, 1999) p. 305 [Romanian]
118. A.P. Purmal, Yu.I. Scurlatov, Priroda [Nature] **10**, 94 (1984) [Russian]

119. Yu.I. Skurlatov, E.V. Shtamm, in *Self-Purification Processes in Natural Waters*, ed. by Gh. Duca (Bulat Art Glob, Chisinau, 1995) pp. 27–37
120. L.S. Ernestova, Yu.I. Scurlatov, L.. Fursina, Zh. Fiz. Khim. [J. Phys. Chem.] **58**, 739 (1984) [Russian]
121. J.W. Moffett, R.G. Zika, Environ. Sci. Technol. **21**, 804 (1987)
122. Yu.I. Skurlatov, Int. J. Chem. Kinet. **12**, 347 (1980)
123. K. Mopper, Z. Zhou, in *Effects of Solar Ultraviolet Radiation on Biogeochemical Dynamics in Aquatic Environments*, ed. by N.V. Blough, R.G. Zepp (Woods Holle Oceanographic, Massachusetts, 1990) p. 151
124. E.V. Shtamm, L.P. Purmal, Yu.I. Scurlatov, Uspekhi khimii [Russian Chem. Rev.] **60**, 2373 (1991) [Russian]
125. Gh. Duca, *Free Radicals in Biology and Environment* (Kluwer, Dordrecht, 1997) pp. 475–489 (NATO ASI Series. 3. High Techn., vol. 27)
126. L.S. Ernestova, .V. Shtamm, L.V. Semeniak, Yu.I. Scurlatov, R. Zepp, in *Pesticides Behavior in the Environment. Book of Abstract Soviet-American Symposium Leningrad*, 1991, pp. 382–395
127. V.V.J. Gonciaruc, Mendeleev Chem. Soc. **35**, 112 (1990) [Russian]
128. I.P. Pozdnyakov, E.M. Glebov, V.F. Plyusnin, V.P. Grivin, E. Bunduki, N.V. Goryacheva, V. Gladki, Gh.G. Duca, in High Energ. Chem. **43**(5), 406–409 (2009)
129. Yu.I. Scurlatov, L.S. Ernestova, .V. Shtamm, Dokl. AN SSSR [Proc. USSR Acad. Sci.] **276**, 1014 (1984) [Russian]
130. Yu.I. Scurlatov, S.O. Travin, L.S. Ernestova, Vdnye Res. [Water Res.] **5**, 66 (1987) [Russian]
131. W.G. Sunda, S.A. Huntsman, Limnol. Oceanogr. **32**, 552 (1987)
132. L.S. Ernestova, Yu.I. Scurlatov, Zh..Fiz. Khim. [J. Phys. Chem.] **58**, 739 (1984) [Russian]
133. A. Zuberbühler, Helv. Chem. Acta **53**, 473 (1970)
134. S. Skipton, D. Hay, *Drinking Water: Nitrate and Methemoglobinemia: Publ.G98–1369* (University of Nebraska Cooperative Extension, Lincoln, NE, 1998)
135. T.R. Steinheimer, D.K. Scoggin, L.A. Kramer, Environ. Sci. Technol. **32**, 1039–1047 (1998)
136. V. Smil, *Enriching the Earth* (MIT, Cambridge, MA, 2001)
137. A.A. Ratushnyak, V.Z. Latypova, M.G. Andreeva, K.I. Abramova, A.Y. Ratushnyak, M.V. Trushin, Fres. Environ. Bull. **18**, 1381–1384 (2009)
138. A.A. Ratushnyak, R. Rifgat, R.R. Shagidullin, M.G. Andreeva, A.Y. Ratushnyak, M.V. Trushin, American-Eurasin J. Agric. Environ. Sci. **6**, 257–261 (2009)
139. O.C. Zafiriou, M.B. True, Mar. Chem. **8**, 33 (1979)
140. American Cancer Society, *Cancer Facts and Figures, 2004* (American Cancer Society, Atlanta, 2004)
141. R. Kumar, M. Siddiqi, Z. Fazili, Cancer Lett. **65**, 139–143 (1992)
142. A.R. Tricker, Eur. J. Cancer Prev. **6**, 226–268 (1997)
143. P.I. Reed, Eur. J. Cancer Prev. **5**(suppl. 1), 137–147 (1996)
144. M. Gonta, in *National Conference Assessment of Environmental Risk for Mother and Child Health. Book of Abstract Chisinau*, 1998, pp. 68–79 [Romanian]
145. S.S. Mirvish, Cancer Lett. **93**, 17–48 (1995)
146. Gh. Duca, M. Gonta, Yu. Subotin, *Science Annals, Series – Chemistry and Biology* (Publishing House of the Moldova State University (MSU), Chisinau, 1998) pp. 88–94
147. Gh. Duca, M. Gonta, V. Matveevici, V. Iambartev, *ACTES du seminaire d'animation regionale, SAR-2004 (Region Europe Centrale et Occidentale). Book of Abstract Chisinau*, (Publishing House of the Moldova State University (MSU), 2004), pp. 43–48 [Romanian]
148. I. Stepanov, S.S. Carmella, S.S. Hecht, Gh. Duca, J. Agric. Food Chem. **50**(10), 2793–2797 (2002)
149. O.C. Zafiriou, M. Mcfarbang, Geophys. Res. Lett. **86**, 3173 (1981)
150. O.C. Zafiriou, M.B. True, Geophys. Res. Lett. **76**, 1126 (1979)
151. V. Balzani, V. Carassity, *Photochemistry of Coordination Compounds* (Academic, New York, 1970)
152. H.G.G. Bates, N. Uri, J. Am. Chem. Soc. **75**, 2754–2759 (1953)

153. J.H. Baxendale, J. Magee, Trans. Faraday Soc. **51**, 205–213 (1955)
154. H. Kawaguchi, A. Inagaki, Chemosphere **28**, 57–62 (1994)
155. P. Mazellier, M. Bolte, Chemosphere **42**, 361–366 (2001)
156. P. Mazellier, M. Sarakaha, M. Bolte, New J. Chem. **23**, 133–137 (1999)
157. P. Mazellier, G. Mailhot, M. Bolte, New J. Chem. **21**, 389–397 (1997)
158. P. Mazellier, P. Jirkovsky, M. Bolte, Pestic. Sci. 259–267 (1997)
159. C. Catastini, M. Sarakha, G. Mailhot, M. Bolte, Sci. Total Environ. **298**, 219–228 (2002)
160. H. Mestankova, G. Mailhot, J. Krysa, P. Jirkovsky, M. Bolte, Water Sci. Technol. **49**, 165–170 (2004)
161. R. Andreozzi, R. Marotta, Water Res. **38**, 1225–1236 (2004)
162. O. Bajt, G. Mailhot, M. Bolte, Appl. Catal. B Environ. **33**, 239–248 (2001)
163. G.G. Duca, L.S. Chiub, Yu.I. Scurlatov, .Ya. Sychev, rd. khim. [Coord. Chem.] **15**, 91 (1989) [Russian]
164. U. Stafford, K.A. Gray, P.V. Kamat, J. Phys. Chem. **98**, 6343–6351 (1994)
165. V.A. Borzylov, N.I. roianova, L.S. Ernestova, in *Forecast of Pesticides Behavior in the Environment. Proceedings of II Soviet-American Symposium* (Gidrometeoizdat, Leningrad, 1984) p. 245
166. R.G. Zika, W.J. Cooper (eds.), *Photochemistry of Environmental Aquatic Systems* (American Chemical Society, Washington, 1987) (ACS Symposium Series, 327)
167. L.S. Ernestova, I.V. Semenova, Vdnye Resursy [Water Resour.] **21**, 161 (1994) [Russian]
168. H. Schwarzer, Wasser. Abwasser. **129**, 484 (1988)
169. K.B. Yatsimirskyi, *Kinetic Methods of Analysis* (Chemistry, Moscow, 1967) p. 22 [Russian]
170. C.E. Housecroft, A.G. Sharpe, *Inorganic Chemistry* (Pearson Education, Harlow, 2005) p. 949
171. O. Covaliova, V. Covaliov, A.-M. Delort, G. Mailhot, A. Cincilei, Patent MD3911. 2010-01-31. in BOPI, 2009, no. 5
172. V. Covaliov, Patent MD2828. 2005-08-31, in BOPI, 2005, no. 8
173. V. Covaliov, O. Covaliova, Patent MD3416. 2007-10-31, in BOPI, 2007, no. 10
174. V. Covaliov, O. Covaliova, Patent MD3513. 2007-08-31, in BOPI, 2008, no. 2
175. V. Covaliov, V. Jalba, O. Covaliova, D. Nenno, D. Ungureanu, Patent MD39Y. 2009-06-30, in BOPI, 2009, no. 6
176. M. Hoffmann, S. Martin, W. Schoi, D. Bahnemann, Chem. Rev. **95**, 69 (1995)
177. S.J. Gopal, P. Muraleedharan, A. Tyagi, B. Raj, Curr. Sci. **90**(10), 1378–1383 (2005)
178. A. Bunescu, P. Besse-Hogan, M. Sanselme, G. Mailhot, A.-M. Delort, Appl. Environ. Microbiol. **74**(10), 2976–2984 (2008)
179. G. Muller, . tto, G. Werner, *Catalytic Methods in Trace Elements Analysis* (ir, Moscow, 1983) p. 53 [Russian]
180. S.U. Creingold, *Catalymetry in Pure Reagents Analysis* (Khimya, Moscow, 1983) [Russian]
181. B.I. Nabivanets, P.N. Linnik, L.V. Kalabina, *Kinetic Methods of Natural Water Analysis* (Naukova Dumka, Kiev, 1981) [Russian]
182. M. Miró, J.M. Estela, V. Cerdà, Talanta **63**(2), 201–223 (2004)
183. H.A. Mottola, D. Pérez-Bendito, Anal. Chem. **68**, 257–289 (1996)
184. V.K. Zinchuk, O.V. Fedak, J. Anal. Chem. **56**(9), 871–874 (2001)
185. T.S. Gill, H. Tewari, J. Pande, Compar. Biochem. Physiol. Part C: Compar. Pharmacol. 1990, 97(2), 287–292.
186. P. Bonchev, *Complex Formation and Catalytic Activity* (Mir, Moscow, 1975) [Russian]
187. V.G. Isac, U. Pfanmuller, A.Ya. Sychev, Zh. Fiz. Khim. [J. Phys. Chem.] **57**, 1969 (1983) [Russian]
188. A.Ya. Sychev, U. Pfanmuller, V.G. Isac, Zh. Fiz. Khim. [J. Phys. Chem.] **57**, 1974 (1983) [Russian]
189. V.G. Isac, A.Ya. Sychev, *Redox Catalysis by Metal Complexes* (Kishinev State University, Kishinev, 1984) [Russian]
190. A.Ya. Sychev, U. Pfanmuller, V.G. Isac, Zh. Fiz. Khim. [J. Phys. Chem.] **57**, 760 (1983) [Russian].

191. A.Ya. Sychev, U. Pfanmuller, V.G. Isac, Zh. Fiz. Khim. [J. Phys. Chem.] **57**, 2217 (1983) [Russian]
192. A.P. Lavrukhina, L.V. Yukina, *Analytical Chemistry of Manganese* (Nauka, scow, 1974) [Russian]
193. A.Ya. Sychev, U. Pfanmuller, V.G. Isac, Zh. Fiz. Khim. [J. Phys. Chem.] **55**, 794 (1981) [Russian].
194. V.G. Isac, A.Ya. Sychev, Ch.T.T, Fyong, Zh. Fiz. Khim. [J. Phys. Chem.] **60**, 2570 (1986) [Russian]
195. I.F. Dolmanova, N.T. Yatsimirskaia, V.I. Poddubienko, V.M. Peshkova, Zh. Anal. Khim. [J. Anal. Chem.] **26**, 1540 (1971) [Russian]
196. CF. Dolmanova, N.T. Yatsimirskaia, V.M. Peshkova, Zh. Anal. Khim. [J. Anal. Chem.] **28**, 112 (1973) [Russian]
197. Ya.D. ighineanu, Redox catalysis with manganese ions in water solutions. Ph.D. thesis, Moscow, 1984, P. 52 [Russian]
198. E.A. Morgen, N.A. Vlasov, L.A. Kozemiakin, Zh. Anal. Khim. [J. Anal. Chem.] **27**, 2064 1972 [Russian]
199. M.L. Babkin, Zh. Anal. Khim. [J. Analyt. Chem.] **17**, 256 (1962) [Russian]
200. R.-G. Yu, G.-M. Zeng, R.-G. Lin Acta Chim. Sin. **41**, 910 (1983)
201. F. Grasses, R. Forteza, J.G. March, V. Geradu, Anal. Chim. Acta **155**, 299 (1983).
202. S.J. Rao, G.S. Reddy, Y.K. Reddy, Proc. Indian Acad. Sci. Chem. Sci. **95**, 557 (1986)
203. A.Ya Sychev, V.G. Isac, D.V. Lap, Zh. Fiz. Khim. [J. Phys. Chem.] **51**, 363 (1977) [Russian]

Chapter 7
Homogeneous Catalysis with Metal Complexes in the Chemical Industry and Foodstuffs Chemistry

7.1 Catalytic Processes in the Chemical Industry

7.1.1 Catalytic Oxidation of Organic Compounds in Model Systems and Their Possible Industrial Applications

Out of the numerous reactions of homogeneous catalysis (oxidation, polymerization, cracking, alkylation, oligomerization, isomerization, etc.), only some oxidation reactions (with dioxygen, hydrogen peroxide, hydroperoxides, and some oxygen-containing oxidants), related to industrial processes or those that could be used in this field, will be briefly considered here. This limitation is connected with the specifics of this book, and with the fact that a number of reviews and books have been published [1–10], devoted more generally aspect to homogeneous catalysis with metal complexes used in industry.

Examples of such industrial catalytic reactions of oxidation with molecular oxygen are [7]:

1. $H_2C = CH_2 \xrightarrow[PdCl_2+CuCl_2]{O_2} CH_3CHO$

2. $CH_3 - CH_2 = CH_2 \xrightarrow[\text{Mo salts}]{ROOH} CH_3 - \underset{\diagdown O \diagup}{CH - CH_2} + ROH$

3.

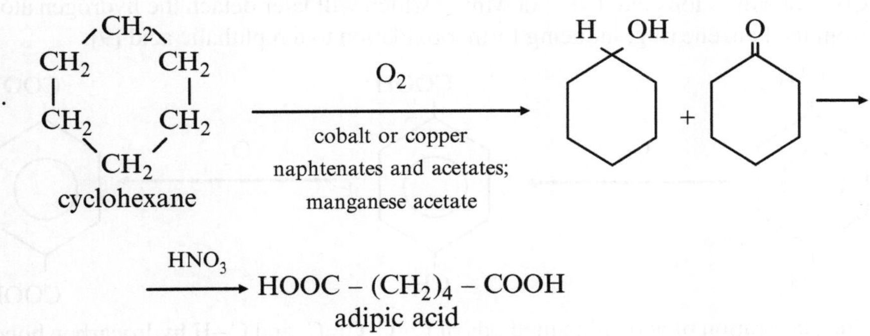

G. Duca, *Homogeneous Catalysis with Metal Complexes*, Springer Series in Chemical Physics 102, DOI 10.1007/978-3-642-24629-6_7,

4. $H - C_4H_{10} \xrightarrow[\text{CO(II) acteate}]{O_2} CH_3COOH$

5. $CH_3CHO \xrightarrow[\text{Mn(II) salts}]{O_2} CH_3COOH$

6. 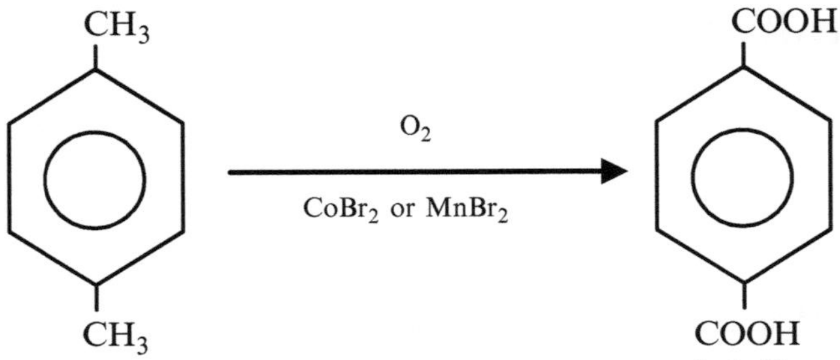

п-xylene terephthalic acid

Oxidation reactions with transition metal complexes involvement usually proceed under relatively mild conditions (<200°C and normal pressure). Advances in catalytic systems development, modeling different enzymes reactions with O_2, H_2O_2, ROOH, C_6H_5IO, and other oxidants participation (see Chap. 3), give grounds for hope that in the near future such "mild" and selective enough processes will find their use in industry, and those used will be applied on a larger scale.

The reactions of hydrocarbons oxidation with molecular oxygen, hydrogen peroxide, and hydroperoxides are used on the largest scale in the field of homogeneous catalysis with metal complexes [11, 12]. From 12 metals (Ti, V, Mn, Fe, Co, Ni, Cu, Zn, Mo, Rh, Pd, and W), whose compounds are most used in industry as catalysts [8], only the systems with iron, manganese, copper, and cobalt participation will be considered here.

Usually, because of low selectivity and high energy consumption, chemical the industry uses mainly, from saturated hydrocarbons, such primary substrates as ethane, propane, and from unsaturated ones—butadiene, benzene, and xylenes. For example, large-scale productions are such processes as cyclohexane oxidation to adipic acid, butane to vinegar, propylene to propylene oxide, *p*-xylene to terephthalic acid. Thus, on *p*-xylene oxidation, hydroperoxide forming will turn Co^{2+} or Mn^{2+} ions into Co^{3+} or Mn^{3+}, which will later detach the hydrogen atom from the benzene ring, inducing further oxidation to terephthalic acid [9]:

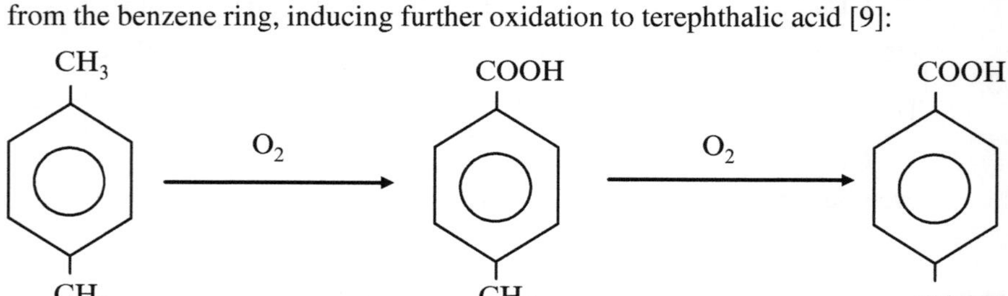

The elaboration of activation methods of C–C, C=C, and C–H hydrocarbon bonds is today and will be in near future the most important fundamental and applied problem. This problem has found its industrial solution, for example, in such processes as alkanes' catalytic cracking, making it possible to split hydrocarbons (C–C bonds) of various fractions of crude oil, into products with lower molecular mass [13]. Cracking of these hydrocarbons usually proceeds on metallic, zeolite, or acid catalyst at more extreme conditions and temperature 400–600 C.

Usually hydrocarbons splitting proceeds through carbocation intermediate formation. In such a way lighter hydrocarbons of petroleum type are obtained. Various catalysts are used. The use of superacids, such as HSO_3F:SbF_5 (its commercial name is "magic acid" because of its wonderful capacity to turn high-molecular hydrocarbons into low-molecular substances, C_4—and other branching isomers) has found wide application. Similar processes, proceeding in more extreme conditions (high temperatures and pressures) are usually radical and thus low-selective (especially low is stereoselectivity). The same refers to the bonds C=C and C–H activation. Really, in a few (for today) industrial processes of alkanes' activation with air oxygen, a very large amount of reaction products is obtained due to the radical processes. For example, on cyclohexane catalytic activation with air participation (process of nylon production by "Dupon" American firm) more than 100 products will be obtained [14].

Today, the solving of hydrocarbons selective oxidation problems is one of the most important, since its practical consequences can be more efficient using natural resources (cheapening of target product) and essential reducing of by-products (environmental improvement). The difficulties with the alkanes selective oxidation problem resolution are dealing with the high breakage energy of bonds C–C and C–H. Thus, bond energy C–H, depending on its placement in the hydrocarbon, will be distributed in the following way: primary ($\sim$410 kJ/mol) > secondary ($\sim$394 kJ/mol) > tertiary ($\sim$381 kJ/mol). The most easily broken allyl and benzyl bonds ($\sim$356 kJ/mol) will be broken, the least easily—the C–H bond in methane ($\sim$416 kJ/mol).

These difficulties are easily overcome by living organisms by using enzymes. Imitation of enzymatic activity and selectivity by model systems creation (mimetic way), similar to the enzyme's active center, is discussed in Sect. 3.2.1.2. All these models to a more or less degree are characterized by relatively low stability and selectivity compared to enzymes.

To create efficient systems, which could have been used in industrial production with the purpose to activate the bonds in alkanes, one has to use the experience and main results obtained on model enzymes systems creation, to learn more about the factors responsible for the stability and selectivity of catalytic systems, and to find the systems, not necessarily similar to enzymes' active centers. Nature has created enzymatic systems, able to work in hard non-changeable conditions of the living organism existence (temperature, pH, water medium, etc.). On bonds C–H activating industrially, all these limitations or some of them may become unnecessary. Revealation of totally new systems, such as polyoxometallic ones, indicates such a possibility. Thus, it is important to use the cheapest oxidants—molecular oxygen, hydrogen peroxide, and hydro peroxides (especially *t*-BuOOH).

7.1.2 Catalytic Hydroxylation and Epoxidation in Industrial Processes

General considerations regarding the reactivity and selectivity of C–H bond oxygenation result in many cases from the results obtained on model systems imitating monooxygenases (Sect. 3.2.1.2). The scheme of alkane oxygenation is given in Fig. 7.1 in rather simplified form.

It follows from Fig. 7.1 that the rate of overall process can be determined either with the rate of catalyst oxygenation (reaction to the right) or oxygen atom transfer from oxometal on substrate (reaction to the left). Both cases can be met in model systems.

For the reaction of metal complex oxygenation, either homolytic or heterolytic breakage of the D–O (or O–O for O_2, H_2O_2, ROOH) bond is characteristic [14]:

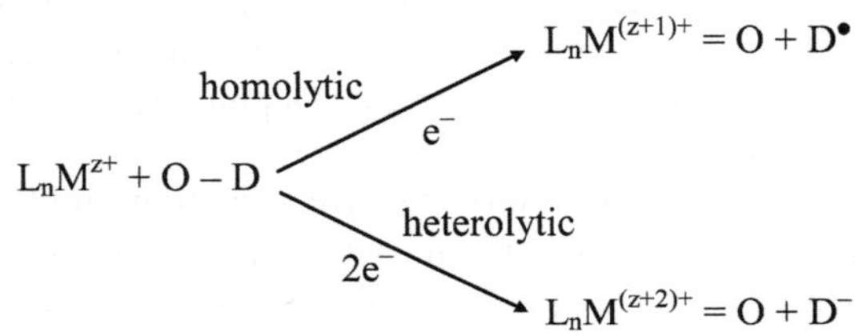

In the first case oxocomplex will be formed, in which the metal oxidation degree is one unit higher, while in the second case it is two units higher.

For alkane selective oxidation the heterolytic process is more preferable, since the heterolytic way will result in free radical $D^\bullet$ formation, reacting with alkanes and thus reducing the selectivity. In the case of hydrocarbons' industrial oxidation, used nowadays, the process proceeds mainly in accordance with the following radical-chain mechanism:

$$L_nM^{z+} + ROOH \longrightarrow L_nM^{(z+1)+} + RO^\bullet + OH^-$$
$$L_nM^{(z+1)+} + ROOH \longrightarrow L_nM^{(z+2)+} + ROO^\bullet + H^-$$

Because of insignificant changes in reaction conditions, it can be directed either in a homolytic or heterolytic way, i.e., the mechanism of the overall hydroxylation

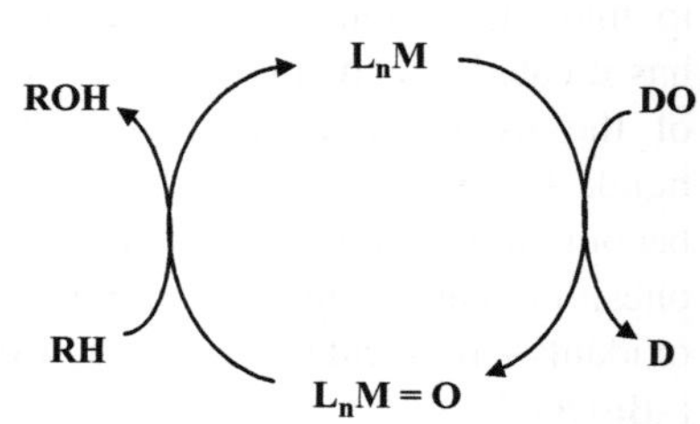

Fig. 7.1 Catalytic cycle of alkane RH oxygenation with oxygen donor DO

process can be changed. This is important for finding the optimal conditions of hydroxylation reaction for use on the industrial scale.

More often the rate and selectivity of alkane hydroxylation will be determined mainly by oxygen transfer from metal oxocomplex on the substrate. These two parameters are mainly dependent both on steric obstacles around the oxocomplex, and on its electron characteristics. The results of the model systems study considered above testify that the composition of products (substrate, regio- and stereoselectivity) depends mostly on steric obstacles.

Reactivity of metal oxocomplex with alkane must be determined not only by its redox potential, but electron state of oxygen in oxocomplex as well. The most reactive must be oxocomplexes which possess high spin density on oxygen boundary orbitals (i.e., it is strong electrophile), able to detach H atoms from C–H bonds of alkane by a "detachment–recombination" mechanism [14]. It is supposed that the most hydroxylation systems work by this mechanism. In its turn, the electron state of oxygen itself in oxocomplex is dependent on the central metal atom and the ligand's nature.

The third parameter characterizing the efficiency of the catalytic system is its stability, which is especially important for industrial processes of hydrocarbons hydroxylation. In the ideal case, it is necessary to exclude the destruction of catalyst ligand. It is with the aim of metal tetraaril porphyrin complexes stability increase, that volumic substitutes were introduced in them in *ortho*-positions of *meso*-aryl groups (for instance, TDCPP-tetrakis-2,6-dichlorophenylporphyrin) which increased complex stability about 100 times with regard to the destruction. Additional introduction of bromium atoms into all β-pyrrole positions of the porphyrin ring sensitive to oxidation, (octabrom octachlor tetraaril porphyrin) resulted in higher stability of the appropriate iron complex on hydrocarbon oxygenation [15].

However, this can only partly resolve the stability problem, since all the organic ligands are thermodynamically unstable in relation to oxidants. A more suitable approach to the stabilization problem is catalyst creation without the organic ligand. Such catalysts were created on the basis of polyoxocomplex use, in which certain peripheral atoms are substituted by transition metal atoms [16–20]. They are soluble and stable to oxidation. These complexes have the composition $(TM)PW_{11}O_{39}{}^{5-}$, where TM = Co^{II}, Mn^{II}, etc. They are written as abbreviation TMSP and are often called "inorganic metalporphyrins." Their peculiarity compared with metalporphyrins is their higher solubility and thermal stability, together with strong retention of transition metal in the active center (prevention of metal sedimentation). Besides, the inorganic ligand possesses certain redox properties, there is free coordination place in the "active center," and its properties can be varied by the axial ligand (heteroatom). The inorganic ligand surrounding is thermodynamically stable to oxidation, i.e., the catalyst is stable [14].

Complexes TMSP ($Co^{II}PW_{11}O_{39}{}^{5-}$, $Mn^{II}PW_{11}O_{39}{}^{5-}$) catalyze olefin epoxidation and alkane hydroxylation. On olefin epoxidation the rates, selectivities, and stabilities for TMSP complexes exceeded any other of the model systems considered above, i.e., they are efficient "inorganic metalporphyrins." The amount of catalytic cycles before deterioration amounted to 10^5 [14].

As mentioned above, for alkane selective oxidation more advantageous (preferable) is heterolytic breakage of D–O bond with metal oxocomplex formation $L_nM^{(z+2)+}=O$. By this mechanism further oxotransfer from this complex to alkane proceeds. Various mechanisms of oxygen atom transfer to substrate were proposed, the review and critical consideration of them being done by Hill [14]. The most probable of them, proceeding from experimental data, is the mechanism of hydrogen atom detachment by the oxygen atom of the oxocomplex. Thus, radical [$L_nM^{(z+1)+}$–OH, $C^{\bullet}$–] will be formed, and then its recombination proceeds with the formation of the hydroxylated product—C–OH. This mechanism is called "detachment-recombination." Formation of such an intermediate radical is confirmed by the number of experimental data [21–26]. Further transformation of radical pair in the product will depend on the electron structure of metal hydroxocomplex $L_nM^{(z+1)+}$–OH.

Numerous studies in this field [24,27–30] have shown that there are many various ways of radical pair recombination depending on the metal hydroxocomplex nature and reaction conditions. A short review of these works [14] shows that the choice of radical pair transformation ways is determined mainly by redox potential of metal hydroxocomplex, its d-orbitals filling, and other (little revealed at present) electron factors. Revealing the connection between such intermediates structures and their further transformations could make it possible, purposely modifying these intermediates, to change the direction of the final stage of hydroxylation and obtain the final products. This could seem to provide the possibility of increasing, in the near future, alkanes' hydroxylation selectivity in industrial processes as well.

New systems of high-selective fictionalization of alkanes with Barton's group have been discovered [31], called Giff or Giff-Orsay systems, which are unlike any of the other aforesaid systems. Primarily Giff system consisted of pyridine, iron powder, oxygen (or air), and carbonic acid. Then it has been modified—zinc was added as reducer (Giff-4 system). These systems oxidized hydrocarbons, and tertiary bonds C–H were not selectively attacked (as in all the preceding model systems) but secondary ones. Besides, alkenes' epoxidation was not observed. Cyclohexene, for instance, will be oxidized in the allyl position. The optimal process temperature is 30 °C. Pyridine could act as ligand, base, and good solvent. From the reaction mixture iron(III) binuclear complex ($Fe^{II}Fe(III)_2O(OAc)_6(C_6H_5N)_{3,5}$) was extracted, and it was supposed that it, or a similar compound, is a catalyst.

Instead of zinc powder as reducer, electrochemical reduction (Giff-Orsay system) was also used, and in Giff and Giff-Orsay systems the main products of hydrocarbon oxidation (alkanes, adamantane, cyclohexane, methylcyclopentane, *cis*- and *trans*-decaline, etc.) were ketones, but not alcohols (unlike the model systems considered above). Pyridine substitution on acetone in the Giff-Orsay system resulted in considerable changes in alkane oxidation products distribution. Ketones were no longer main products—essential amount of appropriate alcohol appears in the system. On addition of certain substances to Giff systems ("traps" such as PPh_3, $CBrCl_3$, CCl_4, etc.), instead of ketones, monosubstituted alkylderivatives were obtained with the same yield (respectively, C_6H_5OH, C_6H_5Br, C_6H_5Cl) [32–34].

Thus, the main differences of the Giff system on cytochrome P-450 model systems are the ketones' formation in it, but not alcohols. Also, unusual selectivity is observed—C–H bonds will be mostly attacked, but not tertiary ones and olefins will not be epoxidized.

To explain ketone generation, intermediate formation was supposed, in which the iron–carbon σ-bond is present [35]. The formation of several other intermediate compounds, among them metal alkylhydroperoxide, was also proved [31]. On alkanes activation in these systems, free radical formation does not take place in most cases, and the active oxidizing particle is not a strong electrophile, i.e., is not the hydroxyl radical $OH^{\bullet}$ or particle similar to that with iron(V) hydroxocomplex from the cytochrome P-450 model systems. It was first supposed that this active particle is a two-nuclear iron complex of $Fe(III)–O–Fe^{V} = O$ type, which, combining with alkane CR_4 via iron–carbon σ-bond, will be turned into OH:

$$Fe(III){-}O{-}Fe^{V}{-}R_3.$$

The latter will then be recombined with ketone (or other products) with initial catalyst formation. Active particle = two-nuclear iron complex have similarity with non-hemocontaining enzymes, and in this sense can be regarded as methamonooxidase model [31]. It was supposed that μ-oxodimers of Fe(III) are initially obtained in solution and then undergo further transformations.

However, later research carried out by Barton and Sawer and their staff [36–40] made it possible to get more detailed ideas. The modern state of the art of the processes proceeding in Giff-systems can be given by the scheme in Fig. 7.2 [41]. Within the squares initial compounds, reaction product, and intermediate compound which can be separated in the free state. In the scheme of alkane oxidation to ketone, iron(III) interaction with hydrogen peroxide proceeds in the following way:

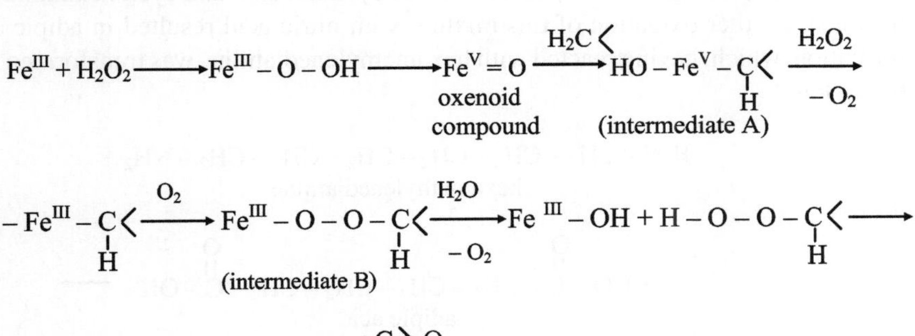

Thus, Fe–C bond formation in the intermediate A is postulated. Fe^{V}, contained in the intermediate A, seems to be rapidly reduced by hydrogen peroxide to the state Fe(III). On interaction with O_2 intermediate B will be formed—this is peroxide which can also be separated. It can be seen from this scheme that free radicals will not be formed within a Giff system involving Fe(III), H_2O_2, and alkane (although in some Giff systems free radicals will also be generated [42]).

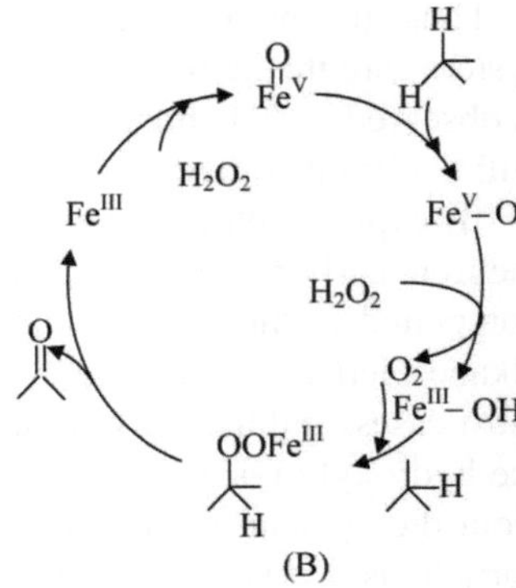

Fig. 7.2 Scheme of alkane oxidation mechanism to ketone on Fe(III), H_2O_2 and substrate interaction within the Giff system

In a similar Giff system hydrogen peroxide can be substituted with *tert*-butylhydroperoxide (TBHP). Thus, the rate of the alkane oxidation reaction will be slow, and the oxidation selectivity and value of kinetic isotope effect will be changed, though the reaction scheme including intermediates A and B will be maintained. These Giff and Giff-Orsay schemes are in perspective for the industrial processes elaboration of low-temperature selective oxidation of alkanes.

The importance of preliminary theoretical consideration of hydrocarbons (alkanes, alkenes) oxidation mechanisms (hydroxylation and epoxidation) with various oxygen-containing oxidants using hemic and non-hemic enzymes, especially of their model systems, can be shown in the cyclohexane hydroxylation process having great practical meaning [43, 44]. Cyclohexane oxidation with air oxygen is carried out on a large scale in industry. Up to the beginning of the 1990s, cyclohexane oxidation by the Du Pont firm only (USA) was done to the amount of about 500,000 tons/year. Thus, the mixture of cyclohexanol and cyclohexanone was obtained. Further oxidation of this mixture with nitric acid resulted in adipic acid formation, which, having reacted with hexamethylenediamine, was transformed into 6,6-nylon:

$$\underset{\text{hexamethylenediamine}}{H_2N-CH_2-CH_2-CH_2-CH_2-CH_2-CH_2-NH_2} +$$

$$+ \underset{\text{adipic acid}}{HO-\overset{\overset{\displaystyle O}{\|}}{C}-CH_2-CH_2-CH_2-CH_2-\overset{\overset{\displaystyle O}{\|}}{C}-OH} \longrightarrow$$

$$\longrightarrow \underset{\text{nylon-6,6}}{H-[NH-CH_2-CH_2-CH_2-CH_2-CH_2-CH_2-NH-\overset{\overset{\displaystyle O}{\|}}{C}-CH_2-CH_2-CH_2-CH_2-\overset{\overset{\displaystyle O}{\|}}{C}]_x-OH^{\bullet}}$$

Nylon-6,6 was so called because it is obtained from diamine containing six carbon atoms, and from dicarbonic acid also containing six carbon atoms. Practical using of nylon-6,6 (artificial silk) is well known.

To provide an acceptable rate of cyclohexane oxidation in industrial processes, the temperature has to be high enough, about 150–175°C. Direct interaction of cyclohexane (CyH) with dioxygen by the reaction

$$\mathrm{Cy{-}H + O_2 \longrightarrow Cy^{\bullet} + HO_2{}^{\bullet}}$$

is strongly endothermic (197 kJ/mol) and, therefore, practically does not proceed. The main role in the industrial process of cyclohexane oxidation is played not only by the direct oxygen interaction with C–H bond, but peroxides' (hydroperoxides') catalytic formation and (further) breakage of the less strong O–O bond in them [32]. Thus, various radicals of $R^{\bullet}$, $RO_2{}^{\bullet}$, $OH^{\bullet}$ type, cyclohexane hydroperoxides (CyOOH), etc. will be formed. The mechanism of high-temperature cyclohexane oxidation is catalytic, radical-chain, and may be represented rather simplified (M–Co, Mn compounds and other catalysts) [45]:

1. $\mathrm{CyO_2{}^{\bullet} + CyH \longrightarrow CyOOH + Cy^{\bullet}}$
2. $\mathrm{M(II) + CyOOH \longrightarrow M(III)OH + CyO^{\bullet}}$
3. $\mathrm{M(III)OH + CyOOH \longrightarrow M(II) + H_2O + CyO_2{}^{\bullet}}$
4. $\mathrm{CyO^{\bullet} + CyH \longrightarrow CyOH + Cy^{\bullet}}$
5. $\mathrm{CyO_2{}^{\bullet} + CyH \longrightarrow CyOOH + Cy^{\bullet}}$
6. $\mathrm{Cy^{\bullet} + O_2 \longrightarrow CyO_2{}^{\bullet}}$
7. $\mathrm{CyO_2{}^{\bullet} + CyO_2{}^{\bullet} \longrightarrow}$ cyclohexanol(CyOH) + cyclohexanone(CyO) + O_2

The catalyst essentially diminishes the activation energy of hydroperoxide decomposition (CyOOH) [stages (2) and (3)]. Then, according to the stages of chain initiation (4) and continuation (5), cyclohexyl radical $Cy^{\bullet}$ will be formed which will rapidly react with oxygen [stage (6)] yielding cyclohexylperoxyradical $CyO_2{}^{\bullet}$. These radicals will be ruined in the breakage reaction [stage (7)].

As can be seen from the mechanism scheme, cyclohexyl hydroperoxides will be formed within the radical-chain process of cyclohexane oxidation. Such catalysts as, for instance, cobalt (and manganese) compounds play an important role in these hydroperoxides' transformations into cyclohexanol and cyclohexanone, though the nature of cobalt(III) intermediates, generated in certain industrial processes, is not exactly known. Further radical-initiated oxidation of cyclohexanone will result in C–C bond breakage and adipic acid formation.

It is clear from the presented mechanism of high-temperature oxidation of cyclohexane that this industrial process is low-selective, since the products forming (cyclohexanol and cyclohexanone) are much easier subjects for radical attack than the initial cyclohexane. Thus, a large amount of various (undesirable) by-products is formed, and the yield of main products is low.

Using the model systems based on metal porphyrin complexes for industrial purposes is also not very hopeful, as this process soon, to a great degree, like that for high-temperature oxidation, has a radical character accompanied by low yield of target products and low selectivity. However, the effort taken on cytochrome P-450 study, and especially its model systems, turned out to be useful from the application aspect. It became clear that spacious geometry and dimensions of enzyme protein surroundings determine the hydrocarbons' oxidation selectivity.

The group of researchers from the "Du Pont" firm has used these considerations for creating new model systems of alkanes' oxidation in which specific (protein) surroundings of an active iron-containing center was replaced with a zeolite structure. The reason behind zeolites usage is that they have canals and cavities ranging from 3 to 13 Å [46], and their alumosilicate matrix is very strong and inert in relation to the oxidation processes. On the basis of zeolites, phthalocyanic iron complexes were synthesized by metal ion and dicyanobenzene (ligand component) successive introduction into zeolite pores. Inside zeolite pores the iron phthalocyanide complex was formed:

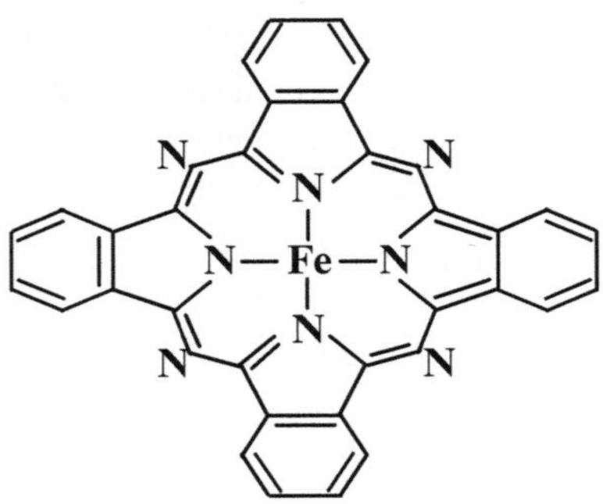

Though its size (16 Å) is too large for zeolite pores, it can be placed within them, since its configuration is distorted and becomes saddle-like, and thus the edge of this complex can enter the adjacent pores. Such a structure was called "ship in the bottle" [45] (ship is iron phthalocyanide complex and bottle is zeolite). Such a catalyst can provide alkanes oxidation, but the oxidation rate is too slow because of the difficulty of oxidizer and substrate diffusion towards the catalyst, as the zeolite's pores are already to some degree occupied with catalyst. For the same reason the number of catalytic cycles and the selectivity is low.

The second, more perfect, system was also created on the zeolite basis; inside of the latter Pd and Fe were placed. They can easily be prepared on the basis of any zeolites despite their pore size, and therefore they must have higher selectivity. Besides, this system used dioxygen at room temperature and pressure, and the reducer (H_2/Pd), in which presence hydrogen peroxide will be catalytically formed. The nature of the oxidizing particle in this system is still unknown—it is either $OH^{\bullet}$ radical or, rather, oxocompounds of high-valency iron. This oxidizing particle, being in a zeolite environment, will provide high selectivity regarding alkane size and form [45]. As a matter of fact, an inorganic model of monooxygenases was created.

In spite of obvious successes in creating monooxygenases model systems, these results used for creating low-temperature and selective industrial processes faces a number of the aforesaid unresolved problems.

With this aim, the other ways of alkanes' mild oxidation were tested. Thus, it appeared that iron complexes with porphyrin are able to oxidize catalytically these substrates to alcohols and ketones at 80 °C under 10 atm dioxygen pressure [47].

One of these ways for this process to work in mild conditions is iron and other metals porphinat complexes photoexcitation [48–52]. Iron(III) porphyrinates irradiation with light ($\lambda = 350$–$400\,\text{nm}$) can result in inner-molecular electron transfer from axial ligand L to iron(III) with its reduction to iron(II) [48, 49, 53] and subsequent interaction with O_2, yielding iron(II) porphyrinate intermediates, and then to the products of electron transfer ($O_2^{\bullet -}$ and $L^\bullet$) [54]:

$$\begin{bmatrix} L \\ | \\ Fe^{III}P \end{bmatrix} \underset{}{\overset{h\nu}{\rightleftarrows}} \begin{bmatrix} L^{\bullet} \\ \\ Fe^{II}P \end{bmatrix} \xrightarrow{O_2} \begin{bmatrix} O_2 \\ |\ L^{\bullet} \\ Fe^{II}P \end{bmatrix} \xrightarrow{\ \ L\ \ } \begin{matrix} L \\ | \\ Fe^{III}P \end{matrix} + O_2^{\bullet -} + L^{\bullet}$$

In such a way the catalytic activation of O_2 proceeds.

The photocatalytic process of cyclohexane oxidation with dioxygen, with iron(III) porphyrin complexes (Fe(III)(TDCPP)(OH^-), Fe(III)($TDCPCl_8P$)(OH^-), Fe(III)($TDCPBr_8P$)(OH^-), where TDCPP is meso-tetrakis(dichlorophenyl)porphyrin, $TDCPCl_8P$ meso-tetrakis(dichlorphenyl)octa-β-chloroporphyrin, and $TDCPBr_8P$ meso-tetrakis(dichlorphenyl)-octa-β-bromporphyrin) was carried out in pure non-polar cyclohexane or in more polar mixture dichlormethane–cyclohexane under 22 °C and irradiation at 365 nm. Chloro- and bromo-substitutes provided steric protection of the porphyrin ring against free radicals influence on it during the catalytic process. Thus, it appeared that the yield of oxidation products (cyclohexanone and cyclohexanol) can be changed depending on the solvent nature and radicals trap presence. Possible mechanism of cyclohexane photocatalytic oxidation is shown in Fig. 7.3 [54].

In the case of pure cyclohexane photooxidation with oxygen in the presence of catalytic quantities of Fe(III)(TDCPP)(OH^-), Fe(III)($TDCPCl_8P$) (OH^-), and Fe(III)($TDCPBr_8P$)(OH^-) under mild conditions (22 °C, 760 Topp O_2 and without additional reducer), $OH^\bullet$ radical will be generated on Fe(III)–OH bond photosplitting (stage *a*), and only cyclohexanone will be selectively formed (stages *a*–*d*, Fig. 7.3). $OH^\bullet$ radical will readily detach the hydrogen atom from cyclohexane, yielding Fe^{II}-cyclohexyl radical pair formation (stage *b*), and then on very rapid

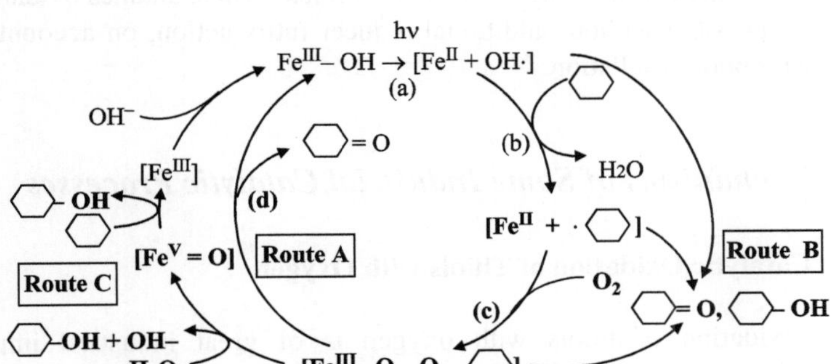

Fig. 7.3 Possible mechanism of cyclohexane photooxidation

(less than 5 ns) O_2 interaction with Fe^{II}, complex Fe(III)-peroxyalkyl will be formed (stage *c*). This complex will undergo inner-molecular transformation in non-polar solvent (cyclohexane), with hydrogen transfer from cyclohexyl group to oxygen atom from the group Fe–O, with rapid and selective formation of cyclohexanone and initial complex (stage *d*).

In a more polar solvent (dichlormethane–cyclohexane mixture), the main oxidation product will be cyclohexanol (way C). In the presence of free radicals traps (phenyl-*tert*-butylnitron)

$$\overset{H}{C_6H_5} - \overset{O^-}{C} = N^+ - C(CH_3)_3 + X^\bullet \longrightarrow \underset{X}{\overset{H}{C_6H_5}} - \overset{O^\bullet}{C} - N - C(CH_3)_3$$

nitroxide formation is detected by the EPR method, and radical-chain self-oxidation of substrate will be suppressed, which increases the yield of cyclohexanol. Medium polarity and proton presence favors the heterolytic splitting of O–O bond in the complex Fe(III)-peroxylalkyl with the formation of iron(V) oxocomplex (which is formally equivalent to the particle $Fe^V = O^+$):

$$(P)Fe^{III} - O - O - CH\langle \;+ H^+ \longrightarrow \rangle CH - OH + (P)Fe^V = O^+$$

Such particles formation also proceeds in catalytic cycles of catalase, peroxidase, cytochrome P-450.

On solvent polarity increasing, iron provision with OH^--ligand will be increased, which will contribute to free radical coming out of the cell and heterolytic splitting of O–O bond in Fe(III)–OOR. On this basis two ways of substrate oxidation may be realized [54]: (1) free radical cyclohexane oxidation resulting in cyclohexanol and cyclohexanone formation (way B) and (2) cyclohexane selective oxidation with $Fe^V = O^+$ particles with the exclusive formation of cyclohexanole only (way C).

In such free radicals traps introduction as phenyl-*tert*-butylnitron, naturally, only way B was inhibited, while way C was not inhibited. Thus, alkanes oxidation with dioxygen is possible without additional reducer introduction, on account of iron porphyrinates photoirradiation.

7.1.3 Mechanisms of Some Industrial Catalytic Processes

7.1.3.1 Catalytic Oxidation of Thiols with Oxygen

Catalytic oxidation of thiols with oxygen is of great industrial importance in the field of crude oil and oil products purification from sulfur-containing compounds (mercaptan oxidation with air oxygen). A large number of papers and patents was published in this field. The necessity for methods elaboration of vulcanization accelerators synthesis, such as dithiazolyl disulfides, thiuram sukfides and sulfenamides related to the thiazol series, also stimulates research in this field.

The researchers agree that two main mechanisms of thiol catalytic oxidation can be distinguished, namely free radical and ion-molecular—by substrate and oxidant transition complex formation within the coordination sphere of metal-catalyst. Thus, both in the first and second mechanism, the electron transfer proceeds by metal ion from thiol to oxygen [55].

Free radical mechanism implies the transition metal compounds involvement as catalysts, which will be oxidized with molecular oxygen and then regenerated by thiolate-ion (RS^-) oxidation by them into thiyl radical ($RS^\bullet$) with further dimerization into disulfide [55]:

$$2M^{z+} + O_2 \longrightarrow 2M^{(z+1)+} + O_2{}^{2-}$$
$$2RS^- + 2M^{(z+1)+} \longrightarrow 2M^{z+} + 2RS^\bullet$$
$$2RS^\bullet \longrightarrow RSSR$$
$$O_2{}^{2-} + H_2O \longrightarrow 2OH^- + 1/2O_2$$

It is possible that hydrogen peroxide forming in the last stage

$$O_2{}^{2-} + 2H_2O \longrightarrow H_2O_2 + 2OH^-$$

will at once be catalytically decomposed by transition metal ions.

An example of such thiol oxidation mechanism is cysteine (RSH) oxidation into cystine, where catalysts serve as cobalt substituted phthalocyanines $Co^{III}PcX_n$ (X represents substitutes in the benzene nuclei of phthalocyanine) [56–58]:

$$RSH \rightleftarrows RS^- + H^+$$
$$Co^{III}Pc(X)_n + RS^- \longrightarrow Co^{II}PcX_n + RS^\bullet$$
$$Co^{II}PcX_n + RS^- \longrightarrow Co^{I}PcX_n + RS^\bullet$$
$$2Co^{I}PcX_n + 1/2O_2 + 2H^+ \longrightarrow 2Co^{II}PcX_n + H_2O$$
$$2Co^{II}PcX_n + 1/2O_2 + 2H^+ \longrightarrow 2Co^{III}PcX_n + H_2O$$
$$2RS^\bullet \longrightarrow RSSR$$

The mechanism of such a process is alternating metal reduction with cysteine and metal regeneration with oxygen, as well as free thiol radicals generation, dimerizing into disulfide (RSSR).

The second mechanism is an ion-molecular one with inner-complex electrons transfer, which can be demonstrated in the example of cysteine (Cys) catalytic oxidation with oxygen in the presence of cobalt tetrasulfophtalocyanine ($Co^{II}TSPc$).

Thus, alternating $Co^{II}TSP$ reducing into $Co^{I}TSP$ will proceed and the oxidation of the latter by O_2 [59]:

$$Co^{II}TSPc + Cys^- \rightleftarrows (Cys^-)Co^{II}TSPc \rightleftarrows (Cys^-)Co^{I}TSPc$$

The formation of such adducts which form with O_2 triple complexes was proved spectrophotometrically [60]. The appropriate mechanism of such process is as follows:

$$Co^{II}TSPc + Cys^- \rightleftharpoons Co^{I}TSPc(Cys^-)$$
$$Co^{I}TSPc(Cys^-) + Cys^- \rightleftharpoons Co^{I}TSPc(Cys^-)_2$$
$$Co^{I}TSPc(Cys^-) + O_2 \rightleftharpoons Co^{II}TSPc(Cys^-)(O_2)$$
$$Co^{I}TSPc(Cys^-)_2 + O_2 \rightleftharpoons Co^{II}TSPc(Cys^-)_2(O_2)$$
$$Co^{II}TSPc(Cys^-) + Cys^- \rightleftharpoons Co^{II}TSPc(Cys^-)_2(O_2)$$
$$Co^{II}TSPc(Cys^-)_2(O_2) \rightleftharpoons \text{products}$$

It is supposed that the last stage of $Co^{II}TSPc(Cys^-)_2(O_2)$ decomposition is accompanied by reversible electron transfer onto oxygen molecule:

$$Co^{II}TSPc(Cys^-)_2(O_2) \rightleftarrows Co^{II}TSPc(Cys^\bullet)_2(O_2^{2-}) \xrightarrow{2H^+}$$
$$\longrightarrow \underset{(RSSR)}{Cys-Cys} + Co^{II}TSPc + 1/2O_2 + H_2O$$

with inner-complex thiol oxidation to disulfide.

Such catalytic oxidation of thiols is usually carried out both in water medium and in organic solvents, and the solvent nature renders great influence on the reation rate, often changing the latter thousands of times. Various compounds of transition metals, especially copper, iron, cobalt, and nickel pyrophosphates were used as catalysts [61, 62].

Methods of dithiazolyl disulfides obtaining from 2-mercaptobenzothiazole (2-MBT), using as catalysts metallic iron, cobalt, zinc, copper, manganese, molybdenum, etc., as well as these metals' complex compounds, are described and patented [55].

Copper compounds are active catalysts on 2-MBT oxidative condensation with amines in the process of obtaining sulfonamides.

Though many ligands are used on thiols oxidation with molecular oxygen, transition metal (cobalt, manganese, iron, etc.) complex compounds with phthalocyanines are especially efficient:

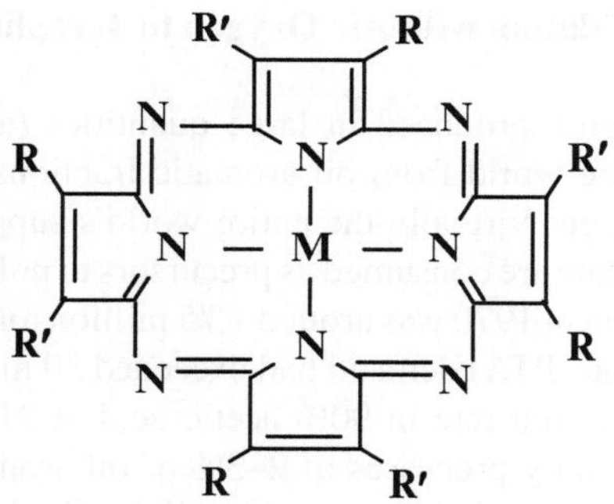

where R and R′ form benzene ring condensed with the pyrrole cycle. Such catalysts are used for mercaptan oxidation when contained in oil, in dithiazolyldisulphides (vulcanization accelerators) synthesis, and in preparation of sulfenamides (patents are given in [55]).

The leading firms in this field, such as Rhône Poulenc SA (France), AKZO NV (Germany), and American Cyanamide Company (USA), are dealing with obtaining vulcanization accelerators (dithiazolyl disulfides and thiuram disulfides). All these firms produce dithiazolyl disulfides by mercaptothiazols oxidation with molecular oxygen at temperatures of ≈0–150 °C in organic solvents (sometimes also in water medium) with disulfide yield equal to about 96–99% [55].

Sulfenamide accelerators of vulcanization are obtained by thiols' oxidation by molecular oxygen in the presence of primary or secondary amines:

$$R'SH + HN\begin{matrix}\diagup R'' \\ \diagdown R'''\end{matrix} + 1/2O_2 \longrightarrow R'-S-N\begin{matrix}\diagup R'' \\ \diagdown R'''\end{matrix} + H_2O^{\bullet}$$

It is supposed that this reaction proceeds by thiol initial oxidation with dioxygen to disulfide, and then the latter interaction with amine [63, 64]:

$$2ArS^{-} \xrightarrow[\text{Catalyst}]{O_2} ArSSAr \xrightarrow[-ArS^{-}]{NR''R'''} ArSHNR''R''', \text{ where } Ar = C_6H_5.$$

These reactions proceed at 50–80 °C in organic solvents. Cobalt and manganese phthalocyanines are especially efficient as catalysts. Sulfenamides yield reaches ≈80–90% (and even 99%).

This research has laid a good basis for the proposed mild methods of thiols oxidation with molecular oxygen used practically in industry for obtaining disulfides and sulfenamides.

7.1.3.2 *p*-Xylene Oxidation with Air Oxygen to Terephthalic Acid

Terephthalic acid is being produced in large quantities (about ten millions tons annually) throughout the world from oil aromatic fractions by *p*-xylene catalytic oxidation with air oxygen. Virtually the entire world's supply of terephthalic acid and dimethyl terephthalate are consumed as precursors to polyethylene terephthalate (PET). World production in 1970 was around 1.75 million tons [65]. By 2006, global purified terephthalic acid (PTA) demand had exceeded 30 million tons. The process involved proceeds with high rate in 90% acetic acid at 210 °C and at a pressure of 21 atm. Thus, secondary processes of 2–3% of oil aromatic fraction oxidation and 1–2% of acetic acid oxidation to yield carbon dioxide and water, proceed. To reduce the unnecessary losses of initial raw material, amethod of redox potentials measuring in the course of these processes was proposed, which makes it possible to propose measures for reducing the undesirable oxidation with oxygen reactionary particles [66].

Establishing an optimal oxidation regime under the aforesaid conditions was carried out in the technical model titane reactor. Catalysts involved cobalt(II) acetic salts jointly with bromide-ions. During the course of the process the initial mixture changes its colouration from cobalt-blue, through violet-blue, then black-green, then violet, and once again through black-green to the colouration of kerosene type. At last, from this reacted mixture, white crystals of terephthalic acid $R{-}(COOH)_2$ (where $R{=}C_6H_4$) will be precipitated. Thus, redox potential will be changed, and the process course of initial *p*-xylene ($CH_3{-}R{-}CH_3$), intermediate *p*-toluic acid ($CH_3{-}R{-}COOH$) and other products consumption, can be followed by this change.

The model reaction mechanism of *p*-xylene catalytic oxidation to terephthalic acid with air oxygen is rather complex, and proceeds through free radicals' formation, their interaction with catalyst, aldehydes, and other intermediate compounds generation, that can be clearly seen in Fig. 7.1 [66].

Proceeding from the scheme of *p*-xylene oxidation mechanism, redox potential change and its correlation with CO_2 evolving from solvent—acetic acid, the method has been elaborated for by-products minimization in the technological process [66].

A few catalytic processes with metal complexes participation used on industrial scale, are given here as examples (Scheme 7.1).

7.2 Foodstuffs Chemistry and Catalysis

7.2.1 Catalytic Lipids Peroxidation in Foodstuffs

In conditions of demographic explosion at the end of the twentieth century when, during about a 10-year period, Earth's population increases by about one billion people, the problem of enough foodstuffs production becomes more and more real. Equally with food production, a no less important component of this problem is the preservation of already produced foodstuffs. This is a complex problem. One of

$$Co^{2+} + O_2 \rightarrow Co^{3+} - OO^{\bullet -}$$

$$Co^{3+} - OO^{\bullet -} + CH_3 - R - CH_3 \rightarrow CH_3 - R - CH_2OO^{\bullet} + Co^{2+} + H^+$$

$$Co^{3+} - OO^{\bullet -} + HOOC - R - CH_3 \rightarrow HOOC - R - CH_2OO^{\bullet} + Co^{2+} + H^+$$

$$CH_3 - R - CH_2OO^{\bullet} + Co^{2+} + H^+ \rightarrow CH_3 - R - CHO + H_2O + Co^{3+}$$

$$HOOC - R - CH_2OO^{\bullet} + Co^{2+} + H^+ \rightarrow HOOC - R - CHO + H_2O + Co^{3+}$$

$$Co^{3+} + CH_3 - R - CH_3 \rightarrow CH_3 - R - CH_2^{\bullet} + Co^{2+} + H^+$$

$$Co^{3+} + CH_3 - R - CHO \rightarrow CH_3 - R - CO^{\bullet} + Co^{2+} + H^+$$

$$Co^{3+} + HOOC - R - CH_3 \rightarrow HOOC - R - CH_2^{\bullet} + Co^{2+} + H^+$$

$$Co^{3+} + HOOC - R - CHO \rightarrow HOOC - R - CO^{\bullet} + Co^{2+} + H^+$$

$$CH_3 - R - CH_2^{\bullet} + O_2 \rightarrow CH_3 - R - CH_2OO^{\bullet}$$

$$CH_3 - R - CO^{\bullet} + O_2 \rightarrow CH_3 - R - COOO^{\bullet}$$

$$HOOC - R - CH_2^{\bullet} + O_2 \rightarrow HOOC - R - CH_2OO^{\bullet}$$

$$HOOC - R - CO^{\bullet} + O_2 \rightarrow HOOC - R - COOO^{\bullet}$$

$$CH_3 - R - COOO^{\bullet} + CH_3 - R - CHO + Co^{2+} + H^+ \rightarrow 2CH_3 - R - COOH + Co^{3+}$$

$$CH_3 - R - COOO^{\bullet} + HOOC - R - CHO + Co^{2+} + H^+ \rightarrow R - (COOH)_2 + \\ + CH_3 - R - COOH + Co^{3+}$$

$$HOOC - R - COOO^{\bullet} + HOOC - R - CHO + Co^{2+} + H^+ \rightarrow 2R\text{–}(COOH)_2 + Co^{3+}$$

Scheme 7.1 Mechanism of catalytical disintegration of *n*-xylene by air oxygen

its component parts involves the procedures for foodstuffs storage. Such processes involve, in particular, a food lipids peroxidation process [67]. Several of reviews, a number of international conferences, and a lot of special books have been devoted to this process [67–69]. Therefore, only catalytic processes of lipids peroxidation with participation of iron ions and compounds, mechanisms of this process, their influence on foodstuffs deterioration, and the possibility of preventing foodstuffs deterioration will be briefly and in generalized form considered here.

Foodstuffs quality problems on storage are connected with oxidation catalytic processes, influencing foodstuffs condition, quality, odor, and many other properties. Negative consequences of storage such as, for example, the bitter taste of oil, can be reduced by temperature reduction. Thus, ham storage at 4°C will preserve its qualityfor weeks. However, storage of fried mackerel under the same conditions is possible for 2 days only.

A safety problem appears connected with oxidation products appearing in foodstuffs (hydroperoxides, epoxides, aldehydes, carbonyl compounds, etc.) and their influence on numerous diseases threatening people.

One of the main processes causing foodstuffs spoilage is lipids' catalytic peroxidation, of which the mechanism in general form (regardless of foodstuffs preservation) has been considered in Sect. 3.1. This process, as a rule, is radical-chain, being initiated with endogenic substances of foodstuffs (H_2O_2, ROOH) and free radicals [70, 71] ($O_2^{\bullet-}$, $OH^{\bullet}$, $ROO^{\bullet}$, $GS^{\bullet}$) or exogenic molecules ($^1O_2$, O_3), free radicals (NO_x, $SO_3^{\bullet-}$), as well as with light, heat, and other influences. These processes in foodstuffs can start with oxygen interaction with iron(II) ions and compounds, including hemes and heme-containing particles (myoglobin, Mb), as well as with certain enzymes (xanthine oxidase, XO) causing pathological conditions in different tissue and resulting in oxidative deterioration in muscle foods [72–75].

$$O_2 \xrightarrow{\text{Fe(II)}} O_2^{\bullet-} + \text{Fe(III)}$$
$$O_2 \xrightarrow{\text{Heme (II)}} O_2^{\bullet-} + \text{Fe(III)}$$
$$O_2 \xrightarrow{\text{Mb(II)}} O_2^{\bullet-} + \text{Mb(III)}$$
$$O_2 \xrightarrow{\text{XO}} nO_2^{\bullet-} + (n-1)H_2O_2 + \text{urea.}$$

Significant amounts of $O_2^{\bullet-}$ forming in one way or another in biological conditions act as nucleophilic reagents (for example, they promote phospholipid hydrolysis inside membranes). Superoxide anion is able to detach hydrogen atoms from organic substrates and initiate lipid peroxidation indirectly (by Haber–Weiss reaction, yielding $OH^{\bullet}$ radical generating).

$O_2^{\bullet-}$ will undergo slow dismutation in a non-catalytic way, and rapid dismutation on superoxide-dismutase participation:

$$2O_2^{\bullet-} + 2H^+ \xrightarrow{\kappa = 0{,}35\ M^{-1}\cdot s^{-1}} H_2O_2 + O_2$$
$$2O_2^{\bullet-} + 2H^+ \xrightarrow[\text{SOD}]{\kappa = 2\cdot 10^9\ M^{-1}\cdot s^{-1}} H_2O_2 + O_2$$

A number of foodstuffs' muscular tissue (aldehyde oxidase, etc.), mitochondria, microsomes, peroxisomes are active generators of H_2O_2 [76]. For example, H_2O_2 is generated in the muscular tissue of turkey and in other foodstuffs.

Table 7.1 Rate constants of free radicals interaction with some biologically important compounds in solution at room temperature [77]

Nr.	Substrate, S	$K_{R\bullet+S}$, $(M^{-1}s^{-1})$				
		$OH^\bullet$	$RO^\bullet$	$HO_2^\bullet$	$ROO^\bullet$	$O_2^{\bullet-}$
1	Stearic acid	$\sim10^9$	2.3×10^6	Low	10^{-3}–10^{-4}	Low
2	Oleic acid	$\sim10^9$	3.3×10^6	Low	0.1–1	Low
3	Linolic acid	9×10^9	8.8×10^6	1.2×10^3	$\sim6 \times 10^1$	Low
4	Linoleic acid	7.3×10^9	1.3×10^7	1.7×10^3	$\sim1.2 \times 10^2$	Low
5	Arachidonic acid	$\sim10^{10}$	2×10^7	3×10^3	$\sim1.8 \times 10^2$	Low
6	Aldehyde	$\sim10^9$	–	5×10^1	2.7×10^3	–
7	α-Tocoferol	$\sim10^{10}$	–	2.6×10^5	5.7×10^6	6
8	Ascorbate (AH^-)	1×10^{10}	–	2.2×10^6		5×10^4

Hydrogen peroxide formed by Fenton reaction

$$M^{z+} + H_2O_2 \longrightarrow M^{(z+1)+} + OH^- + OH^\bullet$$

can be turned into high reactionary $OH^\bullet$ radical which is a source of many other radicals:

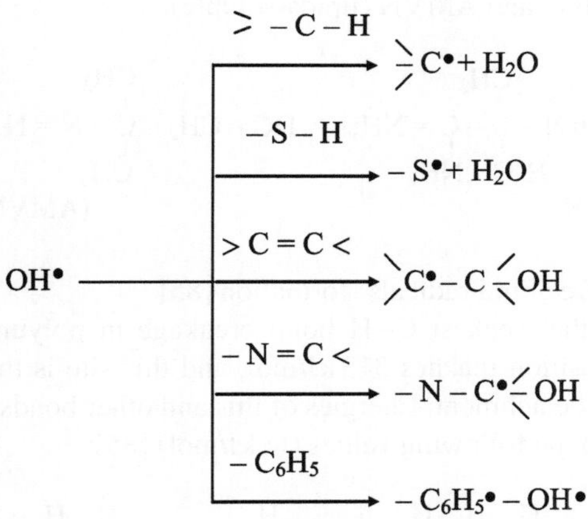

Rate constants of free radicals interaction with some organic compounds making parts of foodstuffs are given in Table 7.1.

Singlet oxygen, interacting with foodstuffs' lipids, will form lipids hydroperoxides (LipOOH), which in subsequent reaction with such catalysts as iron(III) will generate alkoxyl radicals of lipids ($LipO^\bullet$), somewhat similar to $OH^\bullet$, though less reactive:

$$^1O_2 \xrightarrow{RH} LipOOH \xrightarrow{Fe(II)} LipO^\bullet + OH^- + Fe(III)$$

By the reaction with O_2, peroxyl radicals may be formed:

$$LipO^\bullet + O_2 \longrightarrow LipOO^\bullet$$

Rate constants of $^1O_2$ interaction with fatty acids (oleic, linoleic acid, arachidonic) are four orders less that those with $OH^{\bullet}$ radicals.

Methhemoglobin and methmyoglobin (containing iron in 3+ oxidation degree) on interaction with H_2O_2 will generate short-living intermediates—oxy-ferryl hemoglobin radicals [78], which are not completely identical to compounds I or II (see Chap. 3), but are rather similar to both of them [79]. Thus, initiation of membranes' lipids peroxidation proceeds [78]. It was shown that methyl-forms of hemoglobin and myoglobin [80], in the presence of low H_2O_2 concentrations, will form ferryl species, but not $OH^{\bullet}$ radicals [81]. The presence of a free coordination place in Fe(III)-hemoproteins is a necessary condition of ferryl species formation [82]. However, in most cases, for example in the muscular tissue of turkey, the main role in lipids peroxidation can be played by the reversible cycle Fe(II)–Fe(III), and not by ferryl particle [76]. This particle can take part in pathological processes of membranes lipid peroxidation in heart diseases, in oxidized hemoglobin, in erythrocytes on thalassemia, and in crescent-shape cell anaemia [83, 84].

A number of causes (temperature increase, ionizing radiation, etc.) can result in free radicals appearance in unsaturated aliphatic acids of foodstuffs. Even at the relatively low temperature of 37 °C some azoderivatives such as, for example, AAPH (water-soluble) and AMVN (lipids-soluble)

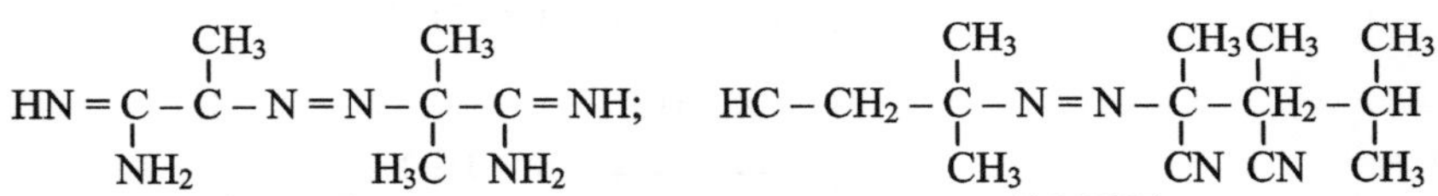

(AAPH) (AMVN)

are able to induce free lipid radicals' formation [85].

The energy of the weakest C–H bond breakage in polyunsaturated aliphatic acids in bisallyl position reaches 313 kJ/mol, and this site is the most reactionary for hydrogen atom detachment. Energies of this and other bonds in polyunsaturated acids correspond to the following values (in kJ/mol) [85]:

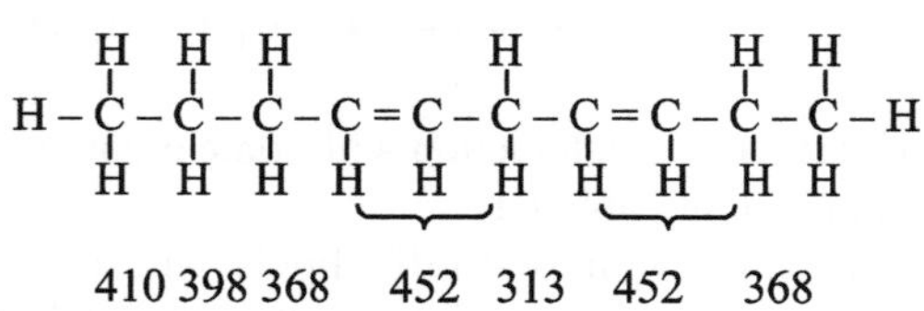

In Fig. 7.4 the mechanism of linolic acid (H_2Lip) peroxidation is presented, characteristic of other unsaturated acids as well, from which it can be seen that first of all the weakest C–H bonds are attacked, especially C_{11} [85].

In the presence of metal compounds, hydroperoxides of linolic and other unsaturated acids will generate alkoxy- and allyl radicals:

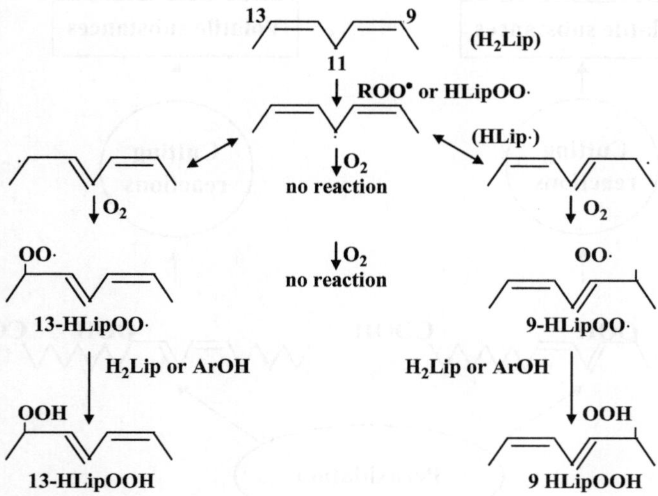

Fig. 7.4 Mechanism of linolic acid peroxidation

$$HLipOOH + Fe(II) \longrightarrow HLipO^{\bullet} + OH^{-} + Fe(III)$$

$$HLipO^{\bullet} + H_2Lip \longrightarrow HLipOH + HLip$$

Alcoxy radicals interaction with H_2Lip results in formation of final products (alcohols, aldehydes, epoxides, ketones, etc.).

On autooxidation of linolic acid, for example, peroxy radical formed, being a good oxidant, will form acyclic hydroperoxide on its reduction. This hydroperoxide is itself the substrate for subsequent reactions, yielding hydrocarbons, carbonyls, and alcohols, many of which represent volatile components with extremely low threshold of odor appearance [86]. The specific feature of such autooxidation is relative non-specificity of oxygen adding place. In linolic acid places 9 and 13 will be attacked by oxygen with the same frequency, which results in racemic mixture formation of linolic acid peroxidation [87] (Scheme 7.2 [88]) and in the following formation of numerous odor-possessing volatile substances on account of hydroperoxide cutting reactions.

Reactions of monohydroperoxides β-cutting are given in Scheme 7.3 [68].

In the water-free foodstuffs greases and oils, homolytic splitting of hydroperoxides predominates. Thus, this is the way A is expelled, and the reaction will result in hexanal ($CH_3(CH_2)_4CHO$) and other aldehydes' formation on oxidation of linolic acid, for example.

On lipids structure complication (amount of double bonds increases), the amount of possible places for oxygen adding will be enhanced too. Thus, in the case of arachidonic acid peroxidation (four double bonds), six main equal places of oxygen adding appear, each of which will result in different hydroperoxide formation.

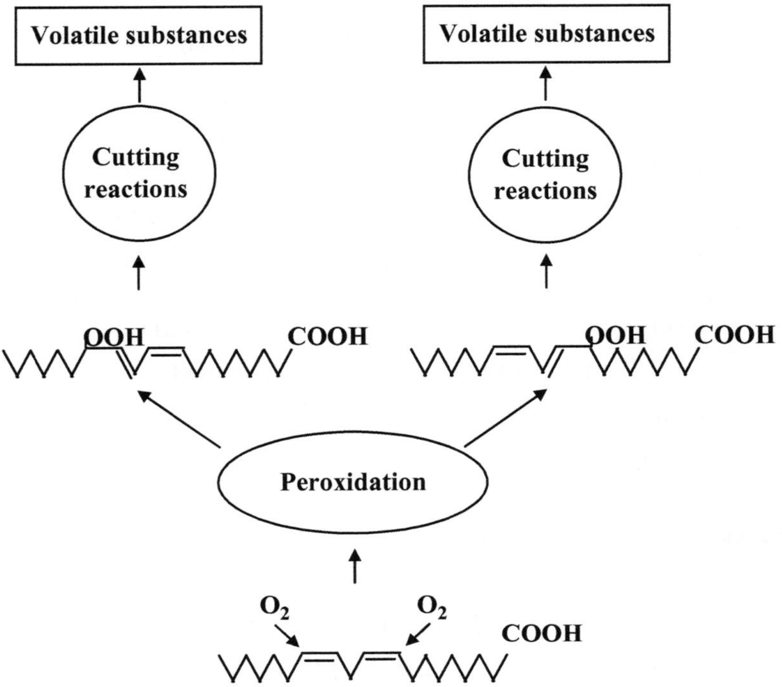

Scheme 7.2 Autooxidation of linoleic acid

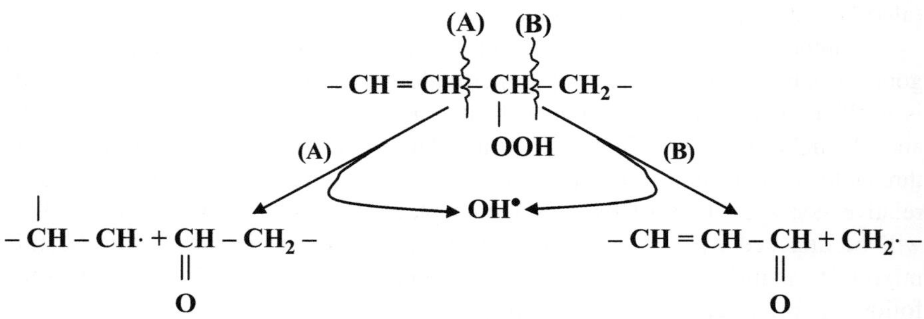

Scheme 7.3 Reactions of β-cutting of monohydroperoxides

Subsequent reactions of their cutting will yield the formation of numerous volatile substances, forming the odors of foodstuffs. Besides, with three and more double-bonds presence in aliphatic acid, peroxy radicals formed can interact with double bonds inside the molecule, with the formation of some hydroxy-epoxides, as, for instance, in the case of linoleic acid (Scheme 7.4) [87].

Lipids oxidation is the main cause of food quality deterioration. Foodstuffs contain a mixture of many polysaturated aliphatic acids, which are the initial

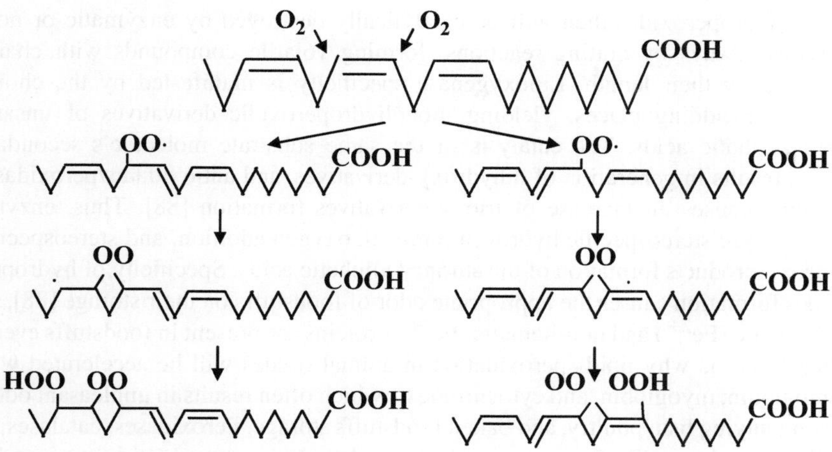

Scheme 7.4 Intramolecular interactions of peroxyradicals with double bonds

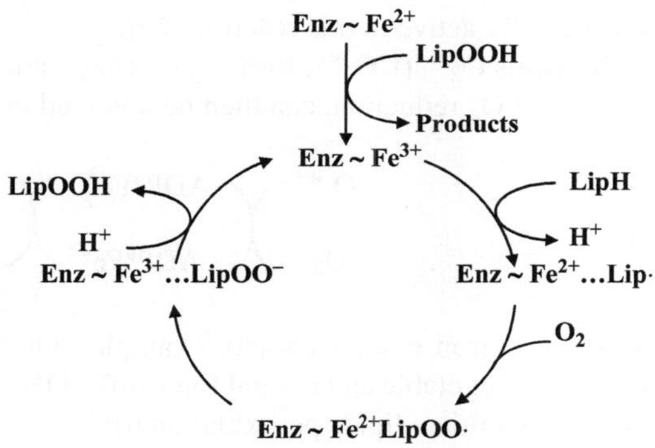

Fig. 7.5 Supposed mechanism of linolic acid oxidation to hydroperoxide and its decomposition to products

components of subsequent autocatalytic peroxidation, resulting in a variety of volatile odor-possessing substances formation. About 1,000 volatile substances were identified (aldehydes, alcohols, phenols, ketones, ethers, carboxyl acids, etc.), which form on cooking. Some of them possess an odor. Lipoxygenases contained in many plants, as well as in erythrocytes and leucocytes, will catalyze unsaturated aliphatic acids oxidation to the appropriate monohydroperoxides. The supposed mechanism of this process is shown in Fig. 7.5 [68].

Activation energy of linolic acid peroxidation is rather low (17 kJ/mol). Under the influence of hydroperoxides, lipoxygenase (LipOOH) may be activated (Enz–Fe^{2+}). Thus, the Fe^{2+}-center will be turned into an Fe^{3+}-center, and the reaction of hydroperoxide formation begins, which will proceed with high stereo- and regioselectivity with the so-called type I lipoxygenase. Type II lipoxygenases also exist, acting much less specifically [68].

Lipoxygenases participate in catalytic reaction of oxygen adding to *cis*, *cis*-4-pentadiene of unsaturated aliphatic acid, generating acyclic hydroperoxide [88].

These hydroperoxides then will be catalytically destroyed by enzymatic or non-enzymatic ways by cutting reactions, forming volatile compounds with chains specified by their length. Lipoxygenase specificity is manifested by the choice of oxygen adding places, yielding monohydroperoxidic derivatives of unsaturated aliphatic acids. The catalysis of the same substrate molecule's secondary places results in generation of dihydroxy-derivatives, and antioxidants/peroxidases reducing causes an increase of trioxy-derivatives formation [88]. Thus, enzyme will catalyze stereospecific hydrogen removal, oxygen addition, and stereospecific hydroxy-products formation of unsaturated aliphatic acids. Specificity of hydroperoxides formation causes the appropriate odor of foodstuffs on their storage [88].

Hematic (Fe^{2+}) and non-hematic (Fe^{3+}) proteins are present in foodstuffs everywhere. That is why lipids peroxidation in animal tissues will be accelerated with hemoglobin, myoglobin, and cytochrome C, which often results in unpleasant odors accompanying fish, poultry, and baked foodstuffs storage. Peroxidases, catalases, as well as cytochrome P-450 play an important role in lipids peroxidation in vegetable foodstuffs. These hematic or non-hematic enzymes catalyze the decomposition of lipids hydroperoxides much faster (by order) than iron ions. Lipoxygenase enzyme is especially active in this reaction [68].

Radicals $O_2^{\bullet -}(HO_2^{\bullet})$, forming by enzymatic of non-enzymatic ways during the process of O_2 reduction, can then be involved in the reactions of Gaber-Weiss-type:

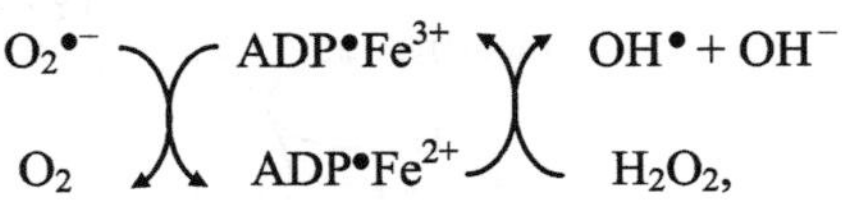

as soon as iron complexes (for example, with adenosine diphosphate, ADP) are present in vegetable and animal foodstuffs. $OH^{\bullet}$ radicals formed in this process will efficiently initiate lipids peroxidation [68].

7.2.2 *Distribution of Iron Compounds in Foodstuffs and Their Influence on Lipids Peroxidation*

Catalytic oxidation of lipids, as shown above, is one way of spoiling the quality (i.e., odor, color, condition, energetic value, toxic compounds appearance) of stored foodstuffs [89–93]. Besides, products of lipids peroxidation (in particular, cholesterol oxides) can result in arterial damage and atherosclerosis [89]. The most important catalytic role in lipids peroxidation belongs to iron compounds. Kinds of such formations are the compounds with low molecular mass—iron (II, III) complexes with organic acids (HADPH, ATP, etc.) and inorganic phosphates, aminoacids, carboxyl [94] and oxyacids (citric acid) [95], as well as with other organic molecules.

Both molecular weight and iron compounds concentration are strongly dependent on muscular tissue type. Iron compounds with low molecular weight make about 3.5–5.9% of all dissolvable iron, and 2.4–3.9% of total iron in the muscular tissues of beef, chicken, mutton, and pork [96]. In fish, for example in mackerel, low-molecular iron concentration (<10,000 Da) will be changed within the limits 6.7–13.9% of total iron [97, 98].

Storage conditions influence iron distribution in muscular foodstuffs. For instance, in non-freezed mackerel storage at 4°C for 7 days, the contents of low-molecular iron will be increased about 1.4 times [98], and for chicken and turkey 2–3 times [99]. Many researchers confirm that heating of beef, pork, mutton, and other foodstuffs cooking will increase non-heme iron contents [100–102], this process, in its turn, being accompanied by oxidative rancidity appearance.

The amount of soluble iron in muscular tissues of beef, pork, chicken, flat-fish, mackerel (light), mackerel (dark) reaches 20.0, 12.3, 13.6, 3.4, 0.3, 0.6, 2.6 μg Fe/g of muscular tissue, respectively. Thus, ~90% of dissolved iron is connected with the block: in myoglobin, hemoglobin, ferritin, transferrin, hemosiderine [103]. Beef, mutton and pork usually contain more myoglobin, while chicken contains a great amount of hemoglobin [104]. On food cooking or on hydrogen peroxide addition to the foodstuffs, hematic iron concentration will be reduced due to iron release from hematic protein [100–102].

In its turn, ferritin will emit Fe^{2+} ions in the presence of such reducers as superoxide-anion, ascorbate, and thiols [104]. A really high molecular fraction (>150,000 Da) of normal mackerel fillet, containing ferritin, in the presence of ascorbate is an active catalyst of lipid oxidation [105]. Ferritin, extracted from turkey muscular tissue, in the presence of ascorbate, will initiate lipid oxidation [106]. Although ferritin heating will not result in direct iron release, an increase the ascorbate's ability to release iron connected within ferritin, and thus, to catalyze lipid oxidation is achieved [107, 108].

Transferrins seem not to be active catalysts of foodstuffs lipids oxidation [109, 110]. Conditions of iron release from another iron-containing proteins of the foodstuffs (cytochrome C, iron-sulfo-proteins of electron-transport chain, peroxidase, lipoxygenase) have not been studied enough [103]. As to hemosiderin, iron connected with it will be released in the presence of ascorbate, dithionate, and superoxide-anion with $OH^{\bullet}$ radicals formation, initiating lipid oxidation and peroxidation [111, 112].

Copper, iron ions, as well as hemin will accelerate lipids oxidation of fresh fish meat [101]. Relative activity of these metals and hemin at 40 °C is the following:

$$Fe^{2+} > \text{hemin} > Cu^{2+} > Fe^{3+}$$

Iron and copper ions catalyze lipid oxidation of mackerel foodstuffs [113]. Enhanced rate of lipids oxidation in cooked food is caused by iron release in this process [101]. Cu^{+} and Fe^{2+} ions are able to accelerate lipid oxidation, catalyzing H_2O_2 and LipOOH decomposition to $OH^{\bullet}$ radicals [92]. Myoglobin and Fe^{2+} ions will accelerate lipid oxidation of mackerel foodstuffs [114], while EDTA, mixture of nitrites with ascorbate addition, will inhibit this process.

Influence of chelate-forming ligands on lipids oxidation is not predictable. It often depends both on ratios between ligand and iron ion, and on other metals presence—especially copper ions. EDTA and desferrioxamine can non-specifically inhibit lipids oxidation, acting as donors at high concentrations.However, EDTA, at low ratio EDTA:Fe, can accelerate catalysis of lipids oxidation. Ceruloplasmin addition (copper-containing protein) transforms iron into non-active condition Fe(III) and thus inhibits lipids' catalytic oxidation with iron(II) [103].

Although $O_2^{\bullet -}$ itself cannot initiate lipids oxidation, its undissociated form $HO_2^{\bullet}$ can do it, for example, in the case of linolic or arachidonic acids. Besides, superoxide-anion uses the most important initiator of Gaber-Weiss reaction, on one hand turning Fe(III) ions into Fe(II), and on the other, being dismutated (superoxide-dismutases are rather widely spread in foodstuffs), will deliver hydrogen peroxide into the system—another most important element of this reaction. Sources of superoxide-anion appearance in foodstuffs are mainly known. These are such enzymes as xanthine oxidase, peroxidases, dehydrogenase flavoprotein, and various organelles—chloroplasts, mitochondria as well. They can also appear in foodstuffs within leuco flavins, hydro quinones oxidation [103], hemoglobin, and myoglobin reduction.

Ascorbate is an even more important reducer (compared to superoxide-ion), influencing lipids oxidation, since its concentration is much higher (from $\sim 5\,\mu mol/100\,g$ in normal mackerel [98] to $\sim 14\,\mu mol/100\,g$ in dark turkey [115]). Muscular tissue of usual mackerel storage within 7 days results in an ascorbate concentration reduction from ~ 8 to $0.2\,\mu mol/100\,g$ of foodstuff.

The role of ascorbate is dual and depends on its concentration and that of iron. Low ascorbate concentration usually promotes lipids oxidation (causing Fe(III) reduction), its low concentration inhibits this process by free radicals inactivate with hydrogen atom transferred to them [103]. Other reducers such as NADH, NADPH, glutathione, and cysteine are much less active than ascorbate and superoxide-anion in foodstuffs lipids catalytic oxidation [116, 117].

Influence of certain enzymes (liver microsomal enzymes dependent on NADH) on lipids oxidation consists mainly in Fe(III) being reduced by them into Fe(II), and, therefore, this process is promoting. Ceruloplasmine added to turkey meat, prevents the transfer Fe(III) $\longrightarrow$ Fe(II), inhibits lipids peroxidation [118]. However, such an enzyme as phospholipase will inhibit lipids catalytic oxidation and peroxidation due to cells membrane structure change or iron inactive complexes formation.

Histidine and ADP stimulate foodstuffs lipids oxidation by iron complexes with histidine and ADP complexes formation 1:1:1, histidine concentration being higher than 7 mM/100 g of muscular tissue [103].

To prevent iron ions' catalytic influence on lipids oxidation, it is important to control their concentration, for example, by complex formation. Thus, it is necessary that all the coordination sites in iron chelates formed be occupied, that irreversibility of redox-cycle $Fe^{II} \longrightarrow Fe(III)$ be reached, and interaction with lipids be prevented by the formation of steric inaccessible or insoluble complexes [119]. For this EDTA is usually used. However, as was mentioned above, EDTA can also promote lipids oxidation as soon as iron complex with EDTA possesses higher reduction potential than iron ions [120]. If EDTA concentration is higher than that of iron ions, complete occupation of coordination places at iron proceeds, which prevents its interaction with peroxides [103]. Phosphates usage to prevent lipids oxidation catalyzed by iron ions is dependent on phosphate type. Thus, antioxidant activity in fried pork changed the sequence [121]: sodium pyrophosphate > sodium tripolyphosphate > hexamethaphosphate > disodium phosphate > sodium phosphate.

For lipids oxidation inhibition, catalyzed with iron ions, citric acid is also used which is efficient in odor inhibition on heating, in pork [121] and in beef [122]. Phytates act as natural chelate-forming agents in cereals and oil seeds [119], and they are also efficient antioxidants in chicken [123], promoting the reaction $Fe^{II} \longrightarrow Fe(III)$ [119]. The same purpose can be reached on ceruloplasmine addition to foodstuffs, which results in lipids oxidation diminishing on, for example, turkey storage in refrigerators [118] or pork storage in frozen form (−15 °C) [124, 125].

Another way of catalysis reduction on lipids oxidation is iron contents diminishing in foodstuffs by preventing pollution with it during technological processes or using a special diet for cattle feeding [126]. Activity diminishing of enzymes, which can reduce Fe(III) to Fe(II) or generate $O_2^{\bullet -}$ and H_2O_2, needs heat treatment and genetic change in these enzymes [103].

7.2.3 Role of Antioxidants in Foodstuffs Preservation

Process inhibition of lipid peroxidation can be provided to some degree by oxygen removal from foodstuffs, their storage at low temperature, or adding antioxidants to them. Antioxidants are purifiers against forming peroxy- and oxy radicals of lipids. Thus, antioxidants, containing the phenol group play the main role in inhibition of oxidation processes in foodstuffs, since they form the radicals stabilized with an aromatic ring. This radicals are unable to detach hydrogen atoms from unsaturated aliphatic acids and thus are unable to initiate lipids peroxidation. In this case antioxidant (ArOH) interaction proceeds in two stages through deprotonation [127]:

$$\mathrm{LipOO^{\bullet} + ArOH \longrightarrow LipOO^{-} + ArOH^{\bullet +}}$$
$$\mathrm{ArOH^{\bullet +} + H_2O \longrightarrow ArO^{\bullet} + H_3O^{+}}$$

$ArO^{\bullet}$ radicals can exist in mesomer forms depending on their substituents[85]:

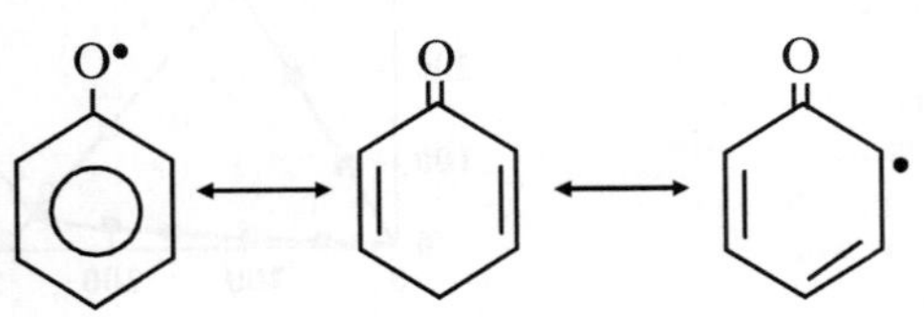

These phenoxyl radicals $ArO^{\bullet}$ usually do not interact with O_2 [128].

Generally, the rate constant of peroxyl radical interaction with antioxidant

$$XLipOO^{\bullet} + ZAH \longrightarrow XLipOO^{-} + ZAH^{\bullet+}$$

depends on the electrophilic character of substitutes X and Z. Thus, electron-attractive (such as –Cl) and electron-giving (such as, OCH_3) substitutes of the reaction partners, will increase, often tens times, the redox potential difference ΔE (i.e., ΔG) and the rate of the reaction [85].

As can be seen from Table 7.1, α-tocopherol (α-Too) is one of the most reactionary radicals in relation to peroxyl radicals. Formation prevention of saturated aliphatic acids hydroperoxides is shown in Fig. 7.6.

The lack of α-tocopherol (vitamin E) in the human organism causes a number of pathologic effects: muscular dystrophy, nerve disturbances, diathesis, liver necrosis, etc. Mitochondria are the most sensitive to E-avitaminosis, in which enzymes of oxidative phosphorylation are placed. On E-vitamin lack of peroxidation will rapidly develop [69].

Antioxidants (AH), being purifiers of lipids free radicals, can realize the following reactions:

$$LipOO^{\bullet} + AH \longrightarrow LipOOH + A^{\bullet} \tag{7.1}$$

$$LipO^{\bullet} + AH \longrightarrow LipOH + A^{\bullet} \tag{7.2}$$

$$LipOO^{\bullet} + A^{\bullet} \longrightarrow LipOOA \tag{7.3}$$

$$LipO^{\bullet} + A^{\bullet} \longrightarrow LipOA \tag{7.4}$$

Final products formed by reactions (7.3) and (7.4), are rather stable, which breaks the chain process of lipids peroxidation.

With the required amount of antioxidants and certain enzymes (superoxide dismutase, glutathione peroxidase) present in living tissues, efficient protection of lipids against peroxidation will be provided.

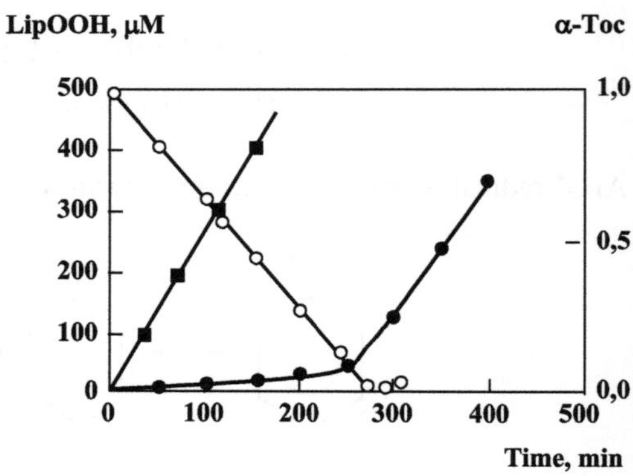

Fig. 7.6 Methyllinoleate oxidation (453 mM) in hexane, induced with (see Sect. 7.2.1) AMVN (0.20 μM) in the absence (*filled squares*) and presence (*filled circles*) of α-tocopherol (3 μM) at 37 °C in air, as well as α-Toc consumption (*circle*) [85]

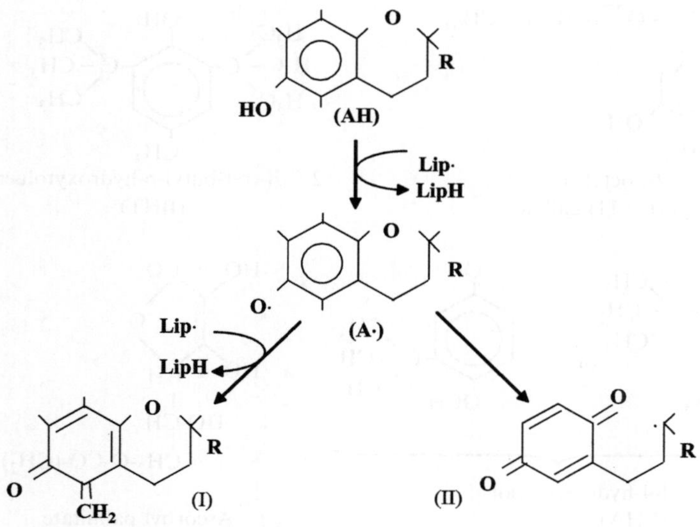

Fig. 7.7 α-Tocopherol (AH) influence on linolic acid (LipH) oxidation delay

During the process of foodstuffs preparation, such as obtaining oil from plants, although certain isolation of the latter from such antioxidants as tocopherols is made, their level will still remain. This provides relative stability even for refined oils.

The retarding of linoleic acid oxidation (LipH) or its substitutes on α-tocopherol (AH) adding (Fig. 7.7) is caused by a whole cascade of reactions inhibiting the lipid radicals and resulting in stable product (I) formation, or that of less active radical (II) [68], which will inhibit substrate oxidation.

The concentration of α-tocopherol increasing from 0.04% to 4% (in relation to linolic acid weight) will result in the fact that this antioxidant will be turned into pro-oxidant, due to the interaction with gaseous oxygen and free radicals' formation. In the case of γ- and δ-tocopherols such a change in antioxidant activity was not discovered (even on higher concentration the antioxidation activity will be preserved) [68].

In aqueous medium an active antioxidant of lipids peroxidation is ascorbic acid, although only at a high enough concentration ($\sim 10^{-3}$ M). At lower concentration ($\sim 10^{-5}$ M) it is a pro-oxidant, especially in the presence of metal ions [68]. In order to use vitamin C as antioxidant in non-water medium (in oil, butter) ascorbylpalmitate is used [129].

Natural antioxidants are flavones and flavonoles—phenolic compounds (natural catechins and other substances) widespread in vegetable tissues. Alongside with them synthetic antioxidants are used [68]:

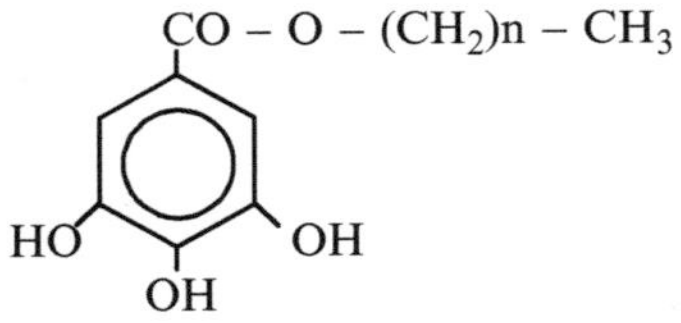

Propyl (n = 2), octyl (n = 7)
and dodecyl (n =11) gallate

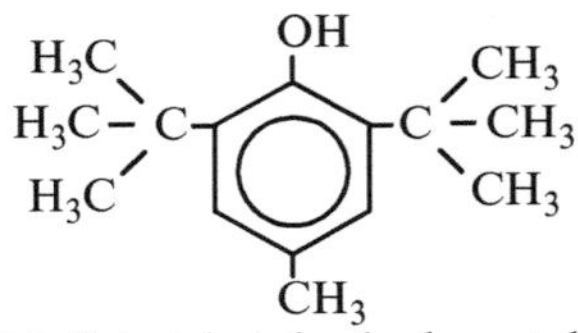

2,6-di-tret-butyl-p-hydroxytoluene
(BHT)

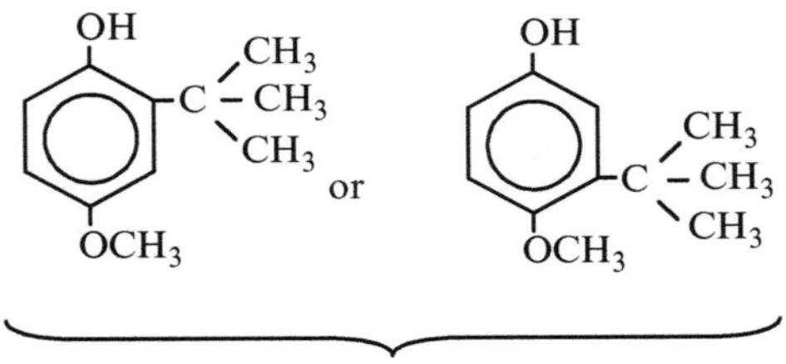

Tret-buthyl-4-hydroxoanisol
(BHA)

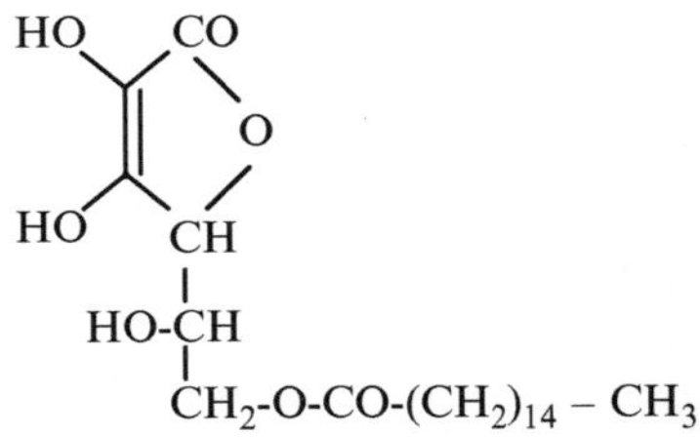

Ascorbyl palmitate

Besides, *tert*-butyl-hydroquinone (TBHQ), being a strong antioxidant, is used as soya oil stabilizer. Such antioxidants are usually used in amounts of 0.01–0.02% of the foodstuff weight. The mixture of BHT and BHA is a more efficient antioxidant of oil or butter than either of them separately. Addition of citric, phosphoric, fumaric, and some other acids to the aforementioned antioxidants results in synergetic effect due to transition metal ions joining by these chelate-forming ligands.

The most complicated problem of foodstuffs preservation from spoiling is the control over its oxidation, accompanied by the characteristic unpleasant odors. Attempts to solve this problem is mainly by additives of synthetic and natural antioxidants being introduced into the food. Among such antioxidants groups, which may be used for certain foodstuffs, are substances forming by Maillard's reaction. This reaction proceeds in many foodstuffs. Maillard's reaction (non-enzymatic browning) represents an interaction of carbonyl compounds and amines yielding, by a series of complex reactions of a number of starting compounds, changes to odor, taste, antioxidants, as well as polymerized brown pigments (mealnoidin) [130].

The second stage of Maillard's reaction is Amadori's re-grouping, in which intermediate products will be formed (1-deoxyozone, 1-deoxyreductone, and 3-deoxyozone), and then a whole series of compounds will be obtained from them (such as isomaltol, maltol, diacetyl, puryvaldehyde, hydroxyacetone, 2-furfural, etc.), possessing odor oreventually leading to it.

Some of them are good antioxidants, for example, various reductones and aminoreductones. Aminoreductones obtained from hexoses and secondary amines [131] will inhibit peroxides' formation in a number of animal oils and butter on their addition at less than 0.02% [130]. These antioxidants are more stable and efficient than their enediol partners, which is caused, in particular, by catalytic metal ions (such as iron and copper) combining into inactive chelate complexes. The same role is played by various forms of reductones in melanoidin polymers, combining iron ions and thus reducing their catalytic activity in oxidation reactions [130]. That is why Maillard's reaction of browning is used in food processing as a weapon against oxidation. In particular, a number of aminoacids (histidine, lysine, arginine) will form with sugars the most efficient antioxidants, by Maillard's reaction products, which protect baked food, milk, and cereals. Glucose and the aforesaid aminoacids being added will enhance the stability of baked products to oxidation [130] (Fig. 7.8).

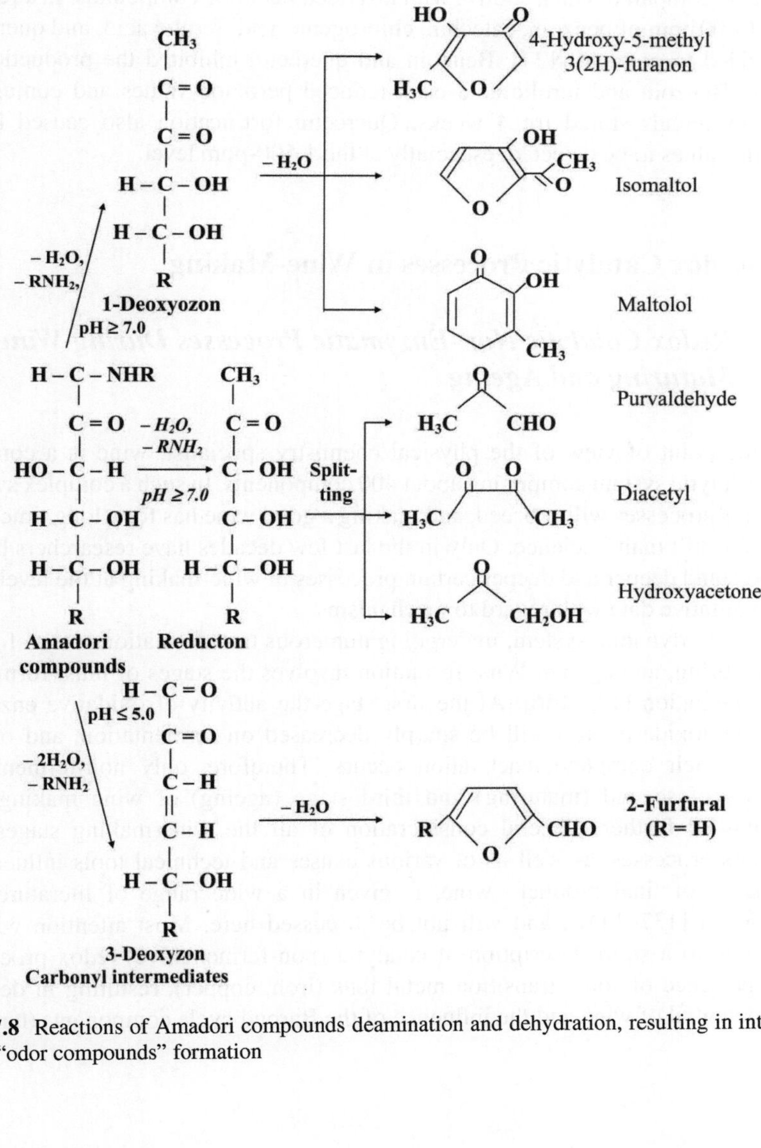

Fig. 7.8 Reactions of Amadori compounds deamination and dehydration, resulting in intermediates -"odor compounds" formation

Phenolic compounds act as free radical terminators, chelators of metal catalysts, and singlet oxygen quenchers. Synthetic antioxidants may be steam-distilled during extrusion and some consumers prefer "natural" products. Antioxidant-rich plant materials could provide protection against oxidation, but their efficacy is not well established. Cinnamic acid and vanillin were better antioxidants than BHT in extruded corn snacks that were fried postextrusion [132]. Simulated ready-to-eat breakfast cereals consisting of 60% rolled oats, 30% oat flour, and 10% sucrose were formulated to contain 200 ppm of BHA, vanillin, and caffeic, cinnamic, ferulic, gallic, *p*-coumaric, or vanillic acids, then were twin-screw extruded and stored at 35 °C for 6 months[133]. No difference in antioxidant activity was found among the treatments compared with a control with no added phenolic compounds. In a related study, 1,000 ppm of benzoin, catechin, chlorogenic acid, ferulic acid, and quercetin were added to oatmeal [134]. Benzoin and quercetin inhibited the production of hexanal. Benzoin and ferulic acid each reduced peroxide values and conjugated dienes in cereals stored for 4 weeks. Quercetin fortification also caused lower peroxide values to be detected, especially at the 1,500-ppm level.

7.3 Redox Catalytic Processes in Wine-Making

7.3.1 Redox Catalytic Non-Enzymatic Processes During Wine Maturing and Ageing

From the point of view of the physical chemistry specialist, wine is a complex redox catalytic system comprising about 400 components. In such a complex system numerous processes will proceed, and making a good wine has for a long time been more of a skill than a science. Only in the last few decades have researchers begun to understand deeper and deeper certain processes in wine-making at the level of at least qualitative data with regard to mechanisms.

Wine is a dynamic system, undergoing numerous transformations on its formation, maturing, and ageing. Wine formation involves the stages of must-formation and fermentation [135, 136]. At the first stage the activity of oxidative enzymes (*o*-diphenoloxidase, etc.) will be sharply decreased on fermentation, and on the fifth day their complete inactivation occurs. Therefore, only non-fermentative processes of second (maturing) and third stage (ageing) of wine making will be discussed further. General consideration of all the wine-making stages and numerous processes, as well as of various causes and technical tools influencing the quality of final product—wine, is given in a wide range of literature (for example, in [137–144]), and will not be discussed here. Most attention will be given only to a short description of catalytic (non-fermentative) redox processes in the presence of some transition metal ions (iron, copper), resulting in definite redox potential of wine and the influence of the Baroud cycle components (tartaric, dihydroxyfumaric acid, etc.) together with the problem of wine stabilization against colloidal muddiness.

The period of wine maturing and ageing is the most time-consuming in its production. This period is characterized by complex physical–chemical processes, among which redox reactions are to be mentioned especially [136]. The latter are caused by oxygen absorbed by wine from the air. The intensity of oxygen absorbtion by wine is subject to the influence of phenolic substances, organic acids (first of all those involved in the Baroud cycle), some metals (see Chap. 5), and also the sulfurous acid contained in wine.

Wine maturing proceeds on limited air oxygen access, which will form hydrogen peroxide and hydroperoxides with some of the wine components. Further, on wine ageing, oxidative processes will proceed without air oxygen access on account of hydrogen peroxide and hydroperoxides. On wine maturing, the content of oxygen will gradually decrease due to its involvement in redox processes. Depth and direction of these processes can be characterized by the value of redox potential E [138, 145]:

$$E = E_o + \frac{0.058}{n} \log \frac{[O_x]}{[Red]}$$

where $[O_x]$ and [Red] are the concentrations of oxidized and reduced forms, n is the number of electrons passing, and E_o the normal potential of system corresponding to the equality $[O_x] = [Red]$ within this system.

The results of much research, for example, in [137, 138], have shown that on wine storage in bottles without oxygen access, its redox potential will be decreased, and its low value will contribute to the wine bouquet developing. The value E for table wine being stored in the bottles reaches 150–250 mV, for Champagne wines—200–350 mV. In old high-quality wine in bottles $E \approx 170$–300 mV, in young wine $E \approx 340$–355 mV. However, good wine bouquet does not always mean low redox potential. Meanwhile, most researchers [137] consider that the low redox potential value influences not only its stability (preventing colloidal and other muddiness), but also the taste and bouquet. For example, Champagne with a more pleasant taste has a lower potential. With the help of redox potential value one may characterize the degree of wine oxidation and to control the production of various wine types. The value of wine redox potential which is influenced by a variety of factors, and is determined by the ratio of oxygen and reducers concentrations:

$$E = f\frac{[Ox]}{[Red]},$$

though different influences on wine (such as SO_2 introduction) will cause simultaneously both O_2 and reducer concentration changes. Therefore, it is not always possible to find the relation between redox potential value and wine quality. Redox processes in wine-making are diverse and rather complex. They are as a rule catalytic. Catalysts are, first of all, iron, copper, and maybe manganese ions and their complex compounds. There are always remaining amounts of iron contained in wine, which will form iron (II) tartaric complexes with tartaric acid, oxidizing spontaneously on contact with oxygen. Practically all the components of Baroud cycle (see Sect. 5.1, Scheme 5.1), natural polyphenols (see Sect. 5.3), as well as many other substances contained in wine, serve as iron, copper, and other metal ions ligands. Interaction of iron tartaric complexes with oxygen will result in tartaric acid oxidation, which is contained in large amounts in wine (up to 5 g/L), and in hydrogen peroxide formation. Thus, typical Fenton's system will be created, yielding the generation of $OH^{\bullet}$ radicals and tartaric acid or other ligands oxidation with them:

$$Fe^{II}L_n + H_2O_2 \longrightarrow Fe(III)L_n + OH^- + OH^{\bullet}$$

$$OH^{\bullet} + L \longrightarrow \text{Products}$$

Initial products of such a process is dihydroxyfumaric acid and triose reductone as well (reductones are substances having enidiolic H–C=O

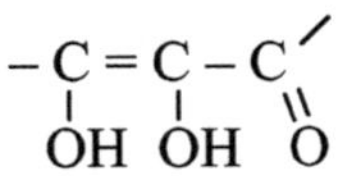

or keto-enolic and –C=O groups

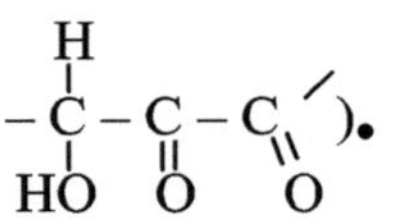

These substances possess strong reducing and antiseptic properties, on one hand establishing rather low wine redox potential, and on the other reducing (protecting of oxidation) such important wine components as natural polyphenols (such as catechins, etc.), thus preventing muddiness. The main processes and mechanisms of wine formation and ageing have been shown in model systems of M^{z+}–L–$O_2(H_2O_2)$ and M^{z+}–L–$O_2(H_2O_2)$—S type (where $M^{z+}=Fe^{2+}$, Cu^{2+}, L—oxyacids of Baroud's type, minoacids, etc., S—substrates, first of all—natural polyphenols and alcohols) and are present and discussed in Chap. 5. In the same chapter in Fig. 5.7 the generalized scheme of polyphenols redox catalytic and photolytic transformation is presented.

The main conclusions that can be made regarding catechins oxidation in model systems, are as follows:

1. Catechin (polyphenols) oxidation in simplified model wine systems proceeds by a radical-chain mechanism.
2. Hydrogen peroxide participates in polyphenols oxidation initiation (it is formed from air O_2), but in the oxidation process an essential role is played by molecular oxygen which can also be directly involved in the chain continuation reactions.

3. The oxidation process is efficiently catalysed with iron (copper) ions. This process is intensified by the reducers' presence in the reaction medium, capable to turn iron(III) into iron(II), due to the formation of redox cycle generating $OH^{\bullet}$ radicals which will oxidize polyphenols.
4. Catechin oxidation in the presence of ethanol proceeds less efficiently, because of both substrates competition for $OH^{\bullet}$ radicals. Hence, ethyl alcohol to a certain degree protects catechin oxidation.

It follows from the aforesaid that to make catechins (polyphenols in general) oxidation slower, it is necessary to do the following:

(a) To withdraw oxygen from dry wine and to isolate wine against contact with oxygen during its storage, a fact which is well-known to wine-makers and to some degree is taken into account in practice
(b) To remove or completely inactivate iron (copper) contained in wine, with the help of suitable food reagents able to form complexes
(c) To introduce food antioxidants able to interact more efficiently with $OH^{\bullet}$ radicals (forming in wine), than polyphenols, and thus to make polyphenols oxidation more difficult. It should also be considered that polyphenols are to certain degree antioxidants

Products of catechins condensation can react with proteins, resulting in tanno-protein compounds gradually precipitating. This reaction is catalyzed by Fe^{3+} ions. As a result wine will turn muddy, and colloidal precipitate which contains protein substances, phenolic compounds, iron, copper ions, and other mineral substances will be formed [138]. Thus, organoleptic properties of wine and its commodity properties will be missed.

7.3.2 *Methods of Wine Stabilization and Improving of Its Organoleptic Properties*

During grapes processing and storage, a series of oxidative-reducing processes occur, which determine the quality of the final product. With the purpose of regulating the catalytic redox processes during wine-making, various preservatives are used possessing antioxidant and antiseptic properties [146].

7.3.2.1 Use of Sulfurous Anhydride

Among the preserving agents applied in wine-making, an outstanding place is occupied by sulfurous anhydride (SO_2), which has been used in wine-making since distant times. Its role as preservative has been recognized as early as in the beginning of the nineteenth century. Until now SO_2 in doses up to 200 mg/L has been used in the production and storage of different grape processing products, due to its antiseptic and antioxidant action.

In spite of its strong antioxidant and antiseptic properties, sulfurous anhydride also possesses disadvantages: on its use in high doses wine aroma will be neutralized—it produces a sharp irritating odor of sulfurous anhydride with characteristic bitter taste. Increased doses of SO_2 will retard wine maturing and the normal maleic-lactic fermentation [147]. An obstacle to its use is also the toxicity—the use should be no more than 25 mg/L of wine. Besides, it does not provide any guarantee of wine stability against colloidal turbidity for longer than 3–4 months, and against mustiness. The effect of SO_2 on oxidative processes can be explained by oxidases inhibition.

Due to these disadvantages of SO_2, as early as 1978 at the International Winemakers Congress in Athens the task was posed to search for efficient and non-toxic preservatives that could completely or at least partly substitute SO_2 (reduce its doses) in wine-making. Test results on model catalytic systems and possible mechanisms with dihydroxyfumaric acid (DFH_4) participation as preservative are described in Chap. 5.

7.3.2.2 Use of Dihydroxyfumaric Acid

To prolong the the wine stability period and improve its hygienic properties, we have used dihydroxyfumaric acid (DFH_4) as a preserving agent [148].

Being an endogen component of grape beverages, DFH_4 possesses the reducing properties though lower than those of SO_2. In grape processing products it can appear in amounts up to 1 mg/L as a result of tartaric acid oxidation.

Due to the pronounced reducing properties, DFH_4 transforms the oxidized components of grape beverages into the reduced forms.

In the case of dihydroxyfumaric acid addition to "Aligote" wine material (and to other wine materials), the most important polyphenols of which are catechins: (−)-catechin, (+)-catechin, (−)-epicatechin, (−)-epigallocatechin, etc., it will be consumed due to the interaction with oxidized polyphenols (Fig. 7.9). Dihydroxyfumaric acid possesses stronger reducing properties ($\varphi^{DFH_4} = 0.34$ V) than catechin ($\varphi^{CT} = 0.43$ V) [149]. Thus, it will be partly transformed into diketosuccinic acid [150] and will be decarboxylized [151]; natural polyphenols will be reduced, i.e., polyphenols' oxidation process is reversible. This fact diminishes the possibility of colloidal muddiness formation in wine, and dihydroxyfumaric acid can be used in certain amounts as preservative [152]. Wine made with dihydroxyfumaric acid additive, compared with wine produced by the usual technology, favorably differs by its better taste and much higher stability. Mechanisms of Fe^{2+} and Fe^{3+} participation in similar catalytic processes of natural polyphenols polymerization are considered in Sect. 5.3.

Our tests with adding DFH_4 into the "Aligote" wine materials have shown (Fig. 7.10) that the decrease in oxidation rate of DFH_4 in the sample (1.7×10^{-8} M/s) is higher than in the bidistilled water (8×10^{-9} M/s). As in both cases the

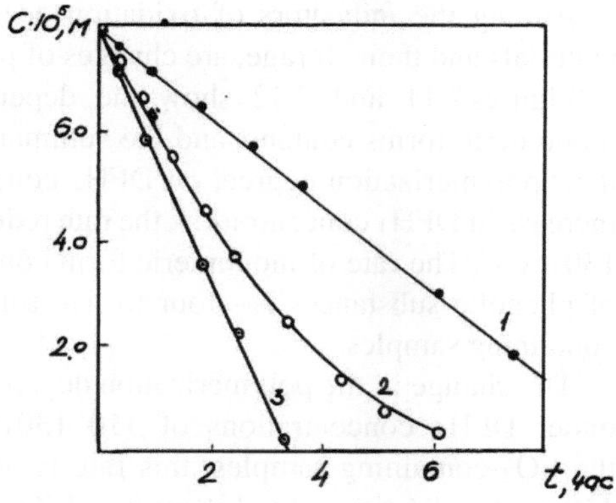

Fig. 7.9 Change in DFH_4 concentration with time: 1—in bidistilled water; 2—in ten-times diluted "Aligote" wine with DFH_4 used as preserving agent; 3—the same with SO_2 as preserving agent

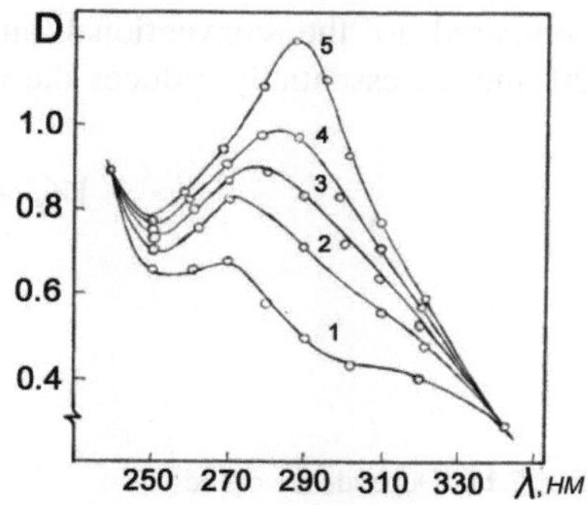

Fig. 7.10 Absorption spectra of "Aligote" wine samples: 1—ten-times diluted; 2—wine + DFH_4; 3—the same in 40 min; 4—the same in 90 min; 5—the same in 150 min. $[DFH_4]_0 = 8 \times 10^{-5}$ M

experiments were made under aerobic conditions, the difference in the DFH_4 oxidation rate can be explained with its interaction with the beverage components. The rate of DFH_4 consumption increases in the case of the drink, prepared by conventional technology (using SO_2 as a preservation agent). This means that once SO_2 is used as a preservation agent, the rate of polyphenol oxidation grows.

The results of these experiments demonstrate that the polyphenols' oxidation is a reversible process.

Another series of tests with DFH_4 as a preservation agent was carried out using the grape material "Onitcani white," following conventional technology, although using DFH_4 instead of SO_2. Different DFH_4 concentrations were taken: 40, 80, 100, 110, 120, 130, 140, 150, 200 mg/L. Organoleptic indicator after 12 months of wine storing in bottles was not worse than that of wine material with SO_2, but at the same time having higher stability [152].

The main physical–chemical indicators of wine materials with DFH_4 and SO_2 were studied.

The contents in the wine quality determining components depend on both the amount of DFH_4 and storage time. The lowest alcohol losses for the oxidation processes were marked with addition of DFH_4 (110–130 mg/L) [153].

Among the indicators of oxidation occurring during the preparation of wine materials and their storage, are changes of phenolic substances.

Figures 7.11 and 7.12 show the dependence of the rate in the change of monomeric forms contents and the summary contents in phenolic substances, and their polymerization degree, on DFH_4 concentration. It can be seen that with the increase in DFH_4 concentration, the rate reduces, reaching the minimum under 110–130 mg/L. The rate of monomeric form concentration change and the total amount of phenolic substances is—four to five times lower, compared to the anhydride-containing samples.

The change in the polymerization degree of phenolic compounds (n) is minimal under DFH_4 concentrations of 110–130 mg/L in wine material. Compared to the SO_2-containing samples, this rate is also somewhat lower: for concentration 110 mg/L—1.6 times, for 130 mg/L—1.2 times lower.

The data obtained demonstrate that DFH_4 in the amounts of 110–130 mg/L, compared to the conventional amount of added sulfurous anhydride (110–200 mg/L), essentially reduces the oxidation rate of polyphenol substances. This

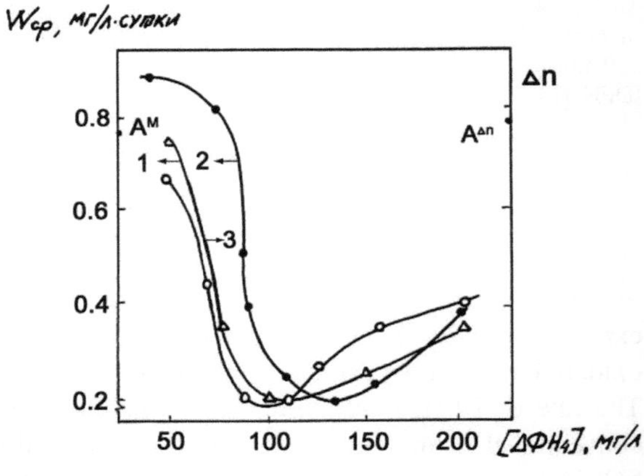

Fig. 7.11 Dependence of the change of monomeric form average rate (1), total amount (2), and polymerization degree (3) of phenolic substances in "Onitcani white" wine material on $[DFH_4]_0$

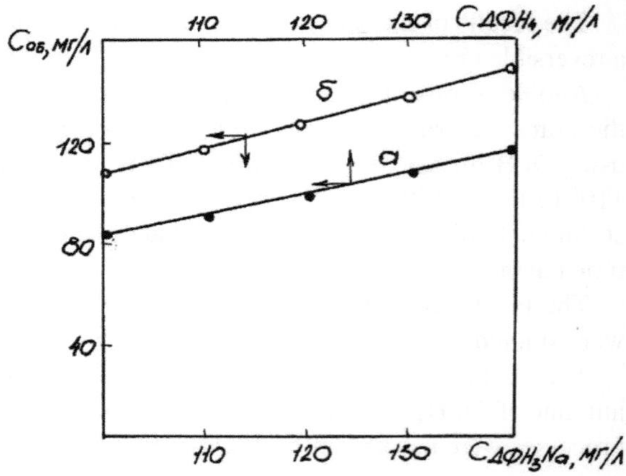

Fig. 7.12 Dependence of total amount of phenolic substances concentration on $[DFH_4]_0$ (**a**) and $[DFH_3Na]$ (**b**) in "Onitcani white" wine material

prevents the accumulation of quinones, which affects negatively the quality of the ready drink. Moreover, introduction of DFH_4 in the aforementioned doses, strongly inhibits polyphenols' condensation (specifically, catechin's), and thus prevents the colloidal clouds appearance. In this way, the stable keeping time period of the bottled grape drinks is—two to four times higher, compared to the samples obtained by the conventional procedure with SO_2 usage.

Experiments have been carried out on the wine material of the "Aligote" type, using the must produced at the winery. The research results have shown that the optimal DFH_4 doses also reaches 110–130 mg/L.

The wine material obtained was used, after 10 months, for experiments at the Cricova Winery with the purpose of studying the effect of SO_2 and DFH_4 on the quality of Champaign wine. The used grape material with the proposed preserving agent DFH_4 was rated as 8.7 by the degustation commission, compared to 8.4 for the control sample. This preserving agent can be obtained from the different secondary products, formed as a result of technological processes in wine-making [154].

From the practical aspect, the results of the performed tests mean that in order to keep the wine quality and to slow down catechin oxidation (polyphenols in general), it is necessary to do the following:

(a) To remove oxygen from the dry wine and to isolate wine from the contact with oxygen during its storing. This fact is well known to wine makers and is more or less is realized practically
(b) To remove the completely deactivated iron (copper), contained in wine, with the appropriate alimentary complex-forming agent
(c) To introduce in wine alimentary antioxidants, able to interact with OH^* radicals (formed in wine) more efficiently than polyphenols, and therefore to hamper polyphenols oxidation. It is also to be noted that polyphenols themselves are to a certain degree antioxidants

The catechins condensation products can react with proteins, which results in the formation of mixed protein compounds, gradually precipitating. This reaction is catalyzed by Fe^{3+} ions. As a result wine grows turbid, and the colloidal precipitate formed contains proteic substances, phenolic compounds, iron, copper ions, and other mineral substances. Organoleptic properties of wine and its marketability will thus be lost.

References

1. M.L. Khidekel, V.E. Wasserberg, J. Mendeleev Chem. Soc. **22**, 73 (1977) [Russian]
2. I. Pasquon, La Chim. a l'Industria **59**, 184 (1977)
3. G.W. Parshall, J. Mol. Catal. **4**, 243 (1978)
4. G.W. Parshall, *Homogeneous Catalysis* (Wiley, New York, 1980)
5. R.A. Sheldon, J.K. Kochi, *Metal-Catalyzed Oxidations of Organic Compounds* (Academic, New York, 1981)
6. W. Keim, Chem. Ind. **36**, 397 (1984)

7. W. Keim, in *Fundamental Research in Homogeneous Catalysis*, vol. 4 (Plenum Press, London, 1984), pp. 131–144
8. W. Keim, in *Industrial Applications of Homogeneous Catalysis*. ed. by A. Mortreux, F. Petit, (D. Reidel, Dordrecht, 1988), pp. 335–347
9. J. Falbe, H. Bahrmann, J. Chem. Educ. **61**, 961 (1984)
10. H. Bönnemann, Österr. Chem. Z. **85**(2), 29 (1984)
11. G.W. Parshall, S.D. Ittel, *Homogeneous Catalysis: The Applications and Chemistry of Catalysis by Soluble Transition Metal Complexes* (Wiley, New York, 1992) p. 342
12. B. Cornils, W.A. Herrmann, *Applied Homogeneous Catalysis with Organometallic Compounds: A Comprehensive Handbook in Two Volumes* (VCH, Weinheim, 1996)
13. S. Bhaduri, D. Mukesh, *Homogeneous Catalysis: Mechanisms and Industrial Applications* (Wiley, New York, 2000)
14. κ.L. Hill, in *Activation and Catalytic Reactions of Alkanes*, ed. by κ.L. Hill (Mir, Moscow, 1992), pp. 265–307
15. T.G. Traylor, S. Tsuchiya, Inorg. Chem. **26**, 1338 (1987)
16. C.L. Hill, R.B. Brown Jr, J. Am. Chem. Soc. **108**, 536 (1986)
17. R.B. Brown, R. Renneke, C.L. Hill, Prepr. Am. Chem. Soc., Div. Pet. Chem. **32**, 205 (1987)
18. M. Faraj, C.L. Hill, J. Chem. Soc. Chem. Commun. 1497 (1987)
19. C.L. Hill, R.F. Renneke, M. Faraj, R.B. Brown Jr, in *The Role of Oxygen in Chemistry and Biology*, ed. by W. Ando, Y. Morooka (Elsevier, Amsterdam, 1988), p. 185
20. M. Faraj, C.-H. Lin, C.L. Hill, New J. Chem. **12**, 745 (1988)
21. J.T. Groves, W.J. Kruper Jr, R.C. Haushalter, J. Am. Chem. Soc. **102**, 6375 (1980)
22. J.T. Groves, T.E. Nemo, J. Am. Chem. Soc. **105**, 6243 (1983)
23. B.C. Schardt, C.L. Hill, J. Am. Chem. Soc. **102**, 6374 (1980)
24. J.A. Smegal, C.L. Hill, J. Am. Chem. Soc. **105**, 2920 (1983)
25. R.B. Brown Jr, M.M. Williamson, C.L. Hill, Inorg. Chem. **26**, 1602 (1987)
26. J.T. Groves, D.V. Subramanian, J. Am. Chem. Soc. **106**, 2177 (1984)
27. J. Halpern, Inorg. Chim. Acta. **100**, 41 (1985)
28. J. Halpern, in *Fundamental Research in Homogeneous Catalysis*, vol. 1, ed. by A.E. Shilov, (Gordon and Brench, New York, 1986), p. 393
29. C.L. Hill, J.A. Smegal, Nouv. J. Chem. **6**, 287 (1982)
30. C.L. Hill, J.A. Smegal, T.J. Henly, J. Org. Chem. **48**, 3277 (1983)
31. D.H.R. Barton, N. Ozbalik, in *Activation and Catalytic Reactions of Alkanes,* ed. by κ.L. Hill (Mir, Moscow, 1992), pp. 302–329
32. T.J. McMurry, J.T. Groves, in *Cytochrome P-450: Structure Mechanism and Biochemistry*, ed. by P. Ortiz de Mantellano (Plenum Press, New York, 1985)
33. D.H.R. Barton, M.J. Gastiger, W.B. Montherwell, J. Chem. Soc. Chem. Commun. 41 (1983)
34. D.H.R. Barton, J. Boivin, W.B. Montherwell, N. Ozbalik, K.M. Schwartzentruber, K. Jankowski, Nouv. J. Chim. **10**, 387 (1986)
35. D.H.R. Barton, J. Boivin, P.L.E. Coupanec, J. Chem. Soc. Chem. Commun. 1379 (1987)
36. C. Sheu, A. Sobkowiak, L. Zhang, N. Ozbalik, D.H.R. Barton, D.T. Sawyer, J. Am. Chem. Soc. **111**, 8030 (1989)
37. C. Sheu, S.A. Richert, P. Cofré, B. Ross, Jr., A. Sobkowiak, D.T. Sawyer, J.A. Kanofsky, J. Am. Chem. Soc. **112**, 1936 (1990)
38. C. Sheu, D.T. Sawyer, J. Am. Chem. Soc. **112**, 8212 (1990)
39. H.-C. Tung, C. Kang, D.T. Sawyer, J. Am. Chem. Soc. **114**, 3445 (1992)
40. U. Schuchardt, C.E.Z. Krähembühl, W.A. Carvalho, New J. Chem. **15**, 955 (1991)
41. D.H.R. Barton, D.K. Taylor, Izv. AN Rossii. Ser. Khim. [Rev. Acad. Sci. Rus. Chem. Ser.] 595 (1995)
42. F. Minisci, F. Fontana, S. Araneo, F. Recupero, S. Banfi, S. Quici, J. Am. Chem. Soc. **117**, 226 (1995)
43. A.K. Suresh, M.M. Sharma, T. Sridhar, Ind. Eng. Chem. Res. **39**(11), 3958–3997 (2000)
44. W. Kanjina, W. Trakarnpruk, J. Met. Mater. Miner. **20**(2), 29–34 (2010)

45. Ch.A. Tolman, J.D. Druminer, M.J. Nappa, N. Herron, in *Activation and Catalytic Reactions of Alkanes*, ed. by κ.L. Hill (Mir, Moscow, 1992), pp. 330–405
46. D.W Breck, *Zeolite Molecular Sieves: Structure, Chemistry, and Use* (R.E. Krieger, Malabar, 1984)
47. P.E. Ellis, J.E. Lyons, Coord. Chem. Rev. **105**, 181 (1990)
48. A. Malldotti, R. Amadelli, C. Bartocci, V. Carassiti, E. Polo, G. Varani, Coord. Chem. Rev. **125**, 143 (1993)
49. K.S. Suslick, R.A. Watson, New J. Chem. **16**, 633 (1992)
50. A. Maldotti, C. Bartocci, R. Amadelli, E. Polo, P. Battioni, D. Mansuy, J. Chem. Soc. Chem. Commun. 1487 (1991)
51. L. Weber, I. Imiolezyk, G. Haufe, D. Rehorek, H. Hennig, J. Chem. Soc. Chem. Commun. 301 (1992)
52. L. Weber, R. Hommel, J. Behling, G. Haufe, H. Hennig, J. Am. Chem. Soc. **116**, 2400 (1994)
53. M. Hoshino, K. Ueda, M. Takahashi, M. Yamaji, Y. Hama, J. Chem. Soc. Faraday Trans. **88**, 405 (1992)
54. A. Maldotti, C. Bartocci, G. Varani, A. Molinari, P. Battioni, D. Mansuy, Inorg. Chem. **35**, 1126 (1996)
55. B.N. Gorbunov, L.S. Ilina, M.A. Sheinin, V.V. Konov, *Chemical Industry. Ser. Production and Treatment of Plastic Masses. Chemicals for Polymers Production.* Review (NIITEKhIM, Moscow, 1990), pp. 1–27 [Russian]
56. N.N. Kundo, N.P. Keyer, Zh. Fiz. Khim. [J. Phys. Chem.] **42**, 1352 (1968) [Russian]
57. N.N. Kundo, N.P. Keyer, G.V. Glazneva, Kinetika I kataliz [Kinetics Catal.] **8**, 1325 (1967) [Russian]
58. A.D. Simonov, N.P. Keyer, N.N. Kundo, Kinetika I kataliz [Kinetics Catal.] **14**, 988 (1973) [Russian]
59. A.K. Yatsimirskii, B.I. Kozlyak, A.S. Erokhin, Kinetika I kataliz [Kinetics Catal.] **29**, 352 (1988) [Russian]
60. B.I. Kozlyak, A.S. Krokhin, A.K. Yatsimirskii, Izv. AN SSSR [Rev. Acad. Sci. USSR] **4**, 815 (1986) [Russian]
61. O. Sygeru *Chemistry of Sulphur Organic Compounds* (Chemistry Publications, Moscow, 1975), p. 98. [Russian transl.]
62. T.J. Wallace, A. Schriesheim, H.B. Jonassen, Chem. Ind. **18**, 734 (1963)
63. B.I. Kozlyak, A.S. Erokhin, V.F. Borodkin, Izv. VUZov. Khim i khim. Tekhnol. [Univ. Rev. Chem. and Chem. Technol.] **23**, 39 (1980) [Russian]
64. A.N. Lazovenko, V.A. Ignatov, V.B. Maizlish, Izv. VUZov. Khim i khim. Tekhnol. [Univ. Rev. Chem. and Chem.Technol.] **24**, 685 (1981) [Russian]
65. R.J. Sheehan, *Terephthalic acid, dimethyl terephthalate, and isophthalic acid*, in *Ullmann's Encyclopedia of Industrial Chemistry* (Wiley-VCH, Weinheim, 2002)
66. B. Rohland, T. Pyl, Wiss. Z. Ernst-Moritz-Arndt-Univ. Greifswald. Math.-Nat. wiss. Reihe. **35**(1), 55 (1986)
67. P.B. Cheah, D.A. Ledward, J. Food Sci. **62**(6), 1135–1139 (1997)
68. H.D. Belitz, W. Grosch, *Food Chemistry* (Springer, Berlin, 1987)
69. V.N. Ushakova, *Stability of Foodstuffs Lipids* (Agropromizdat, Moscow, 1988) [Russian]
70. J. Kanner, Meat Sci. **36**, 169–189 (1994)
71. J.M.C. Gutteridge, B. Halliwell, Trends Biochem. Sci. **15**(4), 129–135 (1990)
72. J. Kanner, in *Lipid Oxidation in Food*, ed. by A.J.St. Angelo (American Chemical Society, Washington, 1992), pp. 55–73
73. E.A. Decker, H.O. Hultin, in *Lipid Oxidation in Food* (American Chemical Society, Washington, 1992), pp. 33–54
74. J.I. Gray, R.L. Crackel, in *The Chemistry of Muscle Based Foods,* ed. by D.E. Johnston, M.K. Knight, D.A. Ledward (Royal Society of Chemistry, Cambridge, 1992), pp. 145–168
75. J.I. Gray, E.A. Gomaa, D.J. Buckley, Meat Sci. **43**, 111–123 (1996)

76. J. Kanner, in *Lipid Oxidation in Food*, ed. by A.J.St. Angelo (American Chemical Society, Washington, 1992), pp. 55–73
77. M.G. Simic, J. Environ. Sci. Health **9**, 113 (1991)
78. J. Kanner, S. Harel, Arch. Biochem. Biophys. **237**, 314 (1985)
79. M. Chance, L.C. Kirmar, M. Power, B. Chance, Biochemistry **215**, 1259 (1986)
80. C.P. Baron, H.J. Andersen, Myoglobin-induced lipid oxidation. A review. J. Agric. Food Chem. **50**, 3887–3897 (2002)
81. S. Harel, J. Kanner, Free Radic. Res. Commun. **5**, 21 (1988)
82. J. Kanner, J.B. German, J.E. Kinsella, Crit. Rev. Food Sci. Nutr. **25**, 317 (1987)
83. D. Galaris, E. Cadenas, P. Hochstein, Biochem. Biophys. Res. Commun. **160**, 1162 (1989)
84. I. Solar, J. Dulitzky, N. Shaklai, Arch. Biochem. Biophys. **283**, 81 (1990)
85. M.G. Simic, S.V. Jovanovic, E. Niki, in *Lipid Oxidation in Food,* ed. by A.J.St. Angelo (American Chemical Society, Washington, 1992), pp. 14–32
86. R.G. Berger, F. Drawert, H. Kollmansberger, S. Nitz, B. Schraufstetter, J. Agric. Food. Chem. **33**, 232 (1985)
87. E. Frankel, Prog. Lipid Res. **23**, 197 (1984)
88. J.B. German, H. Zhang, R. Berger, in *Lipid Oxidation in Food*, ed. by A.J.St. Angelo (American Chemical Society, Washington, 1992), pp. 74–92
89. M.G. Simic, M. Karel (eds), *Autoxidation in Food and Biological Systems* (Plenum Press, New York, 1980), p. 659
90. A.J.St. Angelo, M.E. Bailey (eds), *Warmed-Over Flavor of Meat* (Academic, Orlando, 1987), p. 294
91. A. Asghar, J.I. Gray, D.J. Buckley, A.M. Pearson, A.M. Boosen, Food Technol. **42**(6), 102 (1988)
92. J. Kanner, J.B. German, J.E. Kinsella, Crit. Rev. Food Sci. Nutr. **25**, 317 (1987)
93. P.B. Addis, S.W. Park, in *Food Toxicology: A Perspective on the Relative Risk*, ed. by S.L. Taylor, R.A. Scanlan (Marcel Dekker, New York, 1989), pp. 297–330
94. H.B. Dunford, Free Radic. Biol. Med. **3**, 405 (1987)
95. D.L. Bakkeren, C.M.H. Jeu Jaspars, C. Van Der Heul, H.G. Van Eijk, Int. J. Biochem. **17**, 925 (1985)
96. T. Hazell, J. Sci. Food Agric. **33**, 1049 (1982)
97. E.A. Decker, C.-H. Huang, J.E. Osinchek, H.O. Hultin, J. Food Biochem. **13**, 179 (1989)
98. E.A. Decker, H.O. Hultin, J. Food Sci. **55**, 947 (1990)
99. J. Kanner, B. Hazan, L. Doll, J. Agric. Food Chem. **36**, 412 (1988)
100. B.R. Schricker, D.D. Miller, J.R. Stouffer, J. Food Sci. **47**, 740 (1982)
101. J.O. Igene, J.A. King, A.M. Pearson, J.I. Gray, J. Agric. Food Chem. **27**, 838 (1979)
102. B.R. Schricker, D.D. Miller, J. Food Sci. **48**, 1340, 1349 (1983)
103. E.A. Decker, H.O. Hultin, in *Lipid Oxidation in Food*, ed. by A.J.St. Angelo (American Chemical Society, Washington, 1992), pp. 33–54
104. R.F. Boyer, C.J. Mccleary, Free Radic. Biol. Med. **3**, 389 (1987)
105. E.A. Decker, H.O. Hultin, J. Food Sci. **55**, 951 (1990)
106. J. Kanner, L. Doll, J. Agric. Food Chem. **39**, 247 (1991)
107. E.A. Decker, B. Welch, J. Agric. Food Chem. **38**, 674 (1990)
108. S. Apte, P.A. Morrissey, Food Chem. **25**, 127 (1987)
109. C.C. Winterbourn, J. Biochem. **182**, 625 (1979)
110. D.A. Baldwin, E.R. Jenny, P. Aisen, J. Biol. Chem. **259**, 13391 (1984)
111. M. Ozaki, T. Kawabata, M. Awai, Biochem. J. **250**, 589 (1988)
112. M.J. O'Connell, R.J. Ward, H. Baum, T.J. Peters, Biochem. J. **229**, 135 (1985)
113. S.H. Yong, N. Karel, J. Am. Oil Chem. Soc. **55**, 352 (1978)
114. T. Ohshima, S. Wada, C. Koizumi, Nippon Suisan Gakkaishi **54**, 2165 (1988)
115. J. Kanner, M.A. Salan, S. Harel, J. Shegalovich, J. Agric. Food Chem. **39**, 242 (1991)
116. C.C. Winterbourn, Biochem. J. **210**, 15 (1983)
117. J. Kanner, S. Harel, B. Hazan, J. Agric. Food Chem. **34**, 506 (1986)
118. J. Kanner, F. Sofer, S. Harel, L. Doll, J. Agric. Food Chem. **36**, 415 (1988)
119. E. Graf, J.W. Eaton, Free Radic. Biol. Med. **8**, 61 (1990)

120. J.R. Mahoney, E. Graf, J. Food Sci. **51**, 1293 (1986)
121. F. Shahidi, L.J. Rubin, L.L. Diosady, N. Kassam, J.C.-L.-S. Fong, Food Chem. **21**, 145 (1986)
122. J.P. Roozen, Food Chem. **24**, 167 (1987)
123. K.L. Empson, T.P. Labuza, E. Graf, J. Food Sci. **56**, 560 (1991)
124. H. Faraji, E.A. Decker, J. Food Sci. **56**, 1038 (1991)
125. H. Faraji, E.A. Decker, D.K. Aaron, J. Agric. Food Chem. **39**, 1288 (1991)
126. J. Kanner, I. Bartov, M. Salan, L. Doll, J. Agric. Food Chem. **38**, 601 (1990)
127. S.V. Jovanovic, M.G. Simic, in *Oxygen Radicals in Biology and Medicine*, ed. by M.G. Simic, K.A. Taylor, J.F. Ward, C. Von Sonntag (Plenum Press, New York, 1988), pp. 115–122
128. E.P.L. Hunter, M.F. Desrosiers, M.G. Simic, Free Radic. Biol. Med. **6**, 581 (1989)
129. J.B. German, Adv. Exp. Med. Biol. **459**, 23–50 (1999)
130. M.E. Bailey, K.W. Um, in *Lipid Oxidation in Food.* ed. by A.J.St. Angelo (American Chemical Society, Washington, 1992), pp. 122–139
131. C.D. Evans, H.A. Moser, P.M. Cooney, J.E. Hodge, J. Am. Oil Chem. Soc. **35**, 84 (1958)
132. M.E. Camire, M.P. Dougherty, J. Food Sci. **63**, 516–518 (1998)
133. M.E. Camire, M.P. Dougherty, H. Osborn, D.K. Avis-Dentici, J. Briggs. *Simple Phenolic Compounds as Antioxidants for Extruded Oat Cereals. Abstract 50A-38* (IFT, Chicago, 1999)
134. V.K. Viscidi, M.E. Camire, M.P. Dougherty, J. Briggs, Complex phenolic compounds reduce lipid oxidation in extruded oat cereals. Lebensm.-Wiss. Technol. **37**, 789–796 (2004)
135. T. Garde-Cerdán, C. Ancín-Azpilicueta, Trends Food Sci. Technol. **17**(8), 438–447 (2006)
136. M.R. Salinas, G.L. Alonso, G. Navarro, F. Pardo, J. Jimeno, M.D. Huerta, Am. J. Enol. Viticult. **47**(2), 134–144 (1996)
137. A.K. Rodopulo, *Biochemistry of Wine-Making* (Food Industry, Moscow, 1971) [Russian]
138. Z.N. Kishkovskii, I.M. Skurikhin, *Chemistry of Wine* (Food Industry, Moscow, 1976) [Russian]
139. J. Ribereau-Gayon, E. Peynaud, P. Sudraud, P. Ribereau-Gayon, *Traité d'oenologie. Sciences et techniques du vin.* T.1. (Dunod, Paris, 1972)
140. M.N. Zaprometov, *Principles of Phenol Compounds Biochemistry* (Vysshaya shkola, Moscow, 1974) [Russian]
141. G.G. Valuiko, *Biochemistry and Technology of Red Wine* (Food. Industry, Moscow, 1973) [Russian]
142. V.L. Kretovich, M.A. Metlitskii, N.I. Bokuchaeva, N.I. Skobeleva, Z.N. Kishkovskii, G.S. Ilyin, R.V. Pheniksova, *Technical Biochemistry* (Vysshaya shkola, Moscow, 1973)
143. M.A. Amerine, H.W. Berg, W.V. Cruess, *The Technology of Wine Making* (Avi Publishing, Westport, 1972), p. 802
144. V. Radovanovic, *Technologija vina* (Beograd, Gradevinska knjiga, 1970)
145. M.V. Moreno-Arribas, P.M. Carmen (eds), *Wine Chemistry and Biochemistry* (Springer, New York, 2009)
146. P. Ribéreau-Gayon, Y. Glories, A. Maujean, D. Dubourdieu, *Handbook of Enology: The Chemistry of Wine Stabilization and Treatments,* vol. 2, 2nd edn, (Wiley, Chichester, 2006)
147. G.G. Valuiko. *Biochemistry and technology of red wine making*, (Food Industry Publishing, Moscow, 1973), p. 296, [Russian]
148. A.Ya. Sychev, G.G. Duca, *Wine-Making and Wine-Growing in Moldova* (Kishinev, 1985), 12, p. 38. [Russian]
149. A.Ya Sychev, S.O. Travin, G.G. Duca, Yu.I. Scurlatov, *Catalytic Reactions and Environmental Protection* (Kishinev, Stiinta, 1983), p. 272
150. G.G. Duca, B.S. Gaina, O.V. Covaliova, V.V. Covaliov, M.V. Gonta, *Ecologically Pure Wine Production* (Stiinta Publishing, Chisinau, 2004), p. 432 [Russian]
151. A.Ya. Sychev, Yu.V. Scutaru, G.G. Duca, Zh. Fiz. Khim. [J. Phys. Chem.] **61**, 2264 (1987) [Russian]
152. A.Ya. Sychev, T.A. Postolatii, G.G. Duca, M.V. Gonta, Patent SU1159946. 1985–06–07
153. M. Gonta, in *Bioantioxidant.* Book of Abstr. Chernogolovka, (1983), p. 12
154. Gh. Duca, A. Mereuta, M. Gonta, C. Cojocaru, in *Proc. of 2rd Int. Conf. on Environmental Engineering and Management.* Iasi, 2004, pp. 771–780

Index

G. Duca, *Homogeneous Catalysis with Metal Complexes*, Springer Series in Chemical Physics 102, DOI 10.1007/978-3-642-24629-6,

About the Author

Acad., Prof. Gheorghe Duca, President of the Academy of Sciences of Moldova

Hability Doctor in Chemistry (1989), University Professor (1992), Correspondent Member (1992) and Full Member (2000) of the Academy of Sciences of Moldova.

Research areas: Physical Chemistry and Ecological Chemistry.

As a founder of the Scientific School in Ecological Chemistry, has developed together with Prof. Yu. Scurlatov the theory of complex formation with partial charge transfer during reduction–oxidation processes, and the concept of the redox state of the environment, has established the oxidation mechanisms of chemical components in aquatic systems, has determined the quality formation mechanisms of food and wine products, has elaborated new processes of biologically active substances preparation, has improved the technologies of physical–chemical and

G. Duca, *Homogeneous Catalysis with Metal Complexes*, Springer Series in Chemical Physics 102, DOI 10.1007/978-3-642-24629-6,

biochemical wastewaters treatment and reduction of toxic air emissions, has assessed the risk of chemical substances for human health.

Prof. Duca is the Head of the Editor's Offices of the following journals: *Chemistry Journal of Moldova*; *Mediul ambiant* (Moldova); *Akademos* (Moldova), member of the Editor's Office of the journals (Ukraine), *Environmental Engineering and Management Journal* (Romania), CLEAN—*Soil, Air, Water* (Germany).

The results of his research are presented in more than 700 works, including 67 monographs and manuals. He is author of 136 patents, many of them being awarded gold and silver medals and diplomas at the International Salons and Exhibitions in Brussels, Geneva, Budapest, Iassy, Moscow, London, Sofia, Beijing, Pittsburgh, etc.

Member of the American Chemical Society (1997), Member of Russian Academy of Education and Social Sciences (1999), Honorary Member of the Academy of Sciences and Arts of Montenegro (2006), Foreign Honorary Member of the Romanian Academy (2007), Member of the Central-European Academy of Sciences and Arts (2008), Foreign Member of the National Academy of Sciences of Ukraine (2009).

MIX
Papier aus verantwortungsvollen Quellen
Paper from responsible sources
FSC® C105338

If you have any concerns about our products,
you can contact us on
ProductSafety@springernature.com

In case Publisher is established outside the EU,
the EU authorized representative is:
Springer Nature Customer Service Center GmbH
Europaplatz 3, 69115 Heidelberg, Germany

Printed by Libri Plureos GmbH
in Hamburg, Germany